# Methods in Enzymology

Volume 398

UBIQUITIN AND PROTEIN DEGRADATION

Part A

# METHODS IN ENZYMOLOGY

EDITORS-IN-CHIEF

John N. Abelson    Melvin I. Simon

DIVISION OF BIOLOGY
CALIFORNIA INSTITUTE OF TECHNOLOGY
PASADENA, CALIFORNIA

FOUNDING EDITORS

Sidney P. Colowick and Nathan O. Kaplan

*Methods in Enzymology*

*Volume 398*

# *Ubiquitin and Protein Degradation*

*Part A*

EDITED BY

*Raymond J. Deshaies*

HOWARD HUGHES MEDICAL INSTITUTE
DIVISION OF BIOLOGY
CALIFORNIA INSTITUTE OF TECHNOLOGY
PASADENA, CALIFORNIA

ELSEVIER
ACADEMIC
PRESS

AMSTERDAM • BOSTON • HEIDELBERG • LONDON
NEW YORK • OXFORD • PARIS • SAN DIEGO
SAN FRANCISCO • SINGAPORE • SYDNEY • TOKYO

Elsevier Academic Press
525 B Street, Suite 1900, San Diego, California 92101-4495, USA
84 Theobald's Road, London WC1X 8RR, UK

This book is printed on acid-free paper. ♾



For all information on all Academic Press publications
visit our Web site at www.books.elsevier.com

ISBN-13: 978-0-12-182803-5
ISBN-10: 0-12-182803-4

PRINTED IN THE UNITED STATES OF AMERICA
05 06 07 08 09 9 8 7 6 5 4 3 2 1

# Table of Contents



## Section I. E1 Enzyme



## Section II. E2 Enzyme

## Section III. E3 Enzyme

## Section IV. Proteasome

## Section V. Isopeptidases

# Contributors to Volume 398

Article numbers are in parentheses following the names of contributors.
Affiliations listed are current.


OK SUN BANG (41), *School of Biological Sciences, Seoul National University, Seoul 151-742, Korea*

ROHAN T. BAKER (44), *Molecular Genetics Group, Division of Molecular Medicine, John Curtin School of Medical Research, Australian National University, Canberra ACT 0200, Australia*

DAVID BARFORD (17), *Section of Structural Biology, Institute for Cancer Research, London SW3 6JB, United Kingdom*

SYLVIE BEAUDENON* (1, 4, 11), *Institute for Cellular and Molecular Biology, University of Texas at Austin, Austin, Texas 78712-1095*

DAWADSCHARGAL BECH-OTSCHIR (39), *MRC Human Genetics Unit, Western General Hospital, Edinburgh EH4 2XU, United Kingdom*

RENEE BERNARDS (45), *Division of Molecular Carcinogenesis and Center for Biomedical Genetics, The Netherlands Cancer Institute, 1066 CX Amsterdam, Netherlands*

PHILIP G. BOARD (44), *Molecular Genetics Group, Division of Molecular Medicine, John Curtin School of Medical Research, Australian National University, Canberra ACT 0200, Australia*

GUILAUME BOSSIS (3), *Max Planck Institute for Biochemistry, 82131 Martinsried, Germany*

THIJN R. BRUMMELKAMP (45), *Division of Molecular Carcinogenesis and Center for Biomedical Genetics, The Netherlands Cancer Institute, 1066 CX Amsterdam, Netherlands*

CHRISTOPHER W. CARROLL (18), *Department of Physiology, University of California, San Francisco, California 94143-2200*

PAOLO CASCIO (28), *Department of Morphophysiology, School of Veterinary Medicine, University of Turin, 10095 Grugliasco, Italy*

ANN-MAREE CATANZARITI† (44), *Molecular Genetics Group, Division of Molecular Medicine, John Curtin School of Medical Research, Australian National University, Canberra ACT 0200, Australia*

SUMIT K. CHANDA (23), *Department of Genomics, Genomics Institute of the Novartis Research Foundation (GNF), San Diego, California 92121*

TOMOKI CHIBA (7), *Laboratory of Frontier Science, The Tokyo Metropolitan Institute of Medical Science, Tokyo 113-8613, Japan*

KATARZYNA CHMIELARSKA (3), *Department of Biochemie I, Georg-August-University Göttingen, 37073 Göttingen, Germany*

CHIN HA CHUNG (41), *School of Biological Sciences, Seoul National University, Seoul 151-742, Korea*

* *Current Affiliation: Ambion Diagnostics, Austin, Texas 78744-1837.*

† *Current Affiliation: Division of Plant Industry, CSIRO, Canberra ACT 2601, Australia.*

Sung Soo Chung (41), *School of Biological Sciences, Seoul National University, Seoul 151-742, Korea*

Philip Coffino (33), *Department of Microbiology and Immunology, University of California, San Francisco, California 94143-0414*

Robert E. Cohen (43), *Department of Biochemistry, University of Iowa, Iowa City, Iowa 52242-1109*

Anahita Dastur (11), *Molecular Genetics and Microbiology, Institute for Cellular and Molecular Biology, University of Texas at Austin, Austin, Texas 78712-1095*

George N. DeMartino (24), *Department of Physiology, University of Texas Southwestern Medical Center, Dallas, Texas 75390-9040*

Xing-Wang Deng (38), *Department of Molecular, Cellular, and Development Biology, Yale University, New Haven, Connecticut 06520-8104*

Raymond J. Deshaies (13, 32), *Howard Hughes Medical Institute, Division of Biology, California Institute of Technology, Pasadena, California 91125*

Annette M. G. Dirac (45), *Division of Molecular Carcinogenesis and Center for Biomedical Genetics, The Netherlands Cancer Institute, 1066 CX Amsterdam, Netherlands*

Wolfgang Dubiel (39), *Department of Surgery, Division of Molecular Biology, Medical Faculty Charité, Humboldt University, 10117 Berlin, Germany*

Michael Ellison (5), *Institute for Biomolecular Design, University of Alberta, Edmonton, Alberta, Canada T6G 2H7*

Suzanne Elsasser (29), *Department of Cell Biology, Harvard Medical School, Boston, Massachusetts 02115*

Daniel Finley (29), *Department of Cell Biology, Harvard Medical School, Boston, Massachusetts 02115*

Andreas Förster (25), *Department of Biochemistry, University of Utah, Salt Lake City, Utah 84103*

Maria Gaczynska (34, 35), *Department of Molecular Medicine, University of Texas Health Science Center at San Antonio, San Antonio, Texas 78245-3207*

Ulrike Gärtner (3), *Max Planck Institute for Biochemistry, 82131 Martinsried, Germany*

Alfred L. Goldberg (28, 30), *Department of Cell Biology, Harvard Medical School, Boston, Massachusetts 02115*

Carlos Gorbea (26), *Department of Biochemistry, University of Utah School of Medicine, Salt Lake City, Utah 84132-3201*

Michael Groll (27), *Institute fuer Physiologische Chemie, Ludwig Maximilians Universitat, Muenchen, 81377 Muenchen, Germany*

William Hankey (37), *Department of Molecular Biophysics and Biochemistry, Yale University, New Haven, Connecticut 06520-8114*

J. Wade Harper (17), *Department of Pathology, Harvard Medical School, Boston, Massachusetts 02115*

Patrick Hauser (9), *Novartis Institutes for BioMedical Research, Novartis Pharma AG, 4002 Basel, Switzerland*

Klavs B. Hendil (36), *Institute of Molecular Biology and Physiology, University of Copenhagen, DK-2100 Copenhagen, Denmark*

Avram Hershko (15), *Unit of Biochemistry, Technion-Israel Institute of Technology, Haifa 31096, Israel*

FRANZ HERZOG (16, 19), *Research Institute of Molecular Pathology (IMP), 1030 Vienna, Austria*

BETTINA K. J. HETFELD (39), *Department of Surgery, Division of Molecular Biology, Medical Faculty Charité, Humboldt University, 10117 Berlin, Germany*

CHRISTOPHER P. HILL (25), *Department of Biochemistry, University of Utah, Salt Lake City, Utah 84103*

MARK HOCHSTRASSER (37), *Department of Molecular Biophysics and Biochemistry, Yale University, New Haven, Connecticut 06520-8114*

FRANCESCO HOFMAN (9), *Novartis Institutes for BioMedical Research, Novartis Pharma AG, 4002 Basel, Switzerland*

MARTIN A. HOYT (33), *Department of Microbiology and Immunology, University of California, San Francisco, California 94143-0414*

DANNY T. HUANG (2), *Department of Structural Biology and Genetics/Tumor Cell Biology, St. Jude's Children's Research Hospital, Memphis, Tennessee 38105*

JEN-WEI HUANG (31), *Department of Biochemistry, Weill Medical College of Cornell University, New York, New York 10021*

ROBERT HUBER (27), *Max-Planck Institute for Biochemistry, Abteilung Strukturforschung, D-82152 Martinsried, Germany*

JON M. HUIBREGTSE (1, 11), *Institute for Cellular and Molecular Biology, University of Texas at Austin, Austin, Texas 78712-1095*

HARUTO ISHIKAWA (21), *Department of Molecular Cell Biology, Graduate School of Medicine, Osaka City University and Core Research for Evolutional Science and Technology, Osaka 545-8585, Japan*

KAZUHIRO IWAI (21), *Department of Molecular Cell Biology, Graduate School of Medicine, Osaka City University and Core Research for Evolutional Science and Technology, Japan Science and Technology Agency, Osaka 545-8585, Japan*

JANE P. JENSEN (6, 10), *Laboratory of Protein Dynamics and Signaling, Center for Cancer Research, National Cancer Institute, Frederick, Maryland 21702*

CLAUDIO A. P. JOAZEIRO (23), *Department of Cell and Molecular Biology, Genomics Institute of the Novartis Research Foundation (GNF), San Diego, California 92121*

SUNG HWAN KANG (41), *School of Biological Sciences, Seoul National University, Seoul 151-742, Korea*

YAMUNA KARUNASEKARA (44), *Molecular Genetics Group, Division of Molecular Medicine, John Curtin School of Medical Research, Australian National University, Canberra ACT 0200, Australia*

KEUN IL KIM (40), *Department of Molecular and Experimental Medicine (MEM-L51), The Scripps Research Institute, La Jolla, California 92037*

TAKAYOSHI KIRISAKO (21), *Department of Molecular Cell Biology, Graduate School of Medicine, Osaka City University and Core Research for Evolutional Science and Technology, Japan Science and Technology Agency, Osaka 545-8585, Japan*

ALEXEI F. KISSELEV (30), *Department of Pharmacology and Toxicology & Norris Cotton Cancer Center, Dartmouth Medical School, Lebanon, New Hampshire 03756; and Department of Cell Biology, Harvard Medical School, Boston, Massachusetts 02115*

ROBERT M. KRUG (4), *Institute for Cellular and Molecular Biology, University of Texas at Austin, Austin, Texas 78712*

Y. AMY LAM (31), *Department of Biochemistry, Weill Medical College of Cornell University, New York, New York 10021*

SHYR-JIANN LI (37), *Celera, South San Francsico, California 94020*

TI LI (12), *Department of Pharmacology, University of Washington, Seattle, Washington 98195*

WEI LI (23), *Department of Cell and Molecular Biology, Genomics Institute of the Novartis Research Foundation (GNF), San Diego, California 92121*

CHRISTOPHER D. LIMA (8), *Structural Biology Program, Sloan-Kettering Institute, New York, New York 10021*

KEVIN L. LORICK (6, 10), *Laboratory of Protein Dynamics and Signaling, Center for Cancer Research, National Cancer Institute, Frederick, Maryland 21702*

XUELIAN LUO (20), *Department of Pharmacology, University of Texas, Southwestern Medical Center, Dallas, Texas 75390-9041*

EUGENE I. MASTERS (25), *Department of Biochemistry, University of Utah, Salt Lake City, Utah 84103*

SEAN MCKENNA* (5), *Department of Biochemistry, University of Alberta, Edmonton, Alberta, Canada T6G 2H7*

KARL MECHTLER (19), *Research Institute of Molecular Pathology (IMP), 1030 Vienna, Austria*

FRAUKE MELCHIOR (3), *Department of Biochemie I, Georg-August-University Göttingen, 37073 Gottingen, Germany*

SUCHITHRA MENON (38), *Department of Molecular, Cellular, and Developmental Biology, Yale University, New Haven, Connecticut 06520-8104*

IVANA MICIK (23), *Vala Sciences, La Jolla, California 92037*

MICHIKO MINAMI (22), *Department of Natural and Environmental Science, Faculty of Education, Tokyo Gakugei University, Tokyo 184-8501, Japan*

YASUFUMI MINAMI (22), *Department of Biophysics and Biochemistry, and Undergraduate Program for Bioinformatics and Systems Biology, Tokyo 153-8902, Japan*

TREVOR MORAES† (5), *Department of Biochemistry, University of Alberta, Edmonton, Alberta, Canada T6G 2H7*

DAVID MORGAN (18), *Department of Physiology, University of California, San Francisco, California 94143-2200*

SHIGEO MURATA (22), *Laboratory of Frontier Science, Core Technology and Research Center, Tokyo Metropolitan Institute of Medical Science, Precursory Research for Embryonic Science and Technology (PRESTO); Japan Science and Technology Agency (JST), Tokyo 113-8613, Japan*

SEBASTIAN M. B. NIJMAN (45), *Division of Molecular Carcinogenesis and Center for Biomedical Genetics, The Netherlands Cancer Institute, 1066 CX Amsterdam, Netherlands*

* *Current Affiliation: Department of Structural Biology, Stanford University School of Medicine, Stanford, California 94305-5126.*

† *Current Affiliation: Department of Biochemistry, University of British Columbia, Vancouver, British Columbia V6T 1Z3.*

PAWEL A. OSMULSKI (34, 35), *Department of Molecular Medicine, University of Texas Health Science Center at San Antonio, San Antonio, Texas 78245-3207*

ZHEN-QIANG PAN (42), *Department of Oncological Sciences, The Mount Sinai School of Medicine, New York, New York 10029-6574*

JUNG JUN PARK (41), *School of Biological Sciences, Seoul National University, Seoul 151-742, Korea*

LORI A. PASSMORE (17), *Section of Structural Biology, Institute for Cancer Research, London SW3 6JB, United Kingdom*

NIKOLA P. PAVLETICH (12), *Cellular Biochemistry and Biophysics, Howard Hughes Medical Institute, Memorial Sloan-Kettering Cancer Center, New York, New York 10021*

JAN-MICHAEL PETERS (16, 19), *Research Institute of Molecular Pathology (IMP), 1030 Vienna, Austria*

MATTHEW D. PETROSKI (13), *Howard Hughes Medical Institute, Division of Biology, California Institute of Technology, Pasadena, California 91125*

ANDREA PICHLER (3), *Department of Biochemie I, George-August-University, 37073 Göttingen, Germany*

GREGORY PRATT (25, 26), *Department of Biochemistry, University of Utah, Salt Lake City, Utah 84103*

CHRISTOPHER PTAK (5), *Department of Biochemistry, University of Alberta, Edmonton, Alberta, Canada T6G 2H7*

MARTIN RECHSTEINER (26), *Department of Biochemistry, University of Utah School of Medicine, Salt Lake City, Utah 84132-3201*

VICENTE RUBIO (38), *Department of Molecular, Cellular, and Developmental Biology, Yale University, New Haven, Connecticut 06520-8104*

MARION SCHMIDT (29), *Department of Cell Biology, Harvard Medical School, Boston, Massachusetts 02115*

BRENDA A. SCHULMAN (2, 12), *Departments of Structural Biology and Genetics and Tumor Cell Biology, St. Jude Children's Research Hospital, Memphis, Tennessee 38105*

ROBERT SHARWOOD (44), *Molecular Plant Physiology Group, Australian National University, Canberra ACT 0200, Australia*

OLUWAFEMI SHOWOLE (31), *Department of Biochemistry, Weill Medical College of Cornell University, New York, New York 10021*

TATIANA A. SOBOLEVA* (44), *Molecular Genetics Group, Division of Molecular Medicine, John Curtin School of Medical Research, Australian National University, Canberra ACT 2601, Australia*

EVELYN STIEGER (3), *Max Planck Institute for Biochemistry, 82131 Martinsried, Germany*

VICENCA USTRELL (26), *Department of Biochemistry, University of Utah School of Medicine, Salt Lake City, Utah 84132-3201*

XARALABOS VARELAS[†] (5), *Department of Biochemistry, University of Alberta, Edmonton, Alberta, Canada T6G 2H7*

* *Current Affiliation: Cytokine Molecular Biology and Signalling Group, John Curtain School of Medical Research, Australian National University, Canberra ACT 0200, Australia.*

[†] *Current Affiliation: Samuel Lunenfeld Research Institute, Mount Sinai Hospital, Toronto, Ontario, Canada M5G 1X5.*

RATI VERMA (32), *Division of Biology, California Institute of Technology, Pasadena, California 91125*

XIPING WANG (38), *Department of Molecular, Cellular, and Developmental Biology, Yale University, New Haven, Connecticut 06520-8104*

NING WEI (38), *Department of Molecular, Cellular, and Developmental Biology, Yale University, New Haven, Connecticut 06520-8104*

ALLAN M. WEISSMAN (6, 10), *Laboratory of Protein Dynamics and Signaling, Center for Cancer Research, National Cancer Institute, Frederick, Maryland 21702*

SPENCER WHITNEY (44), *Molecular Plant Physiology Group, Australian National University, Canberra ACT 0200, Australia*

KENNETH WU (42), *Department of Oncological Sciences, The Mount Sinai School of Medicine, New York, New York 10029-6574*

KOSJ YAMOAH (42), *Department of Oncological Sciences, The Mount Sinai School of Medicine, New York, New York 10029-6574*

YILI YANG (10), *Dynamics and Signaling, National Cancer Institute at Frederick, Frederick, Maryland 21702*

TINGTING YAO (43), *Stowers Institute for Medical Research, Kansas City, Missouri 64110*

YUKIKO YOSHIDA (14), *Laboratory of Frontier Science, The Tokyo Metropolitan Institute of Medical Science, Honkomagome, Bunkyo-ku, Tokyo 113-8613, Japan*

HONGTAO YU (20), *Department of Pharmacology, University of Texas Southwestern Medical Center, Dallas, Texas 75390-9041*

ALI A. YUNUS (8), *Structural Biology Program, Sloan-Kettering Institute, New York, New York 10021*

DONG-ER ZHANG (40), *Department of Molecular and Experimental Medicine, Scripps Research Institute, La Jolla, California 92037*

MINGSHENG ZHANG* (33), *Department of Microbiology and Immunology, University of California, San Francisco, California 94143-0414*

CHEN ZHAO (4), *Institute for Cellular and Molecular Biology, University of Texas at Austin, Austin, Texas 78712*

NING ZHENG (12), *Department of Pharmacology, University of Washington, Seattle, Washington 98195*

* *Current Affiliation: Department of Biology, Massachusetts Institute of Technology, Cambridge, Massachusetts 02139.*

# Acknowledgments

These volumes, which I hope will be valuable to both rookie and veteran researchers in the ubiquitin field, would not exist were it not for the the many authors who invested considerable time and energy. I wish to thank them for their excellent contributions and for their patience in dealing with the editor. I also wish to acknowledge the many people who assisted in the preparation of these volumes, including Drs. Avram Hershko, Stefan Jentsch, Cecile Pickart, and Keiji Tanaka for their help in selecting chapters, my assistant Daphne Shimoda, for her help with soliciting and processing manuscripts, Gracy Noelle and Cindy Minor at Elsevier for their overall help in pulling this project together and keeping it on track, and Alan Palmer for his efforts on the page proofs. Finally, I wish to thank Linda Silveira, who inevitably bears greater burdens as a result of projects such as this.

RAY DESHAIES

# METHODS IN ENZYMOLOGY

VOLUME I. Preparation and Assay of Enzymes
*Edited by* SIDNEY P. COLOWICK AND NATHAN O. KAPLAN

VOLUME II. Preparation and Assay of Enzymes
*Edited by* SIDNEY P. COLOWICK AND NATHAN O. KAPLAN

VOLUME III. Preparation and Assay of Substrates
*Edited by* SIDNEY P. COLOWICK AND NATHAN O. KAPLAN

VOLUME IV. Special Techniques for the Enzymologist
*Edited by* SIDNEY P. COLOWICK AND NATHAN O. KAPLAN

VOLUME V. Preparation and Assay of Enzymes
*Edited by* SIDNEY P. COLOWICK AND NATHAN O. KAPLAN

VOLUME VI. Preparation and Assay of Enzymes *(Continued)*
Preparation and Assay of Substrates
Special Techniques
*Edited by* SIDNEY P. COLOWICK AND NATHAN O. KAPLAN

VOLUME VII. Cumulative Subject Index
*Edited by* SIDNEY P. COLOWICK AND NATHAN O. KAPLAN

VOLUME VIII. Complex Carbohydrates
*Edited by* ELIZABETH F. NEUFELD AND VICTOR GINSBURG

VOLUME IX. Carbohydrate Metabolism
*Edited by* WILLIS A. WOOD

VOLUME X. Oxidation and Phosphorylation
*Edited by* RONALD W. ESTABROOK AND MAYNARD E. PULLMAN

VOLUME XI. Enzyme Structure
*Edited by* C. H. W. HIRS

VOLUME XII. Nucleic Acids (Parts A and B)
*Edited by* LAWRENCE GROSSMAN AND KIVIE MOLDAVE

VOLUME XIII. Citric Acid Cycle
*Edited by* J. M. LOWENSTEIN

VOLUME XIV. Lipids
*Edited by* J. M. LOWENSTEIN

VOLUME XV. Steroids and Terpenoids
*Edited by* RAYMOND B. CLAYTON

VOLUME LIV. Biomembranes (Part E: Biological Oxidations)
*Edited by* SIDNEY FLEISCHER AND LESTER PACKER

VOLUME LV. Biomembranes (Part F: Bioenergetics)
*Edited by* SIDNEY FLEISCHER AND LESTER PACKER

VOLUME LVI. Biomembranes (Part G: Bioenergetics)
*Edited by* SIDNEY FLEISCHER AND LESTER PACKER

VOLUME LVII. Bioluminescence and Chemiluminescence
*Edited by* MARLENE A. DELUCA

VOLUME LVIII. Cell Culture
*Edited by* WILLIAM B. JAKOBY AND IRA PASTAN

VOLUME LIX. Nucleic Acids and Protein Synthesis (Part G)
*Edited by* KIVIE MOLDAVE AND LAWRENCE GROSSMAN

VOLUME LX. Nucleic Acids and Protein Synthesis (Part H)
*Edited by* KIVIE MOLDAVE AND LAWRENCE GROSSMAN

VOLUME 61. Enzyme Structure (Part H)
*Edited by* C. H. W. HIRS AND SERGE N. TIMASHEFF

VOLUME 62. Vitamins and Coenzymes (Part D)
*Edited by* DONALD B. MCCORMICK AND LEMUEL D. WRIGHT

VOLUME 63. Enzyme Kinetics and Mechanism (Part A: Initial Rate and Inhibitor Methods)
*Edited by* DANIEL L. PURICH

VOLUME 64. Enzyme Kinetics and Mechanism (Part B: Isotopic Probes and Complex Enzyme Systems)
*Edited by* DANIEL L. PURICH

VOLUME 65. Nucleic Acids (Part I)
*Edited by* LAWRENCE GROSSMAN AND KIVIE MOLDAVE

VOLUME 66. Vitamins and Coenzymes (Part E)
*Edited by* DONALD B. MCCORMICK AND LEMUEL D. WRIGHT

VOLUME 67. Vitamins and Coenzymes (Part F)
*Edited by* DONALD B. MCCORMICK AND LEMUEL D. WRIGHT

VOLUME 68. Recombinant DNA
*Edited by* RAY WU

VOLUME 69. Photosynthesis and Nitrogen Fixation (Part C)
*Edited by* ANTHONY SAN PIETRO

VOLUME 70. Immunochemical Techniques (Part A)
*Edited by* HELEN VAN VUNAKIS AND JOHN J. LANGONE

VOLUME 71. Lipids (Part C)
*Edited by* JOHN M. LOWENSTEIN

VOLUME 72. Lipids (Part D)
*Edited by* JOHN M. LOWENSTEIN

VOLUME 73. Immunochemical Techniques (Part B)
*Edited by* JOHN J. LANGONE AND HELEN VAN VUNAKIS

VOLUME 74. Immunochemical Techniques (Part C)
*Edited by* JOHN J. LANGONE AND HELEN VAN VUNAKIS

VOLUME 75. Cumulative Subject Index Volumes XXXI, XXXII, XXXIV–LX
*Edited by* EDWARD A. DENNIS AND MARTHA G. DENNIS

VOLUME 76. Hemoglobins
*Edited by* ERALDO ANTONINI, LUIGI ROSSI-BERNARDI, AND EMILIA CHIANCONE

VOLUME 77. Detoxication and Drug Metabolism
*Edited by* WILLIAM B. JAKOBY

VOLUME 78. Interferons (Part A)
*Edited by* SIDNEY PESTKA

VOLUME 79. Interferons (Part B)
*Edited by* SIDNEY PESTKA

VOLUME 80. Proteolytic Enzymes (Part C)
*Edited by* LASZLO LORAND

VOLUME 81. Biomembranes (Part H: Visual Pigments and Purple Membranes, I)
*Edited by* LESTER PACKER

VOLUME 82. Structural and Contractile Proteins (Part A: Extracellular Matrix)
*Edited by* LEON W. CUNNINGHAM AND DIXIE W. FREDERIKSEN

VOLUME 83. Complex Carbohydrates (Part D)
*Edited by* VICTOR GINSBURG

VOLUME 84. Immunochemical Techniques (Part D: Selected Immunoassays)
*Edited by* JOHN J. LANGONE AND HELEN VAN VUNAKIS

VOLUME 85. Structural and Contractile Proteins (Part B: The Contractile Apparatus and the Cytoskeleton)
*Edited by* DIXIE W. FREDERIKSEN AND LEON W. CUNNINGHAM

VOLUME 86. Prostaglandins and Arachidonate Metabolites
*Edited by* WILLIAM E. M. LANDS AND WILLIAM L. SMITH

VOLUME 87. Enzyme Kinetics and Mechanism (Part C: Intermediates, Stereo-chemistry, and Rate Studies)
*Edited by* DANIEL L. PURICH

VOLUME 88. Biomembranes (Part I: Visual Pigments and Purple Membranes, II)
*Edited by* LESTER PACKER

VOLUME 89. Carbohydrate Metabolism (Part D)
*Edited by* WILLIS A. WOOD

VOLUME 90. Carbohydrate Metabolism (Part E)
*Edited by* WILLIS A. WOOD

VOLUME 91. Enzyme Structure (Part I)
*Edited by* C. H. W. HIRS AND SERGE N. TIMASHEFF

VOLUME 92. Immunochemical Techniques (Part E: Monoclonal Antibodies and General Immunoassay Methods)
*Edited by* JOHN J. LANGONE AND HELEN VAN VUNAKIS

VOLUME 93. Immunochemical Techniques (Part F: Conventional Antibodies, Fc Receptors, and Cytotoxicity)
*Edited by* JOHN J. LANGONE AND HELEN VAN VUNAKIS

VOLUME 94. Polyamines
*Edited by* HERBERT TABOR AND CELIA WHITE TABOR

VOLUME 95. Cumulative Subject Index Volumes 61–74, 76–80
*Edited by* EDWARD A. DENNIS AND MARTHA G. DENNIS

VOLUME 96. Biomembranes [Part J: Membrane Biogenesis: Assembly and Targeting (General Methods; Eukaryotes)]
*Edited by* SIDNEY FLEISCHER AND BECCA FLEISCHER

VOLUME 97. Biomembranes [Part K: Membrane Biogenesis: Assembly and Targeting (Prokaryotes, Mitochondria, and Chloroplasts)]
*Edited by* SIDNEY FLEISCHER AND BECCA FLEISCHER

VOLUME 98. Biomembranes (Part L: Membrane Biogenesis: Processing and Recycling)
*Edited by* SIDNEY FLEISCHER AND BECCA FLEISCHER

VOLUME 99. Hormone Action (Part F: Protein Kinases)
*Edited by* JACKIE D. CORBIN AND JOEL G. HARDMAN

VOLUME 100. Recombinant DNA (Part B)
*Edited by* RAY WU, LAWRENCE GROSSMAN, AND KIVIE MOLDAVE

VOLUME 101. Recombinant DNA (Part C)
*Edited by* RAY WU, LAWRENCE GROSSMAN, AND KIVIE MOLDAVE

VOLUME 102. Hormone Action (Part G: Calmodulin and Calcium-Binding Proteins)
*Edited by* ANTHONY R. MEANS AND BERT W. O'MALLEY

VOLUME 103. Hormone Action (Part H: Neuroendocrine Peptides)
*Edited by* P. MICHAEL CONN

VOLUME 104. Enzyme Purification and Related Techniques (Part C)
*Edited by* WILLIAM B. JAKOBY

VOLUME 105. Oxygen Radicals in Biological Systems
*Edited by* LESTER PACKER

VOLUME 106. Posttranslational Modifications (Part A)
*Edited by* FINN WOLD AND KIVIE MOLDAVE

VOLUME 107. Posttranslational Modifications (Part B)
*Edited by* FINN WOLD AND KIVIE MOLDAVE

VOLUME 108. Immunochemical Techniques (Part G: Separation and Characterization of Lymphoid Cells)
*Edited by* GIOVANNI DI SABATO, JOHN J. LANGONE, AND HELEN VAN VUNAKIS

VOLUME 109. Hormone Action (Part I: Peptide Hormones)
*Edited by* LUTZ BIRNBAUMER AND BERT W. O'MALLEY

VOLUME 110. Steroids and Isoprenoids (Part A)
*Edited by* JOHN H. LAW AND HANS C. RILLING

VOLUME 111. Steroids and Isoprenoids (Part B)
*Edited by* JOHN H. LAW AND HANS C. RILLING

VOLUME 112. Drug and Enzyme Targeting (Part A)
*Edited by* KENNETH J. WIDDER AND RALPH GREEN

VOLUME 113. Glutamate, Glutamine, Glutathione, and Related Compounds
*Edited by* ALTON MEISTER

VOLUME 114. Diffraction Methods for Biological Macromolecules (Part A)
*Edited by* HAROLD W. WYCKOFF, C. H. W. HIRS, AND SERGE N. TIMASHEFF

VOLUME 115. Diffraction Methods for Biological Macromolecules (Part B)
*Edited by* HAROLD W. WYCKOFF, C. H. W. HIRS, AND SERGE N. TIMASHEFF

VOLUME 116. Immunochemical Techniques (Part H: Effectors and Mediators of Lymphoid Cell Functions)
*Edited by* GIOVANNI DI SABATO, JOHN J. LANGONE, AND HELEN VAN VUNAKIS

VOLUME 117. Enzyme Structure (Part J)
*Edited by* C. H. W. HIRS AND SERGE N. TIMASHEFF

VOLUME 118. Plant Molecular Biology
*Edited by* ARTHUR WEISSBACH AND HERBERT WEISSBACH

VOLUME 119. Interferons (Part C)
*Edited by* SIDNEY PESTKA

VOLUME 120. Cumulative Subject Index Volumes 81–94, 96–101

VOLUME 121. Immunochemical Techniques (Part I: Hybridoma Technology and Monoclonal Antibodies)
*Edited by* JOHN J. LANGONE AND HELEN VAN VUNAKIS

VOLUME 122. Vitamins and Coenzymes (Part G)
*Edited by* FRANK CHYTIL AND DONALD B. MCCORMICK

VOLUME 123. Vitamins and Coenzymes (Part H)
*Edited by* FRANK CHYTIL AND DONALD B. MCCORMICK

VOLUME 124. Hormone Action (Part J: Neuroendocrine Peptides)
*Edited by* P. MICHAEL CONN

VOLUME 141. Cellular Regulators (Part B: Calcium and Lipids)
*Edited by* P. MICHAEL CONN AND ANTHONY R. MEANS

VOLUME 142. Metabolism of Aromatic Amino Acids and Amines
*Edited by* SEYMOUR KAUFMAN

VOLUME 143. Sulfur and Sulfur Amino Acids
*Edited by* WILLIAM B. JAKOBY AND OWEN GRIFFITH

VOLUME 144. Structural and Contractile Proteins (Part D: Extracellular Matrix)
*Edited by* LEON W. CUNNINGHAM

VOLUME 145. Structural and Contractile Proteins (Part E: Extracellular Matrix)
*Edited by* LEON W. CUNNINGHAM

VOLUME 146. Peptide Growth Factors (Part A)
*Edited by* DAVID BARNES AND DAVID A. SIRBASKU

VOLUME 147. Peptide Growth Factors (Part B)
*Edited by* DAVID BARNES AND DAVID A. SIRBASKU

VOLUME 148. Plant Cell Membranes
*Edited by* LESTER PACKER AND ROLAND DOUCE

VOLUME 149. Drug and Enzyme Targeting (Part B)
*Edited by* RALPH GREEN AND KENNETH J. WIDDER

VOLUME 150. Immunochemical Techniques (Part K: *In Vitro* Models of B and T Cell Functions and Lymphoid Cell Receptors)
*Edited by* GIOVANNI DI SABATO

VOLUME 151. Molecular Genetics of Mammalian Cells
*Edited by* MICHAEL M. GOTTESMAN

VOLUME 152. Guide to Molecular Cloning Techniques
*Edited by* SHELBY L. BERGER AND ALAN R. KIMMEL

VOLUME 153. Recombinant DNA (Part D)
*Edited by* RAY WU AND LAWRENCE GROSSMAN

VOLUME 154. Recombinant DNA (Part E)
*Edited by* RAY WU AND LAWRENCE GROSSMAN

VOLUME 155. Recombinant DNA (Part F)
*Edited by* RAY WU

VOLUME 156. Biomembranes (Part P: ATP-Driven Pumps and Related Transport: The Na, K-Pump)
*Edited by* SIDNEY FLEISCHER AND BECCA FLEISCHER

VOLUME 157. Biomembranes (Part Q: ATP-Driven Pumps and Related Transport: Calcium, Proton, and Potassium Pumps)
*Edited by* SIDNEY FLEISCHER AND BECCA FLEISCHER

VOLUME 158. Metalloproteins (Part A)
*Edited by* JAMES F. RIORDAN AND BERT L. VALLEE

VOLUME 159. Initiation and Termination of Cyclic Nucleotide Action
*Edited by* JACKIE D. CORBIN AND ROGER A. JOHNSON

VOLUME 160. Biomass (Part A: Cellulose and Hemicellulose)
*Edited by* WILLIS A. WOOD AND SCOTT T. KELLOGG

VOLUME 161. Biomass (Part B: Lignin, Pectin, and Chitin)
*Edited by* WILLIS A. WOOD AND SCOTT T. KELLOGG

VOLUME 162. Immunochemical Techniques (Part L: Chemotaxis and Inflammation)
*Edited by* GIOVANNI DI SABATO

VOLUME 163. Immunochemical Techniques (Part M: Chemotaxis and Inflammation)
*Edited by* GIOVANNI DI SABATO

VOLUME 164. Ribosomes
*Edited by* HARRY F. NOLLER, JR., AND KIVIE MOLDAVE

VOLUME 165. Microbial Toxins: Tools for Enzymology
*Edited by* SIDNEY HARSHMAN

VOLUME 166. Branched-Chain Amino Acids
*Edited by* ROBERT HARRIS AND JOHN R. SOKATCH

VOLUME 167. Cyanobacteria
*Edited by* LESTER PACKER AND ALEXANDER N. GLAZER

VOLUME 168. Hormone Action (Part K: Neuroendocrine Peptides)
*Edited by* P. MICHAEL CONN

VOLUME 169. Platelets: Receptors, Adhesion, Secretion (Part A)
*Edited by* JACEK HAWIGER

VOLUME 170. Nucleosomes
*Edited by* PAUL M. WASSARMAN AND ROGER D. KORNBERG

VOLUME 171. Biomembranes (Part R: Transport Theory: Cells and Model Membranes)
*Edited by* SIDNEY FLEISCHER AND BECCA FLEISCHER

VOLUME 172. Biomembranes (Part S: Transport: Membrane Isolation and Characterization)
*Edited by* SIDNEY FLEISCHER AND BECCA FLEISCHER

VOLUME 173. Biomembranes [Part T: Cellular and Subcellular Transport: Eukaryotic (Nonepithelial) Cells]
*Edited by* SIDNEY FLEISCHER AND BECCA FLEISCHER

VOLUME 174. Biomembranes [Part U: Cellular and Subcellular Transport: Eukaryotic (Nonepithelial) Cells]
*Edited by* SIDNEY FLEISCHER AND BECCA FLEISCHER

VOLUME 175. Cumulative Subject Index Volumes 135–139, 141–167

VOLUME 176. Nuclear Magnetic Resonance (Part A: Spectral Techniques and Dynamics)
*Edited by* NORMAN J. OPPENHEIMER AND THOMAS L. JAMES

VOLUME 177. Nuclear Magnetic Resonance (Part B: Structure and Mechanism)
*Edited by* NORMAN J. OPPENHEIMER AND THOMAS L. JAMES

VOLUME 178. Antibodies, Antigens, and Molecular Mimicry
*Edited by* JOHN J. LANGONE

VOLUME 179. Complex Carbohydrates (Part F)
*Edited by* VICTOR GINSBURG

VOLUME 180. RNA Processing (Part A: General Methods)
*Edited by* JAMES E. DAHLBERG AND JOHN N. ABELSON

VOLUME 181. RNA Processing (Part B: Specific Methods)
*Edited by* JAMES E. DAHLBERG AND JOHN N. ABELSON

VOLUME 182. Guide to Protein Purification
*Edited by* MURRAY P. DEUTSCHER

VOLUME 183. Molecular Evolution: Computer Analysis of Protein and Nucleic Acid Sequences
*Edited by* RUSSELL F. DOOLITTLE

VOLUME 184. Avidin-Biotin Technology
*Edited by* MEIR WILCHEK AND EDWARD A. BAYER

VOLUME 185. Gene Expression Technology
*Edited by* DAVID V. GOEDDEL

VOLUME 186. Oxygen Radicals in Biological Systems (Part B: Oxygen Radicals and Antioxidants)
*Edited by* LESTER PACKER AND ALEXANDER N. GLAZER

VOLUME 187. Arachidonate Related Lipid Mediators
*Edited by* ROBERT C. MURPHY AND FRANK A. FITZPATRICK

VOLUME 188. Hydrocarbons and Methylotrophy
*Edited by* MARY E. LIDSTROM

VOLUME 189. Retinoids (Part A: Molecular and Metabolic Aspects)
*Edited by* LESTER PACKER

VOLUME 190. Retinoids (Part B: Cell Differentiation and Clinical Applications)
*Edited by* LESTER PACKER

VOLUME 191. Biomembranes (Part V: Cellular and Subcellular Transport: Epithelial Cells)
*Edited by* SIDNEY FLEISCHER AND BECCA FLEISCHER

VOLUME 192. Biomembranes (Part W: Cellular and Subcellular Transport: Epithelial Cells)
*Edited by* SIDNEY FLEISCHER AND BECCA FLEISCHER

VOLUME 193. Mass Spectrometry
*Edited by* JAMES A. MCCLOSKEY

VOLUME 194. Guide to Yeast Genetics and Molecular Biology
*Edited by* CHRISTINE GUTHRIE AND GERALD R. FINK

VOLUME 195. Adenylyl Cyclase, G Proteins, and Guanylyl Cyclase
*Edited by* ROGER A. JOHNSON AND JACKIE D. CORBIN

VOLUME 196. Molecular Motors and the Cytoskeleton
*Edited by* RICHARD B. VALLEE

VOLUME 197. Phospholipases
*Edited by* EDWARD A. DENNIS

VOLUME 198. Peptide Growth Factors (Part C)
*Edited by* DAVID BARNES, J. P. MATHER, AND GORDON H. SATO

VOLUME 199. Cumulative Subject Index Volumes 168–174, 176–194

VOLUME 200. Protein Phosphorylation (Part A: Protein Kinases: Assays, Purification, Antibodies, Functional Analysis, Cloning, and Expression)
*Edited by* TONY HUNTER AND BARTHOLOMEW M. SEFTON

VOLUME 201. Protein Phosphorylation (Part B: Analysis of Protein Phosphorylation, Protein Kinase Inhibitors, and Protein Phosphatases)
*Edited by* TONY HUNTER AND BARTHOLOMEW M. SEFTON

VOLUME 202. Molecular Design and Modeling: Concepts and Applications (Part A: Proteins, Peptides, and Enzymes)
*Edited by* JOHN J. LANGONE

VOLUME 203. Molecular Design and Modeling: Concepts and Applications (Part B: Antibodies and Antigens, Nucleic Acids, Polysaccharides, and Drugs)
*Edited by* JOHN J. LANGONE

VOLUME 204. Bacterial Genetic Systems
*Edited by* JEFFREY H. MILLER

VOLUME 205. Metallobiochemistry (Part B: Metallothionein and Related Molecules)
*Edited by* JAMES F. RIORDAN AND BERT L. VALLEE

VOLUME 206. Cytochrome P450
*Edited by* MICHAEL R. WATERMAN AND ERIC F. JOHNSON

VOLUME 207. Ion Channels
*Edited by* BERNARDO RUDY AND LINDA E. IVERSON

VOLUME 208. Protein–DNA Interactions
*Edited by* ROBERT T. SAUER

VOLUME 209. Phospholipid Biosynthesis
*Edited by* EDWARD A. DENNIS AND DENNIS E. VANCE

VOLUME 210. Numerical Computer Methods
*Edited by* LUDWIG BRAND AND MICHAEL L. JOHNSON

VOLUME 211. DNA Structures (Part A: Synthesis and Physical Analysis of DNA)
*Edited by* DAVID M. J. LILLEY AND JAMES E. DAHLBERG

VOLUME 212. DNA Structures (Part B: Chemical and Electrophoretic Analysis of DNA)
*Edited by* DAVID M. J. LILLEY AND JAMES E. DAHLBERG

VOLUME 213. Carotenoids (Part A: Chemistry, Separation, Quantitation, and Antioxidation)
*Edited by* LESTER PACKER

VOLUME 214. Carotenoids (Part B: Metabolism, Genetics, and Biosynthesis)
*Edited by* LESTER PACKER

VOLUME 215. Platelets: Receptors, Adhesion, Secretion (Part B)
*Edited by* JACEK J. HAWIGER

VOLUME 216. Recombinant DNA (Part G)
*Edited by* RAY WU

VOLUME 217. Recombinant DNA (Part H)
*Edited by* RAY WU

VOLUME 218. Recombinant DNA (Part I)
*Edited by* RAY WU

VOLUME 219. Reconstitution of Intracellular Transport
*Edited by* JAMES E. ROTHMAN

VOLUME 220. Membrane Fusion Techniques (Part A)
*Edited by* NEJAT DÜZGÜNEŞ

VOLUME 221. Membrane Fusion Techniques (Part B)
*Edited by* NEJAT DÜZGÜNEŞ

VOLUME 222. Proteolytic Enzymes in Coagulation, Fibrinolysis, and Complement Activation (Part A: Mammalian Blood Coagulation Factors and Inhibitors)
*Edited by* LASZLO LORAND AND KENNETH G. MANN

VOLUME 223. Proteolytic Enzymes in Coagulation, Fibrinolysis, and Complement Activation (Part B: Complement Activation, Fibrinolysis, and Nonmammalian Blood Coagulation Factors)
*Edited by* LASZLO LORAND AND KENNETH G. MANN

VOLUME 224. Molecular Evolution: Producing the Biochemical Data
*Edited by* ELIZABETH ANNE ZIMMER, THOMAS J. WHITE, REBECCA L. CANN, AND ALLAN C. WILSON

VOLUME 225. Guide to Techniques in Mouse Development
*Edited by* PAUL M. WASSARMAN AND MELVIN L. DEPAMPHILIS

VOLUME 226. Metallobiochemistry (Part C: Spectroscopic and Physical Methods for Probing Metal Ion Environments in Metalloenzymes and Metalloproteins)
*Edited by* JAMES F. RIORDAN AND BERT L. VALLEE

VOLUME 227. Metallobiochemistry (Part D: Physical and Spectroscopic Methods for Probing Metal Ion Environments in Metalloproteins)
*Edited by* JAMES F. RIORDAN AND BERT L. VALLEE

VOLUME 228. Aqueous Two-Phase Systems
*Edited by* HARRY WALTER AND GÖTE JOHANSSON

VOLUME 229. Cumulative Subject Index Volumes 195–198, 200–227

VOLUME 230. Guide to Techniques in Glycobiology
*Edited by* WILLIAM J. LENNARZ AND GERALD W. HART

VOLUME 231. Hemoglobins (Part B: Biochemical and Analytical Methods)
*Edited by* JOHANNES EVERSE, KIM D. VANDEGRIFF, AND ROBERT M. WINSLOW

VOLUME 232. Hemoglobins (Part C: Biophysical Methods)
*Edited by* JOHANNES EVERSE, KIM D. VANDEGRIFF, AND ROBERT M. WINSLOW

VOLUME 233. Oxygen Radicals in Biological Systems (Part C)
*Edited by* LESTER PACKER

VOLUME 234. Oxygen Radicals in Biological Systems (Part D)
*Edited by* LESTER PACKER

VOLUME 235. Bacterial Pathogenesis (Part A: Identification and Regulation of Virulence Factors)
*Edited by* VIRGINIA L. CLARK AND PATRIK M. BAVOIL

VOLUME 236. Bacterial Pathogenesis (Part B: Integration of Pathogenic Bacteria with Host Cells)
*Edited by* VIRGINIA L. CLARK AND PATRIK M. BAVOIL

VOLUME 237. Heterotrimeric G Proteins
*Edited by* RAVI IYENGAR

VOLUME 238. Heterotrimeric G-Protein Effectors
*Edited by* RAVI IYENGAR

VOLUME 239. Nuclear Magnetic Resonance (Part C)
*Edited by* THOMAS L. JAMES AND NORMAN J. OPPENHEIMER

VOLUME 240. Numerical Computer Methods (Part B)
*Edited by* MICHAEL L. JOHNSON AND LUDWIG BRAND

VOLUME 241. Retroviral Proteases
*Edited by* LAWRENCE C. KUO AND JULES A. SHAFER

VOLUME 242. Neoglycoconjugates (Part A)
*Edited by* Y. C. LEE AND REIKO T. LEE

VOLUME 243. Inorganic Microbial Sulfur Metabolism
*Edited by* HARRY D. PECK, JR., AND JEAN LEGALL

VOLUME 244. Proteolytic Enzymes: Serine and Cysteine Peptidases
*Edited by* ALAN J. BARRETT

VOLUME 245. Extracellular Matrix Components
*Edited by* E. RUOSLAHTI AND E. ENGVALL

VOLUME 246. Biochemical Spectroscopy
*Edited by* KENNETH SAUER

VOLUME 247. Neoglycoconjugates (Part B: Biomedical Applications)
*Edited by* Y. C. LEE AND REIKO T. LEE

VOLUME 248. Proteolytic Enzymes: Aspartic and Metallo Peptidases
*Edited by* ALAN J. BARRETT

VOLUME 249. Enzyme Kinetics and Mechanism (Part D: Developments in Enzyme Dynamics)
*Edited by* DANIEL L. PURICH

VOLUME 250. Lipid Modifications of Proteins
*Edited by* PATRICK J. CASEY AND JANICE E. BUSS

VOLUME 251. Biothiols (Part A: Monothiols and Dithiols, Protein Thiols, and Thiyl Radicals)
*Edited by* LESTER PACKER

VOLUME 252. Biothiols (Part B: Glutathione and Thioredoxin; Thiols in Signal Transduction and Gene Regulation)
*Edited by* LESTER PACKER

VOLUME 253. Adhesion of Microbial Pathogens
*Edited by* RON J. DOYLE AND ITZHAK OFEK

VOLUME 254. Oncogene Techniques
*Edited by* PETER K. VOGT AND INDER M. VERMA

VOLUME 255. Small GTPases and Their Regulators (Part A: Ras Family)
*Edited by* W. E. BALCH, CHANNING J. DER, AND ALAN HALL

VOLUME 256. Small GTPases and Their Regulators (Part B: Rho Family)
*Edited by* W. E. BALCH, CHANNING J. DER, AND ALAN HALL

VOLUME 257. Small GTPases and Their Regulators (Part C: Proteins Involved in Transport)
*Edited by* W. E. BALCH, CHANNING J. DER, AND ALAN HALL

VOLUME 258. Redox-Active Amino Acids in Biology
*Edited by* JUDITH P. KLINMAN

VOLUME 259. Energetics of Biological Macromolecules
*Edited by* MICHAEL L. JOHNSON AND GARY K. ACKERS

VOLUME 260. Mitochondrial Biogenesis and Genetics (Part A)
*Edited by* GIUSEPPE M. ATTARDI AND ANNE CHOMYN

VOLUME 261. Nuclear Magnetic Resonance and Nucleic Acids
*Edited by* THOMAS L. JAMES

VOLUME 280. Vitamins and Coenzymes (Part J)
*Edited by* DONALD B. MCCORMICK, JOHN W. SUTTIE, AND CONRAD WAGNER

VOLUME 281. Vitamins and Coenzymes (Part K)
*Edited by* DONALD B. MCCORMICK, JOHN W. SUTTIE, AND CONRAD WAGNER

VOLUME 282. Vitamins and Coenzymes (Part L)
*Edited by* DONALD B. MCCORMICK, JOHN W. SUTTIE, AND CONRAD WAGNER

VOLUME 283. Cell Cycle Control
*Edited by* WILLIAM G. DUNPHY

VOLUME 284. Lipases (Part A: Biotechnology)
*Edited by* BYRON RUBIN AND EDWARD A. DENNIS

VOLUME 285. Cumulative Subject Index Volumes 263, 264, 266–284, 286–289

VOLUME 286. Lipases (Part B: Enzyme Characterization and Utilization)
*Edited by* BYRON RUBIN AND EDWARD A. DENNIS

VOLUME 287. Chemokines
*Edited by* RICHARD HORUK

VOLUME 288. Chemokine Receptors
*Edited by* RICHARD HORUK

VOLUME 289. Solid Phase Peptide Synthesis
*Edited by* GREGG B. FIELDS

VOLUME 290. Molecular Chaperones
*Edited by* GEORGE H. LORIMER AND THOMAS BALDWIN

VOLUME 291. Caged Compounds
*Edited by* GERARD MARRIOTT

VOLUME 292. ABC Transporters: Biochemical, Cellular, and Molecular Aspects
*Edited by* SURESH V. AMBUDKAR AND MICHAEL M. GOTTESMAN

VOLUME 293. Ion Channels (Part B)
*Edited by* P. MICHAEL CONN

VOLUME 294. Ion Channels (Part C)
*Edited by* P. MICHAEL CONN

VOLUME 295. Energetics of Biological Macromolecules (Part B)
*Edited by* GARY K. ACKERS AND MICHAEL L. JOHNSON

VOLUME 296. Neurotransmitter Transporters
*Edited by* SUSAN G. AMARA

VOLUME 297. Photosynthesis: Molecular Biology of Energy Capture
*Edited by* LEE MCINTOSH

VOLUME 298. Molecular Motors and the Cytoskeleton (Part B)
*Edited by* RICHARD B. VALLEE

Volume 299. Oxidants and Antioxidants (Part A)
*Edited by* Lester Packer

Volume 300. Oxidants and Antioxidants (Part B)
*Edited by* Lester Packer

Volume 301. Nitric Oxide: Biological and Antioxidant Activities (Part C)
*Edited by* Lester Packer

Volume 302. Green Fluorescent Protein
*Edited by* P. Michael Conn

Volume 303. cDNA Preparation and Display
*Edited by* Sherman M. Weissman

Volume 304. Chromatin
*Edited by* Paul M. Wassarman and Alan P. Wolffe

Volume 305. Bioluminescence and Chemiluminescence (Part C)
*Edited by* Thomas O. Baldwin and Miriam M. Ziegler

Volume 306. Expression of Recombinant Genes in Eukaryotic Systems
*Edited by* Joseph C. Glorioso and Martin C. Schmidt

Volume 307. Confocal Microscopy
*Edited by* P. Michael Conn

Volume 308. Enzyme Kinetics and Mechanism (Part E: Energetics of Enzyme Catalysis)
*Edited by* Daniel L. Purich and Vern L. Schramm

Volume 309. Amyloid, Prions, and Other Protein Aggregates
*Edited by* Ronald Wetzel

Volume 310. Biofilms
*Edited by* Ron J. Doyle

Volume 311. Sphingolipid Metabolism and Cell Signaling (Part A)
*Edited by* Alfred H. Merrill, Jr., and Yusuf A. Hannun

Volume 312. Sphingolipid Metabolism and Cell Signaling (Part B)
*Edited by* Alfred H. Merrill, Jr., and Yusuf A. Hannun

Volume 313. Antisense Technology (Part A: General Methods, Methods of Delivery, and RNA Studies)
*Edited by* M. Ian Phillips

Volume 314. Antisense Technology (Part B: Applications)
*Edited by* M. Ian Phillips

Volume 315. Vertebrate Phototransduction and the Visual Cycle (Part A)
*Edited by* Krzysztof Palczewski

Volume 316. Vertebrate Phototransduction and the Visual Cycle (Part B)
*Edited by* Krzysztof Palczewski

Volume 335. Flavonoids and Other Polyphenols
*Edited by* Lester Packer

Volume 336. Microbial Growth in Biofilms (Part A: Developmental and Molecular Biological Aspects)
*Edited by* Ron J. Doyle

Volume 337. Microbial Growth in Biofilms (Part B: Special Environments and Physicochemical Aspects)
*Edited by* Ron J. Doyle

Volume 338. Nuclear Magnetic Resonance of Biological Macromolecules (Part A)
*Edited by* Thomas L. James, Volker Dötsch, and Uli Schmitz

Volume 339. Nuclear Magnetic Resonance of Biological Macromolecules (Part B)
*Edited by* Thomas L. James, Volker Dötsch, and Uli Schmitz

Volume 340. Drug–Nucleic Acid Interactions
*Edited by* Jonathan B. Chaires and Michael J. Waring

Volume 341. Ribonucleases (Part A)
*Edited by* Allen W. Nicholson

Volume 342. Ribonucleases (Part B)
*Edited by* Allen W. Nicholson

Volume 343. G Protein Pathways (Part A: Receptors)
*Edited by* Ravi Iyengar and John D. Hildebrandt

Volume 344. G Protein Pathways (Part B: G Proteins and Their Regulators)
*Edited by* Ravi Iyengar and John D. Hildebrandt

Volume 345. G Protein Pathways (Part C: Effector Mechanisms)
*Edited by* Ravi Iyengar and John D. Hildebrandt

Volume 346. Gene Therapy Methods
*Edited by* M. Ian Phillips

Volume 347. Protein Sensors and Reactive Oxygen Species (Part A: Selenoproteins and Thioredoxin)
*Edited by* Helmut Sies and Lester Packer

Volume 348. Protein Sensors and Reactive Oxygen Species (Part B: Thiol Enzymes and Proteins)
*Edited by* Helmut Sies and Lester Packer

Volume 349. Superoxide Dismutase
*Edited by* Lester Packer

Volume 350. Guide to Yeast Genetics and Molecular and Cell Biology (Part B)
*Edited by* Christine Guthrie and Gerald R. Fink

Volume 351. Guide to Yeast Genetics and Molecular and Cell Biology (Part C)
*Edited by* Christine Guthrie and Gerald R. Fink

Volume 352. Redox Cell Biology and Genetics (Part A)
*Edited by* Chandan K. Sen and Lester Packer

Volume 353. Redox Cell Biology and Genetics (Part B)
*Edited by* Chandan K. Sen and Lester Packer

Volume 354. Enzyme Kinetics and Mechanisms (Part F: Detection and Characterization of Enzyme Reaction Intermediates)
*Edited by* Daniel L. Purich

Volume 355. Cumulative Subject Index Volumes 321–354

Volume 356. Laser Capture Microscopy and Microdissection
*Edited by* P. Michael Conn

Volume 357. Cytochrome P450, Part C
*Edited by* Eric F. Johnson and Michael R. Waterman

Volume 358. Bacterial Pathogenesis (Part C: Identification, Regulation, and Function of Virulence Factors)
*Edited by* Virginia L. Clark and Patrik M. Bavoil

Volume 359. Nitric Oxide (Part D)
*Edited by* Enrique Cadenas and Lester Packer

Volume 360. Biophotonics (Part A)
*Edited by* Gerard Marriott and Ian Parker

Volume 361. Biophotonics (Part B)
*Edited by* Gerard Marriott and Ian Parker

Volume 362. Recognition of Carbohydrates in Biological Systems (Part A)
*Edited by* Yuan C. Lee and Reiko T. Lee

Volume 363. Recognition of Carbohydrates in Biological Systems (Part B)
*Edited by* Yuan C. Lee and Reiko T. Lee

Volume 364. Nuclear Receptors
*Edited by* David W. Russell and David J. Mangelsdorf

Volume 365. Differentiation of Embryonic Stem Cells
*Edited by* Paul M. Wassauman and Gordon M. Keller

Volume 366. Protein Phosphatases
*Edited by* Susanne Klumpp and Josef Krieglstein

Volume 367. Liposomes (Part A)
*Edited by* Nejat Düzgüneş

Volume 368. Macromolecular Crystallography (Part C)
*Edited by* Charles W. Carter, Jr., and Robert M. Sweet

Volume 369. Combinational Chemistry (Part B)
*Edited by* Guillermo A. Morales and Barry A. Bunin

Volume 370. RNA Polymerases and Associated Factors (Part C)
*Edited by* Sankar L. Adhya and Susan Garges

Volume 390. Regulators of G-protein Sgnalling (Part B)
*Edited by* David P. Siderovski

Volume 391. Liposomes (Part E)
*Edited by* Nejat Düzgüneş

Volume 392. RNA Interference
*Edited by* Engelke Rossi

Volume 393. Circadian Rhythms
*Edited by* Michael W. Young

Volume 394. Nuclear Magnetic Resonance of Biological Macromolecules (Part C)
*Edited by* Thomas L. James

Volume 395. Producing the Biochemical Data (Part B)
*Edited by* Elizabeth A. Zimmer and Eric H. Roalson

Volume 396. Nitric Oxide (Part E)
*Edited by* Lester Packer and Enrique Cadenas

Volume 397. Environmental Microbiology (in preparation)
*Edited by* Jared R. Leadbetter

Volume 398. Ubiquitin and Protein Degradation (Part A)
*Edited by* Raymond J. Deshaies

Volume 399. Ubiquitin and Protein Degradation (Part B) (in preparation)
*Edited by* Raymond J. Deshaies

# Section I

# E1 Enzyme

# [1] High-Level Expression and Purification of Recombinant E1 Enzyme

*By* Sylvie Beaudenon and Jon M. Huibregtse


## Abstract

The ubiquitin E1 enzyme is an ATP-dependent enzyme that activates ubiquitin for use in all ubiquitin conjugation pathways. This chapter describes the expression and purification of human E1 enzyme for use in *in vitro* ubiquitination reactions.


## Introduction

Most of the known enzymes involved in ubiquitin conjugation fall into what are known as E1, E2, and E3 families of proteins. These families of proteins are part of a pyramid-like organization of enzymes, with a lone E1 enzyme at the top, a limited number of E2s in the middle, and many E3s at the bottom. The E1 enzyme activates all E2 enzymes, and each E2 enzyme is likely to function with multiple E3 enzymes. Assuming each E3 recognizes multiple substrates, this organization is thought to at least partially account for the wide substrate diversity of the conjugation system. There are likely to be many complexities to this model and there are certainly additional protein components involved in some cases that do not fall into one of the three categories [e.g., E4 proteins (Hoppe *et al.,* 2004; Koegl *et al.,* 1999)]. Nevertheless, the overall model has been supported by many studies that have examined the fate of individual substrate proteins, including biochemical experiments that have reconstituted substrate ubiquitination with purified E1, E2, and E3 enzymes *in vitro*. This chapter focuses on the preparation of the E1 enzyme for such *in vitro* experiments.

Most organisms have a single gene encoding the ubiquitin E1 enzyme, including humans (UBE1) (Handley *et al.,* 1991) and yeast (UBA1) (McGrath *et al.,* 1991). This is almost certainly an essential gene in all eukaryotes. Protein isoforms of mammalian E1 enzymes may be present in some cell types (Cook and Chock, 1992; Shang *et al.,* 2001), as well as multiple phosphorylated states of E1 (Cook and Chock, 1995; Stephen *et al.,* 1996); however, biochemical characteristics have not been shown to differ among any of these alternative products. Phosphorylation has been reported to regulate localization of E1 to the nucleus (Stephen *et al.,* 1996, 1997).

METHODS IN ENZYMOLOGY, VOL. 398

0076-6879/05 $35.00
DOI: 10.1016/S0076-6879(05)98001-4

The enzymatic mechanism of the E1 enzyme is well understood and ha been reviewed by Huang *et al.* (2004). Briefly, E1 first binds ATP, followed by ubiquitin, and catalyzes the adenylation of the carboxyl terminus of ubiquitin, with release of inorganic pyrophosphate (PPi). The active-site cysteine thiolate then attacks the carboxyl group of the adenylate intermediate, forming a ubiquitin-thioester. The E1 adenylates another molecule of ubiquitin before it is capable of transferring ubiquitin to an E2 enzyme. The double ubiquitin-loaded E1 can interact with the full spectrum of ubiquitin E2 enzymes, and in a transthiolation reaction, ubiquitin is transferred from the active-site cysteine of the E1 to the active-site cysteine of the E2. While a structure of the ubiquitin E1 enzyme has proven elusive, further insights into the E1 mechanism have come from structural and biochemical analysis of the E1-like enzyme for Nedd8 (Huang *et al.,* 2004; Waldren *et al.,* 2003a,b), as well as analogies with the bacterial MoeB and ThiF enzymes. MoeB and ThiF function with structural homologs of ubiqutin (MoaD and ThiS) and play roles in bacterial molybdopterin and thiamin biosynthesis, respectively (Lake *et al.,* 2001; Rudolph *et al.,* 2001; Wang *et al.,* 2001).

## Expression and Purification of E1

Mechanistic studies of E1 have, for the most part, been performed on the E1 enzyme purified from rabbit reticulocyte lysate. Hershko and colleagues (1983) described the covalent affinity purification of the enzyme from reticulocytes on a ubiquitin-Sepharose column in the presence of ATP, starting with "fraction II," a crude DEAE high-salt fraction. Elution of the E1 enzyme from the column was with AMP and pyrophosphate. While this chapter describes options for the preparation of recombinant human E1 enzyme, the classical method of purification is relatively simple and remains a viable option for investigators who require E1 for *in vitro* studies. Ubiquitin-Sepharose for affinity purification is available commercially (Boston Biochem) and rabbit reticulocytes are an inexpensive and abundant source of the enzyme.

The first widely used recombinant source of ubiquitin-activating enzyme was the bacterially expressed wheat E1 enzyme (Hatfield *et al.,* 1990). This is an acceptable source of E1 and covalent affinity chromatography of this enzyme has been described (Tongaonkar and Madura, 1998). In addition, preparation of a high salt DEAE fraction from crude bacterial lysate-expressing wheat E1 yields acceptable E1 activity for many types of experiments (Scheffner *et al.,* 1993). We developed two baculovirus expression vectors for human E1 that offer some advantages over the wheat E1 expression system in terms of both yield and activity. The simplest system expresses human E1 as a glutathione *S*-transferase (GST) fusion protein in insect cells. The protein is affinity purified on glutathione-Sepharose, followed by cleavage and release of the E1 protein with a site-specific protease. The GST–E1 baculovirus was created by cloning the full-length human E1 (UBE1) open reading frame (ORF) into a pVL1393 baculovirus transfer vector that had been modified previously by insertion of the GST ORF and the multicloning region of pGEX-6p-1 (Amersham Pharmacia). This plasmid was used for the production of recombinant virus using the BaculGold cotransfection system (Pharmingen). High five insect cells (Invitrogen) are used for protein expression, and a typical preparation will utilize 5 to 10 15-cm plates of subconfluent cells, grown in Grace's insect cell media with 5% fetal bovine serum. The amount of virus to use per plate is determined empirically for each batch of amplified virus by titration in six-well plates, determining the minimum volume of virus stock necessary to achieve maximum protein yield. Cells are collected 48 h post-infection and lysed in Nonidet P-40 (NP-40) lysis buffer [100 m*M* Tris, pH 7.9, 100 m*M* NaCl, 0.1% NP-40, 1 m*M* dithiothreitol (DTT), 100 $\mu M$ phenylmethylsulfonyl fluoride], using 1 ml buffer per 15-cm plate. Debris is removed by centrifugation at 10,000*g*, 10 min, at 4°. The supernatant is rotated at 4° for 2 h in the presence of glutathione-Sepharose (25 $\mu$l bead volume per 15-cm plate). The beads are collected by centrifugation and washed four times with NP-40 lysis buffer. E1 protein is released from the beads by cleavage with PreScission protease (4 units, Pharmacia) for 5 h at 4° in a buffer containing 10 m*M* Tris–HCl, 50 m*M* NaCl, 0.01% Triton, and 1 m*M* DTT. The beads are removed by centrifugation, glycerol is added to a final concentration of 5%, and the purified E1 protein is aliquoted and stored at −80°. As PreScission protease is, itself, a GST fusion protein, it also binds to the glutathione beads and does not significantly contaminate the final soluble E1 protein. Figure 1A shows a Coomassie blue-stained SDS–PAGE gel of purified GST–E1 and the E1 protein present in the supernatant following protease digestion. The typical yield for this preparation is between 25 and 50 $\mu$g of E1 protein.

We have also created a recombinant virus that expresses native human E1 protein as a nonfusion protein. This is useful in two ways. First, if highly purified E1 is not required, a crude DEAE high salt fraction can be prepared from infected insect cells and used directly as a source of E1 enzyme. This fraction will contain both recombinant E1 and endogenous insect cell E1 enzyme, in addition to many other components of the ubiquitin system. Nevertheless, this crude fraction can be useful for certain types of *in vitro* assays, given appropriate controls. The DEAE fraction can be used for further purification on a ubiquitin affinity column in the presence of ATP, if desired (Hershko *et al.,* 1983). Again, it is expected that the final purified protein will contain both recombinant human E1 and

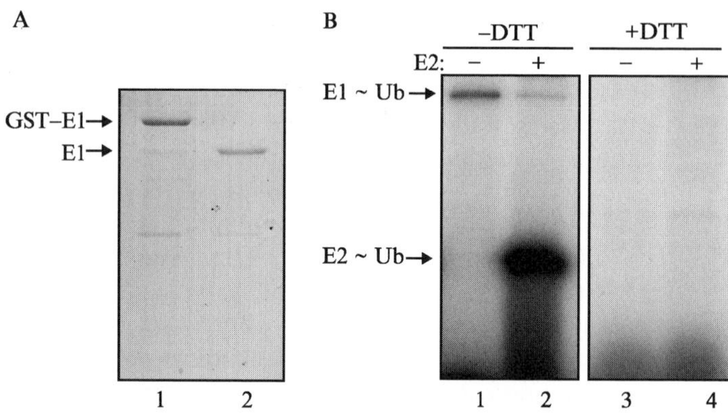


FIG. 1. (A) Coomassie blue-stained SDS–PAGE gel of purified GST–E1 (lane 1) and free E1 after protease digestion (lane 2). (B) Ubiquitin thioester assay with purified E1 enzyme (lanes 1 and 3) and E1 in the presence of the UbcH7 E2 enzyme (lanes 2 and 4). Ubiquitin was $^{32}$P labeled and adducts were detected by autoradiography. Reactions in lanes 1 and 2 were run in SDS–PAGE loading buffer without reducing agent, and lanes 3 and 4 were run in standard SDS–PAGE loading buffer with DTT.

insect cell E1 enzyme. For this reason, and to simplify the purification, we prefer to use the GST–E1 expression system for most *in vitro* reconstitution experiments.

## Assay of E1 Activity

Formation of a ubiquitin–thioester adduct at the active-site cysteine and its ability to transfer ubiquitin to E2 enzymes are the most direct tests of E1 activity. Figure 1B shows formation of the ubiquitin–thioester adduct on E1 using $^{32}$P-labeled ubiquitin. A $^{32}$P-labeled adduct is seen at the appropriate size for an E1–ubiquitin adduct when incubated in the presence of ATP, and this adduct disappears when treated with high concentration of DTT, as expected for a thioester-linked adduct. Figure 1B also shows an assay for E1-dependent E2 activation. In the presence of purified UbcH7 E2 protein, a DTT-sensitive adduct is formed at the appropriate size for E2–ubiquitin adducts, with a concomitant decrease in the amount of the E1 thioester. The labeled ubiquitin used in the aforementioned experiments was prepared from GST–ubiquitin that is expressed in pGEX6p with the addition of a heart muscle kinase (HMK) recognition site downstream of the PreScission protease cleavage site. The labeling reaction is performed according to previously published methods (Kaelin *et al.*, 1992), and the labeled ubiquitin is cleaved from GST with PreScission protease (Pharmacia). As an alternative to radioactively labeled ubiquitin, we prepare FLAG-tagged ubiquitin from a GST-6p-FLAG-ubiquitin protein, with detection by immunoblotting with the anti-FLAG antibody. Reactions in Fig. 1B contained 100 ng E1, 400 ng $^{32}$P-labeled ubiquitin ($\sim 10^6$ cpm), with or without 200 ng UbcH7 in 40 $\mu$l buffer containing 25 m$M$ Tris (pH 7.5), 50 m$M$ NaCl, 10 m$M$ $MgCl_2$, 5 m$M$ ATP, and 0.1 m$M$ DTT. Reactions are incubated for 10 min at room temperature and stopped with standard (DTT containing) SDS-PAGE loading buffer.

## Conclusions

Preparation of the active E1 enzyme is necessary for *in vitro* reconstitution of all ubiquitination reactions. The traditional method of covalent affinity chromatography, starting with either a cell extract from a mammalian cell extract or an extract from E1-expressing bacteria, has been described previously (Hershko *et al.*, 1983; Tongaonkar and Madura, 1998). The generation of E1 from a baculovirus-expressed GST–E1 described here represents a quick and simple alternative method for obtaining a highly purified E1 enzyme.

## Acknowledgment

This work was supported by a grant from the National Institutes of Health to J. M. H. (CA072943).

## References

Cook, J. C., and Chock, P. B. (1992). Isoforms of mammalian ubiquitin-activating enzyme. *J. Biol. Chem.* **267,** 24315–24321.

Cook, J. C., and Chock, P. B. (1995). Phosphorylation of ubiquitin-activating enzyme in cultured cells. *Proc. Natl. Acad. Sci. USA* **92,** 3454–3457.

Handley, P. M., Mueckler, M., Siegel, N. R., Ciechanover, A., and Schwartz, A. L. (1991). Molecular cloning, sequence, and tissue distribution of the human ubiquitin-activating enzyme E1. *Proc. Natl. Acad. Sci. USA* **88,** 258–262 [published erratum appears in *Proc. Natl. Acad. Sci. USA* **15** 88(16), 7456(1991)].

Hatfield, P. M., Callis, J., and Vierstra, R. D. (1990). Cloning of ubiquitin activating enzyme from wheat and expression of a functional protein in *Escherichia coli*. *J. Biol. Chem.* **265,** 15813–15817.

Hershko, A., Heller, H., Elias, S., and Ciechanover, A. (1983). Components of ubiquitin-protein ligase system: Resolution, affinity purification, and role in protein breakdown. *J. Biol. Chem.* **258,** 8206–8214.

Hoppe, T., Cassata, G., Barral, J. M., Springer, W., Hutagalung, A. H., Epstein, H. F., and Baumeister, R. (2004). Regulation of the myosin-directed chaperone UNC–45 by a novel E3/E4-multiubiquitylation complex in *C. elegans*. *Cell* **118,** 337–349.

Huang, D. T., Miller, D. W., Mathew, R., Cassell, R., Holton, J. M., Roussel, M. F., and Schulman, B. A. (2004). A unique E1-E2 interaction required for optimal conjugation of the ubiquitin-like protein NEDD8. *Nature Struct. Mol. Biol.* **11,** 927–935.

Huang, D. T., Walden, H., Duda, D., and Schulman, B. A.. (2004). Ubiquitin-like protein activation. *Oncogene* **23,** 1958–1971.

Kaelin, W. G., Jr., Krek, W., Sellers, W. R., De Caprio, J. A., Ajchenbaum, F., Fuchs, C. S., Chittenden, T., Li, Y., Farnham, P. J., Blanar, M. A., Livingston, D. M., and Flemington, E. K. (1992). Expression cloning of a cDNA encoding a retinoblastoma-binding protein with E2F-like properties. *Cell* **70,** 351–364.

Koegl, M., Hoppe, T., Schlenker, S., Ulrich, H. D., Mayer, T. U., and Jentsch, S. (1999). A novel ubiquitination factor, E4, is involved in multiubiquitin chain assembly. *Cell* **96,** 635–644.

Lake, M. W., Wuebbens, M. M., Rajagopalan, K. V., and Schindelin, H. (2001). Mechanism of ubiquitin activation revealed by the structure of a bacterial MoeB–MoaD complex. *Nature* **414,** 325–329.

McGrath, J. P., Jentsch, S., and Varshavsky, A. (1991). UBA 1: An essential yeast gene encoding ubiquitin-activating enzyme. *EMBO J.* **10,** 227–236.

Rudolph, M. J., Wuebbens, M. M., Rajagopalan, K. V., and Schindelin, H. (2001). Crystal structure of molybdopterin synthase and its evolutionary relationship to ubiquitin activation. *Nature Struct. Biol.* **8,** 42–46.

Scheffner, M., Huibregtse, J. M., Vierstra, R. D., and Howley, P. M. (1993). The HPV–16 E6 and E6-AP complex functions as a ubiquitin-protein ligase in the ubiquitination of p53. *Cell* **75,** 495–505.

Shang, F., Deng, G., Obin, M., Wu, C. C., Gong, X., Smith, D., Laursen, R. A., Andley, U. P., Reddan, J. R., and Taylor, A. (2001). Ubiquitin-activating enzyme (E1) isoforms in lens epithelial cells: Origin of translation, E2 specificity and cellular localization determined with novel site-specific antibodies. *Exp. Eye Res.* **73,** 827–836.

Stephen, A. G., Trausch-Azar, J. S., Ciechanover, A., and Schwartz, A. L. (1996). The ubiquitin-activating enzyme E1 is phosphorylated and localized to the nucleus in a cell cycle-dependent manner. *J. Biol. Chem.* **271,** 15608–15614.

Stephen, A. G., Trausch-Azar, J. S., Handley-Gearhart, P. M., Ciechanover, A., and Schwartz, A. L. (1997). Identification of a region within the ubiquitin-activating enzyme required for nuclear targeting and phosphorylation. *J. Biol. Chem.* **272,** 10895–10903.

Tongaonkar, P., and Madura, K. (1998). Reconstituting ubiquitination reactions with affinity-purified components and 32P-ubiquitin. *Anal. Biochem.* **260,** 135–141.

Walden, H., Podgorski, M. S., Huang, D. T., Miller, D. W., Howard, R. J., Minor, D. L., Jr., Holton, J. M., and Schulman, B. A. (2003a). The structure of the APPBP1-UBA3-NEDD8-ATP complex reveals the basis for selective ubiquitin-like protein activation by an E1. *Mol. Cell* **12,** 1427–1437.

Walden, H., Podgorski, M. S., and Schulman, B. A. (2003b). Insights into the ubiquitin transfer cascade from the structure of the activating enzyme for NEDD8. *Nature* **422,** 330–334.

Wang, C., Xi, J., Begley, T. P., and Nicholson, L. K. (2001). Solution structure of ThiS and implications for the evolutionary roots of ubiquitin. *Nature Struct. Biol.* **8,** 47–51.

# [2] Expression, Purification, and Characterization of the E1 for Human NEDD8, the Heterodimeric APPBP1–UBA3 Complex

*By* DANNY T. HUANG and BRENDA A. SCHULMAN

## Abstract

The NEDD8 pathway is important for numerous biological processes, including cell proliferation, signal transduction, and development. The heterodimeric activating enzyme of NEDD8, APPBP1–UBA3, plays an essential role in NEDD8 conjugation. Not surprisingly, mutations in APPBP1 and UBA3 lead to defects in many cellular functions. The APPBP1–UBA3 complex initiates NEDD8 conjugation by first catalyzing adenylation of the C terminus of NEDD8 and ultimately catalyzing transfer of NEDD8 to the downstream enzyme in the pathway, Ubc12. This chapter describes methods for expressing and purifying APPBP1–UBA3 for *in vitro* studies of NEDD8 conjugation.

## Introduction

NEDD8 is an ubiquitin-like protein with ~60% sequence identity to ubiquitin (Kumar *et al.*, 1993). The best characterized targets of NEDD8 are members of the cullin family of proteins (Lammer *et al.*, 1998; Liakopoulos *et al.*, 1998; Osaka *et al.*, 1998). Cullin family members are subunits of SCF (Skp1-Cul1-F-box) and SCF-like E3 ubiquitin ligase complexes. SCF complexes regulate many important cellular events by polyubiquitinating key regulatory proteins to target them for degradation by the proteasome. For example, different SCF complexes regulate the G1–S transition of the cell cycle, activation of transcription factors, and signal transduction pathways (Bai *et al.*, 1996; Feldman *et al.*, 1997; Jiang and Struhl, 1998; Skowyra *et al.*, 1997; Yaron *et al.*, 1998). While SCF complexes contain the cullin family member Cul1, other related ubiquitin E3 ligases contain other cullin family members, such as Cul2, Cul3, Cul4A, Cul4B, or Cul5 (Hori *et al.*, 1999; Pan *et al.*, 2004). NEDD8 modification of Cul1 and other cullins serves to activate SCF-like E3s in two ways. First, NEDD8 displaces the SCF inhibitor CAND1 (Liu *et al.*, 2002; Zheng *et al.*, 2002). Second, NEDD8 increases the E3 ubiquitin ligase activity of SCF complexes (Furukawa *et al.*, 2000; Morimoto *et al.*, 2000; Read *et al.*, 2000;

METHODS IN ENZYMOLOGY, VOL. 398

0076-6879/05 $35.00
DOI: 10.1016/S0076-6879(05)98002-6

Wu *et al.*, 2000). NEDD8 has also been found to modify and negatively regulate the activities of MDM2 and p53 (Xirodimas *et al.*, 2004). Given the broad functions of NEDD8 modification, it is not surprising that the NEDD8 pathway is essential for viability in organisms ranging from fission yeast to mammals (Osaka *et al.*, 2000; Tateishi *et al.*, 2001).

Posttranslational modification by NEDD8, like ubiquitin, requires sequential action of an E1-activating enzyme, an E2-conjugating enzyme, and an E3 ligase (Gong and Yeh, 1999; Kamura *et al.*, 1999; Liakopoulos *et al.*, 1998, 1999; Osaka *et al.*, 1998). The E1 for human NEDD8 is a heterodimeric complex, consisting of APPBP1 (molecular mass ≈60 kDa) and UBA3 (molecular mass ≈49 kDa). APPBP1 and UBA3 are homologous to the amino and carboxyl regions of the ubiquitin-activating enzyme, respectively. APPBP1–UBA3 first catalyzes the adenylation of C-terminal Gly76 of NEDD8 in the presence of $Mg^{2+}$ and ATP and subsequently forms a thioester intermediate between its catalytic cysteine (Cys216 on UBA3) and the C terminus of NEDD8. The APPBP1–UBA3 complex ultimately binds the E2 of NEDD8, Ubc12, and promotes the transfer of NEDD8 onto the catalytic cysteine of Ubc12.

The biological functions of APPBP1 and UBA3, and their orthologs, have been revealed by genetic studies. RNA interference experiments knocking down levels of the *Caenorhabditis elegans* orthologs of APPBP1 and UBA3, and targeted deletion of the UBA3 gene in mice, lead to embryonic lethality (Jones *et al.*, 2002; Tateishi *et al.*, 2001). Mutations in the AXR1 and ECR1 genes in *Arabidopsis thaliana*, the orthologs of human APPBP1 and UBA3, respectively, result in defects in auxin signaling (Leyser *et al.*, 1993; Pozo *et al.*, 1998). Moreover, hamster cells expressing temperature-sensitive APPBP1 are defective in coupling DNA synthesis and mitosis at the nonpermissive temperature (Chen *et al.*, 2000; Handeli and Weintraub, 1992), and *C. elegans* embryos expressing temperature-sensitive UBA3 have myriad defects in cytokinesis at the nonpermissive temperature (Kurz *et al.*, 2002). Because of the widespread functional importance of APPBP1, UBA3, and the NEDD8 pathway, purified proteins for reconstituting NEDD8 activation would be useful.

This chapter describes (1) a cloning strategy that allows coexpression and affinity purification of APPBP1 and UBA3 from *Escherichia coli*; (2) a detailed procedure for protein expression; (3) a step-by-step method for purification that yields approximately 200 $\mu$g of pure APPBP1–UBA3 complex from 1 liter of *E. coli* culture; and (4) a method for assaying APPBP1–UBA3 activity.

## Plasmid Construction for Expressing Recombinant Human APPBP1–UBA3

### *Generation of pGSTHsAPPBP1rbsUBA3*

APPBP1 and UBA3 are coexpressed in bacteria from a bicistronic plasmid, named pGSTHsAPPBP1rbsUBA3 (Fig. 1), where both APPBP1 and UBA3 are expressed from the same message. pGSTHsAPPBP1rbs-UBA3 can be derived from combining parts of two independent vectors, one expressing APPBP1 as a GST fusion and the other expressing UBA3 in untagged form, using a procedure described here.

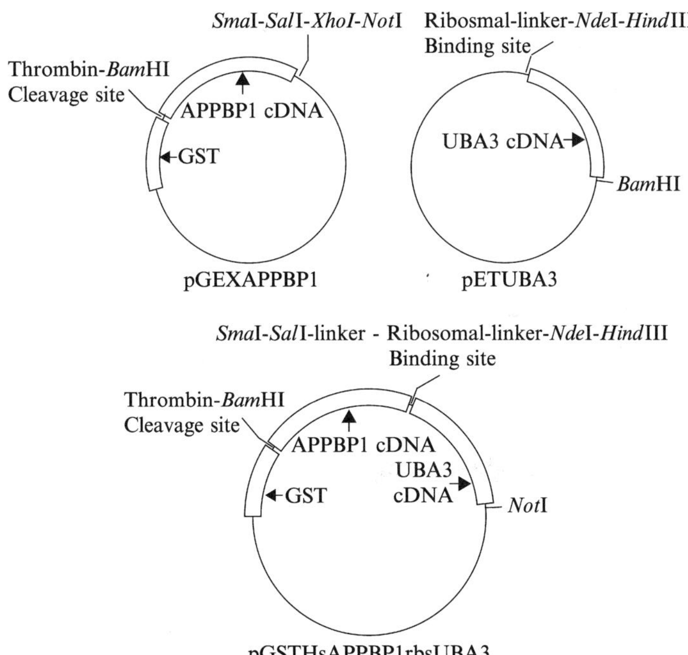


FIG. 1. Generation of pGSTHsAPPBP1rbsUBA3 bicistronic coexpression vector for APPBP1–UBA3. (Top left) Generate pGEXAPPBP1 by subcloning APPBP1 cDNA between the *Bam*HI and the *Sma*I sites of pGEX4T3. (Top right) Generate pETUBA3 by subcloning UBA3 cDNA between the *Nde*I and the *Bam*HI sites of pET3a. Note that it is important to insert a *Hind*III site between the *Nde*I site and the UBA3 cDNA to allow swapping the UBA3 cDNA in and out of the final vector. (Bottom) The final pGSTHsAPPBP1rbsUBA3 vector is generated by subcloning a PCR product containing the ribosomal binding site (rbs) and UBA3 cDNA amplified from pET3a between the *Sal*I and the *Not*I sites of pGEXAPPBP1.

*Materials*

1. Expression vectors
   a. pGEX4T1, pGEX4T2 or pGEX4T3 (Amersham Biosciences), or a similar variant
   b. pET3a (Novagen)
2. cDNAs
   a. Human APPBP1: we use IMAGE Clone 2964638 obtained from Research Genetics/Incyte Genomics
   b. Human UBA3: we use IMAGE Clone 2371074 obtained from Research Genetics/Incyte Genomics
3. Oligonucleotides for polymerase chain reaction (PCR) amplification of inserts for subcloning
   a. APPBP1 left *Bam*HI: cgcggatccatggcgcagctgggaaagctgctcaag
   b. APPBP1 right *Sma*I: tcccccgggctacaactggaaagttgctgaagtttg
   c. UBA3 left *Nde1Hind*III: ggaattccatatgaagcttatggctgttgatggtggg-tgtgggg
   d. UBA3 right *Bam*HI: cgcggatccttaagaagtaaaatgaagtttgaatag
   e. Bicistronic left *Sal*I: acgcgtcgactcgagtagaaataattttgtttaactttaagaag
   f. Bicistronic right *Not*I: atagtttagcggccgcttaagaagtaaaatgaagtttgaatag

*Procedure for Generating Bicistronic APPBP1–UBA3 Expression Vector*

1. pGEXAPPBP1: PCR amplify APPBP1 cDNA using oligonucleotides a and b and subclone into the *Bam*HI and *Sma*I restriction sites of the pGEX4T3 vector.

2. pETUBA3: PCR amplify UBA3 cDNA using oligonucleotides c and d and subclone into the *Nde*I and *Bam*HI restriction sites of the pET3a vector. Note that a *Hind*III site is inserted between the *Nde*I site and UBA3 to allow swapping mutant versions of UBA3 in and out of the final vector.

3. pGSTHsAPPBP1rbsUBA3: PCR amplify the ribosomal binding site (rbs) and UBA3 cDNA from pETUBA3 using oligonucleotides e and f and subclone into the *Sal*I and *Not*I sites of pGEXAPPBP1. In this final vector, the APPBP1 cDNA is flanked by the unique restriction sites *Bam*HI and *Sal*I and UBA3 by *Hind*III and *Not*I.

## Expression of APPBP1–UBA3

The expression and purification described here are for 1 liter of *E. coli* culture and can be readily scaled-up to obtain larger amounts of APPBP1–UBA3.

1. Transform pGSTHsAPPBP1rbsUBA3 into BL21-Gold(DE3) *E. coli* (Stratagene). Follow manufacturer's instructions and plate on LB agar containing 200 μg/ml ampicillin. Allow colony formation at 37°.
2. Grow overnight culture by inoculating 4 ml of LB containing 200 μg/ml ampicillin with a single colony. Shake culture at 37° for 10–16 h.
3. Grow large culture in a 2-liter flask containing 1 liter LB with 200 μg/ml ampicillin. Inoculate with 4 ml of overnight culture. Grow cells at 37° and monitor closely by measuring the absorbance at 600 nm. Add isopropyl-β-D-thiogalactoside (IPTG) at a final concentration of 0.6 m*M* when the $OD_{600}$ reaches 0.8–1.0. After adding IPTG, transfer cells to a 16° shaking incubator and grow for 16 h.

While the APPBP1–UBA3 complex can be induced at $OD_{600}$ up to 1.5 and can be expressed at temperatures up to 37°, we find that induction at this $OD_{600}$ and expression at 16° consistently gives a better level of protein expression.

## Purification of APPBP1–UBA3

All purification steps are performed in a 4° cold room. Protein and buffers are kept at 4° or on ice.

### *Preparation of Cell Lysate*

1. Harvest cells by centrifugation at 4500 rpm for 15 min at 4°.
2. Resuspend cell pellet in 8 ml of lysis buffer [50 m*M* Tris–HCl, 0.2 *M* NaCl, 2.5 m*M* phenylmethysulfonyl fluoride (PMSF), 5 m*M* dithiothreitol (DTT), pH 7.6]. This should give a final volume of about 12 ml, including the cells.
3. Lyse cells by sonication, 5 × 12 s with intermittent cooling on ice.
4. Clear lysate by centrifugation at 15,000 rpm for 25 min at 4°. Pour the lysate into a new centrifuge tube and spin a second time at 15,000 rpm for 25 min at 4°. Keep lysate on ice.

### *Glutathione-Affinity Chromatography*

1. Prepare 400 μl of glutathione (GSH)-Sepharose 4B beads (Amersham Biosciences) by washing in 14 ml of wash buffer (50 m*M* Tris–HCl, 0.2 *M* NaCl, 5 m*M* DTT, pH 7.6) in a 15-ml tube. Centrifuge at 1500 rpm for 5 min at 4° and then decant the wash buffer. Repeat so that the beads are equilibrated in wash buffer.

2. Incubate lysate with the washed GSH-Sepharose beads in a 15-ml tube. Mix gently by inversion and then rock gently at 4° for 1 h.
3. Make a small affinity column by pouring the lysate and beads into a Poly-Prep chromatography column (0.8 × 4.0 cm; Bio-Rad). Rinse the 15-ml tube with wash buffer and pour rinse into the column to recover all the beads. Allow lysate to flow through the column over 1–2 h.
4. Wash the column with 2 ml of wash buffer. Repeat three times.
5. Elution buffer is 50 m*M* Tris–HCl, 0.2 *M* NaCl, 5 m*M* DTT, 10 m*M* reduced glutathione, adjusted to a final pH between 7.6 and 8.0. We elute in three fractions by pipetting a fixed volume of elution buffer onto the column while simultaneously collecting the fraction in a microfuge tube. Add 300 $\mu$l elution buffer and collect fraction 1. Add 450 $\mu$l elution buffer and collect fraction 2. Add 450 $\mu$l elution buffer and collect fraction 3. The majority of GST–APPBP1–UBA3 elutes in fraction 2 with a small amount in fraction 3 (Fig. 2, lanes 1–3).
6. Pool fractions 2 and 3. Determine the total protein content using the Bio-Rad protein assay, with bovine serum albumin as standard. The total protein content is usually in the range of 2–3 mg/liter of culture.

### *Thrombin Cleavage of GST–APPBP1–UBA3*

1. Add $CaCl_2$ to the pooled fraction of GST–APPBP1–UBA3 to a final concentration of 2.5 m*M*.

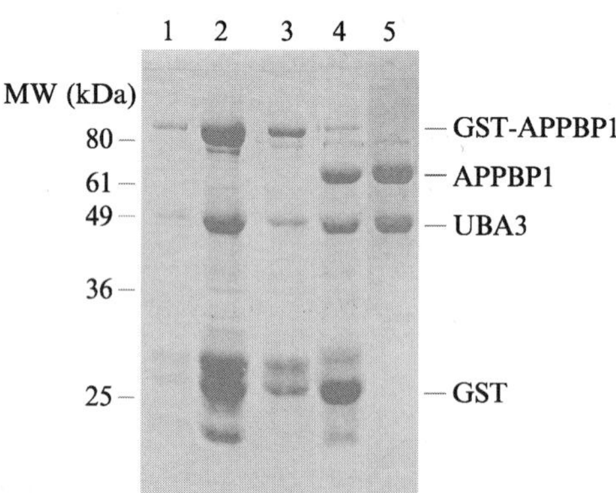


FIG. 2. A 15% SDS–PAGE showing APPBP1–UBA3 after each purification step. Lanes 1–3 correspond to fractions 1–3 eluted from the initial GSH-Sepharose affinity purification. Lane 4 shows the result of thrombin cleavage of combined fractions 2 and 3 from initial GSH-Sepharose affinity purification after a ~16-h incubation at 4°. Lane 5 contains the final purified product: Peak B from gel filtration chromatography using a Superdex 200 column was passed through a second GSH-Sepharose column and was concentrated prior to loading on the gel.

2. Add thrombin at a ratio of 1 mg of thrombin per 100 mg of total protein in the pooled fraction. Allow the cleavage to proceed on ice for 12–24 h. We use thrombin from Sigma (Cat. No. T 4648) dissolved in 25 m*M* Tris–HCl, 0.2 *M* NaCl, 5 m*M* DTT, pH 7.6, at 1–2 mg/ml. Other brands of thrombin should work as well.
3. Check the thrombin cleavage by SDS–PAGE with a 15% acrylamide gel. As shown in Fig. 2 (lane 4), GST–APPBP1 should be cleaved to yield GST and APPBP1 as separate polypeptides.
4. Inhibit thrombin activity with PMSF (dissolved at 0.2 *M* in ethanol) at a final concentration of 1 m*M*.

### *Gel Filtration Chromatography*

Further purification is achieved by gel filtration chromatography using a Superdex 200 column (10 $\times$ 300 mm; Amersham Biosciences).

1. Equilibrate the Superdex 200 column on an AKTA FPLC system (Amersham Biosciences) with storage buffer (25 m*M* HEPES, 0.15 *M* NaCl, 1 m*M* DTT, pH 7.0).
2. Clarify protein prior to loading on gel filtration column by microcentrifugation at 13,000 rpm for 10 min at 4°.
3. Load entire protein sample ($\sim$950 $\mu$l) onto Superdex 200 column and run column in storage buffer, collecting 0.5-ml fractions. Aggregates elute as peak A, APPBP1–UBA3 as peak B, GST as peak C, and buffer components in peak D in the elution profile shown in Fig. 3.
4. Pool fractions corresponding to peak B. At this point, the concentration of APPBP1–UBA3 is so low that detection is difficult. Therefore, we analyze concentration and purity at the end of the preparation (see later).

### *Remove GST Contaminant with a Second Glutathione-Affinity Chromatography Step*

The elution profiles (Fig. 3) of APPBP1–UBA3 (peak B) and GST (peak C) partially overlap during gel filtration chromatography, but the proteins are readily separated from reduced glutathione (peak D). Therefore, any remaining GST contaminants can be removed from APPBP1–UBA3 by a second round of glutathione-affinity chromatography, during which GST binds the column and the APPBP1–UBA3 complex flows through.

1. Regenerate the used GSH-Sepharose column by washing with 10 ml of elution buffer. Then equilibrate with 10 ml of storage buffer.

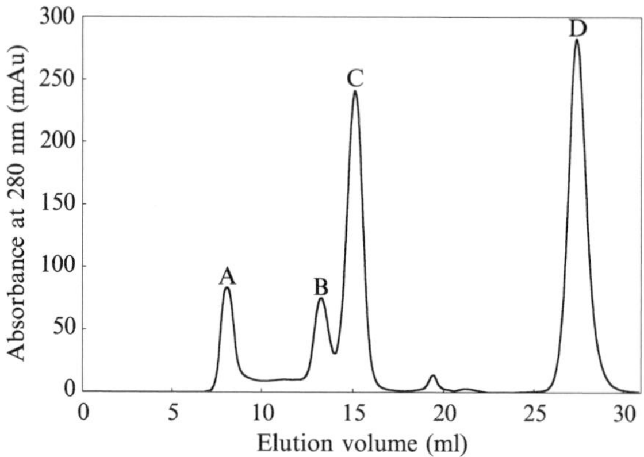


FIG. 3. Superdex 200 elution profile after thrombin cleavage. Peak A, void volume; peak B, APPBP1–UBA3; peak C, GST; and peak D, small molecules including reduced glutathione.

2. Load pooled fractions corresponding to peak B onto the GSH-Sepharose column and collect the entire flow through. Add another 500 $\mu$l of storage buffer and collect to wash out all the APPBP1–UBA3. GST and any uncleaved GST–APPBP1–UBA3 are retained by the GSH-Sepharose column.

### *Concentrating and Storing APPBP1–UBA3*

Concentrate the purified APPBP1–UBA3 to 1.0 mg/ml with an Amicon Centricon centrifugal filter device (10,000-Da molecular mass cutoff) (Millipore) according to the manufacturer's instructions. The protein is now sufficiently concentrated to be analyzed by SDS–PAGE and is sufficiently pure for enzymatic and X-ray crystallographic studies (Fig. 2, lane 5). Store APPBP1–UBA3 in 10- to 15-$\mu$l aliquots at $-80°$ after flash-freezing in liquid nitrogen. Thaw on ice prior to use.

## APPBP1–UBA3–Ubc12 Transthiolation Assay

Following purification, the activity of APPBP1-UBA3 can be assayed readily through a variety of approaches (Bohnsack and Haas, 2003; Huang *et al.*, 2004; Walden *et al.*, 2003a,b). This section outlines a convenient assay for monitoring APPBP1–UBA3-mediated transfer of NEDD8 to Ubc12.

*Materials*

*$^{32}$P-Labeled NEDD8*

a. Subclone NEDD8 cDNA into pGEX-2TK (Amersham Biosciences), which adds a protein kinase A phosphorylation site (GSRRASV) sequence at the N terminus.
b. Express and purify 2TK-NEDD8 in roughly the same manner as APPBP1–UBA3 described here, but with a significantly higher yield (>5 mg/liter *E. coli* culture).
c. Phosphorylate 2TK-NEDD8 by incubating 15 $\mu$g NEDD8, 30 units protein kinase A (Sigma, P2645), 3 $\mu$l [$\gamma$-$^{32}$P]ATP (6000 Ci/mmol) in 15 m*M* Tris–HCl, 0.1 *M* NaCl, 12 m*M* $MgCl_2$, 1 m*M* DTT, pH 7.6, in a total volume of 25 $\mu$l for 2–3 h at room temperature.
d. Store at 4°.

*Ubc12.* Prepare Ubc12 according to the protocol by Chiba (2005) or by an analogous method to that described for APPBP1–UBA3 or NEDD8.

*Procedure for E1-E2 Transthiolation Assay*

1. Perform the reaction at 18° in a 10-$\mu$l volume by mixing 1 n*M* APPBP1–UBA3, 5 $\mu M$ $^{32}$P-labeled NEDD8, and 50 $\mu M$ Ubc12 in 50 m*M* Tris–HCl, 50 m*M* NaCl, 10 m*M* $MgCl_2$, 5 m*M* ATP, 1 m*M* DTT, 0.3 U/ml inorganic pyrophosphatase, 0.3 U/ml creatine phosphatase, 5 m*M* creatine phosphate, and 2 mg/ml ovalbumin, pH 7.6. Quench the reaction with an equal volume of 2× SDS sample buffer after 30 s.
2. Resolve proteins by SDS–PAGE. We use 15% acrylamide gels.
3. Dry the gel and visualize the $^{32}$P-labeled NEDD8 and Ubc12~NEDD8 thioester by autoradiography. We typically observe a strong band after exposing the gel to film for 10 h at room temperature.

## Concluding Remarks

The small-scale (1 liter) expression and purification of APPBP1–UBA3 described here yields approximately 200 $\mu$g of pure enzyme complex. The preparation can be scaled up to 48 liters to obtain milligram quantities to be concentrated to 15–30 mg/ml for X-ray crystallographic studies.

We find that aliquots of APPBP1–UBA3 stored at −80° are stable for up to a year, giving consistent results in NEDD8 adenylation assays, Ubc12~NEDD8 thioester formation assays, and in crystallization trials (Huang *et al.*, 2004; Walden *et al.*, 2003a,b).

## Acknowledgments

Work in the Schulman laboratory is supported by American Lebanese Syrian Associated Charities (ALSAC), the NIH (P30CA21765 NCI Cancer Center Core grant to St. Jude, R01GM69530 to BAS), the Phillip and Elizabeth Gross Foundation, a Pew Scholar in Biomedical Sciences Award, and a Beckman Young Investigator Award.

## References

Bai, C., Sen, P., Hofmann, K., Ma, L., Goebl, M., Harper, J. W., and Elledge, S. J. (1996). SKP1 connects cell cycle regulators to the ubiquitin proteolysis machinery through a novel motif, the F-box. *Cell* **86,** 263–274.

Bohnsack, R. N., and Haas, A. L. (2003). Conservation in the mechanism of Nedd8 activation by the human AppBp1-Uba3 heterodimer. *J. Biol. Chem.* **278,** 26823–26830.

Chen, Y., McPhie, D. L., Hirschberg, J., and Neve, R. L. (2000). The amyloid precursor protein-binding protein APP-BP1 drives the cell cycle through the S-M checkpoint and causes apoptosis in neurons. *J. Biol. Chem.* **275,** 8929–8935.

Chiba, T. (2005). *In vitro* systems for Nedd8 conjugation by Ubc12. *Methods Enzymol.* **398**(7), 2005 (this volume).

Feldman, R. M., Correll, C. C., Kaplan, K. B., and Deshaies, R. J. (1997). A complex of Cdc4p, Skp1p, and Cdc53p/cullin catalyzes ubiquitination of the phosphorylated CDK inhibitor Sic1p. *Cell* **91,** 221–230.

Furukawa, M., Zhang, Y., McCarville, J. T. O., and Xiong, Y. (2000). The CUL1 C-terminal sequence and ROC1 are required for efficient nuclear accumulation, NEDD8 modification, and ubiquitin ligase activity of CUL1. *Mol. Cell. Biol.* **20,** 8185–8197.

Gong, L., and Yeh, E. T. (1999). Identification of the activating and conjugating enzymes of the NEDD8 conjugation pathway. *J. Biol. Chem.* **274,** 12036–12042.

Handeli, S., and Weintraub, H. (1992). The ts41 mutation in Chinese hamster cells leads to successive S phases in the absence of intervening G2, M, and G1. *Cell* **71,** 599–611.

Hori, T., Osaka, F., Chiba, T., Miyamoto, C., Okabayashi, K., Shimbara, N., Kato, S., and Tanaka, K. (1999). Covalent modification of all members of human cullin family proteins by NEDD8. *Oncogene* **18,** 6829–6834.

Huang, D. T., Miller, D. W., Mathew, R., Cassell, R., Holton, J. M., Roussel, M. F., and Schulman, B. A. (2004). A unique E1-E2 interaction required for optimal conjugation of the ubiquitin-like protein NEDD8. *Nature Struct. Mol. Biol.* **11,** 927–935.

Jiang, J., and Struhl, G. (1998). Regulation of the Hedgehog and Wingless signalling pathways by the F-box/WD40-repeat protein Slimb. *Nature* **391,** 493–496.

Jones, D., Crowe, E., Stevens, T. A., and Candido, E. P. (2002). Functional and phylogenetic analysis of the ubiquitylation system in *Caenorhabditis elegans*: Ubiquitin-conjugating enzymes, ubiquitin-activating enzymes, and ubiquitin-like proteins. *Genome Biol.* **3,** 0002.1–0002.15.

Kamura, T., Conrad, M. N., Yan, Q., Conaway, R. C., and Conaway, J. W. (1999). The Rbx1 subunit of SCF and VHL E3 ubiquitin ligase activates Rub1 modification of cullins Cdc53 and Cul2. *Genes Dev.* **13,** 2928–2933.

Kumar, S., Yoshida, Y., and Noda, M. (1993). Cloning of a cDNA which encodes a novel ubiquitin-like protein. *Biochem. Biophys. Res. Commun.* **195,** 393–399.

Kurz, T., Pintard, L., Willis, J. H., Hamill, D. R., Gonczy, P., Peter, M., and Bowerman, B. (2002). Cytoskeletal regulation by the Nedd8 ubiquitin-like protein modification pathway. *Science* **295,** 1294–1298.

Lammer, D., Mathias, N., Laplaza, J. M., Jiang, W., Liu, Y., Callis, J., Goebl, M., and Estelle, M. (1998). Modification of yeast Cdc53p by the ubiquitin-related protein rub1p affects function of the SCFCdc4 complex. *Genes Dev.* **12,** 914–926.

Leyser, H. M., Lincoln, C. A., Timpte, C., Lammer, D., Turner, J., and Estelle, M. (1993). Arabidopsis auxin-resistance gene AXR1 encodes a protein related to ubiquitin-activating enzyme E1. *Nature* **364,** 161–164.

Liakopoulos, D., Busgen, T., Brychzy, A., Jentsch, S., and Pause, A. (1999). Conjugation of the ubiquitin-like protein NEDD8 to cullin–2 is linked to von Hippel-Lindau tumor suppressor function. *Proc. Natl. Acad. Sci. USA* **96,** 5510–5515.

Liakopoulos, D., Doenges, G., Matuschewski, K., and Jentsch, S. (1998). A novel protein modification pathway related to the ubiquitin system. *EMBO J.* **17,** 2208–2214.

Liu, J., Furukawa, M., Matsumoto, T., and Xiong, Y. (2002). NEDD8 modification of CUL1 dissociates p120(CAND1), an inhibitor of CUL1-SKP1 binding and SCF ligases. *Mol. Cell* **10,** 1511–1518.

Morimoto, M., Nishida, T., Honda, R., and Yasuda, H. (2000). Modification of cullin–1 by ubiquitin-like protein Nedd8 enhances the activity of SCF(skp2) toward p27(kip1). *Biochem. Biophys. Res. Commun.* **270,** 1093–1096.

Osaka, F., Kawasaki, H., Aida, N., Saeki, M., Chiba, T., Kawashima, S., Tanaka, K., and Kato, S. (1998). A new NEDD8-ligating system for cullin–4A. *Genes Dev.* **12,** 2263–2268.

Osaka, F., Saeki, M., Katayama, S., Aida, N., Toh, E. A., Kominami, K., Toda, T., Suzuki, T., Chiba, T., Tanaka, K., and Kato, S. (2000). Covalent modifier NEDD8 is essential for SCF ubiquitin-ligase in fission yeast. *EMBO J.* **19,** 3475–3484.

Pan, Z. Q., Kentsis, A., Dias, D. C., Yamoah, K., and Wu, K. (2004). Nedd8 on cullin: Building an expressway to protein destruction. *Oncogene* **23,** 1985–1997.

Pozo, J. C., Timpte, C., Tan, S., Callis, J., and Estelle, M. (1998). The ubiquitin-related protein RUB1 and auxin response in Arabidopsis. *Science* **280,** 1760–1763.

Read, M. A., Brownell, J. E., Gladysheva, T. B., Hottelet, M., Parent, L. A., Coggins, M. B., Pierce, J. W., Podust, V. N., Luo, R. S., Chau, V., and Palombella, V. J. (2000). Nedd8 modification of cul–1 activates SCF(beta(TrCP))-dependent ubiquitination of Ikappa-Balpha. *Mol. Cell. Biol.* **20,** 2326–2333.

Skowyra, D., Craig, K. L., Tyers, M., Elledge, S. J., and Harper, J. W. (1997). F-box proteins are receptors that recruit phosphorylated substrates to the SCF ubiquitin-ligase complex. *Cell* **91,** 209–219.

Tateishi, K., Omata, M., Tanaka, K., and Chiba, T. (2001). The NEDD8 system is essential for cell cycle progression and morphogenetic pathway in mice. *J. Cell Biol.* **155,** 571–579.

Walden, H., Podgorski, M. S., Huang, D. T., Miller, D. W., Howard, R. J., Minor, D. L., Holton, J. M., and Schulman, B. A. (2003a). The structure of the APPBP1-UBA3-NEDD8-ATP complex reveals the basis for selective ubiquitin-like protein activation by an E1. *Mol. Cell* **12,** 1427–1437.

Walden, H., Podgorski, M. S., and Schulman, B. A. (2003b). Insights into the ubiquitin transfer cascade from the structure of the E1 for NEDD8. *Nature* **422,** 330–334.

Wu, K., Chen, A., and Pan, Z. Q. (2000). Conjugation of Nedd8 to CUL1 enhances the ability of the ROC1-CUL1 complex to promote ubiquitin polymerization. *J. Biol. Chem.* **275,** 32317–32324.

Xirodimas, D. P., Saville, M. K., Bourdon, J. C., Hay, R. T., and Lane, D. P. (2004). Mdm2-mediated NEDD8 conjugation of p53 inhibits its transcriptional activity. *Cell* **118,** 83–97.

Yaron, A., Hatzubai, A., Davis, M., Lavon, I., Amit, S., Manning, A. M., Andersen, J. S., Mann, M., Mercurio, F., and Ben-Neriah, Y. (1998). Identification of the receptor component of the IkappaBalpha-ubiquitin ligase. *Nature* **396,** 590–594.

Zheng, J., Yang, X., Harrell, J. M., Ryzhikov, S., Shim, E. H., Lykke-Andersen, K., Wei, N., Sun, H., Kobayashi, R., and Zhang, H. (2002). CAND1 binds to unneddylated CUL1 and regulates the formation of SCF ubiquitin E3 ligase complex. *Mol. Cell* **10,** 1519–1526.

# [3] A Fluorescence Resonance Energy Transfer-Based Assay to Study SUMO Modification in Solution

*By* GUILLAUME BOSSIS, KATARZYNA CHMIELARSKA, ULRIKE GÄRTNER, ANDREA PICHLER, EVELYN STIEGER, and FRAUKE MELCHIOR

## Abstract

Analysis of posttranslational modifications with ubiquitin and ubiquitin-related proteins (Ubl) generally involves detection of the modified species by immunoblotting or autoradiography, techniques that are not easily applicable for kinetic, quantitative, or high-throughput assays. To circumvent these limitations for studies on ubiquitin-related proteins of the SUMO family, we have developed a fluorescence resonance energy transfer (FRET)-based assay system using yellow fluorescent protein (YFP)-tagged mature SUMO1 (amino acids 1–97) and cyan fluorescent protein (CFP)-tagged RanGAP1 (amino acids 400–589) as model substrates. Reactions are set up in 384-well microtiter plates and are followed online using a fluorescence microtiter plate reader. Applications may involve identification and analysis of SUMO-modifying enzymes and isopeptidases, comparison of enzyme and substrate mutants, and screens for small molecular weight inhibitors. The principal outline of the assay should be applicable to other Ubl conjugation systems as well.

## Introduction

Sumoylation results in the formation of an isopeptide bond between the C-terminal carboxy group of mature SUMO and the $\varepsilon$-amino group of a lysine residue in the target protein. While the mechanism of sumoylation resembles ubiquitination, it involves SUMO-specific enzymes (Johnson, 2004; Melchior, 2000). The SUMO E1-activating enzyme heterodimer Aos1/Uba2 (Sae1/Sae2) activates SUMO through ATP-dependent thioester bond formation. SUMO is then transferred to the single E2-conjugating enzyme Ubc9. SUMO-loaded Ubc9 recognizes and modifies

METHODS IN ENZYMOLOGY, VOL. 398

0076-6879/05 $35.00
DOI: 10.1016/S0076-6879(05)98003-8

a specific target, often with the help of SUMO E3 ligases. SUMOylation usually leads to the addition of single SUMO moieties rather than SUMO chains.

Modification with SUMO has been reconstituted *in vitro* for many known SUMO targets. Some are modified efficiently just with Aos1/Uba2 and Ubc9, most prominently the Ran GTPase-activating protein RanGAP1 (Pichler *et al.*, 2002), whereas others require E3 ligases for efficient modification (e.g., Johnson and Gupta, 2001; Pichler *et al.*, 2002; Sachdev *et al.*, 2001). Assays usually involve incubation of bacterially expressed or *in vitro*-translated targets with recombinant or purified enzymes, SUMO, and ATP. As modification results in a target mobility shift of around 20 kDa, it can be detected by Coomassie staining, immunoblotting, or autoradiography (e.g., Desterro *et al.*, 1998; Johnson and Gupta, 2001; Pichler *et al.*, 2002; Terui *et al.*, 2004). Some studies employed $^{125}$I-labeled SUMO, which results in a radioactive signal at the size of the sumoylated target (Desterro *et al.*, 1999). These types of assays all involve SDS–PAGE, and are unsatisfactory for applications that require simultaneous analysis of many samples. We therefore developed a GFP-based fluorescence resonance energy transfer assay for sumoylation. FRET is a process by which the excited state energy of a fluorescent donor molecule is transferred emission free to an acceptor molecule. The consequence of this is a significant reduction in donor–and a concomitant appearance of acceptor–emission. Efficient energy transfer not only requires overlapping emission and excitation spectra, it also requires very close proximity (less than 10 nm) of the donor and acceptor molecules. FRET is therefore widely used as an indicator for inter- or intramolecular protein interactions (the phenomenon and applications are reviewed in Herman *et al.*, 2004; Jares-Erijman and Jovin, 2003; Sekar and Periasamy, 2003). FRET-based ubiquitination assays have already been described, but these are still rather complicated as they require secondary reagents for detection (Boisclair *et al.*, 2000; Hong *et al.*, 2003).

The principle of our FRET-based sumoylation assay is depicted in Fig. 1A. As a model target, we have chosen RanGAP1, as it is the most efficient SUMO target known to date. Fusion proteins of YFP with SUMO1 and CFP with the 20-kDa C-terminal domain of RanGAP1 (GAPtail) are used for conjugation. CFP and YFP represent a well-characterized donor and acceptor pair for *in vivo* and *in vitro* FRET applications (Pollok and Heim, 1999; Tsien, 1998). Isopeptide bond formation between these components enables FRET directly, hence secondary reagents for analysis are not required. YFP-SUMO1, CFP-GAPtail, and the required enzymes Aos1/Uba2 and Ubc9 can be expressed and purified well from bacteria (Fig. 1B, protocols described later) and are functional in ATP-dependent isopeptide

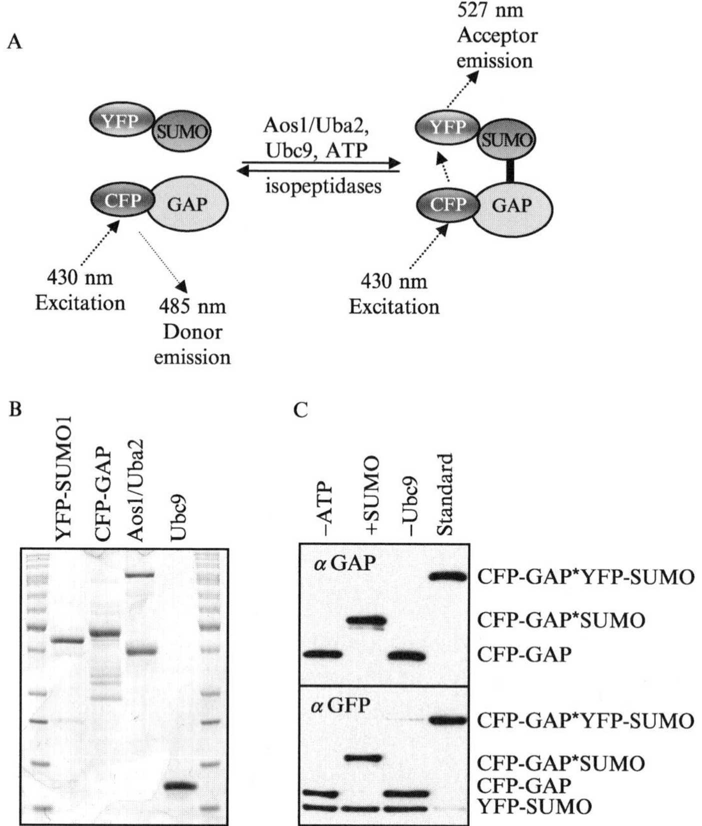


FIG. 1. A FRET-based sumoylation assay: Principle and components. (A) Principle of the FRET assay. When YFP-SUMO and CFP-GAPtail are not conjugated, excitation of CFP at 430 nm results in strong emission at 485 nm. Upon addition of the conjugating enzymes (Aos1/Uba2 and Ubc9) and ATP, YFP-SUMO is covalently conjugated to CFP-GAPtail. This allows FRET to take place. As a consequence, emission at 485 nm is reduced and emission at 527 nm increases. (B) Purified components used in the FRET-based assay. Coomassie staining of recombinant YFP-SUMO1, CFP-GAPtail, Aos1/Uba2, and Ubc9. (C) The four recombinant proteins function in sumoylation. Standard reactions contain all four components and ATP. Covalent modification of YFP-SUMO1 and CFP-GAPtail is revealed by the expected size shift in immunoblotting with αRanGAP1 or αGFP antibodies. The reaction is ATP and Ubc9 dependent and can be completed with untagged SUMO1.

bond formation (Fig. 1C). The clear FRET signal that can be observed upon YFP-SUMO1 and CFP-GAPtail conjugation makes this assay a very useful platform for kinetic analysis of the basic sumoylation machinery.

## Expression and Purification of Recombinant Proteins

### *Expression Constructs*

*pET23a-Ubc9.* The coding region for mouse Ubc9 was obtained by polymerase chain reation (PCR) amplification from EST clone No. IMAGp998A061122 using primers that introduced an *Nde*I site at the start codon (5′-*CATATG*TCGGGGATCGCCCTCAGCCGC-3′) and a *Bam*HI site after the stop codon (5′-*GGATCC*TTATGAGGGGGCAAACT-TCTTCGC-3′). The PCR product was cloned into pCR11, cut out with *Nde*I/*Bam*HI, and ligated into the *Nde*I/*Bam*HI sites of pET23a.

*pET28a-Aos1.* The coding region for human Aos1 was obtained by amplification from clone DKFZp434J0913 (obtained from the RZPD) using primers that introduced a *Nhe*I site at the start codon (5′-G*GCTAG-C*ATGGTGGAGAAGGAGGAGGCTGG-3′) and a *Bam*HI site after the stop codon (5′-G*GGATCC*CGGGCCAATGACTTCAGTTTTCC-3′). The PCR product was cloned into pBluescript, cut out with *Nhe*I and *Bam*HI, and ligated into the respective sites of pET28a.

*pET11d-Uba2.* The coding region for human Uba2 was obtained by amplification from clone DKFZp434O1810 using primers that introduced an *Nco*I site at the start codon (5′-GGCTAGCG*CCATGG*CACTGTCG-CGGGGGCTGCCCC-3′) and a *Bgl*II site after the stop codon (5′-G*A-GATCT*GGCATTTCTGTTCAATCTAATGC-3′). The PCR product was cloned into pBluescript, cut out with *Nco*I and *Bgl*II, and ligated into the *Nco*I and *Bam*HI sites of pET11d.

*pET-YFP-SUMO1.* The coding sequence for human mature SUMO1 (amino acids 1–97) was obtained by PCR amplification from pET11SUMO1ΔC4 (Pichler *et al.*, 2002) using primers introducing a *Kpn*I site 5′ of the SUMO1 sequence (5′-GGTTCCGCGT*GGTACC*ATGTCT-GACCAGGAG-3′) and a *Bam*HI site 3′ of the stop codon (5′-AGA*G-GATCC*TAACCCCCCGTTTGTTCCTG-3′). The PCR product was cloned into the *Kpn*I and *Bam*HI sites of pEYFP-C1 (Clontech). pEYFP-SUMO1 was digested with *Nco*I and *Bam*HI, and the inframe YFP-SUMO fragment was cloned into the *Nco*I and *Bam*HI sites of pET11d.

*pET-CFP-GAPtail.* The C-terminal domain of mouse RanGAP1 (referred to as GAPtail; encoding amino acids 400–589) was obtained by digesting pHHS10B GAPtail (Mahajan *et al.*, 1998) with *Bgl*II and *Eco*RI and was cloned into the same sites of pECFP-C1 (Clontech). pECFP-GAPtail was then digested with *Nco*I and *Bam*HI, and the inframe CFP-GAPtail fragment was cloned into the respective sites of pET11d.

*Expression and Purification*

*Expression and Purification of Recombinant Ubc9.* The following protocol was designed for the purification of untagged Ubc9 from *Escherichia coli*. Isolation is normally carried out in 2 days from a frozen bacterial cell pellet and leads to approximately 5–10 mg of Ubc9 per liter of bacterial culture.

- Buffer 1 (B1): 50 m*M* Na-phosphate, pH 6.5, 50 m*M* NaCl, 1 m*M* dithiothreitol (DTT), and 1 $\mu$g/ml each of aprotinin, leupeptin, and pepstatin
- Buffer 2 (B2): 50 m*M* Na-phosphate, pH 6.5, 300 m*M* NaCl, 1 m*M* DTT, and 1 $\mu$g/ml each of aprotinin, leupeptin, and pepstatin
- Transport buffer (TB): 110 m*M* KOAc, 20 m*M* HEPES, pH 7.3, 2 m*M* Mg(OAc)2, 1 m*M* EGTA, 1 m*M* DTT, and 1 $\mu$g/ml each of leupeptin, pepstatin, and aprotinin

For protein purification, pET23a-Ubc9 is transformed into *E. coli* strain BL21(DE3). A single colony is used to inoculate a 20-ml overnight culture. Bacteria are harvested by centrifugation at 4000 rpm in a Beckman JS5.2 rotor, resuspended in fresh LB/Amp, and used to inoculate 2 liters of LB/Amp medium. At an OD600 of 0.6 (after approximately 2 h at 37°, 250 rpm), 1 m*M* isopropyl-$\beta$-D-thiogalactoside (IPTG) is added and the cells are grown for a further 4 h at 37° before harvesting. The cells are resuspended in 60 ml B1 and subjected to one freeze–thaw cycle (−80°). Once thawed, 0.1 m*M* phenylmethylsulfonyl fluoride (PMSF) is added, and the bacterial debris is removed by centrifugation at 1 h 100,000*g*, 4°, in a 45Ti rotor. Of note, neither lysozyme nor sonication is required, as Ubc9 leaks out of the bacteria upon simple freeze/thawing. The 100,000*g* supernatant is applied to a 10-ml SP Sepharose column equilibrated in B1 and flow through (gravity flow) is discarded. After washing the column with 30 ml B1, Ubc9 is eluted with B2. Here, 15 fractions of 2 ml are collected and analyzed by SDS–PAGE for the presence of Ubc9. Ubc9-containing fractions are combined and concentrated to approximately 2 ml using a centrifugal concentrator (5K-Centriprep). Molecular sieving over a preparative S200 FPLC column (Pharmacia Biotech Inc., Piscataway, NJ) is the final step in the preparation. The column is equilibrated in TB, 2 ml of the protein solution is injected per run, and 5-ml fractions are collected. Ubc9 runs on the column with an apparent molecular mass of 20 kDa and can be detected readily in the absorption profile at 280 nm. Fractions containing Ubc9 are checked for purity by SDS–PAGE, pooled, aliquoted, flash frozen in liquid nitrogen, and stored at −80°. Under these conditions, Ubc9 is stable for years and can be thawed and refrozen several times (for reproducibility, we prefer to use each aliquot only once). Dilutions are routinely done in TB buffer containing 0.05% Tween and 0.2 mg/ml ovalbumin.

*Expression and Purification of Recombinant SUMO E1 Enzyme.* The following protocol uses coexpression of His-tagged Aos1 with untagged Uba2. Although the protocol is a bit laborious, we find it worthwhile, as the resulting enzyme is much more active than, e.g., separately expressed His-Aos1 and His-Uba2. Isolation is normally carried out in 4 days from a frozen bacterial cell pellet and leads to approximately 0.5–1 mg of SUMO E1 enzyme per liter of bacterial culture.

Lysis buffer: 50 m*M* Na-phosphate, pH 8.0, 300 m*M* NaCl, 10 m*M* imidazol

Wash buffer: 50 m*M* Na-phosphate, pH 8.0, 300 m*M* NaCl, 20 m*M* imidazol, 1 m*M* $\beta$-mercaptoethanol, and 1 $\mu$g/ml each of aprotinin, leupeptin, and pepstatin

Elution buffer: 50 m*M* Na-phosphate, pH 8.0, 300 m*M* NaCl, 300 m*M* imidazol, 1 m*M* $\beta$-mercaptoethanol, and 1 $\mu$g/ml each of aprotinin, leupeptin, and pepstatin

S200 buffer: 50 m*M* Tris, pH 7.5, 50 m*M* NaCl, 1 m*M* DTT, and 1 $\mu$g/ml each of aprotinin, leupeptin, and pepstatin

Q buffer 1: 50 m*M* Tris, pH 7.5, 50 m*M* NaCl, 1 m*M* DTT, and 1 $\mu$g/ml each of aprotinin, leupeptin, and pepstatin

Q buffer 2: 50 m*M* Tris, pH 7.5, 1 *M* NaCl, 1 m*M* DTT, and 1 $\mu$g/ml each of aprotinin, leupeptin, and pepstatin

Transport buffer (TB): 110 m*M* KOAc, 20 m*M* HEPES, pH 7.3, 2 m*M* Mg(OAc)2, 1 m*M* EGTA, 1 m*M* DTT, and 1 $\mu$g/ml each of leupeptin, pepstatin, and aprotinin

For protein purification, pET 28a-Aos1 and pET11d-Uba2 are transformed simultaneously into *E. coli* strain BL21(DE3) and used to directly inoculate 500 ml of LB with 50 $\mu$g/ml ampicillin and 30 $\mu$g/ml kanamycin (due to poor growth of the bacteria, we omit selection for single colonies on plates). After growth for 18 h at 37°, bacteria are harvested by centrifugation, resuspended in 2 liter fresh medium, and directly induced for protein expression with 1 m*M* IPTG. The cells are grown for 6 h at 25° before harvesting by centifugation at 4000 rpm in a Beckman JS5.2 rotor. After resuspension in 50 ml buffer A, the cells are subjected to one freeze–thaw cycle (−80°). Once thawed, 1 m*M* $\beta$-mercaptoethanol, 0.1 m*M* PMSF, 1 $\mu$g/ml each of aprotinin, leupeptin, and pepstatin, and 50 mg lysozyme (Sigma) are added, and the bacterial suspension is incubated on ice for 1 h. Bacterial debris is removed by centrifugation at 1 h 100,000*g*, 4°, in a 45Ti rotor. His-Aos1 and associated Uba2 are enriched from the supernatant by batch incubation for 1 h at 4° (slow rotation) with 6 ml Probond resin (Invitrogen) equilibrated in lysis buffer, including protein inhibitors and

$\beta$-mercaptoethanol. After harvesting by centrifugation (5 min 500 rpm, 4°, Beckman J6B), resin is transferred into a column and washed extensively with cold wash buffer until no more protein elutes from the column (protein detection by $OD_{280}$ or by Ponceau staining on a nitrocellulose membrane). Proteins are eluted with 3 volumes elution buffer (gravity flow), and 2-ml fractions are collected. Protein containing fractions are combined and concentrated to 2–5 ml using a centrifugal device (e.g., 30K-Millipore-concentrator). After filtration of the concentrate through a 0.2-$\mu$m low protein binding filter, it is applied to an FPLC S200 preparative gel filtration column equilibrated in S200 buffer. Five-milliliter fractions are collected and analyzed by SDS–PAGE. His-Aos1 is expressed in large excess and smears over many column fractions. Only fractions that contain both His-Aos1 (migrates at 40 kDa) and Uba2 (note that it migrates at 90 kDa despite a predicted size of 72 kDa) should be combined and applied for further purification on a 1-ml MonoQ anion-exchange column (Pharmacia FPLC, 1 ml). Elution from the MonoQ column involves a linear gradient from 50 to 500 m*M* NaCl (generated from Q buffers 1 and 2). Fractions (0.5 ml) are collected and analyzed by SDS–PAGE. Fractions containing approximately equimolar levels of His-Aos1 and Uba2 are combined and dialyzed against TB. The enzyme is aliquoted (at 5 $\mu$l aliquots), flash frozen in liquid nitrogen, and stored at −80°. Under these conditions, it is stable for years. As it loses some activity upon freeze thawing, we use each aliquot only once. Dilutions are routinely done in TB buffer containing 0.05% Tween 20 and 0.2 mg/ml ovalbumin.

*Expression and Purification of Recombinant YFP-SUMO and CFP-Gaptail.* The overall scheme for YFP-SUMO and CFP-GAPtail purifications is so similar that their purification can be described in parallel. The key difference is the use of slightly different lysis buffers due to their different abilities to bind to an anion exchanger. The yield for YFP-SUMO and CFP-GAPtail is approximately 0.5–1 mg /liter culture.

- Lysis buffer for YFP-SUMO: 50 m*M* Tris-HCl, pH 8.0, 50 m*M* NaCl, and 1 m*M* EDTA
- Lysis buffer for CFP-GAPtail: 50 m*M* Tris-HCl, pH 8.0, 20 m*M* NaCl, and 1 m*M* EDTA
- Buffer A: 50 m*M* Tris, pH 8.0, 1 m*M* DTT, and 1 $\mu$g/ml each of aprotinin, leupeptin, and pepstatin
- Buffer B: 50 m*M* Tris, pH 8.0, 1 *M* NaCl, 1 m*M* DTT, and 1 $\mu$g/ml each of aprotinin, leupeptin, and pepstatin
- Transport buffer (TB): 110 m*M* KOAc, 20 m*M* HEPES, pH 7.3, 2 m*M* Mg(OAc)2, 1 m*M* EGTA, 1 m*M* DTT, and 1 $\mu$g/ml each of leupeptin, pepstatin, and aprotinin

For protein purification, pET11d-YFP-SUMO1 or pET11d-CFP-GAPtail is transformed into the *E. coli* strain BL21(DE3) and used to directly inoculate 500 ml of LB with 50 $\mu$g/ml ampicillin. After growth overnight at 37°, the culture is diluted with fresh medium to 2 liter and is induced directly for protein expression with 1 m*M* IPTG. The cells are grown for a further 6 h at 20° before harvesting by centifugation at 4000 rpm in a Beckman JS5.2 rotor (the bacterial pellet is slightly yellow). After resuspension in 50 ml lysis buffer, the cells are subjected to one freeze–thaw cycle (−80°). Once thawed, 1 m*M* DTT, 0.1 m*M* PMSF, 1 $\mu$g/ml each of aprotinin, leupeptin, and pepstatin, and 50 mg lysozyme (Sigma) are added, and the bacterial suspension is incubated on ice for 1 h. Bacterial debris is removed by centrifugation at 1 h 100,000*g*, 4°, in a 45Ti rotor. The supernatant, which is visibly yellow, is filtered through 0.2-$\mu$m low protein binding filters prior to loading on a Hightrap Q Sepharose column connected to a chromatography system (e.g., Aekta prime, Amersham Pharmacia). A 25-ml supernatant is loaded per run on a 5-ml Hightrap Q Sepharose column, equilibrated to 50 m*M* NaCl for YFP-SUMO or 20 m*M* NaCl for CFP-GAPtail using buffers A and B. After sample application and washing of the column with 20 ml loading buffer, YFP-SUMO or CFP-GAPtail is eluted using a linear salt gradient (up to 0.5 *M* NaCl) in a total of 20 ml. Yellow-stained fractions (5 ml each) are analyzed for the presence of the desired full-length protein by SDS–PAGE (of note, color alone is not enough as fluorescent cleavage fragments are also present). The cleanest fractions are combined, concentrated, and purified further by gel filtration on a preparative S200 column in TB. Yellow fractions are analyzed by SDS–PAGE, and appropriate fractions are combined. YFP-SUMO or CFP-GAPtail is aliquoted a 20 $\mu$l without prior concentration, flash frozen in liquid nitrogen, and stored at −80°.

## The FRET Assay

### *Buffers and Proteins*

Transport buffer: 110 m*M* KOAc, 20 m*M* HEPES, pH 7.3, 2 m*M* Mg (OAc)2, 1 m*M* EGTA, 1 m*M* DTT, and 1 $\mu$g/ml each of leupeptin, pepstatin, and aprotinin

FRET buffer: TB + 0.2 mg/ml ovalbumin and 0.05% Tween 20

ATP: 5 m*M* ATP in TB

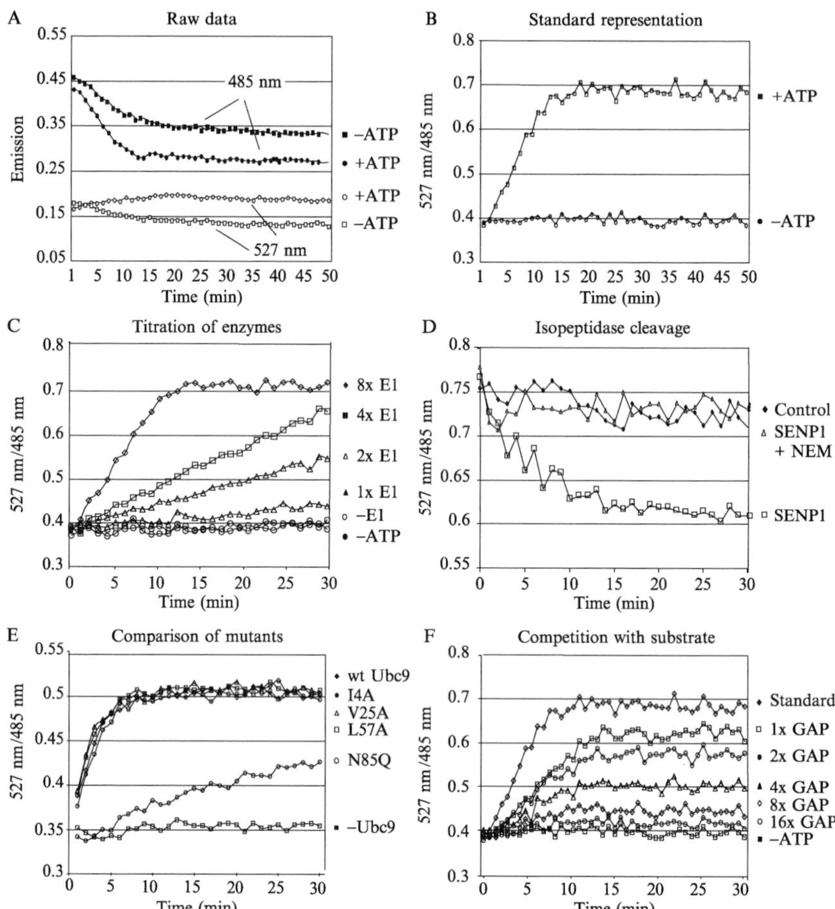


FIG. 2. Applications of the FRET-based sumoylation assay. (A) Unprocessed emission data. Twenty-five-microliter reaction mixes containing YFP-SUMO1, CFP-GAPtail, Aos1/Uba2, and Ubc9 were incubated in a 384-well plate at 30° in the absence or presence of 1 m*M* ATP. Fluorescence after excitation at 430 nm was measured every minute at 485 and 527 nm. (B) Processed data. The rate of conjugation correlates directly with the change in the ratio of emissions (527 nm/485 nm). The value observed in the absence of ATP is due to the fluorescence of CFP at 485 and 527 nm (YFP is not excited to significant levels at 430 nm) and reflects 0% modification. After the addition of ATP, the ratio of emissions increases linearly until it reaches a stable plateau. This reflects 100% modification (demonstrated by subsequent immunoblotting, see Fig. 1C). (C) Titration of the E1 activating enzyme Aos1/Uba2. Reactions included YFP-SUMO1, CFP-GAPtail, Ubc9, and increasing concentrations of Aos1/Uba2 (1–8×). The reaction was started by the addition of ATP and was monitored over 30 min. An ATP control was included to provide a baseline. (D) Deconjugation reaction with SENP1. First, a complete conjugation reaction was carried out as in A. After 30 min, apyrase (1 unit) was added to remove ATP and to prevent further conjugation. Recording started

Proteins: aliquots of YFP-SUMO1, CFP-GAPtail, Ubc9, and His-Aos1/Uba2 in TB (if desired, untagged Ubc9 can be replaced with His-Ubc9, which works with similar efficiency). Where applicable, additional factors, such as isopeptidases or nonfluorescent proteins, are included.

### *Instrumentation*

A fluorescence microtiter plate reader (Fluoroskan Ascent; Labsystems) equipped with an automated sample dispenser and temperature control is used. Parameter settings and data processing are carried out using the software provided with the fluometer (Ascent). Filters: 430 nm for excitation, 485 nm and 527 nm for emission. Microtiter plates: black 384-well plates, e.g., Cliniplate 384 (Labsystems).

### *The Assay*

Twenty-microliter reaction mixes without ATP are set up directly in the 384-well plates (96-well plates can also be used, but require 100 $\mu$l volumes). Due to the efficient modification and good signal-to-noise ratio, our fluorescent substrates can be used in a wide concentration range (we tested 40 n*M* to 2 $\mu$*M*). Typical reactions use equimolar amounts of CFP-GAPtail and YFP-SUMO1 (200 n*M* each) and variable enzyme concentrations, ranging from 1 to 100 n*M*. FRET buffer, in which all proteins are diluted, contains Tween 20 and ovalbumin to prevent nonspecific adsorption of the proteins. Adding these components strongly improved reproducibility of the assay. Special care needs to be taken to avoid air bubbles, as these obscure the reading. After a 5-min preincubation at 30°, reactions are started automatically by the addition of ATP (5 $\mu$l of a 5 m*M* solution, supplied by the dispenser). At desired time points, samples are excited at 430 nm, and fluorescence emissions at 485 and 527 nm are recorded with an integration time of 20 ms. We usually take measurements every minute over the course of 30 min, but reactions can also be followed for longer periods. With this setup, up to 40 samples can be analyzed in parallel. For larger sample sizes the measurement intervals need to be increased. Figure 2A shows raw data recorded for two reactions: one with and one without ATP. Some overall

upon addition of the catalytic domain of the SUMO isopeptidase SENP1. One control was carried out in the absence of SENP1 and a second control included SENP1 pretreated with *N*-ethylmaleimide. (E) Comparison of Ubc9 mutants. The reaction was set up as in A except that Ubc9 wild type or indicated point mutants were used in the reaction. (F) Competition reaction with nonfluorescent RanGAP1. The reaction was set up as in A except that increasing concentrations (1 to 16×) of full-length RanGAP1 were added to the reactions.

loss of fluorescence at 485 nm is generally observed; this is most likely due to photobleaching and/or denaturation of the recombinant proteins. As expected, ATP addition results in a stronger decline of emission at 485 nm, and a slight increase in emission at 527 nm, compared to the ATP control. For easier evaluation, data are usually represented as the emission ratio (527 nm/485 nm). In the absence of ATP, this ratio is unaltered over the entire time course (Fig. 2B). The value of 0.3–0.4 reflects emission of CFP at both wavelengths; YFP excitation at 430 nm is negligible. After the addition of ATP, the ratio of emissions increases linearly over time until it reaches a stable plateau. The maximal value (0.6–0.7) is reached when nearly 100% of the substrates are conjugated (as shown in Fig. 1C, this can be verified easily by immunoblotting of the samples after the time course). The change in the emission ratio over time is a direct readout for the rate of the conjugation reaction.

## Applications of the SUMO FRET Assay

This assay can be used for a large number of different applications dealing with the basic sumoylation machinery. First, it can be used to characterize enzymes involved in the pathway. For example, Fig. 2C compares modification rates in dependence of increasing concentrations of the E1 activating enzyme Aos1/Uba2. As expected, linear rates over the whole time of the experiment are observed with low enzyme concentrations. Doubling the E1 enzyme concentration approximately doubles the reaction rate. Figure 2D shows the converse reaction—cleavage of the conjugate by an isopeptidase. Here, a preformed conjugate was incubated with a recombinant catalytic domain (amino acids 419–644) of the isopeptidase SENP1 [cloned from a full-length construct kindly provided by Bailey and O'Hare (2004)]. As expected, a time-dependent decrease of the FRET signal was observed. If SENP1 was inhibited by *N*-ethylmaleimide prior to addition, no loss of signal was observed. Obviously, one could use similar setups for the identification of stimulatory or inhibitory factors from cell extracts or use this assay as a platform for small compound inhibitors of modifying enzymes or isopeptidases. Comparison of different SUMO family members for modification and demodification rates is achieved by including their respective YFP fusion proteins in the analysis. YFP-SUMO2 and YFP-SUMO3 are expressed and purified like YFP-SUMO1 (not shown). Figure 2E provides an example for mutant enzyme analysis. Assays were set up with wt His-Ubc9 or several single point mutants (Pichler *et al.*, 2004) and were compared for modification rates. Only the N85Q mutant is impaired for RanGAP1 modification. Finally, the FRET assay can be used to compare different nonfluorescent targets or target mutants for their ability to compete with YFP-SUMO1 or CFP-GAPtail. Figure 2F shows the effect of increasing amounts of recombinant RanGAP1 on the readout. As this protein competes with CFP-GAPtail for modification, and as YFP-SUMO1 is limiting in the reaction (present at equimolar levels), one observes a concentration-dependent decrease in the maximum FRET level, as well as a decrease in the rate of CFP-GAPtail modification. A RanGAP1 mutant that lacks the SUMO modification site does not compete even at very high concentrations (data not shown).

The two major advantages of this assay over other available techniques are the reduced time and material requirements. Every set of experiments shown in Fig. 2 was completed in less than 2 h, and a single 25-$\mu$l reaction provided data for 30 time points. The possibility to analyze up to 384 reactions at the same time makes this assay particularly suitable for screening of chemical or biological compounds. Replacement of CFP-GAPtail by other target proteins will be required to allow analysis of specific SUMO E3 ligases, but that should be quite feasible. At least in principle, similar assays could be set up for every pair of ubiquitin-like protein (including ubiquitin) and target protein, as long as chain formation does not take place.

## Acknowledgments

Our special thanks go to the "Tuebinger students" Sven Kröning and David Ermert for generating constructs and recombinant proteins and to Andrea Klanner and Jenni Vordemann for excellent technical assistance. Dr. Daniel Bailey and Dr. Peter O'Hare are gratefully acknowledged for providing a Senp1 plasmid. Funding was obtained from the BMBF (BioFUTURE 0311869), the Engelhorn Stiftung, and a Marie Curie intra-European fellowship to G. B.

## References

Bailey, D., and O'Hare, P. (2004). Characterization of the localization and proteolytic activity of the SUMO-specific protease, SENP1. *J. Biol. Chem.* **279,** 692–703.

Boisclair, M. D., McClure, C., Josiah, S., Glass, S., Bottomley, S., Kamerkar, S., and Hemmila, I. (2000). Development of a ubiquitin transfer assay for high throughput screening by fluorescence resonance energy transfer. *J. Biomol. Screen* **5,** 319–328.

Desterro, J. M., Rodriguez, M. S., and Hay, R. T. (1998). SUMO–1 modification of IkappaBalpha inhibits NF-kappaB activation. *Mol. Cell* **2,** 233–239.

Desterro, J. M., Rodriguez, M. S., Kemp, G. D., and Hay, R. T. (1999). Identification of the enzyme required for activation of the small ubiquitin-like protein SUMO–1. *J. Biol. Chem.* **274,** 10618–10624.

Herman, B., Krishnan, R. V., and Centonze, V. E. (2004). Microscopic analysis of fluorescence resonance energy transfer (FRET). *Methods Mol. Biol.* **261,** 351–370.

Hong, C. A., Swearingen, E., Mallari, R., Gao, X., Cao, Z., North, A., Young, S. W., and Huang, S. G. (2003). Development of a high throughput time-resolved fluorescence resonance energy transfer assay for TRAF6 ubiquitin polymerization. *Assay Drug Dev. Technol.* **1,** 175–180.

Jares-Erijman, E. A., and Jovin, T. M. (2003). FRET imaging. *Nature Biotechnol.* **21,** 1387–1395.

Johnson, E. S. (2004). Protein modification by SUMO. *Annu. Rev. Biochem.* **73,** 355–382.

Johnson, E. S., and Gupta, A. A. (2001). An E3-like factor that promotes SUMO conjugation to the yeast septins. *Cell* **106,** 735–744.

Mahajan, R., Gerace, L., and Melchior, F. (1998). Molecular characterization of the SUMO-1 modification of RanGAP1 and its role in nuclear envelope association. *J. Cell Biol.* **140,** 259–270.

Melchior, F. (2000). SUMO: Nonclassical ubiquitin. *Annu. Rev. Cell. Dev. Biol.* **16,** 591–626.

Pichler, A., Gast, A., Seeler, J. S., Dejean, A., and Melchior, F. (2002). The nucleoporin RanBP2 has SUMO1 E3 ligase activity. *Cell* **108,** 109–120.

Pichler, A., Knipscheer, P., Saitoh, H., Sixma, T. K., and Melchior, F. (2004). The RanBP2 SUMO E3 ligase is neither HECT- nor RING-type. *Nature Struct. Mol. Biol.* **11,** 984–991.

Pollok, B. A., and Heim, R. (1999). Using GFP in FRET-based applications. *Trends Cell Biol.* **9,** 57–60.

Sachdev, S., Bruhn, L., Sieber, H., Pichler, A., Melchior, F., and Grosschedl, R. (2001). PIASy, a nuclear matrix-associated SUMO E3 ligase, represses LEF1 activity by sequestration into nuclear bodies. *Genes Dev.* **15,** 3088–3103.

Sekar, R. B., and Periasamy, A. (2003). Fluorescence resonance energy transfer (FRET) microscopy imaging of live cell protein localizations. *J. Cell Biol.* **160,** 629–633.

Terui, Y., Saad, N., Jia, S., McKeon, F., and Yuan, J. (2004). Dual role of sumoylation in the nuclear localization and transcriptional activation of NFAT1. *J. Biol. Chem.* **279,** 28257–28265.

Tsien, R. Y. (1998). The green fluorescent protein. *Annu. Rev. Biochem.* **67,** 509–544.

# [4] Properties of the ISG15 E1 Enzyme UbE1L

*By* ROBERT M. KRUG, CHEN ZHAO, and SYLVIE BEAUDENON

ISG15 is an ubiquitin-like protein that is induced by treating mammalian cells with interferon-$\alpha/\beta$. UbE1L was identified as the E1-activating enzyme for ISG15 in the course of experiments that demonstrated that the nonstructural or NS1 protein of influenza B virus inhibits ISG15 conjugation. UbE1L is a monomeric protein of 120 kDa. Its most distinctive property is its selectivity in transferring its ISG15 substrate to a single E2 enzyme, UbcH8, an E2 that has been reported to also function in Ub conjugation. This chapter describes the methods for expressing and purifying UbE1L using a baculovirus vector and for assaying UbE1L activity using a $^{32}$P-labeled ISG15 substrate.

METHODS IN ENZYMOLOGY, VOL. 398

0076-6879/05 $35.00
DOI: 10.1016/S0076-6879(05)98004-X

## Introduction

A large number of genes are transcriptionally induced in mammalian cells by IFN-$\alpha/\beta$, including the gene encoding ISG15, a 15-kDa ubiquitin (Ub)-like protein, or Ubl (Haas *et al.*, 1987; Stark *et al.*, 1998). The ISG15 protein consists of two Ub-related domains, approximately 30% (N-terminal domain) and 36% (C-terminal domain) identical to Ub (Bloomstrom *et al.*, 1986; Haas *et al.*, 1987). Following IFN-$\alpha/\beta$ stimulation, ISG15 is conjugated to cellular proteins (Loeb and Haas, 1992). Like Ub and other Ubls, the ISG15 protein is attached covalently via its C-terminal glycine residue to its target proteins (Narasimhan *et al.*., 1996; Potter *et al.*, 1999). Although ISG15 was the first Ubl identified, the function of ISG15 conjugation remains unknown. In contrast, we have identified two of the enzymes involved in ISG15 conjugation: the E1-activating enzyme (UbE1L/E1$^{ISG15}$) and the specific E2 enzyme that functions with ISG15 (Yuan and Krug, 2001; Zhao *et al.*, 2004). Surprisingly, this E2 is UbcH8 (Zhao *et al.*, 2004), an E2 that has been reported to also function in Ub conjugation (Chin *et al.*, 2002; Kumar *et al.*, 1997; Urano *et al.*, 2002; Zhang *et al.*, 2000), suggesting that the ISG15 conjugation pathway may intersect or converge with a branch of the Ub conjugation system. As is the case for the ISG15 gene (Haas *et al.*, 1987), genes encoding UbE1L and UbcH8 are induced transcriptionally by IFN-$\alpha/\beta$ (Nyman *et al.*, 2000; Yuan and Krug, 2001; Zhao *et al.*, 2004), indicating coordinate regulation of these components of the ISG15 conjugation system.

UbE1L was identified as the ISG15 E1-activating enzyme (E1$^{ISG15}$ ) in the course of experiments to elucidate the functions of a viral protein, the nonstructural protein of influenza B virus (NS1B protein) (Yuan and Krug, 2001). This viral protein was shown to bind ISG15 specifically. To determine whether this binding resulted in the inhibition of ISG15 conjugation, we developed an *in vitro* assay for E1$^{ISG15}$ using extracts from IFN-$\beta$-treated A549 cells, a human lung cancer cell line. Using radiolabeled ISG15, we detected a putative E1$^{ISG15}$ –ISG15 complex of high molecular mass (~140 kDa), which was sensitive to reducing agents (dithiothreitol, DTT), as expected for a thioester-linked complex. To purify the E1$^{ISG15}$ , extracts from IFN-$\beta$-treated A549 cells were bound to GST-ISG15 immobilized on glutathione agarose beads (Yuan and Krug, 2001). The E1$^{ISG15}$ , which was expected to form a thioester-linked ISG15 complex on the column, was eluted with DTT. Gel electrophoresis of the eluate demonstrated a single major band of approximately 120 kDa (Fig. 1). Edman degradation sequencing of one of the trypsin-derived peptides established that this protein is UbE1L (Yuan and Krug, 2001), which was identified previously as a potential tumor suppressor because its gene is deleted in

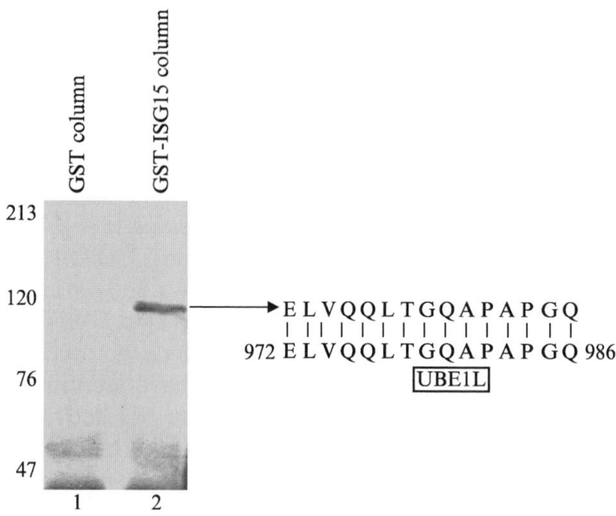


FIG. 1. Identification of the E1 enzyme that activates ISG15. Cell extracts from IFN-$\beta$-treated A549 cells were subjected to affinity selection on a glutathione Sepharose 4B column containing either GST (lane 1) or the GST-ISG15 fusion protein (lane 2). The eluate was resolved by SDS–polyacrylamide (7.5%) gel electrophoresis, followed by staining with Coomassie blue. The 110-kDa protein species from lane 2 was analyzed by MALDI-MS, and one of the tryptic peptides was microsequenced by automated Edman degradation. The sequence of this peptide is compared to the amino acid sequence of UBE1L (amino acids 972–986). From Yuan and Krug (2001).

many lung cancer cells (Kok *et al.*, 1993, 1995). The UbE1L protein was expressed using a baculovirus vector, and the recombinant protein was shown to form an ATP-dependent thioester adduct with ISG15, but not Ub (Yuan and Krug, 2001), demonstrating conclusively that UbE1L is the E1 enzyme for ISG15. The NS1B protein of influenza B virus inhibited the UbE1L-catalyzed activation of ISG15 (Yuan and Krug, 2001). This finding indicates that IFN-$\alpha/\beta$-induced ISG15 conjugation is likely to be an important aspect of the cellular antiviral response and that influenza B virus has evolved a mechanism to specifically counter this response.

## Expression, Purification, and Assay of UbE1L

GST-UbE1L can be expressed using various types of baculovirus vectors (Zhao *et al.*, 2004). The expression system currently in use in our laboratory uses a modified pFastbac1 vector (Invitrogen) in which the GST open reading frame (ORF), the PreScission protease cleavage site, and the polylinker from the pGEX 6p–1 vector had been inserted adjacent to the polyhedrin promoter (S. Beaudenon and J. Huibregtse, unpublished result). The UbE1L ORF is cloned between the *Eco*RI and the *Not*I restriction sites in the polylinker. The pFastbac1-GST-UbE1L recombinant plasmid is transformed into *Escherichia coli* strain DH10Bac for transposition into bacmids following the procedure recommended by the manufacturer. Purified UbE1L recombinant bacmid DNA is transfected into insect cells (High Five, Invitrogen), and virus is collected, which is subjected to two additional rounds of amplification to obtain a high titer stock. For protein expression, High Five cells are infected with the recombinant virus, and 48 h postinfection, cells are extracted using NP–40 lysis buffer (0.1 *M* Tris-HCl, pH 7.5; 0.1 *M* NaCl; 1% NP-40). GST-UbE1L protein is purified by incubating the cell extract with glutathione Sepharose beads for 2 h at 4°. The beads are washed three times with 0.2 ml of phosphate-buffered saline (PBS) containing 1% Triton. The UbE1L protein is released from the beads by cleavage of the GST-UbE1L bond with PreScission protease (Promega) for 5 h at 4° in a buffer containing 10 m*M* Tris-HCl, pH 7.5, 50 m*M* NaCl, 0.01% Triton, and 1 m*M* DTT. Figure 2 shows the Coomassie-blue stained gel of both GST-UbE1L and purified UbE1L. Based on current experiments, it is best to use this purified UbE1L as soon as possible because it is relatively unstable. Experiments indicate that using the pGEX 6p–1 vector for expression of UbE1L in *E. coli* (DH10B) grown at 27° results in higher yields of purified UbE1L (1.5–2.0 mg/liter) that is significantly more stable.

The assay for UbE1L activity uses a $^{32}$P-labeled ISG15 substrate (Yuan and Krug, 2001). To prepare this substrate, the ISG15 ORF is cloned into pGEX–2TK, thereby attaching a cAMP-dependent kinase domain onto the N terminus of ISG15 (Fig. 3A). The GST–2KT-ISG15 fusion protein is purified using glutathione Sepharose beads. For each labeling reaction, 1 $\mu$g of purified GST–2KT-ISG15 is diluted in a final volume of 500 $\mu$l PBS and incubated with 50 $\mu$l glutathione Sepharose beads for 3 h at 4°. The glutathione beads are spun down by centrifugation, washed once with 1 ml of kinase buffer (40 m*M* Tris-HCl, pH 7.5, 20 m*M* $MgCl_2$), and then resuspended in 60 $\mu$l of kinase buffer together with 0.05 mCi ($\gamma$-$^{32}$P ATP) and 1 unit of cAMP-dependent protein kinase catalytic subunit (Promega). The kinase reaction is carried out at room temperature for 30 min and is stopped by spinning down the beads and removing the supernatant. After washing the Sepharose beads twice with 1 ml of PBS, the $^{32}$P-labeled GST–2KT-ISG15 fusion protein can be eluted with glutathione and used in the UbE1L thioester assays described later. Alternatively, the $^{32}$P-labeled 2TK-ISG15 protein can be cleaved from the GST using thrombin. For this cleavage reaction, the PBS-washed glutathione beads are mixed with 100 $\mu$l thrombin buffer (50 m*M* Tris-HCl, pH 8.0, 150 m*M* NaCl, 2.5 m*M* $CaCl_2$,

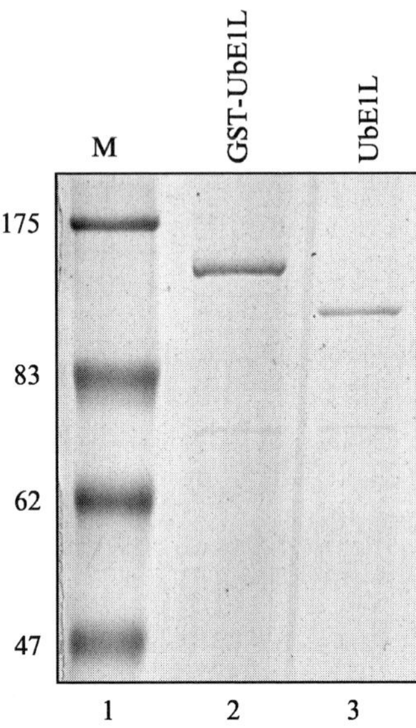


FIG. 2. Purification of the baculovirus-expressed UbE1L enzyme. The baculovirus expressing GST-UbE1L (constructed as described in the text) was used to infect six 150-mm plates of High Five cells. Forty-eight hours postinfection, a cell extract was prepared and incubated with glutathione Sepharose beads for 2 h at 4°. The beads were washed three times with 0.2 ml of PBS containing 1% Triton. A small aliquot (10 $\mu$l of the 0.2 ml) of the beads was diluted in SDS-containing gel-loading buffer and electrophoresed on a 8% SDS–PAGE gel (lane 2). The UbE1L protein was released from the rest of the washed beads by cleavage of the GST-UbE1L bond with PreScission protease as described in the text, and a 5-$\mu$l aliquot of the 0.1-ml eluant was subjected to SDS–PAGE electrophoresis (lane 3). Lane 1: M, molecular weight marker proteins whose molecular weights are denoted on the left side of the gel.

5 m*M* $MgCl_2$) containing 0.5 units of biotin-labeled thrombin (Novagen), and the mixture is incubated for 30 min at room temperature. After centrifugation, the supernatant containing the labeled ISG15 protein is transferred to a new tube, and thrombin is removed using the thrombin cleavage capture kit (Novagen). It is essential to remove the thrombin. The resulting $^{32}$P-labeled ISG15 protein is aliquoted and stored at 4°. Full activity is maintained for 2 or 3 days.

To assay for UbE1L-ISG15 thioester formation, 0.10 $\mu$g of purified UbE1L and 4 $\mu$l of $^{32}$P-labeled ISG15 ($\sim 10^6$ cpm) are incubated for 15 min at room temperature in a 40-$\mu$l reaction containing 50 m*M* Tris-HCl, pH 7.5, 3.75 m*M* ATP (Roche), and 10 m*M* $MgCl_2$ (Zhao *et al.*, 2004). The reaction is terminated by adding an equal volume of SDS gel-loading buffer without DTT, followed by incubation at room temperature for

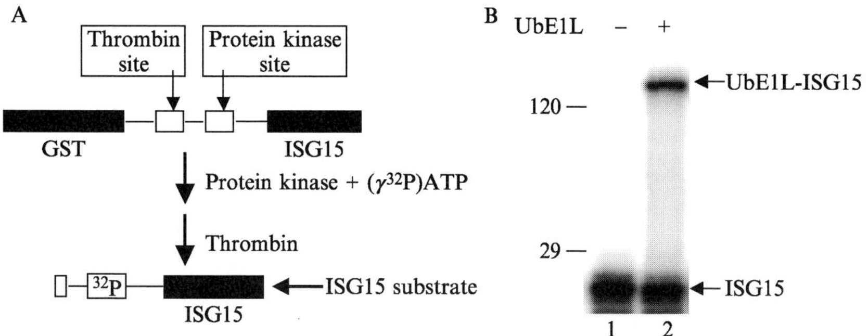


Fig. 3. (A) Schematic diagram of the strategy for preparing the $^{32}$P-labeled ISG15 substrate. From Yuan and Krug (2001). (B) Assay for UbE1L-ISG15 thioester formation. $^{32}$P-labeled ISG15 was incubated with buffer (lane 1) or purified UbE1L (lane 2) as described in the text, and the reaction products were analyzed by SDS–PAGE under nonreducing conditions.

10 min. An aliquot of the mixture (20 $\mu$l) is analyzed by SDS–PAGE under nonreducing conditions. Figure 3B shows the typical result of an assay carried out under these conditions: approximately 10% of the labeled ISG15 is linked to UbE1L. The UbE1L-ISG15 species is sensitive to DTT, as expected for a thioester.

## Distinctive Properties of UbE1L

UbE1L, the ISG15 E1 enzyme ($E1^{ISG15}$), is a monomeric protein of 120 kDa (Yuan and Krug, 2001), similar in size to the monomeric E1 Ub enzyme ($E1^{Ub}$) (Hershko and Ciechanover, 1998; Schwartz and Hochstrasser, 2003). In contrast, the E1 enzymes of several other Ubls, e.g., SUMO and NEDD8, are heterodimers (Johnson *et al.*, 1997). Perhaps the most distinctive feature of UbE1L is its selectivity in transferring its ISG15 substrate to a single E2 enzyme, UbcH8, both *in vitro* and *in vivo* (Zhao *et al.*, 2004), whereas $E1^{Ub}$ can transfer its Ub substrate to all of the approximately 20 Ub E2 enzymes, including UbcH8 (Pickart, 2001). Figure 4A shows the *in vitro* experiments to determine whether other E2s that are closely related to UbcH8 can also function as E2s for ISG15 (Zhao *et al.*, 2004). No E2-ISG15 thioester formation was detected with UbcH5b, which is 56% similar to UbcH8 (compare lanes 8 and 10), and a low amount of thioester formation was observed with UbcH7, which is 72% similar to UbcH8 (~5% of that of UbcH8, as determined by a long exposure of the gel of Fig. 4A). RNA interference experiments in IFN-$\beta$-treated HeLa cells showed that UbcH8 serves as a major E2 enzyme for ISG15 conjugation

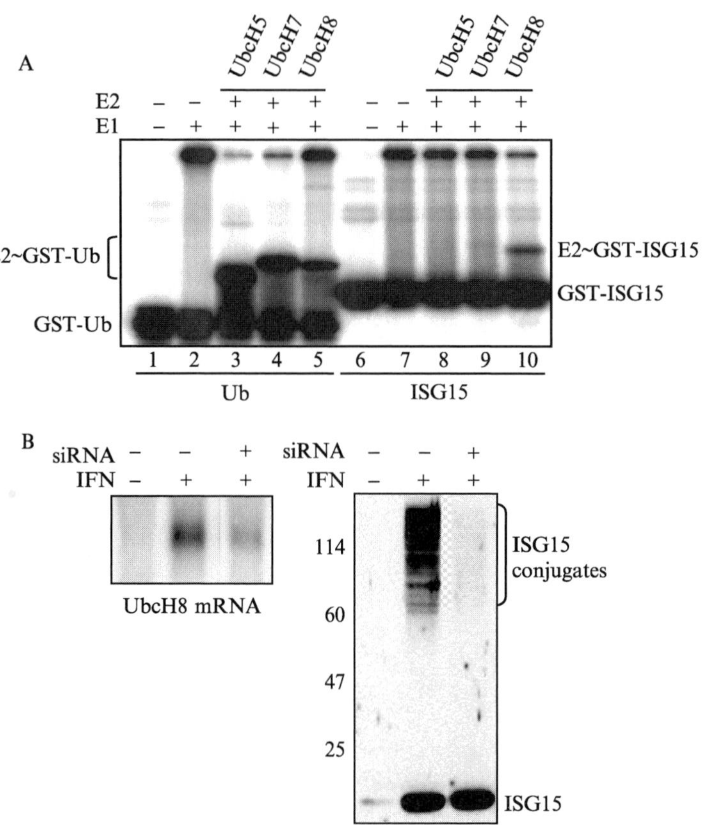


FIG. 4. (A) UbcH8 functions as a major E2 enzyme for ISG15 *in vitro*. Either $^{32}$P-GST-Ub (lanes 1–5) or $^{32}$P-GST-ISG15 (lanes 6–10) was incubated in the absence (lanes 1 and 6) or presence of E1$^{Ub}$ (lanes 2–5) or E1$^{ISG15}$ (lanes 7–10) and in the absence or presence of the indicated E2 proteins. Reactions were stopped with SDS–PAGE loading buffer lacking DTT and were analyzed by SDS–PAGE under nonreducing conditions. (B) UbcH8 functions as a major E2 enzyme for ISG15 *in vivo*. Cells were either mock transfected (–siRNA) or transfected with an siRNA (20 n*M* final concentration) directed against bases 28–49 of the open reading frame of human UbcH8 mRNA (+ siRNA lanes). After 24 h the cells were left untreated (–IFN lanes) or were treated with IFN-$\beta$ (1000 units/ml; + IFN lanes). After another 24 h, the cells were collected. RNA was analyzed for UbcH8 mRNA by Northern analysis (left), and proteins were analyzed by immunoblotting with ISG15 antiserum (right). Each lane of the Northern blot contained 12 $\mu$g total RNA, and the presence of equal amounts of RNA in each lane was confirmed by ethidium bromide staining of 28S ribosomal RNA (not shown). The same results were obtained using a second siRNA that was directed against bases 239–258 of the open reading frame of UbcH8 mRNA, and an siRNA directed against a sequence in the mRNA for green fluorescent protein did not decrease either UbcH8 mRNA or ISG15 conjugation (not shown). The Northern blot also verified that UbcH8 mRNA is induced by IFN-$\beta$. From Zhao *et al.* (2004).

*in vivo* (Zhao *et al.*, 2004). Thus, a UbcH8-specific double-stranded short interfering RNA (siRNA) reduced the amount of UbcH8 mRNA by 80–90% and inhibited IFN-$\beta$-induced ISG15 conjugation to a similar extent (Fig. 4B). Another siRNA against UbcH8 had the same effect on UbcH8 mRNA and ISG15 conjugation, and Northern and RT-PCR analyses showed that UbcH7 mRNA was not affected by either siRNA.

## References


Bloomstrom, D. C., Fahey, D., Kutny, R., Korant, B. D., and Knight, E. (1986). Molecular characterization of the interferon-induced 15-kDa protein: Molecular cloning and nucleotide and amino acid sequence. *J. Biol. Chem.* **261,** 8811–8816.

Chin, L. S., Vavalle, J. P., and Li, L. (2002). Staring, a novel E3 ubiquitin-protein ligase that targets syntaxin I for degradation. *J. Biol. Chem.* **277,** 35071–35079.

Haas, A. L., Ahrens, P., Bright, P. M., and Ankel, H. (1987). Interferon induces a 15-kilodalton protein exhibiting marked homology to ubiquitin. *J. Biol. Chem.* **262,** 11315–11323.

Hershko, A., and Ciechanover, A. (1998). The ubiquitin system. *Annu. Rev. Biochem.* **67,** 425–479.

Johnson, E. S., Schwienhorst, I., Dohmen, R. J., and Blobel, G. (1997). The ubiquitin-like protein Smt3p is activated for conjugation to other proteins by an Aos1p/Uba2p heterodimer. *EMBO J.* **16,** 5509–5519.

Kok, K., Hofstra, R., Pilz, A., van den Berg, A., Terpstra, P., Buys, C. H., and Carritt, B. (1993). A gene in the chromosomal region 3p21 with greatly reduced expression in lung cancer is similar to the gene for ubiquitin-activating enzyme. *Proc. Natl. Acad. Sci. USA* **90,** 6071–6075.

Kok, K., Van den Berg, A., Veldhuis, P. M., Franke, M., Terpstra, P., and Buys, C. H. (1995). The genomic structure of the human UBE1L gene. *Gene Expr.* **4,** 163–175.

Kumar, S., Kao, W. H., and Howley, P. M. (1997). Physical interaction between specific E2 and Hect E3 enzymes determines functional cooperativity. *J. Biol. Chem.* **272,** 13548–13554.

Loeb, K. R., and Haas, A. L. (1992). The interferon-inducible 15-kDa ubiquitin homolog conjugates to intracellular proteins. *J. Biol. Chem.* **267,** 7806–7813.

Narasimhan, J., Potter, J. L., and Haas, A. L. (1996). Conjugation of the 15-kDa interferon-induced ubiquitin homolog is distinct from that of ubiquitin. *J. Biol. Chem.* **271,** 324–330.

Nyman, T. A., Matikainen, S., Sareneva, T., Julkunen, I.., and Kalkkinen, N. (2000). Proreome analysis reveals ubiquitin-conjugating enzymes to be a new family of interferon-alpha-regulated genes. *Eur. J. Biochem.* **267,** 4011–4019.

Pickart, C. M. (2001). Mechanisms underlying ubiquitination. *Annu. Rev. Biochem.* **70,** 503–533.

Potter, J. L., Narasimhan, J., Mende-Mueller, L., and Haas, A. L. (1999). Precursor processing of pro-ISG15/UCRP, an interferon-$\beta$-induced ubiquitin-like protein. *J. Biol. Chem.* **274,** 25061–25068.

Schwartz, D. C., and Hochstrasser, M. (2003). A superfamily of protein tags: Ubiquitin, SUMO and related modifiers. *Trends Biochem. Sci.* **28,** 321–328.

Stark, G. R., Kerr, I. M., Williams, B. R., Silverman, R. H., and Schreiber, R. D. (1998). How cells respond to interferons. *Annu. Rev. Biochem.* **67,** 227–264.

Urano, T., Saito, T., Tsukui, T., Fujita, M., Hosoi, T., Muramatsu, M., Ouchi, Y., and Inoue, S. (2002). Efp targets 14–3–3 sigma for proteolysis and promotes breast tumour growth. *Nature* **417,** 871–875.

Yuan, W., and Krug, R. M. (2001). Influenza B virus NS1 protein inhibits conjugation of the interferon (IFN)-induced ubiquitin-like ISG15 protein. *EMBO J.* **20,** 362–371.

Zhang, Y., Gao, Y., Chung, K. K., Huang, H., Dawson, V. L., and Dawson, T. M. (2000). Parkin functions as an E2-dependent ubiquitin-protein ligase and promotes the degradation of the synatic vesicle-associated protein, CDCrel–1. *Proc. Natl. Acad. Sci. USA* **97,** 13354–13359.

Zhao, C., Beaudenon, S. L., Kelley, M. L., Waddell, M. B., Yuan, W., Schulman, B. A., Huibregtse, J. M., and Krug, R. M. (2004). The UbcH8 ubiquitin enzyme is also the E2 enzyme for ISG15, an IFN-$\alpha/\beta$-induced ubiquitin-like protein. *Proc. Natl. Acad. Sci. USA* **101,** 7578–7582.

# Section II

# E2 Enzyme

# [5] Purification and Properties of the Ubiquitin-Conjugating Enzymes Cdc34 and Ubc13·Mms2

*By* CHRISTOPHER PTAK, XARALABOS VARELAS, TREVOR MORAES, SEAN MCKENNA, and MICHAEL J. ELLISON

## Abstract

A prerequisite for structure/function studies on the ubiquitin-conjugating enzymes (Ubc) Cdc34 and Ubc13·Mms2 has been the ability to express and purify recombinant derivatives of each. This chapter describes the methods used in the expression and purification of these proteins from *Escherichia coli,* including variations of these protocols used to generate $^{35}$S, $^{15}$N, $^{13}$C/$^{15}$N, and seleno-L-methionine derivatives. Assays used to measure the Ub thiolester and Ub conjugation activities of these Ubcs are also described.

## Expression and Purification of Cdc34

Expression of Cdc34 from *Saccharomyces cerevisiae* in *E. coli* used a standard two plasmid heat-inducible system (Ptak *et al.*, 1994). One of the plasmids, pET3a, is ampicillin selectable and possesses bacteriophage T7 promoter and terminator sequences that flank the open reading frame of the protein to be expressed (Studier *et al.*, 1990). To facilitate cloning the oligonucleotides, 5′-CATATGAGCTCTCCCGGGTACCGATCC-3′ was inserted between the *Nde*I and the *Bam*HI sites of pET3a, introducing an *Sst*I site that incorporates a start codon, and a *Kpn*I site. The Cdc34 coding region was modified accordingly by polymerase chain (PCR) such that an *Sst*I site was introduced at its 5′ end and the *Kpn*I site was introduced after the stop codon. These changes had no effect on the Cdc34 amino acid sequence. This approach has also been used to generate pET3a expression plasmids for other Ubcs, including Ubc1 (Hodgins *et al.*, 1996), Ubc4 (Gwozd *et al.*, 1995), and Rad6 (Ptak *et al.*, 2001). In addition, these changes allow for the transfer of these coding regions between these pET3a expression plasmids and a set of yeast overexpression plasmids that places these coding sequences under control of the *Cup1* promoter and the *Cyc1* terminator (Ptak *et al.*, 1994). The second plasmid, pGP1–2, contains the coding region for the bacteriophage T7 polymerase under control of the

METHODS IN ENZYMOLOGY, VOL. 398

0076-6879/05 $35.00
DOI: 10.1016/S0076-6879(05)98005-1

$\lambda p_L$ promoter, the coding region for a temperature-sensitive derivative of the $\lambda$ repressor under control of the $p_{lac}$ promoter and the $kan^r$ selectable marker (Tabor and Richardson, 1985).

For the expression of Cdc34:

1. Cotransform the pET3a-*Cdc34* and pGP1–2 plasmids into *E. coli* BL21 cells. Plate the transformed cells onto LB media (10 g Tryptone, 5 g yeast extract, 5 g NaCl, 20% agar per liter) containing 50 $\mu$g/ml ampicillin and 75 $\mu$g/ml kanamycin (LB-AK). Incubate the plate overnight at 30°.
2. For a starter culture, inoculate 25 ml LB-AK liquid media in a 250-ml Erlenmeyer flask with a single colony and grow the culture overnight at 30° with shaking. For culture growth we use a New Brunswick Scientific temperature-controlled incubator shaker that is run at 250 rpm.
3. Using this starter culture, inoculate 1 liter of LB-AK liquid media in a 2-liter Erlenmeyer flask and incubate at 30° with shaking to a final absorbance of 0.4 at 590 n*M*.
4. Shift the culture to a 42° water bath for 1 h. Although incubations in liquid culture employ shaking, we found that the 42° induction step could be carried out in a water bath with only periodic shaking, leading to equivalent levels of protein expression.
5. Shift the culture to a 37° shaking incubator for an additional 2 h. After these incubations Cdc34 should be the most abundant protein within the cells (Fig. 1A, lanes 1 and 2).
6. Harvest the cells by centrifugation (we pellet the cells at 3000 rpm for 20 min at 4° using a Damon DPR–6000 centrifuge) and resuspend the resulting cell pellet in 25 ml of chilled buffer A [50 m*M* Tris-HCl (pH 7.5), 1 m*M* EDTA, 1 m*M* dithiothreitol (DTT)].
7. Cells may then be lysed using a number of different methods, but for this volume, use of a French press was simplest. Two to three passes through a French press are sufficient for efficient lysis.
8. Clarify the lysate by ultracentrifugation. We centrifuge the lysate at 40,000 rpm for 1 h at 4° using a Beckman L7–65 ultracentrifuge with a Ti70 rotor.

Subsequent purification of Cdc34 from the lysate exploits the relatively low pI of this enzyme (~3.9) resulting from a high content of acidic amino acids present in its carboxy-terminal extension. This simplifies purification, requiring only two chromatographic steps performed at 4° using an FPLC system (Amersham Biosciences).

1. Load the clarified lysate (~25 ml) onto a HiLoad 26/10 Q Sepharose (Amersham Biosciences) anion-exchange column that has been equilibrated with several column volumes of buffer A.

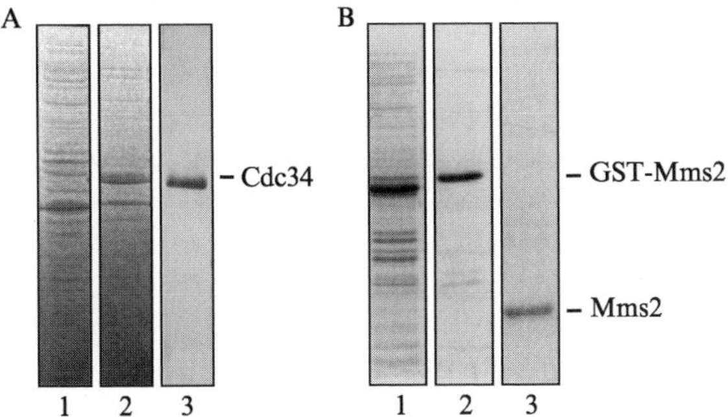


FIG. 1. Expression and purification of Cdc34 and Mms2 from *E. coli*. (A) Coomassie-stained SDS-polyacrylamide gels of cell lysate from uninduced BL21 cells harboring the pET3a-*Cdc34* and pGP1-2 plasmids (lane 1), cell lysate from induced cells (lane 2), and a sample of Cdc34 after column purification (lane 3). (B) Coomassie-stained SDS–polyacrylamide gels of cell lysate from BL21-RIL cells harboring the pGEX–6P1-*Mms2* plasmid after IPTG induction (lane 1), sample from the preparation after running the lysate over a glutathione Sepharose 4B column (lane 2), and a sample of the preparation after protease treatment and column purification (lane 3).

2. Wash the flow through off of the column using buffer A until the $OD_{280}$ approaches baseline. This requires about 2 column volumes of buffer A.

3. Elute the bound protein from the column using a salt gradient ranging from 0 to 2 *M* NaCl in buffer A over 700 ml. Cdc34 elutes from the column at ~470 n*M* NaCl as a well-resolved peak.

4. Pool the peak fractions (~25 ml), and filter concentrate (we use Amicon centriprep YM-10 units) at 4° to a final volume of ~5 ml.

5. Load the concentrate onto a Superdex 75 HR 16/60 gel filtration column equilibrated with 50 m*M* HEPES, pH 7.5, 150 m*M* NaCl, 1 m*M* EDTA, and 1 m*M* DTT. Buffer A with NaCl and DTT may also be used, but the HEPES buffer was employed as it was used in subsequent *in vitro* ubiquitination assays (see later). Given the highly asymmetric structure of Cdc34, it possesses a large Stoke's radius, causing it to elute from the gel filtration column at a high apparent molecular mass (~80 kDa on the Superdex 75 column).

6. Pool the peak fractions (~5–10 ml), and filter concentrate to desired final volume. Add glycerol (5%, v/v), aliquot the preparation, flash freeze in liquid nitrogen, and store at −80°.

7. The purity of the Cdc34 preparation is checked using SDS–polyacrylamide gel electrophoresis (SDS-PAGE), and the Cdc34 concentration is determined using the BCA assay (Pierce). This preparation

yields approximately 5–10 mg of Cdc34 with greater than 90% purity (Fig. 1A, lane 3).

These methods are readily adapted to smaller or larger scale preparations requiring appropriate volume adjustments as well as the use of appropriately sized columns.

## Expression and Purification of Human Ubc13 and Mms2

For reasons that were not explored, the heat-inducible expression system employed for Cdc34 and other Ubcs failed to express human Ubc13 and Mms2 to significant levels. An alternative system was used where both proteins are expressed as N-terminal GST fusions (McKenna *et al.*, 2001). In addition to facilitating expression, the presence of GST also facilitates their purification. Expression of both was chemically induced from the pGEX-6P1 plasmid (Amersham Biosciences) that places the GST coding sequence under control of the $p_{tac}$ promoter. The 3′ flanking region encodes for an amino acid linker sequence recognized by the PreScission protease (Amersham Biosciences) and a multiple cloning sequence (MCS) to the 3′ side of the linker. pGEX–6P1 also contains the $lacI^q$ gene encoding for the lac repressor and the $amp^r$ selectable marker. The coding regions for both human Ubc13 and Mms2 were amplified by PCR from plasmids containing each of their coding sequences (Xiao *et al.*, 1998; Yamaguchi *et al.*, 1996). The oligonucleotides used for amplification introduced a *Bam*HI site at the 5′ end and a *Sal*I site at the 3′ end of these coding sequences, facilitating cloning into the pGEX-6P1 MCS in frame with the GST coding sequence.

For the expression of both GST-Ubc13 and GST-Mms2:

1. Transform the pGEX-6P1 derivatives into the *E. coli* BL21 ($DE_3$)-RIL (Stratagene) strain and plate onto LB media containing 50 $\mu$g/ml ampicillin (LB-A).
2. For a starter culture, inoculate 50 ml LB-A liquid media in a 500-ml Erlenmeyer flask with a single colony and grow the culture overnight at 37° with shaking.
3. Using the 50-ml starter culture, inoculate 2 liter of fresh LB-A liquid media in a 4-liter Erlenmeyer flask and incubate at 37° with shaking to a final absorbance of 0.4 at 590 n*M*.
4. Induce expression of the GST fusion by adding isopropyl-$\beta$-D-galactosidase (IPTG) to final concentration of 0.4 m*M* and incubate the culture at 30° with shaking for an additional 10 h.
5. Harvest the cells by centrifugation (we pellet the cells at 3000 rpm for 20 min at 4° using a Damon DPR–6000 centrifuge), and resuspended the pellet in 25 ml of lysis buffer [20 m*M* Tris, pH 7.9, 10 m*M* $MgCl_2$, 1 m*M*

EDTA, 5% glycerol, 1 m$M$ DTT, 0.3 $M$ ammonium sulfate, and 1 m$M$ phenylmethysulfonyl fluoride (PMSF); 20 mg/ml PMSF in isopropanol is prepared fresh prior to addition to the lysis buffer].

6. Lyse cells using a French press and clarify by ultracentrifugation at 40,000 rpm for 1 h (Beckman L7-65 ultracentrifuge with a Ti70 rotor). Expression of the GST fusion may not be apparent in the cell lysate as shown for GST-Mms2 (Fig. 2B, lane 1).

7. Dialyze lysates (~25 ml) against 4 liter of PBS (140 m$M$ NaCl, 2.7 m$M$ KCl, 10 m$M$ $HPO_4$, 1.8 m$M$ $KH_2PO_4$, pH 7.3) overnight at 4° and then clarify further by syringe filtration using a 0.45-$\mu$m low protein binding filter (MilliPore). This dialysis step ensures that the GST fusion protein is in a buffer that is optimal for subsequent column purification.

Purification is achieved by (1) binding the GST fusions to a glutathione column, (2) elution of the GST fusions from the column using glutathione, (3) cleavage of GST using the PreScission protease, and (4) separation of Ubc13 or Mms2 from GST and PreScission protease by column chromatography.

1. Load the clarified lysate (~25 ml) onto a 5-ml glutathione Sepharose 4B column (Amersham Biosciences) equilibrated with 50 ml of PBS buffer (140 m$M$ NaCl, 2.7 m$M$ KCl, 10 m$M$ $Na_2HPO_4$, 1.8 m$M$ $KH_2PO_4$, pH 7.3).

2. Wash the column with 90 ml PBS to elute unbound protein.

3. Elute the bound GST fusions with 15 ml of elution buffer (50 m$M$ Tris-Cl, pH 8.0, and 10 m$M$ reduced glutathione). The eluant should consist primarily of the GST fusion protein (Fig. 1B, lane 2).

4. Dialyze the 15 ml of eluant against 4 liter of PreScission cleavage buffer (50 m$M$ Tris-Cl, pH 7.0, 150 m$M$ NaCl, 1 m$M$ EDTA, 1 m$M$ DTT) for 4 h at 4°. This step also removes the glutathione used in the elution buffer.

5. Add 40 units of PreScission protease (Amersham Biosciences) to the dialysate and then incubate at 4° for 10 h. Based on SDS–PAGE analysis, GST cleavage is complete under these conditions with no detectable levels of nonspecific cleavage by the protease.

6. Remove the cleaved GST by loading the sample onto a 5-ml glutathione Sepharose 4B column (Amersham Biosciences) and elute with 15 ml PBS. Collect the flow through and filter concentrate to approximately 5 ml. The column can be regenerated by washing with 15 ml of elution buffer followed by equilibration with 50 ml of PBS.

7. Load the concentrated sample onto a Superdex 75 HR 16/60 gel exclusion column equilibrated with 50 m$M$ HEPES, pH 7.5, 150 m$M$ NaCl, 1 m$M$ EDTA, and 1 m$M$ DTT.

8. Pool peak fractions, and filter concentrate to desired volume. Add glycerol (5%, v/v), aliquot the preparation, flash freeze in liquid nitrogen,

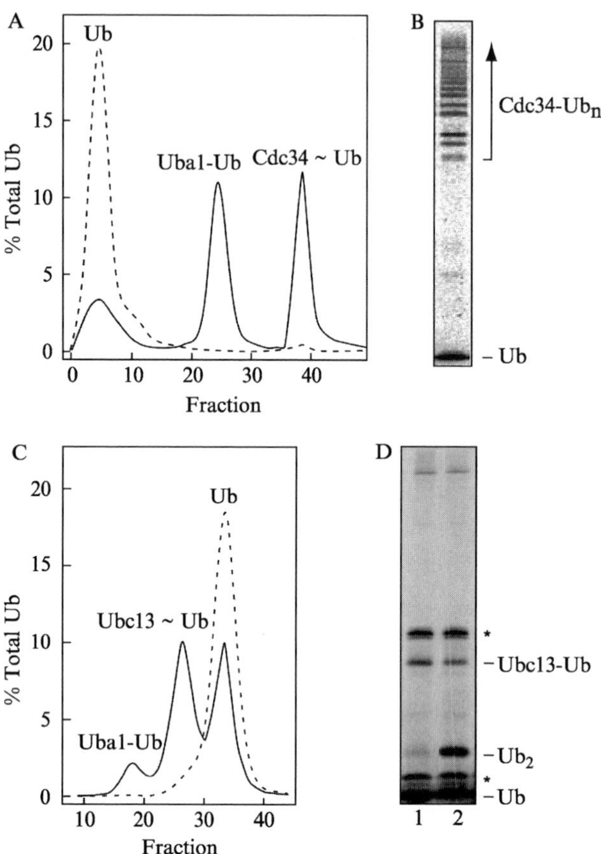


FIG. 2. Assaying Cdc34 and Ubc13·Mms2 activity *in vitro*. (A) Cdc34 $\sim$ Ub thiolester synthesis. Elution profiles of a Cdc34 $\sim$ ($^{35}$S-Ub) thiolester reaction either untreated (solid line) or treated (dotted line) with 100 m*M* DTT prior to loading onto a HiTrap Q-Sepharose HP anion-exchange column (Amersham Biosciences) are shown. (B) Cdc34 autoubiquitination. The same reaction conditions as used for the thiolester assay were employed except that the reaction time was increased to 2 h. The reaction was then treated with 100 m*M* DTT to hydrolyze Ub thiolester species and was loaded onto a 12% SDS–polyacrylamide gel. Reaction products containing $^{35}$S-Ub were visualized by autoradiography. (C) Ubc13 $\sim$ Ub thiolester synthesis. Elution profiles of a Ubc13 $\sim$ Ub thiolester reaction either untreated (solid line) or treated (dotted line) with 100 m*M* DTT prior to loading onto a Superdex 75 HR 10/30 gel filtration column (Amersham Biosciences) are shown. (D) Ubc13 autoubiquitination and Ub$_2$ synthesis by the Ubc13·Mms2 complex. The same reaction conditions as used for the Ubc13 $\sim$ Ub thiolester assay were employed except that Mms2 (100 n*M*) was included in the reaction shown in lane 2. Both reactions were incubated for 4 h, treated with 100 m*M* DTT to hydrolyze Ub thiolester species, and loaded onto an 18% SDS–polyacrylamide gel and reaction products containing $^{35}$S-Ub visualized by autoradiography. * indicates contaminants present in the $^{35}$S-Ub preparation.

and store at −80°. Check purity of the preparation by SDS–PAGE and determine the concentration using the BCA assay (Pierce). These preparations yield approximately 5–10 mg of Ubc13 and 10–20 mg of Mms2 with greater than 90% purity (Fig. 1B, lane 3).

## Preparation of Labeled Cdc34, Ubc13, and Mms2

Various studies required the use of labeled forms of these and other Ubcs as well as Ub, including $^{35}$S labeled for biochemical analysis (Gwozd *et al.*, 1995; Hodgins *et al.*, 1996; McKenna *et al.*, 2001; Ptak *et al.*, 1994; Varelas *et al.*, 2003), $^{13}$C/$^{15}$N labeled for nuclear magnetic resonance (NMR) chemical shift assignments, $^{15}$N labeled for NMR chemical shift perturbation studies (Hamilton *et al.*, 2001; McKenna *et al.*, 2001, 2003), and seleno-L-methionine derivatives for crystal structure determinations (Moraes *et al.*, 2001). Expression of these labeled forms was generated using variations of the expression protocols described earlier.

### *$^{35}$S-Labeled Cdc34*

1. Follow steps 1–3 from the expression protocol described earlier for Cdc34, except inoculate 100 ml of LB-AK media in a 500-ml Erlenmeyer flask with 2.5 ml of starter culture.
2. Pellet cells by centrifugation (we pellet the cells at 5000 rpm for 10 min. using a Sorvall RC-5B refrigerated centrifuge with a GSA rotor), wash with 50 ml M9 media (5 g $Na_2HPO_4$, 3 g $KH_2PO_4$, 1 g $NH_4Cl$, 0.5 g NaCl per liter), and repeat.
3. Pellet the cells again, resuspend in 100 ml M9 media supplemented with 1 m*M* $MgSO_4$, 0.1 m*M* $CaCl_2$, 12 m*M* glucose, 18 $\mu$g/ml thiamine, all amino acids (each at 40 $\mu$g/ml final) with the exception of cysteine and methionine, 50 $\mu$g/ml ampicillin, and 75 $\mu$g/ml kanamycin, and then incubate at 30° for 1 h with shaking.
4. Shift the culture to a 42° water bath for 20 min, add rifampicin (20 mg/ml in methanol) to a final concentration of 200 $\mu$g/ml, and incubate at 42° for an additional 10 min.
5. Shift the culture to 37° for 1 h, add [*trans*-$^{35}$S]methionine (25 $\mu$Ci), and incubate with shaking for an additional 10 min.
6. Pellet the cells by centrifugation as described earlier and lyse. Given the smaller volume and use of radioactive material, cells are lysed using lysozyme (Gonda *et al.*, 1989) instead of a French press.
7. Purify $^{35}$S-Cdc34 as outlined previously for Cdc34 using appropriately sized columns, such as the Mono Q HR 5/10 anion-exchange column

and the Superdex 75 HR 10/30 gel filtration column (Amersham Biosciences).

8. Determine the purity of $^{35}S$-labeled Cdc34 by SDS–PAGE followed by autoradiography and the Cdc34 concentration using the BCA assay (Pierce). The specific activity of purified $^{35}S$-Cdc34 using this method is approximately 1–2 $\times$ $10^4$ cpm/$\mu$g.

### *$^{15}N$, $^{13}C/^{15}N$, and Seleno-L-Methionine-Labeled Ubc13 and Mms2*

Expression of $^{15}N$ and $^{13}C/^{15}N$-labeled Mms2 and Ubc13 followed the same protocol outlined earlier with the exception that the starter culture is used to inoculate 2 liter of M9 media supplemented with 2 m*M* MgSO4, 0.2 m*M* $CaCl_2$, 1% glucose, and 50 mg/ml ampicillin. For $^{15}N$ labeling, replace $NH_4Cl$ in the M9 media with $^{15}$NH4Cl and for $^{13}C$ labeling replace glucose with $^{13}$C6-glucose. Grow cultures at 25° to an absorbance of 0.3 at 590 n*M*. To induce expression, add IPTG to a final concentration of 0.4 m*M* and incubate for 24 h at 25° with shaking.

For the expression of seleno-L-methionine derivatives of Mms2 and Ubc13, use the starter culture to inoculate 2 liter of M9 media supplemented with 2 m*M* $MgSO_4$, 0.2 m*M* $CaCl_2$, 1% glucose, and 50 $\mu$g/ml ampicillin. Incubate at 37° with shaking to an absorbance of 0.4 at 590 n*M* and then add 100 mg seleno-L-methionine, 200 mg lysine-HCl, 200 mg threonine, 100 mg leucine, 200 mg phenylalanine, 100 mg isoleucine, and 100 mg valine. Incubate at 25° for 30 min and then add IPTG to a final concentration of 0.4 m*M*. Incubate for 10 h at 25°. Purify labeled Ubc13 and Mms2 following the same procedures described previously for their unlabeled counterparts.

## Measuring Cdc34 and Ubc13·Mms2 Activity

The Ub thiolester formation and Ub conjugation activities of purified Cdc34 and Ubc13·Mms2 may be tested using a simple *in vitro* assay. The same reaction is used to analyze both of these activities with differences in incubation times and in the methods used to analyze reaction products.

### *Thiolester Assays*

Components of the thiolester assay include reaction buffer [50 m*M* HEPES, pH 7.5 (we have found that HEPES concentrations can be varied between 10 and 50 m*M* with no observable affect on the reaction), 40 m*M*

NaCl, 5 m*M* ATP, 5 m*M* $MgCl_2$, 0.6 units/ml inorganic pyrophosphatase, protease inhibitors (antipain, chymostatin, leupeptin, pepstatin A all at 20 $\mu$g/ml and PMSF at 180 $\mu$g/ml)], purified Ubc (100 n*M*), purified Ub (200 n*M*), and purified ubiquitin activating enzyme (10 n*M*). We use recombinant Uba1-$His_6$ purified from yeast (Varelas *et al.*, 2003). The reaction is incubated for 5 min at 30° and is then quenched by the addition of EDTA to final concentration of 50 m*M*. EDTA inhibits the reaction by chelating $Mg^{2+}$, thereby inhibiting Uba1 function. The short reaction time favors quantitative Ub thiolester formation while minimizing Ub conjugation.

To monitor the production of either Cdc34 ~ Ub or Ubc13 ~ Ub thiolester, $^{35}$S-Ub is used in the reaction. Expression and purification of $^{35}$S-Ub follows the same procedures described for $^{35}$S-Cdc34, keeping in mind that Ub elutes from ion-exchange columns in the flow through. Thiolester synthesis has been traditionally analyzed by coupling SDS–PAGE and autoradiography to visualize the Ubc ~ Ub product (Goebl *et al.*, 1988; Wu *et al.*, 2003). In our hands, this method of analysis tends to result in the smearing of $^{35}$S-Ub during electrophoresis, likely stemming from hydrolysis of the Ubc ~ ($^{35}$S-Ub) thiolester bond upon SDS denaturation. As an alternative, we analyze thiolester synthesis by column chromatography under native conditions. This chromatographic step also allows for the quantitation (Varelas *et al.*, 2003) and purification of the Ubc ~ Ub thiolester for mechanistic studies.

Figure 2 shows examples of column-based Cdc34 and Ubc13 thiolester assays. In both cases, 1-ml reactions were set up using the conditions outlined earlier. After the 5-min incubation, the reaction was split into two 500-$\mu$l samples. One sample was immediately subjected to column chromatography and the second was treated with 100 m*M* DTT prior to column chromatography. For Ubc13 ~ ($^{35}$S-Ub) the samples were loaded onto a Superdex 75 HR 10/30 gel filtration column equilibrated with 50 m*M* HEPES, 150 m*M* NaCl, 1 m*M* EDTA (50 $\mu$g/ml of bovine serum albumin may also be included in this buffer to minimize nonspecific interactions between reaction components and the column matrix during elution). Fractions were collected (0.5 ml), and 50 $\mu$l from each fraction was used for scintillation counting. Using these counts, an elution profile was generated by plotting the amount of $^{35}$S-Ub present in each fraction as a percentage of the total $^{35}$S-Ub that eluted from the column (Fig. 2C). We found that greater than 90% of the $^{35}$S-Ub added to the reaction was recovered after the column run. The resulting elution profile showed three peaks corresponding to Uba1 charged with Ub (Uba-Ub), Ubc13 ~ Ub thiolester, and free Ub. The DTT-treated sample was analyzed in the same fashion. The resulting elution profile showed only free Ub, indicating that Uba-Ub and Ubc13 ~ Ub peaks are DTT sensitive and so contain thiolester. To probe Ubc13 ~ Ub thiolester synthesis within the context of the Ubc13·Mms2 complex, the same procedures just described were employed except that Mms2 was added to the reaction at an equal molar concentration to Ubc13 (McKenna *et al.*, 2003).

An alternative chromatographic step was required for the Cdc34 ~ Ub thiolester assay given that its elution from a gel filtration column overlaps that of Uba-Ub. To facilitate separation of these species, anion-exchange chromatography was employed exploiting the different pI values of Ub, Uba1, and Cdc34. Thus, the Cdc34 ~ Ub thiolester reaction was loaded onto a HiTrap Q-Sepharose HP anion-exchange column (Amersham Biosciences) equilibrated with 50 m*M* Tris-Cl, pH 7.5, and eluted with a 0–800 m*M* NaCl gradient over 70 ml. An elution profile was generated in the same manner as described previously for the Ubc13 ~ Ub assay (Fig. 2A). This profile shows the elution of $^{35}$S-Ub in the flow through, Uba1-Ub at ~250 m*M* NaCl, and Cdc34 ~ Ub at ~470 m*M* NaCl.

We have also scaled the thiolester assay for NMR chemical shift perturbation studies to map surface residues on Ub, Ubc13, and Mms2 that participate in the formation of the (Ubc13 ~ Ub)·Mms2·Ub ubiquitination complex. These reactions required the use of $^{15}$N-labeled protein at much higher concentrations than that used in the standard reaction outlined earlier. For example, to identify Ubc13 surface residues that interact with Ub in the Ubc13 ~ Ub thiolester complex, the reaction included $^{15}$N-Ubc13 and Ub at 300 $\mu M$ and Uba1 at 0.3 $\mu M$. NMR spectra were acquired between 10 and 120 min after the addition of Uba1. Comparison of spectra for $^{15}$N-Ubc13 ~ Ub and $^{15}$N-Ubc13 alone identified chemical shift perturbations stemming from thiolester formation identifying amino acid residues on the surface of Ubc13 that interact with Ub upon thiolester formation (McKenna *et al.*, 2003).

### *Ub Conjugation Assays*

Both purified Cdc34 and Ubc13·Mms2 exhibit *in vitro* Ub conjugation activities in the absence of a ubiquitin protein ligase. Cdc34 has been shown to autoubiquitinate itself with polyubiquitin chains situated on lysine residues found near its carboxy terminus (Banerjee *et al.*, 1993). Ubc13 monoubiquitinates itself on K92, whereas the Ubc13·Mms2 complex synthesizes a K63-linked $Ub_2$ chain (McKenna *et al.*, 2001). To assay for these Ub conjugation activities, the same reaction conditions described for Ub thiolester synthesis were used over extended reaction times.

Figure 2B shows a Cdc34 autoubiquitination reaction in which 100 n*M* Cdc34, 200 n*M* $^{35}$S-Ub, and 10 n*M* Uba1 were incubated at 30° for 2 h. The reaction was quenched by the addition of 100 m*M* DTT to hydrolyze the remaining Ub thiolester. To precipitate reaction components, TCA (10%, v/v) was added and the sample was incubated on ice for 10 min and then centrifuged for 10 min at 14,000 rpm using a tabletop microfuge. The resulting pellet was dried and resuspended in SDS load mix containing 100 m*M* DTT. The sample was loaded and run over a 12% SDS–polyacrylamide gel, and reaction products containing $^{35}$S-Ub were visualized by autoradiography.

Figure 2D shows reactions in where Ubc13 (100 n*M*) was incubated with $^{35}$S-Ub (200 n*M*) and Uba1 (10 n*M*) in the absence (Fig. 2D, lane 1) or presence (Fig. 2D, lane 1) of Mms2. The same procedures described earlier for Cdc34 Ub conjugation activity were employed here except that the reactions were incubated for 4 h and the samples were loaded and run over an 18% SDS–polyacrylamide gel.

Together, the Ub thiolester and Ub conjugation assays provide a means by which the activity of purified Cdc34 and Ubc13·Mms2 may be measured *in vitro*. In addition, these assays have proven to be useful tools for the mechanistic characterization of these Ubcs.

## References


Banerjee, A., Gregori, L., Xu, Y., and Chau, V. (1993). The bacterially expressed yeast CDC34 gene product can undergo autoubiquitination to form a multiubiquitin chain-linked protein. *J. Biol. Chem.* **268,** 5668–5675.

Goebl, M. G., Yochem, J., Jentsch, S., McGrath, J. P., Varshavsky, A., and Byers, B. (1988). The yeast cell cycle gene *CDC34* encodes a ubiquitin-conjugating enzyme. *Science* **241,** 1331–1335.

Gonda, D. K., Bachmair, A., Wunning, I., Tobias, J. W., Lane, S. W., and Varshavsky, A. (1989). Universality and structure of the N-end rule. *J. Biol. Chem.* **264,** 16700–16712.

Gwozd, C. S., Arnason, T., Cook, W. J., Chau, V., and Ellison, M. J. (1995). The yeast Ubc4 ubiquitin conjugating enzyme monoubiquitinates itself *in vivo*: Evidence for an E2-E2 homointeraction. *Biochemistry* **34,** 6296–6302.

Hamilton, K. S., Ellison, M. J., Barber, K. R., Williams, R. S., Huzil, J. T., McKenna, S., Ptak, C., Glover, M., and Shaw, G. S. (2001). Structure of a conjugating enzyme-ubiquitin thiolester intermediate reveals a novel role for the ubiquitin tail. *Structure* **9,** 897–904.

Hodgins, R., Gwozd, C., Arnason, T., Cummings, M., and Ellison, M. J. (1996). The tail of ubiquitin-conjugating enzyme redirects multi-ubiquitin chain synthesis from the lysine 48-linked configuration to a novel nonlysine-linked form. *J. Biol. Chem.* **271,** 28766–28771.

McKenna, S., Moraes, T., Pastushok, L., Ptak, C., Xiao, W., Spyracopoulos, L., and Ellison, M. J. (2003). An NMR-based model of the ubiquitin-bound human ubiquitin conjugation complex Mms2·Ubc13. *J. Biol. Chem.* **278,** 13151–13158.

McKenna, S., Spyracopoulos, L., Moraes, T., Pastushok, L., Ptak, C., Xiao, W., and Ellison, M. J. (2001). Noncovalent interaction between ubiquitin and the human DNA repair protein Mms2 is required for Ubc13-mediated polyubiquitination. *J. Biol. Chem.* **276,** 40120–40126.

Moraes, T., Edwards, R. A., McKenna, S., Pastushok, L., Xiao, W., Glover, J. N. M., and Ellison, M. J. (2001). Crystal structure of the human ubiquitin conjugating enzyme complex, hMms2-hUbc13. *Nat. Struct. Biol.* **8,** 669–673.

Ptak, C., Gwozd, C., Huzil, J. T., Gwozd, T. J., Garen, G., and Ellison, M. J. (2001). Creation of a pluripotent ubiquitin-conjugating enzyme. *Mol. Cell. Biol.* **21,** 6537–6548.

Ptak, C., Prendergast, J. A., Hodgins, R., Kay, C. M., Chau, V., and Ellison, M. J. (1994). Functional and physical characterization of the cell cycle ubiquitin-conjugating enzyme CDC34 (UBC3): Identification of a functional determinant within the tail that facilitates CDC34 self-association. *J. Biol. Chem.* **42,** 26539–26545.

Studier, F. W., Rosenberg, A. H., Dunn, J. J., and Dubendorff, J. W. (1990). Use of bacteriophage T7 RNA polymerase to direct selective high-level expression of cloned genes. *Methods Enzymol.* **185,** 60–89.

Tabor, S., and Richardson, C. C. (1985). A bacteriophage T7 RNA poymerase/promoter system for controlled exclusive expression of specific genes. *Proc. Natl. Acad. Sci. USA* **82,** 1074–1078.

Varelas, X., Ptak, C., and Ellison, M. J. (2003). Cdc34 self-association is facilitated by ubiquitin thiolester formation and is required for its catalytic activity. *Mol. Cell. Biol.* **23,** 5388–5400.

Wu, P.-Y., Hanlon, M., Eddins, M., Tsui, C., Rogers, R. S., Jensen, J. P., Matunis, M. J., Weissman, A. M., Wolberger, C. P., and Pickart, C. M. (2003). A conserved catalytic residue in the ubiquitin-conjugating enzyme family. *EMBO J.* **22,** 5241–5250.

Xiao, W., Lin, S. L., Broomfield, S., Chow, B. L., and Wei, Y. F. (1998). The products of the yeast *Mms2* and two human homologs (*hMms2* and *CROC-1*) define a structurally and functionally conserved Ubc-like protein family. *Nucleic Acids Res.* **26,** 3908–3914.

Yamaguchi, T., Kim, N. S., Sekine, S., Seino, H., Osaka, F., Yamao, F., and Kato, S. (1996). Cloning and expression of cDNA encoding a human ubiquitin-conjugating enzyme similar to *Drosophila bendless* gene product. *J. Biochem. (Tokyo)* **120,** 494–497.

# [6] Expression, Purification, and Properties of the Ubc4/5 Family of E2 Enzymes

*By* KEVIN L. LORICK, JANE P. JENSEN, and ALLAN M. WEISSMAN

## Abstract

Ubiquitin-conjugating enzymes (E2s) play a central role in ubiquitylation. They function to bridge the first, nonspecific step of ubiquitin activation by E1 with the transfer of activated ubiquitin to substrates by substrate-specific E3s. While sharing a common core UBC domain, members of this family exhibit significant specificity in their physical and functional interactions with E3s. Among the families of E2s, members of the


0076-6879/05 $35.00
DOI: 10.1016/S0076-6879(05)98006-3

yeast Ubc4/5 family are particularly well conserved in higher metazoans. In humans, these are represented by the UbcH5 family. Members of this ubiqutiously expressed family show a capacity to interact with a wide range of E3s from both HECT and RING finger families, making them particularly useful tools in the laboratory. Using the UbcH5 family as a prototype, this chapter describes methods for the expression, purification, and characterization of E2 enzymes *in vitro* and some of the basics for their use in experiments in cells.

## Introduction

Ubiquitylation occurs as the result of a multienzyme process that utilizes a cascade of three enzyme classes referred to as E1, E2, and E3. A single E1 interacts with all ubiquitin-specific E2s. There are 11 such E2s in yeast and at least 33 in humans. In humans each E2 interacts with a subset of over 500 different ubiquitin ligases. E3 catalytic domains include HECT domains, RING and PHD fingers, and U-boxes. Variations in the conserved core ubiquitin-conjugating (UBC) domain define the various families of E2s. Several families of E2 are further distinguished by the presence of regions either N- or C-terminal to the core UBC domain. The *sine qua non* of active E2s is a conserved catalytic cysteine within the UBC domain that forms a thiolester linkage with the C terminus of ubiquitin. Some E2s function with ubiquitin-like molecules instead of with ubiquitin, e.g., UbcH9/Ube2i for Sumo1 or UbcH12/Ube2M for Nedd8. UbcH8/Ube2L6 is unique in its capacity to function both with ubiquitin and with the interferon-inducible ubiquitin-like modifier ISG15/UCRP (Zhao *et al.*, 2004). However, there is also a class of E2-like proteins that retain the core UBC domain but lack the canonical active site cysteine. Some of these ubiquitin enzyme variants (UEVs) cooperate with active E3s to form specific ubiquitin linkages. This is exemplified by Uev1/MMS2, which functions together with UbcH13 and the RING finger E3, Traf6, to specifically mediate K63 polyubiquitin chains linkages (see Hofmann and Pickart, 1999).

The E2s with highest homology to yeast Ubc4p and Ubc5p are represented by the UbcH5 family in humans (Jensen *et al.*, 1995; Scheffner *et al.*, 1994). This group of proteins is the most evolutionarily conserved family of E2s with sequence identity and similarity between yeast and human of $\sim$ 80 and 90%, respectively. Not only are these E2s highly conserved, but they appear to have structural characteristics that allow them to function productively, at least *in vitro*, with a wide range of E3s of various classes. In fact, with the exception of E3s that function with E2-Uev1 dimers, few active ubiquitin ligases have been described for which at least one member of the UbcH5 family is not a productive enzymatic partner.

The UbcH5 family of ubiquitin-conjugating enzymes (E2s) consists of four genes and at least four pseudogenes. The coding sequences collectively produce 13 alternatively spliced RNA sequences. The resulting proteins exist in 8 isoforms. These are described as UbcH5A (E2D1); UbcH5B (E2D2) isoforms 1 and 2; UbcH5C (E2D3) isoforms 1, 2, and 3; and two forms of a brain-expressed ubiquitin-conjugating enzyme, hBUCE, which should be referred more properly to as UbcH5D or E2D4. The most commonly studied members of this family are the full-length, catalytically active isoforms (isoform1) of UbcH5A-C. These human proteins have at least 88% identity and 95% similarity to each other. All of these polypeptides contain an active site cysteine and all but UbcH5B isoform 2 and UbcH5D/hBUCE1 isoform 2 are full length.

The names of E2s of the same family are highly varied between species, even when the amino acid sequences are identical. This has resulted in an extremely confusing and potentially misleading nomenclature. Table I is an attempt to clarify some of the confusion. We have used current GenBank reference sequence nomenclature for the human ortholog of each E2 (i.e., Ube2 for ubiquitin pathway enzyme 2). Accession numbers provided are current GenBank reference sequences where available.

The following sections provide some commonly used protocols to express, purify, and evaluate the activity of E2 using members of the UbcH5 family as the prototype. Such protocols should be easily adapted to most other E2s. Also provided in this chapter is a section that should help in the design of experiments to assess the activity of E2s in cells. Before embarking on E2 production and purification it is worth considering the cost-effectiveness and applicability of the growing number of commercially available E2s.

## Expression of E2s in *Escherichia coli* without a Purification Tag

Due to their generally small size, E2s are amenable to expression in bacteria with the attendant advantages of rapid production and low cost. For most applications, a "crude bacterial lysate" containing expressed protein is sufficient for use in *in vitro* ubiquitylation assays. As bacteria lack a ubiquitin-conjugating system, this eliminates the need to consider effects of contamination with other components of the ubiquitin-conjugating system. Thus, for many purposes, tags that allow for purification are not a necessity. Typically, plasmids for protein production use a lac promoter and their expression is controlled by the addition of the lac inducer, isopropyl-$\beta$-D-thiogalactoside (IPTG). Alternatively, they can be expressed from a phage T7 promoter, with expression controlled by IPTG in bacterial strains containing inducible T7 RNA polymerase, i.e., a DE3 plasmid. The

TABLE I
CURRENT E2 NOMENCLATURE

| Common name(s) of human E2/E2 variant | "Official" name | Refseq/GenBank ID | Closest *S. cerevisiae* E2 homolog |
|---|---|---|---|
| **A. E2s for ubiquitin or ubiquitin-like molecules** | | | |
| Apollon, BRUCE, BIR6 | BIRC6 | NP_057336 | Ubc7p/Qri8p |
| HHR6A, RAD6A | Ube2a | NP_003327 | Rad6p/Ubc2p |
| HHR6B, RAD6B | Ube2b | NP_003328 | Rad6p/Ubc2p |
| UbcH10 | Ube2c | NP_008950 | Ubc11p |
| UbcH5A | Ube2d1 | NP_003329 | Ubc4p or Ubc5p |
| UbcH5B | Ube2d2 | NP_003330 | Ubc4p or Ubc5p |
| UbcH5c | Ube2d3 | NP_003331 | Ubc4p or Ubc5p |
| UbcH5D, hBUCE1 | Ube2d4 | NP_057067 | Ubc4p or Ubc5p |
| UbcH6 | Ube2e1 | NP_003332 | Ubc5p or Ubc4p |
| UbcH8 | Ube2e2 | NP_689866 | Ubc5p or Ubc4p |
| UbcH9, UbcM2, E2-23K | Ube2e3 | NP_872619, NP_006348 | Ubc4p or Ubc5p |
| UbcM2 | Ube2e4 | NP_997236 | Ubc4p or Ubc5p |
| Ubc7, E2-17K, HH *C. elegans* Ubc7 | Ube2g1 | NP_003333 | Ubc7p/Qri8p |
| HH mouse Ubc7, MmUbc7 | Ube2g2 | NP_003334 | Ubc7p/Qri8p |
| UbcH2, E2–20K, ubcH8 | Ube2h | NP_003335 | Ubc8p |
| ubcH9/sumo1-conjugating enzyme | Ube2i | NP_003336, NP_919235, NP_919236, NP_919237 | Ubc9p |
| NCUBE1, UBC6 homolog E, HSUBC6e, CGI-76, HSPC153/HSPC205 | Ube2j1 | NP_057420, NP_057105 | Ubc6p, some 3' homology to IRA2 |
| NCUBE2, UBC6 homolog, MmUbc6 | Ube2j2 | NP_919296 | Ubc6p |

*(continued)*

TABLE I *(continued)*

| Common name(s) of human E2/E2 variant | "Official" name | Refseq/GenBank ID | Closest *S. cerevisiae* E2 homolog |
|---|---|---|---|
| UbcH7, E2-18K, UbcM4, E2-F1, L-UBC | Ube2l3 | NP_003338 | Ubc4p or Ubc5p |
| UbcH8, RIG-B | Ube2l6 | NP_004214 | Ubc4p or Ubc5p |
| UbcH12, Nedd8-conjugating enzyme | Ube2m1 | NP_003960 | Ubc12p |
| Similar to E2M, UbcH12, Nedd8-conjugating enzyme | Ube2m2 | XP_497504 | Ubc12p |
| Nedd-conjugating enzyme | Ube2m3 | NP_542409 | Ubc12p or Ubc5p or Ubc4p or Ubc13p |
| UbcH13, bendless homolog | Ube2n1 | NP_003339 | Ubc13p |
| UBE2NL, similar to ubcH13 | Ube2n2 | XP_372257 | Ubc13p |
| Ube2Q | Ube2q | NP_060052 | Ubc11p or Ubc3/Cdc34p |
| cdc34, E2-32K | Ube2r1 | NP_004350 | Ubc3p/Cdc34 or Ubc7p |
| UBC3B | Ube2r2 | NP_060281 | Ubc3p/Cdc34 |
| E2-EPF5, E2-24K | Ube2s | NP_055316 | Ubc13p |
| E2-24K-2 | Ube2s2 | XP_496186 | Ubc13p |
| E2-25K, HsUBC1 | HIP2 | NP_005330 | Ubc1p or Ubc-X |
| KIAA1734, ortholog of mouse E2-230K | E2–230K | NP_071349 | Rad6p |
| HOYS7 | FLJ13855 | NP_075567 | Rad6p |
| Similar to *Drosophila* CG4502 | CG4502 | XP_059689 | Ubc3p/Cdc34 |
| Hypothetical | FLJ11011 | NP_001001481 | Ubc7p/Qri8p |
| Hypothetical | HSPC150 | NP_054895 | Ubc13p |
| Hypothetical | MGC31530 | NP_689702 | Rad6p |

| | | | |
|---|---|---|---|
| **B. E2 variants/UEVs** | | | |
| MMS2, CROC-1, IKK activator, UEV1a | Ube2v1 | NP_068823 | Mms2p |
| MMS2, EDAF/EDPF-1, DDVit 1, enterocyte differentiation associated/ promoting factor, vitamin $D_3$ inducible protein | Ube2v2 | NP_003341 | Mms2p |
| UEV3 | Ube2v3 | NP_060784 | STP22p |
| LOC441372, similar to E2V1 | Ube2v4 | XP_496988 | Mms2p |
| LOC92912 | Ube2q2 | NP_775740 | Ubc3p/Cdc34 |
| Fts, Ftl, fused toes homolog | FTH | NP_071921 | Ubc8p |
| TSG101 | TSG101 | NP_006283 | STP22p |

[a]See also http://home.ccr.cancer.gov/lpds/weissman.

BL21 or BL21(DE3) strains and their variants have reduced protease activity and are particularly useful for protein expression.

To produce UbcH5 proteins in bacteria, start by transforming *E. coli* with the E2-encoding plasmid, use a single colony to inoculate a 2-ml LB culture, and grow overnight at 37°. Transfer the culture to 200 ml sterile 2× YT (16 g Bacto-tryptone, 10 g Bacto yeast extract, 5 g NaCl/ liter) or other rich medium until the $OD_{600}$ is 0.6–1.0. Induce expression with 0.2 m*M* IPTG and grow an additional 1–2 h. *E. coli* are pelleted at 3000*g* for 15 min. The cell pellet may be frozen and stored at −80° or sonicated immediately.

To sonicate, add 8 ml of cold lysis buffer [50 m*M* Tris, pH 7.4–8.0, 1 m*M* EDTA, 1% Triton-X 100 supplemented with freshly made 5 m*M* dithiothreitol (DTT) and 2 m*M* phenylmethylsulfonyl fluoride (PMSF)] for each 200 ml of culture. Sonicate on ice for 30 s in 2 s pulses using a needle sonicator with a microtip probe set at the maximum level. These settings may vary with different sonicators. For culture volumes below 250 ml, sonication times may be reduced, with a change in pitch or the presence of foaming indicating complete lysis. Cellular debris is then pelleted at 20,000*g* at 4° for 15 min. The supernatant is a crude lysate containing the E2. This should be divided into small aliquots to avoid freeze–thaw cycles and stored at −80°. From a 100-ml culture of a 17-kDa protein, such as the UbcH5 family members, expect as much as 250 $\mu$g. For a larger protein, e.g., CDC34/Ube2R, a 100-ml culture may produce closer to 2.5 $\mu$g. Function will need to be assessed in a thiolester assay.

*Notes*: Other E2s vary in conditions required for maximal activity relative to UbcH5 family members. Therefore, consider varying the pH of the sonication buffer between 6.8 and 8.0 and, perhaps more importantly, varying the level of DTT (or other reducing agents) from between 0 and 10 m*M* (see section on thiolester bond formation).

The level of inducing agent and bacterial density, as well as time and temperature of induction, may all be varied to maximize protein yield and complete protein translation. IPTG may be varied from 0.1 to 1 m*M*, with smaller E2s, such as members of the UbcH5 family, generally requiring less. Bacteria should be in logarithmic growth phase at the time of induction with the optimal $OD_{600}$ of 0.6–1.0. The optimal induction time can vary from 30 min to 3 h. Longer incubation generally does not increase expression once the culture has reached the stationary phase. In general, larger proteins should be induced earlier and for a longer period of time. Finally, most *E. coli* machinery works best at 37°; however, because this also includes proteases, larger proteins may be produced with less degradation if their cultures are induced at 22–30° and, in some cases, as low as 16°.

## Purification of Tagged UbcH5 Family E2s

Plasmids encoding a number of tagged E2s, including members of the UbcH5 family, already exist that allow purification and that facilitate detection. When these plasmids are not available, tags may be added by cloning into commercially available vectors. E2s can be used with the tags intact or may be cleaved when appropriate protease recognition sites, such as thrombin, tobacco etch virus (TEV), enterokinase, or factor Xa, are present. All tags, especially larger ones, such as green fluorescent protein (GFP) or glutathione S-transferase (GST), have to be tested for specific applications to ensure that they do not alter or eliminate activity of the E2. In general, we tend to place tags at the N termini of E2s and have found that smaller tags such as $His_6$, Myc, Flag, or hemaglutinin (HA) have minimal effects on function. However, C-terminal tags have the advantage of allowing purification of the full-length protein away from incomplete translation products. One of the most commonly used tags for E2 constructs is $His_6$. The purification of $His_6$-tagged proteins using a "batch method" is described later. If a protease cleavage site has been engineered into the cloning vector, purification using other tags may be carried out using, for example, GST (see later) or an agarose-conjugated antitag antibody followed by cleavage with protease.

### *Purification of His-Tagged E2s from* E. coli

To purify $His_6$-tagged E2, start by rinsing 25–50 $\mu$l NiNTA beads (Qiagen) or other affinity matrix twice with 50 m*M* NaPO4, pH 7.0. Add 1.0 ml crude bacterial lysate (up to 500 $\mu$g, depending on the capacity of the matrix) to the beads and adjust the concentration of the solution to 8 m*M* imidazole by adding 8 $\mu$l of 1 *M* imidazole (this serves to reduce nonspecific binding). Allow the protein to bind to the beads by tumbling 2 h at 4°. Pellet beads and remove the depleted supernatant, saving an aliquot to later check the efficiency of binding by electrophoresis. Wash beads three times with 1 ml wash buffer (50 m*M* $NaPO_4$, pH 7.0, 300 m*M* NaCl, 8 m*M* imidazole). Elute the protein from the beads by adding 500 $\mu$l elution buffer (50 m*M* $NaPO_4$, pH 7.0, 300 m*M* NaCl, 250 m*M* imidazole) and incubating 10 min at room temperature. Collect the supernatant into a fresh tube on ice. Repeat the elution and pool supernatants. At this point imidazole should be removed by dialyzing the eluate at 4° against 50 m*M* $NaPO_4$, pH 7.5, with 2 m*M* DTT. After dialysis, add an equal volume of glycerol and mix thoroughly. The protein in 50% glycerol can be stored at −20°. Before using the E2 to evaluate an E3 or ubiquitylate a substrate its activity should be evaluated by E1-dependent thiolester bond formation, particularly if there is an intention to compare multiple E2s.

*Note:* When highest purity is required, it helps to determine a narrow "window" of imidizole concentrations to use in the wash and elution steps, i.e., find the highest concentration of imidizole in the wash that does not elute the protein and the lowest concentration in the elution step that effectively elutes the protein. DTT is not included in the bead-washing steps because of effects on the NiNTA beads. Some reactions with E2s, particularly those involving RING finger proteins, may be sensitive to phosphate buffers. If this is the case, dialysis may be performed using a buffer containing 50 m*M* Tris-HCl, pH 7.5, in place of 50 m*M* $NaPO_4$. Again, consider varying the pH of the dialysis buffer with each different E2 to maximize activity.

### *Purification of E2s from* E. coli *Using Glutathione-Sepharose and Thrombin Cleavage*

Plasmids are available from several sources that utilize a GST tag for protein purification. These are available with multiple reading frames and a choice of cleavage sites to remove the tag, e.g., thrombin, factor Xa, enterokinase, or TEV protease. This protocol utilizes a plasmid construct carrying an N-terminal GST tag followed by a thrombin cleavage site in front of E2. The protein can be used as a GST fusion or can be cleaved following purification to remove the tag. Complete purification of the cleaved protein is possible by removing the thrombin with benzamidine-Sepharose (Amersham Biosciences). The protocol for generating GST-E2 is the same as for untagged E2. The crude lysate is the starting material for this protocol.

To prepare purified GST-E2, add 100 $\mu$l glutathione-Sepharose bead slurry to a 1.5-ml microfuge tube. Wash three times with 50 m*M* Tris, pH 7.5. Add 750 $\mu$l crude bacterial lysate to the beads (up to 100 $\mu$g). Allow the GST protein to bind at room temperature for 30–60 min by tumbling. Pellet the beads and remove the depleted supernatant, saving an aliquot for later evaluation of binding efficiency. Wash the beads four times with 50 m*M* Tris, pH 7.5. At this point, the protein may be eluted with glutathione and used directly as a GST fusion or the E2 may be cleaved away from the GST.

To generate purified protein without the GST tag, add three times the bead volume of thrombin in PBS (50 units/ml final concentration). Tumble at room temperature for 2–3 h or overnight at 4°. If the cleavage appears incomplete (commercially available thrombin can be variable in activity) the length of incubation can be extended to as long as 16 h at room temperature. Pellet the beads and add 100 $\mu$l washed benzamidine-Sepharose beads to the supernatant to remove thrombin. Tumble the samples again at room temperature for 30–60 min. Pellet the beads and transfer the supernatant to a new tube. The purified E2 is in the supernatant. Expect to recover 90% of the cleaved E2 (up to 34 $\mu$g from 100 $\mu$g of GST UbcH5B). This may be stored in aliquots at $-80^{\circ}$ or mixed with an equal volume of glycerol and stored at $-20^{\circ}$. Test the E2 for thiolester formation before use.

*Alternatives.* If removing the GST tag is not necessary, the protein may be eluted from the glutathione-Sepharose beads after washing and used directly. However, in some cases it is necessary to remove excess reduced glutathione by dialysis because of effects on ubiquitylation.

Elute the labeled protein with 150 $\mu$l PBS containing 20 m*M* reduced glutathione by incubating for 10 min at $4^{\circ}$. Pellet the beads rapidly and transfer the eluate to a new tube. Repeat the elution step three times and combine the eluates. Remove any remaining beads by centrifugation and transfer the supernatant to a new tube. Test the E2 in a thiolester assay before using it in ubiquitylation reactions.

## Expression of E2s with *In Vitro* Transcription and Translation Systems

*In vitro* translation of $^{35}$S-labeled E2 in eukaryotic systems such as rabbit reticulocyte lysate or wheat germ extract provides a source of easily detectable material produced in a milieu that includes eukaryotic chaperones and the machinery necessary for many posttranslational modifications. E2s produced in these systems can be used for binding studies and evaluation of structure–function relationships, as well as other functions. There are several commercially available kits that allow for coupled transcription and translation that generally function quite well for the production of UbcH5 family E2s. In other cases it may be preferable for either technical or other reasons to first produce RNA and then carry out translation using commercial systems. The following are selected protocols for sequential *in vitro* transcription and translation. The translation described here uses commercially available *in vitro* translation kits (Promega).

### *Transcription of RNA*

In a 1.5-ml tube, mix

- 5 $\mu$l 10× transcription buffer (0.4 *M* Tris-HCl, pH 7.5, 0.1 m*M* $MgCl_2$, 50 m*M* DTT, 0.5 mg/ml BSA)
- 2 $\mu$l RNAsin (Promega)
- 4 $\mu$l rNTP mixture
- 2 $\mu$g plasmid DNA

the appropriate RNA polymerase for the vector
q.s. 50 $\mu$l water

Incubate at 30° for 1.5 h. Clean up the reaction by placing on commercially available RNA-binding columns or extract the RNA with phenol-chloroform followed by ethanol precipitation. Measure the RNA content by $OD_{260}$ and by gel electrophoresis. Use 100 ng–1 $\mu$g for translation. The remainder may be frozen down in aliquots at −80° for later use.

*Translation of Protein*

To a 1.5-ml tube add
5 $\mu$l 10× IVT buffer
25 $\mu$l reticulocyte lysate
4 $\mu$l amino acid mixture (–methionine or -cysteine)
2 $\mu$l [$^{35}$S]methionine (1000 Ci/mmol, Amersham Biosciences)
or
[$^{35}$S]cysteine (1000 Ci/mmol, Amersham Biosciences)
1 $\mu$l RNAsin
100 ng–1 $\mu$g *in vitro*-translated E2 mRNA
1 $\mu$l MG132 or other proteasome inhibitor (see *Notes*)
q.s. 50 $\mu$l water

Spin briefly to mix and incubate 1.5 h at 30°. Incorporation of $^{35}$S into proteins should be evaluated by TCA precipitation. Proteins should be directly evaluated immediately after synthesis by SDS–PAGE.

*Note:* Reticulocyte lysates are highly enriched in ubiquitylation machinery and proteasomes. On the one hand, these lysates have the added benefit of potentially providing all one needs to assess the activity of the E2. On the other hand, there is a possibility that E2s and other proteins endogenous to reticulocytes may mask effects of the translated E2 of interest. The high activity of proteasomes in reticulocytes may make inhibition of degradation problematic—high concentrations of proteasome inhibitor may be necessary to prevent this. A proteasome inhibitor is included in the protocol above for this reason; however, if the goal of an experiment is to detect degradation of E2, E3, or substrate, this should be omitted. While there is endogenous E1 in wheat germ, for some mammalian E2s there is the potential for lack of compatibility between the plant E1 and the mammalian E2—this should not be a problem for members of the UbcH5 family. *In vitro*-translated E2s may also be expressed with affinity tags and purified as described earlier. As there is a high concentration of protein in lysates used for *in vitro* translation, proteins may aggregate when boiled in SDS–PAGE sample buffer. To avoid this, we recommend heating to only 70°. If issues of aggregation persist, as evidenced by labeled material at the interface of the stacking and resolving gels that cannot be accounted for by ubiquitylated protein, consider incubating in SDS–PAGE sample buffer at 37° for 10–30 minutes before electrophoresis.

## Assessing Thiolester Formation between E2 and Ubiquitin

Whether comparing multiple E2s in E3 assays or evaluating other properties it is important to demonstrate that the E2s have activity and to compare this activity among E2s. To assess this, begin by evaluating the capacity of E2s to form E1-dependent thiolester bonds with ubiquitin. This generally results in adducts that migate 8 kDa above the E2 itself and disappear with addition of a reducing agent. There are multiple ways to evaluate thiolester bond formation, either by looking at an upward shift in the migration of the E2 or, particularly when issues of relative activity are concerned, to evaluate the E2 indirectly by looking at bound ubiquitin. The following is one such protocol:

Dilute E1 to 50 ng/$\mu$l in 100 m$M$ HEPES, pH 7.5. Set up two identical reactions for each E2 in 1.5-ml tubes as follows:

1.2 $\mu$l E1
1.2 $\mu$l E2 or appropriate negative control
1.2 $\mu$l 10× thiolester buffer (200 m$M$ Tris-HCl, pH 7.6, 500 m$M$ NaCl, 50 m$M$ ATP, 50 m$M$ $MgCl_2$)
1.0 $\mu$l $^{32}$P-labeled ubiquitin (10–40,000 cpm/$\mu$l) (see Yang *et al.*, 2005) or unlabeled ubiquitin (10 mg/ml )
q.s. 12 $\mu$l water

Incubate at room temperature for 5 min. Add 4 $\mu$l of 4× nonreducing SDS–PAGE sample buffer to one tube and 4 $\mu$l 4× reducing sample buffer to the second. After heating the samples, analyze by SDS–PAGE. It is crucial that the reduced and nonreduced samples are adequately separated on gels to prevent diffusion of reducing agent to the nonreduced samples. If radiolabeled ubiquitin is used (see Yang *et al.*, 2005), gels should be fixed in an appropriate acetic acid/methanol mixture prior to drying the gel. The radioactive signal is then detected by autoradiography. If unlabeled ubiquitin is used, ubiquitin may be detected by anti-ubiquitin immunoblotting.

*Note:* Two different E2s prepared under identical conditions may produce different amounts of ubiquitin thiolester bonds. This may be due in part to the relative activity of the proteins. This may in turn reflect differing requirements for DTT during purification to maintain the E2 at maximal activity. Thus, if there are large differences among E2s it may be necessary to experiment with varying DTT amounts used in production and purification. E1 and E2s lose activity after freeze–thaw cycles from −80°. We avoid this problem in one of two ways, we either purify the protein and store it at −20° in 50% glycerol or we aliquot the crude material in quantities suitable for individual experiments and store at −80°.

To guarantee that the amount of thiolester formation is equal when using various E2s, it is a good idea to test all of those to be used for a particular experiment at the same time. Results from the initial assessment can be utilized to equalize the thiolester formation potential and run a second assay. One problem that can occur with E2 equalization, particularly when an E2 exhibits relatively low activity, is that too much E2 added to the assay can result in a paradoxical decrease of the apparent activity in ubiquitylation assays.

## *In Vivo* Studies with UbcH5 Proteins

Studying E2 activity in cell culture usually consists of assessing specific E2–E3 interactions. This may be achieved partially by immunoprecipitation of a complex of E2 and E3, such as that seen with AO7 and UbcH5B/E2D2 (Lorick *et al.*, 1999); determining the localization of the E2, as with E2s involved in endoplasmic reticulum-associated degradation, such as MmUbc6/Ube2j2 and MmUbc7/Ube2g2 (Tiwari and Weissman, 2001) or transport to the nucleus, such as UbcM2 (Plafker and Macara, 2000); or evaluating the effects of mutation or knock down of E2, as with UbcH8/Ube2L6 (Zhao *et al.*, 2004).

Despite several instances of demonstrated stable E2–E3 interactions, in many cases where a particular E2–E3 pair has strong ubiquitin-modifying ability, no physical interaction is readily detectable by coimmunoprecipitation. Thus, a lack of demonstrable physical interaction either *in vitro* or by coimmunoprecipitation should not be taken as an indication that no functionally significant interaction exists. Conversely, there are instances where E2–E3 interactions have been demonstrated where the potential for a particular pair to functionally interact is questionable (Brzovic *et al.*, 2003; Lorick *et al.*, 2005).

Localization studies may provide hints as to the function of an E2. For instance, the potential role of the mammalian ortholog of Ubc7p, MmUBC7/Ube2g2, in ERAD is reinforced by the immunolocalization of Myc-tagged version of this E2 to the ER (Fang *et al.*, 2001; Tiwari and Weissman, 2001). CDC34/Ube2R1 is found on the mitotic spindle, as might be expected for a protein that degrades cyclins, but is both nuclear and cytoplasmic in interphase (Reymond *et al.*, 2000). One of the simplest ways to follow the localization of an E2 is using fluorescent tags and following the protein either in real time or in fixed tissues. An alternative to this is either direct immunofluorescence of the E2 or immunostaining of an epitope-tagged version of the protein. A caveat with the use of fluorescent proteins and epitope tags is that these may influence the normal function or localization of the proteins. For example, certain common epitope tags, such as Flag, contain lysine residues that, in some circumstances, may be substrates for ubiquitin modification.

The most useful way to study if a particular E2 may be involved in the control of specific substrate modification is to overexpress a "dominant-negative" mutant form of the protein or to use small interfering RNAs (siRNAs) to reduce the amount of active E2 available. Generally, potential dominant negatives are generated by mutation of the active site cysteine. A number of considerations need to be taken into account in designing and assessing such studies, including the specificity of the siRNA and the potential for an E2 to function with multiple different E3s and vice versa. For example, UbcH5B functions, at least *in vitro*, with multiple different HECT and RING finger E3s (Lorick *et al.*, 1999), and the RING finger E3s Siah1 and Snurf both have the capacity to utilize multiple E2s (Hakli *et al.*, 2004; Wheeler *et al.*, 2002).

## References


Brzovic, P. S., Keeffe, J. R., Nishikawa, H., Miyamoto, K., Fox, D., 3rd, Fukuda, M., Ohta, T., and Klevit, R. (2003). Binding and recognition in the assembly of an active BRCA1/BARD1 ubiquitin-ligase complex. *Proc. Natl. Acad. Sci. USA* **100,** 5646–5651.

Fang, S., Ferrone, M., Yang, C., Jensen, J. P., Tiwari, S., and Weissman, A. M. (2001). The tumor autocrine motility factor receptor, gp78, is a ubiquitin protein ligase implicated in degradation from the endoplasmic reticulum. *Proc. Natl. Acad. Sci. USA* **98,** 14422–14427.

Hakli, M., Lorick, K. L., Weissman, A. M., Janne, O. A., and Palvimo, J. J. (2004). Transcriptional coregulator SNURF (RNF4) possesses ubiquitin E3 ligase activity. *FEBS Lett.* **560,** 56–62.

Hofmann, R. M., and Pickart, C. M. (1999). Noncanonical MMS2-encoded ubiquitin-conjugating enzyme functions in assembly of novel polyubiquitin chains for DNA repair. *Cell* **96,** 645–653.

Jensen, J. P., Bates, P. W., Yang, M., Vierstra, R. D., and Weissman, A. M. (1995). Identification of a family of closely related human ubiquitin conjugating enzymes. *J. Biol. Chem.* **270,** 30408–30414.

Lorick, K. L., Jensen, J. P., Fang, S., Ong, A. M., Hatakeyama, S., and Weissman, A. M. (1999). RING fingers mediate ubiquitin-conjugating enzyme (E2)-dependent ubiquitination. *Proc. Natl. Acad. Sci. USA* **96,** 11364–11369.

Lorick, K. L., Tsai, Y.-C., Yang, Y., and Weissman, A. M. (2005). RING fingers and relatives: Determination of protein fate. *In* "Protein Degradation: Vol. 1: Ubiquitin and the Chemistry of Life" (R. J. Mayer, A. J. Ciechanover, and M. Rechsteiner, eds.), pp. 44–101. VCH Verlag, Weinheim, Germany.

Plafker, S. M., and Macara, I. G. (2000). Importin–11, a nuclear import receptor for the ubiquitin-conjugating enzyme, UbcM2. *EMBO J.* **19,** 5502–5513.

Reymond, F., Wirbelauer, C., and Krek, W. (2000). Association of human ubiquitin-conjugating enzyme CDC34 with the mitotic spindle in anaphase. *J. Cell Sci.* **113**(Pt. 10), 1687–1694.

Scheffner, M., Huibregtse, J. M., and Howley, P. M. (1994). Identification of a human ubiquitin-conjugating enzyme that mediates the E6-AP-dependent ubiquitination of p53. *Proc. Natl. Acad. Sci. USA* **91,** 8797–8801.

Tiwari, S., and Weissman, A. M. (2001). Endoplasmic reticulum (ER)-associated degradation of T cell receptor subunits: Involvement of ER-associated ubiquitin-conjugating enzymes (E2s). *J. Biol. Chem.* **276,** 16193–16200.

Wheeler, T. C., Chin, L. S., Li, Y., Roudabush, F. L., and Li, L. (2002). Regulation of synaptophysin degradation by mammalian homologues of seven in absentia. *J. Biol. Chem.* **277,** 10273–10282.

Yang, Y., Lorick, K. L., Jensen, J. P., and Weissman, A. M. (2005). Expression and evaluation of RING finger proteins. *Methods Enzymol.* **398**[10] 2005 (this volume).

Zhao, C., Beaudenon, S. L., Kelley, M. L., Waddell, M. B., Yuan, W., Schulman, B. A., Huibregtse, J. M., and Krug, R. M. (2004). The UbcH8 ubiquitin E2 enzyme is also the E2 enzyme for ISG15, an IFN-alpha/beta-induced ubiquitin-like protein. *Proc. Natl. Acad. Sci. USA* **101,** 7578–7582.

# [7] *In Vitro* Systems for NEDD8 Conjugation by Ubc12

*By* TOMOKI CHIBA

## Abstract

Nedd8 is a ubiquitin-like molecule that is highly conserved in eukaryotes. Similar to ubiquitin, Nedd8 attaches to target proteins through an enzymatic cascade composed of Nedd8-specific E1 (activating)- and E2 (conjugating)-enzymes. The E1 for Nedd8 is a heterodimer of APP-BP1 and Uba3, while the E2 is Ubc12. The most well-characterized targets of Nedd8 are proteins of the Cullin family, a core component of SCF (Skp1/Cullin1/F-box proteins) and/or SCF-like ubiquitin ligase complexes. The Nedd8 modification of Cullin (Cul) family proteins is evolutionarily conserved, and genetic analyses in various organisms suggest a positive role of the NEDD8 for the function of Cul family proteins. Further biochemical analysis reveals that NEDD8 modification augments the ubiquitin ligase activity of Cullin-based complexes through the recruitment of ubiquitin-charged E2 to the complex. This chapter describes methods for the purification of NEDD8 conjugation enzymes and *in vitro* Nedd8 conjugation.

## Introduction

Many ubiquitin-like molecules have been identified in eukaryotes and most of them function as covalent modifiers similar to ubiquitin. Among them, NEDD8 has the highest homology with ubiquitin and conjugates to

METHODS IN ENZYMOLOGY, VOL. 398 0076-6879/05 $35.00

DOI: 10.1016/S0076-6879(05)98007-5

target proteins in a reaction similar to ubiquitination (Gong and Yeh, 1999; Lammer *et al.*, 1998; Liakopoulos *et al.*, 1998; Osaka *et al.*, 1998). The most well-characterized targets of NEDD8 are Cul family proteins (reviewed in Deshaies 1999; Chiba and Tanaka 2004). Each Cul family protein serves as the scaffold of a ubiquitin ligase complex by recruiting a broad range of substrate recognition subunits to its N-terminal domain and the RING finger protein ROC1/Rbx1 to its C-terminal domain (Ohta *et al.*, 1999). ROC1/Rbx1, in turn, recruits E2 enzyme to the ubiquitin ligase complex. Unlike ubiquitin, NEDD8 does not target Cul family proteins for degradation, but acts as an activation signal (Kawakami *et al.*, 2001; Morimoto *et al.*, 2000; Podust *et al.*, 2000; Read *et al.*, 2000).

NEDD8 is conjugated to target proteins as follows: (1) nascent NEDD8 is first cleaved by Nedd8-specific C-terminal hydrolase, which exposes its conserved C-terminal diglycine motif. (2) The processed mature form of NEDD8 forms a high-energy thioester linkage between its C-terminal glycine residue and the active site cysteine residue of Uba3, a catalytic subunit of the Nedd8-activating enzyme (heterodimer of APP-BP1 and Uba3), in an ATP-dependent manner. (3) The activated Nedd8 is then transferred to the active site cysteine residue of Ubc12 in a thioester linkage. (4) Finally, the C-terminal glycine residue of NEDD8 forms an isopeptide bond with a lysine residue of the target protein. In the case of Cul family proteins, the association of ROC1/Rbx1 with the C-terminal domain is essential for NEDD8 modification (Gray *et al.*, 2002; Kamura *et al.*, 1999), suggesting that ROC1/Rbx1 is essential not only for ubiquitin ligase activity, but for NEDD8 ligase activity as well.

This chapter describes methods for the *in vitro* NEDD8 modification of the Cul-1-based E3 complex and how to measure its effect on ubiquitin ligase activity of the Cul-1 complex. Proteins used in the NEDD8 conjugation experiment are NEDD8, NEDD8-specific E1 (heterodimer of APP-BP1 and Uba3), E2 (Ubc12) enzymes, and the Cul–1 complex ($SCF^{\beta TrCP1}$ complex; comprising Cullin1, Skp1, F-box protein $\beta$TrCP1, and ROC1/Rbx1). The activity of the Cul-1 complex was measured by assessing the ubiquitination of I$\kappa$B$\alpha$, one of the substrates of the $SCF^{\beta TrCP1}$ complex (Winston *et al.*, 1999; Yaron *et al.*, 1998). All of the materials were recombinant proteins purified from bacteria, insect cells, or, in some cases, human HEK293 cells.

## Expression and Purification of Recombinant Proteins in *Escherichia coli*

Recombinant proteins purified from bacteria are as follow: NEDD8, Ubc12, ubiquitin, and Ubc4. For Ubc12 and Ubc4, each human cDNA is subcloned into a bacterial expression vector (e.g., pET28; Novagen), such that the coding sequence is preceded by an N-terminal His6 tag. For ubiquitin and NEDD8, each open reading frame is modified to produce the processed, mature form by introducing a stop codon after the diglycine motif. This modification bypasses the requirement for the C-terminal processing enzymes for ubiquitin and NEDD8. Modified cDNAs are then subcloned into bacterial expression vectors to yield clones fused N-terminal glutathione *S*-transferase or His6 tags. These bacterial expression vectors are transformed into the *lon* protease-deficient *E. coli* strain BL21 containing the $\lambda$DE3 lysogen and plated onto LB agar containing appropriate antibiotics. A single colony is isolated and grown overnight in 2 ml of LB supplemented with antibiotics. One milliliter of this overnight culture is used to inoculate 500 ml of LB with antibiotics and is incubated until reaching a density of $OD_{600\ nm} = 0.4$. Isopropyl-$\beta$-D-thiogalactopyranoside is added at final concentration of 1 m*M* to induce the expression of fusion protein and is incubated for another 2 h. Cells are harvested by centrifugation at 8000 rpm (4°, 20 min) and washed once in ice-cold phosphate-buffered saline (PBS). The cell pellet is resuspend in 10 ml PBS containing 1% Triton X (PBS-T) and lysed by sonication. The cell lysate is clarified by centrifugation at 5800*g* for 20 min, and the resulting cell debris is resuspended in 10 ml PBS-T for additional sonication. Cell lysates cleared by centrifugation are pooled and filtrated before applying to metal affinity chromatography. Chelating columns are charged according to the manufacture's instruction. The chelating column is rotated gently with the cell lysate in a cold room for 2 h and is then washed twice with PBS-T. The chelating column is then transferred to a column, and bound proteins are eluted by stepwise elution with 2 column bed volumes of PBS containing increasing concentrations of imidazole (10 to 500 m*M*). The purity of the proteins is checked by Coomassie brilliant blue staining following SDS–PAGE, and the proteins are dialyzed in 20 m*M* Tris-HCl, 2 m*M* EDTA (pH 8.0).

## Expression and Purification of Recombinant Proteins in Insect Cells

Proteins purified from insect cells are as follow: APP-BP1/Uba3 heterodimer and $SCF^{\beta TrCP1}$ (Skp1, Cullin–1, ROC1, F-box protein $\beta$TrCP1) complex. To produce the APP-BP1/Uba3 heterodimer, baculoviruses encoding His-tagged APP-BP1 and T7-tagged Uba3 are used simultaneously to infect the insect cells (High Five cells). For the $SCF^{\beta TrCP1}$ complex, baculoviruses encoding His-tagged $\beta$TrCP1, FLAG-tagged Skp1, HA-tagged Cul-1, and T7-tagged ROC1/Rbx1 are used for simultaneous infection. The APP-BP1/Uba3 heterodimer and the $SCF^{\beta TrCP1}$ complex are purified from insect cell lysates by one-step metal affinity chromatography.

In brief, human cDNAs encoding APP-BP1, Uba3, Skp1, Cullin-1, ROC1, and F-box protein $\beta$TrCP1 are tagged as described earlier and subcloned into the pVL1393 vector (BD bioscience). Each virus is isolated following plaque assay and amplified. In general, $5 \times 10^7$ cells are infected with a total $5 \times 10^7$ pfu of viruses (m.o.i. = 1), and the optimum amounts of each virus are determined to acquire the maximum yields of the protein complexes. Cells are harvested 48 to 72 h postinfection, lysed with lysis buffer (20 m*M* Tris-HCl, pH 7.5, 150 m*M* NaCl, 0.5% Nonidet P-40), and subjected to metal affinity chromatography.

## Expression and Purification of Recombinant Proteins from Mammalian Cells

I$\kappa$B kinase is purified from HEK293 cells. HEK293 cells are transfected with the mammalian expression vector encoding FLAG-tagged IKK$\beta$. Two days after transfection, cells are harvested in lysis buffer [20 m*M* Tris-HCl, pH 7.4, 150 m*M* NaCl, 1 m*M* EDTA, 1 m*M* EGTA, 0.5% Nonidet P-40, 2 m*M* dithiothreitol (DTT), 25 m*M* glycerophosphate, with protease inhibitor cocktail], and FLAG-tagged IKK$\beta$ is immunoprecipitated with the agarose-conjugated anti-FLAG antibody (Sigma). After washing the agarose with the lysis buffer, the IKK$\beta$ complex is processed for autophosphorylation to augment its kinase activity. The immunoprecipitated IKK$\beta$ complex is washed twice in kinase buffer (20 m*M* Tris-HCl, pH 7.4, 150 m*M* NaCl, 1 m*M* EDTA, 0.05% Triton X-100, 2 m*M* DTT, 25 m*M* glycerophosphate, protease inhibitor cocktail) and incubated for 30 min at 25° after the addition of 10× ATP-regenerating buffer (20 m*M* ATP, 100 m*M* creatine phosphate, 100 IU/ml creatine kinase, 10 IU/ml inorganic pyrophosphatase). After the reaction, the beads are washed twice with kinase buffer, and the activated IKK$\beta$ complex is eluted by addition of the FLAG peptide (200 $\mu$g/ml).

## Methods for *In Vitro* NEDD8 Modification

For NEDD8 modification of Cul-1 *in vitro*, APP-BP1/Uba3 (0.5 $\mu$g), Ubc12 (0.5 $\mu$g), NEDD8 (3 $\mu$g), and SCF$^{\beta\mathrm{TrCP1}}$ complex (1 $\mu$g) are mixed in a final volume of a 20-$\mu$l reaction mixture (20 m*M* Tris-HCl, pH7.5, 2 m*M* ATP, 1 m*M* DTT) and incubated for 30 min at 30°. After terminating the reaction by the addition of 3× SDS sample buffer, the sample is boiled and separated by 4–20% SDS–PAGE. The NEDD8 modification of HA-Cul-1 is detected by immunoblotting with an anti-HA antibody (Fig. 1).

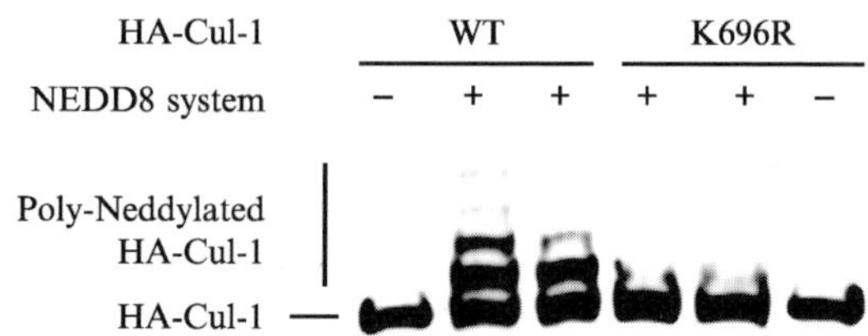


FIG. 1. NEDD8 modification of Cullin1 *in vitro*. The purified recombinant SCF$^{\beta TrCP1}$ complex (FLAG-Skp1/HA-Cullin1/His-$\beta$TrCP/T7-ROC1) was incubated with APP-BP1/Uba3, Ubc12, and NEDD8 *in vitro*. Western blots of HA-Cullin1 show the NEDD8 modification of wild-type Cullin1, but not its K696R mutant, whose essential lysine residue at position 696 is mutated to arginine. Poly-NEDD8 formation occurs *in vitro*, but whether it occurs at physiological condition or its biological role remains to be clarified.

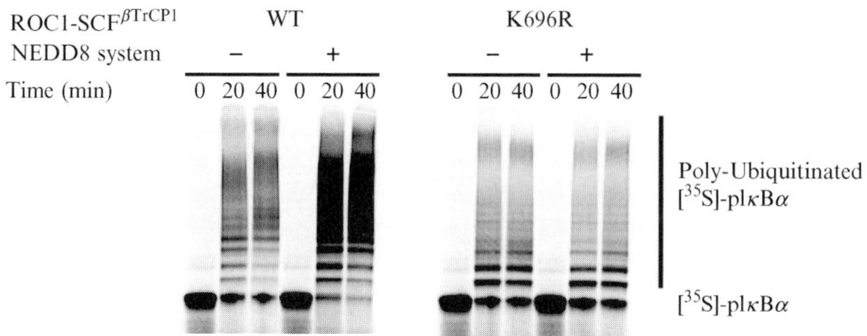


FIG. 2. NEDD8 modification augments the ubiquitin ligase activity of the SCF complex. Phosphorylated I$\kappa$B$\alpha$ was *in vitro* ubiquitinated by the SCF$^{\beta TrCP1}$ complex. Addition of the NEDD8 conjugation system significantly augments ubiquitination of I$\kappa$B$\alpha$ by the wild-type SCF$^{\beta TrCP1}$ complex but not its K696R mutant.

## *In Vitro* Ubiquitination Assay of I$\kappa$B$\alpha$ by the SCF$^{\beta TrCP1}$ Complex

[$^{35}$S]Methionine-labeled I$\kappa$B$\alpha$ is synthesized by the *in vitro* transcription and translation system using reticulocytes. Two microliter lysates of the labeled I$\kappa$B$\alpha$ are phosphorylated by IKK$\beta$ (50 ng) in a final 12-$\mu$l reaction mixture (50 m*M* Tris, pH 7.5, 10 m*M* $MgCl_2$, 0.5 m*M* DTT, 0.5 $\mu M$ ocadaic acid, 2 m*M* ATP, 10 m*M* creatine phosphate, 10 IU/ml creatine kinase, 1 IU/ml inorganic pyrophosphatase) for 30 min at room temperature. Phosphorylated I$\kappa$B$\alpha$ is incubated further with E1 (0.5 $\mu$g), Ubc4 (1 $\mu$g), SCF$^{\beta TrCP1}$ (1 $\mu$g), and ubiquitin (10 $\mu$g) in the presence or absence of the NEDD8 conjugation system consisting of NEDD8 (10 $\mu$g), APP-BP1/Uba3 (0.5 $\mu$g), and Ubc12 (0.5 $\mu$g) at a final 25-$\mu$l reaction volume. After terminating the reaction with 5 $\mu$l of 10% SDS and 10 $\mu$l 4$\times$ SDS sample buffer, 10 $\mu$l of the boiled samples is separated by 10% SDS–PAGE. The gels are fixed, dried, and exposed to X-ray film or Image Screen (Fuji BAS2500) (Fig. 2).

## References


Chiba, T., and Tanaka, K. (2004). Cullin-based ubiquitin ligase and its control by NEDD8-conjugating system. *Curr. Protein Pept. Sci.* **5,** 177–184.

Deshaies, R. J. (1999). SCF and cullin/ring H2-based ubiquitin ligases. *Annu. Rev. Cell Dev. Biol.* **15,** 435–467.

Gong, L., and Yeh, E. T. (1999). Identification of the activating and conjugating enzymes of the NEDD8 conjugation pathway. *J. Biol. Chem.* **274,** 12036–12042.

Gray, W. M., Hellmann, H., Dharmasiri, S., and Estelle, M. (2002). Role of the Arabidopsis RING-H2 protein RBX1 in RUB modification and SCF function. *Plant Cell* **14,** 2137–2144.

Kamura, T., Conrad, M. N., Yan, Q., Conaway, R. C., and Conaway, J. W. (1999). The Rbx1 subunit of SCF and VHL E3 ubiquitin ligase activates Rub1 modification of cullins Cdc53 and Cul2. *Genes Dev.* **13,** 2928–2933.

Kawakami, T., Chiba, T., Suzuki, T., Iwai, K., Yamanaka, K., Minato, N., Suzuki, H., Shimbara, N., Hidaka, Y., Osaka, F., Omata, M., and Tanaka, K. (2001). NEDD8 recruits E2-ubiquitin to SCF E3 ligase. *EMBO J.* **20,** 4003–4012.

Lammer, D., Mathias, N., Laplaza, J. M., Jiang, W., Liu, Y., Callis, J., Goebl, M., and Estelle, M. (1998). Modification of yeast Cdc53p by the ubiquitin-related protein rub1p affects function of the SCFCdc4 complex. *Genes Dev.* **12,** 914–926.

Liakopoulos, D., Doenges, G., Matuschewski, K., and Jentsch, S. (1998). A novel protein modification pathway related to the ubiquitin system. *EMBO J.* **17,** 2208–2214.

Morimoto, M., Nishida, T., Honda, R., and Yasuda, H. (2000). Modification of cullin-1 by ubiquitin-like protein Nedd8 enhances the activity of SCF(skp2) toward p27(kip1). *Biochem. Biophys. Res. Commun.* **270,** 1093–1096.

Ohta, T., Michel, J. J., Schottelius, A. J., and Xiong, Y. (1999). ROC1, a homolog of APC11, represents a family of cullin partners with an associated ubiquitin ligase activity. *Mol. Cell.* **4,** 535–541.

Osaka, F., Kawasaki, H., Aida, N., Saeki, M., Chiba, T., Kawashima, S., Tanaka, K., and Kato, S. (1998). A new NEDD8-ligating system for cullin-4A. *Genes Dev.* **12,** 2263–2268.

Podust, V. N., Brownell, J. E., Gladysheva, T. B., Luo, R. S., Wang, C., Coggins, M. B., Pierce, J. W., Lightcap, E. S., and Chau, V. A (2000). Nedd8 conjugation pathway is essential for proteolytic targeting of p27Kip1 by ubiquitination. *Proc. Natl. Acad. Sci. USA* **97,** 4579–4584.

Read, M. A., Brownell, J. E., Gladysheva, T. B., Hottelet, M., Parent, L. A., Coggins, M. B., Pierce, J. W., Podust, V. N., Luo, R. S., Chau, V., and Palombella, V. J. (2000). Nedd8 modification of cul-1 activates SCF(beta(TrCP))-dependent ubiquitination of IkappaBalpha. *Mol. Cell Biol.* **20,** 2326–2333.

Winston, J. T., Strack, P., Beer-Romero, P., Chu, C. Y., Elledge, S. J., and Harper, J. W. (1999). The SCFbeta-TRCP-ubiquitin ligase complex associates specifically with phosphorylated destruction motifs in IkappaBalpha and beta-catenin and stimulates IkappaBalpha ubiquitination *in vitro*. *Genes Dev.* **13,** 270–283.

Yaron, A., Hatzubai, A., Davis, M., Lavon, I., Amit, S., Manning, A. M., Andersen, J. S., Mann, M., Mercurio, F., and Ben-Neriah, Y. (1998). Identification of the receptor component of the IkappaBalpha-ubiquitin ligase. *Nature* **396,** 590–594.

# [8] Purification and Activity Assays for Ubc9, the Ubiquitin-Conjugating Enzyme for the Small Ubiquitin-like Modifier SUMO

*By* ALI A. YUNUS and CHRISTOPHER D. LIMA

## Abstract

The small ubiquitin-like modifier (SUMO) can be conjugated to lysine residues directly by the ubiquitin-conjugating protein Ubc9. SUMO conjugation can be catalyzed *in vitro* using only E1, Ubc9 (E2), mature SUMO, and ATP because Ubc9 directly recognizes consensus SUMO modification sites found in many identified targets of SUMO conjugation. This article describes methods to prepare Ubc9 and provides details for assay conditions used to evaluate E2 thioester formation and E2-mediated SUMO conjugation under single turnover and multiple turnover conditions. It also briefly describes parameters used to evaluate E3-mediated SUMO conjugation. Conservation of the SUMO conjugation apparatus from yeast to human has enabled *in vivo* assessment of human Ubc9 function through yeast complementation assays.

## Introduction

Ubiquitin (Ub) and ubiquitin-like (Ubl) modifiers are small ~100 amino acid globular proteins that regulate critical cellular pathways such as differentiation, apoptosis, the cell cycle, and responses to stress through posttranslational covalent attachment between a substrate lysine $\varepsilon$ amino group and the Ub/Ubl C terminus (Hershko and Ciechanover, 1998; Hochstrasser, 1996; Johnson, 2004; Laney and Hochstrasser, 1999; Melchior, 2000; Muller *et al.*, 2001). The small ubiquitin-related modifier SUMO-1 shares significant structural similarity to ubiquitin and is a member of the Ub/Ubl family. Despite similarity to ubiquitin, the cellular functions for SUMO remain distinct insomuch as SUMO conjugation does not directly result in protein degradation, but alters the target protein function through changes in activity, cellular localization, or by protecting the substrate from ubiquitination through competition between pathways for a common lysine residue on the protein target (Johnson, 2004; Melchior, 2000; Melchior *et al.*, 2003).

Ub/Ubl conjugation occurs in sequential steps (Hershko and Ciechanover, 1998). Proteases process Ub/Ubls to expose a conserved Ub/Ubl C-terminal

METHODS IN ENZYMOLOGY, VOL. 398

0076-6879/05 $35.00
DOI: 10.1016/S0076-6879(05)98008-7

diglycine (Gly-Gly) motif that is required for E1 adenylation and subsequent transfer to an intramolecular E1-Ub/Ubl thioester adduct. The Ub/Ubl is then transferred from E1 to E2 to form an E2-Ub/Ubl thioester that is now competent for Ub/Ubl transfer to target lysine $\varepsilon$ amino groups either directly or through intermediates mediated by E3s. Ubiquitin conjugation utilizes several E2s and numerous E3s to direct catalysis and to ensure specificity during conjugation. Ubiquitin E3s that belong to either RING or HECT families function by recruiting the substrate and activated E2-Ub/Ubl thioester into a single complex to promote Ub/Ubl conjugation. RING-type E3s utilize a domain containing zinc to bind the activated E2 thioester, enabling direct attack at the E2 thioester by the substrate lysine $\varepsilon$ amino group (Deshaies, 1999). In contrast, HECT E3s recruit activated E2 thioesters to generate a HECT E3 thioester intermediate prior to conjugation (Huibregtse *et al.*, 1995).

The SUMO E2 Ubc9 is unique among conjugating E2s in that it exhibits specificity and interacts directly with many substrates containing the consensus sequence $\Psi$-K-x-D/E, where $\Psi$ is hydrophobic, K is the lysine conjugated to SUMO, x is any amino acid, and D or E is an acidic residue (Bernier-Villamor *et al.*, 2002). While Ubc9 can be sufficient for SUMO conjugation *in vitro* using only E1, Ubc9, mature SUMO, and ATP, a few SUMO E3 factors have been identified that facilitate conjugation *in vivo* and *in vitro*. SP ring SUMO E3s resemble RING-type Ub E3s and include two yeast E3-like ring finger proteins (Siz1 and Siz2) and the PIAS protein family (Hochstrasser, 2001; Johnson and Gupta, 2001; Kahyo *et al.*, 2001; Schmidt and Muller, 2003; Takahashi *et al.*, 2001). In addition, and perhaps more intriguing, is the discovery of SUMO E3 activities identified for RanBP2/NUP358 and Pc2 (Pichler *et al.*, 2002; Kagey *et al.*, 2003). These E3s are unrelated to either RING or HECT E3 ligases, suggesting that the SUMO pathway may utilize mechanisms distinct from the ubiquitin pathway to promote SUMO conjugation.

Functional assessment of Ubc9 has been reported by many groups. Of particular note was the mutagenesis and biochemical characterization of Ubc9 from *Saccharomyces cerevisiae* (Bencsath *et al.*, 2002). This work identified several residues and Ubc9 surfaces involved in E1 thioester transfer and conjugation. Steady-state and half-reaction kinetic analysis of Ubc9 has also been described and provided a detailed analysis of the complicated reaction mechanisms required for Ubc9 association with E1 and its substrates (Tatham *et al.*, 2003a). Further functional analysis for Ubc9 was described in work that assessed the role of an N-terminal Ubc9 site that appeared critical for the Ubc9 interaction with SUMO isoforms and its cognate E1 (Tatham *et al.*, 2003b). Finally, structure-guided mutational and biochemical analyses of human Ubc9 substrate interactions have been described (Bernier-Villamor *et al.*, 2002; Lin *et al.*, 2002). This article describes the expression and purification of Ubc9 and assays used to assess Ubc9 function *in vitro* and *in vivo*.

## Recombinant Protein Expression

Active Ubc9 protein can be obtained from organisms such as *Saccharomyces cerevisiae*, *Homo sapiens*, and *Arabidopsis thaliana* through recombinant protein expression in the *Escherichia coli* bacterium (Bernier-Villamor *et al.*, 2002; Johnson and Blobel, 1997; Lois *et al.*, 2003). Human Ubc9 can be expressed easily at milligram levels per liter of bacterial culture through the use of T7-based expression vectors. We generally utilize N-terminal hexahistidine-tagged Ubc9 fusions followed by thrombin cleavage sites to facilitate removal of the hexahistidine tag during purification. A typical example of our cloning and expression strategy follows for human Ubc9.

The polymerase chain reaction (PCR) is used with a cDNA library to generate a coding region for human Ubc9 flanked at the 5′ end by an *Nde*I restriction site and at the 3′ end with a stop codon followed by a *Xho*I restriction site. This PCR product is digested with the appropriate restriction endonucleases and cloned into the pET-based plasmid (pET28b) to encode an N-terminal thrombin-cleavable hexahistidine-tagged fusion protein. This plasmid is transformed into *E. coli* BL21(DE3) CodonPlus RIL (Novagen) or *E. coli* BL21(DE3) pLysS. A 10-liter culture is grown by fermentation at 37° to an $A_{600}$ of 3, adjusted to 30° and 0.75 m*M* isopropyl-$\beta$-D-thiogalactoside, and fermented for an additional 4 h. Cells are harvested by centrifugation, and cell pellets are weighed, resuspended in 50 m*M* Tris-HCl, pH 8.0, and 20% sucrose at a concentration of 2 ml/g, and distributed into 50-ml conical tubes before flash freezing in liquid nitrogen (one 50-ml tube corresponds to cell pellets obtained from a 2-liter culture).

## Purification of Ubc9

Ubc9 is isolated from the soluble fraction of an *E. coli* lysate. A cell pellet obtained from 2 liter of culture is thawed, and buffer components are added to result in a final concentration of 350 m*M* NaCl, 20 m*M* imidazole, 1 m*M* $\beta$-mercaptoethanol (BME), 20 $\mu$g/ml lysozyme, 1 m*M* phenylmethylsulfonyl fluoride (PMSF), and 0.1% detergent [IGEPAL CA-630 ([octylphenoxy]polyethoxyethanol)] prior to sonication. After sonication, insoluble material is removed by centrifugation. The cell lysate supernatant from a 2-liter culture containing $\sim$300 mg $His_6$-Ubc9 is applied to 30 ml of a metal affinity resin such as Ni-NTA Superflow (Qiagen; capacity $\sim$10 mg/

ml). Batch or gradient elution of the protein is achieved by application of up to 500 m*M* imidazole to elute Ubc9 from the resin. The protein content of the resulting peak is quantified using the Bradford reagent (Bio-Rad) and applied to a gel filtration column (Amersham; Superdex75). Ubc9 migrates on gel filtration media as a monodisperse protein with an apparent molecular mass of ~20 kDa. The $His_6$ tag is removed by proteolysis using a 1:1000 ratio of bovine thrombin (Sigma) to Ubc9. Proteolysis is monitored by SDS–PAGE analysis; after proteolysis is complete (2–4 h at room temperature, overnight at 4°), the solution containing Ubc9 is exchanged or diluted to 50 m*M* NaCl and applied to MonoS cation-exchange chromatography and elutes with a gradient from 50 to 500 m*M* NaCl. Ubc9 elutes from MonoS at approximately 150 m*M* NaCl. The peak is exchanged into buffer containing 10 m*M* Tris-HCl, pH 7.5, 50 m*M* NaCl, and 1 m*M* dithiothreitol (DTT) or BME and concentrated to >5 mg/ml. The protein is then flash frozen in liquid nitrogen and stored at −80° for later use.

## Expression and Isolation of Components for *In Vitro* Conjugation Assays

Heterodimeric SUMO E1 enzymes have been cloned, purified, and characterized in several laboratories, including our own (Lois and Lima, 2005), and are described thoroughly elsewhere (Bossis *et al.*, 2005). We have opted for a coexpression strategy for SUMO E1 enzymes from yeast, plant, and human. The smaller Aos1/Sae1 subunit is cloned into a plasmid that encodes the native protein without an affinity tag. The Uba2/Sae2 subunit is placed on a second plasmid to encode a protein with a N-terminal hexahistidine tag. The two plasmids are cotransformed into *E. coli* and selected with the appropriate antibiotics. Coexpression plasmids have also been developed utilizing the DUET series of vectors available from Novagen. The heterodimeric E1 is coexpressed and copurified by virtue of the single hexahistidine affinity tag located on Uba2/Sae2. Once purified by metal affinity chromatography, the peak is filtered and applied to a Superdex200 gel filtration column (Pharmacia). The SUMO E1 heterodimer migrates on gel filtration media as a monodisperse protein with an apparent molecular mass of ~120 kDa. The resulting peak is diluted or desalted to 100 m*M* NaCl and applied to MonoQ for anion-exchange chromatography and eluted using a gradient of 100 to 500 m*M* NaCl. SUMO E1 elutes from MonoQ at approximately 200–250 m*M* NaCl. The resulting peak is exchanged into buffer containing 75 m*M* NaCl, 20 m*M* Tris-HCl, pH 8, and 1 m*M* DTT, concentrated to >4 mg/ml, flash frozen in liquid nitrogen, and stored at −80°.

Several substrates we use to assay Ubc9 conjugation were described previously and include the C-terminal domain of human and mouse

RanGAP1 (residues 418–587 and residues 420–589, respectively), the C-terminal tetramerization domain of human p53 (residues 320–393), and full-length human I$\kappa$B$\alpha$ (residues 1–317) (Bernier-Villamor *et al.*, 2002). These proteins were chosen as substrates because their respective SUMO conjugation sites are located in different positions within respective polypeptide sequences. I$\kappa$B$\alpha$ contains a SUMO consensus site near its N terminus, p53 near its C terminus, and RanGAP1 within an internal position in the protein. Several additional substrates have been described and are utilized extensively in the literature (reviewed in Gill, 2003; Johnson, 2004; Melchior, 2000). We have also developed assays that utilize substrate-derived peptides that contain a consensus SUMO conjugation site. Peptides are either made synthetically or by utilizing a Smt3 fusion protein expression system to make peptides fused C-terminal to Smt3 (Mossessova and Lima, 2000; licensed and distributed by Invitrogen). After cleavage by Ulp1, Smt3 and the respective peptide were separated by reverse phase chromatography.

Three human isoforms (SUMO-1, SUMO-2, and SUMO-3) have been cloned and expressed in our laboratory (Reverter and Lima, 2004). Each isoform is isolated by PCR from cDNA libraries and cloned into the *Nco*I and *Xho*I restriction sites of pET28b to produce a native N-terminal methionine. The native stop codon is removed to produce SUMO isoforms fused to a C-terminal hexahistidine tag to facilitate purification by metal affinity chromatography (see earlier discussion). After metal affinity chromatography, the resulting protein peak is filtered and applied to gel filtration using Superdex75 (Pharmacia). All SUMO isoforms migrate as monodisperse proteins with apparent molecular masses of ~20 kDa. We generated native SUMO isoforms by proteolysis using Ulp1 or Senp2 (Mossessova and Lima, 2000; Reverter and Lima, 2004). After proteolysis, the solution containing the mature SUMO isoform is equilibrated to less than 100 m*M* NaCl and further purified by anion-exchange chromatography (MonoQ) and eluted using a NaCl gradient from 100 to 500 m*M*. SUMO isoforms elute from MonoQ at approximately 200 m*M* NaCl. The resulting peak is concentrated to >5 mg/ml and flash frozen in liquid nitrogen prior to storage at −80°.

A few E3s have been identified thus far for Ubc9 and the SUMO pathway that include the Sp-RING family, Pc2, and catalytic domains from Nup358/RanBP2 (Johnson and Gupta, 2001; Kagey *et al.*, 2003; Pichler *et al.*, 2002; Schmidt and Muller, 2003; see earlier discussion). We and others have utilized these proteins to assess their ability to catalyze or enhance the ability of Ubc9 to conjugate SUMO to various targets. Because the exact mechanisms utilized by SUMO E3s to promote conjugation remain unclear, it is important to assess the catalytic potential of each E3 by titration with respect to the existing components. It is thought that the PIAS proteins behave in a manner similar to RING-type E3s, recruiting both substrate and activated E2-SUMO-thioester complexes via distinct domains within the respective E3. The mechanisms for Nup358/RanBP2 or Pc2 E3 activity remain unclear at this time, although both interact directly with Ubc9, thereby partially explaining their ability to promote SUMO conjugation.

## Functional Assays for Ubc9

Ubc9 can be assayed directly for substrate conjugation *in vitro* in combination with E1, ATP, and SUMO. Several key factors must be considered before assessing Ubc9 function *in vitro*. Because the Ubc9 E2 must interact with substrate, SUMO, and the SUMO E1 for conjugation to occur, determination of the optimal concentration range for each of these partners is critical before proper interpretation of the results can occur. For example, we have noted that large substrate excess in an *in vitro* reaction can actually inhibit conjugation. In addition, if high concentrations are used for either E1 or E2, the SUMO enzymes actually become the substrate for conjugation and compete with the substrate one is trying to modify. SUMO can also become the substrate, leading to synthesis of SUMO chains in the reaction. As such, we are sometimes forced to individually determine reaction conditions for each of the targets we have assessed through titration of the various reaction components (time, enzyme concentrations, and substrate concentration). This chapter attempts to describe the most typical reaction conditions used to monitor E2 thioester formation and E2-mediated conjugation assays.

### *Assay for Thioester Formation and Single Turnover Thioester Transfer*

Thioester assays for Ubc9 are typically carried out at 37°. A reaction mix includes 50 m*M* NaCl, 20 m*M* HEPES, pH 7.5, 0.1% Tween 20, 5 m*M* $MgCl_2$, 100 n*M* hUbc9, 100 n*M* hE1, and 100 n*M* SUMO. Reactions are initiated by the addition of 100 n*M* to 1 $\mu M$ ATP and are stopped by denaturing the sample in buffer that contains 50 m*M* Tris-HCl, pH 6.8, 2% SDS, 4 *M* urea, and 10% glycerol (Scheffner *et al.*, 1994). Samples are incubated at 37° for 10 min before resolving the products using nonreducing SDS–PAGE. Although thioester adducts can be observed by Sypro or Coomassie blue staining, we have obtained superior results by analyzing the respective adducts by Western analysis using either anti-Ubc9 or anti-SUMO antibodies. Thioester adducts can be specifically cleaved by replacing 4 *M* urea with 100 m*M* DTT in the denaturing buffer in conjunction with a 10-min incubation at 95° before analysis of the samples by reducing SDS–PAGE.

We have also utilized preparative thioester reactions to produce sufficient quantities of E2 thioester SUMO for single turnover conjugation assays. In this instance, we prepared a reaction mix that includes 50 m*M* NaCl, 20 m*M* HEPES, pH 7.5, 0.1% Tween 20, 5 m*M* $MgCl_2$, 1 $\mu M$ hUbc9, 100 n*M* hE1, and 1 $\mu M$ SUMO. The reaction is initiated by the addition of 1 $\mu M$ ATP and is incubated at 37° for up to 8 min to generate approximately 500 n*M* Ubc9-SUMO-thioester (Fig. 1A). The reaction is quenched to prevent further E1-mediated SUMO activation by placing 5 $\mu$l of the E2 thioester reaction mix into 45 $\mu$l of buffer that includes 50 m*M* NaCl, 20 m*M* HEPES, pH 7.5, 0.1% Tween 20, and 5 m*M* EDTA. This buffer also includes substrate at the desired final concentration. Thioester transfer is then monitored by the decay of the E2-SUMO-thioester complex and the appearance of conjugated substrate (Fig. 1B). We typically analyze the results of this assay by Western analysis using anti-SUMO antibodies with

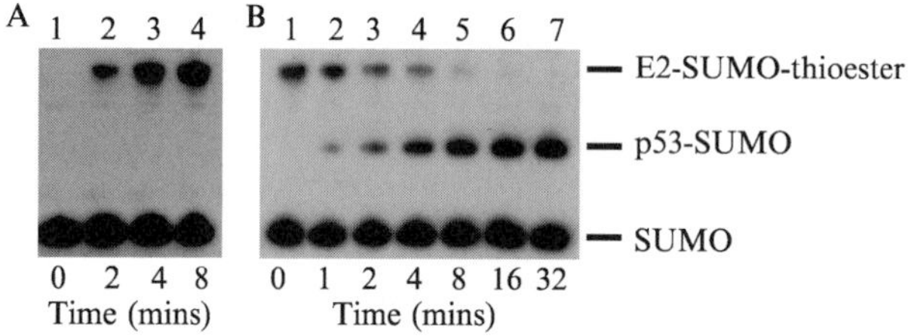


FIG. 1. Time course for E2-SUMO-thioester formation and conjugation. (A) Thioester formation between E2 and SUMO. Reaction details provided in the text. Reactions were stopped using buffer containing 50 m*M* Tris-HCl, pH 6.8, 2% SDS, 4 *M* urea, and 10% glycerol before protein samples were resolved by nonreducing SDS–PAGE, transferred to PVDF, and probed with a polyclonal rabbit antibody raised against human SUMO-1 (Boston Biochem). Lane 1, 0 min; lane 2, 2 min; lane 3, 4 min; and lane 4, 8 min. The approximate positions for E2-SUMO-thioester and SUMO are indicated at the right. (B) Single turnover assay for SUMO transfer from the E2-SUMO-thioester to the C-terminal tetramerization domain of human p53. Reaction details are provided in the text. The assay represented here utilized the p53 tetramerization domain at a concentration of 8 $\mu M$. Protein samples were treated as in A to enable monitoring of both thioester decay and product accumulation. Lane 1, 0 min; lane 2, 1 min; lane 3, 2 min; lane 4, 4 min; lane 5, 8 min; lane 6, 16 min; and lane 7, 32 min. The approximate positions for E2-SUMO-thioester, p53-SUMO, and SUMO are indicated at the right. The higher quantity of p53-SUMO in lane 7 compared to the E2-thioester in lane 1 suggests that the chase is not 100% effective. This seeming inconsistency can be attributed to the poor stability and decay of the E2 thioester compared to the substrate-SUMO conjugate during sample handling at room temperature prior to SDS–PAGE analysis. The rates of thioester decay and product accumulation are nearly equivalent, both of which are complete between 16 and 32 min (lanes 6 and 7).

the SDS–PAGE-resolved complexes. If substrate is provided at saturating conditions, rate constants for the transfer reaction can be calculated from E2 thioester decay or accumulation of the conjugated substrate (not shown).

### *Sypro-Based Conjugation Assays*

In this assay, we describe *in vitro* conjugation of RanGAP1 under conditions that require multiple turnovers of E1 and E2 thioester formation. Other substrates can be assessed using a similar protocol, although the time required to measure conjugation varies depending on the substrate and its stable association with the E2 Ubc9. SUMO conjugation assays are typically carried out at 37°. The reaction mix is composed of 50 m*M* NaCl, 20 m*M* HEPES, pH 7.5, 0.1% Tween 20, 5 m*M* $MgCl_2$, 1 m*M* DTT, 100 n*M* hUbc9, 100 n*M* hE1, 4 $\mu M$ Sumo, and 2 $\mu M$ RanGAP1. Reactions are initiated by the addition of ATP to a final concentration of 1 m*M*. Ten-microliter aliquots are removed at various time points, and the reactions are stopped by the addition of SDS loading buffer and incubation at 95° for 10 min. The products are resolved by SDS–PAGE [4–12% NuPAGE Bis-Tris gel with NuPAGE MES buffer (Invitrogen)], and the gel is stained by the fluorescent dye Sypro Ruby (Bio-Rad) for 4–12 h. The gel is destained for at least 1 h before imaging using UV illumination and a Bio-Rad gel documentation system (Fig. 2A). Data are then analyzed using Quantity One Software (Fig. 2B; Bio-Rad). Proper quantification of the resulting products is ensured by preparation of a standard curve that combines known concentrations of purified RanGAP1, SUMO, and RanGAP1-SUMO on the same gel.

### *Western-Based Conjugation Assays*

This assay also describes Ubc9-dependent SUMO conjugation to RanGAP1, but under conditions where only 0.1 n*M* E2 was used and where the substrate was titrated between nanomolar and micromolar concentrations. In addition, only very small quantities (<10%) of SUMO-conjugated substrate are allowed to accumulate prior to detection by Western analysis. Typically, 1-ml reactions are carried out at 37° and include 50 m*M* NaCl, 20 m*M* HEPES, pH 7.5, 0.1% Tween 20, 5 m*M* $MgCl_2$, 1 m*M* DTT, 0.1 n*M* hUbc9, 10 n*M* hE1, 0.4 $\mu M$ Sumo, and RanGAP1 (concentrations ranging from 15 n*M* to 2 $\mu M$). The reactions are initiated by the addition of ATP to a final concentration of 1 m*M* and allowed to proceed for 120 min. The reactions are terminated by precipitation of the proteins using trichloroacetic acid (TCA).

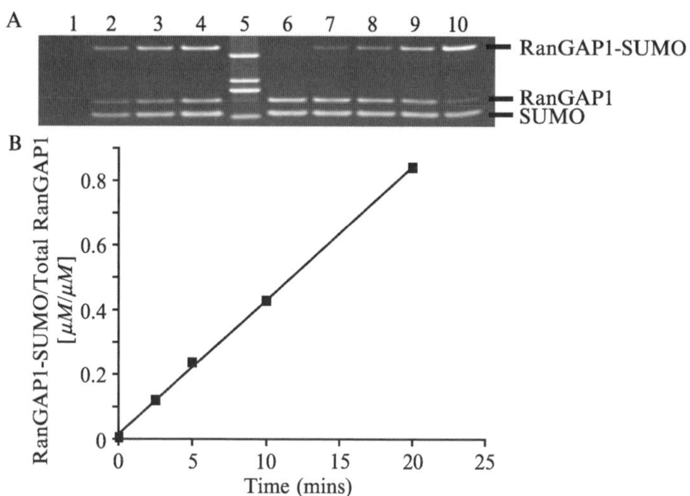


FIG. 2. Time course for SUMO conjugation. (A) Standards and time course for SUMO conjugation to RanGAP1. Samples were resolved by SDS–PAGE and stained with Sypro Ruby (Bio-Rad). Standards include known concentrations of RanGAP1-SUMO, RanGAP1, and SUMO. Lane 1, blank, 0 $\mu M$ RanGAP1-SUMO, 0 $\mu M$ RanGAP1, and 0 $\mu M$ SUMO; lane 2, 0.75 $\mu M$ RanGAP1-SUMO, 0.75 $\mu M$ RanGAP1, and 1.5 $\mu M$ SUMO; lane 3, 1.25 $\mu M$ RanGAP1-SUMO, 1.25 $\mu M$ RanGAP1, and 2.5 $\mu M$ SUMO; lane 4, 2.0 $\mu M$ RanGAP1-SUMO, 2.0 $\mu M$ RanGAP1, and 4 $\mu M$ SUMO; and lane 5, molecular mass markers, 30, 25, 20, and 15 kDa (top to bottom). Ten-microliter aliquots were removed from the SUMOylation assay at indicated time points. Lane 6, 0 min; lane 7, 2.5 min; lane 8, 5 min; lane 9, 10 min; and lane 10, 20 min. (B) RanGAP1 SUMOylation kinetics depicted graphically. The *X* axis denotes the time in minutes. The *Y* axis denotes the fraction of RanGAP1 SUMOylated [$\mu M/\mu M$].

The protein pellet is dissolved in 15 $\mu$l of buffer composed of 1× NuPAGE LDS sample-loading buffer (Invitrogen), 100 m$M$ DTT, 45 m$M$ Tris, pH 7.5, and 10% glycerol. The samples and standards (described later) are subjected to SDS–PAGE [4–12% NuPAGE Bis-Tris gel with NuPAGE MES buffer (Invitrogen)] and transferred to PVDF membranes (Immunoblot; Bio-Rad). The membrane is blocked and probed with anti-RanGAP1 or anti-SUMO primary antibody for 1–2 h. Proteins are detected using the appropriate secondary antibody in combination with ECL-Plus (Amersham Biosciences) and imaged using a Fujifilm LAS-3000 Imager (Fig. 3A). The resulting images are quantified using Image Gauge v4.0 (Fig. 3B; FujiFilm). Standards are prepared using a known concentration of RanGAP1 (0.625 to 10 n$M$) in reactions allowed to proceed to completion.

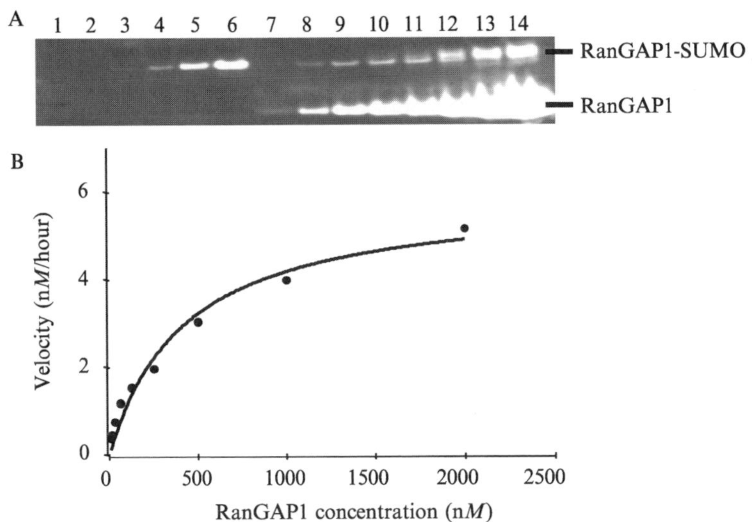


FIG. 3. Substrate titration for SUMO conjugation. (A) SUMO conjugation assay using low E2 concentrations and RanGAP1 titration. Standards for RanGAP1-SUMO were prepared as described in the text. Lane 1, 0 n*M*; lane 2, 0.625 n*M*; lane 3, 1.25 n*M*; lane 4, 2.5 n*M*; lane 5, 5 n*M*; and lane 6, 10 n*M* RanGAP1-SUMO. SUMO conjugation assays were carried out at indicated RanGAP1 concentrations for 2 h, TCA precipitated, and resolved by SDS–PAGE. Proteins were transferred to a PVDF membrane and probed with an antibody raised against the C-terminal domain of RanGAP1 (gift of F. Melchior) (lanes 7–14). Reactions included RanGAP1 concentrations: lane 7, 15.6 n*M*; lane 8, 31.3 n*M*; lane 9, 62.5 n*M*; lane 10, 125 n*M*; lane 11, 250 n*M*; lane 12, 0.5 $\mu M$; lane 13, 1.0 $\mu M$; and lane 14, 2.0 $\mu M$. (B) Data in A depicted graphically. The *X* axis denotes the RanGAP1 concentration [n*M*]. The *Y* axis denotes the reaction velocity [n*M*/h]. Data points are fit to a rectangular hyperbola of the form $y = ax/(b + x)$.

## *Peptide-Based Conjugation Assays*

SUMO conjugation can be measured using peptides that include a consensus SUMO modification site such as that derived from p53 (380-HKKLM*FKTE*GPDSD-393). Reactions are carried out at 37° as described previously. The reaction mix (100 $\mu$l) includes 16 $\mu M$ SUMO, 0.3 $\mu M$ E1, 3 $\mu M$ hUbc9, 0.5 units inorganic pyrophosphatase (Sigma), and up to 500 $\mu M$ p53 peptide (saturating substrate concentration). The inclusion of inorganic pyrophosphatase is not essential, but assists in removing pyrophosphate and its potential inhibitory affects on E1 activation. Reactions are initiated by the addition of 20 $\mu$l of 250 m*M* NaCl, 100 m*M* HEPES, pH 7.5, 0.5% Tween 20, 25 m*M* $MgCl_2$, 5 m*M* DTT, and 5 m*M* ATP to yield a final concentration of 50 m*M* NaCl, 20 m*M* HEPES, pH 7.5, 0.1% Tween

20, 5 m$M$ $MgCl_2$, 1 m$M$ DTT, and 1 m$M$ ATP. A 10-$\mu$l aliquot is removed immediately after initiation of the reaction, and the reaction is stopped by the addition of SDS-loading buffer. Ten-microliter aliquots are removed at various time points during the reaction, and the products are resolved by SDS–PAGE [12% NuPAGE Bis-Tris gel with NuPAGE MES buffer (Invitrogen)]. The gel is stained using Sypro Ruby (Bio-Rad) for 4 to 12 h and is destained for at least 1 h prior to imaging under UV illumination using the Bio-Rad gel documentation system (Fig. 4A). The images are quantified using Quantity One Software (Fig. 4B; Bio-Rad).

## In Vivo *Assessment of Ubc9 Function*

UBC9 is an essential gene in *S. cerevisiae* (Seufert *et al.*, 1995). As such, the *S. cerevisiae* ubc9 null mutation can be used to study mutational effects

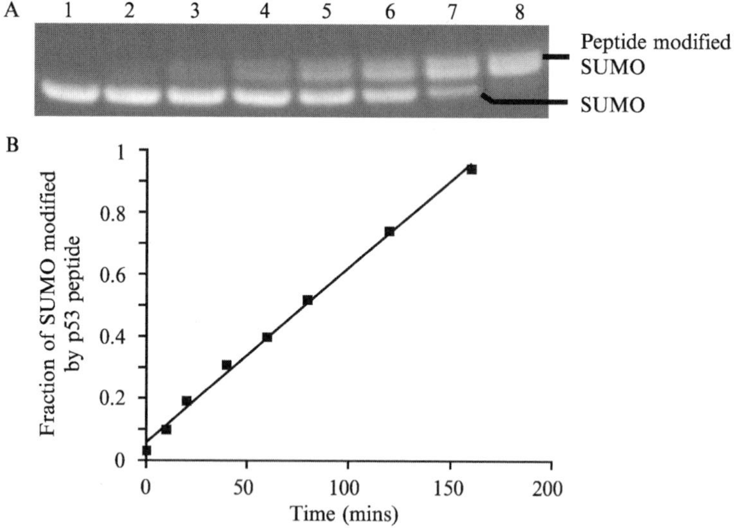


FIG. 4. Time course for SUMO conjugation to a synthetic peptide. (A) Time course for SUMO conjugation to a synthetic peptide derived from C-terminal amino acids of p53 that includes its consensus SUMO modification site (see text for sequence). Samples were resolved by SDS–PAGE and stained with Sypro Ruby (Bio-Rad). Ten-microliter aliquots were removed at the indicated time. Lane 1, 0 min; lane 2, 10 min; lane 3, 20 min; lane 4, 40 min; lane 5, 60 min; lane 6, 80 min; lane 7, 120 min; and lane 8, 160 min. (B) Data in A depicted graphically. Plotted data represent the mean from two SUMO conjugation assays (only one shown in A). The $X$ axis denotes time in minutes, and the the $Y$ axis denotes the fraction SUMO that is modified by the p53 peptide.

on Ubc9 function *in vivo*. This can be accomplished by gene replacement and plasmid shuffle with either extrachromosomal copies of yeast *UBC9* or human *UBC9* using standard techniques for the growth and maintenance of yeast strains (Guthrie and Fink, 1991). We have utilized the haploid strain W303-1A (*MATa ura3-1 ade2-1 trp1-1 his3, -1115, leu2-3, -112 can1-100*) to generate the ubc9 null strain by a one-step gene replacement procedure (Rothstein, 1991). We and others have found the W303–1A strain to be more sensitive than the S288C strain to mutations in the SUMO pathway (Mossessova and Lima, 2000). The *ubc9Δ* strain W303UBC9 (*MATa ura3-1 ade2-1 trp1-1 his3, -1115, leu2-3, -112 can1-100, ubc9Δ::LEU2*, pSE360-*UBC9*) was used for *in vivo* complementation by the plasmid shuffle technique (Boeke *et al.*, 1987). W303UBC9 was transformed with the pSE358 plasmid (*CEN TRP1*) bearing *UBC9* alleles under its natural promoter (Ingenius). $Trp^+$ colonies were selected on 5-fluoroorotic acid (5-FOA) containing media for the loss of pSE360-*UBC9* (*CEN URA3 UBC9* plasmid). To evaluate growth phenotypes, $Trp^+$ colonies were grown in SD-Trp media to $OD_{600}$ 0.1, serially diluted, and 5 $\mu$l was spotted onto YPAD plates.

Human *UBC9* cannot complement the *ubc9Δ* strain using the endogenous yeast *UBC9* promoter. Human *UBC9* must be placed behind a strong promoter such as *TPI1* or *GAL*. We have tested mutational effects on human Ubc9 function *in vivo* by plasmid shuffle in a *S. cerevisiae ubc9Δ* strain containing wild-type *UBC9* on a *CEN URA3* plasmid (described earlier). Wild-type and mutant human *UBC9* alleles are inserted into the vector pYN132 (*CEN TRP1*) in which human *UBC9* expression is driven by the yeast *TPI1* promoter (Fabrega *et al.*, 2003; Schwer *et al.*, 2001). W303UBC9 is then transformed with the pYN-hUBC9 plasmids, and $Trp^+$ colonies are streaked on agar medium containing 5-FOA to select for loss of the *UBC9 URA3* plasmid. Lethal mutations are those that fail to yield 5-FOA-resistant colonies after incubation for 10 days at 23 or 30° . Viable human *UBC9* mutants are assessed for growth defects by growing the respective strains in YPD broth (with 0.03% adenine) to an $A_{600} = 0.5$. Aliquots (3 $\mu$l) of serial 10-fold dilutions ($10^{-1}$, $10^{-2}$, $10^{-3}$) are spotted onto YPAD plates, which were incubated for 3 days at 23° or for 2 days at 30 and 37°.

## Acknowledgments

We thank Dr. Frauke Melchior for the antibody used in Fig. 3. A. A. Y. and C. D. L. are supported in part by a grant from the National Institutes of Health (GM65872). C. D. L. acknowledges additional support from the Rita Allen Foundation.

## References

Bencsath, K. P., Podgorski, M. S., Pagala, V. R., Slaughter, C. A., and Schulman, B. A. (2002). Identification of a multifunctional binding site on Ubc9p required for Smt3p conjugation. *J. Biol. Chem.* **277,** 47938–47945.

Bernier-Villamor, V., Sampson, D. A., Matunis, M. J., and Lima, C. D. (2002). Structural basis for E2-mediated SUMO conjugation revealed by a complex between ubiquitin-conjugating enzyme Ubc9 and RanGAP1. *Cell* **108,** 345–356.

Boeke, J. D., Trueheart, J., Natsoulis, G., and Fink, G. R. (1987). 5-Fluoroorotic acid as a selective agent in yeast molecular genetics. *Methods Enzymol.* **154,** 164–175.

Bossis, G., Chmielarska, K., Gärtner, U., Pichler, A., Stieger, E., and Melchior, F. (2005). A fluorescence resonance energy transfer-based assay to study SUMO modification in solution. *Methods Enzymol.* **398**(3) (this volume).

Deshaies, R. J. (1999). SCF and Cullin/Ring H2-based ubiquitin ligases. *Annu. Rev. Cell Dev. Biol.* **15,** 435–467.

Fabrega, C., Shen, V., Shuman, S., and Lima, C. D. (2003). Structure of an mRNA capping enzyme bound to the phosphorylated carboxy-terminal domain of RNA polymerase II. *Mol. Cell* **11,** 1549–1561.

Gill, G. (2003). Post-translational modification by the small ubiquitin-related modifier SUMO has big effects on transcription factor activity. *Curr. Opin. Genet. Dev.* **13,** 108–113.

Guthrie, C., and Fink, G. R. (1991). Guide to yeast genetics and molecular biology. *Methods Enzymol.* **194**. Academic Press.

Hershko, A., and Ciechanover, A. (1998). The ubiquitin system. *Annu. Rev. Biochem.* **67,** 425–479.

Hochstrasser, M. (1996). Ubiquitin-dependent protein degradation. *Annu. Rev. Genet.* **30,** 405–439.

Hochstrasser, M. (2001). SP-RING for SUMO: New functions bloom for a ubiquitin-like protein. *Cell* **107,** 5–8.

Huibregtse, J. M., Scheffner, M., Beaudenon, S., and Howley, P. M. (1995). A family of proteins structurally and functionally related to the E6-AP ubiquitin-protein ligase. *Proc. Natl. Acad. Sci. USA* **92,** 2563–2567.

Johnson, E. S. (2004). Protein modification by SUMO. *Annu. Rev. Biochem.* **73,** 355–382.

Johnson, E. S., and Blobel, G. (1997). Ubc9p is the conjugating enzyme for the ubiquitin-like protein Smt3p. *J. Biol. Chem.* **272,** 26799–26802.

Johnson, E. S., and Gupta, A. A. (2001). An E3-like factor that promotes SUMO conjugation to the yeast septins. *Cell* **106,** 735–744.

Kagey, M. H., Melhuish, T. A., and Wotton, D. (2003). The polycomb protein Pc2 is a SUMO E3. *Cell* **113,** 127–137.

Kahyo, T., Nishida, T., and Yasuda, H. (2001). Involvement of PIAS1 in the sumoylation of tumor suppressor p53. *Mol. Cell* **8,** 713–718.

Laney, J. D., and Hochstrasser, M. (1999). Substrate targeting in the ubiquitin system. *Cell* **97,** 427–430.

Lin, D., Tatham, M. H., Yu, B., Kim, S., Hay, R. T., and Chen, Y. (2002). Identification of a substrate recognition site on Ubc9. *J. Biol. Chem.* **277,** 21740–21748.

Lois, L. M., and Lima, C. D. (2004). Structures of the Small Ubiquitin-like Modifier E1 activating enzyme suggest conformational changes required for SUMO activation and a general mechanism for E2 recruitment to E1. *EMBO J.* **24,** 439–451.

Lois, L. M., Lima, C. D., and Chua, N.-H. (2003). SUMO modulates abscisic acid signaling in *Arabidopsis*. *Plant Cell* **15,** 1347–1359.

Melchior, F. (2000). SUMO-nonclassical ubiquitin. *Annu. Rev. Cell Dev. Biol.* **16,** 591–626.

Melchior, F., Schergaut, M., and Pichler, A. (2003). SUMO: Ligases, isopeptidases and nuclear pores. *Trends Biochem. Sci.* **28,** 612–618.

Mossessova, E., and Lima, C. D. (2000). Ulp1-SUMO crystal structure and genetic analysis reveal conserved interactions and a regulatory element essential for cell growth in yeast. *Mol. Cell* **5,** 865–876.

Muller, S., Hoege, C., Pyrowolakis, G., and Jentsch, S. (2001). SUMO, ubiquitin's mysterious cousin. *Nature Rev. Mol. Cell. Biol.* **2,** 202–210.

Pichler, A., Gast, A., Seeler, J. S., Dejean, A., and Melchior, F. (2002). The nucleoporin RanBP2 has SUMO1 E3 ligase activity. *Cell* **108,** 109–120.

Reverter, D., and Lima, C. D. (2004). A basis for SUMO protease specificity provided by analysis of human Senp2 and a Senp2-SUMO complex. *Structure* **12,** 1519–1531.

Rothstein, R. (1991). Targeting, disruption, replacement, and allele rescue: Integrative DNA transformation in yeast. *Methods Enzymol.* **194,** 281–301.

Scheffner, M., Huibregtse, J. M., and Howley, P. M. (1994). Identification of a human ubiquitin-conjugating enzyme that mediates the E6-AP-dependent ubiquitination of p53. *Proc. Natl. Acad. Sci. USA* **91,** 8797–8801.

Schmidt, D., and Muller, S. (2003). PIAS/SUMO: New partners in transcriptional regulation. *Cell Mol. Life Sci.* **60,** 2561–2574.

Schwer, B., Lehman, K., Saha, N., and Shuman, S. (2001). Characterization of the mRNA capping apparatus of *Candida albicans*. *J. Biol. Chem.* **276,** 1857–1864.

Seufert, W., Futcher, B., and Jentsch, S. (1995). Role of a ubiquitin-conjugating enzyme in degradation of S- and M-phase cyclins. *Nature* **373,** 78–81.

Takahashi, Y., Toh-e, A., and Kikuchi, Y. (2001). A novel factor required for the SUMO1/Smt3 conjugation of yeast septins. *Gene* **275,** 223–231.

Tatham, M. H., Chen, Y., and Hay, R. T. (2003a). Role of two residues proximal to the active site of Ubc9 in substrate recognition by the Ubc9.SUMO–1 thiolester complex. *Biochemistry* **42,** 3168–3179.

Tatham, M. H., Kim, S., Yu, B., Jaffray, E., Song, J., Zheng, J., Rodriguez, M. S., Hay, R. T., and Chen, Y. (2003b). Role of an N-terminal site of Ubc9 in SUMO-1, -2, and -3 binding and conjugation. *Biochemistry* **42,** 9959–9969.

# [9] High-Throughput Assay to Monitor Formation of the E2-Ubiquitin Thioester Intermediate

*By* PATRICK HAUSER and FRANCESCO HOFMANN

## Abstract

Targeting components of ubiquitination pathways for drug discovery necessitates the development of high-capacity assays that monitor the ubiquitination process at defined steps of the E1–E2–E3 cascade. This chapter describes the development of an assay based on time-resolved fluorescence to monitor formation of the thioester intermediate between ubiquitin-conjugating enzymes (E2s) and ubiquitin. The methodology is exemplified by an assay tailored for the ubiquitin-conjugating enzyme

METHODS IN ENZYMOLOGY, VOL. 398 0076-6879/05 $35.00

DOI: 10.1016/S0076-6879(05)98009-9

Cdc34. This assay setup can be easily adapted to other E2s and is suitable to screen small molecule inhibitors of E2-thioester formation in a high-throughput mode.

## Introduction

The increasing understanding of the ubiquitin-proteasome system, of its importance in the regulation of a myriad of biological processes, and of its implication in a variety of human diseases has attracted the interest of drug discovery toward its components. Following the successful example of targeting the proteolytic activity of the 26S proteasome for the treatment of cancer, the question remains whether other components of the system, such as E1s, E2s, E3s, and deubiquitylating enzymes, will be equally attractive as targets, druggable, and of potential therapeutic value.

Assuming the attractiveness of certain ubiquitination pathway components as potential therapeutic targets, the next important step toward their pharmaceutical exploitation is the availability of sensitive and reliable assays. These assays have to faithfully recapitulate the activity or function of a selected target in a manner that allows rapid screening of large compound libraries in search for small molecules capable of modulating the activity or function of such a target.

A potential source of targets within the ubiquitination system is represented by the ubiquitin-conjugating enzyme (E2) family, encompassing ~50 members in the human genome. Selected family members have been associated with the ubiquitin or ubiquitin-like modification of substrate proteins, whose stability or function is then altered, thereby determining or at least contributing to a particular disease condition. One such example is represented by the ubiquitin-conjugating enzyme Cdc34, which has been described to regulate cell cycle progression (Plon *et al.*, 1993) and to function in conjunction with the $SCF^{Skp2}$ E3 ligase (Lisztwan *et al.*, 1998), which in turn has been implicated in the ubiquitin-dependent degradation of the cyclin-dependent kinase inhibitor $p27^{Kip1}$ (Carrano *et al.*, 1999; Sutterluty *et al.*, 1999). Downregulation of the latter, through enhanced ubiquitin-dependent degradation, has been associated with neoplastic diseases (Sgambato *et al.*, 2000; Slingerland and Pagano, 2000). Thus, Cdc34 could qualify as a potential therapeutic target, whose inhibition should reduce aberrant $p27^{Kip1}$ ubiquitination and degradation, thereby restoring $p27^{Kip1}$ abundance and tumor suppressor function.

Theoretically, the function of an E2 could be inhibited by several different approaches, including impairing its interaction with the E1-Ub donor complex, preventing the formation of the thioester intermediate between the E2 and ubiquitin, inhibiting the release and transfer of ubiquitin to the E3 or to the substrate, and interfering with the interaction of the E2 with its cognate E3. Hereafter, an assay to monitor the formation of the Cdc34-ubiquitin thioester intermediate is described.

## Experimental Procedures

### *Cloning of Human Ubiquitin-Activating Enzyme E1*

E1, isoform a, is cloned from a human uterus Quick Clone cDNA library (Clontech, No. 637237) by polymerase chain reaction (PCR). Primers are designed to cover the whole coding sequence (nucleotide 33-3209, GeneBank accession No. X56976) and to allow subcloning in frame, downstream of GST and a factor Xa cleavage site, into FastBacGSTx3 (Garcia-Echeverria *et al.*, 2004). The 5′ primer carries a *Bam*HI cleavage site: 5′-T CGT GGG ATC CCC ATG TCC AGC TCG CCG CTG TCC AAG AAA-3′. The 3′ primer carries a *Hin*dIII cleavage site: 5′-TCG ACA AGC TTA GCG GAT GGT GTA TCG GAC ATA GGG AAC CTC-3′. The PCR mix contains 1 ng cDNA library, 200 $\mu M$ dNTPs, 200 n$M$ of each primer, and 2.5 U Takara ExTaq DNA polymerase (Takara Mirus Bio, No. RR001A) in 50 $\mu$l of 1× Takara PCR buffer. The PCR is initiated with 10 min denaturing at 94°, followed by 30 amplification cycles (1 min denaturing at 94°, 1 min annealing at 55°, 3 min elongation at 72°) and is terminated with 10 min elongation at 72°.

The PCR product of the expected size is subcloned as a *Bam*HI–*Hin*dIII insert into pFastBacGSTx3 to yield pFastBacGSTx3-E1.

### *Expression and Purification of GST-E1*

GST-E1 is expressed from pFastBacGSTx3-E1 using the baculovirus system according to the protocol supplied by Invitrogen and using High Five cells (Invitrogen, No. B855–02). Cells harvested from 10 P150 plates are resuspended by vortexing in 50 ml lysis buffer [25 m$M$ Tris-HCl, pH 7.5, 2 m$M$ EDTA, 1% NP-40, 1 m$M$ dithiothreitol (DTT), incubated. Complete protease inhibitor cocktail (Roche, No. 1697498)] and incubating on ice for 30 min. After centrifugation at 8000*g* using a SS34 rotor from Sorvall for 20 min, the supernatant is incubated with 2 ml glutathione-Sepharose 4B (Pharmacia, No. 17-0756-01) overnight at 4° on a rotor table. The glutathione-Sepharose 4B is then packed in a column and washed extensively with 25 m$M$ Tris-HCl, pH 7.5, 2 m$M$ EDTA, and 1 m$M$ DTT. The GST-tagged protein is eluted using 50 m$M$ Tris·HCl, pH 7.5, 10 m$M$ reduced glutathione, 1 m$M$ DTT, and 20% glycerol, aliquoted, and stored at −70°. Typically, ~5.0 mg GST-E1 is obtained.

### *Cloning of Human Ubiquitin-Conjugating Enzyme Cdc34*

Cdc34 is cloned from a human fetal brain Quick Clone cDNA library (Clontech, No. 637221) by PCR. Primers are designed to cover the whole coding sequence (nucleotide 187-894, see GeneBank accession No. L22005) and to allow subcloning in frame, downstream of GST and a factor Xa cleavage site, into pFastBacGSTx3 (Garcia-Echeverria *et al.*, 2004). The 5′ primer carries a *Bam*HI cleavage site: 5′-T CGT GGG ATC CCC ATG GCT CGG CCG CTA GTG CCC AGC TCG-3′. The 3′ primer carries a *Eco*RI cleavage site: 5′-CCT TTG AAT TCA GGA CTC CTC CGT GCC AGA GTC ATC CTC-3′. The PCR mix contains 1 ng cDNA library, 200 $\mu M$ dNTPs, 200 n*M* of each primer, and 2.5 U Takara ExTaq DNA polymerase (Takara Mirus Bio, No. RR001A) in 50 $\mu$l of 1× Takara PCR buffer. The PCR is initiated with 10 min denaturing at 94°, followed by 30 amplification cycles (1 min denaturing at 94°, 1 min annealing at 55°, 1 min elongation at 72°) and is terminated with 10 min elongation at 72°.

The PCR product of the expected size is subcloned as a *Bam*HI–*Eco*RI insert into pFastBacGSTx3 to yield pFastBacGSTx3-Cdc34.

Using the clone obtained earlier as a template, a version containing a N-terminal factor Xa cleavage site and a C-terminal flag tag (DYKDDDDK) is constructed by PCR using the following 5′ primer (5′-ATC GAA GGT CGT ATG GTC CGG CCG CTA GTG–3′) and 3′ primer (5′-TCA CTT GTC GTC GTC GTC CTT GTA GTC GGA CTC CTC CGT GCC AGA-3′). Subsequently, attB1 and attB2 sequences are added at both extremities to allow subcloning by recombination into a Gateway pDEST20 vector (Invitrogen, No. 11807-013) to generate pDEST20-GST-FXa-Cdc34-Flag for expression of Cdc34-Flag, as cleavable GST fusion using the baculovirus system. The PCR reactions are performed as described earlier using 2.5 U PFU DNA polymerase (Stratagene).

### *Expression and Purification of Cdc34-Flag*

The GST-Cdc34-Flag is expressed from pDEST20-GST-FXa-Cdc34-Flag using the baculovirus system according to the protocol supplied by Invitrogen and using High Five cells (Invitrogen, No. B855-02). Cells harvested from 10 P150 plates are resuspended by vortexing in 50 ml lysis buffer [25 m*M* Tris-HCl, pH 7.5, 2 m*M* EDTA, 1 m*M* DTT, Complete protease inhibitor cocktail (Roche, # 1697498)] and incubated on ice for 30 min. After centrifugation at 8000*g* for 20 min with a SorvalL SS34 rotor, the supernatant is incubated with 2 ml glutathione-Sepharose 4B (Pharmacia, No. 17-0756-01) overnight at 4° on a rotor table. The glutathione-Sepharose 4B is then packed in a column and washed extensively with 25 m*M*

Tris-HCl, pH 7.5, 100 m*M* NaCl, and 0.5 m*M* DTT. Glutathione-Sepharose 4B is washed with factor Xa buffer (50 m*M* Tris-HCl, pH 8.0, 100 m*M* NaCl, 1 m*M* $CaCl_2$) and resuspended in 2 ml of the same buffer. Cdc34-Flag is subsequently cleaved off the GST fusion by adding 100 $\mu$g of factor Xa (Roche, No. 1585924) to the glutathione-Sepharose 4B suspension and incubating for 2 h at room temperature on a rotor table. The column is fixed again in a vertical position and the flow through, containing Cdc34-Flag, is collected and stored at $-70°$, upon addition of 0.5 m*M* DTT and 10% glycerol. Typically, ~4.0 mg Cdc34-Flag is obtained.

*Labeling of Ubiquitin with Europium*

Bovine red blood cell ubiquitin (Sigma, U-6253) is labeled with Europium using the reagents supplied with the DELFIA Europium labeling kit (Wallac, No. 1244-302) as described by the manufacturer. Typically, 2.5 mg ubiquitin (~300 nmol) is dissolved in 500 $\mu$l labeling buffer (50 m*M* $NaHCO_3$, 150 m*M* NaCl, pH 8.5) and added to 1 vial of DELFIA Europium reagent (0.2 mg, 300 nmol) in order to have an equimolar solution. After overnight incubation at room temperature, the labeled protein is separated from the remaining labeling reagent by gel filtration. For this purpose, a Sephadex G50 medium (Amersham Biosciences, No. 17-0043-02) column (30 $\times$ 0.8 cm) equilibrated in elution buffer (50 m*M* Tris-HCl, pH 7.8, 150 m*M* NaCl) is used and 1-ml fractions are collected. Fractions containing the labeled material are identified by measuring the protein content using the Bradford method (Bio-Rad, No. 500-006) and by analyzing Europium time-resolved fluorescence using a Victor II reader (Wallac). Fractions containing Europium-labeled ubiquitin (Eu-Ub) are pooled, aliquoted, and stored at $-20°$. Analysis of Europium-labeled ubiquitin and comparison with the standard supplied with the labeling kit showed that the pooled fractions (8 ml) contain ~2 mg ubiquitin (~250 nmol) and 175 nmol Europium, or 0.7 mol Europium/mole ubiquitin, corresponding to $3.2 \times 10^9$ counts/nmol ubiquitin.

*Gel-Based E2-Ubiquitin Thioester Intermediate Assay*

Formation of a thioester bond between ubiquitin and an E2, upon activation of ubiquitin by E1, can be visualized by SDS gel electrophorisis (Scheffner *et al.*, 1995). One microgram of purified GST-E1 and 4 $\mu$g of purified Cdc34-Flag are incubated for 10 min at room temperature with 4 $\mu$g ubiquitin in 25 m*M* Tris-HCl, pH 7.5, 50 m*M* NaCl, 5 m*M* $MgCl_2$, 4 m*M* ATP, and 0.25 m*M* DTT in a total volume of 20 $\mu$l. Half the sample is then mixed with an equal volume of stop buffer A (50 m*M* Tris-HCl, pH 6.8, 2% SDS, 4 *M* urea, 10% glycerol, bromphenol blue) and incubated for

15 min at 30°. The other half is mixed with an equal volume of stop buffer B (50 m*M* Tris-HCl, pH 6.8, 2% SDS, 10% glycerol, 100 m*M* DTT, bromphenol blue) and incubated for 5 min at 95°, prior to analysis on a 12% SDS–PAGE. Thioester bond formation is then visualized as a ~8-kDa shift of the Cdc34 band by staining the gel with gelcode blue (Pierce, No. 24590).

*High-Throughput E2-Ubiquitin Thioester Intermediate Assay*

The assay requires Eu-labeled ubiquitin (Eu-Ub), GST-E1, and Cdc34-Flag and because it is aimed at screening for compounds that modulate Cdc34-Ub thioester formation, small molecular weight compounds are added to assess their effect. The buffer required to drive the reaction (thioester buffer, TEB) is prepared 10× concentrated and contains 1 *M* Tris-HCl, pH 7.5, 1 *M* NaCl, 100 m*M* $MgCl_2$, 40 m*M* ATP (Sigma, No. A-2383), and 2.5 m*M* DTT. The same buffer, lacking ATP, is also prepared (TEB - no ATP).

The following protocol refers to conditions optimized to achieve a robust signal to noise (presence versus absence of Cdc34), with a minimal requirement of reagents to maximize high-thoughput screening capacity, in a 96-well plate format.

The final concentration of the key components are

| | |
|---|---|
| Compound | 50, 10, or 1 μ*M* for single point determination, or 50/12.5/3.1/0.8/0.2/0.05/0.012 μ*M* for dose–response curves |
| Eu-Ub | 6.25 n*M* |
| GST-E1 | 25 n*M* |
| Cdc34-Flag | 12.5 n*M* |

1. In nonabsorbing 96-well plates, V-bottom (Catalys AG, # 2605) mix per well, 10 μl 5% dimethyl sulfoxide (DMSO, positive and negative control), or small molecular weight compound 5× concentrated in 5% DMSO and 30 μl Cdc34-Flag (20.8 n*M*) in 1× TEB - no ATP, or 1× TEB - no ATP (negative control). Incubate for 10 min at room temperature.

2. Prepare activating mixture (equivalent of 10 μl/well) containing Eu-Ub (31.2 n*M*) and GST-E1 (125 n*M*) in 1 × TEB and preincubate for 10 min at room temperature.

3. To reaction plate, add 10 μl activating mixture. Incubate for 30 min at room temperature.

4. Transfer 45 μl reaction mix to streptavidin-coated plates (Pierce No. 15124) containing 50 μl/well Superblock-TBS (Pierce No. 37535) + 1 m*M* *N*-ethylmaleimide (Fluka No. 04259) + 100 n*M* biotinylated anti-Flag antibody (Sigma No. F-9291). Incubate for 2 h at room temperature.

5. Wash 6 × 200 μl/well with 1× wash solution (Wallac, No.1244-114).

6. Add 200 μl/well Wallac enhancer solution (Wallac, No.1244-105) and mix for 5 min.

7. Read time-resolved fluorescence on a Victor II reader using the standard Europium protocol (excitation filter D340, emission filter D615, delay 400 μs, window time 400 μs, cycle 1000 μs).

Positive control: Cdc34, no compound (10 μl 5% DMSO)
Negative control: no Cdc34-Flag (30 μl 1 × TEB - no ATP), no compound (10 μl 5% DMSO)

## Results

Analysis of the formation of a thioester intermediate between ubiquitin and an E2, upon activation of ubiquitin by E1, has been described and is based on SDS gel electrophoresis (Scheffner *et al.*, 1995). This chapter described the adaptation of this assay to a suitable high-throughput screening format.

GST-E1 and GST-Cdc34-Flag were expressed using the baculovirus system and affinity purified (Fig. 1); Cdc34-Flag was cleaved from the fused GST moiety using factor Xa because the latter was found to interfere with its activity. The activity of the enzymes was first assessed in a classical gel-based E2 thioester assay, as described by Scheffner (1995); formation of the Cdc34-ubiquitin thioester intermediate was confirmed and visualized under nonreducing conditions as a ~8-kDa mobility shift, which is abolished under reducing conditions (Fig. 2).

For development of a suitable high-throughput assay format, several options were evaluated. The most reproducible activity, combined with a robust signal and the most economic use of ubiquitin, E1, and E2, was obtained with a 96-well-based two-step procedure; the enzymatic reaction was carried out under homogeneous conditions, followed by capturing Cdc34-Flag with a biotinylated anti-Flag antibody on streptavidin-coated plates. Thioester formation is then monitored by Europium time-resolved fluorescence, through Europium-labeled ubiquitin (Fig. 3).

In a first series of experiments, each of the components was tested at concentrations between 1 and 1000 n*M*. Measurable and robust signals were observed readily at concentrations below 100 n*M*. The signal was shown to increase substantially in a Cdc34-driven manner over a range of Eu-ubiquitin concentrations (Fig. 4) and to depend on the presence of E1 (25 n*M*) and Cdc34 (Fig. 5). Furthermore, when Cdc34 was pretreated with 1 m*M* *N*-ethylmaleimide to modify free cysteines, such as the one that accepts ubiquitin at the catalytic site, the signal was reduced to background level (Fig. 5). Similarly, the omission of ATP during ubiquitin activation by

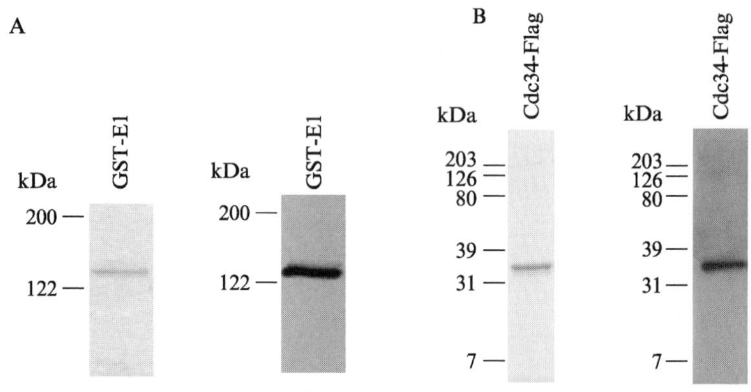


FIG. 1. Purified recombinant GST-E1 and Cdc34-Flag. (A) Affinity-purified GST-E1 (2 μg) was analyzed by SDS–PAGE and Coomassie blue staining. The identity of the purified protein (200 ng) was confirmed by immunostaining of a corresponding Western blot with an anti-E1 (isoform a) antibody (Calbiochem, No. 662102). (B) Affinity-purified and factor Xa-cleaved Cdc34-Flag (2 μg) was analyzed by SDS–PAGE and GelCode staining. The identity of the purified protein (0.5 ng) was confirmed by immunostaining of a corresponding Western blot with a monoclonal anti-Cdc34 antibody (136.1.12; Butz *et al.*, 2005). (See color insert.)

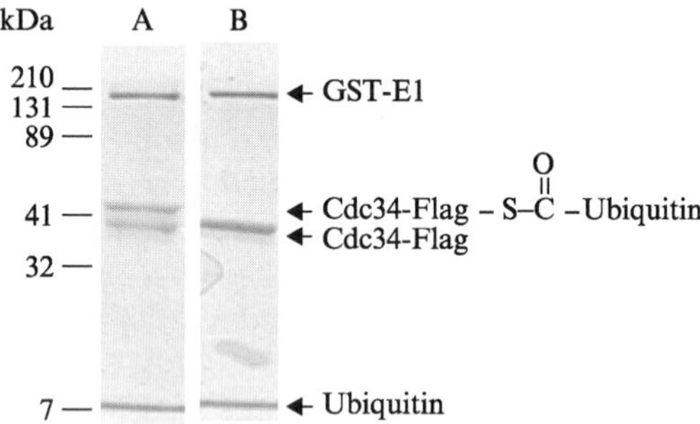


FIG. 2. Activity test. Activity of the purified enzymes was confirmed by assessing formation of a thioester intermediate between Cdc34 and ubiquitin, which can be monitored as a SDS–PAGE mobility shift under nonreducing conditions (A). Upon treatment with DTT, the thioester bond between the active site cysteine of Cdc34 and the carboxy terminus of ubiquitin is cleaved and the mobility shift is abolished (B). (See color insert.)

E1 or the addition of 5 m*M* DTT at the end of the reaction (and prior to transferring it to streptavidin-coated plates to capture the E2) was sufficient to abolish the signal (data not shown). Thus, the time resolved-fluorescence measured upon incubation of all necessary components

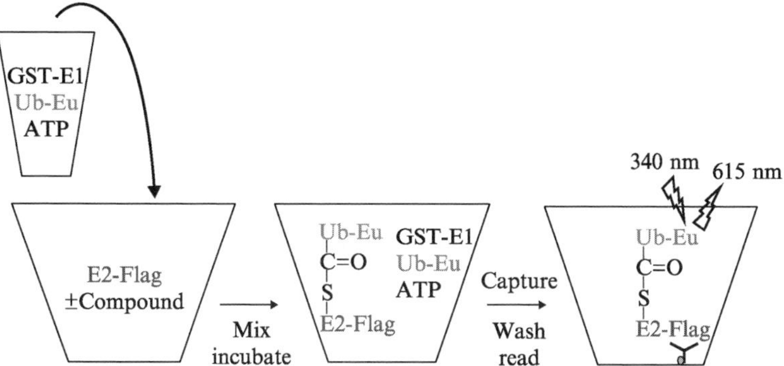


FIG. 3. E2-ubiquitin thioester assay. A 96-well-based assay to monitor formation of the thioester intermediate between a ubiquitin-conjugating enzyme and ubiquitin was developed. Ubiquitin labeled with Europium was activated by E1 in the presence of ATP and added to a flag-tagged E2 in the presence or absence of a small molecular weight compound. Separation of the E2 from the remaining reagents was achieved by capturing the flag-tagged E2 on streptavidin-coated plates via a biotinylated anti-Flag antibody. Formation of the E2-ubiquitin thioester intermediate was then monitored by measuring Europium time-resolved fluorescence (excitation 340 nm; emission 615 nm). (See color insert.)

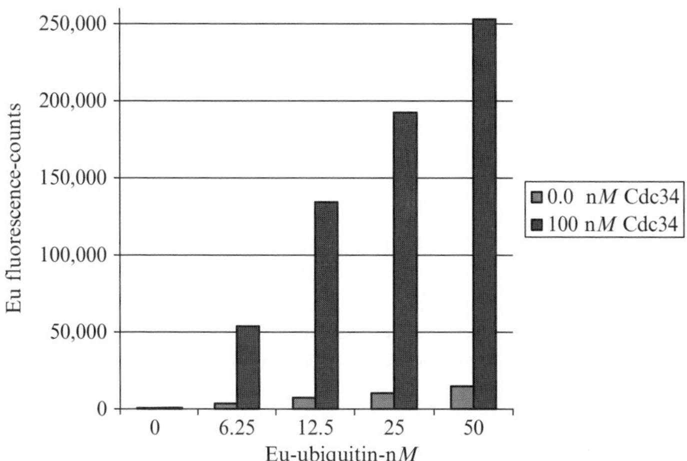


FIG. 4. Europium fluorescence. Europium time-resolved fluorescence increases in a Cdc34-dependent manner over a range of Eu-ubiquitin concentrations.

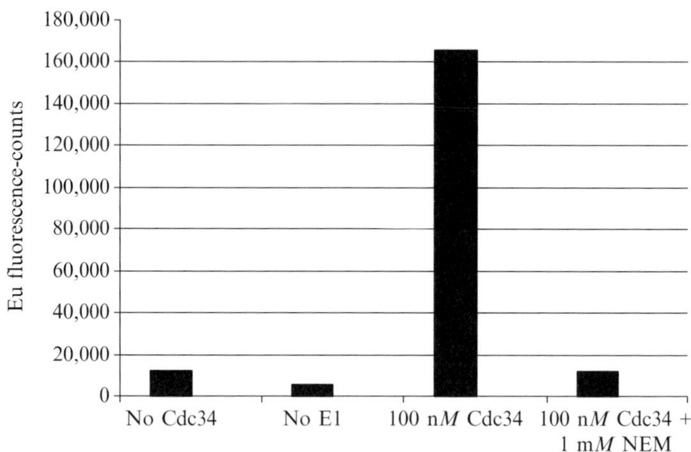


FIG. 5. Europium fluorescence as a measure of E2-Ub thioester formation. The Europium time-resolved fluorescence signal was validated as a measure of formation of the thioester intermediate between Cdc34 and ubiqutin by demonstrating its dependence on E1 and Cdc34, as well as its prevention by modification of the Cdc34 active site cysteine with *N*-ethylmaleimide (NEM).

indeed represents formation of the thioester intermediate between Eu-ubiquitin and Cdc34.

Based on these observations, positive (signal) and negative (noise) readouts were defined as reaction mixtures containing all the necessary components and, respectively, mixtures where the target and ubiquitin acceptor Cdc34 were omitted.

In an attempt to better define conditions where a reproducible activity and a robust signal-to-noise ratio can be measured, increasing amounts of Cdc34 were added into the system, resulting in a dose-dependent response (Fig. 6A) and yielding signal-to-noise ratios between 2 and 18 (Fig. 6B). Among the latter, only those greater than or equal to 10-fold can be recommended for the selection of suitable concentrations of each assay component.

Assay conditions were then optimized, according to the protocol described under Experimental Procedures, to minimize the amount of Eu-Ub, E1, and Cdc34 necessary to run a high-throughput screening, which typically encompasses a few thousand 96-well plates. Under such conditions, a robust and reproducible signal over noise ratio was consistently maintained over hundreds of independent assay plates (Fig. 7). The optimized assay is suitable for single point determinations in a high-throughput mode or for determining dose–response curves and $IC_{50}$ values for selected compounds (Fig. 8).

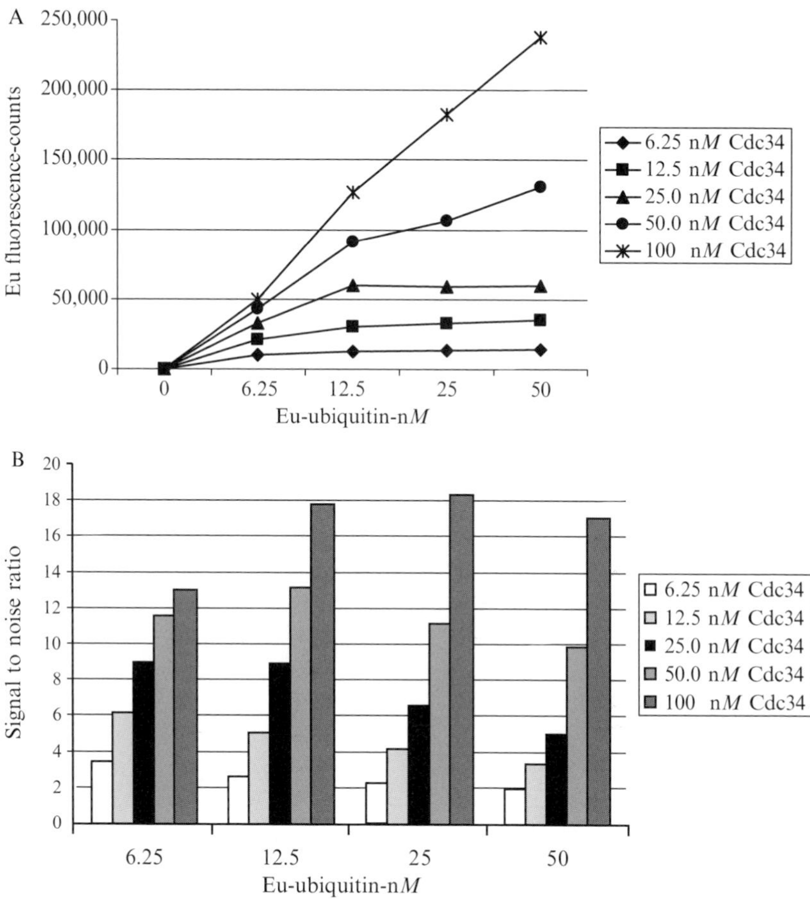


FIG. 6. Signal-to-noise ratio. Positive and negative (noise) signals were defined as presence and absence of Cdc34. (A) Formation of the thioester intermediate was monitored over a range of Cdc34 and Eu-ubiquitin concentrations at a fixed E1 concentration (25 n*M*). (B) The signal-to-noise ratio was then determined over the tested concentration range to identify conditions sustaining a robust signal over noise.

## Discussion

The 96-well-based Cdc34-ubiquitin thioester intermediate assay described here was optimized with the aim of having a reliable and robust signal over noise with a minimal use of components. Lowering the concentrations of each of the reagents also improved the sensitivity of the assay, which benefited the identification and the further profiling of small

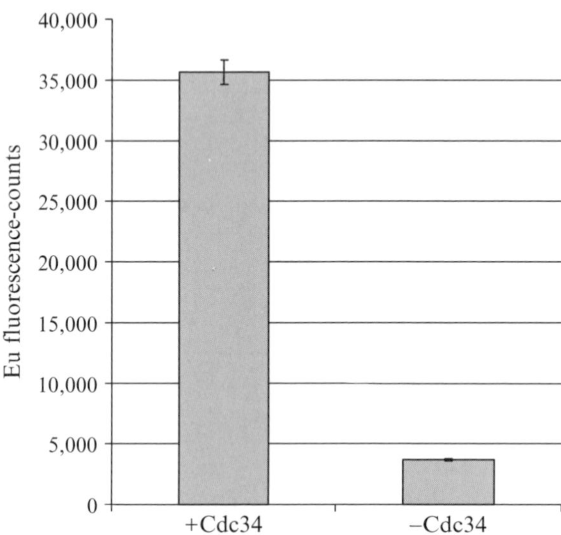


FIG. 7. Interplate signal variation. Positive signal (presence of Cdc34, absence of compound) and negative signal (absence of Cdc34, absence of compound) are determined on each 96-well assay plate. Their average (±SEM) over a representative set of 100 independent assay plates was plotted, yielding an average signal-to-noise ratio of 9.65.

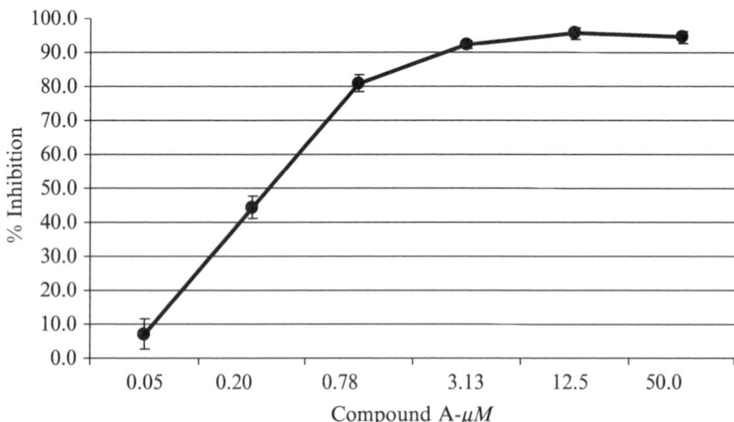


FIG. 8. Dose-dependent inhibition of Cdc34-ubiquitin thioester formation. Compound A was shown to inhibit formation of a thioester intermediate between Cdc34 and ubiquitin in a dose-dependent manner. The average percentage inhibition (±SEM) of seven independent measurements is plotted, allowing determination of an $IC_{50}$ of 0.25 $\mu M$ (±0.026) for compound A.

molecule inhibitors. During the optimization process, we also observed that stopping the reaction prior to capturing Cdc34 on streptavidin-coated plates stabilizes the maximal signal. To this purpose, the addition of *N*-ethylmaleimide to modify and block any remaining free active site cysteine residue on E1 or E2 was introduced.

In order to favor the selection of inhibitors binding to the E2, as opposed to E1, incubation of Cdc34 with compounds and activation of ubiquitin by E1 were initiated separately. Concerning the assessment of the selectivity of the inhibitors toward Cdc34, this assay was adapted successfully to other E2s, maintaining identical experimental conditions. Indeed, using Flag-tagged versions of UbcH5 (GeneBank accession No. X78140), UbcH7 (GeneBank accession No. X92962), UbcH10 (GeneBank accession No. U73379), HHR6B (GeneBank accession No. M74525), and E2-EPF5 (GeneBank accession No. M91670), it was possible to assess the selectivity profile of selected compounds over a limited panel of representative E2s. Moreover, selective inhibition of a particular E2 also indicates that the compound is not interfering with the activation of ubiquitin by E1, as the latter is a common denominator in all assays.

Thus, this experimental setup and this panel of E2 thioester assays are suitable for the screening of compound libraries in a high-throughput mode, as well as for the profiling of compounds with the aim of assessing their potency and selectivity.

## Acknowledgments

We thank Dennis France for valuable advice during the development of the assay and Mark Andrew Pearson for critical reading of the manuscript.

## References

Butz, N., Ruetz, S., Natt, F., Hall, J., Weiler, J., Mestan, J., Ducarre, M., Grossenbacher, R., Hauser, P., Kempf, D., and Hofmann, F. (2005). The human ubiquitin-conjugating enzyme Cdc34 controls cellular proliferation through regulation of p27Kip1 protein levels. Submitted for publication. *Exp. Cell Res.* **303,** 482–493.

Carrano, A. C., Eytan, E., Hershko, A., and Pagano, M. (1999). SKP2 is required for ubiquitin-mediated degradation of the CDK inhibitor p27. *Nature Cell Biol.* **1,** 193–199.

García-Echeverría, C., Pearson, M. A., Marti, A., Meyer, T., Mestan, J., Zimmermann, J., Gao, J., Brueggen, J., Capraro, H.-G., Cozens, R., Evans, D. B., Fabbro, D., Furet, P., Graus Porta, D., Liebetanz, J., Martiny-Baron, G., Ruetz, S., and Hofmann, F. (2004). *In vivo* anti-tumour activity of NVP-AEW541: A novel, potent and selective inhibitor of the IGF-IR kinase. *Cancer Cell* **5,** 231–239.

Lisztwan, J., Marti, A., Sutterluety, H., Gsteiger, M., Wirbelauer, C., and Krek, W. (1998). Association of human CUL–1 and ubiquitin conjugating enzyme CDC34 with the F-box protein p45SKP2: Evidence for evolutionary conservation in the subunit composition of the CDC34-SCF pathway. *EMBO J.* **17,** 368–383.

Plon, S. E., Leppig, K. A., Do, H. N., and Groudine, M. (1993). Cloning of the human homologue of the CDC34 cell cycle gene by complementation in yeast. *Proc. Natl. Acad. Sci. USA* **90,** 10484–10488.

Scheffner, M., Nuber, U., and Huibregtse, J. M. (1995). Protein ubiquitination involving an E1-E2-E3 enzyme ubiquitin thioester cascade. *Nature* **373,** 81–83.

Sgambato, A., Cittadini, A., Faraglia, B., and Weinstein, I. B. (2000). Multiple functions of p27(Kip1) and its alterations in tumor cells: A review. *J. Cell Physiol.* **183,** 18–27.

Slingerland, J., and Pagano, M. (2000). Regulation of the cdk inhibitor p27 and its deregulation in cancer. *J. Cell Physiol.* **183,** 10–17.

Sutterluty, H., Chatelain, E., Marti, A., Wirbelauer, C., Senften, M., Muller, U., and Krek, W. (1999). p45SKP2 promotes p27Kip1 degradation and induces S phase in quiescent cells. *Nature Cell Biol.* **1,** 207–214.

# Section III

# E3 Enzyme

# [10] Expression and Evaluation of RING Finger Proteins

*By* Yili Yang, Kevin L. Lorick, Jane P. Jensen, and Allan M. Weissman


## Abstract

RING finger proteins represent the largest class of potential ubiquitin ligases. This chapter describes methods used to express and assess the activity of proteins containing RING fingers based on our experience with a number of different family members. In addition to general protocols for assessing activity, specific protocols are provided for evaluating the ubiquitylation of p53 by the RING finger E3 Hdm2/Mdm2. Use of these methods may help identify new E3s, dissect factors involved in ubiquitylation of substrates, and screen for molecules that affect ubiquitylation.


## Introduction

The RING finger motif was first identified in the early 1990s in the protein encoded by the *Really Interesting New Gene 1* (Freemont *et al.*, 1991). Like other cysteine-rich motifs, the RING finger binds Zn through its conserved cysteine and histidine residues. However, the RING finger binds two Zn ions in a unique "cross-brace" arrangement, which distinguishes it from tandem zinc fingers and other similar motifs. The canonical RING finger consensus sequence has been defined as Cys-$X_2$-Cys-$X_{9\text{-}39}$-Cys-$X_{1\text{-}3}$-His-$X_{2\text{-}3}$-Cys/His-$X_2$-Cys-$X_{4\text{-}48}$-Cys-$X_2$-Cys, where X can be any amino acid residue. The majority of RING fingers fall into two subclasses: RING-HC and RING-H2. The classification depends on whether there is a Cys or a His in the fifth of the eight Zn coordinating sites. The PHD/LAP domain represents a variation on the RING finger in which a Cys replaces the His in the fourth position and in which the fifth position is an invariant His (RING-C4HC3); another feature of PHD fingers is the presence of an invariant tryptophan two amino acids before the seventh Zn-coordinating residue. In addition, there are active E3s that contain a RING finger with eight coordinating cysteines and other E3s with noncanonical RING finger variants. These include some with coordinating residues other than Cys or His. The U-box is a more distant RING finger relative that has a RING-like conformation but lacks the canonical Zn-coordinating residues. It has been estimated that there are more than 270 genes in the human genome

METHODS IN ENZYMOLOGY, VOL. 398 0076-6879/05 $35.00
DOI: 10.1016/S0076-6879(05)98010-5

encoding RING finger proteins, whereas the numbers for PHD domain and U-box are 86 and 7, respectively (Lorick *et al.*, 2005b; Semple, 2003; von Arnim, 2001; Wong *et al.*, 2003) (see also http://home.ccr.cancer.gov/lpds/weissman).

Many RING finger proteins possess ubiquitin ligase activity by themselves or as parts of multisubunit E3s (Deshaies, 1999; Lorick *et al.*, 1999, 2005b). Unlike the HECT domain that acts as a catalytic intermediate in the transfer of ubiquitin from E2 to substrate, there is no evidence that RING fingers form catalytic intermediates with ubiquitin. One view of RING fingers is that they simply serve as E2-docking sites positioning E2-Ub to enable transfer of Ub to substrate. There is accumulating evidence, however, for a more active role in facilitating the transfer of ubiquitin from E2 to substrates (Lorick *et al.*, 2005b). Many RING finger-containing proteins also promote their own ubiquitylation *in vitro* and *in vivo* in a RING finger-dependent manner. Although the physiological significance of autoubiquitylation remains to be established in many cases, this characteristic is widely used as a sensitive indicator of potential E3 activity of RING finger proteins, particularly *in vitro,* and especially when the nature of heterologous substrates is not well established.

The methods described in this chapter include (1) techniques for expression of GST-RING finger protein fusions *in vitro*; (2) generation of $^{32}$P-labeled ubiquitin as a tool for detecting ubiquitylation; (3) assessment of autoubiquitylation activity *in vitro*; (4) *in vitro* ubiquitylation of p53 by Hdm2 protein; and (5) Hdm2-mediated ubiquitylation of p53 in cells.

## Expressing RING Finger Proteins

To express GST fusion proteins, cDNAs encoding either full-length RING finger proteins or portions of these proteins can be cloned into pGEX family vector (Pharmacia) or similar vectors in the appropriate reading frame and transformed into *Escherichia coli* BL-21(DE3) (Novagen). A single colony is inoculated in 2 ml SuperBroth with ampicillin (100 $\mu$g/ml) and incubated at 37° overnight. The culture is transferred into 200 ml SuperBroth or other rich medium and incubated at room temperature. After an $OD_{600} = 0.6$–1.0 is reached, isopropyl-$\beta$-D-thiogalactoside (0.2 m*M*) is added to induce the expression of fusion protein for 2 h. Bacteria are then collected by centrifugation at 4° and resuspended in cold lysis buffer [50 m*M* Tris, pH 8.0, 5 m*M* dithiothreitol (DTT), and 1% Triton X–100, 2m*M* PMSF]. Bacteria are lysed by sonicating on ice for 15–45 s, depending on the volume of bacteria and power setting of the sonicator (see Lorick *et al.*, 2005a). After removing cellular debris by centrifugation, the supernatant is aliquoted and stored at –80° for further experiments. GST fusion proteins bound to glutathione-Sepharose beads (Amersham) are prepared by incubating the supernatant with beads at room temperature for 15 min and washing extensively with 1× TBS (50 m$M$ Tris, pH 7.5, 150 m$M$ NaCl). Once a protein is prepared, the level of expression and purity of translation products are determined by SDS–PAGE and Coomassie blue staining. However, because of stability issues, the protein level may need to be confirmed for each experiment.

*Notes:* When the primary objective of generating a GST fusion of a RING finger protein is to evaluate its potential activity by autoubiquitylation, one must consider whether to express the full-length protein or a limited region that includes the RING finger. For membrane-spanning proteins, solubility is a factor. For larger RING finger proteins, such as BRCA1, folding and yield of the full-length protein can be problematic. However, expression of a limited region containing the RING finger and surrounding regions can result in a failure to appreciate the role of other domains in E3 activity. Depending on lysine availability in the expressed protein, there is the potential that only limited ubiquitylation will be observed.

While many GST-fusion proteins can be induced at 37°, higher yields of full-length proteins are often obtained when induced at room temperature for 1–4 h or at even lower-temperature for longer time. This may require emperical determination. Zn is critical to proper folding and activity of RING finger E3s. The addition of $ZnCl_2$ to culture medium (up to 250 $\mu M$) or buffers (up to 25 $\mu M$) may facilitate proper folding and recovery of active protein. Phosphate buffers should be avoided due to Zn-phosphate precipitates and use of metal chelators such as EDTA and EGTA should be minimized.

## Generating $^{32}$P-Labeled Ubiquitin

While there are a number of ways of detecting ubiquitin and ubiquitylated proteins, we have found the use of $^{32}$P-labeled ubiquitin, first described by Peter Howley and colleagues (Lorick *et al.*, 1999; Scheffner *et al.*, 1994), to be sensitive, cost effective, and dependable. Radiolabeled ubiquitin is generated from GST-ubiquitin cloned into the pGEX-2TK vector (Pharmacia). Fusion proteins expressed from this plasmid contains a protein kinase A recognition site used to introduce $^{32}$P. The GST moiety is then cleaved, leaving the $^{32}$P just N-terminal to ubiquitin.

Bacterial lysate expressing pGEX-2TK-Ub is made according to the procedures described in the previous section. After incubating 750 $\mu$l supernatant (containing 100–150 $\mu$g GST-Ub) with 100 $\mu$l glutathione-Sepharose beads at room temperature for 15 min, the beads are washed three times with phosphate-buffered saline (PBS) followed by three washes with

× kinase buffer (20 m$M$ Tris, pH 7.5, 100 m$M$ NaCl, 12 m$M$ $MgCl_2$, and 1 m$M$ DTT). To label the GST-Ub with $^{32}P$, the GST-Ub beads are combined with the following components: 5 $\mu$l 10× kinase buffer (200 m$M$ Tris, pH 7.5, 1 $M$ NaCl, 120 m$M$ $MgCl_2$, and 10 m$M$ DTT), 2 $\mu$l protein kinase A (see note), 2 $\mu$l [$\gamma$-$^{32}P$]ATP (Amersham, >5000 Ci/mmol), and 42 $\mu$l $H_2O$. The kinase reaction mixture is incubated on ice for 30 min with occasional mixing. The beads are then washed four to five times with PBS to remove unincorporated $^{32}P$. At this point, the radiolabeled GST-Ub may be eluted by the addition of 150 $\mu$l PBS containing 20 m$M$ reduced glutathione and incubating for 10 min at 4°. The labeled GST-Ub can be used without further purification in applications where the GST moiety will not interfere with activity. We generally cleave the GST moiety. To generate purified $^{32}P$-Ub without the GST tag, add 400 $\mu$l of thrombin (Amersham Biosciences) (50 units/ml final concentration) in cold PBS. Tumble at room temperature for 1–2 h. Pellet beads and remove the supernatant containing the cleaved $^{32}P$-Ub and thrombin to a new tube. Add 100 $\mu$l of washed benzamidine-Sepharose beads (Amersham) to the supernatant (this will remove thrombin). Tumble at room temperature for 30–60 min. Pellet beads and transfer the supernatant containing $^{32}P$-Ub to a new tube. Spot 1–2 $\mu$l of $^{32}P$-Ub onto nitrocellulose disc, dry, and quantify incorporated radioactivity by scintillation counting. Expect 10,000–40,000 cpm/$\mu$l.

*Notes:* The protein kinase A catalytic subunit (Sigma, 250 units / vial) is provided lyophilized and should be resuspended in 25 $\mu$l 40 m$M$ DTT in $dH_2O$ on ice. Freshly resuspended protein kinase A should be used for each labeling. Because the activity of commercial thrombin is variable, if cleavage is incomplete the length of incubation can be extended up to 16 h at room temperature.

### Assaying the E3 Activity of RING Finger Proteins *In Vitro*

The most sensitive means of assessing the potential activity of a RING finger protein *in vitro* is to evaluate its capacity for autoubiquitylation. This can be achieved in a variety of ways that either measure the appearance of higher molecular weight forms of the potential E3 or E3 fusion protein or allow visualization of ubiquitin migrating at higher molecular weights after an ubiquitylation reaction. Two methods are described in the following sections. The first involves bacterially expressed GST fusion proteins of putative E3s and detection of ubiquitylated species. The second approach is based on *in vitro* translation of $^{35}S$-labeled putative RING finger E3s using commercial wheat germ or rabbit reticulocyte systems. Use of GST bacterially expressed fusion proteins allows for evaluation of activity without concern for contaminating components of the ubiquitin system and the capacity to scale up to preparative levels of protein. Advantages of material produced by *in vitro* translation in the presence of eukaryotic chaperones include enhanced protein folding, which is often reflected in a higher percentage of proteins undergoing autoubiquitylation.

### *Autoubiquitylation of GST Fusion Proteins (Lorick* et al., *1999)*

Glutathione-Sepharose beads containing 20 pmol of GST-RING finger protein are washed twice with 50 m*M* Tris (pH 7.4) and resuspended in 27 $\mu$l of ubiquitylation reaction buffer #1 [50 m*M* Tris, pH 7.4, 0.2 m*M* ATP, 0.5 m*M* $MgCl_2$, 0.1 m*M* DTT, 10 m*M* creatine phosphate, 5 units/ml phosphocreatinine kinase (Calbiochem, prepared at 50 unit/ml in 100 m*M* Tris, pH 8.0, and aliquoted)]. The ubiquitylation reaction is carried out by adding approximately 100 ng of purified rabbit, mouse, or wheat E1, 100 ng of UbcH5 family E2 in bacterial lysate (see Lorick *et al.*, 2005a), and $2 \times 10^4$ cpm of $^{32}$P-Ub to the mixture and incubating at 30° for 90 min with agitation. The reaction is terminated by the addition of 10 $\mu$l of 4× sample buffer (250 m*M* Tris, pH 6.8, 20% glycerol, 8% SDS, 4% 2-mercaptoethanol, and 0.2% bromphenol blue) and boiling for 3 min. Samples are then resolved by SDS–PAGE. The optimal percentage gel is determined by the size of the GST fusion. If using radioactive ubiquitin, the ubiquitylated protein is visualized by autoradiography. If unlabeled ubiquitin is employed, the electrophoresed proteins are transferred to a PVDF membrane (Immobilon P). After rinsing with PBS, the membrane is soaked in a solution of 0.5% glutaraldehyde in 0.1 *M* sodium phosphate, pH 7.0, for 20 min at room temperature. Following thorough washing with PBS to remove residual glutaraldehyde, the membrane is blocked with 0.5% bovine serum albumin in TBST (TBS + 0.1% Tween 20) for a minimum of 2 h and then probed with the anti-ubiquitin antibody overnight. To avoid nonspecific background, the blot should be washed extensively (e.g., five times × 5 min with TBST). Anti-ubiquitin is visualized with the HRP-labeled secondary antibody and chemiluminescence (ECL, Amersham) or $^{125}$I-labeled protein A followed by autoradiography. In general, most ubiquitylated protein in the reaction mixture is the fusion protein itself. However, if it is necessary to determine whether this is the case, the supernatant and the bead fractions from the ubiquitylation reaction can be separated before SDS–PAGE. Depending on the efficiency of the ubiquitylation reaction, immunoblotting with antibody directed against the specific RING protein or the GST moiety can also be used to detect ubiquitylated RING finger proteins (Lorick *et al.*, 1999).

*Notes:* We have found that, like E1 and certain E2s, some RING finger proteins, e.g., GST-Hdm2, are sensitive to cycles of freezing and thawing or that their activities are altered by changes in redox potential. Therefore, fusion proteins should be stored in small aliquots. Occasionally, proteins in the lysate do not survive in an active form at –80° for more than 1 week. To circumvent this, bacteria induced to express the GST fusion protein can be aliquoted as pellets in amounts appropriate for individual experiments. Final processing is carried out immediately before use. For autoubiquitylation reactions, E1 from mouse, wheat, and rabbit all work well when used in conjunction with members of the UbcH5 family. Thus, in many cases, a bacterial crude lysate containing UbcH5A, B, or C is sufficient to be used in the reaction. Evaluating other E2s may require changing the E1 used as, for example, wheat E1 does not support thiolester formation with all mammalian E2s. Additionally, if comparisons of E2s are to be made it is important to know their relative activity as assessed by E1-dependent thiolester bond formation (see Lorick *et al.*, 2005a). Negative controls for autoubiquitylation should include samples lacking either E1 or E2. GST fusions containing mutations of Zn coordinating residues should be included if possible.

Original protocols for anti-ubiquitin immunoblotting employed either autoclaving of nitrocellulose filters before immunoblotting or the treatment of PVDF filters with glutaraldehyde as described earlier. We have found that such treatment is often not necessary. Nonetheless, we recommend using such a protocol until a determination is made of whether it is required in your system.

### *Autoubiquitylation Using* In Vitro-*Translated Proteins*

An alternative means of evaluating the capacity of a RING finger protein for autoubiquitylation is to produce it in wheat germ or rabbit reticulocyte lysate. Such *in vitro* translation systems permit labeling the E3 with $^{35}$S-Met or $^{35}$S-Cys and visualizing the proteins directly by autoradiography. Protocols for protein generation by *in vitro* translation are discussed for E2 production in Lorick *et al.* (2005a). A general protocol for ubiquitylation is as follows.

To a 1.5-ml tube add

- up to 5 $\mu$l translation product ($\sim 2 \times 10^4$ cpm)
- 1.5 $\mu$l of 20× ubiquitylation buffer #2 (1.0 *M* Tris, pH 7.5, 40 m*M* DTT, 40 m*M* ATP, and 100 m*M* $MgCl_2$)
- 1 $\mu$l E1 ($\sim$100 ng)
- 1 $\mu$l bacterially expressed UbcH5B or other E2 ($\sim$100 ng)
- 1 $\mu$l (10 $\mu$g) of ubiquitin
- 1 $\mu$l MG132 or other proteasome inhibitor
- q.s. 30 $\mu$l $H_20$

After incubating with agitation at 30° for 90 min, the reaction is terminated by adding 4× sample buffer and heating as appropriate before SDS–PAGE and autoradiography.

*Note:* Reticulocyte lysates are particularly enriched in ubiquitylation machinery and proteasomes. This potentially allows for some autoubiquitylation and proteasomal degradation even without the addition of exogenous E1, E2, and ubiquitin. High concentrations of a proteasome inhibitor may be required to prevent protein degradation (up to 500 $\mu M$ MG132). As there is a high concentration of protein in lysates used for *in vitro* translation, proteins may aggregate when boiled in SDS–PAGE sample buffer. We recommend heating to only 70° or carrying out a 10- to 30-min incubation at 37° in sample buffer prior to resolution on SDS–PAGE. Aggregates may be decreased further by high-speed centrifugation of the translation product before addition of the sample buffer. If no evidence of ubiquitylation is observed, deubiquitylating enzymes (DUBs) intrinsic to the lysate may be responsible. In this case, consider the addition of ubiquitin aldehyde to inhibit DUB activity. Ubiquitin aldehyde (Calbiochem) is prepared as a 30 $\mu M$ stock in TBS and used at a final concentration of 0.5–1 $\mu M$. As usual, appropriate negative controls are required. RING finger proteins may be targets for heterologous E3s contained in the translation system. Therefore, a RING finger mutant version of the protein of interest is an important negative control for any assay of autoubiquitylation. Plant E1 works with some but not all mammalian E2s and might be a consideration in choosing one or the other translation system.

## Ubiquitylation of p53 by Hdm2

The following method can also be used for other substrate–E3 pairs (Fang *et al.*, 2000). The N terminus of Hdm2 interacts with p53 and promotes p53 ubiquitylation in a manner dependent on the C-terminal RING finger of Hdm2. To measure Hdm2-mediated p53 ubiquitylation *in vitro*, [$^{35}$S]methoinine-labeled p53 is generated by *in vitro* translation in reticulocyte using the TnT kit (Promega). The translation product (~2 × $10^4$ cpm) is diluted into binding buffer (25 m$M$ Tris, pH 7.5, 50 m$M$ NaCl, 5 m$M$ DTT, 0.5% NP-40) incubated with glutathione-Sepharose-bound Hdm2 for 2 h at 4°. After washing the beads three times with binding buffer, the ubiquitylation reaction is initiated by resuspending the beads that contain Hdm2/p53 complexes in 30 $\mu$l of ubiquitylation buffer #2 containing E1, E2, and ubiquitin as described earlier. The ubiquitylation reaction is carried out at 30° with agitation. After 90 min, samples are mixed with 10 $\mu$l of 4× sample buffer, heated, and the supernatant resolved on 8% SDS–PAGE, and the migration of p53 and ubiquitylated p53 visualized by autoradiography.

*Note:* Some of the substrate may be ubiquitin modified by proteins endogenous to the reticulocyte lysate. Therefore, use of proteasome inhibitor to prevent degradation of the substrate is often necessary when performing the translation reaction. In addition, appropriate negative controls are needed for the ubiquitylation reaction. These may include a complete reaction incubated on ice, a complete reaction terminated at time 0 by the addition of SDS sample buffer, or incomplete reactions with E1, E2, or E3 omitted.

Hdm2-mediated p53 ubiquitylation can also be examined with p53 expressed in cultured cells. For example, lysate from COS cells treated with the proteasome inhibitor MG132 can be used as a source of p53 (Fang *et al.*, 2000). More recently, we have made use of a SAOS–2 cell line stably transfected with p53 cDNA under the control of the "Tet-on" promoter (Nakano *et al.*, 2000). Cell lysates are made by disrupting the cells with lysis buffer (50 m*M* Tris, pH 7.4, 150 m*M* NaCl, 0.5 % NP-40) at 4° and pelleting insoluble material at 15,000*g* for 15 min in a refrigerated centrifuge. GST-Hdm2 (50–100 ng) that has been prebound to glutathione-Sepharose beads is then incubated with the cell lysate for 2 h at 4°. After washing four times with 50 m*M* Tris to remove unbound material, ubiquitylation reactions are carried out as described earlier. The reaction mixture is separated on 8% SDS–PAGE and immunoblotting is carried out with anti-p53 antibody DO–1 (BD Biosciences), which can recognize both unmodified and ubiquitylated p53.

*Note:* As with the TnT-generated p53, a portion of substrate from whole cell lysates may be ubiquitylated without added E3 or ubiquitylation reaction components. Therefore, appropriate negative controls should be included.

## Hdm2-Mediated Ubiquitylation and Degradation of p53 in Cells

Cultured adherent cells (such as U2OS or SAOS-2) are transfected with vectors expressing p53 (pCB6-p53) and Hdm2 (pCHDM) (Kubbutat *et al.*, 1997) using Lipofectamine 2000 (Invitrogen) according to the manufacturer's instruction. To transfect cells that are 90% confluent in each well of the six-well tissue culture cluster, 1 $\mu$g of pCB6-p53 and 3 $\mu$g of pCHDM are mixed gently in 250 $\mu$l of Opti-MEM I reduced serum medium (Invitrogen) or other medium without serum. After 5 min, the DNA is mixed with 10 $\mu$l Lipofectamine 2000 that has been diluted in 250 $\mu$l Opti-MEM I and incubated at room temperature for 20 min. The mixture (500 $\mu$l) is then added to each well and mixed gently by rocking the plate back and forth.

A proteasome inhibitor such as MG132 (20–100 $\mu M$) may be added to the culture 20 h after transfection to facilitate the accumulation of ubiquitylated p53. The cells are harvested 32 h after transfection and lysed in RIPA buffer (50 m*M* Tris, pH 7.5, 150 m*M* NaCl, 1% NP-40, 0.5% sodium deoxycholate, 0.1% SDS, 10 m*M* iodoacetamide, 1 $\mu$g/ml aprotinin, 100 $\mu$g/ml phenylmethylsulfonyl fluoride, and 5 $\mu$g/ml leupeptin). Following centrifugation at 4° for 20 min to remove cell debris, the supernatant is subjected to SDS–PAGE. The levels of p53 and ubiquitylated p53 are examined by Western blot using anti-p53 antibody DO–1 (BD Biosciences).

*Note:* For many substrates, the level of higher molecular weight ubiquitylated species present at steady state is low and undetectable by direct immunoblotting of the substrate. Therefore, both to prove directly that the substrate is ubiquitylated and to detect low levels of higher molecular weight species, it may be necessary to immunoprecipitate the target protein with a specific antibody. The immunoprecipitated material may then be immunoblotted with an anti-ubiquitin antibody. Alternatively, a plasmid expressing N-terminally tagged ubiquitin (such as HA-ubiquitin) can be cotransfected into cells. This can facilitate the detection of target protein ubiquitylation by using high-quality anti-HA antibodies. There is also evidence that proteins bearing polyubiquitin chains that include N-terminal tags may be less susceptible to degradation and possibly less susceptible to the action of deubiquitylating enzymes. Thus, the use of tagged ubiquitins may increase the detectability of modified species even when immunoblotting is carried out with an anti-ubiquitin antibody (Ellison and Hochstrasser, 1991). Iodoacetamide is used to inhibit cysteine proteases, including deubiquitylating enzymes. However, for maximal inhibition of deubiqutylation, other measures may be required, including the use of other alkylating agents such as *N*-ethylmaleimide, the addition of ubiquitin aldehyde, or the use of lysis buffers containing higher levels of SDS (SDS–PAGE sample buffer, for example).

## References

Deshaies, R. J. (1999). SCF and cullin/Ring H2-based ubiquitin ligases. *Annu. Rev. Cell Dev. Biol.* **15,** 435–467.

Ellison, M. J., and Hochstrasser, M. (1991). Epitope-tagged ubiquitin: A new probe for analyzing ubiquitin function. *J. Biol. Chem.* **266,** 21150–21157.

Fang, S., Jensen, J. P., Ludwig, R. L., Vousden, K. H., and Weissman, A. M. (2000). Mdm2 is a RING finger-dependent ubiquitin protein ligase for itself and p53. *J. Biol. Chem.* **275,** 8945–8951.

Freemont, P. S., Hanson, I. M., and Trowsdale, J. (1991). A novel cysteine-rich sequence motif. *Cell* **64,** 483–484.
Kubbutat, M. H., Jones, S. N., and Vousden, K. H. (1997). Regulation of p53 stability by Mdm2. *Nature* **387,** 299–303.
Lorick, K. L., Jensen, J. P., Fang, S., Ong, A. M., Hatakeyama, S., and Weissman, A. M. (1999). RING fingers mediate ubiquitin-conjugating enzyme (E2)-dependent ubiquitination. *Proc. Natl. Acad. Sci. USA* **96,** 11364–11369.
Lorick, K. L., Jensen, J. P., and Weissman, A. M. (2005a). Expression, purification, and properties of the Ubc4/5 family of E2 enzymes. *Methods Enyzmol.* **398,** 54–68.
Lorick, K. L., Tsai, Y.-C., Yang, Y., and Weissman, A. M. (2005b). RING fingers and relatives: Determination of protein fate. *In* "Protein Degradation: Vol. 1: Ubiquitin and the Chemistry of Life" (R. J. Mayer, A. J. Ciechanover, and M. Rechsteiner, eds.), pp. 44–101. VCH Verlag, Weinheim, Germany.
Nakano, K., Balint, E., Ashcroft, M., and Vousden, K. H. (2000). A ribonucleotide reductase gene is a transcriptional target of p53 and p73. *Oncogene* **19,** 4283–4289.
Scheffner, M., Huibregtse, J. M., and Howley, P. M. (1994). Identification of a human ubiquitin-conjugating enzyme that mediates the E6-AP-dependent ubiquitination of p53. *Proc. Natl. Acad. Sci. USA* **91,** 8797–8801.
Semple, C. A. (2003). The comparative proteomics of ubiquitination in mouse. *Genome Res.* **13,** 1389–1394.
von Arnim, A. G. (2001). A hitchhiker's guide to the proteasome. *Sci. STKE* 2001, PE2.
Wong, B. R., Parlati, F., Qu, K., Demo, S., Pray, T., Huang, J., Payan, D. G., and Bennett, M. K. (2003). Drug discovery in the ubiquitin regulatory pathway. *Drug Discov. Today* **8,** 746–754.

# [11] Expression and Assay of HECT Domain Ligases

*By* SYLVIE BEAUDENON, ANAHITA DASTUR, and JON M. HUIBREGTSE

## Abstract

HECT domain ubiquitin ligases (HECT E3s), typified by human E6AP and yeast Rsp5p, are unique among the several classes of known ubiquitin ligases in that they participate directly in the chemistry of substrate ubiquitination reactions. This chapter discusses strategies for the expression of active HECT E3s and the assays that are available for analyzing E2 interaction, ubiquitin-thioester formation, and substrate ubiquitination.

## Introduction

HECT ubiquitin ligases are monomeric enzymes that are mechanistically distinct from other known families of ubiquitin ligases (Pickart, 2001; Scheffner *et al.*, 1995) and are defined by a carboxy-terminal catalytic

METHODS IN ENZYMOLOGY, VOL. 398

0076-6879/05 $35.00
DOI: 10.1016/S0076-6879(05)98011-7

domain of approximately 350 amino acid (the HECT domain, for homologous to E6AP carboxy terminus) (Huibregtse *et al.*, 1995). The HECT domain contains an active-site cysteine residue that, like the E1 and E2 enzymes, forms a ubiquitin-thioester intermediate during each round of catalysis of ubiquitin-isopeptide bond formation. Thus, unlike other families of E3s, which appear to facilitate ubiquitination by bringing activated E2s and substrates into close proximity, HECT E3s directly participate in the chemistry of the ubiquitination reaction.

There are relatively few HECT E3s compared to other types of E3s, with only 5 encoded in the *Saccharomyces cerevisae* genome and approximately 50 in the human genome. Only a handful of HECT E3s have been characterized with respect to their biologic functions and their substrate specificity. The first to be characterized was human E6AP (E6-associated protein), whose substrate specificity is altered by the E6 protein of the cancer-associated human papillomaviruses (e.g., HPV types 16 and 18) (Huibregtse *et al.*, 1991; Scheffner *et al.*, 1993). The HPV E6 oncoprotein binds to E6AP in the central portion of the protein, amino-terminal to the HECT domain, and allows E6AP to then stably interact with the p53 tumor suppressor protein (Huibregtse *et al.*, 1993b). p53 is then multiubiquitinated by E6AP and degraded by the 26S proteasome. Importantly, p53 is only a target of E6AP in cells that express the viral E6 protein. The natural targets of E6AP remain largely unknown; however, it is known that mutations in the maternal allele of E6AP (UBE3A) are the cause of Angelman syndrome (AS), a severe neurologic disorder (Albrecht *et al.*, 1997; Fang *et al.*, 1999; Jiang *et al.*, 1999; Kishino *et al.*, 1997). AS patients lack a functional maternal allele of UBE3A and therefore lack E6AP protein in subregions of the brain, particularly the hippocampal and Purkinje neurons, as the paternal allele is permanently inactivated in these cells. This leads to the hypothesis that one or more targets of E6AP are not ubiquitinated in subregions of the brain of AS patients, which leads to the severe neurologic phenotype. Further understanding of the basis of AS will require thorough understanding of the enzymatic mechanism of HECT ubiquitin ligases and characterization of the natural substrate specificity of this enzyme. This example therefore emphasizes the importance of establishing conditions for expression and biochemical analysis of HECT ubiquitin ligases.

Following E6AP, the best characterized HECT E3s are the WW domain containing HECT E3s, typified by *S. cerevisiae* Rsp5p (Dunn and Hicke, 2001; Hein *et al.*, 1995; Wang *et al.*, 1999) and mammalian Nedd4 (Ingham *et al.*, 2004) and Smurf proteins (Arora and Warrior, 2001; Suzuki *et al.*, 2002; Zhang *et al.*, 2004). These proteins play an important role in targeting of plasma membrane-associated proteins, such as receptors and permeases, for ubiquitin-dependent endocytosis. WW-HECT E3s also have soluble cytoplasmic and nuclear substrates, and in most cases it is clear that enzyme–substrate interactions are mediated by the WW domains, binding to proline-rich ligands ("PY" motifs or variants) within substrates (Huibregtse *et al.*, 1997; Staub *et al.*, 1996).

HECT E3s range in molecular mass from about 92 kDa (e.g., *S. cerevisiae* Rsp5p) to over 500 kDa (human Herc1), posing challenges for expression and analysis of full-length proteins. We have had success in expressing full-length, biochemically active HECT E3s for those that are in the 100-kDa range (e.g., E6AP, Rsp5p). Larger proteins are more likely to be expressed well in a baculovirus system rather than a bacterial expression system, although in some cases it may not be necessary to express full-length protein in order to address certain biochemical or functional questions. For example, the isolated HECT domain has the ability to interact with the activating E2 enzyme and can form the ubiquitin-thioester intermediate. In addition, while there are a limited number of characterized HECT E3s to make generalizations, we speculate that direct substrate specificity determinants for HECT E3s may lie at a common position relative to the HECT domain. For WW domain HECT E3s, the center of the WW domain region is roughly equivalent in position to the HPV E6-binding domain in E6AP (Huibregtse *et al.*, 1993b). This region is centered approximately 120 amino acids N-terminal to the beginning of the HECT domain. Therefore, if one is interested in studying one of the larger HECT E3s that might not be amenable to expression of the full-length protein, a reasonable alternative might be to express the C-terminal 50–75 kDa, on the possibility that this protein might retain direct substrate interaction determinants as well as a functional catalytic domain. The expression of dominant-negative mutants in some systems, such as yeast, can also be used to indirectly map potential substrate interaction domains (Hoppe *et al.*, 2000). It is important to note that with the exception of the WW domain HECT E3s, many HECT E3s, including E6AP, do not contain obvious protein–protein interaction motifs, so it is difficult in most cases to predict domains that might mediate substrate interactions.

## Expression

We typically establish both bacterial and baculovirus expression systems for the expression of previously uncharacterized HECT E3s. In our experience, even when both systems result in acceptable levels of protein expression, there is often a significant difference in enzymatic activity for proteins made in the two systems. That is the case for the two proteins discussed here, human E6AP and *S. cerevisiae* Rsp5p. E6AP is more active when expressed in the baculovirus system, whereas Rsp5p is more active when expressed in the bacterial system, although both proteins are expressed reasonably well (in terms of yield) in both bacteria and insect cells. In both systems, we typically express HECT E3s as GST fusion proteins, purify the protein on gluthathione-Sepharose, and then separate the protein from GST by site-specific proteolysis. If there is a need for subsequent antibody-based detection, epitope tags can be incorporated into the protein sequence, but the epitope should, for most purposes, be placed at the amino-terminal end of the E3, as the extreme carboxyl terminus plays a critical role in substrate ubiquitination (Salvat *et al.*, 2004). We have shown that even small carboxy-terminal epitopes interfere with the ability to catalyze substrate ubiquitination, although they do not block ubiquitin-thioester formation (Salvat *et al.*, 2004).

For bacterial expression we utilize the commercially available pGEX vector series (Amersham Pharmacia), which are isopropyl-$\beta$-D-thiogalactoside (IPTG)-inducible vectors for the expression of GST fusion proteins. The full-length *RSP5* open reading frame (ORF) is cloned into pGEX-6p, and the yield of GST-Rsp5p protein is approximately 100 $\mu$g per liter. For a typical preparation, a 25-ml overnight culture will be grown to saturation in LB media and diluted 10-fold the next day with fresh media. The culture will be grown to an OD of 0.6 (approximately 2 h) and IPTG will be added to 100 $\mu M$ final concentration. Cells will be collected after another 3–4 h at 37° and resupended in 10 ml of phosphate-buffered saline (PBS) with 1% Triton X-100 containing 1 m*M* phenymethylsulfonyl fluoride (PMSF). The cells are lysed by sonication on ice, and debris is removed by centrifugation at 10,000*g* for 10 min. The supernatant is collected and incubated with 100 $\mu$l (bed volume) of glutathione-Sepharose (Amersham Pharmacia) in a 15-ml conical disposable tube. Incubation is at 4°, rotating, with times varying between 1 h and overnight. If the goal is to isolate biochemically active protein, we use shorter incubation times, and if our goal is high yield (for binding assays or antibody production, for example), we use longer incubation times. Beads are collected by centrifugation and are washed at least four times with PBS–1% Triton X-100. To check expression, an aliquot of the Sepharose beads (5–10 $\mu$l) is mixed with SDS–PAGE loading buffer, heated at 95° for 5 min, and analyzed by SDS–PAGE and Coomassie blue staining. To cleave the HECT E3 from GST, we remove all liquid from the beads and resuspend in 200 $\mu$l of buffer containing 10 m*M* Tris (pH 7.5), 50 m*M* NaCl, 0.01% Triton X-100, 1 m*M* EDTA, 1 m*M* dithiothreitol (DTT) and 4 units PreScission protease (Amersham Pharmacia). The reaction is incubated at 4°, rotating, and the supernatant is collected after a minimum of 2 h, glycerol is added to 5%, and the protein is frozen in aliquots at −80°.

For baculovirus expression of GST-E6AP, we modified the pFastbac vector (Invitrogen) by inserting the GST ORF, the sequence encoding the PreScission protease cleavage site, and the polylinker region from the pGEX 6p-1 vector (Amersham Pharmacia) adjacent to the polyhedrin promoter. The E6AP ORF is then cloned between the *Bam*HI and the *Not*I restriction sites in the polylinker. The pFastbac1-GST-E6AP plasmid is transformed into *Escherichia coli* strain DH10Bac for transposition into bacmids following the procedure recommended by the manufacturer (Invitrogen). Purified recombinant bacmid DNA is transfected into insect cells (High Five cells, Invitrogen), and virus is collected and amplified to obtain a high-titer stock. For protein expression, High Five cells are infected with the recombinant virus, and cell extracts are made 48 h postinfection in NP-40 lysis buffer (0.1 *M* Tris-HCl, pH 7.5; 0.1 *M* NaCl; 1% NP-40). GST-E6AP protein is purified as described earlier for GST-Rsp5p, incubating the insect cell extract with glutathione-Sepharose beads for 2 h at 4°. The beads are washed three times with NP-40 lysis buffer, and the E6AP protein is then released by cleavage with PreScission protease (Promega) for 5 h at 4° in a buffer containing 10 m*M* Tris-HCl, 50 m*M* NaCl, 0.01% Triton, and 1 m*M* DTT. Figure 1 shows a Coomassie blue-stained SDS–PAGE showing purified Rsp5p and E6AP proteins, before and after protease digestion to remove the GST moiety.

## Assays

A minimum of three separable activities can be defined for HECT E3 proteins: E2 binding, ubiquitin-thioester formation, and isopeptide bond catalysis (or substrate ubiquitination). Assays for each of these activities

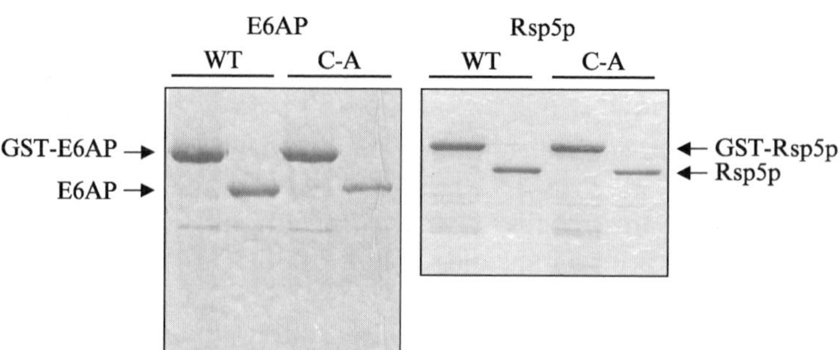


FIG. 1. Purification of E6AP and Rsp5p. Wild-type and Cys-to-Ala (C-A) mutant E6AP proteins were purified from insect cells infected with recombinant baculovirus expressing GST-E6AP. GST fusion proteins are shown before and after protease digestion to remove the GST moiety. GST-Rsp5p and GST-Rsp5p C-A were purified similarly from *E. coli*. Proteins were analyzed on a 10% acrylamide SDS–PAGE gel and detected by Coomassie blue staining.

are described here, followed by a brief discussion of general approaches toward identifying substrates of uncharacterized HECT E3s.

*E2 Binding*

In all cases where E2 specificity has been characterized, the subgroups of E2s most closely related to the yeast Ubc4/5 proteins are those that function with HECT E3s (Huang *et al.*, 1999; Schwarz *et al.*, 1998). In most cases, multiple E2s from this family can function with a single HECT E3. For example, UbcH6, H7, H8, and at least some isoforms of UbcH5 can all function with E6AP *in vitro* (Kumar *et al.*, 1997; Nuber and Scheffner, 1999). A set of three related enzymes in *S. cerevisiae,* Ubc1, Ubc4, and Ubc5, are presumed to function with Rsp5p (de la Fuente *et al.*, 1997; Gitan and Eide, 2000) (A. Dastur and J. M. Huibregtse, unpublished result). A key feature of all of these E2s is a conserved phenylalanine in Loop 1 of the E2 structure (Huang *et al.*, 1999). In the X-ray crystal structure of the E6AP HECT domain in complex with UbcH7, the Phe63 side chain makes extensive contacts in a deep groove at the E2 interaction surface, and mutation of either this residue (Nuber and Scheffner, 1999) or the contact sites on the HECT domain (A. Salvat and J. M. Huibregtse, unpublished result) abrogates all E2-dependent HECT E3 activities. The affinity of the HECT domain for E2 enzymes is quite low, with a $K_d$ of approximately 2 $\mu M$ for the UbcH7–E6AP interaction, as determined by biophysical methods by Brian Kuhlman (University of North Carolina, personal communication). Consistent with this, coimmunoprecipitations or GST pull-down experiments have generally failed to detect a stable interaction between HECT E3s and E2 enzymes. Yeast two-hybrid assays have been successful in some cases for detecting HECT–E2 interactions (Kumar *et al.*, 1997). In addition, we have found that it is possible to detect HECT–E2 interaction by native gel shift assays, when both proteins are at high concentrations. This is dependent on the behavior of both the particular HECT E3 and E2 in native gels, as some proteins simply do not enter native gels (e.g., Rsp5p). We have been able to detect a gel shift with full-length E6AP in the presence of purified AtUbc8 (*Arabidopsis thaliana*), as shown in Fig. 2. This experiment is performed by incubating approximately 2 $\mu$g each of E6AP and Ubc8 in a total volume of 15 $\mu$l in buffer containing 25 m$M$ Tris (pH 7.5), 50 m$M$ NaCl, and 1 $\mu M$ DTT. The mixture is incubated for 10 min at room temperature before loading onto a native 8% polyacrylamide gel. Proteins are detected by Coomassie blue staining.

In summary, given the low affinity for the E2 and the difficulty in detecting the interaction, it is generally best to rely on downstream consequences of E2 binding (thioester formation or substrate ubiquitination) in order to characterize E2 specificity of a given HECT E3.

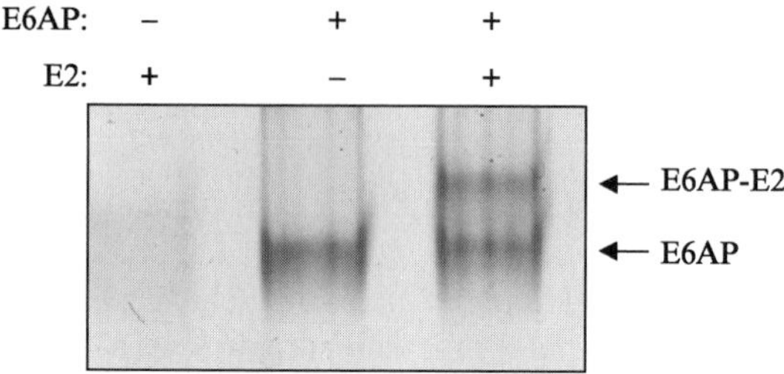


FIG. 2. Detection of the E6AP–E2 interaction by native gel shift. Approximately 2 $\mu$g each of purified E6AP and AtUbc8 were incubated under conditions described in the text and run on a 10% acrylamide gel prepared without SDS in the gel or buffers. Proteins were detected by Coomassie blue staining.

*Ubiquitin-Thioester Formation*

In the absence of known substrates of a given HECT E3, demonstration of a ubiquitin-thioester intermediate is the best indicator of potential ligase activity. Figure 3A shows ubiquitin-thioester formation for E6AP, using purified E1 and E2 enzymes and E6AP, prepared as described earlier. The ubiquitin-thioester adduct is seen with the wild-type protein, but not if the active-site cysteine is mutated to an alanine. The adduct with the wild-type protein disappears when run on an SDS–PAGE gel using standard loading buffer with DTT or $\beta$-mercaptoethanol. Each reaction contains 100 ng of purified FLAG-tagged E6AP, 20 ng of E1 protein, and 100 ng of E2 protein in a total volume of 40 $\mu$l containing 25 m$M$ Tris (pH 7.5), 50 m$M$ NaCl, 10 m$M$ $MgCl_2$, 5 m$M$ ATP, and 0.1 m$M$ DTT. Reactions are incubated at room temperature for 10 min and stopped with SDS–PAGE loading buffer with or without DTT.

With Rsp5p, an additional characteristic of some HECT E3s is observed. Rsp5p will spontaneously catalyze isopeptide-linked multiubiquitination to internal lysine residues, following formation of the thioester adduct (Huibregtse *et al.*, 1995). This self-ubiquitination is dependent on thioester formation, as the active-site Cys-to-Ala mutation does not do this. Interestingly, as noted earlier, alteration of the extreme carboxy-terminal sequences, and the –4Phe, in particular, blocks self-ubiquitination because the protein is no longer able to catalyze substrate ubiquitination. If the −4Phe is altered to an Ala, a DTT-sensitive thioester adduct is then evident, as the reaction is blocked at this step (Fig. 3B).

An alternative approach for detecting the ubiquitin-thioester adduct is to translate the HECT E3 *in vitro* in either a reticulocyte lysate system or a wheat germ extract system in the presence of [$^{35}$S]methionine (Huibregtse *et al.*, 1995). The lysate can be supplemented with recombinant E1 and E2

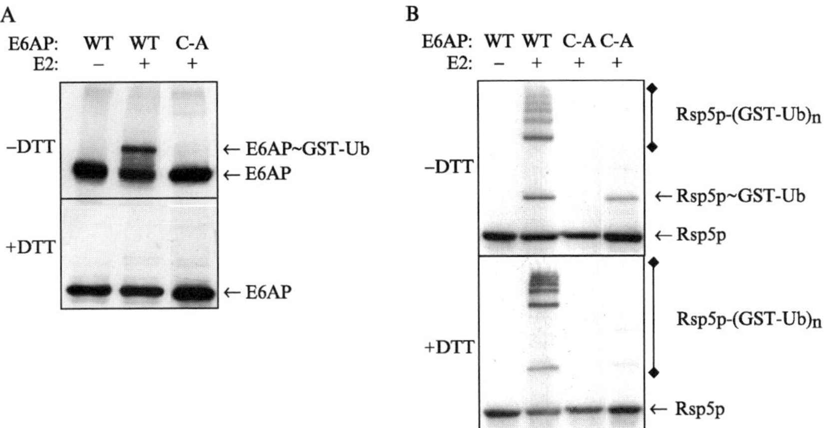


FIG. 3. Ubiquitin-thioester assays. (A) Purified E6AP, wild type (WT), and Cys-to-Ala (C-A) mutant proteins were incubated with purified E1 +/– Ubc8 protein and GST-ubiquitin. The reaction was split and run with or without DTT in the SDS–PAGE loading buffer. E6AP was detected by immunoblotting with the anti-E6AP antibody. (B) Purified Rsp5p proteins, wild type (WT), the C-A active-site mutant, and the F806A mutant (–4F to A) were analyzed as described in A, and Rsp5p was detected by immunoblotting with the anti-Rsp5p antibody.

enzymes, although often the lysate, itself, may have sufficient E1 and E2 activities to support activation of the E3. The reactions are then loaded on SDS–PAGE gels in parallel, with and without reducing agent in the loading buffer, and the thioester intermediate should be evident as a DTT-sensitive adduct of an additional 8 kDa. GST-ubiquitin can also be used as the source of ubiquitin to provide further evidence of the thioester adduct (based on a predictable shift in size of the GST-ubiquitin adduct) that any apparent adducts evident in the –DTT lanes were due to ubiquitin-thioester formation (Huibregtse *et al.*, 1995). The active-site cysteine-to-alanine mutation in the HECT domain is again a necessary control to ensure that additional higher molecular weight bands seen in the absence of reducing agent were due to thioester formation, particularly as some proteins do not run as single sharp bands in the absence of a reducing agent.

### *Substrate Ubiquitination*

Substrate ubiquitination can be reconstituted with all purified components in the case of both E6-dependent ubiquitination of p53 by E6AP and ubiquitination of Rsp5p substrates, such as WBP2. Such assays are shown

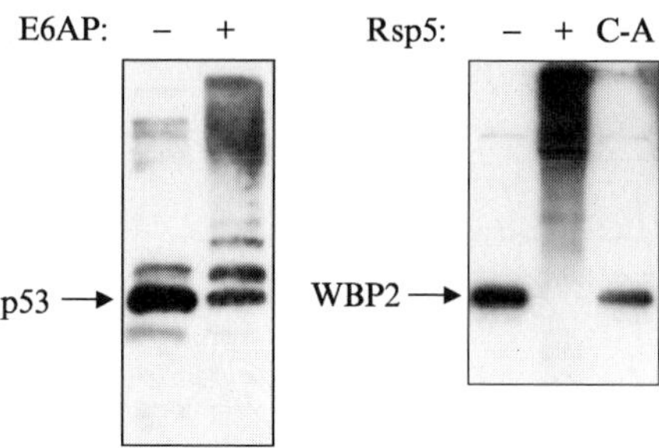


FIG. 4. p53 ubiquitination was reconstituted *in vitro* with all purified components (E1, *At*Ubc8, E6AP, p53, and HPV33 E6 protein and ubiquitin). p53 was detected by immunoblotting with the anti-p53 antibody. Similarly, WBP2 ubiquitination was reconsituted *in vitro* with all purified components (E1, *At*Ubc8, FLAG-WBP2, Rsp5p, and ubiquitin). WBP2 was detected by immunoblotting with the anti-FLAG antibody.

in Fig. 4. These reactions contained 200 ng of Ub, 100 ng E1, 200 ng UbcH8, 100 ng Rsp5p or E6AP, and 1 $\mu$g of purified substrate protein (p53 or WBP2) in 40 $\mu$l buffer containing 25 m$M$ Tris (pH 7.5), 50 m$M$ NaCl, 10 m$M$ $MgCl_2$, 5 m$M$ ATP, and 0.1 m$M$ DTT. Reactions are incubated for 30 min at room temperature and stopped with standard (DTT containing) SDS–PAGE loading buffer. Substrate proteins are detected by immunoblotting with either p53-specific antibody or anti-FLAG antibody.

Substrate ubiquitination can, in many cases, also be performed by translating the substrate *in vitro* in either a wheat germ or reticulocyte lysate system and adding recombinant purified E3. Some E3s may be present in the lysate to begin with, making such an approach unfeasible. For example, E6AP is present in rabbit reticulocyte lysate and therefore E6-dependent ubiquitination of *in vitro*-translated p53 does not require the addition of E6AP (Scheffner *et al.*, 1990). E6AP is not present in wheat germ extract, so E6- and E6AP-dependent p53 ubiquitination can be effectively reconstituted in this system (Huibregtse *et al.*, 1993a). Rsp5p activity, or any apparent Rsp5p homologs, is not present in either reticulocyte or wheat germ extract, and Rsp5p-dependent ubiquitination of WBP2 can be effectively reconstituted in either system.

### *Identification of Substrates*

Only a few HECT E3s have been characterized with respect to their substrate specificity, most notably the WW-HECT E3s. Here we briefly suggest some general methods that have proven useful for identifying substrates. An obvious first choice is a yeast two-hybrid screen, which has proven successful in some cases (Kumar *et al.*, 1999; Lin *et al.*, 2000). It is generally a good strategy to utilize the active-site cysteine-to-alanine mutant of the HECT E3 as the "bait," as it is possible that any positive signal in the assay would otherwise be obliterated by ubiquitin-mediated degradation of the target protein. HECT E3s have also been isolated as interacting proteins in two hybrid screens with several different baits that turned out to be substrates (Mitsui *et al.*, 1999; Traweger *et al.*, 2002), further supporting the two-hybrid system as a reasonable approach for substrate identification.

We previously described a biochemical assay for the isolation of HECT E3 substrates *in vitro* (Huibregtse *et al.*, 1997; Nakagawa and Huibregtse, 2000). In this method, we purified GST-Rsp5p or E6AP on gluthione-Sepharose and used this as an affinity matrix for the isolation of proteins from $^{35}$S-labeled total yeast or human cell extract. Binding and washing conditions were varied with respect to detergent and salt concentrations, and the spectrum of isolated proteins was analyzed by boiling the matrix in loading buffer and analyzing total protein by SDS–PAGE and autoradiography. This spectrum of proteins was compared to the identical pull-down that, following the wash step, had been treated with ATP, ubiquitin, E1 enzyme, and an appropriate E2 enzyme. This activates the E3, and any of the bound proteins that represent potential substrates should then be ubiquitinated. This would be evident as a loss of specific protein bands compared to the straight pull-down. We run these experiments in parallel with the GST-HECT E3 Cys-to-Ala active-site mutation to ensure that any apparent ubiquitination was E3 dependent (and not due to the addition of E1 and E2 enzymes alone). This approach was used successfully to isolated Rpb1p as a substrate of Rsp5p and Scribble as a substrate of the HPV E6/E6AP complex (Huibregtse *et al.*, 1997; Nakagawa and Huibregtse, 2000).

Tandem affinity purification (TAP) affinity approaches (Puig *et al.*, 2001) have also proven useful in our laboratory for the isolation of HECT E3 substrates (Y. Kee and J. M. Huibregtse, in preparation), particularly in yeast where precise chromosomal gene alterations ensure that tagged E3s will not compete with endogenous E3s and the E3 can be expressed from its natural promoter. Finally, "*in vitro* expression cloning" has been used successfully for the identification of APC substrates (King *et al.*, 1997; Lustig *et al.*, 1997; McGarry and Kirschner, 1998). This method screens pools of *in vitro*-translated proteins for the presence of substrates of a given E3. This approach would appear to be well suited to HECT E3s, although to our knowledge there are no published reports that have used this for HECT E3 substrate identification.

## Conclusion

HECT E3s play important roles in many important biological processes and disease states, including uterine cervical cancer (Scheffner, 1998), Angelman syndrome (Jiang *et al.*, 1999), Liddle's syndrome (Abriel *et al.*, 1999; Goulet *et al.*, 1998), the TGF-$\beta$ response (Arora and Warrior, 2001; Lin *et al.*, 2000; Suzuki *et al.*, 2002), budding of enveloped viruses (Harty *et al.*, 2000; Kikonyogo *et al.*, 2001; Sakurai *et al.*, 2004; Yasuda *et al.*, 2003), and trafficking of integral membrane proteins (Dunn and Hicke, 2001; Ingham *et al.*, 2004; Springael and Andre, 1998). They are amenable to biochemical analysis due to their relative simplicity compared to multimeric ubiquitin ligases, and they have provided unique opportunities to study the enzymology of ubiquitin transfer in highly purified reconstituted systems. Several mechanistic problems remain unanswered, including the question of how the gap between the active-site cysteines of the E2 and E3 observed in crystal structures (Huang *et al.*, 1999; Verdecia *et al.*, 2003) is breached, how the growing end of the multiubiquitin chain apparently remains oriented for attachment of the next ubiquitin moiety, and how the substrate-binding domain interfaces with the catalytic HECT domain. Further biochemical characterization of HECT E3s will certainly be aided by characterization and study of additional HECT E3-substrate pairs. In addition, the finding that HECT E3s have the capacity to catalyze conjugation of ISG15 (Zhao *et al.*, 2004), a 15-kDa interferon-inducible Ubl, will surely provide additional opportunities for gaining insight into the biochemical and biological functions of HECT E3s.

## Acknowledgments

We thank Brian Kuhlman for communication of unpublished results. This work was supported by a grant from the National Institutes of Health to J. M. H. (CA072943).

## References

Abriel, H., Loffing, J., Rebhun, J. F., Pratt, J. H., Schild, L., Horisberger, J. D., Rotin, D., and Staub, O. (1999). Defective regulation of the epithelial $Na^+$ channel by Nedd4 in Liddle's syndrome. *J. Clin. Invest.* **103,** 667–673.

Albrecht, U., Sutcliffe, J. S., Cattanach, B. M., Beechey, C. V., Armstrong, D., Eichele, G., and Beaudet, A. L. (1997). Imprinted expression of the murine Angelman syndrome gene, *Ube3a*, in hippocampal and Purkinje neurons. *Nature Genet.* **17,** 75–78.

Arora, K., and Warrior, R. (2001). A new Smurf in the village. *Dev. Cell* **1,** 441–442.

de la Fuente, N., Maldonado, A. M., and Portillo, F. (1997). Glucose activation of the yeast plasma membrane H+-ATPase requires the ubiquitin-proteasome proteolytic pathway. *FEBS Lett.* **411,** 308–312.

Dunn, R., and Hicke, L. (2001). Domains of the Rsp5 ubiquitin-protein ligase required for receptor-mediated and fluid-phase endocytosis. *Mol. Biol. Cell* **12,** 421–435.

Fang, P., Lev-Lehman, E., Tsai, T. F., Matsuura, T., Benton, C. S., Sutcliffe, J. S., Christian, S. L., Kubota, T., Halley, D. J., Meijers-Heijboer, H., Langlois, S., Graham, J. M., Jr., Beuten, J., Willems, P. J., Ledbetter, D. H., and Beaudet, A. L. (1999). The spectrum of mutations in UBE3A causing Angelman syndrome. *Hum. Mol. Genet.* **8,** 129–135.

Gitan, R. S., and Eide, D. J. (2000). Zinc-regulated ubiquitin conjugation signals endocytosis of the yeast ZRT1 zinc transporter. *Biochem. J.* **346**(Pt 2), 329–336.

Goulet, C. C., Volk, K. A., Adams, C. M., Prince, L. S., Stokes, J. B., and Snyder, P. M. (1998). Inhibition of the epithelial $Na^+$ channel by interaction of Nedd4 with a PY motif deleted in Liddle's syndrome. *J. Biol. Chem.* **273,** 30012–30017.

Harty, R. N., Brown, M. E., Wang, G., Huibregtse, J., and Hayes, F. P. (2000). A PPxY motif within the VP40 protein of Ebola virus interacts physically and functionally with a ubiquitin ligase: Implications for filovirus budding. *Proc. Natl. Acad. Sci. USA* **97,** 13871–13876.

Hein, C., Springael, J. Y., Volland, C., Haguenauer-Tsapis, R., and Andre, B. (1995). NPl1, an essential yeast gene involved in induced degradation of Gapl and Fur4 permeases, encodes the Rsp5 ubiquitin-protein ligase. *Mol. Microbiol.* **18,** 77–87.

Hoppe, T., Matuschewski, K., Rape, M., Schlenker, S., Ulrich, H. D., and Jentsch, S. (2000). Activation of a membrane-bound transcription factor by regulated ubiquitin/proteasome-dependent processing. *Cell* **102,** 577–586.

Huang, L., Kinnucan, E., Wang, G., Beaudenon, S., Howley, P. M., Huibregtse, J. M., and Pavletich, N. P. (1999). Structure of an E6AP-UbcH7 complex: Insights into ubiquitination by the E2-E3 enzyme cascade. *Science* **286,** 1321–1326.

Huibregtse, J. M., Scheffner, M., Beaudenon, S., and Howley, P. M. (1995). A family of proteins structurally and functionally related to the E6-AP ubiquitin-protein ligase. *Proc. Natl. Acad. Sci. USA* **92,** 2563–2567.

Huibregtse, J. M., Scheffner, M., and Howley, P. M. (1991). A cellular protein mediates association of p53 with the E6 oncoprotein of human papillomavirus types 16 or 18. *EMBO J.* **10,** 4129–4135.

Huibregtse, J. M., Scheffner, M., and Howley, P. M. (1993a). Cloning and expression of the cDNA for E6-AP, a protein that mediates the interaction of the human papillomavirus E6 oncoprotein with p53. *Mol. Cell. Biol.* **13,** 775–784.

Huibregtse, J. M., Scheffner, M., and Howley, P. M. (1993b). Localization of the E6-AP regions that direct human papillomavirus e6 binding, association with p53, and ubiquitination of associated proteins. *Mol. Cell. Biol.* **13,** 4918–4927.

Huibregtse, J. M., Yang, J. C., and Beaudenon, S. L. (1997). The large subunit of RNA polymerase II is a substrate of the Rsp5 ubiquitin-protein ligase. *Proc. Natl. Acad. Sci. USA* **94,** 3656–3661.

Ingham, R. J., Gish, G., and Pawson, T. (2004). The Nedd4 family of E3 ubiquitin ligases: Functional diversity within a common modular architecture. *Oncogene* **23,** 1972–1984.

Jiang, Y., Lev-Lehman, E., Bressler, J., Tsai, T. F., and Beaudet, A. L. (1999). Genetics of Angelman syndrome. *Am. J. Hum. Genet.* **65,** 1–6.

Kikonyogo, A., Bouamr, F., Vana, M. L., Xiang, Y., Aiyar, A., Carter, C., and Leis, J. (2001). Proteins related to the Nedd4 family of ubiquitin protein ligases interact with the L domain of Rous sarcoma virus and are required for gag budding from cells. *Proc. Natl. Acad. Sci. USA* **98,** 11199–11204.

King, R. W., Lustig, K. D., Stukenberg, P. T., McGarry, T. J., and Kirschner, M. W. (1997). Expression cloning in the test tube. *Science* **277,** 973–974.

Kishino, T., Lalande, M., and Wagstaff, J. (1997). *UBE3A*/E6-AP mutations cause Angelman syndrome. *Nature Genet.* **15,** 70–73.

Kumar, S., Kao, W. H., and Howley, P. M. (1997). Physical interaction between specific E2 and Hect E3 enzymes determines functional cooperativity. *J. Biol. Chem.* **272,** 13548–13554.

Kumar, S., Talis, A. L., and Howley, P. M. (1999). Identification of HHR23A as a substrate for E6-associated protein-mediated ubiquitination. *J. Biol. Chem.* **274,** 18785–18792.

Lin, X., Liang, M., and Feng, X. H. (2000). Smurf2 is a ubiquitin E3 ligase mediating proteasome-dependent degradation of Smad2 in transforming growth factor-beta signaling. *J. Biol. Chem.* **275,** 36818–36822.

Lustig, K. D., Stukenberg, P. T., McGarry, T. J., King, R. W., Cryns, V. L., Mead, P. E., Zon, L. I., Yuan, J., and Kirschner, M. W. (1997). Small pool expression screening: Identification of genes involved in cell cycle control, apoptosis, and early development. *Methods Enzymol.* **283,** 83–99.

McGarry, T. J., and Kirschner, M. W. (1998). Geminin, an inhibitor of DNA replication, is degraded during mitosis. *Cell* **93,** 1043–1053.

Mitsui, K., Nakanishi, M., Ohtsuka, S., Norwood, T. H., Okabayashi, K., Miyamoto, C., Tanaka, K., Yoshimura, A., and Ohtsubo, M. (1999). A novel human gene encoding HECT domain and RCC1-like repeats interacts with cyclins and is potentially regulated by the tumor suppressor proteins. *Biochem. Biophys. Res. Commun.* **266,** 115–122.

Nakagawa, S., and Huibregtse, J. M. (2000). Human scribble (Vartul) is targeted for ubiquitin-mediated degradation by the high-risk papillomavirus E6 proteins and the E6AP ubiquitin-protein ligase. *Mol. Cell. Biol.* **20,** 8244–8253.

Nuber, U., and Scheffner, M. (1999). Identification of determinants in E2 ubiquitin-conjugating enzymes required for hect E3 ubiquitin-protein ligase interaction. *J. Biol. Chem.* **274,** 7576–7582.

Pickart, C. M. (2001). Mechanisms underlying ubiquitination. *Annu. Rev. Biochem.* **70,** 503–533.

Puig, O., Caspary, F., Rigaut, G., Rutz, B., Bouveret, E., Bragado-Nilsson, E., Wilm, M., and Seraphin, B. (2001). The tandem affinity purification (TAP) method: A general procedure of protein complex purification. *Methods* **24,** 218–229.

Sakurai, A., Yasuda, J., Takano, H., Tanaka, Y., Hatakeyama, M., and Shida, H. (2004). Regulation of human T-cell leukemia virus type 1 (HTLV–1) budding by ubiquitin ligase Nedd4. *Microbes Infect.* **6,** 150–156.

Salvat, C., Wang, G., Dastur, A., Lyon, N., and Huibregtse, J. M. (2004). The -4 phenylalanine is required for substrate ubiquitination catalyzed by HECT ubiquitin ligases. *J. Biol. Chem.* **279,** 18935–18943.

Scheffner, M. (1998). Ubiquitin, E6-AP, and their role in p53 inactivation. *Pharmacol. Ther.* **78,** 129–139.

Scheffner, M., Huibregtse, J. M., Vierstra, R. D., and Howley, P. M. (1993). The HPV-16 E6 and E6-AP complex functions as a ubiquitin-protein ligase in the ubiquitination of p53. *Cell* **75,** 495–505.

Scheffner, M., Nuber, U., and Huibregtse, J. M. (1995). Protein ubiquitination involving an E1-E2-E3 enzyme ubiquitin thioester cascade. *Nature* **373,** 81–83.

Scheffner, M., Werness, B. A., Huibregtse, J. M., Levine, A. J., and Howley, P. M. (1990). The E6 oncoprotein encoded by human papillomavirus types 16 and 18 promotes the degradation of p53. *Cell* **63,** 1129–1136.

Schwarz, S. E., Rosa, J. L., and Scheffner, M. (1998). Characterization of human hect domain family members and their interaction with UbcH5 and UbcH7. *J. Biol. Chem.* **273,** 12148–12154.

Springael, J. Y., and Andre, B. (1998). Nitrogen-regulated ubiquitination of the gap1 permease of *Saccharomyces cerevisiae. Mol. Biol. Cell* **9,** 1253–1263.

Staub, O., Dho, S., Henry, P. C., Correa, J., Ishikawa, T., McGlade, J., and Rotin, D. (1996). WW domains of Nedd4 bind to the proline-rich PY motifs in the epithelial $Na^+$ channel deleted in Liddle's syndrome. *EMBO J.* **15,** 2371–2380.

Suzuki, C., Murakami, G., Fukuchi, M., Shimanuki, T., Shikauchi, Y., Imamura, T., and Miyazono, K. (2002). Smurf1 regulates the inhibitory activity of Smad7 by targeting Smad7 to the plasma membrane. *J. Biol. Chem.* **277,** 39919–39925.

Traweger, A., Fang, D., Liu, Y. C., Stelzhammer, W., Krizbai, I. A., Fresser, F., Bauer, H. C., and Bauer, H. (2002). The tight junction-specific protein occludin is a functional target of the E3 ubiquitin-protein ligase itch. *J. Biol. Chem.* **277,** 10201–10208.

Verdecia, M. A., Joazeiro, C. A., Wells, N. J., Ferrer, J. L., Bowman, M. E., Hunter, T., and Noel, J. P. (2003). Conformational flexibility underlies ubiquitin ligation mediated by the WWP1 HECT domain E3 ligase. *Mol. Cell* **11,** 249–259.

Wang, G., Yang, J., and Huibregtse, J. M. (1999). Functional domains of the Rsp5 ubiquitin-protein ligase. *Mol. Cell. Biol.* **19,** 342–352.

Yasuda, J., Nakao, M., Kawaoka, Y., and Shida, H. (2003). Nedd4 regulates egress of Ebola virus-like particles from host cells. *J. Virol.* **77,** 9987–9992.

Zhang, Y., Wang, H. R., and Wrana, J. L. (2004). Smurf1: A link between cell polarity and ubiquitination. *Cell Cycle* **3,** 391–392.

Zhao, C., Beaudenon, S. L., Kelley, M. L., Waddell, M. B., Yuan, W., Schulman, B. A., Huibregtse, J. M., and Krug, R. M. (2004). The UbcH8 ubiquitin E2 enzyme is also the E2 enzyme for ISG15, an IFN-alpha/beta-induced ubiquitin-like protein. *Proc. Natl. Acad. Sci. USA* **101,** 7578–7582.

# [12] High-Level Expression and Purification of Recombinant SCF Ubiquitin Ligases

*By* Ti Li, Nikola P. Pavletich, Brenda A. Schulman, and Ning Zheng

## Abstract

The SCF complexes are the prototype of a superfamily of cullin-dependent ubiquitin ligases, which regulate diverse cellular functions by promoting the ubiquitination of a large number of regulatory and signaling proteins. The SCF complexes are organized by the elongated scaffold protein subunit Cul1, which interacts with the Rbx1 RING finger protein at one end and the Skp1 adaptor protein at the other. By binding to Skp1, members of the F-box protein family are responsible for recruiting specific substrates to the ligase machine. This chapter describes methods that we have developed to achieve high-level expression and purification of two recombinant SCF complexes from both insect cells and bacteria. We emphasize the power of protein coexpression and a novel "Split-n-Coexpress" method in producing soluble and functional recombinant proteins and

METHODS IN ENZYMOLOGY, VOL. 398 0076-6879/05 $35.00

DOI: 10.1016/S0076-6879(05)98012-9

protein complexes. We propose that similar approaches can be used to obtain large quantities of other SCF and SCF-like complexes for biochemical and structural investigations.

## Introduction

The SCF (*S*kp1-*C*ul1-*F*-box protein) complexes represent the largest family of ubiquitin-protein ligases and mediate the ubiquitination of a broad spectrum of regulatory and signaling proteins in diverse cellular pathways (Deshaies, 1999). The SCF consists of three invariant components, Skp1, Cul1, and Rbx1 (also known as Roc1 or Hrt1), and an interchangeable subunit, an F-box protein (Bai *et al.*, 1996; Feldman *et al.*, 1997; Kamura *et al.*, 1999; Ohta *et al.*, 1999; Seol *et al.*, 1999; Skowyra *et al.*, 1997, 1999; Tan *et al.*, 1999). Cul1-Rbx1 forms the catalytic core of the complex and is responsible for recruiting the ubiquitin-conjugating enzyme, whereas Skp1 serves as an adaptor, which is able to bind different F-box proteins. The F-box proteins, bearing a Skp1-interacting F-box motif and a protein–protein interaction domain, can dock different protein substrates, often phosphorylated, to the SCF complex for ubiquitination. The large number of the F-box proteins in eukaryotes, with more than 60 members in mammals, allows a large number of substrates to be specifically ubiquitinated by the SCF.

The SCF is the prototype of an even larger superfamily of cullin-dependent multisubunit E3 ligase complexes. In addition to Cul1, the cullin protein family has five additional closely related paralogues in humans (Cul2, 3, 4A, 4B, and 5) (Kipreos *et al.*, 1996), all of which can bind Rbx1 and form ubiquitin ligases (Furukawa *et al.*, 2002). Similar to Cul1, most, if not all, cullin family members are able to form numerous ligase complexes by recruiting variable substrate-binding subunits. For example, Cul2-Rbx1, together with Elongin-B and the Skp-1 homologous Elongin-C, can assemble with members of the SOCS-box protein family to form SCF-like complexes (Kamura *et al.*, 1998), whereas Cul3-Rbx1 can organize ligase complexes by interacting directly with members of the BTB/POZ domain protein family (Furukawa *et al.*, 2003; Geyer *et al.*, 2003; Pintard *et al.*, 2003; Xu *et al.*, 2003).

Preparations of recombinant SCF and SCF-like complexes are crucial for biochemical and structural characterizations of the multisubunit ubiquitin ligases. This chapter describes methods to express, purify, and assemble two well-studied SCF complexes, $SCF^{Skp2}$ and $SCF^{\beta\text{-TRCP1}}$(the superscript denotes the F-box protein). We expect that these methods can be adapted to the preparation of other SCF and SCF-like ubiquitin ligase complexes.

## General Strategy for Preparation of Recombinant SCF complexes

The SCF complexes adopt a rigid overall structure, which is important for the functions of the ubiquitin ligases (Schulman *et al.*, 2000; Wu *et al.*, 2003; Zheng *et al.*, 2002). In the complex, extensive protein–protein interfaces are found between Cul1 and Rbx1 and between Cul1 and Skp1, as well as between Skp1 and the F-box protein. As these interactions often involve intermolecular hydrophobic packing, recombinant SCF components, when produced individually in the absence of their direct binding partner(s), would have otherwise buried hydrophobic surfaces exposed, making the polypeptides mostly insoluble and inappropriate for biochemical and structural analyses. In our efforts to obtain soluble and functional SCF complexes for structural studies, we found that an effective solution to the protein insolubility problem is to coexpress and copurify two subunits of the protein complex together. Specifically, we have been successful in preparing soluble recombinant Cul1-Rbx1 and Skp1-F-box protein complexes separately in quantities adequate for biochemical and structural analyses. Using these SCF subcomplexes, we have also been able to assemble two active whole SCF complexes.

The largest subunit of SCF is the 776 amino acids scaffold protein Cul1. For reasons to be clarified, *Escherichia coli* often has difficulty overproducing eukaryotic proteins of this size or larger. Even though full-length Cul1 becomes significantly soluble when coexpressed with Rbx1, its low expression level makes it unpractical to prepare the complex from bacteria. To overcome this problem, we have developed a "Split-n-Coexpress" method to make bacteria overexpress a split yet functional form of Cul1 together with Rbx1 in a large quantity (Zheng *et al.*, 2002). Based on the structure of Cul1, we have chosen a Cul1 surface loop, which connects the N-terminal domain (NTD) and the C-terminal domain (CTD) of the protein, to split the polypeptide into two fragments, each being small enough to be efficiently overproduced by *E. coli.* These two Cul1 domains pack against each other through a hydrophobic interface formed among the last two helices of the NTD and the first two helices of the CTD. Because protein folding and protein interaction share the same principles, upon coexpression, the two domains are able to dock onto each other as if they were in the same polypeptide. When Rbx1 is made at the same time, the CTD forms a complex with the NTD as well as Rbx1. Based on previous structural and biochemical studies, this split form of Cul1 prepared from bacteria overexpression has the same structure and ligase activity in the SCF complex as the wild-type intact form (Zheng *et al.*, 2002). A similar approach with an alternative split site within a surface loop in the four helical bundle subdomain of the Cul1 CTD has proved equally effective (Zheng *et al.*, 2002).

## Expression and Purification of Recombinant Cul1-Rbx1 Complex

A complex of the full-length intact Cul1 and Rbx1 proteins can be produced from insect cells using recombinant baculoviruses, whereas high-level production of a similar complex containing the split form of Cul1 can be achieved in *E. coli*. This section provides protocols for both approaches. Although the relatively low yield of the baculovirus system did not prevent us from making enough protein samples to crystallize the Cul1-Rbx1 complex in the early stage of our studies, we took advantage of the simpler and more cost-effective *E. coli* system in all of our subsequent structural studies of the SCF complexes.

### *Expression and Purification of Cul1-Rbx1 from Insect Cells*

#### *Generation of Recombinant Baculoviruses*

1. Subclone human Cul-1 cDNA into the pAcGHLT-A vector (PharMingen) vector and mouse Rbx1 cDNA into the pAcSG2 vector (PharMingen). The pAcGHLT-A vector introduces a gluathione *S*-transferase (GST) tag followed by a thrombin cleavage site to the N terminus of Cul-1.
2. Use the BaculoGold DNA baculovirus expression system (PharMingen) and follow the manufacturer's instructions to make separate recombinant baculoviruses for GST-Cul1 and Rbx1.
3. Use Sf9 insect cells (*Spodoptera frugiperda*, GIBCO) to amplify the viruses on plates with Grace's insect medium (GIBCO) supplemented with 10% fetal bovine serum and 2 m*M* L-glutamine and 100 U/ml penicillin/streptomycin. After three to four generations, high-titer virus stock can be obtained.

#### *Expression of Cul1-Rbx1 in Insect Cells*

1. For large-scale preparation, adapt Hi5 insect cells (High Five, Invitrogen) to serum-free suspension culture in Sf-900 II SFM (GIBCO) medium. Culture 1-liter cells in a 2.4-liter flask at 27°, shaking at 110 rpm.
2. When the cell density reaches $1–2 \times 10^6$ cells/ml, spin down the cells in an autoclaved bottle at 1000 rpm for 3 min. Remove the medium as much as possible.
3. Add 5–10 mL of each high-titer GST-Cul1 and Rbx1 viruses to the cell pellet. Gently resuspend and incubate the cells with the viruses for 1 h at room temperature with an occasional stir.
4. Transfer the infected cells back to the flask with 1-liter fresh medium and return the flask to the shaker. Harvest the infected cells 2 days after the infection.

*Purification of the Cul1-Rbx1 Complex*

1. Harvest 4 liters of infected cells by centrifugation at 2000 rpm for 5 min. Perform all the following purification steps at 4° or on ice.

2. Resuspend cell pellets in 200 ml lysis buffer [20 m*M* Tris-HCl, 200 m*M* NaCl, 5 m*M* dithiothreitol (DTT), pH 8.0] supplemented with a protease inhibitor cocktail [1 m*M* phenylmethylsulfonyl fluoride (PMSF), 1 $\mu$g/ml aprotinin, 1 $\mu$g/ml leupeptin, and 1 $\mu$g/ml pepstatin].

3. Lyse the cells with a Microfluidizer processor (M-110EHI, Microfluidics Corp.) at 10,000 psi twice or by sonication until all the cells are broken as monitored under a microscope.

4. Remove cell debris by a 1-h centrifugation at 15,000 rpm followed by another 1-h high-speed centrifugation at 45,000 rpm in a Beckman Ti 45 rotor.

5. Equilibrate a gravity column of 5 ml glutathione-Sepharose 4B (Amersham Biosciences) with 10 column volume (CV) lysis buffer.

6. Apply the clarified lysate to the column and control the flow rate to be less or equal to 1 ml/min. Wash the column with 20 CV lysis buffer after the lysate has passed through.

7. Elute the protein with 5 CV elution buffer (50 m*M* Tris-HCl, 200 m*M* NaCl, 10 m*M* reduced glutathione, pH 8.0). Determine the protein concentration by the Bio-Rad protein assay and check the purity of the eluted protein by SDS–PAGE. At this step, we routinely obtain 4–8 mg GST-Cul1-Rbx1 from 1 liter of insect cells.

8. Add thrombin to the eluted protein at a ratio of thrombin:fusion-protein = 1:100 (w/w) and incubate the sample overnight at 4°. Check the thrombin cleavage efficiency the next morning by SDS–PAGE (Fig. 1). If high purity is not needed, skip to step 12.

9. Equilibrate a 1-ml Resource-S column on an Akta FPLC system (Amersham Biosciences) with buffer A [50 m*M* 2-(*N*-morpholino)ethanesulfonic acid (MES), 5 m*M* DTT, pH 6.5]. Dilute the thrombin-cleaved GST-Cul1-Rbx1 sample three folds by buffer A and load the sample onto the column. Other cation-exchange columns such as a 1-ml Mono-S column can be used at this step. Cul1-Rbx1 also binds anion-exchange Q columns. We prefer S columns, as GST is in the flow through.

10. After 2 CV wash with buffer A, elute the protein with a gradient of 100–500 m*M* NaCl in 40 CV generated by mixing buffer A and buffer B (50 m*M* MES, 1 *M* NaCl, 5 m*M* DTT, pH 6.5).

11. Check fractions of eluted proteins by SDS–PAGE and pool the fractions containing purified Cul1-Rbx1.

12. Concentrate the protein sample to 5 mg/ml by ultrafiltration with a 30-kDa molecular mass cut-off membrane (Millipore).

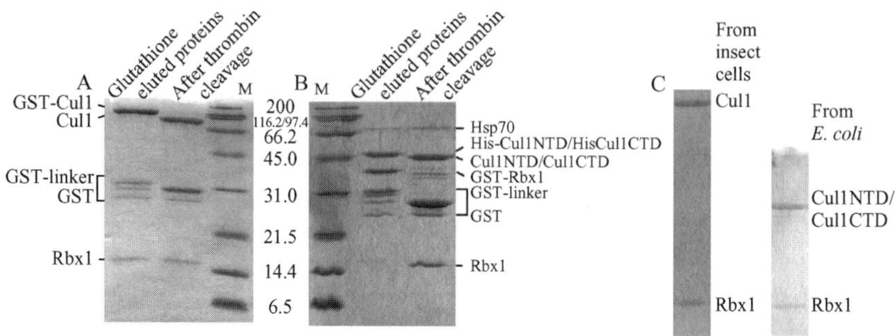


FIG. 1. Affinity purification of the Cul1-Rbx1 complex from insect cells and *E. coli.* (A) Glutathione eluted GST-Cul1 in complex with Rbx1 purified from baculoviruses-infected Hi5 insect cells. After thrombin digestion, GST is removed from Cul1. (B) Glutathione eluted GST-Rbx1 in complex with the split and coexpressed form of Cul1 from *E. coli.* The Cul1-NTD and the Cul1-CTD migrate together as one band on the SDS–PAGE gel. Upon thrombin digestion, GST is removed from Rbx1 and His tags are cleaved off from the two Cul1 domains. Note that the molar ratio of Rbx1 and Cul1NTD/Cul1CTD is probably larger than one at this step. Upon further purification by ion-exchange and/or size-exclusion chromatography, excessive Rbx1 can be removed easily. (C) Purified Cul1-Rbx1 from insect cells and *E. coli.*

13. Equilibrate a 24-ml Superdex 200 column with buffer C (20 m*M* MES, 200 m*M* NaCl, 5 m*M* DTT, pH 6.5) on an Akta FPLC system. Load 5 mg Cul1-Rbx1 sample each time and pool the fractions containing the purified proteins. The protein complex can now be used for biochemical or structural studies or stored in –80° after being aliquoted and snap-frozen in liquid nitrogen.

### *Expression and Purification of Cul1-Rbx1 from* E. coli

#### *Generation of Constructs for Expression*

1. Generate pGEX-Rbx1 by subcloning the full-length mRbx1 gene into the pGEX-4T1 vector (Amersham Biosciences), which introduces an N-terminal GST tag followed by a thrombin cleavage site (Fig. 2).

2. Generate pETCul1NTD and pETCul1CTD by subcloning human Cul1 NTD (1– 410) and CTD (411–776) separately into the pET15b vector, which introduces an N-terminal His tag followed by a thrombin cleavage site.

3. Amplify the ribosome binding site (rbs) and the open reading frame containing the His-tagged Cul1-CTD from pETCul1CTD by polymerase chain reaction (PCR). Insert the PCR product into pGEX-Rbx1 following the Rbx1 gene at about 20–30 bp away from its stop codon.

4. Obtain pALCul1NTD by swapping the *Sph*I–*Hind*III fragments of pETCul1NTD and pACYC184 (New England BioLabs). The pALCul1NTD

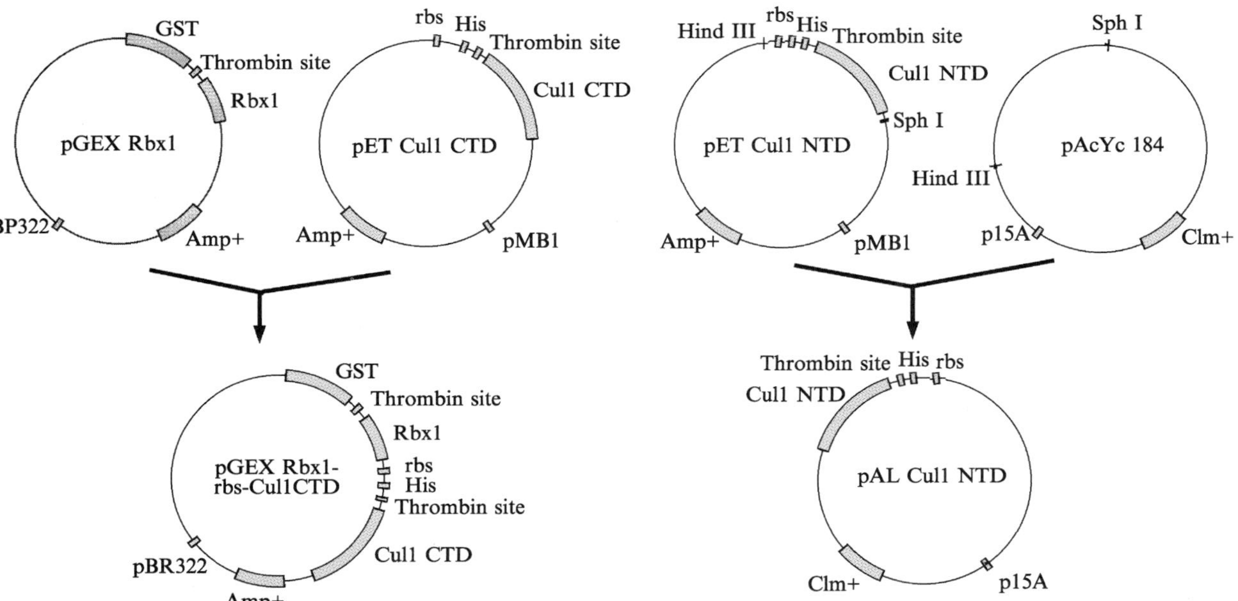


FIG. 2. Plasmids construction for coexpressing the Cul1-NTD, the Cul1-CTD, and Rbx1 together in bacteria. The vector pALCul1NTD has the proper origin and selection marker to coexist with pGEXRbx1-rbs-Cul1CTD. The His tag introduced into the two Cul1 domains makes it easier to confirm copurification of the Cul1 domains without disrupting their packing.

vector should retain the p15A origin and chloramphenicol resistance gene from pACYC184.

*Expression of Cul1-Rbx1 in* E. coli

1. Cotransform the *E. coli* BL21(DE3) strain with pGEXRbx1rbs-Cul1CTD and pALCul1NTD. Desired transformants are selected by an $Amp^R/Clm^R$ LB agar plate. Because the transformation efficiency of BL21(DE3) is very low, a higher DNA concentration (1 $\mu$g/$\mu$l) might be necessary to improve the efficiency of cotransformation.
2. Inoculate 60 ml LB culture containing 50 $\mu$g/ml ampicillin and 25 $\mu$g/ml chloramphenicol from a freshly transformed single colony. Grow the culture overnight at 37° with 250 rpm shaking.
3. Use the 60 ml preculture to inoculate 12 liters of LB medium. After $OD_{600}$ reaches 1.5, decrease the incubation temperature to 16°. Following half an hour of cooling, add isopropyl-$\beta$-D-thiogalactoside (IPTG) to 0.5 m*M* and carry out induction overnight (about 16 h).

*Purification of Cul1-Rbx1.* The aforementioned general procedures for purifying Cul1-Rbx1 from insect cells can be followed with the following exceptions.

1. Harvest cells by centrifugation at 4000 rpm for 15 min at 4°.
2. High-speed centrifugation at 45,000 rpm can be replaced by a second 1-h centrifugation at 15,000 rpm.
3. Use a 10-ml glutathione-Sepharose 4B column for a 12-liter preparation. We routinely obtain 10 mg GST-Rbx1-Cul1CTD-NTD from a 1-liter culture of *E. coli*.
4. Use an 8-ml Resource S column for a 12-liter preparation.

## Expression and Purification of Recombinant Skp1-F Box Protein Complexes

Over the course of our efforts to obtain large quantities of Skp1-F box protein complexes for crystallization studies, we have found a good general strategy for expressing Skp1-F box protein complexes from insect cells that has worked for complexes with all F box proteins tested to date. Our initial studies focused on production of a complex between Skp1 and the F box protein Skp2. We tested coexpression of both proteins in insect cells, with a variety of tags on either Skp1 or Skp2, and found that we were able to obtain complexes with 1:1 stoichiometry with an initial affinity purification for Skp2. In contrast, if the initial affinity purification selected for the tag on Skp1, we always copurified a massive excess of free Skp1. The common approach for obtaining the Skp1-F box protein involves coexpression of a GST-tagged F box protein with untagged Skp1, and initial purification of a 1:1 complex by glutathione affinity purification. We present a protocol for isolation of full-length Skp1-GSTSkp2, Skp1-GST$\beta$-TRCP1, and Skp1-GSTCdc4 complexes from insect cells. We also present protocols for large-scale purification of the Skp1-$\beta$TRCP1 and Skp1-Skp2 complexes used for crystallographic studies from insect cells and *E. coli*, respectively.

### *Expression and Purification of Skp1-GST F Box Protein Complexes from Insect Cells*

#### *Generation of Recombinant Baculoviruses*

1. Subclone cDNAs for the following F box proteins into the pAcGHLT-A vector (PharMingen): human Skp2, human $\beta$-TRCP1, or *Saccharomyces cerevisiae* Cdc4. This introduces a GST tag followed by a thrombin cleavage site to the N terminus of the F-box proteins. Subclone human or *S. cerevisiae* Skp1 cDNA into the pAcSG2 vector (PharMingen).
2. Use the BaculoGold DNA baculovirus expression system (PharMingen) and follow the manufacturer's instructions to make separate recombinant baculoviruses for GST-Skp2, GST-$\beta$-TRCP1, GST-Cdc4p, and human or *S. cerevisiae* Skp1.
3. Use Sf9 insect cells (GIBCO) to amplify the viruses on plates to obtain high-titer virus stock.

#### *Small-Scale Expression of Skp1-Skp2, Skp1-β-TRCP, and Skp1-Cdc4 in Insect Cells*

1. For small-scale preparation, culture Hi5 insect cells (Invitrogen) on 15-cm tissue culture dishes in Hink's insect medium (Cellgro) supplemented with 10% fetal bovine serum, 2 m*M* L-glutamine, and 100 U/mL penicillin/streptomycin at 27°. Use five plates for each complex.
2. Grow the Hi5 cells to 80% confluency, prior to infection. Remove all but 1 ml of the media.
3. To each plate, add 1 ml of each high-titer virus for the desired Skp1-F box protein complex combination. For example, to obtain the Skp1-Skp2 complex, add 1 ml of the virus for Skp1 and 1 ml of the virus for GST-Skp2. Incubate the cells with the viruses for 1 h, with periodic gentle rocking.
4. Add 20 ml fresh medium to each plate and incubate for 2 days.

#### *Initial Glutathione Affinity Purifications of Skp1-GSTSkp2, Skp1-GSTβ-TRCP1, and Skp1-GSTCdc4 Complexes*

1. Resuspend the infected cells by trituration or scraping and harvest the cell pellet by centrifugation at 2000 rpm for 5 min. Perform all the following purification steps at 4° or on ice.

2. Resuspend cell pellets in 10 ml lysis buffer (50 m*M* Tris-HCl, 200 m*M* NaCl, 0.1% NP-40, 5 m*M* DTT, pH 7.6) supplemented with a protease inhibitor cocktail (see earlier discussion).

3. Lyse the cells by sonication.

4. Remove cell debris by centrifugation for 30 min at 15,000 rpm. Transfer the supernatant to a fresh tube and repeat.

5. Equilibrate a 0.2-ml glutathione-Sepharose 4B column with 10 CV of lysis buffer.

6. Apply clarified lysate to the column and control the flow rate to be less or equal to 1 ml/min. Wash the column with 20 CV lysis buffer after the lysate has passed through.

7. Elute the protein with 5 CV elution buffer (50 m*M* Tris-HCl, 200 m*M* NaCl, 10 m*M* reduced glutathione, pH 8.0). Determine the protein concentration by the Bio-Rad protein assay and check the purity of the eluted protein by SDS–PAGE (Fig. 3).

*Deletions in Skp1, Skp2, and β-TRCP1 Prepared in Large Scale for X-Ray Crystallographic Studies.* For crystallization of Skp1-Skp2 and Skp1-β-TRCP1 complexes, it was necessary to remove a number of residues from Skp1, Skp2, and β-TRCP1, which presumably are disordered. The crystallized version of human Skp1 has two internal deletions, which are

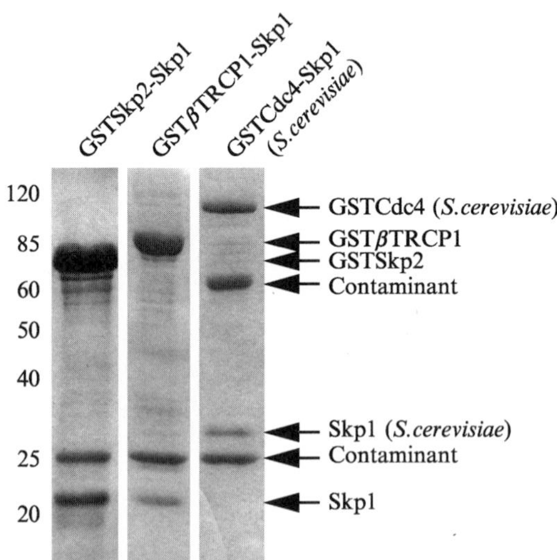


FIG. 3. SDS–PAGE gel slices from small-scale glutathione affinity purifications of Skp1 in complexes with F box proteins. From left to right: GSTSkp2-Skp1 (human), GSTβTRCP1-Skp1 (human), and GSTCdc4-Skp1 (*S. cerevisiae*).

the loops spanning residues 38 to 43 and 71 to 82. This version is referred to hereafter as Skp1ΔΔ. These two internal deletions do not affect expression or solubility of Skp1. Moreover, they did not affect the ability of Skp1 to form complexes with F box proteins or Cul1 in our studies.

The crystallized version of Skp2 lacks the N-terminal 100 residues (referred to hereafter as Skp2ΔN), and the crystallized version of β-TRCP1 lacks the N-terminal 138 residues (referred to hereafter as β-TRCP1ΔN). These N-terminal truncations in both F box proteins resulted in dramatic improvements in expression and solubility in our experiments.

*Large-Scale Expression of Skp1ΔΔ–GSTβ-TRCP1ΔN in Insect Cells and Initial Glutathione Affinity Purification.* The general procedures for expressing Skp1ΔΔ–β-TRCP1ΔN in insect cells, harvesting and lysing the cells, and performing the initial glutathione affinity purification are the same as for expressing Cul1-Rbx1 in insect cells with the following exceptions.

1. Generate cDNAs for Skp1 with two internal deletions, and for β-TRCP1 with the N-terminal deletion by PCR, and generate recombinant baculoviruses using the protocols described earlier for full-length Skp1 and GSTβ-TRCP1.
2. Coinfect cells with 5–10 ml of high-titer virus for each Skp1ΔΔ and GSTβ-TRCP1ΔN.

*Further Large-Scale Purification of Skp1ΔΔ–βTRCP1ΔN by Anion-Exchange, Glutathione Affinity, and Size-Exclusion Chromatography*

1. Following initial glutathione affinity purification, we can routinely obtain 10–20 mg GST-β-TRCP1ΔN-Skp1ΔΔ (Fig. 4) from 1 liter of cell culture.

2. Further purify GST-β-TRCP1ΔN-Skp1ΔΔ by anion-exchange chromatography. Equilibrate a 10-ml Resource-Q column on an Akta FPLC system (Amersham Bioscience) with buffer A (25 m*M* Tris-HCl, 5 m*M* DTT, pH 7.6). Dilute the GST-β-TRCP1ΔN-Skp1ΔΔ 1:4 (v/v) with buffer A and load the sample onto the column. After 2 CV wash with buffer A, elute the protein with a gradient of 30–400 m*M* NaCl in 40 CV generated by mixing buffer A and buffer B (25 m*M* Tris-HCl, 1 *M* NaCl, 5 m*M* DTT, pH 7.6). The GST-β-TRCP1ΔN-Skp1ΔΔ elutes at NaCl concentrations between 150 and 200 m*M*. Pool the purified fractions (Fig. 4) and concentrate to >2 mg/ml by ultrafiltration with a 10-kDa molecular mass cut-off membrane (Millipore).

3. Perform thrombin cleavage by adding thrombin to the eluted protein at thrombin:fusion-protein = 1:100 (w/w). Add $CaCl_2$ to a final concentration of 2.5 m*M*, and incubate the sample 8–16 h at 4°. Check the thrombin cleavage by SDS–PAGE (Fig. 4).

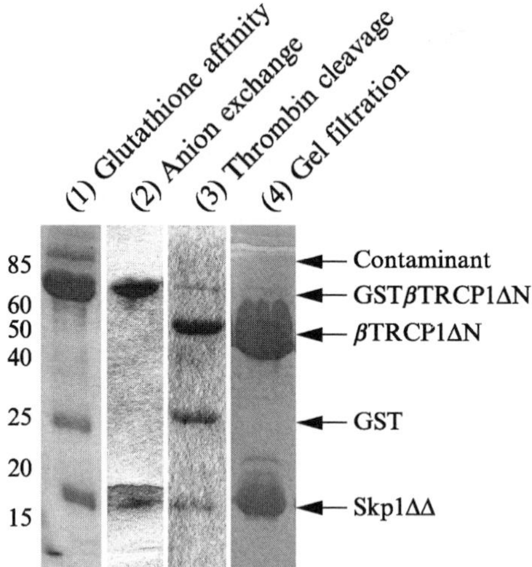


FIG. 4. SDS–PAGE gel slices from large-scale purification of $\beta$TRCP1ΔN-Skp1ΔΔ. Lane 1 corresponds to the initial pool of GST$\beta$TRCP1ΔN-Skp1ΔΔ after glutathione affinity chromatography. Lane 2 corresponds to the pool of GST$\beta$TRCP1ΔN-Skp1ΔΔ after anion-exchange purification with a Resource Q column. Lane 3 shows the result of thrombin cleavage after a ~16-h incubation at 4° to remove the GST tag. This step is followed by another purification by glutathione affinity chromatography to remove GST contaminants. Lane 4 corresponds to the final purified $\beta$TRCP1ΔN-Skp1ΔΔ complex after gel filtration chromatography.

4. Remove free GST and any remaining uncleaved GST-$\beta$-TRCP1ΔN-Skp1ΔΔ with a second glutathione affinity step. Regenerate the used glutathione-Sepharose column by washing with 5 CV of elution buffer. Then equilibrate with 5 CV storage buffer (50 m*M* Tris-HCl, 200 m*M* NaCl, 5 m*M* DTT, pH 7.6). Load protein onto the glutathione-Sepharose column and collect the entire flow through. Add and collect another CV of storage buffer to wash out all the $\beta$-TRCP1ΔN-Skp1ΔΔ. GST-containing contaminants are retained by the glutathione-Sepharose column (Fig. 4). The yield of purified complex at this step should be roughly 5–10 mg/liter of insect cell culture. Concentrate the protein sample to 10 mg/ml by ultrafiltration with a 10-kDa molecular mass cut-off membrane (Millipore).

5. Perform final purification by size-exclusion chromatography. Equilibrate a 24-ml Superdex 200 column with storage buffer on an Akta FPLC system. Load 1.5 ml (roughly 15 mg) $\beta$-TRCP1ΔN-Skp1ΔΔ sample each time and pool the fractions containing the purified protein complex (Fig. 4). The protein complex can now be used for biochemical or structural studies or stored in –80° after being aliquoted and snap-frozen in liquid nitrogen.

*Expression and Purification of Skp1ΔΔ–Skp2ΔN from* E. coli

*Generation of the Bicistronic pGEXSkp2ΔNrbsSkp1ΔΔ Expression Vector*

1. Generate pGEX-Skp2ΔN by subcloning the cDNA encoding human Skp2 lacking the N-terminal 100 residues into pGEX-4T1 or related vector to introduce an N-terminal GST tag followed by a thrombin cleavage site.
2. Generate pETSkp1ΔΔ by subcloning the cDNA encoding human Skp1 lacking residues 38–43 and 71 to 82 into pET3a.
3. Amplify the rbs and the ORF encoding Skp1ΔΔ from pETSkp1ΔΔ by PCR. Generate the plasmid pGEXSkp2ΔNrbsSkp1ΔΔ by subcloning the PCR product into pGEX-Skp2ΔN, about 20–30 bases downstream of the stop codon for the Skp2ΔN gene (Fig. 5).

*Expression of Skp1ΔΔ-Skp2ΔN in* E. coli

1. Transform the *E. coli* BL21(DE3) strain with pGEXSkp2ΔNrbs Skp1ΔΔ.

2. Inoculate 200 ml LB culture containing 150 μg/ml ampicillin from a freshly transformed single colony. Grow the culture overnight at 37°, shaking at 200–250 rpm.

3. Carry out the large-scale expression in twenty-four 2-liter baffle flasks, each containing 1 liter of LB medium with 150 μg/ml ampicillin. Inoculate each liter of media with 5 ml of the preculture. Grow the twenty-four 1-liter cultures at 37°, shaking at 200–250 rpm. After $OD_{600}$ reaches 0.8–1.2, add IPTG to a final concentration of 1 m*M*. After adding IPTG, transfer cells to a 16° shaking incubator and grow for 16–20 h.

*Prepare Cell Lysate for the Purification of Skp1ΔΔ-Skp2ΔN.* All purification steps are performed in a 4° cold room. Protein and buffers are kept at 4° or on ice.

1. Harvest cells by centrifugation at 4500 rpm for 15 min at 4°.
2. Resuspend the cell pellet for each liter of culture in 8 ml of lysis buffer (50 m*M* Tris-HCl, 200 m*M* NaCl, 2.5 m*M* PMSF, 5 m*M* DTT, pH 7.6). This should give a final volume of about 300 ml including the cells.
3. Lyse the cells with a Microfluidizer processor (M–110EHI, Microfluidics Corp.) at 10,000 psi twice or by sonication.
4. Clear lysate by centrifugation at 15,000 rpm for 30 min at 4°. Pour the lysate into a new centrifuge tube and spin a second time at 15,000 rpm for 30 min at 4°. Keep lysate on ice.

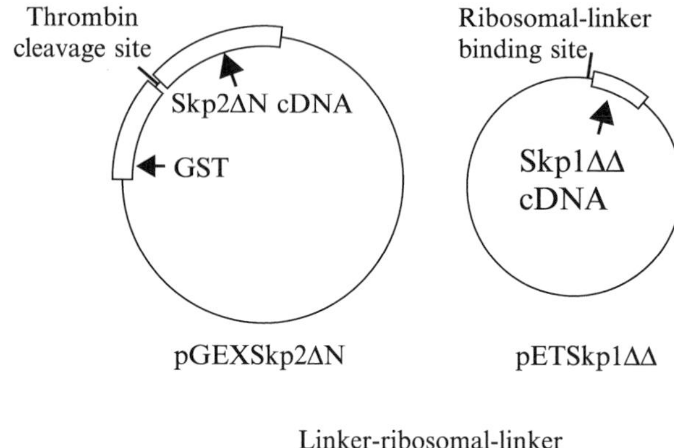


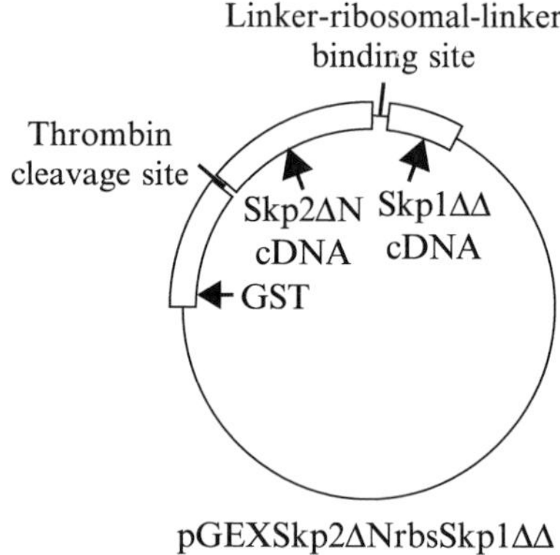


FIG. 5. Generation of pGEXSkp2ΔNrbsSkp1ΔΔ bicistronic coexpression vector for the Skp2ΔN-Skp1ΔΔ complex. (Top left) Generate pGEXSkp2ΔN by subcloning the ORF for human Skp2 lacking the N-terminal 100 residues into pGEX4T1. (Top right) Generate pETSkp1ΔΔ by subcloning the ORF for human Skp1 lacking residues 38–43 and 71 to 82 into pET3a. (Bottom) The final pGEXSkp2ΔNrbsSkp1ΔΔ vector is generated by subcloning a PCR product containing the ribosomal binding site (rbs) and Skp1ΔΔ cDNA amplified from pETSkp1ΔΔ into pGEXSkp2ΔN.

### *Initial Glutathione Affinity and Anion-Exchange Chromatography of GSTSkp2ΔN-Skp1ΔΔ*

1. Equilibrate a 20-ml glutathione-Sepharose 4B gravity column with 10 CV lysis buffer.

2. Apply the clarified lysate to the column and control the flow rate to be less or equal to 1 ml/min. Wash the column with 20 CV lysis buffer after the lysate has gone through.

3. Elute the protein with five fractions, each 1 CV, with elution buffer (50 m*M* Tris-HCl, 200 m*M* NaCl, 5 m*M* DTT, 10 m*M* reduced glutathione, pH 8.0). Determine the protein concentration by the Bio-Rad protein

assay and check the purity of the eluted protein by SDS–PAGE (Fig. 6). Pool fractions containing the most pure GSTSkp2ΔN-Skp1ΔΔ. At this step, we can routinely obtain 10 mg GSTSkp2ΔN-Skp1ΔΔ from a 1-liter culture.

4. Equilibrate a 30-ml Q-Sepharose fast-flow (Sigma) gravity column with 10 CV low-salt buffer (25 m*M* Tris-HCl, 50 m*M* NaCl, 5 m*M* DTT, pH 7.6).

5. Dilute the GSTSkp2ΔN-Skp1ΔΔ 1:4 (v/v) in no-salt dilution buffer (25 m*M* Tris-HCl, 5 m*M* DTT, pH 7.6).

6. Apply the diluted protein to the column and control the flow rate to be less or equal to 5 ml/min. Wash the column with 20 CV low-salt buffer after loading.

7. Elute the protein with three fractions, each 1 CV, with high-salt buffer (25 m*M* Tris-HCl, 400 m*M* NaCl, 5 m*M* DTT, pH 7.6). Determine the protein concentration by the Bio-Rad protein assay and pool fractions containing the most pure GSTSkp2ΔN-Skp1ΔΔ. At this step, we can routinely recover 90–100% of the GSTSkp2ΔN-Skp1ΔΔ. Concentrate the proteins to >5 mg/ml.

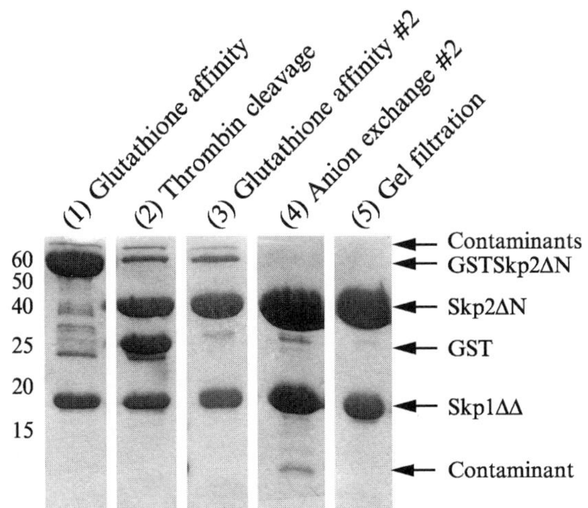


FIG. 6. SDS–PAGE gel slices from large-scale purification of Skp2ΔN-Skp1ΔΔ. Lane 1 corresponds to the initial pool of GSTSkp2ΔN-Skp1ΔΔ after glutathione affinity chromatography. This is followed by anion-exchange chromatography over a gravity Q-Sepharose followed column (not shown). Lane 2 shows the result of thrombin cleavage after a ~16-h incubation at 4° to remove the GST tag. Lane 3 shows the result of further purification by glutathione affinity chromatography to remove GST contaminants. Lane 4 corresponds to the pool following anion-exchange chromatography over a Resource Q column. Lane 5 shows the final purified Skp2ΔN-Skp1ΔΔ complex after gel filtration chromatography.

8. Add thrombin to the eluted protein at thrombin:fusion-protein = 1:300 (w/w). Add $CaCl_2$ to a final concentration of 2.5 m*M*, and incubate the sample 8–16 h at 4°. Check the thrombin cleavage by SDS–PAGE (Fig. 6). When cleavage is complete, inhibit thrombin activity with PMSF (dissolved at 0.2 *M* in ethanol) at a final concentration of 1 m*M*.

*Further Purification by Glutathione Affinity, Anion-Exchange, and Size-Exclusion Chromatography*

1. Free GST and any uncleaved fusion protein can be removed from Skp2ΔN-Skp1ΔΔ by a second round of glutathione affinity chromatography, during which GST binds the column and the Skp2ΔN-Skp1ΔΔ complex flows through. Regenerate the used glutathione-Sepharose column by washing with 5 CV of elution buffer. Then equilibrate with 5 CV high-salt buffer.
2. Load protein onto the glutathione-Sepharose column and collect the entire flow through. Add and collect another CV of high-salt buffer to wash out all the Skp2ΔN-Skp1ΔΔ. GST-containing contaminants are retained by the glutathione-Sepharose column (Fig. 6).
3. The complex can be purified further by anion-exchange chromatography. Equilibrate a 10-ml Resource-Q column on an Akta FPLC system (Amersham Biosciences) with buffer A (25 m*M* Tris-HCl, 5 m*M* DTT, pH 7.6). Dilute the Skp2ΔN-Skp1ΔΔ 1:4 (v/v) with buffer A and load the sample onto the column.
4. After 2 CV wash with buffer A, elute the protein with a gradient of 50–400 m*M* NaCl in 35 CV generated by mixing buffer A and buffer B (25 m*M* Tris-HCl, 1 *M* NaCl, 5 m*M* DTT, pH 7.6). The Skp2ΔN-Skp1ΔΔ elutes at NaCl concentrations between 150 and 200 m*M*.
5. Check fractions with protein by SDS–PAGE and pool fractions containing purified Skp2ΔN-Skp1ΔΔ (Fig. 6). The yield of purified complex at this step should be roughly 5 mg/liter of *E. coli* culture.
6. Concentrate the protein sample to 50 mg/ml by ultrafiltration with a 10-kDa molecular mass cut-off membrane (Millipore).
7. Perform final purification by size-exclusion chromatography. Equilibrate a 120-ml Superdex 200 column with storage buffer (50 m*M* Tris-HCl, 200 m*M* NaCl, 5 m*M* DTT, pH 7.6) on an Akta FPLC system. Load 1.5 ml (roughly 75 mg) Skp2ΔN-Skp1ΔΔ sample each time and pool fractions containing the purified protein complex (Fig. 6).

## Preparation of Recombinant SCF Ubiquitin Ligase Complexes

The Skp1-Skp2 complex can readily bind to the Cul1-Rbx1 complex upon mixing. After careful determination of the protein concentration,

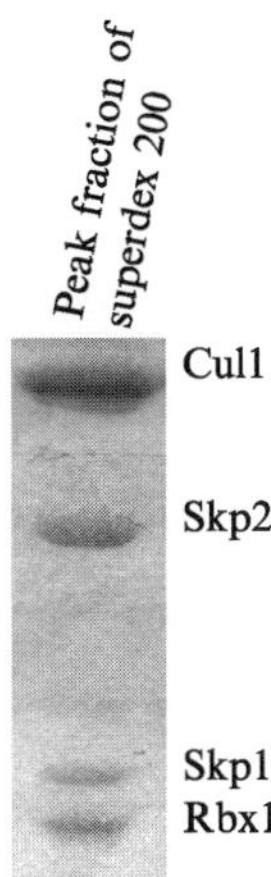


Fig. 7. A Cul1-Rbx1-Skp1-Skp2 complex purified by gel filtration chromatography using separately isolated recombinant Cul1-Rbx1 and Skp1-Skp2 complexes.

we have used this direct mixing method to obtain the full SCF$^{Skp2}$ complex for crystallographic studies. Alternatively, size-exclusion chromatography can be used to prepare whole SCF complexes with a more accurate molar ratio of the subunits. For example, mix the Cul1-Rbx1 and Skp1-Skp2 complexes at a molar ratio of 1:1.2 and apply the sample to a Superdex 200 column. The full SCF$^{Skp2}$ complex can be well separated from the excessive Skp1-Skp2 complex (Fig. 7).

## Acknowledgments

B. A. S. and N. Z. are Pew Scholars. B. A. S. was supported by a Foundation for Advanced Cancer Studies fellowship from the Life Sciences Research Foundation. This work is supported by NIH Grant CA107134 and the Howard Hughes Medical Institute.

## References

Bai, C., Sen, P., Hofmann, K., Ma, L., Goebl, M., Harper, J. W., and Elledge, S. J. (1996). SKP1 connects cell cycle regulators to the ubiquitin proteolysis machinery through a novel motif, the F-box. *Cell* **86,** 263–274.

Deshaies, R. J. (1999). SCF and cullin/Ring H2-based ubiquitin ligases. *Annu. Rev. Cell. Dev. Biol.* **15,** 435–467.

Feldman, R. M., Correll, C. C., Kaplan, K. B., and Deshaies, R. J. (1997). A complex of Cdc4p, Skp1p, and Cdc53p/cullin catalyzes ubiquitination of the phosphorylated CDK inhibitor Sic1p. *Cell* **91,** 221–230.

Furukawa, M., He, Y. J., Borchers, C., and Xiong, Y. (2003). Targeting of protein ubiquitination by BTB-Cullin 3-Roc1 ubiquitin ligases. *Nat. Cell Biol.* **5,** 1001–1007.

Furukawa, M., Ohta, T., and Xiong, Y. (2002). Activation of UBC5 ubiquitin-conjugating enzyme by the RING finger of ROC1 and assembly of active ubiquitin ligases by all cullins. *J. Biol. Chem.* **277,** 15758–15765.

Geyer, R., Wee, S., Anderson, S., Yates, J., and Wolf, D. A. (2003). BTB/POZ domain proteins are putative substrate adaptors for cullin 3 ubiquitin ligases. *Mol. Cell* **12,** 783–790.

Kamura, T., Koepp, D. M., Conrad, M. N., Skowyra, D., Moreland, R. J., Iliopoulos, O., Lane, W. S., Kaelin, W. G., Jr., Elledge, S. J., Conaway, R. C., Harper, J. W., and Conaway, J. W. (1999). Rbx1, a component of the VHL tumor suppressor complex and SCF ubiquitin ligase. *Science* **284,** 657–661.

Kamura, T., Sato, S., Haque, D., Liu, L., Kaelin, W. G., Jr., Conaway, R. C., and Conaway, J. W. (1998). The Elongin BC complex interacts with the conserved SOCS-box motif present in members of the SOCS, ras, WD–40 repeat, and ankyrin repeat families. *Genes Dev.* **12,** 3872–3881.

Kipreos, E. T., Lander, L. E., Wing, J. P., He, W. W., and Hedgecock, E. M. (1996). cul–1 is required for cell cycle exit in *C. elegans* and identifies a novel gene family. *Cell* **85,** 829–839.

Ohta, T., Michel, J. J., Schottelius, A. J., and Xiong, Y. (1999). ROC1, a homolog of APC11, represents a family of cullin partners with an associated ubiquitin ligase activity. *Mol. Cell* **3,** 535–541.

Pintard, L., Willis, J. H., Willems, A., Johnson, J. L., Srayko, M., Kurz, T., Glaser, S., Mains, P. E., Tyers, M., Bowerman, B., and Peter, M. (2003). The BTB protein MEL–26 is a substrate-specific adaptor of the CUL–3 ubiquitin-ligase. *Nature* **425,** 311–316.

Schulman, B. A., Carrano, A. C., Jeffrey, P. D., Bowen, Z., Kinnucan, E. R., Finnin, M. S., Elledge, S. J., Harper, J. W., Pagano, M., and Pavletich, N. P. (2000). Insights into SCF ubiquitin ligases from the structure of the Skp1-Skp2 complex. *Nature* **408,** 381–386.

Seol, J. H., Feldman, R. M., Zachariae, W., Shevchenko, A., Correll, C. C., Lyapina, S., Chi, Y., Galova, M., Claypool, J., Sandmeyer, S., Nasmyth, K., and Deshaies, R. J. (1999). Cdc53/cullin and the essential Hrt1 RING-H2 subunit of SCF define a ubiquitin ligase module that activates the E2 enzyme Cdc34. *Genes Dev.* **13,** 1614–1626.

Skowyra, D., Craig, K. L., Tyers, M., Elledge, S. J., and Harper, J. W. (1997). F-box proteins are receptors that recruit phosphorylated substrates to the SCF ubiquitin-ligase complex. *Cell* **91,** 209–219.

Skowyra, D., Koepp, D. M., Kamura, T., Conrad, M. N., Conaway, R. C., Conaway, J. W., Elledge, S. J., and Harper, J. W. (1999). Reconstitution of G1 cyclin ubiquitination with complexes containing SCFGrr1 and Rbx1. *Science* **284,** 662–665.

Tan, P., Fuchs, S. Y., Chen, A., Wu, K., Gomez, C., Ronai, Z., and Pan, Z. Q. (1999). Recruitment of a ROC1-CUL1 ubiquitin ligase by Skp1 and HOS to catalyze the ubiquitination of I kappa B alpha. *Mol. Cell* **3,** 527–533.

Wu, G., Xu, G., Schulman, B. A., Jeffrey, P. D., Harper, J. W., and Pavletich, N. P. (2003). Structure of a beta-TrCP1-Skp1-beta-catenin complex: Destruction motif binding and lysine specificity of the SCF(beta-TrCP1) ubiquitin ligase. *Mol. Cell* **11,** 1445–1456.

Xu, L., Wei, Y., Reboul, J., Vaglio, P., Shin, T. H., Vidal, M., Elledge, S. J., and Harper, J. W. (2003). BTB proteins are substrate-specific adaptors in an SCF-like modular ubiquitin ligase containing CUL–3. *Nature* **425,** 316–321.

Zheng, N., Schulman, B. A., Song, L., Miller, J. J., Jeffrey, P. D., Wang, P., Chu, C., Koepp, D. M., Elledge, S. J., Pagano, M., Conaway, R. C., Conaway, J. W., Harper, J. W., and Pavletich, N. P. (2002). Structure of the Cul1-Rbx1-Skp1-F boxSkp2 SCF ubiquitin ligase complex. *Nature* **416,** 703–709.

# [13] *In Vitro* Reconstitution of SCF Substrate Ubiquitination with Purified Proteins

*By* MATTHEW D. PETROSKI and RAYMOND J. DESHAIES

## Abstract

The development of *in vitro* systems to monitor ubiquitin ligase activity with highly purified proteins has allowed for new insights into the mechanisms of protein ubiquitination to be uncovered. This chapter describes the methodologies employed to reconstitute ubiquitination of the budding yeast cyclin-dependent kinase inhibitor Sic1 by the evolutionarily conserved ubiquitin ligase $SCF^{Cdc4}$ and its ubiquitin-conjugating enzyme Cdc34. Based on our experience in reconstituting Sic1 ubiquitination, we suggest some parameters to consider that should be generally applicable to the study of different SCF complexes and other ubiquitin ligases.

## Introduction

The development of *in vitro* systems with highly purified proteins allows for the molecular mechanisms of the steps involved in the ubiquitination of a protein to be directly explored and dissected. We have developed such a system to study the ubiquitination of the budding yeast cyclin-dependent kinase (CDK) inhibitor Sic1 by the ubiquitin ligase $SCF^{Cdc4}$ and the ubiquitin-conjugating enzyme Cdc34. Whereas genetic studies initially identified some of the proteins involved in this process, it was not until biochemical approaches were developed and applied that all of the proteins necessary and sufficient for Sic1 ubiquitination were identified (Feldman *et al.*, 1997; Kamura *et al.*, 1999; Seol *et al.*, 1999; Skowyra *et al.*, 1997, 1999; Verma *et al.*, 1997b).

Sic1 binds to and inhibits the activity of the S-phase cyclin-dependent kinase complex (S-CDK) during the G1 phase of the budding yeast cell cycle (Mendenhall, 1993; Schwob *et al.*, 1994). As the level of G1 phase CDK (G1-CDK) activity increases, Sic1 becomes multiply phosphorylated at consensus CDK sites (Nash *et al.*, 2001; Verma *et al.*, 1997a). Phosphorylated Sic1 is recognized by $SCF^{Cdc4}$ and becomes ubiquitinated through the activity of Cdc34 (Verma *et al.*, 1997b). Once ubiquitinated, Sic1 is degraded rapidly by the 26S proteasome, resulting in activation of S-CDK and allowing for progression into the S phase of the cell cycle (Verma *et al.*, 2001).

METHODS IN ENZYMOLOGY, VOL. 398

0076-6879/05 $35.00
DOI: 10.1016/S0076-6879(05)98013-0

Sic1 destruction at the G1–S phase boundary represents an important model system for understanding the mechanisms of protein ubiquitination and degradation. *In vitro* reconstitution of this process has been used successfully to study how SCF recognizes phosphorylated substrates (Nash *et al.*, 2001; Orlicky *et al.*, 2003; Verma *et al.*, 1997a), the mechanisms of ubiquitin transfer through SCF (Deffenbaugh *et al.*, 2003), the minimum set of components that can sustain Sic1 turnover and S-CDK activation (Verma *et al.*, 2001), the number and position of ubiquitin chains that can promote 26S proteasome-mediated degradation (Petroski and Deshaies, 2003), and the role of targeting factors in guiding ubiquitinated Sic1 to the 26S proteasome (Verma *et al.*, 2004). As a result, SCF is perhaps the best understood member of the RING class of ubiquitin ligases. These ligases, in contrast with HECT domain-containing ligases, do not form a catalytic intermediate with ubiquitin, but instead may activate through the RING motif the direct transfer of ubiquitin from the active site cysteine of the ubiquitin-conjugating enzyme onto the lysine of the ubiquitin ligase-bound substrate. RING proteins are probably the most common type of ubiquitin ligase, but the molecular basis of how they work remains largely unknown.

SCF consists of Skp1, Cul1, a variable F-box protein that confers substrate specificity, and the RING-H2 subunit known alternatively as Hrt1, Rbx1, or Roc1 (for a review, see Petroski and Deshaies, 2005). SCF has been implicated in many aspects of cellular and organismal homeostasis, including cell cycle control, transcription, development, and circadian clock regulation. Cul1 is a member of the cullin family of proteins, of which at least five of the mammalian family members have been shown to form similar ubiquitin ligases with an enzymatic core that is targeted to different substrates by variable substrate recognition components. It seems highly likely, therefore, that paradigms established by studying how ubiquitin transfer through SCF works will apply not only to other cullin-based ubiquitin ligases, but also to RING-based ligases in general.

This chapter describes the strategies and methods developed to study the ubiquitination of Sic1 by $SCF^{Cdc4}$. It also discusses efforts to develop substrates containing defined ubiquitin-accepting sites, which are useful for studies aimed at unraveling the mechanism of ubiquitination (see Petroski and Deshaies, 2003). We describe the methods we have developed for the expression and purification of proteins involved in this process and the reaction conditions we use to reconstitute the ubiquitination of Sic1 by $SCF^{Cdc4}$ *in vitro*. We expect that these approaches will be adaptable and applicable to the study of substrates of other ubiquitin ligases.

## General Considerations for the Development of Ubiquitin Ligase Substrates

When developing an *in vitro* assay to examine the ubiquitination of a particular protein by its ubiquitin ligase, it is important to consider several parameters likely to have a significant impact on the probability of success. Many ubiquitin ligase substrates (particularly those of cullin-based ligases) require some sort of posttranslational modification, such as phosphorylation for recognition. This will influence the choice of expression system used to generate sufficient quantities of the substrate of interest. Perhaps the easiest and most robust approach to generate large quantities of protein is by bacterial expression. However, this system has the drawback of lacking eukaryotic posttranslational modifications. While baculovirus expression systems may require slightly more expertise and considerably more time, posttranslational modifications may occur during expression in insect cells, potentially eliminating the requirement for pretreating substrates prior to use in ubiquitination assays. This may be particularly advantageous if the proteins involved in the modification are unknown. However, if degradation-promoting modifications occur in insect cells, the substrate may be degraded if its cognate ubiquitin ligase is present. Thus, in some circumstances, it may be necessary to apply chemical inhibitors of the proteasome several hours prior to harvesting cells. Other eukaryotic systems such as yeast or *in vitro* transcription/translation with reticulocyte lysates (e.g., see Verma *et al.*, 1997a) are possible alternatives to insect cells.

In attempting to elucidate sites of ubiquitin attachment on a substrate, it is important to note that some substrates may be ubiquitinated at the N terminus instead of exclusively at lysine residues (see, for example, Aviel *et al.*, 2000; Bloom *et al.*, 2003; Fajerman *et al.*, 2004; Ikeda *et al.*, 2002; Reinstein *et al.*, 2000). In our efforts to identify the sites of Sic1 ubiquitination, we mutated all 20 lysine residues of Sic1 to arginine (Petroski and Deshaies, 2003). This lysine-less version of Sic1 is stable *in vivo* and cannot be ubiquitinated *in vitro* by $SCF^{Cdc4}$, suggesting that N-terminal ubiquitination of Sic1 does not occur. During the course of these studies, we also determined that it is important to ensure that fusion proteins or epitope tags used to facilitate purification and/or detection do not contain lysines, as these introduced residues may become sites of ubiquitin attachment which can, in turn, complicate analyses.

Traditionally, ubiquitination sites are mapped by mutating lysine residues to arginine. Unfortunately, several of the tRNAs for arginine are exceptionally rare in *Escherichia coli* (see Table I), which may have the effect of reducing expression of the mutated protein or result in an

TABLE I
FREQUENCY OF CODON USAGE FOR LYSINE AND ARGININE tRNAS IN *E. coli*[a]

| Amino acid | Codon | Frequency[b] |
|---|---|---|
| Lysine | AAG | 0.24 |
| | AAA | 0.76 |
| Arginine | AGG | 0.02 |
| | AGA | 0.04 |
| | CGA | 0.06 |
| | CGG | 0.10 |
| | CGU | 0.38 |
| | CGG | 0.40 |

[a] Arginine codons that are utilized infrequently (less than 10% of total usage) are indicated in bold. It is advisable when mutating lysine residues to arginine to introduce CGG or CGT instead of sequences corresponding to the four rare arginine codons or to use bacterial strains that may correct for low-level expression of these tRNAs.
[b] From Nakamura *et al.* (2000).

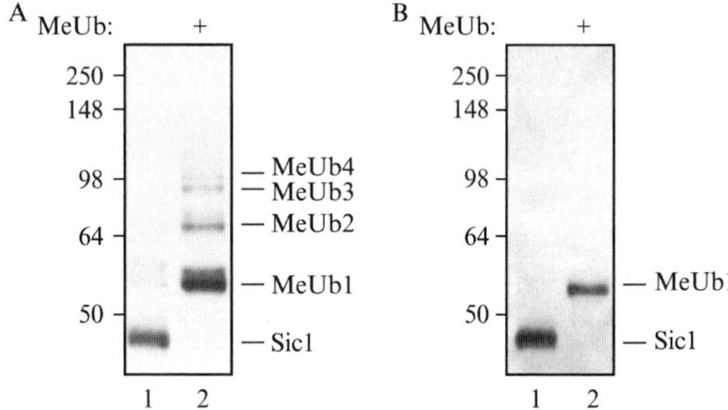


FIG. 1. Influence of rare codon usage on the ubiquitination pattern of *E. coli* expressed Sic1. (A) Sic1 containing a single lysine residue (K32) was expressed and purified from the BL21(DE3)pLyS *E. coli* strain (Novagen), phosphorylated, and utilized in an *in vitro* ubiquitination reaction containing Uba1, Cdc34, $SCF^{Cdc4}$, methyl ubiquitin, and ATP. Lane 1 is the reaction performed without added MeUb and lane 2 is the complete reaction. The number of methyl ubiquitin (MeUb) attachments onto Sic1 is indicated. As lysine residues are required for Sic1 destabilization *in vivo* and this substrate contains only a single lysine (K32) that confers destabilization *in vivo* (Petroski and Deshaies, 2003), secondary sites of MeUb attachment presumably arise from misincorporation of extra (i.e., not genetically encoded) lysines into Sic1 during *E. coli* expression. (B) Sic1 containing a single lysine residue was expressed and purified from BL21-Codon Plus *E. coli* strain (Stratagene) and utilized in an *in vitro* ubiquitination reaction as in A. Note that only a single MeUb was incorporated into this substrate, suggesting that expression of the tRNAs for rare arginine codons in this strain circumvented the misincorporation problem shown in A.

unacceptably high rate of misincorporation of amino acids when using conventional bacterial expression systems (You *et al.*, 1999). The latter problem is highlighted by analysis of a Sic1 substrate containing a single lysine residue that was expressed in BL21(DE3)pLysS and used in ubiquitination assays employing methyl ubiquitin (Fig. 1A). Surprisingly, multiple ubiquitinated species were detected, which indicates the presence of more than one ubiquitin attachment site. Possible solutions to this problem include minimizing protein induction times or utilizing *E. coli* strains (e.g., the Rosetta strains from Novagen or BL21-Codon Plus strains from Stratagene) that compensate for these rare tRNAs by harboring extra copies of these genes. As shown in Fig. 1B, expression of single lysine Sic1 in BL21-Codon Plus-RIL eliminated the problem illustrated in Fig. 1A.

It is also worth noting that preexisting knowledge might reduce the amount of effort required to identify potential targets for ubiquitin conjugation on a substrate. For example, Sic1 is bound to the S-CDK protein complex prior to its degradation (Nugroho and Mendenhall, 1994). While "naked" Sic1 can be ubiquitinated on a large fraction of its 20 lysine residues, this number is reduced significantly when Sic1 is assembled into its "physiological" context prior to its ubiquitination *in vitro* (see Petroski and Deshaies, 2003) (Fig. 2).

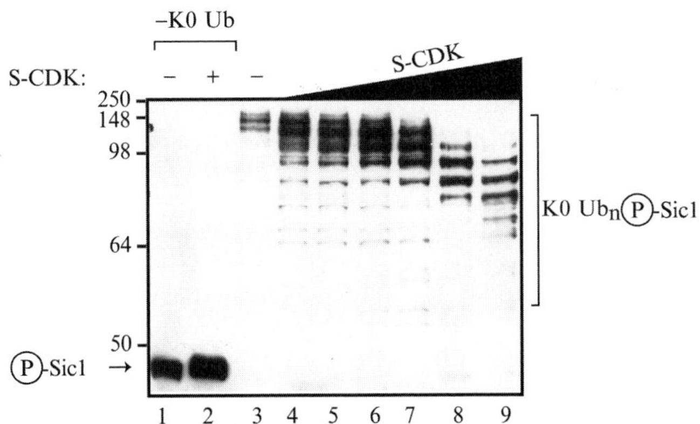


Fig. 2. Suppression of sites of Sic1 ubiquitination by the presence of S-CDK. Sic1 was preincubated in the absence (–S-CDK) or presence (+S-CDK) of purified S-CDK prior to addition to ubiquitination reactions containing Uba1, Cdc34, $SCF^{Cdc4}$, K0 ubiquitin, and ATP. The relative position of Sic1 and the position of K0 ubiquitin (K0 Ub) attachments onto Sic1 are indicated. Note that Sic1 in the presence of saturating amounts of S-CDK is ubiquitinated on up to six or seven sites whereas considerably more sites are utilized in the absence of S-CDK. Reprinted from Petroski and Deshaies (2003), with permission.

## Site-Directed Mutagenesis of the Sic1 Open Reading Frame

Standard cloning procedures are used to generate a bacterial expression construct containing the Sic1 open reading frame in pET11b (Novagen). Sequences for an N-terminal T7 epitope tag (MASMTGGQQMG, Novagen) and a C-terminal hexa-histidine were also included. To generate various lysine to arginine mutations, this plasmid is subjected to site-directed mutagenesis following the Quikchange protocol (Stratagene). In general, oligonucleotides are designed such that the desired mutation is centered on the two complementary oligonucleotides with approximately 15 nucleotides on both sides. Typical polymerase chain reaction (PCR) conditions are used with a 68° extension for 2 min per kilobase of plasmid and a total of 15 cycles.

To ensure high mutagenesis efficiency, it is important to optimize the amount of input plasmid to use as little as possible. It is also possible to introduce multiple mutations simultaneously, either by introducing them on the same oligonucleotides if the sites are sufficiently close together or by using multiple oligonucleotides in the same reaction (Quikchange multisite-directed mutagenesis, Stratagene). Introduction of the desired mutation is confirmed by DNA sequencing.

## Expression and Purification of Proteins

### *Uba1 Expression and Purification from Yeast*

Both yeast and human E1 are available for purchase from Boston Biochem and Affiniti. Alternatively, baculoviruses for insect cell expression or purification from rabbit reticulocyte lysates have been utilized as E1 sources.

1. Grow an overnight culture of yeast strain RJD941 in YPD. This strain contains *Saccharomyces cerevisiae UBA1* encoding a hexa-histidine C-terminal tag under control of the copper-inducible *CUP1* promoter.
2. Inoculate 1 liter of YPD with 4 ml overnight culture.
3. Grow to saturation (at 36 h at 30°).
4. Add copper sulfate to a final concentration of 1 m*M*.
5. Grow an additional 24 h.
6. Pellet cells and wash once with water.
7. Resuspend in 25 ml buffer A (20 m*M* Tris-HCl, pH 7.6, 0.5 *M* NaCl, 10 m*M* imidazole, 5 m*M* $MgCl_2$, 10% glycerol).
8. Lyse cells in Constant Systems One Shot at 25 kpsi (mechanical grinding under liquid nitrogen in a mortar and pestle also works).

9. Pellet insoluble material at 50,000*g* for 30 min.
10. Add supernatant to 3 ml NiNTA agarose (Qiagen) and incubate at 4° for 1 h.
11. Wash four times with buffer A + 0.25% Triton X-100 (usually done in batch format, although column chromatography works as well).
12. Elute three times with 10 ml buffer B (20 m*M* Tris-HCl, pH 7.6, 200 m*M* NaCl, 0.5 *M* imidazole, 5 m*M* $MgCl_2$). Each elution is done in batch for 3 min at room temperature. Usually, the majority of E1 comes off in the first elution.
13. Dialyze eluate against 30 m*M* Tris-HCl, pH 7.6, 100 m*M* NaCl, 5 m*M* $MgCl_2$, 1 m*M* DTT, 15% glycerol (1 liter buffer with two changes, first one usually overnight).
14. Concentrate to desired concentration (greater than 1 mg/ml) in a Biomax (Millipore) centrifugal filter device. (If higher purity is desired, further purification steps such as ubiquitin affinity, gel filtration, or ion-exchange chromatography may be used. However, this single-step procedure generally results in >90% purity.) Typically, 2 mg of E1 is recovered per liter of culture from the NiNTA purification. Store at –80° in small aliquots. A Uba1 preparation is shown in Fig. 3A.

*Cdc34 Expression and Purification from Bacteria*

This protocol is used routinely to express and purify hexa-histidine-tagged proteins from *E. coli.*

The *S. cerevisiae CDC34* open reading frame appended to sequences encoding a C-terminal hexa-histidine tag was cloned by standard techniques into pET11b, transformed into BL21(DE3)pLysS (Novagen), and transformants were selected on LB plates containing ampicillin (50 $\mu$g/ml) and chloramphenicol (25 $\mu$g/ml).

1. Inoculate a 5-ml culture of LB medium containing ampicillin (50 $\mu$g/ml) and chloramphenicol (25 $\mu$g/ml) with a single colony from a fresh transformation and grow overnight at 37° with 200 rpm shaking.
2. Inoculate 1 liter of 2 XYT containing ampicillin (50 $\mu$g/ml) and chloramphenicol (35 $\mu$g/ml) with 5 ml of the overnight culture.
3. Grow to $OD_{600}$ of 1 at 30° with 200 rpm shaking and induce with 0.8 m*M* isopropyl-$\beta$-D-thiogalactoside for 2.5 h.
4. Pellet cells and wash once with 25 m*M* Tris-HCl, pH 7.6.

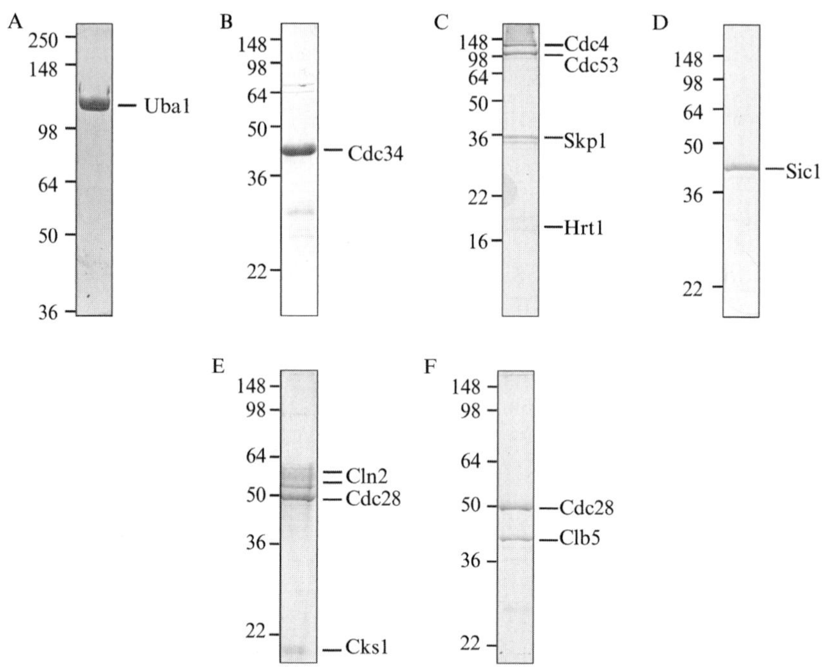


FIG. 3. Purified reaction components for Sic1 ubiquitination assays. (A) Uba1 was purified from yeast strain RJD941 by NiNTA chromatography. This strain expresses hexahistidine-tagged Uba1 under control of a copper-inducible promoter. A total of 2 $\mu$g was loaded and run on an 8% SDS-containing polyacrylamide gel followed by Coomassie staining. (B) Cdc34 was purified from *E. coli* strain BL21(DE3)pLysS transformed with a pET11b plasmid that contains the *CDC34* open reading frame fused to sequences that encode a C-terminal hexahistidine tag. A Coomassie-stained 12% polyacrylamide gel containing SDS loaded with approximately 2 $\mu$g is shown. (C) $SCF^{Cdc4}$ was expressed in Hi5 insect cells by coinfection of baculoviruses encoding Py3HA2-tagged Cdc4, Cdc53, HA-tagged Skp1, and Hrt1. Forty-eight hours postinfection, lysates were prepared and $SCF^{Cdc4}$ was purified on protein A-Sepharose beads that had been coupled with anti-Py antibodies. Approximately 1 $\mu$g of total protein was loaded onto a 15% gel and analyzed by Coomassie staining. (D) After expression in BL21-Codon Plus *E. coli*, a mutant of Sic1 that contains a single lysine (K32) was purified on NiNTA agarose. Approximately 1 $\mu$g from a typical purification that was analyzed on a 12% SDS–polyacrylamide gel by Coomassie staining is shown. (E) G1-CDK was prepared from Hi5 insect cells that were coinfected with baculoviruses encoding GST-Cdc28HA, Myc-tagged Cln2, His6-tagged Cks1, and Cak1. The complex was purified on glutathione-Sepharose from lysates prepared after 48 h of infection. The resulting material (approximately 1 $\mu$g of total protein) was analyzed by SDS–PAGE on a 12% gel followed by Coomassie staining. (F) S-CDK was purified by glutathione-Sepharose chromatography from Hi5 insect cells that were coinfected with baculoviruses encoding GST-Cdc28HA and Clb5. A Coomassie-stained 12% polyacrylamide gel loaded with approximately 1 $\mu$g of total protein is shown.

5. Resuspend pellet in 20 ml of 25 m*M* Tris-HCl, pH 7.6, and add imidazole to 10 m*M*, NaCl to 0.5 *M*, and Triton X-100 to 0.2%.
6. Freeze in liquid nitrogen.
7. Thaw to lyse cells (expression of T7 lysozyme from pLysS leads to cell lysis upon freeze–thaw) and sonicate briefly to reduce lysate viscosity.
8. Clarify lysate by centrifugation.
9. Add to 1 ml NiNTA agarose (Qiagen) and incubate at 4° for 30 min. (All purification steps are done in batch.)
10. Wash resin four times with 25 m*M* Tris-HCl, pH 7.6, 0.5 *M* NaCl, 0.2% Triton X-100, 10 m*M* imidazole, and 10% glycerol.
11. Wash resin four times with 20 m*M* Tris-HCl, pH 7.6, 0.5 *M* NaCl. (An ATP wash step for 5 min at room temperature with mixing of 20 m*M* Tris-HCl, pH 7.6, 2 m*M* ATP, and 5 m*M* $MgCl_2$ may reduce the amount of contaminating heat shock proteins, if desired.)
12. Elute three times for 2 min each with 3 ml 20 m*M* Tris-HCl, pH 7.6, 0.5 *M* NaCl, 0.5 *M* imidazole.
13. Check eluate for protein by SDS–PAGE and pool fractions.
14. Dialyze against 30 m*M* Tris-HCl, pH 7.6, 200 m*M* NaCl, 5 m*M* $MgCl_2$, 1 m*M* dithiothreitol (DTT), 10% glycerol (1 liter buffer, with two changes).

From a 1-liter culture, at least 50 mg of Cdc34 is recovered. A typical Cdc34 preparation is shown in Fig. 3B.

### *SCF Expression and Purification from Baculovirus-Infected Insect Cells*

To generate $SCF^{Cdc4}$ in sufficient quantities for ubiquitination assays, we employ the baculovirus/insect cell expression system, as it allows for multiple proteins to be expressed simultaneously and relative expression levels of SCF subunits can be manipulated by altering the amount of the various viruses added to insect cells.

1. Baculoviruses containing the various open reading frames of *S. cerevisiae* $SCF^{Cdc4}$ components (Skp1, Cdc53, Hrt1, and Cdc4) are prepared according to the manufacturer's instructions (Invitrogen). For ease of purification, *CDC4* contains sequences encoding an N-terminal tag comprising two repeats of the polyoma (Py, EYMPME) epitope followed by three repeats of the hemagglutinin (HA, YPYDVPDYA ) tag. Alternative epitope tags may be used. The baculovirus encoding Skp1 also contains a C-terminal HA tag, although this is not used for purification.

2. Generate high-titer stocks of the individual viruses by infecting 50 ml of Sf9 insect cells (*Spodoptera frugiperda*, $1 \times 10^6$ cells/ml, 200-ml

spinner flasks) with 0.5 ml low titer stock. Grow for 3 to 4 days until cells are approximately 80% lysed as judged by trypan blue exclusion.

3. Seed Hi5 insect cells (Invitrogen) onto 10-cm$^2$ plates at $1 \times 10^7$ and allow cells to adhere.

4. Add high-titer virus stocks in appropriate ratios. (For SCF$^{Cdc4}$, 0.6 ml Py2HA3-Cdc4, 0.6 ml Cdc53, 0.3 ml Skp1-HA, and 0.3 ml Hrt1 are typically used. This was determined empirically to give the best yield of intact SCF$^{Cdc4}$.)

5. Incubate infected cells for 48 h at 27°.

6. Remove cells from plates and wash with phosphate-buffered saline (PBS).

7. Resuspend cell pellets in cold Sf9 lysis buffer (25 m*M* HEPES, pH 7.4, 250 m*M* NaCl, 0.2% Triton X-100, 2 m*M* EDTA, 1 m*M* DTT, 5 m*M* $MgCl_2$, 10% glycerol) using 2 ml buffer per 10-cm$^2$ plate.

8. Sonicate briefly and pellet cellular debris. At this point, the clarified lysates can be stored indefinitely at –80° or used directly to purify SCF$^{Cdc4}$.

9. Add lysates to protein A beads coupled to the anti-Py monoclonal antibody. Typically, 1 ml of anti-Py affinity resin is incubated with 10 ml lysate. Covalently coupled antibody-protein A beads are prepared as described (Harlow and Lane, 1988).

10. Incubate 3 h at 4°.

11. Wash three times with Sf9 lysis buffer. Wash four times with Py wash buffer (20 m*M* HEPES, pH 7.4, 100 m*M* NaCl, 0.5% Igepal CA-630, 1 m*M* EDTA, 1 m*M* DTT, 10% glycerol).

12. Elute 2 h at room temperature with 3 ml 100 $\mu$g/ml Py peptide (EYMPME) in Py wash buffer containing 300 m*M* NaCl, 0.1% *n*-octylglucoside, and 20 $\mu$g/ml arg-insulin. Elution efficiencies are typically 50 to 70%. A second elution may help increase yield. Beads can be reused numerous times (at least three) after stripping with 0.1 *M* glycine, pH 2, and reequilibration with Sf9 lysis buffer. (If soluble SCF is not required or desired, bead-bound SCF can be utilized. In this instance, the protocol can be terminated prior to washing with Py wash buffer and the beads can be equilibrated in ubiquitination reaction buffer prior to use.)

13. Concentrate SCF in a Biomax (Millipore) centrifugal filtration device. As 100 n*M* SCF is typically used in a 20-$\mu$l ubiquitination reaction, 2 $\mu M$ is a desired final concentration. While it is difficult to get an exact concentration of fully assembled SCF due to variations in the stoichiometry of the possible complexes that may form, reasonable approximations can be made by performing a protein concentration assay and by Western blotting of gels containing titrations of the purified SCF to various known concentrations of purified subunits (e.g., individual subunits expressed and purified in *E. coli*). It is reasonable to expect 250 ng to 1 $\mu$g of SCF per 10-cm$^2$ plate by this method.

14. Dialyze eluted material against 30 m*M* Tris-HCl, pH 7.6, 200 m*M* NaCl, 5 m*M* $MgCl_2$, 2 m*M* DTT, and 15% glycerol after concentrating to desired volume and store at –80° in small aliquots.

A typical preparation of eluted $SCF^{Cdc4}$ is shown in Fig. 3C.

*Sic1 Expression and Purification from Bacteria*

1. Transform a pET11b plasmid that encodes hexahistidine-tagged Sic1 into the bacterial strain BL21-Codon Plus-RIL (Stratagene) and plate cells onto LB medium with ampicillin (50 $\mu$g/ml) and chloramphenicol (25 $\mu$g/ml). Alternative strains that also contain genes that encode the rare tRNAs in *E. coli* such as Rosetta (Novagen) can also be used. This is important due to the high number of lysine→arginine replacements introduced into Sic1 mutants that contain a single lysine.
2. For expression and purification, follow the protocol described earlier for Cdc34.
3. Instead of dialysis, purify the eluted protein from NiNTA agarose by gel filtration on a PD10 column equilibrated with kinase buffer (20 m*M* HEPES, pH 8, 200 m*M* NaCl, 5 m*M* $MgCl_2$, 1 m*M* EDTA, and 1 m*M* DTT). Sic1 proteins containing a large number of lysine→arginine mutations tend to precipitate during dialysis. It was determined empirically that using gel filtration for buffer exchange circumvents this problem.

Sic1 purified by this protocol is shown in Fig. 3D. Wild-type Sic1 yields approximately 50 mg per liter of culture, whereas Sic1 containing all lysines mutated to arginine yields approximately 5 mg per liter. Recovery of Sic1 containing other lysine→arginine mutations is within this range. Proteins are stored in small aliquots at –80°.

*G1-CDK and S-CDK Expression and Purification from Baculovirus-Infected Insect Cells*

Baculoviruses are prepared to express the subunits of G1-CDK and S-CDK complexes in insect cells following the manufacturer's instructions (Invitrogen). High-titer stocks of viruses encoding G1-CDK proteins (GST-Cdc28HA, Cln2, Cak1, Cks1-His6) and S-CDK proteins (GST-Cdc28HA and Clb5) are generated as described for $SCF^{Cdc4}$.

1. Seed Hi5 insect cells (Invitrogen) onto 10-$cm^2$ plates at $2 \times 10^7$ and allow cells to adhere.
2. Add high-titer virus stocks in appropriate ratios. (For G1-CDK, 0.6 ml GST-Cdc28HA, 2.4 ml Cln2, 0.3 ml Cak1, and 0.3 ml Cks1-His6 are typically used. For S-CDK, 0.6 ml GST-Cdc28HA and 1.8 ml Clb5 are used.)

3. Incubate infected cells for 48 h at 27°.

4. Remove cells from plates and wash with PBS.

5. Resuspend cell pellets in cold Sf9 lysis buffer (25 m*M* HEPES, pH 7.4, 250 m*M* NaCl, 0.2% Triton X-100, 2 m*M* EDTA, 1 m*M* DTT, 5 m*M* $MgCl_2$, 10% glycerol) using 2 ml buffer per 10-$cm^2$ plate.

6. Sonicate briefly and pellet cellular debris. At this point, the clarified lysates can be stored indefinitely at –80° or used directly to purify G1-CDK or S-CDK.

7. Add 25 μl of glutathione-Sepharose (Pharmacia) to 1 ml of the lysate and incubate for 1 h at 4° with mixing.

8. Wash four times with Sf9 lysis buffer. (At this point, bead-bound complexes can be utilized if desired.)

9. Elute twice with 50 μl 500 m*M* Tris-HCl, pH 8, 100 m*M* glutathione, 250 m*M* NaCl, 0.25% Triton X-100, and 2 mM DTT. (The high concentration of Tris-HCl in the elution buffer is to prevent acidification caused by the high concentration of glutathione used. This glutathione concentration was determined empirically to be optimal for eluting these complexes off of glutathione-Sepharose.)

10. Check eluate and dialyze against 30 m*M* Tris-HCl, pH 8, 150 m*M* NaCl, 5 m*M* $MgCl_2$, 2 m*M* DTT, and 15% glycerol (two changes of 1 liter each). Resulting material can be concentrated if desired in a Biomax (Millipore) centrifugal device. Typically, 0.25 mg of G1-CDK and 2 mg of S-CDK are recovered per 10-$cm^2$ plate.

Purified G1-CDK is shown in Fig. 3E and S-CDK is shown in Fig. 3F. Small aliquots are stored at –80°.

## Phosphorylation of Sic1

Sic1 must be phosphorylated by G1-CDK for its recognition by $SCF^{Cdc4}$ and ubiquitination by Cdc34.

1. Immobilize G1-CDK on glutathione-Sepharose beads. Usually, 10 to 15 μl of glutathione-Sepharose (Pharmacia) is used per 500 μl lysate. Incubate 1 h at 4° with mixing.

2. Wash four times with Sf9 lysis buffer. Wash three times with kinase buffer (20 m*M* HEPES, pH 8, 200 m*M* NaCl, 5 m*M* $MgCl_2$, 1 m*M* EDTA, and 1 m*M* DTT).

3. Assemble kinase reaction. To glutathione beads, add Sic1 diluted in kinase buffer to 88 μl. Several hundred picomoles of Sic1 can be added with little effect on the efficiency of phosphorylation. Add 2 μl of 100 m*M* ATP. (If radio-labeled Sic1 is desired, substitute 2 μl of [$\gamma$-$^{32}$P]ATP (4500 Ci/mmol) and 1 μl of 1 m*M* ATP instead of 2 μl of 100 m*M* ATP.)

4. Incubate 1 h at room temperature with mixing. Add 1 μl 100 m*M* ATP and incubate another hour at room temperature.

5. Pellet beads and keep supernatant. Under these conditions, essentially all of the substrate is converted to a phosphorylated form (as judged by its change in mobility compared to unphosphorylated Sic1 by SDS–PAGE and Western blotting). See Fig. 4 for an example.

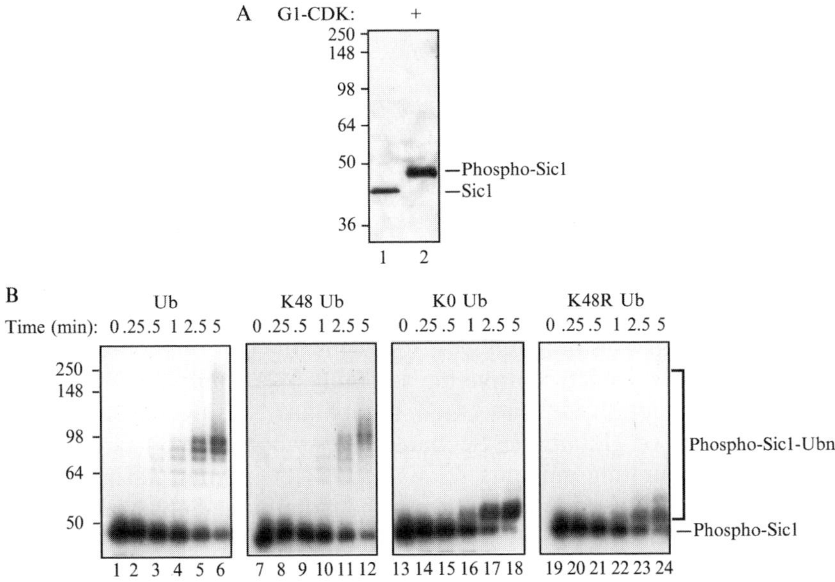


FIG. 4. Ubiquitination of Sic1 with various ubiquitin derivatives. (A) Sic1 was phosphorylated by glutathione-Sepharose-bound G1-CDK in the presence of 2 m*M* ATP. After incubating at room temperature for 2 h, the resulting material was analyzed by SDS–PAGE and Western blotting with antisera against the T7 tag, which is the N-terminal epitope tag on this form of Sic1. The relative position of unphosphorylated Sic1 (lane 1) and phosphorylated Sic1 (lane 2) is shown. (B) *In vitro* ubiquitination assays were performed with purified reaction components and several different ubiquitin derivatives. $^{32}$P-labeled Sic1 (K36 only, 2.5 μ*M*) was premixed with SCF$^{Cdc4}$ (100 n*M*) prior to addition to reactions containing Uba1 (150 n*M*), Cdc34 (800 n*M*), ATP (2 m*M*), and the various ubiquitin derivatives indicated (80 μ*M*). After mixing, aliquots were taken and added to an equal volume of sample buffer at the various times indicated. The resulting reaction products were analyzed by SDS–PAGE, followed by autoradiography. K48 Ub is ubiquitin containing all lysines mutated to arginine except lysine 48, K0 Ub lacks all lysine residues, and K48R ubiquitin has lysine 48 mutated to arginine. The relative positions of phosphorylated Sic1 and its ubiquitinated forms are indicated.

### Sic1 Ubiquitination

After phosphorylation, Sic1 can be added to ubiquitination reactions containing ATP, ubiquitin, Uba1, Cdc34, and $SCF^{Cdc4}$. Perhaps the most difficult aspect of this assay is that multiple proteins are required and all should be titrated individually to determine optimal reaction conditions. Furthermore, buffer conditions and protein purity may have a severe impact on the "quality" of ubiquitination. The conditions described here have been determined empirically to be optimal for Sic1 ubiquitination and may serve as a starting point for substrates of other ubiquitin ligases. It is also important to note that the inclusion of NaCl in these reactions suppresses nonspecific phosphorylation-independent ubiquitination, but may reduce overall reaction efficiency.

#### *Assemble Sic1 into a Complex with S-CDK if Desired*

Binding of S-CDK to the C-terminal domain of Sic1 suppresses spurious ubiquitination on C-terminal lysines. Genetic analysis suggests that ubiquitination of Sic1 on C-terminal lysines is not physiologically relevant, and *in vitro* studies indicate that ubiquitin chains conjugated to the C-terminal domain of Sic1 do not sustain rapid degradation by the proteasome (Petroski and Deshaies, 2003), and thus we have sought to minimize this side reaction. It is probably best to empirically determine the amount of S-CDK that is needed to drive the assembly of Sic1 into heterotrimeric complexes. Alternatively, one could immobilize S-CDK on an affinity support such as on glutathione-Sepharose (described in detail earlier) and add Sic1 in saturating amounts relative to S-CDK. After washing away unbound Sic1, the eluted material from the glutathione-Sepharose should contain stoichiometric complexes of GST-Cdc28, Clb5, and Sic1.

#### *Ubiquitination Assay*

Prepare 10× ubiquitination reaction buffer (300 m*M* Tris-HCl, pH 7.6, 50 m*M* $MgCl_2$, 1 *M* NaCl, 20 m*M* ATP, and 20 m*M* DTT). Usually, reaction components are added to 10× ubiquitination reaction buffer and water to yield 1× concentration at the final volume. A typical 20-$\mu$l reaction consists of the following components added in this order:

- 2 $\mu$l 10× ubiquitination reaction buffer
- 13.2 $\mu$l $H_2O$
- 0.5 $\mu$l ubiquitin (80 $\mu M$)
- 0.3 $\mu$l Uba1 (150 n*M*)
- 1 $\mu$l Cdc34 (variable concentration, >500 n*M* with 100 n*M* SCF will result in rapid and robust ubiquitin conjugation in less than 10 min)

1 μl SCF (variable concentration, typically 100 n*M*)
2 μl phosphorylated Sic1 +/− S-CDK (variable concentration, typically in ~20-fold excess of SCF, 2 μ*M*).

Incubate at room temperature after gentle mixing. The amount of ubiquitinated Sic1 can be monitored over time by removing aliquots and adding to SDS–PAGE sample buffer prior to gel electrophoresis. Depending on how Sic1 is prepared, Western blotting using antisera against Sic1 or its epitope tags or autoradiography of gels containing radio-labeled Sic1 can be used to analyze the reaction products. See Figs. 1 and 2 for examples of Sic1 ubiquitination assays.

Different mutant or chemically modified ubiquitins are available commercially from Boston Biochem and Affiniti. Chain-terminating forms of ubiquitin such as K0 ubiquitin or methyl ubiquitin can be used to determine how many lysine residues are targeted for ubiquitin conjugation (e.g., see Fig. 4).

## Acknowledgments

M. D. P. is an associate and R. J. D. is an assistant investigator of the Howard Hughes Medical Institute. We thank R. Verma, R. Feldman, C. Correll, J. Seol, and members of the Deshaies laboratory for helpful discussions during the course of this work.

## References

Aviel, S., Winberg, G., Massucci, M., and Ciechanover, A. (2000). Degradation of the Epstein-Barr virus latent membrane protein 1 (LMP1) by the ubiquitin-proteasome pathway: Targeting via ubiquitination of the N-terminal residue. *J. Biol. Chem.* **275,** 23491–23499.

Bloom, J., Amador, V., Bartolini, F., De Martino, G., and Pagano, M. (2003). Proteasome-mediated degradation of p21 via N-terminal ubiquitinylation. *Cell* **115,** 71–82.

Deffenbaugh, A. E., Scaglione, K. M., Zhang, L., Moore, J. M., Buranda, T., Sklar, L. A., and Skowyra, D. (2003). Release of ubiquitin-charged Cdc34-S - Ub from the RING domain is essential for ubiquitination of the SCF(Cdc4)-bound substrate Sic1. *Cell* **114,** 611–622.

Fajerman, I., Schwartz, A. L., and Ciechanover, A. (2004). Degradation of the Id2 developmental regulator: Targeting via N-terminal ubiquitination. *Biochem. Biophys. Res. Commun.* **314,** 505–512.

Feldman, R. M., Correll, C. C., Kaplan, K. B., and Deshaies, R. J. (1997). A complex of Cdc4p, Skp1p, and Cdc53p/cullin catalyzes ubiquitination of the phosphorylated CDK inhibitor Sic1p. *Cell* **91,** 221–230.

Harlow, E., and Lane, D. (1988). "Antibodies : A Laboratory Manual." Cold Spring Harbor Laboratory, Cold Spring Harbor, NY.

Ikeda, M., Ikeda, A., and Longnecker, R. (2002). Lysine-independent ubiquitination of Epstein-Barr virus LMP2A. *Virology* **300,** 153–159.

Kamura, T., Koepp, D. M., Conrad, M. N., Skowyra, D., Moreland, R. J., Iliopoulos, O., Lane, W. S., Kaelin, W. G., Jr., Elledge, S. J., Conaway, R. C., Harper, J. W., and Conaway, J. W. (1999). Rbx1, a component of the VHL tumor suppressor complex and SCF ubiquitin ligase. *Science* **284,** 657–661.

Mendenhall, M. D. (1993). An inhibitor of p34CDC28 protein kinase activity from Saccharomyces cerevisiae. *Science* **259,** 216–219.

Nakamura, Y., Gojobori, T., and Ikemura, T. (2000). Codon usage tabulated from international DNA sequence databases: Status for the year 2000. *Nucleic Acids Res.* **29,** 292.

Nash, P., Tang, X., Orlicky, S., Chen, Q., Gertler, F. B., Mendenhall, M. D., Sicheri, F., Pawson, T., and Tyers, M. (2001). Multisite phosphorylation of a CDK inhibitor sets a threshold for the onset of DNA replication. *Nature* **414,** 514–521.

Nugroho, T. T., and Mendenhall, M. D. (1994). An inhibitor of yeast cyclin-dependent protein kinase plays an important role in ensuring the genomic integrity of daughter cells. *Mol. Cell. Biol.* **14,** 3320–3328.

Orlicky, S., Tang, X., Willems, A., Tyers, M., and Sicheri, F. (2003). Structural basis for phosphodependent substrate selection and orientation by the SCFCdc4 ubiquitin ligase. *Cell* **112,** 243–256.

Petroski, M. D., and Deshaies, R. J. (2003). Context of multiubiquitin chain attachment influences the rate of Sic1 degradation. *Mol. Cell* **11,** 1435–1444.

Petroski, M. D., and Deshaies, R. J. (2005). Function and regulation of cullin-RING ubiquitin ligases. *Nat. Rev. Mol. Cell Biol.* **6,** 9–20.

Reinstein, E., Scheffner, M., Oren, M., Ciechanover, A., and Schwartz, A. (2000). Degradation of the E7 human papillomavirus oncoprotein by the ubiquitin-proteasome system: Targeting via ubiquitination of the N-terminal residue. *Oncogene* **19,** 5944–5950.

Schwob, E., Bohm, T., Mendenhall, M. D., and Nasmyth, K. (1994). The B-type cyclin kinase inhibitor p40SIC1 controls the G1 to S transition in *S. cerevisiae. Cell* **79,** 233–244.

Seol, J. H., Feldman, R. M., Zachariae, W., Shevchenko, A., Correll, C. C., Lyapina, S., Chi, Y., Galova, M., Claypool, J., Sandmeyer, S., Nasmyth, K., and Deshaies, R. J. (1999). Cdc53/cullin and the essential Hrt1 RING-H2 subunit of SCF define a ubiquitin ligase module that activates the E2 enzyme Cdc34. *Genes Dev.* **13,** 1614–1626.

Skowyra, D., Craig, K. L., Tyers, M., Elledge, S. J., and Harper, J. W. (1997). F-box proteins are receptors that recruit phosphorylated substrates to the SCF ubiquitin-ligase complex. *Cell* **91,** 209–219.

Skowyra, D., Koepp, D. M., Kamura, T., Conrad, M. N., Conaway, R. C., Conaway, J. W., Elledge, S. J., and Harper, J. W. (1999). Reconstitution of G1 cyclin ubiquitination with complexes containing SCFGrr1 and Rbx1. *Science* **284,** 662–665.

Verma, R., Annan, R. S., Huddleston, M. J., Carr, S. A., Reynard, G., and Deshaies, R. J. (1997a). Phosphorylation of Sic1p by G1 Cdk required for its degradation and entry into S phase. *Science* **278,** 455–460.

Verma, R., Feldman, R. M., and Deshaies, R. J. (1997b). SIC1 is ubiquitinated *in vitro* by a pathway that requires CDC4, CDC34, and cyclin/CDK activities. *Mol. Biol. Cell* **8,** 1427–1437.

Verma, R., McDonald, H., Yates, J. R., 3rd, and Deshaies, R. J. (2001). Selective degradation of ubiquitinated Sic1 by purified 26S proteasome yields active S phase cyclin-Cdk. *Mol. Cell* **8,** 439–448.

Verma, R., Oania, R., Graumann, J., and Deshaies, R. J. (2004). Multiubiquitin chain receptors define a layer of substrate selectivity in the ubiquitin-proteasome system. *Cell* **118,** 99–110.

You, J., Cohen, R. E., and Pickart, C. M. (1999). Construct for high-level expression and low misincorporation of lysine for arginine during expression of pET-encoded eukaryotic proteins in *Escherichia coli. Biotechniques* **27,** 950–954.

# [14] Expression and Assay of Glycoprotein-Specific Ubiquitin Ligases

*By* YUKIKO YOSHIDA


## Abstract

*N*-linked glycosylation of proteins that takes place in the endoplasmic reticulum (ER) plays a key role in protein quality control. Misfolded proteins or unassembled protein complexes that fail to achieve their functional states in the ER are retrotranslocated into the cytosol and degraded by the ubiquitin-proteasome system in a process called ER-associated degradation (ERAD). *N*-linked glycoprotein-specific ubiquitin ligase complexes, $SCF^{Fbs1}$ and $SCF^{Fbs2}$, appear to participate in ERAD for selective elimination of aberrant glycoproteins in the cytosol. This chapter describes methods employed for the isolation and oligosaccharide-binding assay of Fbs proteins that are the substrate-recognition components of the $SCF^{Fbs}$ complex and the *in vitro* ubiquitylation assay of the $SCF^{Fbs}$ ubiquitin ligase complexes.


## Introduction

Eukaryotic cells have an abundant and diverse repertoire of asparagine (*N*)-linked oligosaccharide structures (Kornfeld and Kornfeld, 1985). The *N*-linked oligosaccharides expressed on the surface of most cells and secreted proteins contribute to their multiple biological functions (Varki, 1993). The structural diversification of *N*-glycans is introduced by a series of glycosyltransferases in the Golgi complex. *N*-linked glycosylation of proteins in the secretory pathway initially occurs in the endoplasmic reticulum (ER), and *N*-glycan in the ER represents an evolutionarily conserved high-mannose type structure that occurs from fungal to mammalian cells. Glycans *N*-linked provide a common role in assisting protein folding, quality control, and transport (Helenius and Aebi, 2001). *N*-linked glycoproteins are subjected to protein quality control through which aberrant proteins are distinguished from properly folded proteins and retained in the ER (Ellgaard and Helenius, 2003; Yoshida, 2003). The quality control system includes the calnexin–calreticulin cycle, a unique chaperone system that recognizes $Glc_1Man_{9\text{-}6}GlcNAc_2$ and assists in the refolding of misfolded or unfolded proteins. When the improperly folded or incompletely assembled proteins cannot be restored to their functional states, they are degraded by

METHODS IN ENZYMOLOGY, VOL. 398

0076-6879/05 $35.00
DOI: 10.1016/S0076-6879(05)98014-2

the ER-associated degradation (ERAD) system, which involves the retrograde transfer of proteins from the ER to the cytosol followed by degradation by the proteasome (Tsai *et al.,* 2002). Studies have demonstrated that $Man_8GlcNAc_2$ structures (Man8) serve as part of the signal needed for ERAD and that a lectin for Man8 in the ER accelerates the turnover rate of the misfolded glycoprotein (Hosokawa *et al.,* 2001; Jakob *et al.,* 2001; Nakatsukasa *et al.,* 2001). It has been shown that several ERAD substrates are deglycosylated when proteasomal proteolysis is blocked, indicating that peptide *N*-glycanase (PNGase) in the cytosol acts on the substrates prior to destruction by the proteasome (Halaban *et al.,* 1997; Huppa and Ploegh, 1997; Wiertz *et al.,* 1996a,b). The glycoprotein-specific $SCF^{Fbs}$ ubiquitin ligase complexes that recognize high-mannose oligosaccharides in denatured proteins interact with the retrotranslocated ERAD substrate in the cytosol and ubiquitylate them for degradation prior to deglycosylation (Blom *et al.*, 2004; Yoshida *et al.*, 2002, 2003).

## Purification of Fbs1

Because Fbs1 uniquely binds the innermost chitobiose structure in high-mannose type *N*-glycans (Mizushima *et al.,* 2004), the addition of *N, N'*-diacetylchitobiose allows efficient elution of Fbs1 from glycoproteins with high-mannose oligosaccharides. For example, Fbs1 can bind efficiently to ribonuclease B (RNase B) that contains a high-mannose oligosaccharide and can be released from RNase B by *N, N'*-diacetylchitobiose. Therefore, it can be purified easily by single-step affinity chromatography using RNase B as a ligand and *N, N'*-diacetylchitobiose as an eluent. This method can be applied for purification from tissues and mammalian cells or *Escherichia coli* that overexpress Fbs1 (Fig. 1).

### *Preparation of RNase B-Immobilized Agarose*

Twenty milligrams of RNase B (Sigma) is immobilized to 1 ml of Affi-Gel 10 (Bio-Rad, Richmond, CA) using the procedure supplied by the manufacturer.

### *Purification of Fbs1 from Mouse Brains*

Brains of ICR mice (8 weeks old) are homogenized in 10 volumes of 20 m*M* Tris-HCl (pH 7.5)/150 m*M* NaCl (TBS) containing a cocktail of protease inhibitors (complete EDTA free, Roche Molecular Systems, Inc., NJ). After centrifugation of the homogenate at 30,000*g* for 30 min, the supernatant is incubated with 1/10 volume of Sepharose 4B (Amersham Life Science, Buckinghamshire, UK) under gentle rotation at 4° for 2 h.

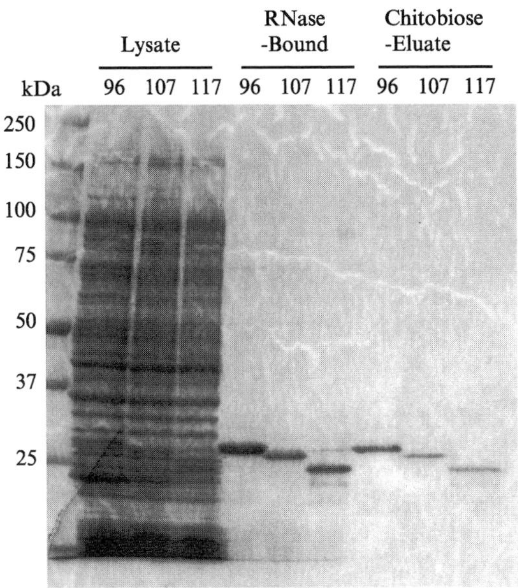


FIG. 1. SDS–PAGE analysis of the affinity-purified recombinant substrate-binding domain (SBD) of Fbs1. The SBD of murine Fbs1 (96: 96–297, 107: 107–297, 117: 117–297) are cloned into pET15b and expressed in *Escherichia coli*. These lysates of *E. coli* are incubated with RNase B agarose beads, and the resulting beads are washed thoroughly. Bound proteins are eluted by boiling with SDS sample buffer (RNase B-bound) or by 0.1 *M* diacetylchitobiose (chitobiose-eluate). Molecular masses of marker proteins are indicated on the left.

After removal of Sepharose and insoluble materials by centrifugation at 30,000*g* for 20 min, TritonX-100 is added to the precleared supernatant at a final concentration of 0.5%, and the mixture is incubated with 0.5 ml of RNase B-immobilized beads for <50 mg total proteins under gentle rotation at 4° for 18 h. The beads are washed in a Poly-Prep column (Bio-Rad) with ice-cold TBS containing 0.5% Triton X-100 (TBS-T), followed with TBS-T containing 0.5 *M* NaCl. After equilibration with TBS without detergent, the proteins adsorbed are eluted by 0.1 *M* *N, N'*-diacetylchitobiose in TBS.

## Glycoprotein Interaction Assays

### *Overlay Assay*

This assay is used to screen for putative Fbs substrates in various cells and tissues (Yoshida *et al.,* 2002).

1. Cultured cells and tissues are homogenized in 10 volumes of TBS containing 0.5% NP-40 (TBS-N) and protease inhibitor cocktail. This homogenate is centrifuged at 20,000*g* for 20 min, and the resulting supernatant is the whole cell extract (WCE). For preparation of fractionated cell extracts, the cell extracts prepared by freezing and thawing in TBS are centrifuged at 24,000*g* for 20 min. The resulting supernatant is used as the cytoplasmic fraction, while the precipitate is solubilized with TBS containing 1% Triton X-100 and the centrifuged supernatant (24,000*g* for 20 min) is used as the membranous fraction.

2. Proteins of WCE or fractionated cell extracts are separated with sodium dodecyl sulfate–polyacrylamide gel electrophoresis (SDS–PAGE) and are blotted onto a membrane (Immobilon, Millipore, Bedford, MA).

3. To prepare [$^{35}$S]methionine-labeled Fbs probes, cDNA of Fbs1 and Fbs2 are cloned into an expression vector derived from the T7 promoter [e.g., pcDNA (Invitrogen, San Diego, CA)] with an additional three methionines appended at their C termini using polymerase chain reaction (PCR) cloning methods. The TNT-coupled reticulocyte lysate system (Promega, Madison, WI) is used for generating [$^{35}$S]methionine-labeled Fbs1 and Fbs2. To reduce background signals, free methionine is removed by desalting on a MicroSpin G25 column (Amersham Pharmacia Biotech, Piscataway, NJ).

4. The blotted membranes are blocked by 5% skim milk in phosphate-buffered saline (PBS), washed, and then incubated with [$^{35}$S]methionine-labeled probes in 1% skim milk in PBS at 4° for 18 h with gentle shaking. Membranes are washed with PBS containing 0.05% Tween 20, air dried, and analyzed by autoradiography.

5. To confirm that these interactions occur through high-mannose type oligosaccharides, the blots are treated with endoglycosidase H (Endo H) before incubation with the probes. In the Endo H treatment step, the blotted membrane is blocked with 3% bovine serum albumin (fraction V; Sigma Chemical Co., St. Louis, MO) in PBS, washed with 0.1 *M* citrate-phosphate buffer, pH 5.5, and then treated with 1 unit of recombinant Endo H (Roche) at 37° for 18 h. The membrane is washed thoroughly with PBS to adjust the pH to 7.5, incubated with 5% skim milk, and then incubated with [$^{35}$S]methionine-labeled probes.

### *Coprecipitation Assay*

This assay is used to screen for Fbs substrates and to examine the oligosaccharide structure of the substrates in mammalian cells (Yoshida *et al.*, 2002, 2003).

1. Using conventional recombinant DNA methods, cDNAs coding for Fbs1 and Fbs2 are isolated. The full-length and F-box domain truncated forms of Fbs1 and Fbs2 are prepared by PCR amplification and cloned into pcDNA3-Flag or the pcDNA3.1-HA vector.

2. Cells ($1 \times 10^6$) of various mammalian cell lines are transfected with Flag- or HA-tagged Fbs expression vectors by lipofection (Lipofectamine Plus reagent, Invitrogen) using the procedure provided by the manufacturer. Cells are treated with 50 $\mu M$ of MG132 (Peptide Institute Inc., Osaka, Japan) for adequate times (293T, 1.5 h; Neuro2a, 4 h; PC12, 4 h) before harvesting cells.

3. Forty-eight hours after transfection, the cultured cells are homogenized on ice in TBS-N containing a protease inhibitor cocktail. This homogenate is centrifuged at 20,000*g* for 20 min, and the resulting supernatant (200 $\mu$l protein solution adjusted to approximately 1 mg/ml) is used for immunoprecipitation or pull-down.

4. For Fbs-coimmunoprecipitate protein analysis by lectin blot, the cell lysates are precleared with 20 $\mu$l protein A-Sepharose beads (Amersham) for 2 h and incubated with 4 $\mu$g of antibody [anti-FLAG (M2, Sigma), anti-HA (HA.11, CRP Inc., Denver, PA)] for 2 h and then 25 $\mu$l protein A-Sepharose beads are added for 4 h at 4°. The resulting immunoprecipitates are washed thoroughly four times with cold TBS-N, separated by SDS–PAGE, and blotted onto membranes (Immobilon, Millipore).

5. To confirm that the overexpressed Fbs proteins are included in the immunoprecipitates, immunoblotting is performed. Membranes are blocked by 5% skim milk in TBS for 1 h, incubated with the required antibody for 1 h at room temperature, followed by horseradish peroxidase (HRP)-conjugated antimouse antibody (Jackson Immunoresearch Laboratories, West Grove, PA), and visualized using the enhanced chemiluminescence (ECL) system.

6. The Fbs1- and Fbs2-binding proteins in the immune complex are analyzed by lectin blotting. Membranes are blocked by 3% bovine serum albumin (BSA) in PBS for 2 h, incubated with HRP-conjugated lectins (e.g., Con A, RCA120, WGA; Seikagaku Corp., Tokyo, Japan, GNA; EY Laboratories, Inc., San Mateo, CA), and visualized using an enhanced ECL. Con A and GNA lectins bind to high-mannose type oligosaccharides, the RCA120 lectin binds to terminal galactose$\beta$1-4GlcNAc, and WGA is specific for terminal GlcNAc or sialic acid.

7. To analyze the ability of Fbs proteins to bind to endogenous glycoproteins that are fractionated by their oligosaccharide structures, the lysates of various cells expressing Flag-tagged Fbs proteins prepared in step 3 are precleared with 20 $\mu$l Sepharose 4B for 2 h and are incubated with 25 $\mu$l of various lectin-conjugated agarose beads (e.g., Con A,

RCA120, WGA; Seikagaku-kogyo, GNA; EY Laboratories, Inc.) for 4 h at 4°. For the negative control experiment, *in vitro*-translational Fbs proteins are synthesized by the TNT-coupled reticulocyte lysate system (Promega) and are used in the pull-down assay. Proteins bound to lectin-agarose are washed thoroughly four times with cold TBS-N, separated by SDS–PAGE, and blotted onto membranes. The blotted membranes are blocked by 5% skim milk in TBS for 1 h, incubated with the antibody against the tag peptide of Fbs for 1 h at room temperature, followed by the HRP-conjugated antimouse antibody (Jackson Laboratory), and visualized using the ECL system.

## *The Pull-Down Assay*

This assay is used for assessing the binding specificities of Fbs proteins for different *N*-glycans (Yoshida *et al.,* 2002, 2003).

1. Preparation of glycoprotein-conjugated beads. Bovine fetuin containing tri-sialylated (tri-antennary complex-type) oligosaccharides, fetuin-derivatives treated with various glycosidases, and other glycoproteins with known oligosaccharide structures are used for this assay. Fetuin and asialofetuin (type II) are obtained from Sigma. GlcNAc-terminated fetuin (GTF) is prepared by incubating 10 mg of asialofetuin with 0.2 unit of β-galactosidase (*Streptococcus* 6646K, Seikagaku-kogyo) in 200 $\mu$l of 50 m*M* citrate buffer (pH 5.5) at 37° for 24 h. For removing the asialofetuin and exchanging the buffer, the β-galactosidase-treated GTF solution is applied to a RCA-120-agarose (Seikagaku-kogyo) column (200 $\mu$l agarose in a 1-ml syringe) equilibrated previously with 0.1 *M* of MOPS buffer (pH 7.5). Mannose-terminated fetuin (MTF) is prepared by incubating 10 mg of GTF with 5 unit of jack bean β-*N*-acetylhexosaminidase (Seikagaku-kogyo) in 200 $\mu$l of 150 m*M* citrate-phosphate buffer (pH 5.0) at 37° for 24 h, and the MTF solution is applied to a WGA-agarose (Seikagaku-kogyo) column equilibrated with 0.1 *M* MOPS (pH 7.5) for removing the GTF. For preparation of deglycosylated fetuin (DGF), 10 mg of asialofetuin is incubated with 200 units of *Flavobacterium meningosepticum* recombinant PNGase F (Roche) in 50 mM phosphate buffer, pH 7.2, at 37° for 24 h, and the enzyme-treated protein is loaded onto successive WGA and RCA-lectin agarose columns. *N*-linked glycoproteins containing high-mannose oligosaccharides (RNase B, thyroglobulin, and ovalbumin) and GlcNAc-terminated tetra-antennary oligosaccharides (ovomucoid) are obtained from Sigma. Each 10 mg of glycoprotein in 2 ml of 0.1 *M* MOPS (pH 7.5) is immobilized to 1 ml of Affi-Gel 10 (RNase B, MTF, DGF) or Affi-Gel 15 (fetuin, ASF, GTF,

thyroglobulin, ovalbumin, ovomucoid) using the procedure provided by the manufacturer.

2. 293T cells ($1 \times 10^6$) are transfected with 5 $\mu$g of the Flag-tagged Fbs1 or Fbs2 expression vector by Lipofectamine plus reagent. Forty-eight hours after transfection, cultured cells are homogenized in TBS-N containing the protease inhibitor cocktail and centrifuged at 20,000*g* for 20 min, and the resulting supernatant is adjusted to a protein concentration of 0.5 mg/ml.

3. The cell lysates are precleared with 1/10 volume of Sepharose 4B beads for 2 h, and each divided 100-$\mu$l aliquot of lysate is incubated overnight at 4° with 10 $\mu$l of various glycoprotein-immobilized beads prepared as described in step 1.

4. The resulting beads are washed thoroughly four times with cold TBS-N, and bound proteins are eluted by boiling with the SDS sample buffer. For detailed analysis of the binding specificities, the bound proteins are eluted by incubation with 15 $\mu$l of various concentrations of synthetic oligosaccharides at room temperature for 10 min. The eluates are separated by spin filtration. *N,N'*-diacetylchitobiose is obtained from Seikagaku-kogyo, $Man_8GlcNAc_2$ is from Glyko (San Leandro, CA), $Man_5GlcNAc_1$ is from Iso sep AB, $GlcNAc_3Man_3GlcNAc_2$ (asialo-, agalacto-, tri-antennary complex) and $GlcNAc_2Man_3GlcNAc_2$ ($Fuc_1$) (asialo-, agalacto-, core-fucosylated biantennary complex) are from Ludger (Oxford, UK), and $Man_3GlcNAc_2$, $Man_5GlcNAc_2$, and $Man_5$-$GlcNAc_2$-Asn are from Sigma.

5. The bound or eluted proteins are analyzed by immunoblotting using the anti-Flag antibody.

## *In Vitro* Ubiquitylation Assay

### *Preparation of Recombinant Proteins of the Ubiquitin System*

N-terminally RGS(His)$_6$-tagged human Ubc4, human Ubc12 (E2 for NEDDylation), and NEDD8 cDNA are subcloned into the pT7-7 bacterial expression vector. The cDNA coding ubiquitin is subcloned into the pGEX6p-1 vector. Recombinant His-tagged Ubc4, GST-tagged ubiquitin, His-tagged Ubc12, and His-tagged NEDD8 are produced in *E. coli*.

The N-terminally His-tagged mouse Uba1 (E1), N-terminally Flag-tagged human Skp1, C-terminally HA-tagged human Cul1, N-terminally His-tagged rat Fbs1, N-terminally T7-tagged human Roc1/Rbx1, N-terminally His-tagged human APP-BP1, and N-terminally T7-tagged human Uba3 are subcloned into pVL1393 (Invitrogen). A recombinant baculovirus is generated using the Bac-PAK6 baculovirus expression system (Clontech Laboratories, Palo Alto, CA). High Five insect cells are cultured in Grace's insect medium supplemented with 10% fetal calf serum at 27° and infected with the appropriate recombinant virus. To generate a $SCF^{Fbs1}$ complex that is competent to function as an E3 for ubiquitylation *N*-linked glycoproteins are simultaneously infected with four baculoviruses (Flag-tagged Skp1/ HA-tagged Cul1/ His-tagged Fbs1/ T7-tagged Roc1). To generate an active E12 enzyme for NEDD8, cells are simultaneously infected with baculoviruses that encode His-tagged APP-BP1 and T7-tagged Uba3. Sixty hours after infection, cells are harvested and lysed in 0.5% NP-40, 10 m*M* 2-mercaptoethanol, 10 m*M* imidazole, 20 m*M* Na-phosphate (pH 7.2), 150 m*M* NaCl, and 1 m*M* phenylmethylsulfonyl fluoride. These proteins are affinity purified by a HiTrap HP column (Amersham), and the purified proteins are dialyzed against 10% glycerol, 10 m*M* 2-mercaptoethanol, and 50 m*M* Tris-HCl (pH 7.4).

### In Vitro *Ubiquitylation of GlcNAc-Terminated Fetuin (GTF) by the $SCF^{Fbs1}$ Ubiquitin-Ligase System*

Fbs1 recognizes high-mannose oligosaccharides, especially their innermost diacetylchitobiose structure, and can interact efficiently with RNase B containing a high-mannose oligosaccharide. $SCF^{Fbs1}$, however, cannot ubiquitylate RNase B in the *in vitro* ubiquitylation system. Structural analysis suggests that the E2 enzyme bound to the SCF complex cannot be accessible to the polypeptide chain of RNase B interacted with Fbs1 due to the too small size of RNase B (data not shown). However, ubiquitylation of GlcNAc-terminated fetuin (GTF) is detected efficiently in the fully reconstituted system with $SCF^{Fbs1}$ (Mizushima *et al.*, 2004; Yoshida *et al.*, 2002).

*Preparation of Substrate Proteins.* GTF is prepared by incubating 0.1 mg of asialofetuin type II (Sigma) with 2 mU $\beta$-galactosidase (*Streptococcus* 6646K, Seikagaku-kogyo) in 50 m*M* citrate buffer (pH 5.5) at 37° for 24 h and removing the terminal galactose-containing proteins using RCA120 lectin agarose. For preparation of deglycosylated fetuin (DGF) as a negative control, 0.1 mg of asialofetuin is incubated with 2 units of recombinant PNGase F (Roche) in 50 m*M* phosphate buffer, pH 7.2, at 37° for 24 h. The enzyme-treated proteins are loaded onto successive RCA-lectin agarose and WGA-lectin agarose columns. The flow-through fraction from both columns is used as DGF.

In Vitro *Ubiquitylation Assay.* Each 1 $\mu$g of GTF or DGF is incubated in 50 $\mu$l of the reaction mixture containing the ATP-regenerating system, 0.5 $\mu$g E1, 1 $\mu$g Ubc4 (E2), 2 $\mu$g $SCF^{Fbs1}$, and 6.5 $\mu$g recombinant GST-ubiquitin in the presence of the NEDD8 system consisting of NEDD8

(10 $\mu$g), APP-BP1/Uba3 (0.5 $\mu$g), and Ubc12 (0.5 $\mu$g) at 30°. After terminating the reaction by the addition of 25 $\mu$l of 3× SDS sample buffer, proteins in 8 $\mu$l of the boiled supernatants are separated by SDS–PAGE on a 4–20% gel, and the high molecular mass-ubiquitylated proteins are detected by immunoblotting with an antifetuin (Chemicon International, Inc., Temecula, CA) or anti-GST (Ab-1, Calbiochem, La Jolla, CA) antibody.

## In Vitro *Ubiquitylation of Integrin β1 by the SCF$^{Fbs1}$ Ubiquitin-Ligase System*

One of the identified *in vivo* Fbs1-substrate proteins is preintegrin $\beta$1, which is modified with high-mannose oligosaccharides and is mainly located in the ER (Yoshida *et al.*, 2002). Most of the integrin $\beta$1 localized on the plasma membrane is modified with complex-type oligosaccharides, which cannot bind to Fbs1. In this *in vitro* ubiquitylation system, the purified recombinant integrin $\beta$1 modified with high-mannose oligosaccharides is used as a substrate of SCF$^{\mathrm{Fbs1}}$.

*Plasmid Construction.* The cDNA coding for human integrin $\beta$1 is amplified by PCR with Pyrobest DNA polymerase (Takara Bio, Japan) from a cDNA clone (ATCC 988953). The PCR primers are as follow: 5′-GGC CCC GAA TTC CGC GCG GAA AAG ATG AAT TTA C–3′ and 5′-TTG CAC GGG CGG CCG CCA TTT TCC CTC ATA CTT CGG ATT–3′. The PCR product is digested with *Eco*RI/*Not*I, subcloned into the pTracer-EF-V5HisA vector (Invitrogen), and sequenced.

*Preparation of Preintegrin β1.* 293T cells (1 × 10$^7$) are transfected with 50 $\mu$g of the C-terminal V5-His-tagged integrin $\beta$1 expression vector by the Lipofectamine plus reagent (Invitrogen). Forty-eight hours after transfection, cultured cells are homogenized in 2 ml of TBS-T containing the protease inhibitor cocktail and centrifuged at 20,000*g* for 20 min, and the resulting supernatant is incubated with 100 $\mu$l of Talon metal affinity resin (Clontech) for 2 h at 4°. The resulting resin is washed thoroughly with cold TBS-T, and bound proteins are eluted three times by 100 $\mu$l of 0.15 *M* imidazole. One-half of the eluate is incubated with 50 $\mu$l of WGA agarose and the other with 50 $\mu$l of Con A agarose for 2 h at 4°. The supernatant that does not bind WGA is used as the preintegrin β1 substrate of SCF$^{\mathrm{Fbs1}}$, and the supernatant that does not bind to Con A is used as mature integrin β1 modified with complex-type oligosaccharides that is used as a negative control of the substrate.

In Vitro *Ubiquitylation Assay.* Each 25 $\mu$l of integrin $\beta$1 modified with high-mannose oligosaccharides (pre-integrin $\beta$1) and complex-type oligosaccharides (mature integrin $\beta$1) is incubated in 50 $\mu$l of the reaction

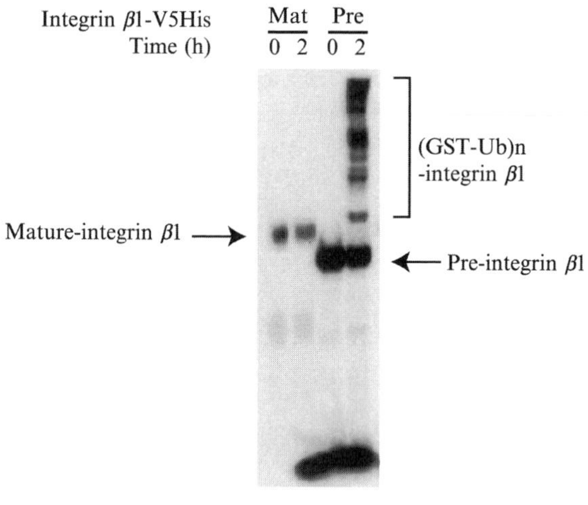


FIG. 2. *In vitro* ubiquitylation of preintegrin β1 by the SCF$^{Fbs1}$ ubiquitin ligase system. *In vitro* ubiquitylation reactions are performed with pre-integrin β1 modified with high-mannose oligosaccharides and mature form integrin β1 with complex-type oligosaccharides. The higher molecular mass-ubiquitylated integrin β1 [(GST-Ub)n-integrin β1] is detected by immunoblotting with an anti-V5 epitope antibody.

mixture containing the ATP-regenerating system, 0.5 μg E1, 1 μg Ubc4 (E2), 2 μg SCF$^{Fbs1}$, and 6.5 μg recombinant GST-ubiquitin in the presence of the NEDD8 system consisting of NEDD8 (10 μg), APP-BP1/Uba3 (0.5 μg), and Ubc12 (0.5 μg) at 30°. After terminating the reaction by the addition of 25 μl of 3× SDS sample buffer, proteins in 8 μl of the boiled supernatants are separated by SDS–PAGE on a 7.5% gel and the high molecular mass-ubiquitylated proteins are detected by immunoblotting with an anti-V5 epitope antibody (Invitrogen). The pattern of ubiquitylation of integrin β1 is shown in Fig. 2.

## References

Blom, D., Hirsch, C., Stern, P., Tortorella, D., and Ploegh, H. L. (2004). A glycosylated type I membrane protein becomes cytosolic when peptide: N-glycanase is compromised. *EMBO J.* **23,** 650–658.

Ellgaard, L., and Helenius, A. (2003). Quality control in the endoplasmic reticulum. *Nature Rev. Mol. Cell. Biol.* **4,** 181–191.

Halaban, R., Cheng, E., Zhang, Y., Moellmann, G., Halon, D., Michalak, M., Setaluri, V., and Hebert, D. N. (1997). Aberrant retention of tyrosinase in the endoplasmic reticulum mediates accelerated degradation of the enzyme and contributes to the dedifferentiated phenotype of amelanotic melanoma cells. *Proc. Natl. Acad. Sci. USA* **94,** 6210–6215.

Helenius, A., and Aebi, M. (2001). Intracellular functions of N-linked glycans. *Science* **291,** 2364–2369.

Hosokawa, N., Wada, I., Hasegawa, K., Yorihuzi, T., Tremblay, L. O., Herscovics, A., and Nagata, K. (2001). A novel ER alpha-mannosidase-like protein accelerates ER-associated degradation. *EMBO Rep.* **2,** 415–422.

Huppa, J. B., and Ploegh, H. L. (1997). The alpha chain of the T cell antigen receptor is degraded in the cytosol. *Immunity* **7,** 113–122.

Jakob, C. A., Bodmer, D., Spirig, U., Battig, P., Marcil, A., Dignard, D., Bergeron, J. J., Thomas, D. Y., and Aebi, M. (2001). Htm1p, a mannosidase-like protein, is involved in glycoprotein degradation in yeast. *EMBO Rep.* **2,** 423–430.

Kornfeld, R., and Kornfeld, S. (1985). Assembly of asparagine-linked oligosaccharides. *Annu. Rev. Biochem.* **54,** 631–664.

Mizushima, T., Hirao, T., Yoshida, Y., Lee, S. J., Chiba, T., Iwai, K., Yamaguchi, Y., Kato, K., Tsukihara, T., and Tanaka, K. (2004). Structural basis of sugar-recognizing ubiquitin ligase. *Nature Struct. Mol. Biol.* **11,** 365–370.

Nakatsukasa, K., Nishikawa, S., Hosokawa, N., Nagata, K., and Endo, T. (2001). Mnl1p, an alpha-mannosidase-like protein in yeast *Saccharomyces cerevisiae*, is required for endoplasmic reticulum-associated degradation of glycoproteins. *J. Biol. Chem.* **276,** 8635–8638.

Tsai, B., Ye, Y., and Rapoport, T. A. (2002). Retro-translocation of proteins from the endoplasmic reticulum into the cytosol. *Nature Rev. Mol. Cell. Biol.* **3,** 246–255.

Varki, A. (1993). Biological roles of oligosaccharides: All of the theories are correct. *Glycobiology* **3,** 97–130.

Wiertz, E. J., Jones, T. R., Sun, L., Bogyo, M., Geuze, H. J., and Ploegh, H. L. (1996a). The human cytomegalovirus US11 gene product dislocates MHC class I heavy chains from the endoplasmic reticulum to the cytosol. *Cell* **84,** 769–779.

Wiertz, E. J., Tortorella, D., Bogyo, M., Yu, J., Mothes, W., Jones, T. R., Rapoport, T. A., and Ploegh, H. L. (1996b). Sec61-mediated transfer of a membrane protein from the endoplasmic reticulum to the proteasome for destruction. *Nature* **384,** 432–438.

Yoshida, Y. (2003). A novel role for N-glycans in the ERAD system. *J. Biochem. (Tokyo)* **134,** 183–190.

Yoshida, Y., Chiba, T., Tokunaga, F., Kawasaki, H., Iwai, K., Suzuki, T., Ito, Y., Matsuoka, K., Yoshida, M., Tanaka, K., and Tai, T. (2002). E3 ubiquitin ligase that recognizes sugar chains. *Nature* **418,** 438–442.

Yoshida, Y., Tokunaga, F., Chiba, T., Iwai, K., Tanaka, K., and Tai, T. (2003). Fbs2 is a new member of the E3 ubiquitin ligase family that recognizes sugar chains. *J. Biol. Chem.* **278,** 43877–43884.

# [15] Affinity Purification of Mitotic Anaphase-Promoting Complex/Cyclosome on p13[Suc1]

*By* AVRAM HERSHKO

## Abstract

A procedure is described for the affinity purification of the mitotic form of anaphase-promoting complex/cyclosome (APC/C) from HeLa cells. It is based on the binding of mitotically phoshorylated APC/C to the phosphate-binding site of p13$^{suc1}$, followed by specific elution with a phosphate-containing compound. The procedure is rapid, simple, and yields 50- to 70-fold purification of soluble APC/C, with a ~30% recovery of activity.

## Introduction

A large multisubunit ubiquitin-protein ligase called the cyclosome or anaphase-promoting complex was first identified as that responsible for targeting cyclin B for degradation in the early embryonic cell cycles of clams (Sudakin *et al.*, 1995) and frogs (King *et al.*, 1995). Subsequent studies have shown that APC/C-mediated destruction of specific cell cycle regulatory proteins is important in eukaryotic cells both for exit from mitosis and for the prevention of inappropriate exit from the G1 phase of the subsequent cell cycle (reviewed in Peters, 2002; Zachariae and Nasmyth, 1999). Most currently used procedures for the purification of APC/C employ immunoprecipitation with specific antibodies, followed by an assay of ubiquitin-protein ligase activity of the complex bound to anti-APC/C beads (Fang *et al.*, 1998; Kramer *et al.*, 1998). This method may cause problems of steric interference in the interaction of APC/C with its substrates of modifiers due to the immobilization of the complex on beads. Thus, the recovery of enzymatic activity of affinity-purified, immobilized APC/C preparations (as compared to that of crude extracts) is usually less than 1%. Such potential problems do not exist with free and soluble preparations of purified APC/C. We have developed an affinity procedure for the purification of soluble mitotic APC/C by affinity chromatography on p13$^{Suc1}$-Sepharose. The method was originally developed to isolate mitotic APC/C from clam oocytes (Sudakin *et al.*, 1997) and was subsequently modified for its purification from cultured human cells (Golan *et al.*, 2002).

METHODS IN ENZYMOLOGY, VOL. 398 0076-6879/05 $35.00

DOI: 10.1016/S0076-6879(05)98015-4

The procedure was based on our serendipitous observation that the mitotic, phosphorylated form of APC/C binds strongly to p13$^{Suc1}$-Sepharose beads (Sudakin *et al.*, 1997). p13$^{suc1}$ is a member of the highly conserved family of Suc1/Cks proteins, all of which bind to cyclin-dependent kinases (Cdks). Because of this property, p13$^{suc1}$-Sepharose beads have been widely used for the purification of Cdk-cyclin complexes. Our original intention was to use p13Suc1-Sepharose for removing Cdk1-cyclin B from partially purified preparations of mitotic clam APC/C, but we were surprised to find that APC/C is also strongly bound to these beads. Further work showed that the binding of mitotic, phosphorylated APC/C to p13$^{suc1}$ was not mediated by Cdk, but was due to interaction with a phosphate-binding site of p13$^{suc1}$. Such phosphate-binding sites exist in all Suc1/Cks proteins, in addition to the Cdk-binding site [see Sitry *et al.* (2002) and references therein].[1] Indeed, mitotically phosphorylated APC/C could be eluted from p13$^{suc1}$-Sepharose with a phosphate-containing compound, such as *p*-nitrophenyl phosphate. These properties were exploited to develop an affinity purification method. By this simple one-step affinity procedure, a ~50- to 70-fold purification of mitotic APC/C from HeLa cells was obtained, with a recovery of ~30% of activity (Golan *et al.*, 2002). Cdks are well separated from APC/C by this procedure because they are not eluted from p13$^{suc1}$-Sepharose by phosphate-containing compounds. Some other mitotically phosphorylated proteins are also enriched in these preparations, but APC/C can be separated from most of these by a further step of fast protein liquid chromatography (FPLC) ion-exchange chromatography on MonoQ (Golan *et al.*, 2002). For many practical purposes, preparations of mitotic APC/C purified only by p13$^{suc1}$-Sepharose are suitable.

## Materials

The following materials are from Sigma Chemical Co: ubiquitin from bovine erythrocytes, bovine serum albumin (BSA, protease free), soybean trypsin inhibitor (STI,), *p*-nitrophenyl phosphate, phosphocreatine, creatine phosphokinase, leupeptin, chymostatin, staurosporine, and glycerol (molecular biology grade). Chymostatin and staurosporine are dissolved in dimethyl sulfoxide (DMSO) in 1000- and 200-fold concentrated stocks, respectively. Nocodazole (Sigma) is dissolved in DMSO at a stock concentration of 1 mg/ml and is stored at –20° in small samples, each of which is used only once. Okadaic acid (Roche) is dissolved in DMSO at a stock

[1] Mammalian Cks1 (but not Cks2) has an additional Skp2-binding site, which is required for the interaction of the SCF$^{Skp2}$ ubiquitin ligase complex with its specific substrates (Sitry *et al.*, 2002).

concentration of 200 $\mu$M and is stored at –70°. E1 (Hershko *et al.*, 1983) and ubiquitin aldehyde (Mayer and Wilkinson, 1989) are prepared as described; these two materials can also be obtained from commercial sources. Recombinant human E2C/UbcH10 is expressed in bacteria and is purified as described (Townsley *et al.*, 1997). The sea urchin cyclin B N-terminal fragment (13–91)-protein A fusion protein is expressed in BL21(DE3) bacteria, purified by affinity chromatography on IgG-Sepharose, and radioiodinated (Glotzer *et al.* 1991). $p13^{suc1}$ is expressed in BL21(DE3)pLys as described (Brizuela *et al.*, 1987) and is purified by gel filtration on a column (3 × 120 cm) of Sephadex G–100 equilibrated with 50 m*M* Tris-HCl (pH 8.0) and 2 m*M* EDTA. By this procedure, $p13^{suc1}$ is well separated from most bacterial proteins, which are of larger size. Prior to its covalent binding to Sepharose beads, $p13^{suc1}$ is concentrated by centrifuge ultrafiltration (Centriprep 3, Amicon) and is extensively dialyzed against water. $p13^{suc1}$ is coupled to cyanogen bromide-activated Sepharose-4B (Sigma) according to the instructions of the manufacturer to a concentration of 11–13 mg protein/ml of swollen beads. The pT7T3 plasmid containing human Cdc20/Fizzy cDNA (Shteinberg *et al.*, 1999) was generously provided by Dr. Michael Brandeis, the Hebrew University, Jerusalem.

## Procedure

HeLa S3 cells (American Type Culture Collection) are grown in suspension culture in Dulbecco's modified Eagles medium supplemented with 10% fetal bovine serum at 37° in the presence of 5% $CO_2$. To arrest cells in mitosis, logarithmically growing cultures (4–6 × $10^5$ cells/ml) are treated with nocodazole (0.2 $\mu$g/ml) for 18 h. Unless specified otherwise, all subsequent operations are carried out at 0–4°. Cells from a 2-liter culture are harvested by centrifugation (400*g*, 10 min) and are washed twice with phosphate-buffered saline (PBS). All cells are collected to a single centrifuge tube with a further wash with PBS, and their amount is estimated by weighing. The cells are suspended in 3 volumes of hypotonic buffer [20 m*M* HEPES-NaOH (pH 7.6), 1.5 m*M* $MgCl_2$, 0.5 m*M* KCl and 1 m*M* dithiothreitol (DTT)] and are collected immediately by centrifugation. The pellet is suspended in 2 volumes (relative to the original volume, prior to the hypotonic wash) of the aforementioned hypotonic buffer that contains leupeptin and chymostatin (10 $\mu$g/ml, each). The sample is allowed to swell on ice for 30 min and then the cells are disrupted by Dounce homogenization (30 strokes). Following centrifugation at 40,000*g* for 30 min, the supernatants are collected, mixed with glycerol [10% (v/v), final concentration], and stored at –70°.

Prior to affinity chromatography, extracts from nocodazole-arrested HeLa cells are incubated with MgATP and okadaic acid to hyperphosphorylate APC/C by endogenous mitotic protein kinases. The reaction mixture contains in a volume of 4 ml: 50 m$M$ HEPES-NaOH (pH 7.2), 5 m$M$ $MgCl_2$, 1 m$M$ DTT, 1 m$M$ ATP, 10 m$M$ phosphocreatine, 100 $\mu$g/ml creatine phosphokinase, ~20 mg of protein extract from nocodazole-arrested HeLa cells, and 1 $\mu M$ okadaic acid. The sample is incubated at 30° for 60 min and is subsequently centrifuged (15,000$g$, 10 min) to remove some insoluble material. The supernatant is mixed with 1 ml (packed volume) of p13$^{suc1}$-Sepharose beads, which were washed twice previously (400$g$, 5 min) with 10-ml portions of 20 m$M$ Tris-HCl (pH 7.2) and 1 m$M$ DTT (buffer A) The sample is mixed with beads by rotation in a 15-ml tube at 60 rpm for 1 h at room temperature and is then transferred to a column (0.7 cm diameter) at 4°. All subsequent operations are at 0–4°. The column is washed with 45 ml of buffer A that contains 300 m$M$ KCl and then with 15 ml buffer A. APC/C is eluted from p13$^{suc1}$-Sepharose beads with 20 ml of buffer A that contains 50 m$M$ $p$-nitrophenyl phosphate and 0.2 mg/ml STI. Elution is at a flow rate of 1–1.5 ml/min. The addition of STI is necessary in order to prevent the adsorption of the dilute enzyme to surfaces. STI was chosen as the carrier protein because it has a relatively low molecular mass (20 kDa) and thus does not interfere with the detection of most larger APC/C subunits by SDS–PAGE. The eluate is concentrated to ~1 ml by centrifuge ultrafiltration (Centriprep–10; Amicon), diluted 10-fold with buffer A that contains 20% (v/v) glycerol, and concentrated again to a volume of ~1 ml. The preparation is divided into small samples in prechilled test tubes and stored at –70°. Because repeated freezing–thawing of APC/C preparations results in loss of activity, each sample should be used only once. The activity of APC/C in such preparations is usually 30–40 units/$\mu$l (see later). Following use, p13$^{suc1}$-Sepharose beads are regenerated by washing with 30 ml of 50 m$M$ Tris-HCl (pH 9.0), followed by a wash with 30 ml of 1 $M$ KCl in 50 m$M$ Tris-HCl (pH 7.2). The beads are stored at 4° in 50 m$M$ Tris-HCl (pH 7.2) containing 0.02% sodium azide. p13$^{suc1}$-Sepharose beads can be reused up to 10 times over a period of 1–2 years.

## Assay of Cyclin-Ubiquitin Ligase Activity of APC/C

Reaction mixtures contain in a volume of 10 $\mu$l: 40 m$M$ Tris-HCl (pH 7.6), 1 mg/ml BSA, 1 m$M$ DTT, 5 m$M$ $MgCl_2$, 10 m$M$ phosphocreatine, 50 $\mu$g/ml creatine phosphokinase, 0.5 m$M$ ATP, 50 $\mu M$ ubiquitin, 1 $\mu M$ ubiquitin aldehyde, 1 pmol E1, 5 pmol E2-C, 1 $\mu M$ okadaic acid, 1–2 pmol of $^{125}$I-labeled cyclin B-(13–91)/protein A (1–2 $\times$ $10^5$ cpm), 0.5–1 $\mu$l of affinity-purified APC/C, and 0.4 $\mu$l of Cdc20/Fizzy produced by *in vitro* translation in reticulocyte lysate as described (Shteinberg *et al.*,

1999). Activity of APC/C in crude mitotic extracts is determined under similar conditions, except that 10–20 $\mu$g protein of extract is added and 10 $\mu M$ staurosporine is also supplemented, to prevent the phosphorylation and inactivation of Cdc20/Fizzy by mitotic protein kinases (Yudkovsky *et al.*, 2000). In parallel control incubations, APC/C or Cdc20 are deleted. Following incubation at 30° for 1 h, samples are subjected to electrophoresis on a 12.5% polyacrylamide–SDS gel. Results are quantified by a phosphorimager and are expressed as the percentage of $^{125}$I-labeled cyclin converted to ubiquitin conjugates. A unit of activity is defined as that converting 1% of $^{125}$I-labeled cyclin to ubiquitinylated derivatives under the aforementioned assay conditions, in the range of assay linear with APC/C concentration.

## References

Brizuela, L., Draetta, G., and Beach, D. (1987). p13$^{suc1}$ acts in the fission yeast cell division cycle as a component of the p34$^{cdc2}$ protein kinase. *EMBO J.* **6,** 3507–3514.

Fang, G., Yu, H., and Kirschner, M. W. (1998). The checkpoint protein Mad2 and the mitotic regulator Cdc20 form a ternary complex with the anaphase-promoting complex to control anaphase initiation. *Genes Dev.* **12,** 1871–1883.

Glotzer, M., Murray, A. W., and Kirschner, M. W. (1991). Cyclin is degraded by the ubiquitin pathway. *Nature* **349,** 132–138.

Golan, A., Yudkovsky, Y., and Hershko, A. (2002). The cyclin-ubiquitin ligase activity of the cyclosome/APC is jointly activated by protein kinases Cdk1/cyclin B and Plk. *J. Biol. Chem.* **277,** 15552–15557.

Hershko, A., Heller, H., Elias, S., and Ciechanover, A. (1983). Components of ubiquitin-protein ligase system: Resolution, affinity purification and role in protein breakdown. *J. Biol. Chem.* **258,** 8206–8214.

King, R. W., Peters, J. M., Tugendreich, S., Rolfe, M., Hieter, P., and Kirschner, M. W. (1995). A 20S complex containing Cdc27 and Cdc16 catalyzes the mitosis-specific conjugation of ubiquitin to cyclin. *Cell* **81,** 279–288.

Kramer, E. R., Gieffers, C., Holzl, G., Hengstschlager, M., and Peters, J. M. (1998). Activation of human anaphase-promoting complex by proteins of the CDC20/Fizzy family. *Curr. Biol.* **8,** 1207–1210.

Mayer, A. N., and Wilkinson, K. D. (1989). Detection, resolution, and nomenclature of multiple ubiquitin carboxyl-terminal esterases from bovine calf thymus. *Biochemistry* **28,** 166–172.

Peters, J. M. (2002). The anaphase-promoting complex: Proteolysis in mitosis and beyond. *Mol. Cell* **9,** 931–943.

Shteinberg, M., Protopopov, Y., Listovsky, T., Brandeis, M., and Hershko, A. (1999). Phosphorylation of the cyclosome is required for its stimulation by Fizzy/Cdc20. *Biochem. Biophys. Res. Commun.* **260,** 193–198.

Sitry, D., Seeliger, M., Ko, T. K., Ganoth, D., Breward, S. E., Itzhaki, L. S., Pagano, M., and Hershko, A. (2002). Three different binding sites of Cks1 are required for p27-ubiquitin ligation. *J. Biol. Chem.* **277,** 42233–42240.

Sudakin, V., Ganoth, D., Dahan, A., Heller, H., Hershko, J., Luca, F. C., Ruderman, J. V., and Hershko, A. (1995). The cyclosome, a large complex containing cyclin-selective ubiquitin ligase activity, targets cyclins for destruction at the end of mitosis. *Mol. Biol. Cell* **6,** 185–197.

Sudakin, V., Shteinberg, M., Ganoth, D., Hershko, J., and Hershko, A. (1997). Binding of activated cyclosome to p13$^{suc1}$: Use for affinity purification. *J. Biol. Chem.* **272,** 18051–18059.

Townsley, F. M., Aristarkhov, A., Beck, S., Hershko, A., and Ruderman, J. V. (1997). Dominant-negative cyclin-selective ubiquitin-carrier protein E2-C/UbcH10 blocks cells in metaphase. *Proc. Natl. Acad. Sci. USA* **94,** 2362–2367.

Yudkovsky, Y., Shteinberg, M., Listovsky, T., Brandeis, M., and Hershko, A. (2000). Phosphorylation of Cdc20/fizzy negatively regulates the mammalian cyclosome/APC in the mitotic checkpoint. *Biochem. Biophys. Res. Commun.* **271,** 299–304.

Zachariae, W., and Nasmyth, K. (1999). Whose end is destruction: Cell division and the anaphase-promoting complex. *Genes Dev.* **13,** 2039–2058.

# [16] Large-Scale Purification of the Vertebrate Anaphase-Promoting Complex/Cyclosome

*By* FRANZ HERZOG and JAN-MICHAEL PETERS

## Abstract

The anaphase-promoting complex or cyclosome (APC/C) is a ubiquitin ligase that controls progression through mitosis and the G1 phase of the cell cycle. The APC/C is a 1.5-MDa complex composed of at least 12 different core subunits. At different stages of mitosis and G1, the APC/C associates with a variety of regulatory proteins, such as the activator proteins Cdc20 and Cdh1 and the mitotic checkpoint complex (MCC), which regulate APC/C activity in a substrate-specific manner. Although APC/C and its regulators have been under intense investigation, it is still poorly understood how substrates are recognized and ubiquitinated by the APC/C, why so many subunits are required for these processes, and how regulators of the APC/C control its ubiquitin ligase activity in a substrate-specific manner. This chapter describes a simple and rapid procedure that allows the isolation of APC/C from vertebrate cells and tissues with reasonable purity and at high concentrations, yielding up to 0.5 mg of APC/C. This procedure should facilitate biochemical, biophysical, and structural analyses of the APC/C that will be needed for a better mechanistic understanding of its function and regulation.

## Introduction

The anaphase-promoting complex/cyclosome is a ubiquitin ligase that controls important transitions during mitosis by targeting regulatory proteins for destruction by the 26S proteasome (reviewed in Harper *et al.*,

METHODS IN ENZYMOLOGY, VOL. 398 0076-6879/05 $35.00

DOI: 10.1016/S0076-6879(05)98016-5

2002; Peters, 2002). The APC/C is a multisubunit complex that contains at least 12 different protein subunits of unknown stoichiometry (Table I). Among these are a cullin protein (Apc2) and a RING finger protein (Apc11), which are thought to recruit the E2 enzymes UbcH10 and UbcH5 to the APC/C, but the function of other subunits is poorly understood. Together, the APC/C subunits form a particle whose sedimentation coefficient in metazoan species is 22S (King *et al.*, 1995) and whose mass is estimated to be 1.5 MDa (Sudakin *et al.*, 1995).

The ubiquitin ligase activity of the APC/C is stimulated in early mitosis by binding to the protein Cdc20 (Fang *et al.*, 1998; Kramer *et al.*, 1998;

TABLE I
APC/C SUBUNITS, ACTIVATORS, AND COLLABORATING E2 ENZYMES[a]

| Protein | Vertebrates | $M_r$ | Domains | Proposed functions |
|---|---|---|---|---|
| APC/C subunit | Apc1 | 200 | PC repeats | |
| | Apc2 | 105 | Cullin | Apc11 binding |
| | Cdc27/Apc3 | 94 | TPRs | Cdh1 binding |
| | Apc4 | 92 | | |
| | Apc5 | 75 | | |
| | Cdc16/Apc6 | 70 | TPRs | |
| | Apc7 | 62 | TPRs | |
| | Cdc23/Apc8 | 60 | TPRs | |
| | Doc1/Apc10 | 24 | DOC | Substrate binding, processivity |
| | Cdc26 | 14 | | |
| | Apc11 | 10 | RING-H2 | UbcH5 binding, *in vitro* ubiquitination activity |
| | Apc13/Swm1[b] | 8.5 | | |
| APC/C activator | Cdc20 | 55 | WD40 | Substrate recruitment |
| | Cdh1 | 55 | WD40 | |
| E2 | UbcH5/Ubc4 | 16 | UBC | Ubiquitin transfer to substrate |
| | UbcH10/UbcX/ E2-C | 21 | UBC | |

[a] In somatic vertebrate cells, active APC/C consists of at least 12 constitutive subunits and either one of two transiently associated activators, Cdc20 and Cdh1. $M_r$, molecular mass; PC repeats, proteasome/cyclosome repeats; TPR, tetratrico peptide repeat; UBC, ubiquitin conjugating enzyme; WD40, Trp-Asp repeats.

[b] Apc13/Swm1 has been identified as an APC/C subunit in yeast (Yoon *et al.*, 2002). Data of Schwickart *et al.* (2004) and tandem mass spectrometric analysis of APC/C immunoprecipitates (F. H., J. Hutchins, K. Mechtler and J.-M. P., unpublished data) indicate that an ortholog of Apc13/Swm1 is also present in human APC/C.

Visintin *et al.*, 1997). This interaction depends on prior phosphorylation of APC/C subunits by mitosis-specific kinases (for details, see Herzog *et al.*, 2005). In prometaphase, the resulting APC/C$^{Cdc20}$ complex mediates destruction of some substrates such as cyclin A, whereas other APC/C$^{Cdc20}$ substrates such as cyclin B and securin are only degraded in metaphase, i.e., once all chromosomes have been attached to both poles of the mitotic spindle (Clute and Pines, 1999; Dawson *et al.*, 1995; Sigrist *et al.*, 1995). Securin destruction results in activation of separase, a protease that dissolves sister chromatid cohesion by cleaving cohesin complexes, whereas cyclin degradation leads to inactivation of cyclin-dependent kinase 1 (Cdk1). Both of these events are required for anaphase and for subsequent exit from mitosis (reviewed by Peters, 2002). The temporal delay of securin and cyclin B destruction relative to cyclin A degradation is caused by the spindle checkpoint (den Elzen and Pines, 2001; Geley *et al.*, 2001; Hagting *et al.*, 2002). This surveillance mechanism detects the presence of kinetochores that are not fully occupied with spindle microtubules, activates inhibitors of APC/C$^{Cdc20}$ such as the Mad2 and BubR1 subunits of the mitotic checkpoint complex (MCC) at these unattached kinetochores, and thereby prevents separase activation by APC/C$^{Cdc20}$ until all chromosomes have been attached to a full complement of microtubules that emanate from opposite spindle poles (reviewed by Musacchio and Hardwick, 2002).

During exit from mitosis, Cdc20 itself is degraded in an APC/C-dependent manner and is replaced by a related activator protein, Cdh1, which keeps APC/C active throughout G1 phase and in postmitotic differentiated cells (reviewed by Harper *et al.*, 2002; Peters, 2002). In cycling cells that enter S phase, APC/C$^{Cdh1}$ is disassembled by the phosphorylation of Cdh1 and possibly by binding of the protein Rca1/Emi1 to Cdh1, allowing the accumulation of APC/C substrates such as mitotic cyclins that are essential for entry into the subsequent mitosis.

Cdc20 and Cdh1 are WD40 repeat proteins that are predicted to form propeller-like structures, as do some of the substrate adaptor proteins of SCF ubiquitin ligases (Orlicky *et al.*, 2003; Wu *et al.*, 2000). This observation, the finding that Cdc20 and Cdh1 can confer some substrate specificity to APC/C-mediated ubiquitination reactions, and the observation that Cdc20 and Cdh1 can bind APC/C substrates, at least *in vitro* or when overexpressed, have led to the hypothesis that these proteins activate the APC/C by recruiting substrates (reviewed by Vodermaier, 2001). A study in *Xenopus* has shown, however, that model substrates of the APC/C can bind stably to the APC/C itself, but not to Cdc20 (Yamano *et al.*, 2004). It remains therefore poorly understood how Cdc20 and Cdh1 activate the ubiquitin ligase activity of the APC/C.

Likewise, it is unknown how the APC/C recognizes and ubiquitinates substrates and how the resulting substrate-ubiquitin conjugates are released and transferred to the 26S proteasome. It is also unknown which of these reaction steps are inhibited by spindle checkpoint proteins and by Rca1/Emi1 and how the spindle checkpoint inhibits APC/C$^{Cdc20}$ in a substrate-specific manner. Answering these questions will require a variety of biochemical, biophysical, and structural approaches. All of these would benefit from an experimental system in which APC/C or subcomplexes could be reconstituted from recombinant proteins. Unfortunately, however, it has not been possible to date to reconstitute functional APC/C or subcomplexes, with the exception of Apc2-Apc11 heterodimers (Tang *et al.*, 2001; Vodermaier *et al.*, 2003). Experiments that aim at a molecular understanding of APC/C function and regulation are therefore presently limited to analyses of endogenous APC/C and subcomplexes that can be isolated from cells or tissues. To facilitate these approaches, we have developed a procedure that allows the rapid purification of APC/C from vertebrate cells and tissues at relatively large scale.

## Preparation of APC/C Antibodies

### *Generation of APC/C Antisera*

We use antibodies raised against a carboxy-terminal peptide derived from human Cdc27 for the immunoaffinity chromatographic purification of APC/C. These antibodies recognize Cdc27 from a variety of vertebrate species and can therefore be used to purify APC/C not only from human cells, but also from bovine brain and *Xenopus laevis* eggs. Synthetic peptides composed of 17 amino acid residues with the sequence CTDADDTQLHAAESDEF are coupled via the amino-terminal cysteine residue to keyhole limpet hemocyanin (KLH; Merck Biosciences, Nottingham, UK) following the manufacturer's instructions. Three milligrams of lyophilized peptide is dissolved in 700 $\mu$l of buffer containing 3 *M* guanidine hydrochloride and 0.5 *M* sodium tetraborate, pH 6.9, and the solution is added to 300 $\mu$l of the same buffer in which 3 mg KLH has been dissolved. After removal of oxygen by aerating with argon, the reaction mixture is incubated for 1 h at room temperature and subsequently stored at 4°. To determine the efficiency of the coupling reaction, 10-$\mu$l aliquots of the peptide-KLH solution are removed at the beginning and the end of the reaction, mixed with 50 $\mu$l of Ellman's reagent [50 m*M* sodium acetate, 2 m*M* 5,5'-dithiobis(2-nitrobenzoic acid); Sigma-Aldrich, St. Louis, MO], 100 $\mu$l solution containing 1 *M* Tris-HCl, pH 8.0, and 840 $\mu$l water, and the absorbance of the samples is measured at 412 nm ("Ellman test"). A decrease in absorbance indicates a reduction in the amount of free thiol groups and is a measure for the efficiency of peptide coupling to KLH (Bulaj *et al.*, 1998; Ellman, 1959). The coupled peptides are used to immunize rabbits at a commercial facility (Gramsch Laboratories, Schwabhausen, Germany). The resulting sera are frozen in 20-ml aliquots in liquid nitrogen and stored at –80°.

### *Generation of Peptide Affinity Columns for the Purification of APC/C Antibodies*

Poros beads (Applied Biosystems, Foster City, CA) are activated with the bifunctional cross-linker sulfo-GMBS (Pierce, Rockford, IL) according to the manufacturer's instructions, which results in the generation of maleimid groups. Five milligrams of lyophilized immunogenic peptide is dissolved in 3 ml buffer containing 3 *M* guanidine hydrochloride and 0.1 *M* HEPES-KOH, pH 7.9, yielding a peptide concentration of ~2.27 $\mu$mol. Activated Poros beads (310 mg) are mixed with the peptide solution and the suspension is incubated on a rotary shaker for 1 h at room temperature. Subsequently, the Poros beads are collected by low-speed centrifugation, and the coupling reaction is stopped by washing with at least 10 bead volumes of a buffer containing 150 m*M* NaCl, 20 m*M* Tris-HCl, pH 7.5. The efficiency of cross-linking is determined by the Ellman test (see earlier discussion). Resin coupled to peptides can be stored in batch at 4°. To obtain homogenously packed affinity columns, the coupled resin is transferred into a HPLC column (0.7 ml; Applied Biosystems) by employing a filling device (Applied Biosystems) that is connected to a BIOCAD HPLC system (Applied Biosystems).

### *Antibody Purification and Concentration*

The affinity column prepared earlier is connected to a BIOCAD HPLC and is equilibrated with 20 column volumes (cv) of HBS buffer (20 m*M* HEPES-KOH, 150 m*M* NaCl) at a flow rate of 7 ml/min. All chromatography steps are carried out at room temperature and are monitored by measuring the absorbance on-line at 280 nm. Aliquots of antiserum prepared as described earlier are thawed rapidly in a water bath at 37°. The thawed serum is extracted once for 2 min with an equal volume of chloroform to remove lipids. To separate the organic fraction from the serum, the emulsion is centrifuged at 2000 rpm in a Varifuge 3.0R (Heraeus; Kendro, Asheville, NC) for 15 min at 4°. Subsequently, the serum is filtered through a 0.45-$\mu$m syringe filter (MILLEX-HA; Millipore, Billerica, MA) and stored on ice. We purify antibodies from 20 ml of serum by four subsequent chromatography runs in each of which 5-ml aliquots are processed. Five milliliters of chloroform-extracted serum is loaded onto the column with a flow rate of 5 ml/min (column pressure: 20–25 bar). The column is washed with 50 cv of HBS. Bound proteins are eluted first with buffer A (1.5 *M* $MgCl_2$, 100 m*M* sodium acetate, pH 5.2) and subsequently by buffer B (100 m*M* glycine-HCl, pH 2.45). The pH of the eluted fractions is immediately neutralized by the addition of 1/10 volume of 2 *M* HEPES-KOH, pH 7.9. Fractions obtained by elution with buffer A or with buffer B are kept separate during all subsequent steps because antibodies eluted by high $MgCl_2$ concentrations (buffer A) or by low pH (buffer B) can differ in their abilities to recognize their antigen under different conditions. The peak fractions are dialyzed overnight at 4° against 2 liters of buffer C [20 m*M* HEPES-KOH, 150 m*M* NaCl, 10% glycerol, 0.5 m*M* dithiothreitol (DTT)] using dialysis tubing with a molecular mass cut-off (MWCO) of 3500 Da (Spectra/Por; Spectrum Laboratories, Rancho Dominguez, CA). Buffer C is exchanged once after 3 h. Subsequently, the antibodies are concentrated by ultrafiltration in a Centriprep centrifugal filter device (YM–30, MWCO 30 kDa, 15 ml, Amicon; Millipore) at 3000 rpm (Varifuge 3.0R) and 4°. The final protein concentration is determined by photometric measurements at a wavelength of 280 nm. An $OD_{280} = 1$ of a mixture of IgG and IgM corresponds to 1.4 mg/ml protein (Harlow and Lane, 1988). The purity of the samples is assessed by SDS–PAGE and silver staining, and 500-$\mu$l aliquots of the purified antibodies are frozen in liquid nitrogen and stored at –80°. A typical purification yields 6 to 9 mg of pure antibodies from 20 ml of antiserum, but depending on the specific serum, higher or lower yields can also be obtained.

### *Coupling and Cross-Linking of Antibodies to Affi-Prep Protein A Beads*

Antibodies purified as described earlier are coupled to Affi-prep protein A beads (Bio-Rad) by using a modified version of the protocol described by Harlow and Lane (1988). Affi-prep beads (0.5 ml) are washed twice with 10 volumes TBS-T [20 m*M* Tris-HCl, pH 7.5, 150 m*M* NaCl, 0.04% Tween 20 (Sigma)] in a 15-ml Falcon tube (BLUE MAX; Becton Dickinson, Franklin Lakes, NJ). The washed beads are resuspended in 10 volumes of TBS-T, 500 $\mu$g antibodies are added, and the resulting suspension is mixed on a rotary shaker for 1 h at room temperature or overnight at 4°. All subsequent steps are performed at room temperature. Unbound protein is removed from the beads by washing three times with 10 volumes TBS-T. To couple the bound antibodies covalently to protein A, the beads are washed twice with 20 volumes 0.2 *M* sodium tetraborate, pH 9.2 (Sigma-Aldrich), and the beads are resuspended in 20 volumes of the same buffer. Solid dimethylpimelimidate (Sigma-Aldrich) is added to a final concentration of 20 m*M*, dissolved immediately by gentle mixing, and the suspension is incubated at room temperature for 30 min on a rotary shaker. The cross-linking reaction is stopped by washing the beads twice for 10 min with 20 volumes of a buffer containing 150 m*M* NaCl and 200 m*M* Tris-HCl, pH 7.5, and twice with TBS-T.

To avoid the elution of antibodies that have not been coupled efficiently to the beads in subsequent APC/C purification steps, the beads can be resuspended twice in 10 volumes of a solution containing 100 m*M* glycine-HCl, pH 2.2. They are subsequently washed three times with TBS-T to restore a neutral pH. The antibody-coupled beads are then stored in TBS-T, containing 0.03% $NaN_3$, at 4°.

## Preparation of Protein Extracts from Mammalian Cells and Tissues

### *Preparation of HeLa Cell Lysates*

For small-scale purification of APC/C, adherent HeLa cells are grown in 145-mm tissue culture dishes (Greiner bio-one, Frickenhausen, Germany) at 37° in the presence of 5% $CO_2$ in high glucose Dulbecco's modified Eagle medium supplemented with 10% (v/v) fetal bovine serum (PAA Laboratories GmbH, Pasching, Austria), 0.3 $\mu$g/ml L-glutamine (Sigma-Aldrich), 100 units/ml penicillin (Sigma-Aldrich), and 100 $\mu$g/ml streptomycin (Sigma-Aldrich). Cells are harvested by scraping with a rubber policeman and by subsequent centrifugation at 1200 rpm for 5 min at 4° in a Biofuge (Heraeus). The cell pellet is washed twice with ice-cold phosphate-buffered saline (PBS; 140 m*M* NaCl, 2.5 m*M* KCl, 10 m*M* $Na_2HPO_4$, 1.4 m*M* $KH_2PO_4$, pH 7.4). A pellet obtained from 20 dishes (approximately 1 ml containing $2 \times 10^8$ cells) is resuspended in one volume (~1 ml) of ice-cold extraction buffer [20 m*M* Tris-HCl, pH 7.5, 150 m*M* NaCl, 20 m*M* $\beta$-glycerophosphate, 5 m*M* $MgCl_2$, 1 m*M* NaF, 1 $\mu M$ okadaic acid, 1 m*M* DTT, 0.2% NP–40, 10% glycerol, 10 $\mu$g/ml each of chymostatin, leupeptin, and pepstatin A (Sigma-Aldrich) and homogenized in a prechilled Potter–Elvehjem glass–Teflon homogenizer (Wheaton Scientific Products, Millville, NJ) for 20 min by striking the piston 10 times every 10 min. To remove the bulk of cell debris and chromatin, lysates are centrifuged at 100,000*g* at 4° for 20 min in a Beckman Optima MAX ultracentrifuge using a Beckman TLA 45 rotor (42,000 rpm) (Beckman Coulter, Krefeld, Germany). Cells from 20 145-mm dishes typically yield 1.5–2 ml of supernatant fraction, which contains 25 mg/ml protein. Aliquots of 0.5 ml are shock frozen in liquid nitrogen and stored at –80°. Because APC/C subunits (especially Cdc27 and Apc7) are quite sensitive to proteolytic degradation, it is important to leave the cells as short as possible at room temperature prior to harvesting and to perform all subsequent steps on ice.

For large-scale purification of APC/C, HeLa cells grown in suspension can be used that are purchased from a commercial supplier as frozen cell pellets ($5 \times 10^9$ cells) in 25-ml aliquot sizes (CILBIOTECH s.a., MONS, Belgium).

### *Preparation of* Xenopus laevis *Egg Extracts*

Female clawed frogs of the species *X. laevis* are obtained from commercial suppliers (Nasco, USA; African Reptile Farm, South Africa) and are kept according to standard procedures (Wu and Gerhart, 1991). For the generation of interphase extracts from laid eggs we use a modified version of the protocol of Murray (1991) for the preparation of "cycling extracts." In brief, laid eggs are first dejellied by incubating them in a solution containing 2% (w/v) cysteine-NaOH, pH 7.8, and subsequently washed four times in MMR (10 m*M* NaCl, 0.2 m*M* KCl, 0.1 m*M* $MgCl_2$, 0.2 m*M* $CaCl_2$, 0.01 m*M* EDTA, 0.5 m*M* HEPES-KOH, pH 7.8). To stimulate exit from meiosis II into interphase, the eggs are activated by incubating them for 3 to 5 min in MMR buffer containing 0.4 $\mu$g/ml ionophore (A23187; Merck Biosciences) until the pigmented animal pole of the eggs begins to contract. Then the eggs are washed three times in XB buffer (10 m*M* HEPES-KOH, pH 7.7, 100 m*M* KCl, 1 m*M* $MgCl_2$, 0.1 m*M* $CaCl_2$) and once in XB containing 10 $\mu$g/ml each of the protease inhibitors chymostatin, pepstatin A, and leupeptin (Sigma-Aldrich). Subsequently, they are transferred with a large-bore pipette into Ultra-Clear centrifugation tubes ($14 \times 89$ mm; Beckman). Fifteen minutes after activation the eggs are tightly packed by centrifugation at 1000 rpm for 1 min in a Central CL2 centrifuge (IEC, DJB labcare, Bucks, UK) and as much buffer as possible is removed from the eggs [in contrast to Murray (1991), we perform this centrifugation step without the addition of mineral oil]. Up until this point all steps are carried out at 19 to 21°. To crush the eggs, they are centrifuged at 10,000 rpm for 10 min at 4° in a HB4 or HB6 rotor in a Sorvall RC 5C PLUS centrifuge (Kendro). To obtain egg extracts that can be stably arrested in an interphase state by the addition of cycloheximide (see later), this centrifugation step should be started 20 min after egg activation has occurred. If the eggs are used later for the generation of mitotic extracts by the addition of recombinant nondegradable cyclin B ($\Delta$90 cyclin B; Glotzer *et al.*, 1991), the eggs can also be left at room temperature for up to 40 min after activation, which allows more cyclin B to be synthesized and facilitates mitotic entry *in vitro*. The straw-colored cytoplasmic layer is collected by side puncture of the centrifugation tube using a 3-ml syringe

(Omnifix, Melsungen, Germany) and a 16-gauge, 1.5-in. needle (EMOSA, Trier, Germany). To prevent cyclin synthesis and actin polymerization, 100 $\mu$g/ml cycloheximide (Sigma-Aldrich) and 10 $\mu$g/ml cytochalasin B (Sigma-Aldrich) are added to the cytoplasmic fraction, respectively. The extract is clarified further by a second centrifugation step, which is carried out as the first one. For storage, sucrose is added to the extract from a 1.5 *M* stock solution in XB to a final concentration of 200 m*M*, and an ATP-regenerating system is added (20$\times$ stock: 150 m*M* creatine phosphate, 20 m*M* ATP, pH 7.4, 2 m*M* EGTA, pH 7.7, 20 m*M* $MgCl_2$). Aliquots of the extract are frozen in liquid nitrogen and stored at $-80°$. Typically, eggs from one frog yield 1–2 ml of extract with a protein concentration (after addition of sucrose and ATP-regenerating system) of 30–50 mg/ml.

### *Preparation and Fractionation of Cow Brain Extracts*

Fresh cow brains are obtained from a local slaughter house, stored on ice for transport, and processed immediately afterward. The tissue is sliced into small pieces using a scalpel and suspended in 2 volumes of ice-cold XB buffer containing 10 $\mu$g/ml each of chymostatin, leupeptin, and pepstatin A (Sigma-Aldrich) and 0.5 m*M* phenylmethysulfonyl fluoride (Pierce). The brain tissue is minced in short intervals in an electrical blender (Kenwood) and subsequently homogenized in a prechilled Potter–Elvehjem glass–Teflon homogenizer for 30 min by striking the piston 10 times every 10 min. The brain extract is precleared by centrifugation in a Sorvall GSA rotor at 8000 rpm for 15 min and 4°. The resulting cow brain extract can be frozen in liquid nitrogen and stored at –80°.

APC/C can be isolated from crude cow brain extracts by immunoaffinity chromatography (see earlier discussion); however, both the yield and the puritiy of APC/C obtained under these conditions are low (data not shown). Before immunoprecipitation, we therefore enrich for APC/C by first fractionating the cow brain extract by anion-exchange chromatography. The extract, typically containing 10 mg/ml protein, has to be further clarified by centrifugation for 1 h in a Beckman SW28 rotor at 24,000 rpm and 4° and subsequent filtration through a 0.45-$\mu$m syringe filter (MILLEX-HA; Millipore). Anion-exchange chromatography is performed on a 50-ml Source 15Q column (Amersham Bioscienes, Uppsala, Sweden) connected to an ÄKTA purifier system (Amersham Biosciences) with a flow rate of 5 ml/min at 4°. The Source 15Q column is first washed with 2 cv of a buffer containing 2 *M* NaCl and 20 m*M* Tris-HCl, pH 7.5, and subsequently equilibrated with 3 cv of buffer containing 150 m*M* NaCl and 20 m*M* Tris-HCl, pH 7.5. Cow brain extract (120 ml) is loaded, and the column is washed with buffer containing 150 m*M* NaCl and 20 m*M*

Tris-HCl, pH 7.5, until the absorbance at 280 nm reaches a value of 0.05. Bound proteins are eluted from the column by a linear salt gradient ranging from 150 to 500 m*M* NaCl with 5 cv of 20 m*M* Tris-HCl, pH 7.5, buffer. Ten-milliter fractions are collected and desalted by dialyzing for 3 h at 4° against 5 liters solution containing 150 m*M* NaCl, 1 m*M* DTT, and 20 m*M* Tris-HCl, pH 7.5, using dialysis tubing with a MWCO of 3500 Da (Spectra/Por; Spectrum Laboratories). The dialyzed fractions are concentrated to 3–5 ml in Centriprep centrifugal filter devices (YM–30, MWCO 30,000 DA, 15 ml, Amicon) at 3000 rpm (Varifuge 3.0R; Heraeus) and 4°. Fractions are frozen in liquid nitrogen and stored at –80°. Aliquots of each fraction are analyzed by SDS–PAGE and immunoblotting for the presence of APC/C. APC/C typically elutes at a concentration of 400–450 m*M* NaCl.

## Immunopurification of APC/C Using Immunoaffinity Chromatography

### *Purification of APC/C at Small Scale*

We have used the following protocol, which is based on a method described in Gieffers *et al.* (2001), to purify APC/C from HeLa lysates, *Xenopus* egg extracts, and cow brain anion-exchange fractions described earlier (collectively called "protein extracts" from now on). We have used this protocol to isolate APC/C from extracts containing between 0.6 and 20 mg protein, which requires between 3 and 100 $\mu$l of Affiprep beads that have been coupled to Cdc27 antibodies. All solutions are precooled and stored on ice. Aliquots of protein extracts are thawed rapidly at 37° and then kept on ice. One volume of Cdc27 antibody beads is preequilibrated once in 20 volumes TBS containing 0.01% Tween 20 (TBS–0.01T) and resuspended in 3 volumes of the same buffer. This suspension is mixed with 10 volumes protein extract (protein concentration: 20 mg/ml) and incubated on a rotary shaker for 1–2 hours at 4°. Subsequently, the extract is removed and the beads are washed. The washing procedure is performed at 4° with 10 to 15 bead volumes and 10-min incubation times per step and comprises two short washing steps with TBS–0.01T and five washing steps with TBS–0.01T containing 400 m*M* NaCl. Finally, the beads are washed twice with TBS without detergent, and as much residual buffer as possible is removed after the final washing step.

Elution of the proteins that are bound to the antibody beads is performed by either one of two different protocols, depending on the planned use of the eluate. To obtain APC/C that can be analyzed by SDS–PAGE, denaturing immunoprecipitation, or mass spectrometry, protein is eluted at 4° under denaturing conditions by resuspending beads with a pipette for

30–60 s in 1.5 volumes of a solution containing 100 m*M* glycine-HCl, pH 2.2. The pH of the eluate is subsequently raised by the addition of 1.5 *M* Tris-HCl, pH 9.2, to a final pH of 8. The samples can be stored at –20 or –80°.

The following procedure is used to obtain native APC/C that can be analyzed in further native fractionation steps, in electron microscopy experiments, in ubiquitination assays, or in protein-binding assays, for example, to measure interactions with activator proteins (Gieffers *et al.*, 2001; Vodermaier *et al.*, 2003). The beads are resuspended with 1.5 volumes of TBS containing 0.5 m*M* DTT and 1 mg/ml of the antigenic Cdc27 peptide. The suspension is incubated for 1–2 h on a rotary shaker at 4°, and subsequently the beads are removed by centrifugation. The purity of APC/C in the supernatant fraction is assessed by SDS–PAGE and staining with either silver (neutral silver staining; Harlow and Lane, 1988) (Figs. 1, 3, and 4) or Coomassie (Coomassie brilliant blue G-250;

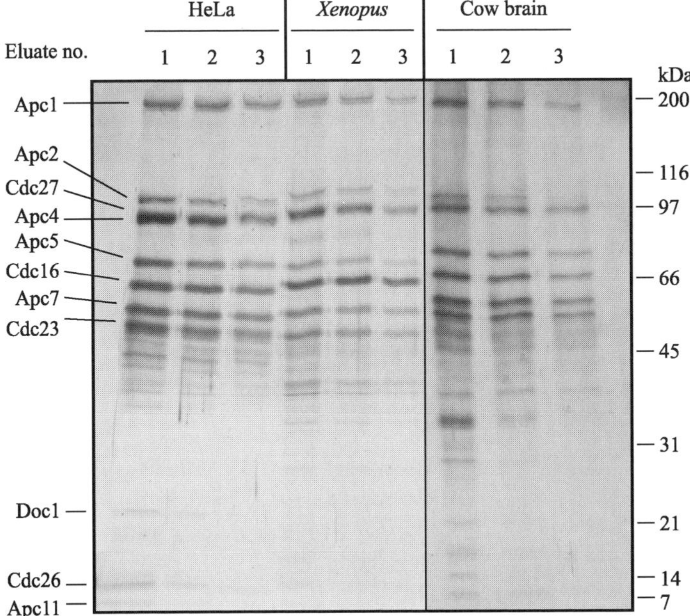


FIG. 1. Small-scale purification of APC/C. APC/C was immunopurified from extracts of logarithmically proliferating HeLa cells, cow brain, and *Xenopus laevis* interphase eggs using Cdc27 antibody beads (see text). Bound proteins were eluted three times with one bead volume of TBS containing 1 mg/ml antigenic Cdc27 peptide and 0.5 m*M* DTT, and 10 $\mu$l of each eluate was analyzed by SDS–PAGE and silver staining. The identity of different APC/C subunits is indicated on the left, and the molecular mass (in kDa) and position of marker proteins are indicated on the right.

Harlow and Lane, 1988) (Fig. 5). In the same experiment the concentration of APC/C is estimated by comparing the staining intensity of APC/C subunits to the staining intensity of known amounts of a marker protein, such as bovine serum albumin (BSA, Sigma-Aldrich) (Fig. 5). From 4 mg of protein extract, we have typically been able to isolate 1.1 $\mu$g APC/C in the case of HeLa lysates, 0.8 $\mu$g APC/C in the case of *Xenopus* egg extracts, and 0.8 $\mu$g APC/C in the case of cow brain fractions (Fig. 1).

## *Purification of APC/C at Large Scale*

We have developed the following protocol to purify native APC/C at a relatively large scale from HeLa lysates and *Xenopus* egg extracts (collectively called "protein extracts" from now on) (Fig. 2). We have used this protocol to isolate APC/C from extracts that contain protein amounts ranging from 0.2 to 4 g, which requires between 1 and 20 ml of Affiprep beads that have been coupled to Cdc27 antibodies.

*Immunoaffinity Chromatography.* APC/C is isolated by imunoaffinity chromatography exactly as described earlier, except that protein extracts are incubated with Affi-prep beads coupled to Cdc27 antibodies in a 50-ml Falcon tube. For optimal recovery of native APC/C from the antibody beads, bound protein is eluted in three steps. In each step the beads are resuspended in one bead volume of TBS containing 1 mg/ml antigenic peptide and 0.5 m*M* DTT, the suspension is mixed on a rotary shaker for 40 min at 4°, and the beads are subsequently removed by centrifugation. The supernatant, containing native APC/C, is immediately processed further as described later. Residual APC/C is removed from the beads by

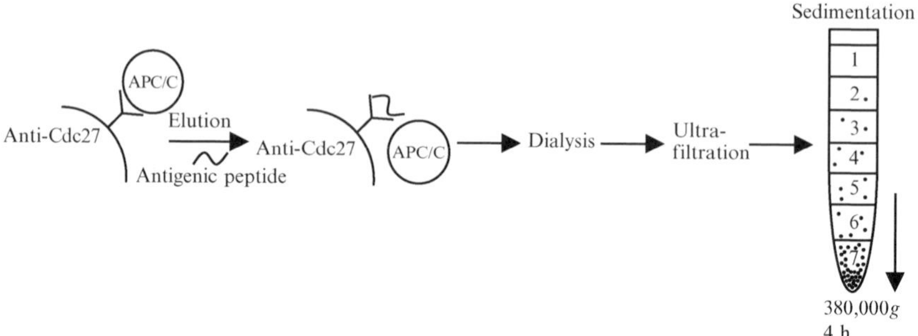


FIG. 2. Schematic illustration of steps used for the large-scale purification of APC/C from HeLa or *Xenopus laevis* egg extracts. Numbers in the centrifugation tube on the right refer to fractions that are removed from the top. APC/C is enriched in fraction number 7 (see Fig. 3).

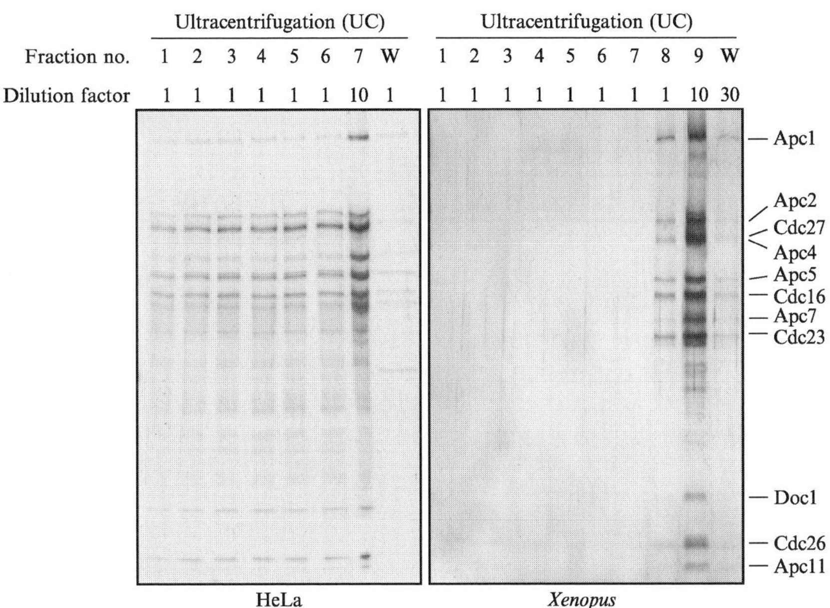


FIG. 3. Concentration of APC/C by ultracentrifugation. APC/C is isolated and concentrated from logarithmically proliferating HeLa cells or *Xenopus* interphase eggs by Cdc27 antibody beads, peptide elution, dialysis, and ultrafiltration (see text). Ultrafiltrate (550–600 $\mu$l) is concentrated further by ultracentrifugation at 380,000*g* for 4 h (HeLa) or 9 h (*Xenopus*). Fractions are collected from the top of the tube and are analyzed by SDS–PAGE and silver staining. APC/C is concentrated in the bottom fraction of the tube. To determine how much APC/C remains bound to the tube wall after removal of the bottom fraction, the tube is washed with 20 $\mu$l sample buffer containing 4% SDS (lane W). Although APC/C is removed more completely from the top fractions if centrifugation is performed for 9 h, the yield of APC/C is higher if centrifugation is only performed for 4 h, presumably because the APC/C binds irreversibly to the tube bottom if centrifugation times are kept longer (data not shown).

elution with a solution containing 100 m*M* glycine-HCl, pH 2.2, as described earlier. The beads are stored in TBS-T containing 0.03% $NaN_3$ at 4° and can be reused at least five times.

*Removal of Peptide by Dialysis.* The excess antigenic peptide, which might interfere with subsequent analyses, is removed from the APC/C eluate by dialysis. The APC/C sample is transferred into one or several Slide-A-Lyzer dialysis cassettes (MWCO 10 kDa, capacity: 3–12 ml; Pierce) and dialyzed for 3 h at 4° against 2 liters of TBS, HBS, or PBS containing 0.5 m*M* DTT. The buffer is exchanged once after 90 min.

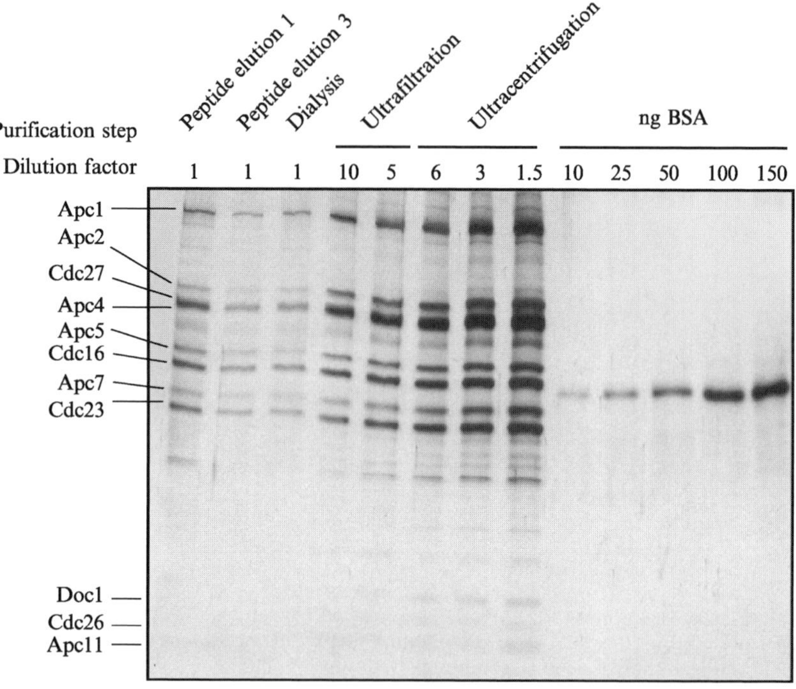


FIG. 4. Large-scale purification of APC/C from *Xenopus* interphase egg extracts. APC/C is immunopurified from a 175-ml interphase egg extract according to the protocol described in the text (Table III). APC/C samples are taken at the individual purification steps and analyzed by SDS–PAGE and silver staining. The concentration of each sample can be roughly estimated by comparing the staining intensity of APC/C subunits with the staining intensity of BSA.

*Concentration of APC/C by Ultrafiltration.* After dialysis, the APC/C sample is concentrated to a final volume of ~650 $\mu$l by ultrafiltration using a Centriprep centrifugal filter device (YM–30, MWCO 30 kDa, 15 ml, Amicon) at 3000 rpm (Varifuge 3.0R; Heraeus) and 4°. In case the volume of the dialyzed sample exceeds the capacity of one Centriprep device, several such devices are used in parallel because refilling of the same device reduces the yield of APC/C. Following this concentration step, all APC/C samples are combined and further concentrated to ~650 $\mu$l in one Centriprep device.

*Further Concentration of APC/C by Ultracentrifugation.* As a second concentration step, APC/C is sedimented by centrifugation with high *g* forces. The APC/C sample is transferred from the Centriprep device into a Beckman centrifuge tube (5 × 41 mm, Ultra-Clear) and is centrifuged in a SW 55 Ti rotor (Beckman) at 52,000 rpm (380,000*g*) for 4 h at 4°. Fractions of ~100 $\mu$l are carefully removed with a Pipetman from the top until 50 to

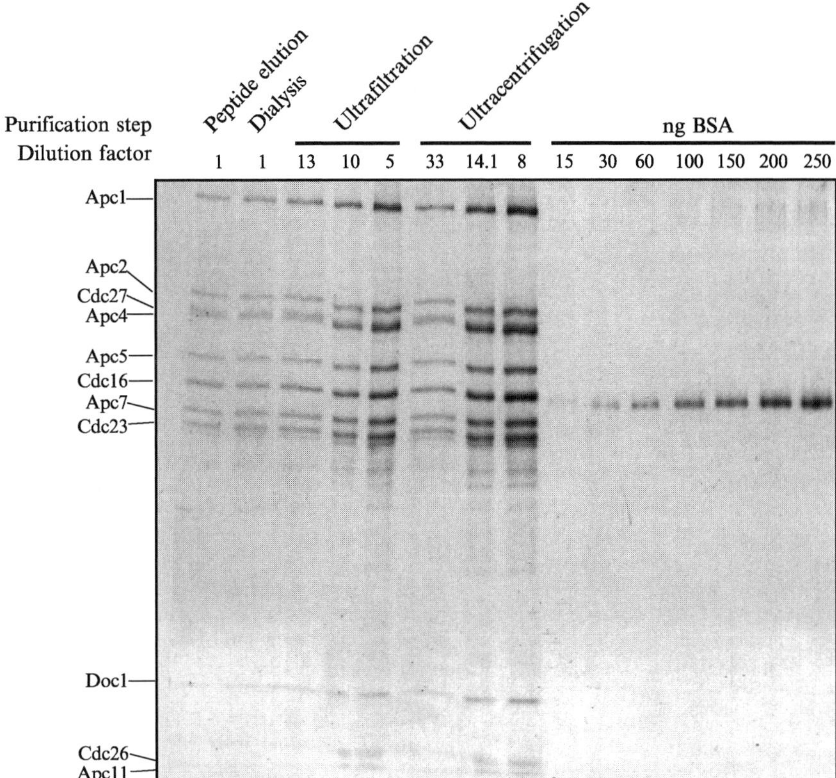


Fig. 5. Large-scale purification of APC/C from HeLa cell extracts. APC/C is purified from a 40-ml lysate of logarithmically grown HeLa cells according to the protocol described in the text (Table II). APC/C samples are taken at the individual purification steps and analyzed by SDS–PAGE and Coomassie staining. To estimate the concentration of APC/C, we compare APC/C samples of the individual purification steps with a BSA dilution series on a Coomassie-stained gel. Silver staining is less suitable for determination of the protein concentration because protein amount and silver staining intensity do no exhibit a linear correlation. The software ImageJ (http://rsb.info.nih.gov/ij/) is used to measure the staining intensities of APC/C subunits and BSA.

80 $\mu$l is left in the tube. APC/C is enriched in this final fraction but many APC/C particles will also have been pelleted on the bottom of the centrifuge tube. To recover as many APC/C particles as possible, the bottom wall of the tube is therefore carefully rinsed several times with the remaining solution in the centrifuge tube.

The purity and concentration of APC/C in the final sample are determined by SDS–PAGE as described earlier. The sample is kept on ice and used for further analyses as soon as possible (see following section). We have noticed that the specific ubiquitination activity of APC/C stored under these conditions does not change significantly for at least 4 days. The use of 4 ml Cdc27 antibody beads and 40 ml of HeLa cell lysate (protein concentration: 25 mg/ml) typically yields 85 $\mu$l of 1.2 mg/ml APC/C (Table II, Fig. 5). Purification of APC/C from 160 ml *Xenopus* interphase egg extracts (protein concentration: 25 mg/ml) using 17.6 ml Cdc27 antibody beads yields 100 $\mu$l of 4.5 mg/ml APC (Table III, Fig. 4).

TABLE II
SUMMARY OF LARGE-SCALE PURIFICATION OF APC/C FROM HeLa CELLS[a]

| Purification step | Volume (ml) | Concentration (mg/ml) | Yield ($\mu$g) | Specific activity [(% substrate/ng APC/C)] |
|---|---|---|---|---|
| Peptide elution | 13.5 | 0.024 | 324 | 0.42 |
| Dialysis | 13 | 0.024 | 312 | 0.45 |
| Ultrafiltration | 0.55 | 0.44 | 242 | 0.39 |
| Ultracentrifugation | 0.085 | 1.20 | 102 | 0.35 |

[a] APC/C is purified from 40 ml lysate of logarithmically proliferating HeLa cells (protein amount: 1 g) according to the protocol described in the text. The ubiquitination activity of APC/C obtained at individual purification steps is measured using an iodinated human cyclin B fragment (1–87 amino acids) as a substrate (see text). Protein concentrations are normalized by diluting the different APC/C samples with dialysis buffer (Fig. 5). We define the specific activity of APC/C, not activated by binding of recombinant Cdh1, as the percentage of ubiquitinated substrate (ubiquitin–substrate conjugates) per nanogram APC/C after 45 min of incubation (see Fig. 6).

TABLE III
SUMMARY OF LARGE-SCALE PURIFICATION OF APC/C FROM *Xenopus laevis* INTERPHASE EGGS[a]

| Purification step | Volume (ml) | Concentration (mg/ml) | Yield ($\mu$g) | Yield (%) |
|---|---|---|---|---|
| Peptide elution | 56 | 0.025 | 1400 | 100 |
| Dialysis | 55 | 0.025 | 1375 | 98 |
| Ultrafiltration 1 | 2.0 | 0.6 | 1200 | 86 |
| Ultrafiltration 2 | 0.6 | 1.8 | 1080 | 77 |
| Ultracentrifugation | 0.1 | 4.5 | 450 | 32 |

[a] APC/C is purified from a 175-ml extract of *X. laevis* interphase eggs (protein amount: 4.3 g) according to the protocol described in the text (Fig. 4). Ultrafiltration is performed in two sequential steps (see text.). During the final concentration step, 60% of APC are lost, presumably due to unspecific binding to the tube wall.

## Assay of *in Vitro* Ubiquitination Activity

To measure ubiquitin ligase activity, the purified APC/C is incubated with ubiquitin, E1, and either human UbcH10 and UbcH5 or with *Xenopus* UbcX and Ubc4 and is analyzed for the ability to form polyubiquitin chains on a substrate protein (see Figs. 6 and 7). A detailed description of methods for the measurement of APC/C activity, including protocols for the generation of recombinant E1 and E2 enzymes, Cdc20 and Cdh1, will be published elsewhere (Kraft *et al.*, 2005). In brief, typically 5 $\mu$l APC-bound antibody beads or 7 $\mu$l of peptide eluate is incubated in TBS (see earlier discussion) in a total reaction volume of 14 $\mu$l containing purified wheat E1 (80 $\mu$g/ml), *X. laevis* Ubc4 and UbcX (50 $\mu$g/ml each), bovine ubiquitin (1.25 mg/ml), and ATP-regenerating system (7.5 m*M* creatine phosphate, 1 m*M* ATP, 1 m*M* $MgCl_2$, 0.1 m*M* EGTA, 30 U/ml rabbit creatine phosphokinase type I; Sigma-Aldrich). As a substrate, we use 3 $\mu$g/ml iodinated human $His_6$-myc-cyclin B fragment (amino acids 1–87; Kraft *et al.*, 2003). The ubiquitination reaction mix is incubated for different periods of time (typically ranging from 5 to 45 min) in an Eppendorf thermomixer and stopped by the addition of SDS–PAGE sample buffer. Mammalian APC/C is assayed at 37°, whereas *Xenopus* APC/C is analyzed at room temperature. Ubiquitin-cyclin B conjugates are separated by SDS–PAGE and detected by phosphorimaging (Fig. 6). The amount of cyclin B that is converted into ubiquitin conjugates can be quantified by using the software ImageQuant (Amersham Biosciences). Alternatively, recombinant epitope-tagged cyclin B and securin can be used as substrates, and

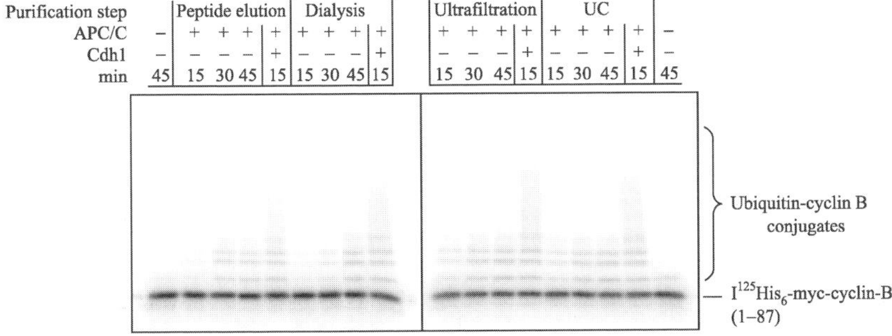


FIG. 6. Ubiquitination activity of HeLa APC/C after different steps in the purification (see Table II). The activity of APC/C at individual purification steps is assessed quantitatively using an iodinated human cyclin B fragment (1–87 amino acids) as a substrate for ubiquitination (see text). The APC/C concentration is normalized by diluting APC samples with dialysis buffer. The ubiquitination activity of APC/C is stimulated greatly by the addition of 20 ng recombinant human $His_6$-Cdh1 per 5 $\mu$l APC/C peptide eluate.

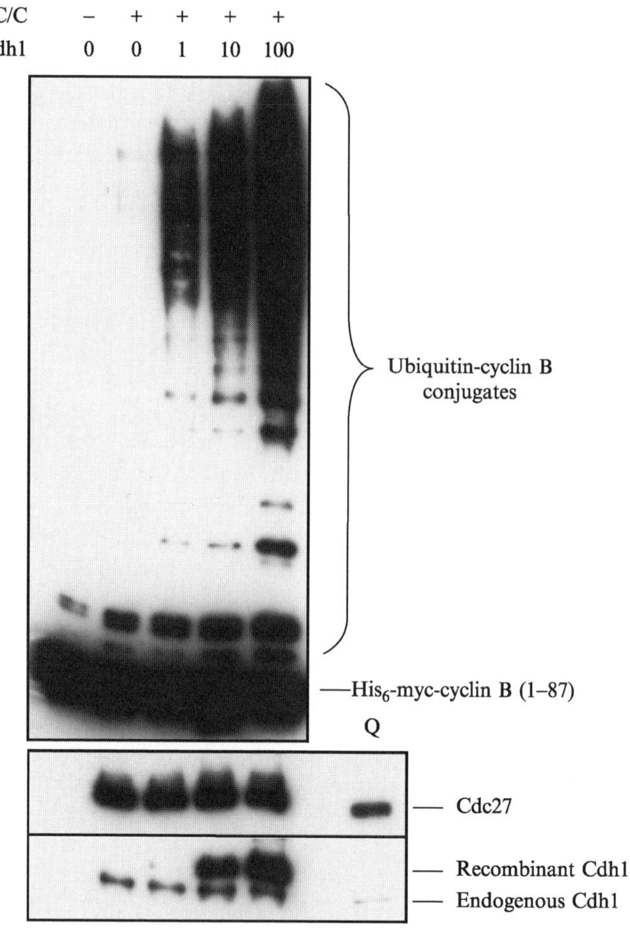


FIG. 7. Ubiquitination activity of APC/C isolated from anion-exchange chromatography fractions of cow brain extracts is stimulated by binding of recombinant human $His_6$-Cdh1. Cow brain APC/C bound to 3.5 $\mu$l Cdc27 antibody beads is incubated with increasing amounts of recombinant human $His_6$-Cdh1 or is mock treated with TBS, followed by peptide elution of the complex from the antibody beads. The eluate is incubated with myc-tagged cyclin B as a substrate for ubiquitination, and cyclin B is detected by immunoblotting using a mouse, monoclonal myc antibody (9E10, Sigma-Aldrich). The peptide-eluted APC/C used in the ubiquitination assay is immunoblotted for the presence of Cdc27 and Cdh1, showing the binding of recombinant Cdh1 to the complex (bottom) (Q, APC/C containing fraction of anion-exchange chromatography, see text).

the resulting ubiquitin conjugates can be detected by Western blotting using the appropriate epitope antibodies (Kraft *et al.*, 2005).

APC/C isolated from extracts of *Xenopus* eggs in interphase, logarithmically proliferating HeLa cells or cow brain has only low ubiquitination activity, but these forms of APC/C can be activated by the addition of recombinant Cdh1. Cdh1 can either be generated by *in vitro* translation in rabbit reticulocyte lysate or by expression in baculovirus-infected insect cells (Fang *et al.*, 1998; Kraft *et al.*, 2005; Kramer *et al.*, 1998, 2000). Figures 6 and 7 show examples of ubiquitination reactions of APC/C from logarithmically proliferating HeLa cells and from cow brain, respectively, where APC/C has been activated by the addition of human $His_6$-Cdh1 purified from baculovirus-infected insect cells (Kraft *et al.*, 2005; Kramer *et al.*, 2000). APC/C can also be activated by recombinant Cdc20, but efficient activation can only occur if APC/C has been phosphorylated previously by mitosis-specific kinases such as Cdk1-cyclin B-p9 (Kraft *et al.*, 2003, 2005; Kramer *et al.*, 1998; Shteinberg *et al.*, 1999).

## References

Bulaj, G., Kortemme, T., and Goldenberg, D. P. (1998). Ionization-reactivity relationships for cysteine thiols in polypeptides. *Biochemistry* **37,** 8965–8972.

Clute, P., and Pines, J. (1999). Temporal and spatial control of cyclin B1 destruction in metaphase. *Nature Cell Biol.* **1,** 82–87.

Dawson, I. A., Roth, S., and Artavanis-Tsakonas, S. (1995). The Drosophila cell cycle gene fizzy is required for normal degradation of cyclins A and B during mitosis and has homology to the CDC20 gene of *Saccharomyces cerevisiae*. *J. Cell Biol.* **129,** 725–737.

den Elzen, N., and Pines, J. (2001). Cyclin A is destroyed in prometaphase and can delay chromosome alignment and anaphase. *J. Cell Biol.* **153,** 121–136.

Ellman, G. L. (1959). Tissue sulfhydryl groups. *Arch. Biochem. Biophys.* **82,** 70–77.

Fang, G., Yu, H., and Kirschner, M. W. (1998). Direct binding of CDC20 protein family members activates the anaphase-promoting complex in mitosis and G1. *Mol. Cell* **2,** 163–171.

Geley, S., Kramer, E., Gieffers, C., Gannon, J., Peters, J. M., and Hunt, T. (2001). Anaphase-promoting complex/cyclosome-dependent proteolysis of human cyclin A starts at the beginning of mitosis and is not subject to the spindle assembly checkpoint. *J. Cell Biol.* **153,** 137–148.

Gieffers, C., Dube, P., Harris, J. R., Stark, H., and Peters, J. M. (2001). Three-dimensional structure of the anaphase-promoting complex. *Mol. Cell* **7,** 907–913.

Glotzer, M., Murray, A. W., and Kirschner, M. W. (1991). Cyclin is degraded by the ubiquitin pathway. *Nature* **349,** 132–138.

Hagting, A., Den Elzen, N., Vodermaier, H. C., Waizenegger, I. C., Peters, J. M., and Pines, J. (2002). Human securin proteolysis is controlled by the spindle checkpoint and reveals when the APC/C switches from activation by Cdc20 to Cdh1. *J. Cell Biol.* **157,** 1125–1137.

Harlow, E., and Lane, D. (1988). Using antibodies: A laboratory manual. CSHL Press.

Harper, J. W., Burton, J. L., and Solomon, M. J. (2002). The anaphase-promoting complex: It's not just for mitosis any more. *Genes Dev.* **16,** 2179–2206.

Herzog, F., Mechtler, K., and Peters, J.-M. (2005). Identification of cell cycle dependent phosphorylation sites on the anaphase-promoting complex/cyclosome by mass spectrometry. *Methods Enzymol.* **398**(19), 2005 (this volume).

King, R. W., Peters, J. M., Tugendreich, S., Rolfe, M., Hieter, P., and Kirschner, M. W. (1995). A 20S complex containing CDC27 and CDC16 catalyzes the mitosis-specific conjugation of ubiquitin to cyclin B. *Cell* **81,** 279–288.

Kraft, C., Gmachl, M., and Peters, J. M. (2005). Methods to measure ubiquitin-dependent proteolysis mediated by the anaphase-promoting complex (APC). *Methods,* in press.

Kraft, C., Herzog, F., Gieffers, C., Mechtler, K., Hagting, A., Pines, J., and Peters, J. M. (2003). Mitotic regulation of the human anaphase-promoting complex by phosphorylation. *EMBO J.* **22,** 6598–6609.

Kramer, E. R., Gieffers, C., Holzl, G., Hengstschlager, M., and Peters, J. M. (1998). Activation of the human anaphase-promoting complex by proteins of the CDC20/Fizzy family. *Curr. Biol.* **8,** 1207–1210.

Kramer, E. R., Scheuringer, N., Podtelejnikov, A. V., Mann, M., and Peters, J. M. (2000). Mitotic regulation of the APC activator proteins CDC20 and CDH1. *Mol. Biol. Cell* **11,** 1555–1569.

Murray, A. W. (1991). Cell cycle extracts. *Methods Cell Biol.* **36,** 581–605.

Musacchio, A., and Hardwick, K. G. (2002). The spindle checkpoint: Structural insights into dynamic signalling. *Nature Rev. Mol. Cell Biol.* **3,** 731–741.

Orlicky, S., Tang, X., Willems, A., Tyers, M., and Sicheri, F. (2003). Structural basis for phosphodependent substrate selection and orientation by the SCFCdc4 ubiquitin ligase. *Cell* **112,** 243–256.

Peters, J. M. (2002). The anaphase-promoting complex: Proteolysis in mitosis and beyond. *Mol. Cell* **9,** 931–943.

Schwickart, M., Havlis, J., Habermann, B., Bogdanova, A., Camasses, A., Oelschlaegel, T., Shevchenko, A., and Zachariae, W. (2004). Swm1/Apc13 is an evolutionarily conserved subunit of the anaphase-promoting complex stabilizing the association of Cdc16 and Cdc27. *Mol. Cell. Biol.* **24,** 3562–3576.

Shteinberg, M., Protopopov, Y., Listovsky, T., Brandeis, M., and Hershko, A. (1999). Phosphorylation of the cyclosome is required for its stimulation by Fizzy/cdc20. *Biochem. Biophys. Res. Commun.* **260,** 193–198.

Sigrist, S., Jacobs, H., Stratmann, R., and Lehner, C. F. (1995). Exit from mitosis is regulated by Drosophila fizzy and the sequential destruction of cyclins A, B and B3. *EMBO J.* **14,** 4827–4838.

Sudakin, V., Ganoth, D., Dahan, A., Heller, H., Hershko, J., Luca, F. C., Ruderman, J. V., and Hershko, A. (1995). The cyclosome, a large complex containing cyclin-selective ubiquitin ligase activity, targets cyclins for destruction at the end of mitosis. *Mol. Biol. Cell* **6,** 185–197.

Tang, Z., Li, B., Bharadwaj, R., Zhu, H., Ozkan, E., Hakala, K., Deisenhofer, J., and Yu, H. (2001). APC2 cullin protein and APC11 RING protein comprise the minimal ubiquitin ligase module of the anaphase-promoting complex. *Mol. Biol. Cell* **12,** 3839–3851.

Visintin, R., Prinz, S., and Amon, A. (1997). CDC20 and CDH1: A family of substrate-specific activators of APC-dependent proteolysis. *Science* **278,** 460–463.

Vodermaier, H. C. (2001). Cell cycle: Waiters serving the Destruction machinery. *Curr. Biol.* **11,** R834–R837.

Vodermaier, H. C., Gieffers, C., Maurer-Stroh, S., Eisenhaber, F., and Peters, J. M. (2003). TPR subunits of the anaphase-promoting complex mediate binding to the activator protein CDH1. *Curr. Biol.* **13,** 1459–1468.

Wu, H., Lan, Z., Li, W., Wu, S., Weinstein, J., Sakamoto, K. M., and Dai, W. (2000). p55CDC/hCDC20 is associated with BUBR1 and may be a downstream target of the spindle checkpoint kinase. *Oncogene* **19,** 4557–4562.

Wu, M., and Gerhart, J. (1991). Raising Xenopus in the laboratory. *Methods Cell Biol.* **36,** 3–18.

Yamano, H., Kominami, K., Harrison, C., Kitamura, K., Katayama, S., Dhut, S., Hunt, T., and Toda, T. (2004). Requirement of the SCFPop1/Pop2 ubiquitin ligase for degradation of the fission yeast S phase cyclin Cig2. *J. Biol. Chem.* **279,** 18974–18980.

Yoon, H. J., Feoktistova, A., Wolfe, B. A., Jennings, J. L., Link, A. J., and Gould, K. L. (2002). Proteomics analysis identifies new components of the fission and budding yeast anaphase-promoting complexes. *Curr. Biol.* **12,** 2048–2054.

# [17] Purification and Assay of the Budding Yeast Anaphase-Promoting Complex

*By* LORI A. PASSMORE, DAVID BARFORD, and J. WADE HARPER

## Abstract

The anaphase-promoting complex (APC) is a central regulator of the eukaryotic cell cycle and functions as an E3 ubiquitin protein ligase to catalyze the ubiquitination of a number of cell cycle regulatory proteins. The APC contains at least 13 subunits in addition to two activator subunits, Cdc20 and Cdh1, that associate with the APC in a cell cycle-dependent manner. This chapter describes methods for preparation and assay of the APC from *Saccharomyces cerevisiae*. Highly active APC is purified from cells expressing Cdc16 fused with a tandem affinity purification (TAP) tag. Enzymatically active APC is achieved upon addition of recombinant Cdc20 or Cdh1 together with E1, Ubc4, ATP, and ubiquitin. Activity assays toward several endogenous substrates, including Clb2 and Pds1, are described. In addition, methods for observation of APC–coactivator and APC–substrate complexes by native gel electrophoresis are described.

## Introduction

Progression through the mitotic phase of the cell cycle is regulated by ubiquitin-mediated proteolysis of regulatory proteins and by the periodic activity of the major mitotic kinase Cdk1/cyclin B. Ubiquitin-mediated proteolysis controls several transitions during mitosis, including chromosomal

METHODS IN ENZYMOLOGY, VOL. 398

0076-6879/05 $35.00
DOI: 10.1016/S0076-6879(05)98017-8

segregation, reorganization of the mitotic spindle, and exit from mitosis, in part through the ubiquitination of B-type cyclins. A major ubiquitin ligase controlling mitosis is the *a*naphase-*p*romoting *c*omplex, which is also referred to as the cyclosome (reviewed in Harper *et al.,* 2002; Peters, 2002). The APC is composed of at least 13 core subunits. Activation of the APC involves phosphorylation by cyclin-dependent kinases (Cdks) (Golan *et al.*, 2002; Kotani *et al.*, 1998; Kramer *et al.*, 2000; Rudner and Murray, 2000), as well as association with two coactivators, Cdc20 and Cdh1, which bind the APC in a cell cycle-dependent manner (Fang *et al.*, 1998; Kramer *et al.*, 2000; Visintin *et al.*, 1997; Zachariae *et al.*, 1998a). Substrate specificity is thought to be largely dictated by Cdc20 and Cdh1 (Burton and Solomon, 2001; Hendrickson *et al.*, 2001; Hilioti *et al.*, 2001; Pfleger and Kirschner, 2000; Pfleger *et al.*, 2001; Visintin *et al.*, 1997). In budding yeast, Cdc20 is largely responsible for ubiquitination of Pds1 at the metaphase–anaphase transition, whereas Cdh1 is largely responsible for ubiquitination of B-type cyclins, Cdc5, Hsl1, and Cdc20 itself (Harper *et al.*, 2002; Peters, 2002). There is some evidence that the coactivators may bind APC substrates directly (Burton and Solomon, 2001; Hilioti *et al.*, 2001; Pfleger and Kirschner, 2000; Pfleger *et al.*, 2001), with the assistance of the Doc1/Apc10 subunit (Passmore *et al.*, 2003). However, the precise function of the coactivators is unclear.

While much of our understanding of the components and function of the APC has derived from budding and fission yeast systems, only recently has budding yeast APC activity toward full-length natural substrates been reconstituted *in vitro*. This is in stark contrast to the situation with *Xenopus* and mammalian APC complexes, where activity assays against multiple substrates have been available for some time (Aristarkhov *et al.*, 1996; King *et al.*, 1995). This chapter describes methods for purification and assay of the budding yeast APC. In addition, it describes methods for visualization of APC complexes with both substrates and coactivator subunits using native gel analysis.

## Overview of Assay Development

Enzymatic analysis of APC activity requires several components: (1) purified APC, (2) coactivator subunits Cdc20 or Cdh1, (3) E1 ubiquitin-activating enzyme, (4) an appropriate E2 ubiquitin-conjugating enzyme (UBC), and (5) appropriate substrates, in addition to ubiquitin and ATP. In the case of vertebrate APC, APC core complexes can be purified by conventional column chromatography to yield a preparation that is active in the presence of added Cdc20 or Cdh1. However, most frequently, APC core complexes are purified using a single-step immunopurification procedure involving anti-Cdc27 or anti-Cdc16 antibodies as an affinity reagent.

The ability to selectively tag yeast genes with epitope tags has made it possible to generate high-purity budding yeast APC using *t*andem *a*ffinity *p*urification (TAP) wherein the epitope tag is fused to the C terminus of Cdc16, a core APC subunit (Carroll and Morgan, 2002; Passmore *et al.*, 2003; Zachariae *et al.*, 1996). As described later, current procedures employ a TAP tag composed of the IgG-binding domain of protein A, one or more TEV protease cleavage sites, and a small peptide capable of binding to calmodulin in a calcium-dependent manner. Purification of Cdc16-TAP and its associated APC subunits involves (1) affinity chromatography on IgG Sepharose beads, (2) cleavage of bound fusion proteins with TEV protease, (3) capture of the Cdc16-calmodulin binding peptide by affinity chromatography on calmodulin beads in the presence of calcium, and (4) elution of the APC complex from the calmodulin beads in the presence of EGTA (Fig. 1B). Such preparations support the ubiquitination of full-length native yeast APC substrates (including Pds1, Hsl1, and Clb2), as well as an artificial substrate composed of residues 12–91 of sea urchin cyclin B, when combined with recombinant Cdc20 or Cdh1 as the activating subunit (Carroll and Morgan, 2002; Passmore *et al.*, 2003). An alternative approach employing a single-step purification of Cdc16-HA using an anti-HA affinity resin has also been reported (Charles *et al.*, 1998). The following section describes the purification and assay of budding yeast APC using the TAP-tag approach.

## Construction of Yeast Strain Expressing Cdc16-TAP

We have used polymerase chain reaction (PCR)-based gene targeting with a modified version of the pFA6a-kanMX6 vector series (Fig. 1A) (Wach *et al.*, 1998) to fuse the coding region for a C-terminal affinity purification tag onto the chromosomal copy of an APC gene (Passmore, 2003; Passmore *et al.*, 2003). Because the gene is tagged at its 3′ end, the promoter region is not disrupted and expression of the fusion protein should be equivalent to that in wild-type yeast. This is important for the modification of APC subunits, as many of them are essential for growth and a change in their endogenous expression levels may be toxic. We use the protease-deficient yeast BJ2168 (*MAT***a** *leu2 trp1 ura3–52 pep4–3 prc1–407 prb1–1122 gal2*; ATCC 208277) (Jones, 1977; Zubenko *et al.*, 1980) for gene targeting and protein purifications. The *CDC16* gene was tagged because, in previous studies, it had been tagged with at least nine Myc epitopes without affecting APC function (Zachariae *et al.*, 1996) and an immunoprecipitation of APC from a *CDC16–9MYC* yeast strain had provided enough protein for analysis of its subunits by mass spectrometry (Zachariae *et al.*, 1998b). These factors indicated that the C terminus of Cdc16p is

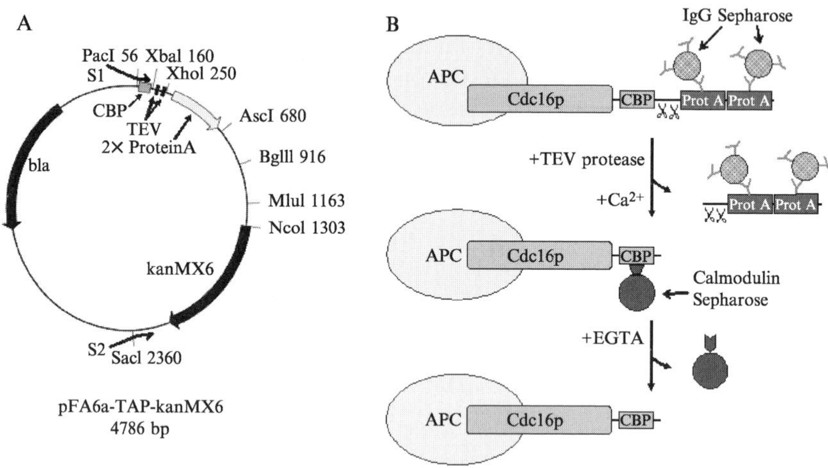


FIG. 1. Tandem affinity purification. (A) Schematic diagram of the pFA6a-TAP-kanMX6 vector. The positions of selected restriction enzyme cleavage sites are shown. The TAP tag was inserted into the *Pac*I/*Asc*I sites of the pFA6a vector and includes the coding region for the calmodulin-binding peptide (CBP), two TEV protease cleavage sites, and two protein A IgG-binding domains. This vector contains an *E. coli* origin of replication and an ampicillin resistance gene (bla) to allow propagation in bacteria and a kanamycin/G418 resistance gene (kanMX6) for selection in yeast. Oligonucleotides used for PCR-based gene targeting (S1 and S2) are shown schematically. (B) Overview of the TAP purification strategy. Clarified yeast lysates are loaded onto an IgG Sepharose column and washed thoroughly. The C-terminal protein A IgG-binding domains (Prot A) bind to the column and are cleaved off using TEV protease. The TEV eluate is supplemented with calcium and loaded onto a calmodulin Sepharose column. The calmodulin-binding peptide (CBP) binds to the column and the column is washed. The bound protein is eluted by the addition of EGTA, which chelates the calcium, thereby disrupting the interaction between calmodulin and the calmodulin-binding peptide.

solvent exposed and not required for APC function. We tried tagging *CDC16* with several different tags (Passmore, 2003) but the TAP tag (Puig *et al.*, 2001; Rigaut *et al.*, 1999) was the most efficient for APC purification.

The TAP tag was synthesized by recursive PCR with 14 overlapping oligonucleotides (Prodromou and Pearl, 1992) and the sequence was modified for optimal *S. cerevisiae* codon usage, avoiding repetitive DNA sequences (Passmore, 2003; Passmore *et al.*, 2003). To increase the efficiency of TEV cleavage during purification, tandem TEV protease recognition sites were inserted within the tag. Thus, the TAP tag in pFA6a-TAP-kanMX6 contains an N-terminal calmodulin-binding peptide (originating from rabbit skeletal muscle myosin light chain kinase) (Blumenthal *et al.*, 1985), followed by two TEV protease cleavage sites and, at its C terminus, two repeats of the *Staphylococcus aureus* protein A IgG-binding domain

(Fig. 1). The following protocol (adapted from Wach *et al.*, 1998) can be used for PCR-based gene targeting and was used to create a yeast strain expressing Cdc16-TAP.

*Buffers and Reagents*

10× LiAc: 1 *M* lithium acetate (Sigma L–6883) adjusted to pH 7.5 with dilute acetic acid and filter sterilized
10× TE: 0.1 *M* Tris, pH 7.5, 10 m*M* EDTA, autoclaved
LiAc/TE: 1× LiAc, 1× TE, made fresh and filter sterilized
50% PEG: 50% (w/v) PEG 3350 (Sigma P–3640) in water (leave overnight on stirplate to allow the PEG to go into solution)
PEG/LiAc: 40% PEG 3350 (8 ml 50% PEG), 1× TE (1 ml 10× TE), 1× LiAc (1 ml 10× LiAc), prepared immediately prior to use and filter sterilized
Dimethyl sulfoxide (DMSO), Sigma D–8779
Salmon sperm DNA, Invitrogen 15632–011
YPD: For 1 liter, autoclave 20 g bacto-peptone and 10 g yeast extract in 950 ml water. Cool to approximately 60° and add sterile-filtered 2 *M* D-glucose to 0.11 *M*
G418/YPD plates: For 1 liter, autoclave 20 g bacto-peptone, 10 g yeast extract, and 20 g bacto-agar in 950 ml water. Cool to approximately 60° and add sterile-filtered 2 *M* D-glucose to 0.11 *M* and sterile-filtered G418 to 300 $\mu$g/ml (from a 25-mg/ml stock solution; Melford G0175)
IP buffer: 50 m*M* Tris-Cl, pH 8.0, 150 m*M* KCl, 10% (w/v) glycerol, 0.2 % (v/v) Triton X–100, and protease inhibitors.

*Preparation of PCR Product*

Oligonucleotide primers should be 5′ phosphorylated and PAGE purified. S1 oligonucleotides are designed to contain 40 bp of the 3′ end of gene to be tagged (up to but not including the stop codon) followed by the 24 bp pFA6a vector sequence, such that the gene is fused, in-frame, with the tag from the vector (Fig. 1A). The S2 oligonucleotide contains 40 bp complementary to the chromosomal region downstream of the targeted gene, followed by a 20 bp pFA6a vector sequence. To make the Cdc16-TAP strain, we used the following oligonucleotides (with vector-specific sequences in italics):

S1-TAP-CDC16: TAATGCCGACGATGATTTTGACGCAGATA TGGAACTGGAA *TCTCACGAAAAGAGAAGATGGAAG*

S2-CDC16: CTTTTACGTGTGGCTGCCTCTAAGAATTAAACT TCTTTTC*CATCG ATGAATTCGAGCTCG*

Perform a 200-$\mu$l PCR reaction using Expand high-fidelity polymerase (Roche Applied Science, 1732641), the S1 and S2 oligonucleotides, and the pFA6a-TAP-kanMX6 template. Use 20 cycles of PCR as follows: 95° 30 s, 54° 30 s, and 72° 3 min. Clean up the PCR products (~2.4 kb) using gel purification and the QIAEX II gel purification kit (Qiagen, 20021).

*Transformation Protocol*

1. Inoculate a 50-ml overnight culture in YPD using one large colony of BJ2168 yeast (less than 2 weeks old). Grow for approximately 24 h at 30°, 220 rpm.
2. Inoculate 300 ml fresh YPD with overnight culture to give an $OD_{600nm}$ of 0.2 and grow at 30°, 220 rpm until the $OD_{600nm}$ is 0.4–0.6.
3. Harvest cells in 50-ml Falcon tubes at 1000*g* for 5 min at room temperature.
4. Combining all pellets, wash once in 30 ml sterile water and once in 1.5 ml LiAc/TE.
5. Resuspend cells in 1.5 ml LiAc/TE and make competent by incubating at 30° for 30 min with gentle shaking (100 rpm).
6. To prepare the DNA for transformation, heat salmon sperm DNA to 100° for 10 min and then chill quickly on ice. For each transformation, mix 100 $\mu$g salmon sperm DNA with 1–4 $\mu$g of PCR product in a 1.5-ml tube. As a negative control, use TE buffer instead of PCR product.
7. Add 100 $\mu$l competent yeast to the DNA and mix well.
8. Add 0.6 ml PEG/LiAc solution to each tube and mix well by flicking the tube.
9. Incubate at 30° for 30 min with shaking (200 rpm).
10. Add 70 $\mu$l DMSO, mix by gentle inversion, heat shock for 15 min at 42°, and then chill on ice for 2 min.
11. Pellet cells at 6000*g* for 1 min, resuspend in 0.5 ml YPD, and recover by incubation at 30° and 200 rpm for 2 h.
12. Finally, pellet cells at 6000*g* for 1 min, resuspend in 0.6 ml YPD, and plate onto six G418/YPD plates (100 $\mu$l/plate) using glass beads to spread the cells. Grow inverted plates for 3 days at 30°.
13. After 3 days, the transformed yeast grow out as large colonies on a thin lawn of background colonies. There should be no colonies on the negative control plates. Streak out several of the large colonies onto fresh G418/YPD plates to obtain single colonies.

*Verification of Integration*

Integration of the purification tag at the 3′ end of the endogenous *CDC16* gene should be verified by colony PCR as described (Wach *et al.*, 1998). Colony PCR confirms that the integration has occurred at the correct locus and that the yeast strain is haploid for the targeted gene. Briefly, design four primers so that a PCR product from one set of primers (V1 and K2) spans the 5′ integration site and a PCR product from a second set of primers (K3 and V4) spans the 3′ integration site (V1 and V4 are gene specific; K2 and K3 are vector specific). The presence of correctly sized PCR products indicates a positive transformant. PCR reactions are performed in tubes containing a small quantity of yeast microwaved on high power for 1 min (Wach *et al.*, 1998). For the CDC16-TAP strain, the following oligonucleotides were used:

V1-CDC16: GCACAAATCATTGTACCTAAAGCC
V4-CDC16: GGAACCTTGAACTTGAACAGCG
K2-kanMX6: CGGATGTGATGTGAGAACTGTATCCTAGC
K3-kanMX6: GCTAGGATACAGTTCTCACATCACATCCG

The PCR products should be 1.6 kb with V1-CDC16 and K2-kanMX6; 1.4 kb with K3-kanMX6 and V4-CDC16; and 3.1 kb with V1-CDC16 and V4-CDC16 (0.7 kb for a wild-type strain).

A small proportion of clones that have integrated tags as determined by colony PCR do not express functional tags, perhaps due to frameshifts or other mutations. Therefore, small-scale immunoprecipitations should also be performed for several clones to test for the presence of functional tags. Inoculate 200 ml YPD with 5 ml of a saturated overnight culture and grow at 30° with shaking (200 rpm) until the $OD_{600\ nm}$ is 0.6–1.0. (Use untagged BJ2168 yeast for a negative control.) Harvest cells by centrifugation at 3000*g* for 12 min at 4°. Rinse pellets once with cold water, flash freeze in a dry ice/ethanol bath, and store at −80°. For immunoprecipitations, add an equal volume of acid-washed glass beads (425–600 $\mu$m; Sigma) to frozen cell pellets and IP buffer to cover the glass beads (~0.4 ml). Lyse yeast by vortexing on maximum speed at 4° for 30 s, followed by a 30-s incubation on ice, for a total of 8 min of vortexing. Transfer the lysate, without glass beads, into 1.5-ml tubes and centrifuge at 20,000*g* for 20 min at 4°. Transfer the supernatant to a new tube and estimate the protein concentration using the Bio-Rad protein assay.

Equilibrate 20 $\mu$l of IgG Sepharose (Amersham Biosciences 17–0969–01) in IP buffer. Add equal amounts of protein extract to the resin and incubate at 4° on a pinwheel rotor for 1 h. Wash IgG Sepharose three times with IP buffer. To elute protein, add 40 $\mu$l 2× SDS–PAGE sample buffer to each sample and vortex. Boil samples for 10 min, vortex, and load directly onto an 8% SDS–PAGE large gel. Western blot with anti-Protein A antibody (SPA–27, Sigma P–2921; 1:2400 dilution). Cdc16-TAP should be visible as a 118-kDa band (99 kDa after TEV cleavage).

## Purification of Yeast APC$^{CDC16\text{-}TAP}$

The TAP purification protocol (Passmore *et al.*, 2003) was modified from Rigaut *et al.* (1999) and is outlined in Fig. 1B. We routinely perform the entire purification in 1 day over ~12 h. However, the purification can be stopped before TEV cleavage and stored overnight at 4°.

### *Yeast Cultures*

To prepare yeast pellets, use a saturated overnight culture, grown from one or two fresh *CDC16-TAP* yeast colonies, to inoculate 10–20 liters of YPD (0.5–2.0 ml overnight culture per liter YPD in a 2-liter flask). Grow yeast cultures overnight with shaking (200 rpm) at 30° to an $OD_{600\ nm}$ of approximately 1.0. The cells must be harvested before they reach stationary phase, as the major yeast proteases become more active, the cell wall becomes thicker, and protein synthesis decreases at stationary phase (Jones, 2002; Werner-Washburne *et al.*, 1993). Harvest the yeast at 3000*g* for 12 min at 4°, rinse once in cold water (to remove extracellular proteases), flash freeze in a dry ice/ethanol bath or liquid nitrogen, and store at –80°.

### *Buffers and Reagents*

Buffers used for large-scale purifications are in italics and are listed in Table I. All purification steps should be performed at 4° unless otherwise indicated. It is important to prechill buffers to 4° before adjusting the pH due to the temperature dependence of Tris buffers. To avoid problems with pH fluctuations, HEPES buffer can be used instead of Tris.

To prepare IgG Sepharose FF (Amersham Biosciences 17–0969–01), pack the resin into a Vantage L column (Amicon) using 0.125 ml resin for every 10 g of yeast pellet. An empty PD10 column (Amersham Biosciences), or similar gravity flow column, may also be used. However, we have observed that lower yields are obtained using gravity flow columns. Wash the column with five bed volumes of *TST* (Table I). To remove unbound IgG, wash with three bed volumes each of *0.5 M HAc, TST,* and *0.5 M HAc*. Wash the column thoroughly with *TST* until the pH returns to 7.6 (monitor the pH using pH strips). Chill it to 4° and equilibrate with *IgG buffer 1* just before use. To prepare calmodulin Sepharose 4B (Amersham

TABLE I
BUFFERS FOR APC PURIFICATION

| Buffer name | Composition |
|---|---|
| TST | 50 m*M* Tris-Cl, pH 7.6, 150 m*M* NaCl, 0.05% (v/v) Tween 20 |
| 0.5 *M* HAc | 0.5 *M* acetic acid, adjusted to pH 3.4 with $NH_4CH_3COOH$ |
| Lysis buffer | 50 m*M* Tris-Cl, pH 8.0, 150 m*M* KCl, 10% (w/v) glycerol |
| DNase I | 1 mg/ml DNase I (Roche 104159) in water |
| Igepal CA-630 | 10% (v/v) Igepal CA-630 (Sigma I-3021) in water |
| IgG buffer 1 | 50 m*M* Tris-Cl, pH 8.0, 150 m*M* KCl, 10% (w/v) glycerol, 0.5 m*M* EDTA, 2 m*M* EGTA, 0.1% (v/v) Igepal CA-630 |
| IgG buffer 2 | 50 m*M* Tris-Cl, pH 8.0, 150 m*M* KCl, 10 % (w/v) glycerol, 0.5 m*M* EDTA, 1 m*M* DTT, 0.1% (v/v) Igepal CA-630 |
| CaM wash buffer | 10 m*M* Tris-Cl, pH 8.0, 150 m*M* NaCl, 10% (w/v) glycerol, 3 m*M* DTT, 1 m*M* Mg-acetate, 2 m*M* $CaCl_2$, 0.1% (v/v) Igepal CA-630 |
| CaM elution buffer | 10 m*M* Tris-Cl, pH 8.0, 150 m*M* NaCl, 10% (w/v) glycerol, 3 m*M* DTT, 1 m*M* Mg-acetate, 2 m*M* EGTA, 0.1% (v/v) Igepal CA-630 |
| Protease inhibitors | Complete EDTA-free protease inhibitor cocktail tablets (Roche Applied Science 1873580). Add one tablet for every 100 ml buffer |
| Phosphatase inhibitors | 50 m*M* NaF, 25 m*M* $\beta$-glycerophosphate, 1 m*M* Na-orthovanadate |

Biosciences 17–0529–01), pack 0.1 ml resin per 10 g yeast pellet into a Bio-Rad polyprep column or empty PD10 column. Wash the column with 10 bed volumes of water and then with 20 bed volumes of *CaM wash buffer* and chill to 4°.

### *Preparation of Yeast Extracts*

1. Thaw cell pellets from 10 to 20 liters yeast culture at room temperature and resuspend in an equal volume of *lysis buffer* with *protease inhibitors* and *phosphatase inhibitors*. Add 1/1000 volume *DNase I.*

2. Lyse cells by passing through an Emusiflex C5 homogenizer (Avestin, Canada) on ice, four times at 20,000 psi. The Emulsiflex C5 homogenizer uses a high-pressure pump to push a continuous flow of cell paste through an adjustable homogenizing valve. Cells are subjected to a shear force due to the rapid change in pressure at the homogenizing valve as well as a physical force due to impact with the homogenizing valve, resulting in cell lysis. We have also successfully used a French press to lyse the cells.

3. Clear lysate by centrifuging in a JA20 Beckman rotor at 19,000 rpm (45,000*g*) for 30 min at 4°. (The lysate remains quite turbid after centrifugation.)

4. Add EDTA to a final concentration of 0.5 m*M* (from 0.5 *M* EDTA stock) and *Igepal CA–630* (a nonionic detergent) to a final concentration of 0.1% (v/v).

### *Purification on IgG Sepharose*

1. Load the cleared lysate onto the IgG Sepharose Vantage column at 0.5–1.0 ml/min using a P1 pump (or slowly pass the lysate over the column, trying not to disturb the bed of resin, if using a gravity flow column).

2. Wash the IgG column with 60 bed volumes of *IgG buffer 1* with *protease inhibitors* and *phosphatase inhibitors*, followed by 30 bed volumes of *IgG buffer 2* with *phosphatase inhibitors*. The EGTA in *IgG buffer 1* chelates calcium to release any calmodulin that may be bound to the calmodulin-binding peptide. We have observed that if EGTA is omitted from *IgG buffer 1*, only ~60% of the APC binds to the calmodulin column. This suggests that endogenous yeast calmodulin binds to the calmodulin-binding peptide of the tagged APC subunit during purification and blocks binding to calmodulin resin. (The purification may be stopped here and stored overnight if necessary.)

3. Carefully remove resin from the IgG column and transfer it into round-bottomed 2-ml tubes. For every 100 $\mu$l of IgG Sepharose, add 100 $\mu$l *IgG buffer 2* and approximately 10 units (or 15 $\mu$g) TEV protease (Invitrogen, 10127–017). Shake the cleavage reactions at 16°, 100 rpm for 2 h.

4. After cleavage, the APC should be present in the supernatant. Spin down the resin (1000*g*, 1 min, 4°). Remove the supernatant and wash the resin twice with a small volume of *IgG buffer 2* (approx. 0.3 bed volumes each wash). Pool the TEV supernatant and washes. In our hands, approximately 95% of Cdc16-TAP is cleaved and eluted with TEV protease, as shown by immunoblotting for the protein A tag.

5. Spin the supernatant at 10,000*g* for 1 min to ensure all IgG beads are removed.

### *Purification on Calmodulin Resin*

1. To the supernatant from the TEV cleavage reaction, add 1 *M* $CaCl_2$ to a final concentration of 5 m*M*. Save 35 $\mu$l for SDS–PAGE analysis.
2. Incubate the TEV supernatant with calmodulin Sepharose in a sealed column on rollers for 1 h at 4°. The calmodulin-binding

peptide of the TAP tag will bind to calmodulin in the presence of calcium.

3. After binding, allow the column to drain by gravity (save 35 $\mu$l of this flow through for SDS–PAGE analysis). Wash with 70 bed volumes *CaM wash buffer* by carefully pipetting several milliliters of buffer onto the resin at a time, trying to maintain an even bed of resin.
4. Elute protein from the calmodulin Sepharose in *CaM elution buffer* in at least 15 fractions of one bed volume each. Be careful not to disturb the bed of resin when pipetting elution buffer. The APC usually elutes in fractions 2–6.

*Analysis and Concentration*

5. Analyze samples by SDS–PAGE on an 18 × 16-cm 8% SDS–polyacrylamide gel and silver stain using the method of Ansorge (1985). Generally, 35 $\mu$l of the TEV eluate or 100 $\mu$l of each calmodulin elution fraction should be loaded onto the gel. We can resolve all of the APC subunits (except Apc4 and Apc5, which run at the same position, Fig. 2) using a modified gel composition of

Resolving gel: 0.75 *M* Tris, pH 9.2, 8% 37.5:1 acrylamide:bis-acrylamide, 0.1% (w/v) SDS

Stacking gel: 0.25 *M* Tris, pH 6.8, 5.1% 37.5:1 acrylamide:bis-acrylamide, 0.1% (w/v) SDS

Running buffer: 25 m*M* Tris, 192 m*M* glycine, pH 8.3, 0.1% SDS

6. Pool the calmodulin elution fractions containing the APC (usually fractions 2–6 contain most of the APC).

7. Concentrate the pooled fractions using YM–50 Centricons or YM–50 Microcons (Millipore). Estimate the protein concentration using the Bio-Rad protein assay and a bovine serum albumin calibration standard made up in *CaM elution buffer*. For *in vitro* assays, a concentration of 0.05–0.1 mg/ml is sufficient.

8. Aliquot the concentrated protein, flash freeze in dry ice/ethanol or liquid nitrogen, and store at –80°. Purified APC loses activity after multiple freeze–thaws.

*Comments*

A silver-stained gel of a typical APC purification is shown in Fig. 2. After the first two steps of the TAP purification (purification on an IgG Sepharose column and elution with TEV protease), a series of bands

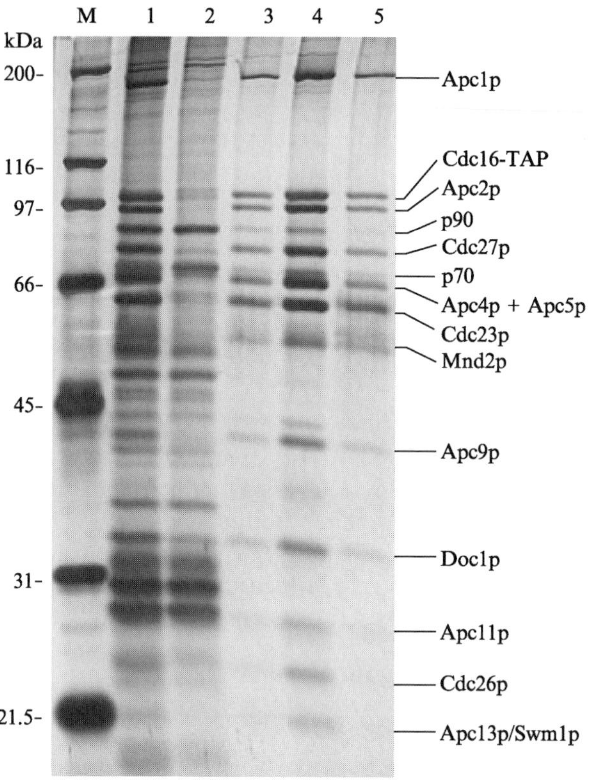


FIG. 2. TAP purification of endogenous APC from *S. cerevisiae*. Analysis of APC purification from *CDC16-TAP* yeast by silver-stained SDS–PAGE. Cdc16-TAP binds to IgG Sepharose and is eluted by cleavage with TEV protease (eluate, lane 1). The eluate is bound to calmodulin Sepharose (flow through, lane 2), washed, and eluted with EGTA (elution fractions, lanes 3–5). Proteins identified by MALDI-TOF mass spectrometry are labeled. p90 is a Cdc16p degradation product, and p70 is the Hsp70 family heat shock protein Ssa2. The diffuse band at 55–60 kDa contains keratins, which originate in contaminated SDS stock solutions. M, molecular weight markers. Reproduced from Passmore *et al.* (2003) with permission.

corresponding to the correct molecular weights for APC subunits is visible (Fig. 2, lane 1). After subsequent purification on calmodulin resin, the APC is ~95% pure (Fig. 2, lanes 3–5) with a yield of ~10–50 $\mu$g APC from 10 liters of yeast culture. In contrast, a negative control purification from untagged BJ2168 yeast yields no major purified proteins (not shown). We have confirmed the identities of all of the 13 APC subunits using MALDI-TOF mass spectrometry (Passmore *et al.*, 2003). We also identified two contaminating proteins: p90 is a 90-kDa degradation product of Cdc16 whereas p70 is the Hsp70 family heat shock protein Ssa2. Hsp70 binds to extended hydrophobic polypeptide segments (Bukau and Horwich, 1998), but its presence does not necessarily indicate unfolding or disassembly of the APC because it is often a contaminant in protein purifications. Ssa2 is a highly expressed protein (Garrels *et al.*, 1997) and it had a high frequency of occurrence in genome-wide mass spectrometric studies in yeast, being present in up to 54% of all protein purifications, as well as in mock purifications from untagged strains (Gavin *et al.*, 2002; Ho *et al.*, 2002). The relative intensities of p90 and p70 on SDS–PAGE decrease after concentration of the APC, probably because they are not integral components of the APC and they pass through the concentrator membrane.

We have examined the stability of the APC in response to pH change (Passmore, 2003). These studies showed that APC is highly sensitive to pH less than 7.5 and that low pH may cause the APC to precipitate or dissociate. The APC is stable at higher pH and it should be purified in a buffer with a pH of 8.0. Because the pH of Tris buffers is temperature dependent, one must ensure that all buffers are kept at 4° and that APC is stored on ice. (As mentioned earlier, we have also used HEPES buffer to avoid this problem.)

The TAP tag was designed specifically to allow for the efficient recovery of proteins present in low quantities (Rigaut *et al.*, 1999). Its success is probably dependent on several factors. First, both the protein A–IgG and calmodulin-binding peptide (CBP)–calmodulin interactions are high affinity. For the CBP–calmodulin interaction, the $K_d$ is less than 10 n$M$ (Blumenthal *et al.*, 1985). The affinity of the protein A–IgG interaction varies with the type of IgG and the pH but it should bind tightly under the conditions used here (human IgG and pH 8.0; Harlow and Lane, 1999). Second, these protein–protein interactions are very specific, unlike those found in many other affinity/epitope tags. Finally, the elution methods (TEV cleavage from IgG Sepharose and EGTA elution from calmodulin Sepharose) are specific and gentle, preventing dissociation of the protein complex.

## Assay of APC Using Native Substrates

As an E3 ubiquitin ligase, the *in vivo* role of the APC is to catalyze the conjugation of polyubiquitin chains onto specific substrate proteins. This activity is regulated both spatially and temporally. We have developed an *in vitro* ubiquitination assay to study APC activity against endogenous yeast targets and therefore in a physiologically relevant context (Passmore,

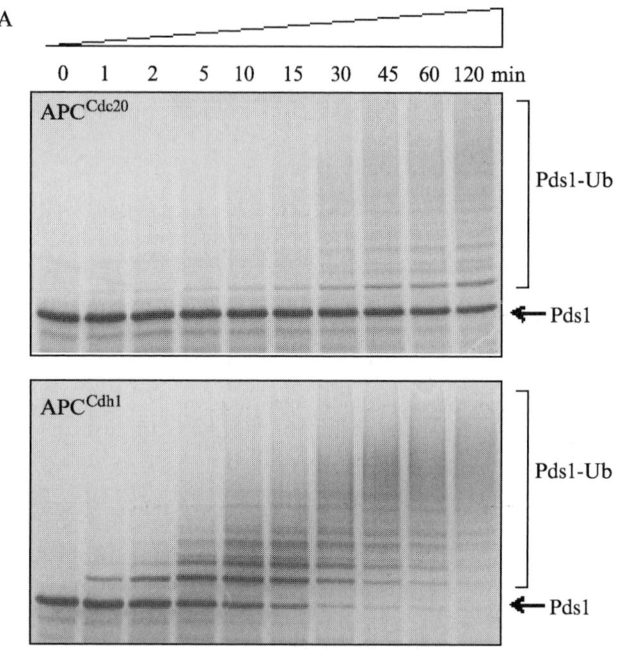
A
0 1 2 5 10 15 30 45 60 120 min
APC^Cdc20
Pds1-Ub
Pds1
APC^Cdh1
Pds1-Ub
Pds1

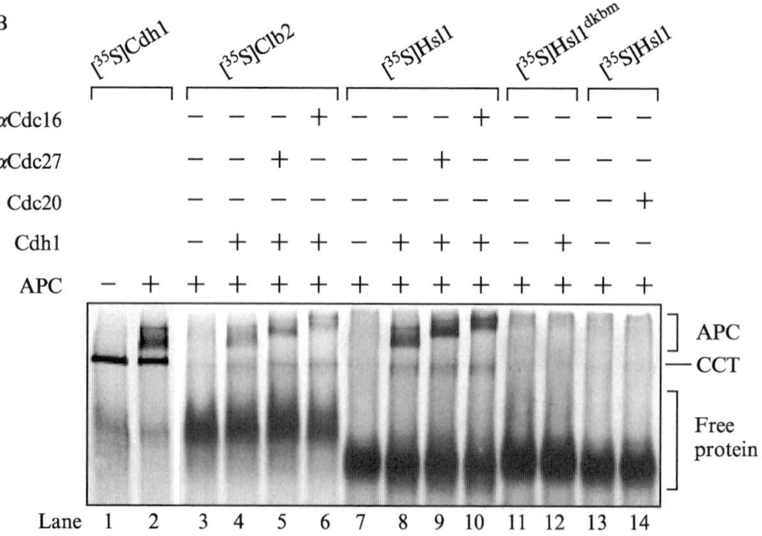
B
[35S]Cdh1
[35S]Clb2
[35S]Hsl1
[35S]Hsl1dkbm
[35S]Hsl1
αCdc16
αCdc27
Cdc20
Cdh1
APC
APC
CCT
Free protein
Lane 1 2 3 4 5 6 7 8 9 10 11 12 13 14

2003; Passmore *et al.*, 2003). Purified APC is incubated with yeast E2 (Ubc4p), ubiquitin, ATP, and $^{35}$S-labeled substrate, and the entire reaction mixture is analyzed by SDS–PAGE followed by autoradiography. Using this assay, purified yeast APC is active as an E3 ubiquitin ligase toward specific yeast substrates, including Pds1, Clb2, and Hsl1, but it is not active toward, nonphysiological substrates such as Cln1 (Fig. 3A) (Passmore *et al.*, 2003). The activity of the APC is dependent on the presence of either the Cdc20 or the Cdh1 coactivator. It catalyzes the formation of polyubiquitin chains onto substrate, resulting in high molecular weight polyubiquitinated products. Thus, the *in vitro* assay reflects the *in vivo* function of the APC.

### *Buffers and Reagents*

Plasmids: All plasmids are constructed using *S. cerevisiae* open reading frames (ORFs). APC substrates (Pds1, Clb2, and Hsl1) and coactivators (Cdc20 and Cdh1) are cloned into vectors containing a T7 promoter (either pET or pRSET) so they can be produced by *in vitro* transcription/translation (IVT). All plasmids for IVT should be highly purified, e.g., using the HiSpeed plasmid midi kit (Qiagen 12643).

20× ubiquitination buffer: 800 m*M* Tris, pH 7.5, 200 m*M* $MgCl_2$, 12 m*M* dithiothreitol (DTT). Aliquot and store at –20°.

0.1 *M* ATP (Amersham Biosciences 27-2056-01): Aliquot and store at –20°. Avoid multiple freeze–thaws.

5 mg/ml ubiquitin: Dissolve ubiquitin (Affiniti, UW8795) to 5 mg/ml in 10 m*M* Tris, pH 7.0. Aliquot and store at –20°.

E2: $His_6$-Ubc4p (E2) can be overexpressed and purified from *Escherichia coli* using standard procedures. Dilute purified Ubc4 to 500 ng/$\mu$l in 10 m*M* Tris, pH 7.5, 150 m*M* NaCl, and 2 m*M* DTT. Aliquot and store at –80°.

---

FIG. 3. *In vitro* APC assays. (A) Time courses of the activities of APC with Cdc20 ($APC^{Cdc20}$) and APC with Cdh1 ($APC^{Cdh1}$) toward the substrate Pds1. APC was incubated with E2 (Ubc4), ubiquitin, ATP, $^{35}$S-labeled substrate, and one of the coactivators, Cdc20 or Cdh1. Samples were taken at the indicated time points and added to SDS–PAGE loading buffer. The addition of polyubiquitin chains onto $^{35}$S-labeled Pds1 results in the appearance of high molecular weight smears correlating with the disappearance of unmodified Pds1 from the bottom of the gel. (B) Analysis of APC–coactivator and APC–substrate interactions. In the first two lanes, $^{35}$S-Cdh1 was incubated with and without APC. In the remaining lanes, $^{35}$S-labeled Clb2, 6His-$Hsl1^{667–872}$, or 6His-Hsl1 containing D- and KEN-box mutations (6His-$Hsl1^{667–872\ dkbm}$) were mixed with APC in the presence or absence of Cdh1 or Cdc20. All samples were run on a native gel and analyzed by autoradiography. Antibodies to Cdc27 or Cdc16, which retard the migration of the APC, were added to some samples. Reproduced from Passmore *et al.* (2003) with permission.

1 $\mu$g/$\mu$l ubiquitin aldehyde: Dissolve ubiquitin aldehyde (Affiniti, UW8450) to 1 $\mu$g/$\mu$l in 10 m*M* Tris, pH 7.0. Aliquot and store at –20°.

200 $\mu M$ LLnL: Dissolve LLnL (*N*-acetyl-Leu-Leu-Norleu-al; Sigma A6185) to 20 m*M* in DMSO and then dilute to 200 $\mu M$ with water. Aliquot and store at –20°.

APC buffer: 10 m*M* Tris, pH 8.0, 150 m*M* NaCl, 10% (w/v) glycerol, 1 m*M* Mg-acetate, 0.01% (v/v) Igepal CA-630, 2 m*M* EGTA, and 3 m*M* DTT. Aliquot and store at –20°.

4× SDS–PAGE loading buffer: 0.2 *M* Tris-Cl, pH 6.8, 8% (w/v) SDS, 40% (v/v) glycerol, 20% (v/v) 2-mercaptoethanol, and bromphenol blue to color.

### In Vitro *Transcription/Translation*

Coactivators and substrates are obtained using a rabbit reticulocyte-coupled *in vitro* transcription/translation (IVT) system, as these proteins are difficult to purify from overexpression systems. The incorporation of [$^{35}$S]methionine into the IVT system also provides a convenient and sensitive mechanism to label substrates for direct visualization on autoradiograms. A problem with using IVT-produced proteins is that other proteins from the IVT mixture might affect the assay. In these ubiquitylation assays, proteins independent of Cdc20 or Cdh1 are not activating the APC because the addition of reticulocyte lysate alone does not stimulate APC activity (Passmore *et al.*, 2003).

Prepare substrates and coactivators by IVT using the TNT T7 Quick coupled *in vitro* transcription/translation kit (Promega L1170). The TNT T7 Quick IVT system contains T7 RNA polymerase, nucleotides, salts, and ribonuclease inhibitor for *in vitro* transcription from a T7 promoter, as well as a rabbit reticulocyte lysate solution and amino acids to allow *in vitro* translation. Therefore, when a plasmid containing a gene under the control of a T7 promoter is incubated in the TNT solution, the gene product is expressed and may undergo posttranslational processing (e.g., chaperone-mediated folding). The IVT reactions should be performed fresh for each assay, as freeze/thaw cycles inactivate Cdc20 and Cdh1. For ubiquitination assays, synthesize substrates using [$^{35}$S]methionine and coactivators using unlabeled methionine. For [$^{35}$S]methionine IVT reactions, use 40 $\mu$l TNT T7 Quick Master Mix, 1 $\mu$g plasmid DNA, 4 $\mu$l [$^{35}$S]methionine (Redivue L-[$^{35}$S]methionine, Amersham AG1594), and water to 50 $\mu$l. For IVT reactions with unlabeled methionine, use 1 $\mu$l 1 m*M* methionine instead of [$^{35}$S]methionine. Incubate the reactions at 30° for 90 min. [$^{35}$S] Methionine-labeled reactions can be checked by running 3 $\mu$l on SDS–PAGE (see later).

*Ubiquitination Assays*

Ubiquitination assays are performed in 10-$\mu$l reaction volumes containing

0.50 $\mu$l 20× ubiquitination buffer
0.27 $\mu$l 0.1 *M* ATP
1.33 $\mu$l 5 mg/ml ubiquitin
2.00 $\mu$l purified E2 (yeast Ubc4, 500 ng/$\mu$l)
0.20 $\mu$l 1 mg/ml ubiquitin aldehyde
0.10 $\mu$l 200 $\mu M$ LLnL
1.00 $\mu$l $^{35}$S-labeled substrate (produced in IVT)
0.67 $\mu$l purified APC (~50 ng/$\mu$l)
2.00 $\mu$l unlabeled Cdc20 or Cdh1 (produced in IVT)
1.93 $\mu$l $H_2O$

Prepare a reaction mixture on ice containing the ubiquitination buffer, ATP, ubiquitin, E2, ubiquitin aldehyde, LLnL, and water and then add substrate, APC, and coactivator as required. Incubate the reaction at room temperature for 45 min and stop it by adding 4 $\mu$l of 4× SDS–PAGE loading buffer. For negative controls, substitute APC with APC buffer, or coactivator with mock IVT reaction (i.e., IVT without plasmid). To control for substrate specificity, D- and/or KEN-box substrate mutants or nonphysiological substrates may be used.

Ubiquitin aldehyde is a specific inhibitor of ubiquitin C-terminal hydrolases (Hershko and Rose, 1987; Melandri *et al.*, 1996; Pickart and Rose, 1986) and is included to prevent the hydrolysis of polyubiquitin chains. The proteasome inhibitor LLnL (Orlowski *et al.*, 1993; Vinitsky *et al.*, 1992) is included to prevent degradation of the polyubiquitinated proteins. The addition of exogenous E1 does not enhance the ubiquitination reaction, as E1 is abundant in reticulocyte lysate (Ciechanover *et al.*, 1982; Haas and Bright, 1988). The optimal amounts of Ubc4 and APC should be determined by titration, as different preparations will have different activities.

*SDS–PAGE and Autoradiography*

For SDS–PAGE, standard gel compositions should be used (Sambrook and Russell, 2001) with the Bio-Rad Mini-PROTEAN 3 system (or similar). For visualizing ubiquitinated products, 8% minigels (8 × 7.3-cm gel plates) with 15-well combs provide good resolution. Run the entire reaction mixture (14 $\mu$l) on the gel at 140 V until the dye front reaches the bottom. Stain the gel with Coomassie blue (50% methanol, 10% acetic acid), destain (20% methanol, 10% acetic acid), dry on Whatman paper at 80° for 1 h, and expose to BioMax MR–1 film (Kodak).

In the presence of E1, E2, ubiquitin, ATP, and $^{35}$S-labeled substrate, active APC conjugates ubiquitin proteins onto the substrate to form a polyubiquitin chain. As successive ubiquitins are added onto the substrate, its molecular mass increases by 8.6-kDa increments (the molecular mass of ubiquitin). On the autoradiogram, a ladder of $^{35}$S-labeled substrate appears, increasing in average molecular weight as the reaction proceeds (Fig. 3A). Whereas other assays have only detected the conjugation of monoubiquitin onto one to three substrate lysines, this assay shows polyubiquitin chain formation. The APC with Cdh1 ($APC^{Cdh1}$) is more active than the APC with Cdc20 ($APC^{Cdc20}$) perhaps due to more efficient processing of Cdh1 relative to Cdc20 in the reticulocyte lysate (Passmore *et al.*, 2003).

## Visualization of APC–Coactivator and APC–Substrate Complexes by Native Gel Electrophoresis

Because the core APC is active in ubiquitination reactions, it must interact with coactivators, E2, and substrates. To examine these interactions further, we have used an APC-binding assay (Passmore *et al.*, 2003). Traditional binding assays on columns could not be used, as IVT-produced Cdc20 and Cdh1 bind nonspecifically to affinity resins such as calmodulin Sepharose and Ni-NTA agarose. Native gels separate proteins based on their size, shape, and charge and therefore can be used to investigate protein–protein interactions that alter one or more of these parameters. A native gel system had been used previously to examine the interactions of the chaperonin containing TCP1 (CCT) with its substrates, one of which is Cdh1 (E. A. McCormack and K. R. Willison, unpublished results) (Liou and Willison, 1997; Valpuesta *et al.*, 2002). A modified version of this native gel assay using purified APC and proteins produced using IVT can be used to examine APC interactions (Passmore *et al.*, 2003).

### *APC–Coactivator Interactions*

To examine APC–coactivator interactions, $^{35}$S-labeled Cdh1 produced using IVT is mixed with purified APC. Interactions with APC can be monitored by changes in the mobility of the labeled coactivator. $^{35}$S-labeled Cdh1 migrates as two species on a native gel (Fig. 3B, lane 1). A diffuse, faster migrating band probably represents free Cdh1, whereas a slower, more discrete band corresponds with the migration of CCT. This CCT comes from the rabbit reticulocyte lysate and is required for proper Cdh1 processing (Camasses *et al.*, 2003). (CCT is a eukaryotic chaperonin that binds to specific substrates and assists in their folding.) Upon addition of APC, two additional bands appear (Fig. 3B, lane2). These bands migrate very slowly in native gels and represent APC–coactivator complexes, as they undergo a band shift upon the addition of antibodies directed against APC subunits (Passmore *et al.*, 2003). In addition, they correspond with the migration position of purified APC (Passmore *et al.*, 2003).

Binding reactions are performed in 14 $\mu$l with the following compositions:

- 2 $\mu$l purified APC (~50 ng) or *CaM elution buffer* (Table I) for negative control
- 2 $\mu$l $^{35}$S-labeled Cdh1 (produced in IVT as described earlier)
- 0.7 $\mu$l 100 m*M* $CaCl_2$

$CaCl_2$ is required for proper migration on the native gels (E. A. McCormack, unpublished results). Adjust the volume to 14 $\mu$l with binding buffer (10 m*M* Tris-Cl, pH 8.0, 150 m*M* NaCl, 3 m*M* DTT, 1 m*M* Mg-acetate, 2 m*M* EGTA). Incubate samples at room temperature for 15 min, add 1.5 $\mu$l native gel sample buffer (see later), and load the entire reaction onto a 5.25% native gel.

We have observed that lower amounts of Cdc20 bind to the APC than Cdh1 in the native gel assay (Passmore *et al.*, 2003). Instead, most Cdc20 remains bound to CCT, suggesting that yeast Cdc20 is processed poorly by rabbit CCT. This Cdc20 interacts poorly with the APC and explains why $APC^{Cdc20}$ has reduced ubiquitin ligase activity compared to $APC^{Cdh1}$ (Fig. 3A). This may be due to species-specific differences between yeast Cdc20 and rabbit CCT. Alternatively, phosphorylation of Cdc20 or additional factors may be required for its efficient release from CCT and subsequent interaction with APC. Because Cdc20 does not appear to be processed properly, we perform most *in vitro* studies with Cdh1.

### *APC–Substrate Interactions*

Similar to the APC–coactivator binding assay described earlier, APC–substrate interactions can be examined using native gels (Passmore *et al.*, 2003). The migration of $^{35}$S-labeled APC substrates (Clb2 and a D- and KEN-box-containing domain of Hsl1, $Hsl1p^{667–872}$) changes upon the addition of APC and Cdh1 (Fig. 3B). Both APC and coactivator are required for this band shift. These bands represent APC–substrate complexes, as their migrations are retarded by the addition of APC antibodies (Fig. 3B, lanes 5, 6, 9, and 10). These interactions are specific and likely to be physiologically relevant because they are dependent on the presence of intact D- and/or KEN-boxes (Fig. 3B, lanes 11 and 12) (Passmore *et al.*, 2003).

APC–substrate-binding reactions are performed in 14 $\mu$l with the following compositions:

- 2 $\mu$l purified APC (~50 ng) or *CaM elution buffer* (Table I) for negative control
- 2 $\mu$l $^{35}$S-labeled substrate (produced in IVT as described earlier)
- 2 $\mu$l unlabeled coactivator (produced in IVT as described earlier) or mock IVT reaction for negative control
- 0.7 $\mu$l 100 m*M* $CaCl_2$

Adjust the volume to 14 $\mu$l with binding buffer (10 m*M* Tris-Cl, pH 8.0, 150 m*M* NaCl, 3 m*M* DTT, 1 m*M* Mg-acetate, 2 m*M* EGTA). Incubate samples at room temperature for 15 min, add 1.5 $\mu$l native gel sample buffer, and load the entire reaction onto a 5.25% native gel.

We have been unable to detect interactions between APC and Pds1 in the native gel assay, presumably due to a lower affinity of $APC^{Cdh1}$ for Pds1. In addition, the interaction of substrates with $APC^{Cdc20}$ could not be detected, probably due to the poor processing of yeast Cdc20 by rabbit CCT as described previously. The pH of the native gels is high (pH 8.8), which may affect some protein–protein interactions negatively. However, our attempts to run native gels at lower pH have been unsuccessful.

### *Antibody Shifts*

Antibodies to APC subunits induce an antibody shift (Fig. 3B), proving that the complexes visualized on native gels contain APC. For antibody shifts, prepare the binding reactions as described earlier, except make them up to 12 $\mu$l instead of 14 $\mu$l. After all of the other components have been mixed together, add 2 $\mu$l antibody and proceed as described earlier.

### *Native Gel Composition*

We perform native gel analysis using the Mini-PROTEAN 3 system (Bio-Rad) with 1.5-mm spacers and 15-well combs. Prepare native gels using the following gel compositions (Liou and Willison, 1997):

- Resolving gel: 0.37 *M* Tris, pH 8.8, 5.25% 37.5:1 acrylamide:bisacrylamide
- Stacking gel: 57 m*M* Tris, pH 8.8, 3.22% 37.5:1 acrylamide:bisacrylamide
- Running buffer: 25 m*M* Tris, 192 m*M* glycine, pH 8.3 (to make a 10× stock solution, dissolve 30 g Tris base and 144.2 g glycine in 1 liter water)

Native gel sample buffer: 125 m$M$ Tris, pH 8.8, 84% (v/v) glycerol, bromphenol blue to color

Both the resolving and the stacking gels are made using a 30% acrylamide solution and a 1.5 $M$ Tris solution (pH 8.8 at room temperature). Chill native gels and running buffer to 4° before use. Run gels at 110 V and 4° for approximately 2 h, until the dye front reaches the bottom of the gel. Fix, stain, and dry the gels as described previously and expose to BioMax MR–1 film.

## Deletion Strains

The functions of individual APC subunits and their interactions within the complex are largely unknown. Because APC subunits are difficult to overexpress and purify, most of them cannot be studied in isolation. Five of the APC subunits are not essential for viability (*DOC1*, *APC9*, *CDC26*, *SWM1*, and *MND2*) and therefore can be deleted in the *CDC16-TAP* yeast strain. Using these deletion strains, the TAP purification approach, the *in vitro* ubiquitination assay, and the APC-binding assay provide excellent tools to delineate the roles of individual APC subunits.

To construct deletion strains, we use PCR-based gene targeting (as described earlier) using the pAG32c vector (Goldstein and McCusker, 1999) instead of pFA6a-TAP-kanMX6. The pAG32c vector contains a hygromycin resistance marker and can be used to transform the *CDC16-TAP* strain, which already has G418 resistance (using 200 $\mu$g/ml hygromycin B and 300 $\mu$g/ml G418 for selection). Strains with deleted APC subunits should be grown at 25°. Deletion of *DOC1, CDC26*, or *SWM1* results in temperature sensitivity, and the *DOC1* deletion strain grows very slowly even at 25°. APC can be purified from the deletion strains to examine its stability, activity, and ability to interact with coactivators or substrates.

## Conclusion

This chapter describes a TAP purification approach to purify endogenous *S. cerevisiae* APC to near homogeneity. In addition, we described an *in vitro* ubiquitination assay and an APC-binding assay. In the ubiquitination assay, the APC forms polyubiquitin chains on yeast substrates (Pds1, Clb2, and Hsl1), but only in the presence of one of the coactivators, Cdc20 or Cdh1. These coactivators confer substrate specificity upon the APC in a D-box and KEN-box-dependent manner, reflecting *in vivo* observations. The native gel APC-binding assay can be used to explore the properties of coactivator and substrate interactions with the APC. In this assay, the association of yeast APC substrates (Clb2 and Hsl1) with the APC is dependent on coactivator and intact substrate D- and KEN-boxes. These properties reflect the *in vivo* requirements of APC-mediated ubiquitination reactions. Finally, we described how these techniques can be used to study the roles of individual APC subunits.

## Acknowledgments

We thank E. A. McCormack and K. R. Willison for assistance in developing the native gel assays. This work is supported by NIH Grant AG11085 to J. W. H. and by Cancer Research UK to D. B.

## References

Ansorge, W. (1985). Fast and sensitive detection of protein and DNA bands by treatment with potassium permanganate. *J. Biochem. Biophys. Methods* **11,** 13–20.

Aristarkhov, A., Eytan, E., Moghe, A., Admon, A., Hershko, A., and Ruderman, J. V. (1996). E2-C, a cyclin-selective ubiquitin carrier protein required for the destruction of mitotic cyclins. *Proc. Natl. Acad. Sci. USA* **93,** 4294–4299.

Blumenthal, D. K., Takio, K., Edelman, A. M., Charbonneau, H., Titani, K., Walsh, K. A., and Krebs, E. G. (1985). Identification of the calmodulin-binding domain of skeletal muscle myosin light chain kinase. *Proc. Natl. Acad. Sci. USA* **82,** 3187–3191.

Bukau, B., and Horwich, A. L. (1998). The Hsp70 and Hsp60 chaperone machines. *Cell* **92,** 351–366.

Burton, J. L., and Solomon, M. J. (2001). D box and KEN box motifs in budding yeast Hsl1p are required for APC-mediated degradation and direct binding to Cdc20p and Cdh1p. *Genes Dev.* **15,** 2381–2395.

Camasses, A., Bogdanova, A., Shevchenko, A., and Zachariae, W. (2003). The CCT chaperonin promotes activation of the anaphase-promoting complex through the generation of functional Cdc20. *Mol. Cell* **12,** 87–100.

Carroll, C. W., and Morgan, D. O. (2002). The Doc1 subunit is a processivity factor for the anaphase-promoting complex. *Nature Cell Biol.* **4,** 880–887.

Charles, J. F., Jaspersen, S. L., Tinker-Kulberg, R. L., Hwang, L., Szidon, A., and Morgan, D. O. (1998). The Polo-related kinase Cdc5 activates and is destroyed by the mitotic cyclin destruction machinery in *S. cerevisiae*. *Curr. Biol.* **8,** 497–507.

Ciechanover, A., Elias, S., Heller, H., and Hershko, A. (1982). "Covalent affinity" purification of ubiquitin-activating enzyme. *J. Biol. Chem.* **257,** 2537–2542.

Fang, G., Yu, H., and Kirschner, M. W. (1998). Direct binding of CDC20 protein family members activates the anaphase-promoting complex in mitosis and G1. *Mol. Cell* **2,** 163–171.

Garrels, J. I., McLaughlin, C. S., Warner, J. R., Futcher, B., Latter, G. I., Kobayashi, R., Schwender, B., Volpe, T., Anderson, D. S., Mesquita-Fuentes, R., and Payne, W. E. (1997). Proteome studies of *Saccharomyces cerevisiae*: Identification and characterization of abundant proteins. *Electrophoresis* **18,** 1347–1360.

Gavin, A. C., Bösche, M., Krause, R., Grandi, P., Marzioch, M., Bauer, A., Schultz, J., Rick, J. M., Michon, A. M., Cruciat, C. M., Remor, M., Höfert, C., Schelder, M., Brajenovic, M., Ruffner, H., Merino, A., Klein, K., Hudak, M., Dickson, D., Rudi, T., *et al.* (2002).

Functional organization of the yeast proteome by systematic analysis of protein complexes. *Nature* **415,** 141–147.

Golan, A., Yudkovsky, Y., and Hershko, A. (2002). The cyclin-ubiquitin ligase activity of cyclosome/APC is jointly activated by protein kinases Cdk1-cyclin B and Plk. *J. Biol. Chem.* **277,** 15552–15557.

Goldstein, A. L., and McCusker, J. H. (1999). Three new dominant drug resistance cassettes for gene disruption in *Saccharomyces cerevisiae*. *Yeast* **15,** 1541–1553.

Haas, A. L., and Bright, P. M. (1988). The resolution and characterization of putative ubiquitin carrier protein isozymes from rabbit reticulocytes. *J. Biol. Chem.* **263,** 13258–13267.

Harlow, E., and Lane, D. (1999). "Using Antibodies: A Laboratory Manual." Cold Spring Harbor Laboratory Press, Cold Spring Harbor, NY.

Harper, J. W., Burton, J. L., and Solomon, M. J. (2002). The anaphase promoting complex: It's not just for mitosis anymore. *Genes Dev.* **16,** 2179–2206.

Hendrickson, C., Meyn, M. A., 3rd, Morabito, L., and Holloway, S. L. (2001). The KEN box regulates Clb2 proteolysis in G1 and at the metaphase-to-anaphase transition. *Curr. Biol.* **11,** 1781–1787.

Hershko, A., and Rose, I. A. (1987). Ubiquitin-aldehyde: A general inhibitor of ubiquitin-recycling processes. *Proc. Natl. Acad. Sci. USA* **84,** 1829–1833.

Hilioti, Z., Chung, Y. S., Mochizuki, Y., Hardy, C. F., and Cohen-Fix, O. (2001). The anaphase inhibitor Pds1 binds to the APC/C-associated protein Cdc20 in a destruction box-dependent manner. *Curr. Biol.* **11,** 1347–1352.

Ho, Y., Gruhler, A., Heilbut, A., Bader, G. D., Moore, L., Adams, S. L., Millar, A., Taylor, P., Bennett, K., Boutilier, K., Yang, L., Wolting, C., Donaldson, I., Schandorff, S., Shewnarane, J., Vo, M., Taggart, J., Goudreault, M., Muskat, B., Alfarano, C., *et al.* (2002). Systematic identification of protein complexes in *Saccharomyces cerevisiae* by mass spectrometry. *Nature* **415,** 180–183.

Jones, E. W. (1977). Proteinase mutants of *Saccharomyces cerevisiae*. *Genetics* **85,** 23–33.

Jones, E. W. (2002). Vacuolar proteases and proteolytic artifacts in *Saccharomyces cerevisiae*. *Methods Enzymol.* **351,** 127–150.

King, R. W., Peters, J. M., Tugendreich, S., Rolfe, M., Hieter, P., and Kirschner, M. W. (1995). A 20S complex containing CDC27 and CDC16 catalyzes the mitosis-specific conjugation of ubiquitin to cyclin B. *Cell* **81,** 279–288.

Kotani, S., Tugendreich, S., Fujii, M., Jorgensen, P. M., Watanabe, N., Hoog, C., Hieter, P., and Todokoro, K. (1998). PKA and MPF-activated Polo-like kinase regulate anaphase-promoting complex activity and mitosis progression. *Mol. Cell* **1,** 371–380.

Kramer, E. R., Scheuringer, N., Podtelejnikov, A. V., Mann, M., and Peters, J. M. (2000). Mitotic regulation of the APC activator proteins CDC20 and CDH1. *Mol. Biol. Cell* **11,** 1555–1569.

Liou, A. K., and Willison, K. R. (1997). Elucidation of the subunit orientation in CCT (chaperonin containing TCP1) from the subunit composition of CCT micro-complexes. *EMBO J.* **16,** 4311–4316.

Melandri, F., Grenier, L., Plamondon, L., Huskey, W. P., and Stein, R. L. (1996). Kinetic studies on the inhibition of isopeptidase T by ubiquitin aldehyde. *Biochemistry* **35,** 12893–12900.

Orlowski, M., Cardozo, C., and Michaud, C. (1993). Evidence for the presence of five distinct proteolytic components in the pituitary multicatalytic proteinase complex: Properties of two components cleaving bonds on the carboxyl side of branched chain and small neutral amino acids. *Biochemistry* **32,** 1563–1572.

Passmore, L. A. (2003). "Structural and Functional Studies of the Anaphase-Promoting Complex (APC)." Ph.D Thesis, University of London.

Passmore, L. A., McCormack, E. A., Au, S. W., Paul, A., Willison, K. R., Harper, J. W., and Barford, D. (2003). Doc1 mediates the activity of the anaphase-promoting complex by contributing to substrate recognition. *EMBO J.* **22,** 786–796.

Peters, J. M. (2002). The anaphase-promoting complex: Proteolysis in mitosis and beyond. *Mol. cell.* **9,** 931–943.

Pfleger, C. M., and Kirschner, M. W. (2000). The KEN box: An APC recognition signal distinct from the D box targeted by Cdh1. *Genes Dev.* **14,** 655–665.

Pfleger, C. M., Lee, E., and Kirschner, M. W. (2001). Substrate recognition by the Cdc20 and Cdh1 components of the anaphase-promoting complex. *Genes Dev.* **15,** 2396–2407.

Pickart, C. M., and Rose, I. A. (1986). Mechanism of ubiquitin carboxyl-terminal hydrolase: Borohydride and hydroxylamine inactivate in the presence of ubiquitin. *J. Biol. Chem.* **261,** 10210–10217.

Prodromou, C., and Pearl, L. H. (1992). Recursive PCR: A novel technique for total gene synthesis. *Protein Eng.* **5,** 827–829.

Puig, O., Caspary, F., Rigaut, G., Rutz, B., Bouveret, E., Bragado-Nilsson, E., Wilm, M., and Séraphin, B. (2001). The tandem affinity purification (TAP) method: A general procedure of protein complex purification. *Methods* **24,** 218–229.

Rigaut, G., Shevchenko, A., Rutz, B., Wilm, M., Mann, M., and Séraphin, B. (1999). A generic protein purification method for protein complex characterization and proteome exploration. *Nature Biotechnol.* **17,** 1030–1032.

Rudner, A. D., and Murray, A. W. (2000). Phosphorylation by Cdc28 activates the Cdc20-dependent activity of the anaphase-promoting complex. *J. Cell Biol.* **149,** 1377–1390.

Sambrook, J., and Russell, D. W. (2001). "Molecular Cloning: A Laboratory Manual," 3rd Ed. Cold Spring Harbor Laboratory Press, Cold Spring Harbor, NY.

Valpuesta, J. M., Martin-Beníto, J., Gómez-Puertas, P., Carrascosa, J. L., and Willison, K. R. (2002). Structure and function of a protein folding machine: The eukaryotic cytosolic chaperonin CCT. *FEBS Lett.* **529,** 11–16.

Vinitsky, A., Michaud, C., Powers, J. C., and Orlowski, M. (1992). Inhibition of the chymotrypsin-like activity of the pituitary multicatalytic proteinase complex. *Biochemistry* **31,** 9421–9428.

Visintin, R., Prinz, S., and Amon, A. (1997). *CDC20* and *CDH1*: A family of substrate-specific activators of APC-dependent proteolysis. *Science* **278,** 460–463.

Wach, A., Brachat, A., Rebischung, C., Steiner, S., Pokorni, K., Hessen, S., and Philippsen, P. (1998). PCR-based gene targetting in *Saccharomyces cerevisiae*. *In* "Yeast Gene Analysis" (A. J. P. Brown and M. Tuite, eds.), Vol. 26, pp. 67–82. Academic Press, London.

Werner-Washburne, M., Braun, E., Johnston, G. C., and Singer, R. A. (1993). Stationary phase in the yeast *Saccharomyces cerevisiae*. *Microbiol. Rev.* **57,** 383–401.

Zachariae, W., Schwab, M., Nasmyth, K., and Seufert, W. (1998a). Control of cyclin ubiquitination by CDK-regulated binding of Hct1 to the anaphase promoting complex. *Science* **282,** 1721–1724.

Zachariae, W., Shevchenko, A., Andrews, P. D., Ciosk, R., Galova, M., Stark, M. J., Mann, M., and Nasmyth, K. (1998b). Mass spectrometric analysis of the anaphase-promoting complex from yeast: Identification of a subunit related to cullins. *Science* **279,** 1216–1219.

Zachariae, W., Shin, T. H., Galova, M., Obermaier, B., and Nasmyth, K. (1996). Identification of subunits of the anaphase-promoting complex of *Saccharomyces cerevisiae*. *Science* **274,** 1201–1204.

Zubenko, G. S., Mitchell, A. P., and Jones, E. W. (1980). Mapping of the proteinase b structural gene *PRB1*, in *Saccharomyces cerevisiae* and identification of nonsense alleles within the locus. *Genetics* **96,** 137–146.

# [18] Enzymology of the Anaphase-Promoting Complex

*By* CHRISTOPHER W. CARROLL and DAVID O. MORGAN

## Abstract

The anaphase-promoting complex (APC) is an ubiquitin-protein ligase that promotes mitotic progression by catalyzing the ubiquitination of numerous proteins, including securin and cyclin. Its complex subunit composition and extensive regulation make the APC an active subject of investigation for both cell biologists and enzymologists. This chapter describes a system for the reconstitution and quantitative analysis of APC activity from budding yeast *in vitro*. We focus in particular on the measurement of processive ubiquitination, which complements traditional analysis of the reaction rate as a means to elucidate the molecular details of substrate recognition and ubiquitination by the APC.

## Introduction

The covalent modification of proteins with ubiquitin is a widespread regulatory mechanism in cell biology. In mitosis, ubiquitin-mediated proteolysis of securin (Pds1 in budding yeast) and mitotic cyclins is required for sister chromatid separation and mitotic exit. The ubiquitination of these proteins is catalyzed by an ubiquitin-protein ligase (or E3 enzyme) called the anaphase-promoting complex.

The APC is a structurally complex E3 containing at least 11–13 subunits, most of which are conserved from yeast to humans (Harper *et al.*, 2002; Peters, 2002). Two subunits, Apc11 and Apc2, contain a RING-H2 domain and a cullin-homology domain, respectively, and are thought to comprise the catalytic core of the enzyme: they form a stable subcomplex, can bind an E2-ubiqutin conjugate, and are sufficient to catalyze some aspects of the ubiquitination reaction (Gmachl *et al.*, 2000; Leverson *et al.*, 2000; Tang *et al.*, 2001). However, the Apc2/11 heterodimer lacks

METHODS IN ENZYMOLOGY, VOL. 398

0076-6879/05 $35.00
DOI: 10.1016/S0076-6879(05)98018-X

the full activity and substrate specificity of the holo-APC. Highly specific and efficient ubiquitination of substrates also requires an additional protein, either Cdc20 or Cdh1, that binds and activates the APC in a cell cycle-dependent manner (Fang *et al.*, 1998; Zachariae *et al.*, 1998). Both the activating protein and core APC subunits contribute to substrate recognition through direct interactions with a conserved sequence element in substrates called the destruction box (Burton and Solomon, 2001; Hilioti *et al.*, 2001; Pfleger *et al.*, 2001; Schwab *et al.*, 2001; Yamano *et al.*, 2004).

Once bound, a substrate can be processively ubiquitinated by the APC, receiving multiple ubiquitin molecules during a single binding event (Carroll and Morgan, 2002). Because each E2 enzyme carries a single ubiquitin, the reaction requires multiple E2 molecules. The processive ubiquitination of a given substrate can occur in two patterns that are not mutually exclusive. First, a substrate can be modified directly on multiple lysine residues, resulting in multiple monoubiquitination. Second, the APC can direct the assembly of a polyubiquitin chain on a substrate. We do not understand what controls the pattern of ubiquitination or how a single active site is able to catalyze these different reactions.

Much of what we know about the regulation of APC activity and the functions of its individual subunits and activating proteins has been learned from detailed biochemical analysis with purified proteins *in vitro*. Although reconstitution of APC activity requires a large number of proteins, the relative ease of purification and the ability to purchase components of the reaction have made the development of an APC activity assay relatively straightforward. This chapter details a method to quantitatively analyze the activity of budding yeast APC *in vitro* using the APC activator Cdh1 and a model substrate containing an N-terminal fragment derived from sea urchin cyclin B (amino acids 13–110).

## Activity of the APC *In Vitro*

Analysis of APC activity *in vitro* requires numerous proteins that can be divided functionally into four groups: (i) the ubiquitin-activating system, (ii) APC, (iii) APC-activating protein, and (iv) substrates. Each is discussed in some detail later, followed by a description of a typical APC assay using these proteins. For quantitative analysis of APC activity, purified Cdh1 is used as the activator and $^{125}$I-labeled sea urchin cyclin B$^{13–110}$ is used as the substrate. With these components, the contents of the reaction are completely defined and the ubiquitinated products are easily detected, interpreted, and quantified. This system has been used in our laboratory to characterize the regulation of Cdh1 binding to the APC and to elucidate the role of E2 dynamics and the APC subunit Doc1 in promoting processive substrate ubiquitination (Carroll and Morgan, 2002; Carroll *et al.*, 2005; Jaspersen *et al.*, 1999).

It is worth noting that alternative means are available for the preparation of both activating proteins and substrates. For example, a coupled *in vitro* transcription and translation system can be used to produce both the activators Cdh1 and Cdc20 and numerous APC substrates (Carroll *et al.*, 2005; Passmore *et al.*, 2003). This approach has been very useful because Cdc20-dependent APC activity had not been analyzed previously with yeast proteins *in vitro*, and to our knowledge no other methods are currently available for obtaining active Cdc20. Furthermore, many substrates of the APC, including the essential target securin, have eluded purification by other means. However, the low amount of protein obtained and the presence of contaminating polyubiquitinating and deubiquitinating enzymes in the transcription/translation extract limit the scope of experiments that can be performed and can obscure the analysis of substrate ubiquitination.

### *Ubiquitin-Activating System*

The ubiquitin-activating system contains E1 (yeast Uba1), E2 (yeast Ubc4), ubiquitin, and ATP. Together these components generate E2-ub conjugates that serve as the ubiquitin donor in the reaction. Protocols for the purification of both E1 and E2 are available upon request, while ubiquitin (Sigma, St. Louis, MO; or Boston Biochem, Boston, MA) is purchased. Under the reaction conditions outlined here, the rate of substrate ubiquitination is linear for extended periods (>1 h, providing that the substrate is not significantly depleted), and increasing the concentration of any individual component of the ubiquitin-activating system does not increase the rate of the reaction (i.e., these components are not rate limiting).

### *Purification of the APC*

APC can be obtained from budding yeast using the TAP-tag purification method. This approach allows for the rapid isolation of protein complexes using two sequential affinity steps, and APC isolated by this method is highly pure and active. The details of this purification scheme have been described elsewhere and therefore are not discussed here (Carroll and Morgan, 2002; Rigaut *et al.*, 1999). In most instances, we have placed the tag at the C terminus of the Cdc16 subunit. Active APC has also been purified successfully using TAP-tagged Apc1 or Cdc27 subunits. TAP-tagged Apc11 is not fully functional and should not be used.

A typical purification yields ~0.1 mg of APC from 2 g of yeast lysate, which corresponds to ~10% of the soluble APC in the extract. The binding of APC to the IgG and calmodulin columns is generally efficient, with the biggest losses attributable to APC that binds but does not elute from the columns. APC purified by this method can be stored in the calmodulin column elution buffer (+10% glycerol) for long periods at −80° and is tolerant of multiple freeze/thaw cycles.

*APC Activating Proteins*

Although the activating proteins Cdc20 and Cdh1 bind the APC directly *in vivo*, they do not copurify with the APC to an appreciable extent. Activation of the APC *in vitro* therefore requires addition of an activating protein. We use Cdh1 containing an N-terminal 6-histidine tag and expressed in insect cells infected with a recombinant baculovirus. Approximately 1–3 mg of highly purified protein can be obtained from the purification scheme described here. This amount is sufficient to perform dose–response experiments that require saturating amounts of Cdh1. Cdc20 is not expressed efficiently in this system.

*Expression of 6His-Cdh1 in Baculovirus-Infected Insect Cells*

a. Sf9 insect cells are cultured in Graces insect medium (GIBCO, Carlsbad, CA) containing 10% fetal bovine serum (GIBCO). The day before baculovirus infection, the culture is diluted to $1 \times 10^6$ cells/ml, allowing the cells to double once by the following day to a final concentration of $1.8–2.2 \times 10^6$ cells/ml. The quality of cells is very important and greater than 99% of the cells should be viable before infecting with virus for protein expression. Cell viability is particularly dependent on good aeration, and suspension cultures should not be deeper than the top of the spinner.

b. Two hundred milliliters of cells is pelleted at 500*g* for 5 min in a tabletop centrifuge and the supernatant is removed.

c. The pellet is resuspended gently in a small volume of media (usually ~20 ml) containing the baculovirus at a multiplicity of infection (MOI) of 5–10 pfu/cell (typically, pass 3 virus at $\sim 10^8$ pfu/ml is used).

d. Cells are incubated for 1 h at 28° while gently swirling the cell suspension every 10 min.

e. The cell suspension is distributed equally to twelve 14-cm plates (Fisher, #12–565–100) containing fresh media (final concentration of cells is $\sim 25 \times 10^6$ cells/plate). The total volume of media in each plate should be ~14 ml.

f. Protein expression is allowed to proceed for 2 days on plates at 28° in a sterile incubator. Infected cells should appear swollen and grainy by standard phase-contrast microscopy. Cells are harvested by gentle scraping with a rubber policeman, after which media and cells are transferred to 50-ml conical tubes. Cells are pelleted at 500*g* for 5 min in a tabletop centrifuge and the supernatant is discarded. Infected cells are extremely fragile and should not be washed. Cell pellets are frozen on liquid $N_2$ and stored at $-80^\circ$ until ready for protein purification.

*Purification of 6His-Cdh1*

a. The frozen cell pellet is thawed on ice and resuspended in 20 ml hypotonic lysis buffer containing protease inhibitors [10 m*M* Tris, pH 7.6, 10 m*M* NaCl, 0.1% Triton X-100, 1 m*M* phenylmethylsulfonyl fluoride (PMSF), and 1 $\mu$g/ml aprotinin, pepstatin, and leupeptin].

b. Cells are homogenized slowly on ice in a prechilled 40-ml Dounce homogenizer (Wheaton, Millville, NJ). Twenty strokes are usually sufficient for cell lysis.

c. NaCl is added to a final concentration of 300 m*M*, and the lysate is centrifuged for 1 h at 4° at 100,000*g*.

d. The supernatant is removed and applied at 0.5 ml/min to a 1-ml Hi-Trap chelating column (Amersham, Uppsala, Sweden) that has been charged previously with $CoCl_2$.

e. The column is washed with 10 ml of lysis buffer, followed by 5 ml of lysis buffer at pH 6.0. After the pH 6.0 wash, the column is reequilibrated with 2 ml lysis buffer, pH 7.6.

f. Cdh1 is eluted from the column in lysis buffer using an imidazole gradient from 0 to 200 m*M*. Cdh1 will elute as a broad peak beginning at ~75 m*M* imidazole.

g. Purified Cdh1 is dialyzed for 4 h into storage buffer (20 m*M* Tris, pH 7.6, 300 m*M* NaCl, 0.1% Triton X-100, 10% glycerol), and insulin is added to 0.1 mg/ml to help stabilize the protein.

## *APC Substrates*

The best substrate for detailed quantitative analyses of APC activity is sea urchin cyclin $B^{13-110}$. It is highly expressed and soluble in bacteria, easily purified to homogeneity, and can be labeled efficiently with $^{125}I$, allowing the detection of trace amounts of ubiquitinated products. Most importantly, the ubiquitinated products are typically the mono-, di-, and triubiquitinated species, which can be resolved easily from one another, facilitating quantification of both the amount of substrate used and the processivity of the reaction. This substrate behaves qualitatively the same as endogenous yeast substrates under all conditions that we have examined, indicating that it is a valid substrate for the analysis of yeast APC activity.

*Expression of Sea Urchin Cyclin $B^{13-110}$*

a. pET 11 plasmid (Novagen, Madison, WI) containing sea urchin cyclin $B^{13-110}$ under control of the inducible lactose promoter is transformed into BL21 bacteria, and transformed cells are selected on LB agar plates containing 75 $\mu$g/ml ampicillin.

b. A single colony is picked and inoculated into 2 ml of LB + ampicillin (75 $\mu$g/ml) and grown overnight at 37°.

c. The following day, the saturated 2-ml starter culture is diluted into 1 liter of LB-amp media and allowed to grow at 37° until reaching an $O.D._{600}$ of ~0.5. The culture is then shifted to 23° for 1 h.

d. Isopropyl-$\beta$-D-thiogalactoside is added to a final concentration of 0.3 m*M* to induce protein expression. After 3 h, cells are harvested by centrifugation (10,000 rpm for 10 min) and washed once in 250 ml ice-cold water. Cells are centrifuged again and the water is decanted, leaving only the pellet in the bottom of the centrifuge flask. The bacteria are then scooped in small pellets directly into liquid $N_2$ with a spatula. Pellets are stored at –80° and should not be thawed and refrozen.

*Purification of Sea Urchin Cyclin $B^{13-110}$*

a. Bacterial pellets are removed from –80° and ground to a fine powder (5–10 min) with a mortar and pestle (Coors, Golden, CO) that have been precooled with liquid $N_2$. Liquid $N_2$ should be added frequently during grinding to ensure that the bacteria do not thaw during this process.

b. The powder is added to a beaker that is precooled with liquid $N_2$ and allowed to sit at room temperature until the powder just begins to thaw (3–5 min). Three volumes of room temperature buffer (20 m*M* HEPES, pH 7.4, 100 m*M* NaCl, 0.1% Triton X-100, 1 m*M* PMSF) are then added. To avoid ice crystal formation, care should be taken not to add the buffer too early. The cell suspension is rapidly placed on ice and stirred vigorously at 4° for 5 minutes.

c. Cells are sonicated on ice four times for 30 s at a medium setting, with 2 min of resting on ice between each sonication pulse.

d. The lysate is spun for 1 h at 4° at 100,000*g* to pellet unlysed cells and insoluble material.

e. The supernatant is transferred to a 50-ml conical tube and placed in a 95°-water bath for 8 min. During this period, a white precipitate will form due to protein aggregation; sea urchin cyclin $B^{13-110}$ is the major soluble protein under these conditions.

f. The boiled lysate is centrifuged again at 50,000*g*, and the supernatant containing sea urchin cyclin $B^{13-110}$ is removed to a fresh tube.

g. While generally pure at this point, we recommend further purification and concentration of the substrate by diluting the supernatant twofold with buffer lacking salt (20 m*M* HEPES, pH 7.4, 1 m*M* PMSF) and applying it to a 1-ml Hi-Trap S column (Amersham, Uppsala, Sweden), followed by elution with a 0–1 *M* NaCl gradient. The protein should elute at ~300 m*M* NaCl. Greater than 10 mg protein at a concentration of over 5 mg/ml can be obtained by this method.

*Labeling Sea Urchin Cyclin $B^{13-110}$ with $Na^{125}I$*

a. Twenty microliters of sea urchin cyclin $B^{13-110}$ (3 mg/ml) is mixed with 70 μl 0.5 *M* $NaPO_4$, pH 7.6, and then mixed with 1 mCi $Na^{125}I$ (10 μl) (Amersham, Uppsala, Sweden). $Na^{125}I$ is extremely volatile, and one should wear protective eyeware, a laboratory coat, and a lead apron when handling. All steps should be performed in a fume hood certified for protein iodination.

b. The coupling reaction is initiated by adding 50 μl of 3.6 mg/ml chloramine T (Fluka Chemical Corp., Ronkonkona, NY) and incubating for 25 s. The reaction is stopped by adding 50 μl of 10 mg/ml sodium pyrosulfite (Fluka Chemical Corp., Ronkonkona, NY). The time of incubation is very important; incubations longer than 25 s can result in breakdown of the substrate.

c. Once the coupling reaction has been stopped, labeled protein is separated from free $Na^{125}I$ by gel filtration on a 10-ml PD–10 column (Amersham, Uppsala, Sweden) that has been preequilibrated with QA buffer (20 m*M* Tris, pH 7.6, 100 m*M* NaCl, 1 m*M* $MgCl_2$). The entire 200 μl is applied to the column and allowed to absorb. Eight 500-μl fractions are collected, and the amount of radioactivity in each is counted to determine the peak fractions. Typically, ~75% of the protein will appear in fractions 6 and 7. Peak fractions are pooled, resulting in a final concentration of ~10 μ*M* (with a specific activity of over $10^5$ cpm/μg).

d. Pooled fractions are aliquoted and frozen at –80° for later use. Cyclin $B^{13-110}$ is stable at –80° for long periods and is tolerant of multiple freeze–thaw cycles.

*A Typical APC Assay*

APC assays are carried out in a total volume of 15 μl. The stock concentrations of purified components can vary in different preparations; thus, the volume of each that is added to the reaction may be adjusted to reach the appropriate final concentration. QA buffer (20 m*M* Tris, pH 7.6, 100 m*M* NaCl, 1 m*M* $MgCl_2$) is used to bring the final volume up to 15 μl.

a. The components of the ubiquitin-activating system are mixed and incubated together for 10 min at room temperature.

| Component | Volume |
|---|---|
| E1 | 0.5 $\mu$l (of a 1-mg/ml stock; 300 n$M$ final concentration) |
| E2 | 1.0 $\mu$l (of a 4-mg/ml stock; 20 $\mu M$ final concentration) |
| ATP | 1.5 $\mu$l (of a 10 m$M$ stock; 1 m$M$ final concentration) |
| Ubiquitin | 1.5 $\mu$l (of a 10-mg/ml stock; 150 $\mu M$ final concentration) |
| QA buffer | 3.0 $\mu$l |
| | 7.5 $\mu$l total |

b. During the 10-min preincubation of the ubiquitin-activating system, APC, Cdh1, and the substrate are mixed and allowed to bind at room temperature.

| Component | Volume |
|---|---|
| APC | 0.5 $\mu$l (of a 30 n$M$ stock; 1 n$M$ final concentration) |
| Cdh1 | 1.0 $\mu$l (of a 1.5 $\mu M$ stock; 100 n$M$ final concentration) |
| $^{125}$I cyclin | 0.5 $\mu$l (of a 10 $\mu M$ stock; 300 n$M$ final concentration) |
| QA buffer | 5.5 $\mu$l |
| | 7.5 $\mu$l total |

c. Reactions are initiated by adding 7.5 $\mu$l of the ubiquitin-activating system to 7.5 $\mu$l of APC mix, followed by incubation for 10–30 min at room temperature. The separate preincubation of the ubiquitin-activating system and APC, Cdh1, and substrate ensure that no lag in the initial kinetics of substrate ubiquitination is observed.

d. Reactions are stopped by the addition of 5 $\mu$l of 4$\times$ SDS sample buffer (260 m$M$ Tris, pH 6.8, 10% SDS, 40% glycerol, 10% $\beta$-mercaptoethanol [BME], and trace bromophenol blue), followed by a brief incubation at 65°. Reaction products are resolved on a 17 $\times$ 11-cm (1.5 mm thick) 15% polyacrylamide gel. The gel should be stopped when the dye front is ~1 in. from the bottom to ensure that the unmodified substrate does not run off the bottom.

e. The gel is stained with Coomassie blue, dried for 1.5 h at 80°, and exposed for 2 h (or overnight) to a Phosphorimager screen (Molecular Dynamics, Sunnyvale, CA). Results of the experiment are visualized by scanning the Phosphorimager screen on a Storm 840 Phosphorimager (Molecular Dynamics, Sunnyvale, CA). The major products of the reaction are the mono-, di-, and triubiquitinated species. A method for the analysis of results is presented.

## Analysis of Ubiquitination Reactions

Once a gel containing APC reactions has been scanned into the Phosphorimager, results of the experiment can be viewed and quantified using Imagequant software (Molecular Dynamics, Sunnyvale, CA). Both the rate of the reaction (expressed as substrate converted to product per unit time) and the processivity of ubiquitination (expressed as the ratio of ubiquitin to substrate in the reaction products) can be analyzed. The accuracy of both measurements requires that each product is modified by the APC only once, and therefore care should be taken to ensure that no more than 5% of starting substrate has been converted to product.

The cyclin $B^{13-110}$ substrate will exist in four major forms after a typical ubiquitination reaction (Fig. 1). The unubiquitinated substrate (equivalent to the starting material) should be the predominant form, with the less abundant mono-, di-, and triubiquitinated species each migrating more slowly on the gel. In some cases, a small amount of ubiquitinated substrate above the tri-modified species is observed, but this typically represents a minor fraction of the total ubiquitinated product.

The rate and processivity of the reaction are quantified as follows. A small box is drawn around the monoubiquitinated product using tools supplied in the Imagequant program. The box is copied to ensure that the

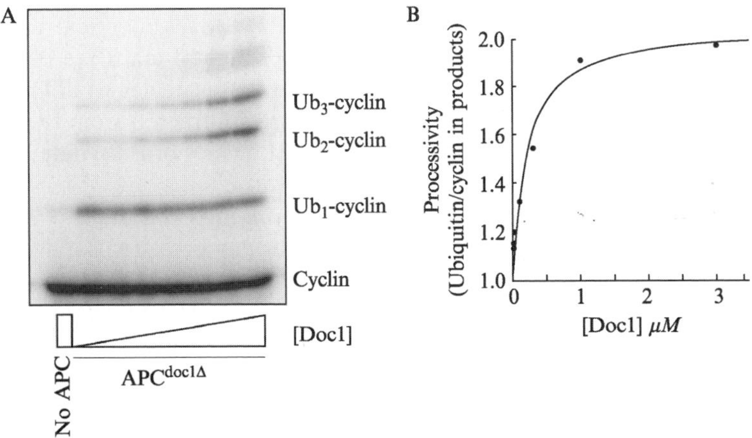


FIG. 1. Quantifying processive substrate ubiquitination as a function of Doc1 concentration. (A) Increasing concentrations of recombinant Doc1 were added to $APC^{doc1\Delta}$ and the effect on processivity was determined. The first lane is a negative control that contains no APC and 3 $\mu M$ Doc1. Other lanes contain ~1 n$M$ $APC^{doc1\Delta}$ and 0, 0.01, 0.03, 0.1, 0.3, 1, and 3 $\mu M$ Doc1, respectively. (B) Data from A are quantified as described in the text and plotted to determine the concentration of Doc1 required for half-maximal stimulation of APC activity.

same size area is quantified for each individual product and is pasted around the diubiquitinated product. This process is then repeated for the triubiquitinated species. The intensity of the different modified substrates is determined independently and is transferred to a work sheet (e.g., Microsoft Excel) for manipulation. The sum of the intensities of the mono-, di-, and triubiquitinated species represents the total amount of substrate converted to product and can be used to calculate the rate of the reaction in arbitrary units per unit time. If one is interested in converting intensity measurements into absolute abundance of cyclin $B^{13-110}$ in each reaction product, known amounts of cyclin $B^{13-110}$ can be run on the same gel for comparison.

To determine the processivity of the reaction, the intensity of each modified species is multiplied by the number of ubiquitin molecules contained in that species; that is, the intensity of the diubiquitinated product is multiplied by two and that of the triubiquitinated species by three. The sum of these adjusted intensities is equal to the total amount of ubiquitin transferred to substrate during the reaction, and processivity is calculated as the ratio of ubiquitin to cyclin in the three major reaction products. The processivity of the reaction does not have a time component and is therefore constant over time under defined conditions, as long as the substrate is not depleted. An ubiquitin to cyclin ratio of ~2 is typically observed using standard reaction conditions.

To control for the background associated with each species, all experiments should include a reaction lacking APC. Boxes identical to those surrounding the ubiquitinated species in experimental lanes are placed in positions corresponding to the position where these species would migrate in the APC-free control lane. These values are subtracted from the intensity of the corresponding species in experimental lanes prior to determining the rate or processivity of the reaction.

For all experiments, we recommend quantifying both the rate and the processivity of the reaction, as both quantities can provide useful insights into APC function. For example, we used the rate of the reaction to determine the concentration of Cdh1 required for half-maximal stimulation of the APC and for determining the $K_M$ of cyclin $B^{13-110}$ for APC (Carroll and Morgan, 2002). In both experiments, the processivity of the reaction was constant over all concentrations of cyclin $B^{13-110}$ and Cdh1. These results are expected if processivity is determined simply by a balance of the intrinsic catalytic activity and substrate dissociation rate, neither of which is affected by the concentration of substrate or Cdh1.

In contrast, dose–response experiments with the E2 enzyme Ubc4 revealed that both the rate and the processivity of the reaction increase with higher E2 concentration and that both can be used to accurately determine the $K_M$ of E2 for APC. These results are consistent with the idea that multiple E2-ub conjugates are required for the processive ubiquitination of a single substrate. At higher E2 concentrations, a higher rate of E2 association increases the number of E2-ub conjugates that bind the APC during a single substrate binding event, thereby increasing processivity as well as the overall reaction rate

Alterations in substrate binding can also be revealed by measurements of processivity (Carroll and Morgan, 2002). This concept is illustrated by studies of the Doc1 subunit of the APC, which enhances substrate binding by reducing the rate of substrate dissociation. APC lacking Doc1 ($APC^{doc1\Delta}$) has an increased $K_M$ for cyclin $B^{13-110}$ when compared to wild-type APC. At low substrate concentrations (at or below $K_M$), $APC^{doc1\Delta}$ displays defects in both overall rate and processivity of substrate ubiquitination (see Fig. 1). At high (saturating) substrate concentrations, however, wild-type APC and $APC^{doc1\Delta}$ convert substrate to product at the same overall rate, but $APC^{doc1\Delta}$ still displays reduced processivity because each substrate-binding event is terminated more rapidly than it is in the wild-type enzyme. Under such conditions, quantifying processivity is the only way to detect and measure the contribution of the Doc1 subunit to substrate binding and ubiquitination.

## Summary

A simple approach is outlined for quantifying the activity and processivity of budding yeast APC. While some details of reagent preparation may be unique to this system, this method of analysis can be applied to APC from other systems and to other ubiquitin-protein ligases, provided the various ubiquitinated reaction products are easily distinguished from one another.

## Acknowledgments

This work was supported by a grant from the National Institutes of Health (GM053270).

## References

Burton, J. L., and Solomon, M. J. (2001). D box and KEN box motifs in budding yeast Hsl1p are required for APC-mediated degradation and direct binding to Cdc20p and Cdh1p. *Genes Dev.* **15,** 2381–2395.

Carroll, C. W., Enquist-Newman, M., and Morgan, D. O. (2005). The APC subunit Doc1 promotes recognition of the substrate destruction box. *Curr. Biol.* **15,** 11–18.

Carroll, C. W., and Morgan, D. O. (2002). The Doc1 subunit is a processivity factor for the anaphase-promoting complex. *Nature Cell Biol.* **4,** 880–887.

Fang, G., Yu, H., and Kirschner, M. W. (1998). Direct binding of CDC20 protein family members activates the anaphase-promoting complex in mitosis and G1. *Mol. Cell* **2,** 163–171.

Gmachl, M., Gieffers, C., Podtelejnikov, A. V., Mann, M., and Peters, J. M. (2000). The RING-H2 finger protein APC11 and the E2 enzyme UBC4 are sufficient to ubiquitinate substrates of the anaphase-promoting complex. *Proc. Natl. Acad. Sci. USA* **97,** 8973–8978.

Harper, J. W., Burton, J. L., and Solomon, M. J. (2002). The anaphase-promoting complex: It's not just for mitosis any more. *Genes Dev.* **16,** 2179–2206.

Hilioti, Z., Chung, Y. S., Mochizuki, Y., Hardy, C. F., and Cohen-Fix, O. (2001). The anaphase inhibitor Pds1 binds to the APC/C-associated protein Cdc20 in a destruction box-dependent manner. *Curr. Biol.* **11,** 1347–1352.

Jaspersen, S. L., Charles, J. F., and Morgan, D. O. (1999). Inhibitory phosphorylation of the APC regulator Hct1 is controlled by the kinase Cdc28 and the phosphatase Cdc14. *Curr. Biol.* **9,** 227–236.

Leverson, J. D., Joazeiro, C. A., Page, A. M., Huang, H., Hieter, P., and Hunter, T. (2000). The APC11 RING-H2 finger mediates E2-dependent ubiquitination. *Mol. Biol. Cell* **11,** 2315–2325.

Passmore, L. A., McCormack, E. A., Au, S. W., Paul, A., Willison, K. R., Harper, J. W., and Barford, D. (2003). Doc1 mediates the activity of the anaphase-promoting complex by contributing to substrate recognition. *EMBO J.* **22,** 786–796.

Peters, J. M. (2002). The anaphase-promoting complex: Proteolysis in mitosis and beyond. *Mol. Cell* **9,** 931–943.

Pfleger, C. M., Lee, E., and Kirschner, M. W. (2001). Substrate recognition by the Cdc20 and Cdh1 components of the anaphase-promoting complex. *Genes Dev.* **15,** 2396–2407.

Rigaut, G., Shevchenko, A., Rutz, B., Wilm, M., Mann, M., and Seraphin, B. (1999). A generic protein purification method for protein complex characterization and proteome exploration. *Nature Biotechnol.* **17,** 1030–1032.

Schwab, M., Neutzner, M., Mocker, D., and Seufert, W. (2001). Yeast Hct1 recognizes the mitotic cyclin Clb2 and other substrates of the ubiquitin ligase APC. *EMBO J.* **20,** 5165–5175.

Tang, Z., Li, B., Bharadwaj, R., Zhu, H., Ozkan, E., Hakala, K., Deisenhofer, J., and Yu, H. (2001). APC2 cullin protein and APC11 RING protein comprise the minimal ubiquitin ligase module of the anaphase-promoting complex. *Mol. Biol. Cell* **12,** 3839–3851.

Yamano, H., Gannon, J., Mahbubani, H., and Hunt, T. (2004). Cell cycle-regulated recognition of the destruction box of cyclin B by the APC/C in Xenopus egg extracts. *Mol. Cell* **13,** 137–147.

Zachariae, W., Schwab, M., Nasmyth, K., and Seufert, W. (1998). Control of cyclin ubiquitination by CDK-regulated binding of Hct1 to the anaphase promoting complex. *Science* **282,** 1721–1724.

# [19] Identification of Cell Cycle-Dependent Phosphorylation Sites on the Anaphase-Promoting Complex/Cyclosome by Mass Spectrometry

*By* FRANZ HERZOG, KARL MECHTLER, and JAN-MICHAEL PETERS

## Abstract

Phosphorylation has an almost universal role in controlling the properties of proteins that govern progression through mitosis and meiosis. The ubiquitin ligase anaphase-promoting complex/cyclosome (APC/C) and its cofactors are no exception to this rule. However, it is poorly understood how APC/C pathway components are regulated by phosphorylation, i.e., little is known about which amino acid residues on subunits and regulators of the APC/C are phosphorylated by which kinase, when during the cell cycle, where in the cell, and with which functional consequence. As a first step toward answering these questions we have established a procedure for the sensitive and relatively rapid identification of phosphorylation sites on small microgram amounts of the APC/C and on associated regulatory proteins. This procedure will enable studies on the dynamic changes of APC/C phosphorylation during the cell cycle and, in conjunction with chemical biology approaches, will allow one to determine which phosphorylation sites depend on the presence of which kinase activity in living cells.

## Introduction

Ubiquitin-dependent proteolysis mediated by the anaphase-promoting complex/cyclosome occurs only during specific periods of mitosis and during the G1 phase of the cell cycle, and during these periods of APC/C activity different APC/C substrates are ubiquitinated and degraded at different times (reviewed in Peters, 2002). For example, cyclin A degradation is initiated in prometaphase once APC/C has been activated by Cdc20 (den Elzen and Pines, 2001; Geley *et al.*, 2001), cyclin B destruction begins in metaphase once substrate-specific inhibition of $APC/C^{Cdc20}$ by the spindle checkpoint has been relieved (Clute and Pines, 1999), and Plk1 is degraded during anaphase once Cdc20 has been replaced by the related APC/C activator protein Cdh1 (Lindon and Pines, 2004).

This complex temporal control of APC/C activity is thought to depend in part on specific phosphorylation events. Several subunits of the APC/C

METHODS IN ENZYMOLOGY, VOL. 398

0076-6879/05 $35.00
DOI: 10.1016/S0076-6879(05)98019-1

are phosphorylated on multiple sites at the beginning of mitosis (Peters *et al.*, 1996), presumably by cyclin-dependent kinase 1 (Cdk1) and, to a lesser extent, by Polo-like kinase 1 (Plk1), but possibly also by other protein kinases (Golan *et al.*, 2002; Kraft *et al.*, 2003; Patra and Dunphy, 1998). Initially it was believed that phosphorylation of APC/C subunits by Plk1 can directly stimulate the ubiquitin ligase activity of the APC/C, and it was further reported that phosphorylation by protein kinase A (PKA) can prevent this activation (Kotani *et al.*, 1998, 1999). In subsequent studies it has been found, however, that phosphorylation of APC/C subunits facilitates the binding of Cdc20 to the APC/C and thereby activates the APC/C through a poorly understood mechanism (Kraft *et al.*, 2003; Kramer *et al.*, 2000; Rudner and Murray, 2000; Shteinberg *et al.*, 1999), perhaps by Cdc20-mediated recruitment of substrates to the APC/C (reviewed in Vodermaier, 2001).

The properties of the APC/C activators Cdc20 and Cdh1 are also controlled by phosphorylation. Phosphorylation of Cdh1 by Cdk1 (Jaspersen *et al.*, 1999; Kramer *et al.*, 2000; Zachariae *et al.*, 1998) or by the related cyclin-dependent kinase 2 (Cdk2; Lukas *et al.*, 1999) inhibits the ability of Cdh1 to bind to the APC/C, and this mechanism is required to keep the APC/C inactive during the S and G2 phases of the cell cycle. For Cdc20, it was initially reported that phosphorylation by Cdk1 increases the ability of Cdc20 to activate the APC/C (Kotani *et al.*, 1999), but since then several other studies have come to the opposite conclusion that phosphorylation inhibits Cdc20 (Chung and Chen, 2003; D'Angiolella *et al.*, 2003; Searle *et al.*, 2004; Tang and Yu, 2004; Yudkovsky *et al.*, 2000). In vertebrates, inhibitory Cdc20 phosphorylation is thought to be required for efficient APC/C inactivation by the spindle checkpoint, although different studies have implicated different kinases in this process, ranging from Cdk1 (D'Angiolella *et al.*, 2003) to mitogen-activated protein kinase (MAP kinase; Chung and Chen, 2003) to the spindle checkpoint kinase Bub1 (Tang and Yu, 2004). In budding yeast, Cdc20 is phosphorylated in response to a different checkpoint, which is activated by DNA damage and depends on the Mec1 kinase, an ortholog of human Atr (Searle *et al.*, 2004). Cdc20 phosphorylation in this pathway is thought to be mediated by PKA and decreases the ability of Cdc20 to bind the mitotic cyclin Clb2 (Searle *et al.*, 2004), consistent with the possibility that Cdc20 phosphorylation reduces substrate recognition by the APC/C$^{Cdc20}$.

Also, many substrates of the APC/C are phosphorylated during mitosis when their degradation is initiated, but it appears that most APC/C substrates can be recognized and ubiquitinated without these modifications, in contrast to SCF substrates that typically have to be phosphorylated to allow their recognition by specific substrate adaptor proteins (reviewed in

King *et al.*, 1996). However, in the case of the APC/C substrate Aurora A there is evidence to suggest that substrate phosphorylation may inhibit proteolysis (Littlepage and Ruderman, 2002). Substrate phosphorylation could therefore be a mechanism that delays the recognition of some APC/C substrates as long as their presence is needed, in this case until the end of mitosis when most mitosis-specific phosphorylation sites are removed by phosphatases and when Aurora A is known to be degraded.

Finally, the properties of APC/C inhibitors are also controlled by phosphorylation. The spindle checkpoint protein Mad2 inhibits the ability of APC/C$^{Cdc20}$ to ubiquitinate cyclin B and specific other substrates until metaphase. Mad2 phosphorylation appears to antagonize this inhibitory activity by preventing the association of Mad2 with its activator protein Mad1 and with the APC/C (Wassmann *et al.*, 2003). The APC/C inhibitor Emi1 is also inactivated by phosphorylation, in this case because its phosphorylation by Plk1 results in ubiquitination by the ubiquitin ligase SCF$^{\beta Trcp}$ and subsequent degradation of Emi1 by the 26S proteasome (Hansen *et al.*, 2004; Moshe *et al.*, 2004; ).

In summary, there is plenty of evidence that phosphorylation has a key role in determining when during the cell cycle specific substrates of the APC/C are degraded. However, many open questions remain, in part because numerous details are unknown and in part because several of the reports in the literature have come to conflicting conclusions. This situation is presumably a reflection of the complexity of APC/C regulation by phosphorylation. The variance of results in the literature may also be due to the use of *in vitro* systems in which the effects of purified protein kinases on APC/C pathway components were analyzed. At least for some kinases such as Plk1 it is known that under such conditions sites can be phosphorylated that are not normally phosphorylated *in vivo* (Kraft *et al.*, 2003; Shou *et al.*, 2002). Results from *in vitro* systems employing purified kinases need therefore to be interpreted with caution and should ideally include a comparison of the phosphorylation sites that were generated *in vitro* with the sites that exist *in vivo*.

To obtain a more complete and detailed understanding of how phosphorylation controls the APC/C pathway, several questions will have to be answered: Which amino acid residues on subunits and regulators of the APC/C are phosphorylated by which kinase, when during the cell cycle, where in the cell, and with which functional consequence? As a first step toward this goal we have established a sensitive method that allows the relatively rapid identification of phosphorylation sites in small microgram amounts of APC/C and associated proteins.

In brief, APC is immunoprecipitated from cultured human HeLa cells that are arrested at different stages of the cell cycle, the immunopurified proteins are eluted from the antibody beads by buffers of low pH, the proteins are reduced and alkylated to avoid the formation of disulfide bridges, and subsequently the proteins are digested into peptides by a combination of different proteases, without prior separation of APC/C subunits by gel electrophoresis. The peptides are separated on a nano-HPLC, transferred into gas phase by electron spray ionization (ESI), and analyzed by tandem mass spectrometry in a Thermo Finnigan LCQ Classic ion trap mass spectrometer. The sequence of each peptide is determined through automated computational analysis using the Sequest program (Eng *et al.*, 1994), and the mass spectra of all potentially phosphorylated peptides are subsequently analyzed manually.

The same technique can be used to analyze which sites on APC/C can be phosphorylated by purified protein kinases *in vitro*, allowing a comparison between phosphorylation sites that can principally be generated by a kinase and those sites that really exist *in vivo* (Kraft *et al.*, 2003). Such information is useful for identifying or excluding potential candidate kinases that could phosphorylate specific sites, and it yields information about the specificity with which a kinase phosphorylates proteins *in vitro*. In the future it will also be possible to use the methods described here to analyze the phosphorylation state of APC/C isolated from cells that had been treated with chemical inhibitors to specific kinases to determine which phosphorylation site depends, directly or indirectly, on the activity of which kinase.

## Purification of Human APC/C for the Identification of Phosphorylation Sites by Mass Spectrometry

### *Preparation of HeLa Cell Lysates*

Adherent HeLa cells are cultured as described in Herzog and Peters (2005). To obtain cell populations that are synchronized with respect to their cell cycle stage, a variety of methods can be used (Johnson *et al.*, 1993; O'Connor and Jackman, 1996). To arrest cells in S phase, we usually grow logarithmically proliferating cells to 60–80% confluency and subsequently culture them for 17 to 18 h in the presence of 2 m*M* hydroxyurea (HU; Sigma-Aldrich). HU inhibits the enzyme ribonucleotide reductase (Elford, 1968), thereby preventing the synthesis of deoxyribonucleotides and activating a DNA replication checkpoint mechanism that prevents cell cycle progression (Elledge, 1996; Koc *et al.*, 2004). To arrest cells in mitosis, we treat logarithmically proliferating cells at 60–80% confluency for 17 to 18 h with 330 n*M* nocodazole (Sigma-Aldrich, St. Louis, MO). Nocodazole depolymerizes microtubules and therefore prevents the attachment of microtubules to kinetochores on mitotic chromosomes. The resulting unattached kinetochores activate the spindle checkpoint, which arrests cells in a prometaphase-like state (Musacchio and Hardwick, 2002). For one mass spectrometry experiment, HeLa cells from ten 145-mm tissue culture dishes are needed. Lysates of the arrested HeLa cells are prepared as described in Herzog and Peters (2005).

### *Purification of APC/C*

To detect phosphorylation sites on all APC/C subunits, we immunoprecipitate APC/C from HeLa cell lysates and digest the eluted protein in solution without prior separation of individual polypeptides by gel electrophoresis. The complete peptide mixture of all APC/C subunits is subsequently analyzed in one mass spectrometry run. This experimental setup requires only 1.2 to 2 $\mu$g of purified APC/C and also allows the analysis of polypeptides that are hardly detectable after electrophoretic separation and silver staining. This amount of APC/C is isolated from HeLa cell extracts containing 12 to 20 mg protein, which requires 60 to 100 $\mu$l Affi-prep protein A beads coupled to Cdc27 antibodies.

The generation of reagents and a detailed purification protocol are described in Herzog and Peters (2005). As described in this chapter, APC/C is isolated in the presence of phosphatase inhibitors to make sure that phosphate groups are not removed from the APC/C subunits during the isolation. The same phosphatase inhibitors are used when APC/C is purified from cells arrested in S phase or in mitosis. To rule out the possibility that protein kinases can phosphorylate amino acid residues during the isolation procedure that were not yet phosphorylated *in vivo*, EDTA can be added to the purification buffers to chelate $Mg^{2+}$, without which kinases cannot utilize ATP. However, we could not detect differences in phosphorylation by mass spectrometry when we compared mitotic APC/C samples that had been isolated in either the presence or the absence of EDTA (data not shown). We therefore routinely purify APC/C in the absence of EDTA.

We control the purity of our APC/C samples by analyzing aliquots by SDS–PAGE and silver staining (Fig. 1). Because mitotic phosphorylation of the APC/C subunits Apc1, Cdc27, Cdc17, and Cdc23 causes electrophoretic mobility shifts of these proteins (Peters *et al.*, 1996), the silver-stained gels can, in the case of mitotic samples, also be used to roughly estimate to which degree the phosphorylation state of these subunits was preserved during the isolation procedure.

Special care has to be taken during the following purification steps. First, before APC/C is eluted from the antibody beads, all detergents have

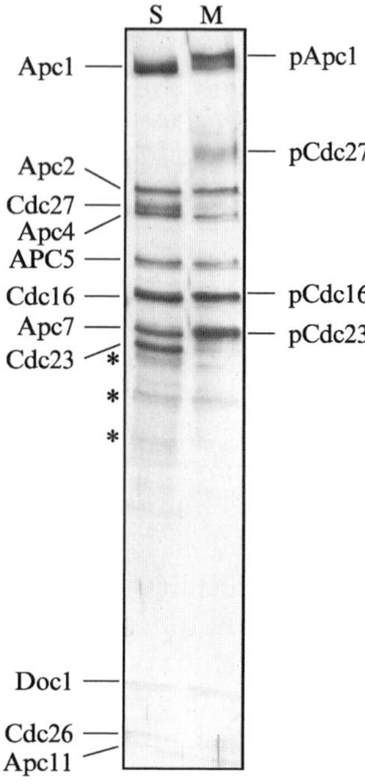


FIG. 1. APC was immunoprecipitated from extracts of HeLa cells arrested with nocodazole (M) or hydroxyurea (S) and subjected to SDS–PAGE and silver staining. The positions of subunits phosphorylated in mitosis are indicated (pApc1, pCdc27, pCdc16, and pCdc23). Stars indicate degradation products of Cdc27 and Apc7.

to be removed quantitatively by washing the beads five times with 10 to 15 bead volumes of TBS, as signals from detergents that are present in the eluate can strongly decrease the detectability of peptides in the mass spectrometer. Second, APC/C has to be eluted from the antibody beads by resuspending the beads in 1.5 volumes of a solution containing 100 m*M* glycine-HCl, pH 2.2, or 60 m*M* HCl. Samples that have been obtained by eluting APC/C with the antigenic Cdc27 peptide cannot be used because the large excess of Cdc27 peptide would obscure the signals of peptides derived from APC/C. Third, the pH of the eluate has to be adjusted to 7.5 to 8.0 by adding 1.5 *M* Tris-HCl, pH 9.2, or 2 *M* $(NH_4)_2CO_3$ to a final concentration of 150 to 200 m*M*. This pH range allows the subsequent digestion of APC/C subunits by various proteases (see later). The samples can be frozen in liquid nitrogen and stored at –80°.

## Protein Reduction, Alkylation, and Proteolytic Digestion in Solution

APC/C samples are processed for mass spectrometric analysis according to the following protocol.

### *Reduction and Alkylation*

The protein sample is reduced and alkylated to dissolve potentially existing disulfide bridges and to avoid the formation of new ones that could affect the subsequent proteolytic digest and mass spectrometric analysis. First, 2 $\mu$g of APC/C (~100 $\mu$l APC/C eluate) is reduced by the addition of 1 $\mu$g dithiothreitol (Roche, Monza, Italy) and incubation at 37° for 1 h. Subsequently, APC/C is alkylated by the addition of 5 $\mu$g iodacetamide (Sigma-Aldrich) and incubation for 30 min at room temperature in the dark.

### *Proteolytic Digest*

To fragment APC/C subunits into peptides whose sequence and phosphorylation state can be analyzed by mass spectrometry, the reduced and alkylated APC/C subunits are digested with proteases. Digestion with one protease can, depending on the cleavage specificity of the enzyme and on the amino acid sequence of the substrate, result in the formation of some very long or very short peptides that are difficult to detect in the mass spectrometer. To be able to analyze all parts of a protein we therefore use four different proteases, either singly (trypsin, chymotrypsin, and subtilisin) or in combination (trypsin and Glu-C). The analysis of all four peptide mixtures generated by these digestion reactions can yield sequence coverages of more than 90% for the large subunits of the APC/C (Table II).

Trypsin, Glu-C, and chymotrypsin (sequencing grade; Roche, Monza, Italy) are dissolved in 1 m*M* HCl to a final concentration of 200 ng/$\mu$l. These enzyme stocks can be stored at –80° for up to 3 months. Subtilisin (Sigma-Aldrich) has to be freshly prepared for every experiment. It is first dissolved in 1 m*M* HCl at a concentration of 10 $\mu$g/$\mu$l and further diluted in a solution containing 5.6 *M* urea (Sigma-Aldrich) and 100 m*M* Tris-HCl, pH 8.5, to a final concentration of 300 ng/ $\mu$l.

One hundred-microliter samples containing 2 $\mu$g reduced and alkylated APC/C are digested proteolytically under different conditions as listed in Table I. The reactions are stopped by the addition of trifluoroacetic acid (Pierce, Rockford, IL) to a final concentration of 0.1% (v/v) and the resulting peptide mixtures are either analyzed immediately by mass spectrometry or stored at –80°.

TABLE I
PROTEOLYTIC DIGEST OF APC/C

| Protease | Quantity of protease (ng) | Incubation | |
|---|---|---|---|
| | | Temperature | Time (h) |
| 1. Trypsin | 200 | 37° | 4 |
| 2. Trypsin | 200 | 37° | 4 |
| 1. Trypsin | 400 | 37° | 12 |
| 2. Glu-C | 400 | 25° | 8 |
| Chymotrypsin | 400 | 25° | 4 |
| Subtilisin | 300 | 37° | 0.5 |

## Nano-HPLC and Mass Spectrometry

Following reduction, alkylation, and proteolytic digest, the samples are applied to an UltiMate nano-HPLC system (LC Packings, Amsterdam, The Netherlands), which is used with the following mobile phases: solvent A, 95% (v/v) water (HPLC grade, Supra-Gradient, Biosolve B.V., Valkenswaard, The Netherlands), 5% (v/v) acetonitrile (HPLC grade, Supra-Gradient, Biosolve B.V), 0.1% (v/v) trifluoroacetic acid; solvent B, 30% (v/v) water, 70% (v/v) acetonitrile, 0.1% (v/v) formic acid (Fluka, Buchs, Switzerland). The samples are applied to a precolumn to concentrate the proteolytic peptides and to remove buffer components and other perturbing substances, and subsequently the peptides are separated according to their hydrophobicity on an analytical column. Peptide solution (100 $\mu$l) is loaded via a 250-$\mu$l sample loop onto a PepMAP C18 precolumn (0.3 $\times$ 5 mm; DIONEX, Sunnyvale, CA) using a loading pump with solvent A at a flow rate of 20 $\mu$l/min. After loading the precolumn is switched in-line with the nano separation column (PepMAP C18, 75 $\mu$m $\times$ 150 mm; DIONEX) and the sample is eluted in back flush mode using the $\mu$HPLC pump with a linear gradient ramping from 0 to 40% solvent B in 4 h at a flow rate of 200 nl/min.

Eluting peptides are introduced via a heated capillary (Pico Tip, FS360–20–10; New Objective, Cambridge, USA) of a nanospray ion source interface (Protana, Toronto, Canada) into an LCQ Classic ion trap mass spectrometer (Thermo Finnigan, Waltham, MA). Electrospray ionization (ESI) is performed by applying the following conditions: spray voltage, 1.8 kV; capillary temperature, 185°; and voltage of the electron multiplier, –1050 V. The collision energy is adjusted automatically depending on the mass of the parent ion, and gain control is set to $5 \times 10^7$. Data are collected in the controid mode using one MS experiment (Full-MS) followed by MS/MS experiments of the two most intensive ions (minimal intensity required $4 \times 10^5$). The repeated recording of tandem mass spectra of peptides, which elute over a long period of time, is limited by using a dynamic exclusion list with an exclusion duration of 1 min and an exclusion mass width of $\pm 3$ Da.

### Data Analysis

All tandem mass spectra are searched against a human nonredundant protein database obtained from NCBI (Betheseda, MD) using the Sequest algorithm (Eng *et al.*, 1994). The sequence of each APC/C peptide, which has been identified by the Sequest algorithm as a candidate phosphopeptide, is compared manually with the corresponding tandem mass spectrum to ensure that the assignment is correct (Fig. 2). We use the following criteria proposed by Yates *et al.* (2000) to validate the sequences of phosphopeptides. First, the mass spectrum of a peptide has to contain a fragment ion representing the mass of the peptide fragment plus the mass of a phosphate residue (80 Da). Second, the MS/MS spectrum must be of good quality with fragment ions clearly above baseline noise. Third, there must be some continuity to the *b* or *y* ion series, i.e., at least four to six ions that differ by the presence of only one amino acid must be detectable in each series. Fourth, because proline-x bonds (where x can represent any amino acid residue) are weaker and therefore fragment more easily than all other peptide bonds between amino acid residues, the signals from *y* ions that correspond to a proline residue should be intense.

### Possible Applications

To obtain a complete phosphorylation site map of APC/C subunits, all potentially phosphorylatable amino acid residues in the protein sequence have to be covered by mass spectrometric analysis. By using the methods described here we have been able to obtain sequence coverages of more than 90% for seven APC/C subunits isolated from mitotic cells (Table II; Kraft *et al.*, 2003). This has allowed the identification of 43 phosphorylation sites on the APC/C and two on Cdc20. By comparing data from APC/C isolated from cells in S phase with data from mitotic APC/C we have found that at least 34 of these phosphorylation sites are mitosis specific. Of these, 32 sites are clustered in parts of Apc1 and the tetratrico peptide repeat (TPR) subunits Cdc27, Cdc16, Apc7, and Cdc23, consistent with the electrophoretic mobility shift that is seen for several of these subunits in mitosis (Fig. 1). We have raised phospho-specific antibodies against five of the

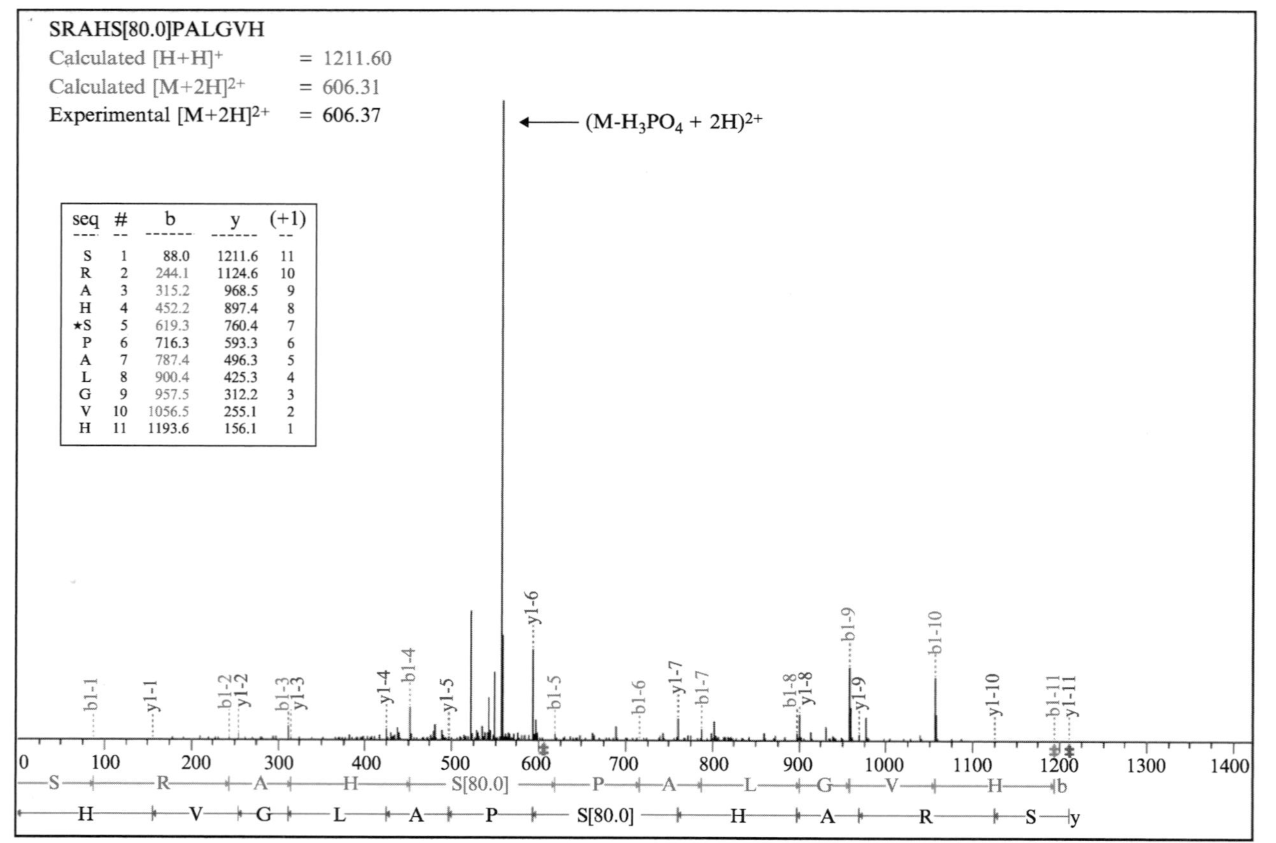
SRAHS[80.0]PALGVH
Calculated [H+H]+ = 1211.60
Calculated [M+2H]2+ = 606.31
Experimental [M+2H]2+ = 606.37
seq # b y (+1)
S 1 88.0 1211.6 11
R 2 244.1 1124.6 10
A 3 315.2 968.5 9
H 4 452.2 897.4 8
*S 5 619.3 760.4 7
P 6 716.3 593.3 6
A 7 787.4 496.3 5
L 8 900.4 425.3 4
G 9 957.5 312.2 3
V 10 1056.5 255.1 2
H 11 1193.6 156.1 1
(M-H3PO4 + 2H)2+
b1-1
y1-1
b1-2
y1-2
b1-3
y1-3
y1-4
b1-4
y1-5
y1-6
b1-5
b1-6
y1-7
b1-7
b1-8
y1-8
b1-9
y1-9
b1-10
y1-10
b1-11
y1-11
0 100 200 300 400 500 600 700 800 900 1000 1100 1200 1300 1400
S R A H S[80.0] P A L G V H b
H V G L A P S[80.0] A H A R S y

TABLE II
SEQUENCE COVERAGE OF APC/C SUBUNITS AND Cdc20[a]

| APC/C | APC/C source | Sequence coverage | No. of samples |
|---|---|---|---|
| Apc1 | M | 73.0 | 6 |
| | S | 54.8 | 2 |
| Apc2 | M | 92.0 | 6 |
| | S | 74.8 | 2 |
| Cdc27 | M | 95.8 | 6 |
| | S | 89.9 | 2 |
| Apc4 | M | 99.0 | 6 |
| | S | 68.7 | 2 |
| Apc5 | M | 97.9 | 6 |
| | S | 87.3 | 2 |
| Cdc16 | M | 92.9 | 6 |
| | S | 87.6 | 2 |
| Apc7 | M | 94.5 | 6 |
| | S | 84.8 | 2 |
| Cdc23 | M | 92.1 | 6 |
| | S | 96.3 | 2 |
| Cdc20 | M | 62.7 | 6 |
| | S | n.d. | 2 |

[a] Sequence coverage represents the percentage of sequences of human APC/C subunits and Cdc20 that is represented by peptides found in the mass spectrometric analyses of different proteolytic digests. M, APC/C immunopurified from nocodazole-arrested HeLa cells; S, APC/C immunopurifed from hydroxyurea-arrested HeLa cells; No. of samples, number of samples obtained by digestion with different proteases; n.d., not detected.

mitosis-specific sites and have confirmed by immunofluorescence microscopy and Western blotting experiments that all five of these sites are generated in HeLa cells specifically in mitosis (Kraft *et al.*, 2003). We are therefore confident that the methodology described here identifies modifications that exist on the proteins *in vivo* and does not lead to false-positive assignment of phosphorylation sites (Table III).

FIG. 2. Tandem mass (MS/MS) spectrum of a phosphopeptide of human Apc1 generated by collision-induced dissociation of the $(M + 2H)^{2+}$ precursor ion with a mass-to-charge ratio ($m/z$) of 606.37. Fragment ions in the spectrum were mainly generated by single-event cleavages of the peptide bonds, resulting in sequence information recorded simultaneously from both N and C termini of the peptide ($b$ and $y$ ions, respectively). This spectrum was searched against the human nonredundant protein database using the Sequest program and was matched to an Apc1 peptide with an additional mass from a phosphate residue. The phosphate mass was unambiguously assigned to serine–355 based on the presence of an ion that was generated by cleavage at the histidine–serine peptide bond. This resulted in a $y$1–7 fragment ion with $m/z = 760.4$, 680.4[peptide] + 80[phosphate].

TABLE III
SUMMARY OF PHOSPHORYLATION SITES FOUND ON HUMAN APC/C$^{Cdc20}$

| APC/C subunit | Length (amino acid residues) | Phosphorylation sites | |
|---|---|---|---|
| | | Mitosis specific | Total |
| Apc1 | 1944 | 9 | 15 |
| Apc2 | 822 | 1 | 2 |
| Cdc27 | 824 | 11 | 11 |
| Apc4 | 808 | 1 | 2 |
| Apc5 | 755 | 0 | 1 |
| Cdc16 | 619 | 6 | 6 |
| Apc7 | 565 | 2 | 2 |
| Cdc23 | 591 | 4 | 4 |
| Cdc20 | 499 | 2 | 2 |
| Sum | | 36 | 45 |

We have also used the same technology to begin to map phosphorylation sites that are generated if APC/C purified from cells in interphase is incubated with recombinant Cdk1 and Plk1 (Kraft *et al.*, 2003). Such comparisons help identify or exclude candidate kinases for specific phosphorylation sites and shed light on the *in vitro* specificity of the kinases tested. Under the purification conditions described here, several APC/C regulators copurify with the APC/C, such as Cdc20, BubR1, and Plk1 (Kraft *et al.*, 2003; data not shown). Information about the cell cycle-dependent phosphorylation state of these molecules can therefore be obtained in the same experiments.

In the future, it should further be possible to use the methods described here to map phosphorylation sites on APC/C that have been purified from mitotic cells treated with chemical inhibitors to specific kinases [see Hauf *et al.* (2003) for an example of such an inhibitor]. Such analyses should allow the identification of *in vivo* dependencies between specific phosphorylation sites and kinase activities, i.e., it should be possible to ask which phosphorylation site can only be generated in the presence of which kinase activity. However, this approach will not be able to distinguish between direct and indirect requirements for a kinase in substrate phosphorylation. In particular, care will have to be taken to rule out cell cycle effects in which a kinase inhibitor would block APC/C phosphorylation indirectly by preventing entry into the cell cycle state in which these modifications are normally generated.

## Acknowledgments

We are grateful to Elisabeth Roitinger for helpful comments on the manuscript. Research in the laboratory of J.-M. P. is supported by Boehringer Ingelheim, the 6th Framework Programme of the European Union via the Integrated Project MitoCheck, the Wiener Wirtschaftsfoerderungsfonds (WWFF), and the European Molecular Biology Organization (EMBO).

## References

Chung, E., and Chen, R. H. (2003). Phosphorylation of Cdc20 is required for its inhibition by the spindle checkpoint. *Nature Cell Biol.* **5,** 748–753.

Clute, P., and Pines, J. (1999). Temporal and spatial control of cyclin B1 destruction in metaphase. *Nature Cell Biol.* **1,** 82–87.

D'Angiolella, V., Mari, C., Nocera, D., Rametti, L., and Grieco, D. (2003). The spindle checkpoint requires cyclin-dependent kinase activity. *Genes Dev.* **17,** 2520–2525.

den Elzen, N., and Pines, J. (2001). Cyclin A is destroyed in prometaphase and can delay chromosome alignment and anaphase. *J. Cell Biol.* **153,** 121–136.

Elford, H. L. (1968). Effect of hydroxyurea on ribonucleotide reductase. *Biochem. Biophys. Res. Commun.* **33,** 129–135.

Elledge, S. J. (1996). Cell cycle checkpoints: Preventing an identity crisis. *Science* **274,** 1664–1672.

Eng, J. K., McCormack, A. L., and Yates, I. J. R. (1994). An approach to correlate tandem mass spectral data of peptides with amino acid sequences in a protein database. *J. Am. Soc. Mass Spectrom.* **5,** 976–989.

Geley, S., Kramer, E., Gieffers, C., Gannon, J., Peters, J. M., and Hunt, T. (2001). Anaphase-promoting complex/cyclosome-dependent proteolysis of human cyclin A starts at the beginning of mitosis and is not subject to the spindle assembly checkpoint. *J. Cell Biol.* **153,** 137–148.

Golan, A., Yudkovsky, Y., and Hershko, A. (2002). The cyclin-ubiquitin ligase activity of cyclosome/APC is jointly activated by protein kinases Cdk1-cyclin B and Plk. *J. Biol. Chem.* **277,** 15552–15557.

Hansen, D. V., Loktev, A. V., Ban, K. H., and Jackson, P. K. (2004). Plk1 regulates activation of the anaphase promoting complex by phosphorylating and triggering SCFbetaTrCP-dependent destruction of the APC inhibitor Emi1. *Mol. Biol. Cell* **15,** 5623–5634.

Hauf, S., Cole, R. W., La Terra, S., Zimmer, C., Schnapp, G., Walter, R., Heckel, A., van Meel, J., Rieder, C. L., and Peters, J. M. (2003). The small molecule Hesperadin reveals a role for Aurora B in correcting kinetochore-microtubule attachment and in maintaining the spindle assembly checkpoint. *J. Cell Biol.* **161,** 281–294.

Herzog, F., and Peters, J. M. (2005). Large scale purification of the vertebrate anaphase-promoting complex/cyclosome. *Methods Enzymol.* **398**(16), 2005 (this volume).

Jaspersen, S. L., Charles, J. F., and Morgan, D. O. (1999). Inhibitory phosphorylation of the APC regulator Hct1 is controlled by the kinase Cdc28 and the phosphatase Cdc14. *Curr. Biol.* **9,** 227–236.

Johnson, R. T., Downes, C. S., and Meyn, R. E. (1993). The synchronisation of mammalian cells. *In* "The Cell Cycle: A Practical Approach" (P. Fantes and R. Brooks, eds.). Oxford Univ. Press, New York.

King, R. W., Deshaies, R. J., Peters, J. M., and Kirschner, M. W. (1996). How proteolysis drives the cell cycle. *Science* **274,** 1652–1659.

Koc, A., Wheeler, L. J., Mathews, C. K., and Merrill, G. F. (2004). Hydroxyurea arrests DNA replication by a mechanism that preserves basal dNTP pools. *J. Biol. Chem.* **279,** 223–230.

Kotani, S., Tugendreich, S., Fujii, M., Jorgensen, P. M., Watanabe, N., Hoog, C., Hieter, P., and Todokoro, K. (1998). PKA and MPF-activated polo-like kinase regulate anaphase-promoting complex activity and mitosis progression. *Mol. Cell* **1,** 371–380.

Kotani, S., Tanaka, H., Yasuda, H., and Todokoro, K. (1999). Regulation of APC activity by phosphorylation and regulatory factors. *J. Cell Biol.* **146,** 791–800.

Kraft, C., Herzog, F., Gieffers, C., Mechtler, K., Hagting, A., Pines, J., and Peters, J. M. (2003). Mitotic regulation of the human anaphase-promoting complex by phosphorylation. *EMBO J.* **22,** 6598–6609.

Kramer, E. R., Scheuringer, N., Podtelejnikov, A. V., Mann, M., and Peters, J. M. (2000). Mitotic regulation of the APC activator proteins CDC20 and CDH1. *Mol. Biol. Cell* **11,** 1555–1569.

Lindon, C., and Pines, J. (2004). Ordered proteolysis in anaphase inactivates Plk1 to contribute to proper mitotic exit in human cells. *J. Cell Biol.* **164,** 233–241.

Littlepage, L. E., and Ruderman, J. V. (2002). Identification of a new APC/C recognition domain, the A box, which is required for the Cdh1-dependent destruction of the kinase Aurora-A during mitotic exit. *Genes Dev.* **16,** 2274–2285.

Lukas, C., Sorensen, C. S., Kramer, E., Santoni-Rugiu, E., Lindeneg, C., Peters, J. M., Bartek, J., and Lukas, J. (1999). Accumulation of cyclin B1 requires E2F and cyclin-A-dependent rearrangement of the anaphase-promoting complex. *Nature* **401,** 815–818.

Moshe, Y., Boulaire, J., Pagano, M., and Hershko, A. (2004). Role of Polo-like kinase in the degradation of early mitotic inhibitor 1, a regulator of the anaphase promoting complex/cyclosome. *Proc. Natl. Acad. Sci. USA* **101,** 7937–7942.

Musacchio, A., and Hardwick, K. G. (2002). The spindle checkpoint: Structural insights into dynamic signalling. *Nature Rev. Mol. Cell. Biol.* **3,** 731–741.

O'Connor, P. M., and Jackman, J. (1996). Synchronisation of mammalian cells. *In* "Cell Cycle: Materials and Methods" (M. Pagano, ed.). Springer-Verlag, Berlin.

Patra, D., and Dunphy, W. G. (1998). Xe-p9, a Xenopus Suc1/Cks protein, is essential for the Cdc2-dependent phosphorylation of the anaphase-promoting complex at mitosis. *Genes Dev.* **12,** 2549–2559.

Peters, J. M., King, R. W., Hoog, C., and Kirschner, M. W. (1996). Identification of BIME as a subunit of the anaphase-promoting complex. *Science* **274,** 1199–1201.

Peters, J. M. (2002). The anaphase-promoting complex: Proteolysis in mitosis and beyond. *Mol. Cell* **9,** 931–943.

Rudner, A. D., and Murray, A. W. (2000). Phosphorylation by Cdc28 activates the Cdc20-dependent activity of the anaphase-promoting complex. *J. Cell Biol.* **149,** 1377–1390.

Searle, J. S., Schollaert, K. L., Wilkins, B. J., and Sanchez, Y. (2004). The DNA damage checkpoint and PKA pathways converge on APC substrates and Cdc20 to regulate mitotic progression. *Nature Cell Biol.* **6,** 138–145.

Shou, W., Azzam, R., Chen, S. L., Huddleston, M. J., Baskerville, C., Charbonneau, H., Annan, R. S., Carr, S. A., and Deshaies, R. J. (2002). Cdc5 influences phosphorylation of Net1 and disassembly of the RENT complex. *BMC Mol. Biol.* **3,** 3.

Shteinberg, M., Protopopov, Y., Listovsky, T., Brandeis, M., and Hershko, A. (1999). Phosphorylation of the cyclosome is required for its stimulation by Fizzy/cdc20. *Biochem. Biophys. Res. Commun.* **260,** 193–198.

Tang, Z., and Yu, H. (2004). Functional analysis of the spindle-checkpoint proteins using an *in vitro* ubiquitination assay. *Methods Mol. Biol.* **281,** 227–242.

Vodermaier, H. C. (2001). Cell cycle: Waiters serving the Destruction machinery. *Curr. Biol.* **11,** R834–R837.

Wassmann, K., Liberal, V., and Benezra, R. (2003). Mad2 phosphorylation regulates its association with Mad1 and the APC/C. *EMBO J.* **22,** 797–806.

Yates, J. R., 3rd, Link, A. J., and Schieltz, D. (2000). Direct analysis of proteins in mixtures. Application to protein complexes. *Methods Mol. Biol.* **146,** 17–46.

Yudkovsky, Y., Shteinberg, M., Listovsky, T., Brandeis, M., and Hershko, A. (2000). Phosphorylation of Cdc20/fizzy negatively regulates the mammalian cyclosome/APC in the mitotic checkpoint. *Biochem. Biophys. Res. Commun.* **271,** 299–304.

Zachariae, W., Schwab, M., Nasmyth, K., and Seufert, W. (1998). Control of cyclin ubiquitination by CDK-regulated binding of Hct1 to the anaphase promoting complex. *Science* **282,** 1721–1724.

# [20] Purification and Assay of Mad2: A Two-State Inhibitor of Anaphase-Promoting Complex/Cyclosome

*By* XUELIAN LUO and HONGTAO YU


## Abstract

To maintain the fidelity of chromosome inheritance, cells utilize a surveillance mechanism called the spindle checkpoint to sense improper attachment of sister chromatids to the mitotic spindle prior to chromosome segregation. The target of the spindle checkpoint is a ubiquitin ligase called the anaphase-promoting complex or cyclosome (APC/C). The spindle checkpoint protein Mad2 inhibits the activity of APC/C through direct binding to its activator Cdc20. Studies have shown that Mad2 has two distinct natively folded conformations and that the unusual two-state behavior of Mad2 plays a crucial role in checkpoint signaling. This article describes methods for the purification of the two Mad2 conformers and for the analysis of their activities in APC/C inhibition in *Xenopus* egg extracts.


## Introduction

During a normal cell division cycle, the chromosomes are duplicated precisely once and then distributed evenly into two daughter cells (Nasmyth, 2002). The accuracy of this process plays a pivotal role in maintaining the genetic stability of the organism. Errors in chromosome segregation might promote aneuploidy in cells and lead to cancer formation (Jallepalli and Lengauer, 2001). Prior to anaphase, the sister chromatids are kept together by a protein complex termed cohesin (Nasmyth, 2002). After all the chromosomes are attached to the microtubules and are aligned at the cell equator, the anaphase-promoting complex or cyclosome, a multisubunit E3 ubiquitin ligase, tags the securin protein with polyubiquitin chains and results in its degradation by the proteasome (Harper *et al.*, 2002; Peters, 2002). The destruction of securin releases its inhibitory effect on separase, a cysteine protease of the CD clan family (Nasmyth, 2002). The activated separase then proteolyzes one of the cohesin subunits, Scc1, thus destroying the cohesion between the sister chromatids and allows the onset of anaphase (Nasmyth, 2002). APC/C is thus required indirectly for the initiation of sister chromatid separation.

METHODS IN ENZYMOLOGY, VOL. 398

0076-6879/05 $35.00
DOI: 10.1016/S0076-6879(05)98020-8

To ensure the fidelity of chromosome segregation, cells employ a surveillance mechanism called the spindle checkpoint to monitor mistakes in the attachment of sister chromatids to the mitotic spindle before their separation (Bharadwaj and Yu, 2004; Musacchio and Hardwick, 2002). The molecular components of the spindle checkpoint include Bub1–3 and Mad1–3. APC/C is an important molecular target of the checkpoint (Yu, 2002). A single kinetochore that has not attached to the mitotic spindle in the cell can activate the spindle checkpoint (Rieder *et al.*, 1995). This suggests that an inhibitory signal is generated by this kinetochore to inhibit APC/C throughout the cell and block chromosome segregation (Yu, 2002). Although the nature of this diffusible "wait anaphase" signal is still unclear, the mitotic checkpoint complex (MCC) that contains BubR1 (Mad3), Bub3, Mad2, and Cdc20, and the subcomplexes of MCC, are likely candidates for this signal (Yu, 2002). In particular, it has been noted that the binding between Mad2 and Cdc20 is absolutely required for the proper function of the checkpoint and is enhanced greatly upon checkpoint activation. Mad2 also interacts with another checkpoint protein, Mad1, that recruits Mad2 to the kinetochore and is required for Mad2 binding to Cdc20 (Chen *et al.*, 1998, 1999). Moreover, $p31^{comet}$, a Mad2-binding protein identified through a yeast two-hybrid screen, has been implicated in the silencing of the checkpoint (Habu *et al.*, 2002; Xia *et al.*, 2004). Finally, phosphorylation of the C terminus of Mad2 also seems to regulate its function negatively (Wassmann *et al.*, 2003).

Recent structural studies have provided new insights into the interactions between Mad2 and its aforementioned binding partners (Musacchio and Hardwick, 2002). Mad2 recognizes similar short peptide motifs in Mad1 (its upstream regulator) and Cdc20 (its downstream target) (Luo *et al.*, 2002). Using nuclear magnetic resonance (NMR) techniques, we have determined the solution structures of free Mad2 and Mad2 in complex with MBP1, a peptide ligand identified using phage display that mimics the Mad2-binding sequences of both Mad1 and Cdc20 (Luo *et al.*, 2000, 2002). Based on our NMR structures, Mad2 undergoes a similarly dramatic conformational change upon binding to Mad1 or Cdc20. The Mad2-binding peptide of Mad1 or Cdc20 inserts as a central strand into the main $\beta$ sheet of Mad2. The C-terminal region of Mad2 then moves across from one side of the $\beta$ sheet to the other side to form two new strands. This unusual conformational change of Mad2 has also been observed in the crystal structure of Mad2 in complex with a 120 residue fragment of Mad1 (Sironi *et al.*, 2002).

Mad1 and Cdc20 bind to the same site on Mad2. Binding of Mad1 to Mad2 also induces the same conformational changes in Mad2 as does Cdc20. Thus, Cdc20 and Mad1 binding to Mad2 are mutually exclusive.

Consistently, overexpression of Mad1 in cells inhibits the function of Mad2 (Luo *et al.*, 2002). However, Mad1 is also required for the proper function of Mad2, as genetic deletion or RNAi-mediated depletion of Mad1 causes checkpoint defects in yeast or human cells, respectively. To explain this paradox of Mad1, we have proposed that Mad2 dissociated from Mad1 must transiently retain an activated conformation that is more suitable for subsequent Cdc20 binding (Yu, 2002). This hypothesis suggests that Mad2 might have a second conformation. Consistent with this notion, previous studies have shown that the bacterially expressed Mad2 protein exists as both monomeric and dimeric forms (Fang *et al.*, 1998). The dimeric Mad2 is more potent in inhibiting APC/C in *Xenopus* egg extracts than the monomeric form, suggesting that the dimeric Mad2 might have an activated conformation (Fang *et al.*, 1998).

To further study this activated conformation of Mad2 by NMR, we decided to use the Mad2 R133A mutant that was exclusively monomeric *in vitro* and yet retained the full biological activity of the wild-type Mad2 (Sironi *et al.*, 2001). To obtain the activated conformer of Mad2 R133A, we first expressed the protein in bacteria and showed that it indeed existed only as a monomer. However, when the Mad2 R133A protein was fractionated on an anion-exchange column, there were two well-separated peaks in the chromatograph (Fig. 1). We named the first low-salt peak N1-Mad2 R133A and the second high-salt peak N2-Mad2 R133A (Luo *et al.*, 2004). N1 and N2 stood for native fold 1 and 2, respectively. Based on two-dimensional NMR experiments, the structure of N1-Mad2 R133A was identical to that of the free Mad2 determined previously, whereas the structure of N2-Mad2 R133A resembled that of the conformation of Mad2 when bound to Mad1 or Cdc20 (Luo *et al.*, 2004). This was confirmed when we solved the structure of the N2-Mad2 R133A by NMR (Fig. 2) (Luo *et al.*, 2004). Indeed, N2-Mad2 R133A had a fold similar to that of Mad2 in complex with Cdc20, except with a vacant peptide-binding site. More surprisingly, we showed that N1-Mad2 R133A could spontaneously convert into N2-Mad2 R133A after an overnight incubation at room temperature (Luo *et al.*, 2004). Thus, the two conformers of Mad2 R133A exist in equilibrium with a large energetic barrier, and N2-Mad2 R133A is more stable than N1-Mad2 R133A *in vitro* (Luo *et al.*, 2004). The wild-type Mad2 also has N1 and N2 conformations and undergoes similar N1 to N2 conversion, except that the wild-type N2-Mad2 is a dimer (Luo *et al.*, 2004). Finally, we showed that the conversion from N1-Mad2 to N2-Mad2 is accelerated greatly by the Mad2-binding motif of Mad1, suggesting that Mad1 might facilitate the N1-N2 structural rearrangement of Mad2 (Luo *et al.*, 2004).

To determine the biological relevance of the two conformations of Mad2, we have tested the two Mad2 R133A conformers in an *in vitro*

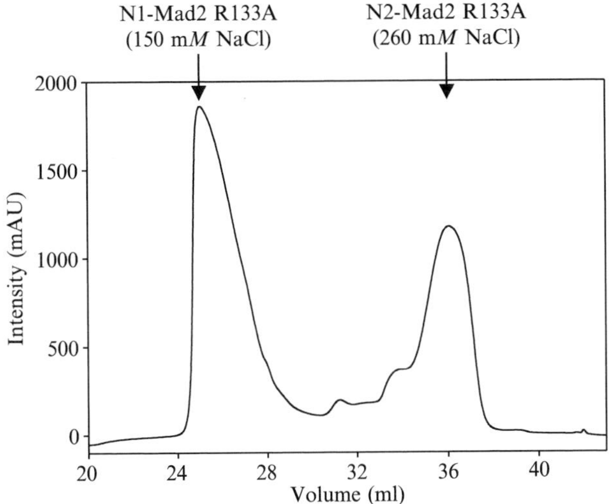


FIG. 1. The chromatograph of Mad2 R133A eluted from a Mono-Q column. The salt concentrations of the two peaks are labeled.

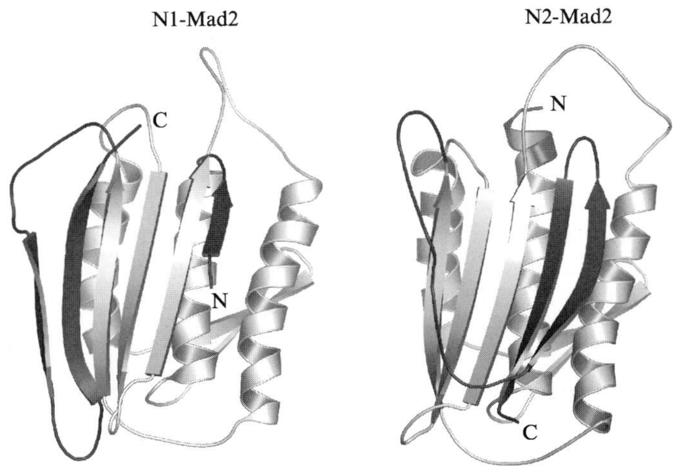


FIG. 2. Structures of N1-Mad2 (left) and N2-Mad2 (right). The C-terminal region of Mad2 that undergoes dramatic conformational changes between the N1 and the N2 conformers is shown as dark gray. N and C termini are labeled.

cyclin B1 degradation assay (Luo *et al.*, 2004). N2-Mad2 was more potent than N1-Mad2 in inhibiting APC/C and cyclin B degradation in *Xenopus* egg extracts (Luo *et al.*, 2004). The wild-type dimeric N2-Mad2 was also more active in blocking cyclin B1 degradation than the wild-type N1-Mad2 monomer (Fang *et al.*, 1998; Luo *et al.*, 2004). Interestingly, the wild-type N2-Mad2 dimer was only slightly more active than the N2-Mad2 R133A monomer, suggesting that dimerization itself is not required for efficient APC/C inhibition (Luo *et al.*, 2004). Thus, N2-Mad2 is very likely the activated form of Mad2. Our data are consistent with the notion that Mad1 facilitates the binding of Mad2 to Cdc20 by catalyzing formation of the N2-Mad2 conformer, which is more compatible for Cdc20 binding. This model explains the paradox that Mad1 is required for binding of Mad2 to Cdc20 *in vivo*, yet Mad1 acts as a competitive inhibitor of Cdc20 binding by Mad2. Thus, the ratio of Mad1/Mad2 seems crucial for proper spindle checkpoint signaling *in vivo*. Without Mad1, N2-Mad2 does not form efficiently, causing defects in the spindle checkpoint. However, overexpression of Mad1 results in the sequestration of Mad2, preventing its interaction with Cdc20.

In summary, our results suggest that the unusual two-state behavior of Mad2 is critical for spindle checkpoint signaling. The unique properties of Mad2 provide a clear example of how the conformational malleability of a protein might be used as a signaling mechanism for an important biological process. This chapter describes the detailed procedure for obtaining the two conformers of the recombinant human Mad2 protein and for assaying the APC/C inhibitory activities of these two conformers in crude *Xenopus* egg extracts.

## Expression and Purification of N1- and N2-Mad2

The human Mad2 protein is overexpressed in *Escherichia coli* and is purified as an N-terminally $His_6$-tagged fusion protein. Due to the direct involvement of the C-terminal region of Mad2 in the conformational change, addition of any tags at the C terminus of Mad2 prevents the formation of N2-Mad2 and inhibits the biochemical function of Mad2. Using conventional recombinant DNA techniques, the coding region of the wild-type Mad2 protein was cloned into the *Bam*HI and *Hin*dIII sites of pQE30 (Qiagen, Valencia, CA) with a 5′ primer encoding a tobacco etch virus (TEV) protease cleavage site. The pQE30-Mad2 R133A was made with the QuikChange site-directed mutagenesis kit (Stratagene, La Jolla, CA). The vectors were validated by DNA sequencing and then transformed into *E. coli* strain M15[pREP4] (Qiagen). The purification procedures for the wild-type Mad2 and the Mad2 R133A mutant are similar except that the wild-type N2-Mad2 dimer elutes earlier from the gel-filtration column than the wild-type N1-Mad2 and Mad2 R133A monomers.

*Materials*

LB medium (GIBCO)
Carbenicillin (Sigma) stock solution (50 mg/ml)
Kanamycin (Sigma) stock solution (50 mg/ml)
Ampicillin (Sigma) stock solution (100 mg/ml)
Isopropyl-$\beta$-D-thiogalactoside (IPTG) (Sigma) (1 *M*)
AEBSF (Research Products International) (100 m*M*)
Protease inhibitor cocktail (P–2714, Sigma), stock concentration (100×): 10 m*M* AEBSF, 5 m*M* EDTA, 0.65 m*M* Bestatin, 7 $\mu M$ E–64, 5 $\mu M$ leupeptin, and 1.5 $\mu M$ aprotinin
$Ni^{2+}$-NTA agarose (Qiagen)
Empty columns (Bio-Rad)
PD–10 column (Amersham)
Sonication buffer: 50 m*M* sodium phosphate (pH 7.8) + 300 m*M* NaCl
Wash buffer: 50 m*M* sodium phosphate (pH 6.0) + 300 m*M* NaCl + 10% glycerol
Elution buffer: 50 m*M* sodium phosphate (pH 6.0) + 300 m*M* NaCl + 10% glycerol + 150 m*M* imidazole
TEV cleavage buffer: 50 m*M* Tris (pH 8.0) + 100 m*M* NaCl + 1 m*M* dithiothreitol (DTT)
QA buffer: 20 m*M* Tris (pH 8.0) + 50 m*M* NaCl + 1 m*M* DTT
QB buffer: 20 m*M* Tris (pH 8.0) + 1 *M* NaCl + 1 m*M* DTT
Superdex–75 buffer: 50 m*M* sodium phosphate (pH 6.8) + 0.3 *M* KCl + 1 m*M* DTT

*Procedures*

1. Spread the M15[pREP4] cells that are freshly transformed with pQE30-Mad2 onto a LB agar plate containing 50 $\mu$g/ml carbenicillin (Carb) and 50 $\mu$g/ml kanamycin (KAN). Incubate the plate overnight at 37°.

2. Inoculate 100 ml of LB medium containing 100 $\mu$g/ml ampicillin and 50 $\mu$g/ml kanamycin in a 250-ml flask with a single colony from the LB/Carb/Kan agar plate. Grow the culture for 12 to 16 h at 37° with vigorous shaking (250 rpm).

3. Inoculate 6 liter of LB medium containing 100 $\mu$g/ml ampicillin and 50 $\mu$g/ml kanamycin (1 liter each in six 2.8-liter flasks) by adding 10 ml of overnight culture to each flask. Incubate the six flasks of culture at 37° with constant shaking (250 rpm) for about 2 h until reaching $OD_{600nm}$ of 0.5–0.6.

Reduce the temperature to 16° and grow the bacterial culture for another 30 min.

4. Induce the expression of the fusion protein by adding IPTG to a final concentration of 0.2 m*M*. Grow the culture for additional 6–8 h at 16°.

5. Harvest the cells in 1-liter centrifuge bottles by centrifugation at 4000*g* at 4° for 20 min.

6. Discard supernatant and resuspend the cell pellets in 15 ml cold sonication buffer (pH 8.0) per liter of culture. Freeze sample in liquid $N_2$ and store at –80° until protein purification.

7. For protein purification, thaw cells in cold water for about 30 min. Add AEBSF, protease inhibitor cocktail, and $\beta$-mercaptomethanol to final concentrations of 0.2 m*M*, 1% (v/v), and 10 m*M*, respectively.

8. Lyse cells by passing the suspensions through an Emulsiflex-C5 homogenizer (Avestin, Canada) at a pressure of 75–150 MPa two times.

9. Pool the bacterial cell lysate into 50-ml centrifuge tubes and centrifuge at 30,000*g* at 4° for 45 min.

10. Collect and filter the supernatant through a 0.45-$\mu$m syringe filter. Add 12 ml of 50% slurry of $Ni^{2+}$-NTA resin (preequilibrated with sonication buffer) to the cleared lysate. Rotate end over end at 4° for 90 min.

11. Load the mixture of the lysate and the $Ni^{2+}$-NTA resin onto an empty column and collect the column flow through.

12. Wash the resins three times with 20 ml sonication buffer in each wash.

13. Wash the resins three times with 20 ml wash buffer each time.

14. Wash twice with 15 ml wash buffer containing 20 m*M* imidazole.

15. Elute the proteins from the $Ni^{2+}$-NTA resins four times with 6 ml elution buffer each time. Collect the four elutions separately in 15-ml conical tubes.

16. Exchange the sample into TEV cleavage buffer by passing through PD–10 columns that have been preequilibrated with the TEV cleavage buffer.

17. Add TEV protease (1 $OD_{280}$ of TEV per 100 $OD_{280}$ of Mad2 fusion protein) and incubate the protein sample overnight at 4°.

18. Purify the cleaved Mad2 protein using an anion-exchange Mono-Q column with the Akta FPLC system (Amersham). The Mono-Q column is first equilibrated with 5 column volume (CV) of QA buffer. The Mad2 protein in QA buffer is loaded onto the column at a flow rate of 1 ml/min. Elution is performed with a 20 CV linear gradient of 0–40% (v/v) QB buffer.

19. Analyze the fractions by SDS–PAGE followed by Coomassie blue staining. Fractions containing N1-Mad2 and N2-Mad2 are pooled separately and concentrated to 5 ml.

20. Equilibrate a HiLoad 16/60 Superdex–75 column (Amersham) with two CV of Superdex–75 buffer. Load the two Mad2 conformers separately onto the column with a flow rate of 1 ml/min. Collect 3-ml fractions. Analyze the fractions by SDS–PAGE followed by Coomassie blue staining. Pool the fractions containing N1- and N2-Mad2 separately.

21. Concentrate the purified proteins to the desired concentration using Centriprep–10 (Millipore) at <3000*g* at 4°. Freeze the samples in liquid nitrogen and store at –80° in small aliquots for further experiments.

## Cyclin B Degradation Assay in *Xenopus* Egg Extracts

In this assay, we use *in vitro*-translated $^{35}$S-labeled full-length human cyclin B1 as the APC/C substrate. To eliminate the possibility that some undetectable substances are copurified with N2-Mad2, but not with N1-Mad2, we have used N2-Mad2 converted from N1-Mad2 by incubating N1-Mad2 at 30° overnight in test tubes. However, for most applications, N2-Mad2 purified using the protocol described earlier can be used directly in the cyclin B degradation assay. All the experiments are done at room temperature. The amount of N1-Mad2 that has transformed into N2-Mad2 during the assays is negligible.

### *Materials*

- Mitotic Δ90 *Xenopus* egg extracts
- Bovine ubiquitin (Sigma). Dissolve in $H_2O$ at a final concentration of 10 mg/ml. Aliquot and store at –80°.
- Energy mix (20×), 100 ml: 3.827 g phosphocreatine (150 m*M* final concentration) (Sigma), 1.102 g ATP (20 m*M*) (Sigma), 0.4 ml of 0.5 *M* EGTA (pH 7.7) (2 m*M*), and 2 ml of 1 *M* $MgCl_2$ (20 m*M*), adjust pH to 7.7, aliquot, and store at –80°.
- XB buffer: 10 m*M* HEPES (pH 7.7), 100 m*M* KCl, 0.1 m*M* $CaCl_2$, 1 m*M* $MgCl_2$, and 50 m*M* sucrose
- $^{35}$S-labeled human cyclin B1: *in vitro* translated in SP6-coupled rabbit reticulocyte lysate (Promega) in the presence of [$^{35}$S]methionine (Amersham) per manufacturer's protocols
- SDS sample buffer

### *Procedures*

1. Prepare mitotic Δ90 *Xenopus* egg extracts as described (Murray, 1991).

2. Thaw an aliquot of the N1-Mad2 protein in cold water. Divide the sample evenly into two Eppendorf tubes. Freeze one tube of N1-Mad2

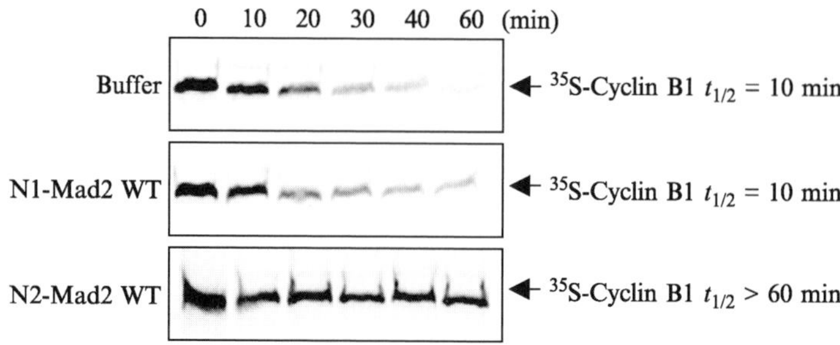


FIG. 3. Cyclin B degradation assay. The $^{35}$S-labeled human cyclin B1 protein in reticulocyte lysate was added to mitotic Δ90 *Xenopus* egg extracts containing XB buffer, N1-Mad2 (0.5 mg/ml final concentration), or N2-Mad2 (0.5 mg/ml final concentration). Samples were taken at the indicated time points and separated by SDS–PAGE followed by Phosphoimager analysis.

with liquid nitrogen and store at –80°. This tube of N1-Mad2 will be used in the cyclin B degradation assay as the negative control. Incubate the other tube of N1-Mad2 in a 30° incubator for 18 h to allow it to convert into N2-Mad2.

3. Thaw the N1-Mad2 sample described in step 2 in cold water. Adjust the concentrations of N1-Mad2 and N2-Mad2 proteins to 5 mg/ml with XB buffer. Set up three 500-μl Eppendorf tubes. Add 37 μl of mitotic Δ90 *Xenopus* egg extracts to each tube. Add 5 μl XB to one tube, 5 μl N1-Mad2 protein to the second tube, and 5 μl N2-Mad2 protein to the third tube. Incubate at room temperature for 20 min.

4. Add 2.5 μl energy mix and 2.5 μl ubiquitin to each of the three reaction tubes. Add 3 μl $^{35}$S-labeled cyclin B1 to each tube. Mix thoroughly. Incubate at room temperature. The total volume of each reaction is 50 μl.

5. Remove 3 μl of reaction mixture from each tube at 0, 10, 20, 30, 40, and 60 min. Mix immediately with 40 μl of SDS sample buffer.

6. After all time points have been completed, boil the samples for 5 min. Load 10 μl of each sample on a 12% SDS–PAGE gel. Dry the gel with a gel dryer. Expose the gel to a Phosphoimager plate and analyze the results with a phosphoimager (Fig. 3).

## Acknowledgments

This work is supported by the National Institutes of Health (5 K01 CA100292 to X. L. and GM61542 to H. Y.), the Packard Foundation, the Burroughs Wellcome Fund, and the Robert A. Welch Foundation (I–1441).

## References


Bharadwaj, R., and Yu, H. (2004). The spindle checkpoint, aneuploidy, and cancer. *Oncogene* **23,** 2016–2027.

Chen, R. H., Brady, D. M., Smith, D., Murray, A. W., and Hardwick, K. G. (1999). The spindle checkpoint of budding yeast depends on a tight complex between the Mad1 and Mad2 proteins. *Mol. Biol. Cell* **10,** 2607–2618.

Chen, R. H., Shevchenko, A., Mann, M., and Murray, A. W. (1998). Spindle checkpoint protein Xmad1 recruits Xmad2 to unattached kinetochores. *J. Cell Biol.* **143,** 283–295.

Fang, G., Yu, H., and Kirschner, M. W. (1998). The checkpoint protein MAD2 and the mitotic regulator CDC20 form a ternary complex with the anaphase-promoting complex to control anaphase initiation. *Genes Dev.* **12,** 1871–1883.

Habu, T., Kim, S. H., Weinstein, J., and Matsumoto, T. (2002). Identification of a MAD2-binding protein, CMT2, and its role in mitosis. *EMBO J.* **21,** 6419–6428.

Harper, J. W., Burton, J. L., and Solomon, M. J. (2002). The anaphase-promoting complex: It's not just for mitosis any more. *Genes Dev.* **16,** 2179–2206.

Jallepalli, P. V., and Lengauer, C. (2001). Chromosome segregation and cancer: Cutting through the mystery. *Nature Rev. Cancer* **1,** 109–117.

Luo, X., Fang, G., Coldiron, M., Lin, Y., Yu, H., Kirschner, M. W., and Wagner, G. (2000). Structure of the Mad2 spindle assembly checkpoint protein and its interaction with Cdc20. *Nature Struct. Biol.* **7,** 224–229.

Luo, X., Tang, Z., Rizo, J., and Yu, H. (2002). The Mad2 spindle checkpoint protein undergoes similar major conformational changes upon binding to either Mad1 or Cdc20. *Mol. Cell* **9,** 59–71.

Luo, X., Tang, Z., Xia, G., Wassmann, K., Matsumoto, T., Rizo, J., and Yu, H. (2004). The Mad2 spindle checkpoint protein has two distinct natively folded states. *Nature. Struct. Mol. Biol.* **11,** 338–345.

Murray, A. W. (1991). Cell cycle extracts. *Methods Cell Biol.* **36,** 581–605.

Musacchio, A., and Hardwick, K. G. (2002). The spindle checkpoint: Structural insights into dynamic signalling. *Nature Rev. Mol. Cell Biol.* **3,** 731–741.

Nasmyth, K. (2002). Segregating sister genomes: The molecular biology of chromosome separation. *Science* **297,** 559–565.

Peters, J. M. (2002). The anaphase-promoting complex: Proteolysis in mitosis and beyond. *Mol. Cell* **9,** 931–943.

Rieder, C. L., Cole, R. W., Khodjakov, A., and Sluder, G. (1995). The checkpoint delaying anaphase in response to chromosome monoorientation is mediated by an inhibitory signal produced by unattached kinetochores. *J. Cell Biol.* **130,** 941–948.

Sironi, L., Mapelli, M., Knapp, S., Antoni, A. D., Jeang, K. T., and Musacchio, A. (2002). Crystal structure of the tetrameric Mad1-Mad2 core complex: Implications of a 'safety belt' binding mechanism for the spindle checkpoint. *EMBO J.* **21,** 2496–2506.

Sironi, L., Melixetian, M., Faretta, M., Prosperini, E., Helin, K., and Musacchio, A. (2001). Mad2 binding to Mad1 and Cdc20, rather than oligomerization, is required for the spindle checkpoint. *EMBO J.* **20,** 6371–6382.

Wassmann, K., Liberal, V., and Benezra, R. (2003). Mad2 phosphorylation regulates its association with Mad1 and the APC/C. *EMBO J.* **22,** 797–806.

Xia, G., Luo, X., Habu, T., Rizo, J., Matsumoto, T., and Yu, H. (2004). Conformation-specific binding of p31(comet) antagonizes the function of Mad2 in the spindle checkpoint. *EMBO J.* **23,** 3133–3143.

Yu, H. (2002). Regulation of APC-Cdc20 by the spindle checkpoint. *Curr. Opin. Cell Biol.* **14,** 706–714.

# [21] Identification, Expression, and Assay of an Oxidation-Specific Ubiquitin Ligase, HOIL-1

*By* KAZUHIRO IWAI, HARUTO ISHIKAWA, and TAKAYOSHI KIRISAKO

## Abstract

The ubiquitin system plays important roles in the regulation of numerous cellular processes. It is well established that ubiquitin ligases (E3s) are key components in determining the specificity of the system and that the modification of substrates such as phosphorylation often plays a critical role in selective substrate recognition by E3s. Through studies analyzing iron-mediated degradation of iron regulatory protein 2 (IRP2), a central regulator of iron metabolism in mammalian cells, we have identified a RING finger protein, HOIL-1, as an ubiquitin ligase recognizing IRP2 through a signal created by heme-mediated oxidative modification of the protein. We have utilized several types of *in vitro* ubiquitination assays that detect IRP2 ubiquitination and a differential yeast two-hybrid screen in which yeast cells were cultured either in the presence or in the absence of oxygen to control the oxidation state of the bait in the cells in our studies. This chapter describes the detailed methods used for the identification and functional analysis of the HOIL-1 ligase.

## Introduction

Most organisms on the earth depend on aerobic growth, but at the same time are exposed to damaging oxygen radicals generated by products of aerobic metabolism, primarily by the reaction of iron with molecular oxygen. Oxygen radicals can oxidatively modify amino acid residues of proteins, causing these proteins to lose their activity. Moreover, oxidized proteins tend to form protein aggregates and are often toxic to cells. In fact, such aggregates participate in the pathogenesis of several neurodegenerative disorders, and oxidation has been suggested to be involved in the formation of these aggregates. Organisms have developed pathways to remove oxidized proteins by rapid protein degradation, which thus serve as an important component of cellular quality control mechanisms (Grune *et al.*, 2003; Iwai, 2003).

Most studies agree that oxidized proteins are degraded by the proteasome (Benaroudj *et al.*, 2001; Grune *et al.*, 2003). However, it remains unclear whether ubiquitination is involved in this process. Some studies

METHODS IN ENZYMOLOGY, VOL. 398

0076-6879/05 $35.00
DOI: 10.1016/S0076-6879(05)98021-X

have demonstrated that oxidized proteins are ubiquitinated (Hershko *et al.*, 1986). However, other researchers have shown that oxidized proteins are degraded by the 20S proteasome and no ubiquitination is necessary for their degradation (Shringarpure *et al.*, 2003). Thus, whether ubiquitination is necessary for the degradation of oxidized proteins remains a controversial issue. We described the HOIL-1 ubiquitin ligase that specifically recognizes oxidized IRP2, a central regulator of iron metabolism in mammalian cells (Yamanaka *et al.*, 2003).

Iron is an essential nutrient in almost all organisms, but it is also toxic because of its high reactivity to molecular oxygen. Therefore, iron metabolism must be tightly regulated, and regulation is achieved, in mammalian cells, at the posttranscriptional level through the interaction between transacting factors named iron regulatory proteins (IRPs) and an RNA stem–loop structure known as the iron responsive element (IRE), which is found in the transcripts of the molecules involved in iron metabolism (Hentze *et al.*, 2004). Two IRPs, IRP1 and IRP2, have been reported. IRP1 and IRP2 have largely redundant roles in posttranscriptional gene regulation. However, IRP2 null mice, but not IRP1 null mice, develop adult-onset neurodegenerative disease (LaVaute *et al.*, 2001; Meyron-Holtz *et al.*, 2004). Although both IRP1 and IRP2 bind to IREs only under conditions of iron depletion, the mode of regulation by iron is different between the two proteins (Hentze *et al.*, 2004). IRP1 is a stable bifunctional protein and its activity is regulated by the iron-dependent assembly/disassembly of an iron–sulfur cluster center (Klausner *et al.*, 1993). IRP2 is regulated by the rapid degradation by the ubiquitin-proteasome system in the presence of iron. An IRP2-specific domain consisting of 73 amino acids, called the iron-dependent degradation (IDD) domain (Yamanaka *et al.*, 2003), is required for iron-dependent degradation of the protein (Iwai *et al.*, 1995). We have demonstrated that IRP2 is oxidized and ubiquitinated in cells prior to its degradation by the proteasome and that oxidative modification of the protein is involved in the selective ubiquitination of IRP2 (Iwai *et al.*, 1998). The oxidative modification of IRP2 is generated by heme, which binds to IRP2, in the presence of oxygen. The HOIL-1 (heme-oxidized IRP2 ubiquitin ligase 1) ubiquitin ligase recognizes oxidized IRP2 specifically (Yamanaka *et al.*, 2003).

This chapter reviews the methods used for analysis of the molecular mechanisms underlying oxidation-induced degradation of IRP2, including identification of HOIL-1, expression and purification of substrates for ubiquitination, and preparation of enzymes needed for *in vitro* ubiquitination assays.

## Preparation of IRP2 Substrate for *In Vitro* Ubiquitination Assay

IRP2 is a cytosolic 105-kDa protein. We used recombinant IRP2 protein as the substrate in all the *in vitro* ubiquitination assays performed. Recombinant IRP2 was expressed and purified using the baculovirus expression system.

## Preparation of IRP2 Using the Baculovirus Expression System

See Iwai *et al.* (1998) for details.

### *Reagents and Instruments*

Recombinant baculovirus expressing C-terminally (His)6-*myc*-tagged IRP2 [C-terminally (His)6-*myc*-tagged IRP2 cDNA was subcloned into pVL1393; recombinant baculovirus was generated using the Bac-PAK6 baculovirus expression system as instructed by the manufacturer (BD Biosciences, Palo Alto, CA)]
Sf21 cells
Grace's insect cell medium supplemented with 10% heat-inactivated fetal bovine serum (FBS)
Phosphate-buffered saline (PBS)
Lysis buffer (20 m*M* Tris-Cl, pH 7.5, protease inhibitors, and 0.1 m*M* deferoxamine)
Wash buffer (20 m*M* Tris-Cl, pH 7.5, containing indicated concentration of imidazole)
Elution buffer (20 m*M* Tris-Cl, pH 7.5, 300 m*M* imidazole)
Dialysis buffer (20 m*M* Tris-Cl, pH 7.5)
Ni-NTA agarose (Quiagen, Valencia, CA)
Dounce homogenizer
Cell culture incubator, 27°

### *Procedure*

1. Seed $3 \times 10^6$ cells of Sf21 in a 100-mm culture dish the day before infection.
2. Pipette off medium.
3. Add 1ml of diluted virus supernatant onto dishes (MOI = 0.1–0.5).
4. Cover the dish completely with virus supernatant by gentle swirling every 15 min.
5. Incubate at 27° for 60 min.
6. Pipette off supernatant.
7. Add 10 ml of Grace's insect cell medium supplemented with 10% heat-inactivated FBS.

8. Culture in 27° incubator.
9. Harvest cells by pipetting at 60 h postinfection.
10. Wash twice with ice-cold PBS.
11. Suspend cells with 10 volumes of lysis buffer.
12. Incubate on ice 10 min.
13. Disrupt cells with a Dounce homogenizer (30 strokes).
14. Incubate the homogenate on ice 10 min.
15. Centrifuge at 15,000 rpm for 15 min at 4°.
16. Transfer supernatant to a new tube.
17. Add imidazole to bring the solution to final concentration of 0.2 m*M* imidazole.
18. Add 0.5–0.7 ml of Ni-NTA agarose (for twenty 100-mm culture dishes).
19. Rotate the tube at 4° for 60 min.
20. Wash six times with 20 m*M* Tris-Cl, pH 7.5, and 5 m*M* imidazole.
21. Wash twice with 20 m*M* Tris-Cl, pH 7.5, and 10 m*M* imidazole.
22. Elute protein by adding 0.6 ml of elution buffer.
23. Rotate the tube at 4° for 10 min.
24. Recover supernatant.
25. Repeat the elution step (steps 22–24) once.
26. Dialyze the supernatant against 20 m*M* Tris-Cl, pH 7.5.
27. Store at –80°.
28. Forty 100-mm dishes will purify 0.4–0.6 mg of the protein.

## IRP2 Oxidation

IRP2 oxidation (Iwai *et al.*, 1998) should be performed immediately prior to the *in vitro* ubiquitination reaction because it is very difficult to control the oxidation state of IRP2. IRP2 oxidation proceeds when samples are kept at –80°; conducting the oxidation reaction immediately before using IRP2 in the assay ensures consistency in the assays.

### *Reagents*

Purified IRP2
$FeCl_3$ (Sigma, St. Louis, MO)
Dithiothreitol (DTT; Wako, Osaka, Japan)

### *Procedure*

1. Prepare a 20-$\mu$l reaction mixture of 20 m*M* Tris-Cl, pH 7.5, 5 $\mu M$ $FeCl_3$, and 10 m*M* DTT.

2. Add 0.25 μg/μl of *myc*-tagged IRP2.
3. Incubate at 37° for 5 min.
4. Stop the reaction by adding deferoxamine (0.1 m*M*) and cooling to 4°.

## Heme Loading of IRP2

Because heme-loaded IRP2 is oxidized readily *in vitro*, all the procedures should be performed on ice or at 4°. Heme-loaded IRP2 can be stored at –80° after quick freezing, but it must be used within 2–3 weeks because it will oxidize, even at –80° (Yamanaka *et al.*, 2003).

### *Reagents*

Hemin (Sigma, St. Louis, MO)
*N,N*-Dimethylformamide (DMF; Nacalai Tesque, Kyoto, Japan)
PD10 (Amersham Biosciences, Piscataway, NJ)
Equilibration buffer (20 m*M* Tris-Cl, pH 7.5)

### *Procedure*

1. Dissolve hemin in DMF (0.5 m*M*).
2. Mix equimolar amounts of IRP2 and hemin.
3. Incubate on ice for 10 min.
4. Remove excess hemin by gel filtration using a PD10 column equilibrated with equilibration buffer.

## Preparation of Lysates for Enzyme Source and Purified Enzymes for *In Vitro* Ubiquitination Assays

### *Preparation of Cell Lysates as Enzyme Sources*

#### *S100 Lysate of HeLa S3 Cells*

REAGENTS AND INSTRUMENTS

Spinner culture flasks
SMEM medium (Invitrogen, Carlsbad, CA) supplemented with 10% FBS
PBS
Lysis buffer (20 m*M* Tris-Cl, pH 7.5, 1 m*M* DTT)
Dounce homogenizer
Ultracentrifuge and angle rotors

Procedure

1. Culture HeLa S3 cells in spinner culture flasks with SMEM medium supplemented with 10% FBS. (The culture is initiated at $1.0 \times 10^5$ cells/ml and cells are harvested when confluent.)
2. Collect HeLa S3 cells by centrifugation (400*g* at 4°) for tissue culture.
3. Wash cells three times with PBS.
4. Lyse cells by adding 3 volumes of lysis buffer.
5. Incubate on ice for 10 min.
6. Disrupt cells with a Dounce homogenizer (30 strokes).
7. Incubate on ice for 10 min.
8. Centrifuge the homogenate at 10,000*g* for 20 min at 4°.
9. Transfer the supernatant to a new tube.
10. Centrifuge supernatants at 100,000*g* for 60 min at 4°.
11. Store at –80°.

*Fraction I (Fr. I) and Fraction II (Fr. II) of HeLa S100 Lysate.* Fr. I and Fr. II are prepared as demonstrated by Ciechanover *et al.* (1980) with slight modifications.

Reagents

HeLa S100 lysate
DEAE-cellulose (DE–52, Whatman, Middlesex, UK)
Equilibration buffer (20 m*M* Tris-Cl, pH 7.5 1, m*M* DTT)
Wash buffer (20 m*M* Tris-Cl, pH 7.5, 30 m*M* NaCl, 1 m*M* DTT)
Elution buffer (20 m*M* Tris-HCl, pH 7.5, 1 *M* NaCl, 1 m*M* DTT)
Saturated ammonium sulfate solution

Procedure

1. Equilibrate DE–52 with equilibration buffer.
2. Load HeLa S100 lysate onto DE–52 column (ratio of lysate to resin is 1.5:1).
3. Collect unbound materials (Fr. I).
4. Wash column with 3 column volumes of wash buffer.
5. Elute DEAE-bound materials with 3 column volumes of elution buffer (Fr. II).
6. Precipitate Fr. I and Fr. II with ammonium sulfate at a final concentration of 80%.
7. Dissolve in equilibration buffer.
8. Dialyze with equilibration buffer to remove ammonium sulfate.
9. Store Fr. I and Fr. II at –80°.

### *Preparation of Purified Enzymes*

*E1 and E2.* Recombinant N-terminally $(His)_6$-tagged mouse E1 generated by the baculovirus expression system and recombinant N-terminally $(His)_6$-tagged human E2s generated by the bacterial expression system (Iwai *et al.*, 1999) are used in the assays. Details for the production and purification of these enzymes are not described, as the procedures are described elsewhere in this volume.

## HOIL-1 Ubiquitin Ligase

HOIL-1 is a RING finger ubiquitin ligase, which specifically ubiquitinates oxidized IRP2. It was identified by a differential yeast two-hybrid screening using the IDD domain as bait. The procedure for the differential two-hybrid screening is described at the end of this section.

We prepare recombinant C-terminally $(His)_6$-tagged HOIL-1 using the baculovirus expression system (Yamanaka *et al.*, 2003). The reagents and procedures are identical to those of IRP2 purification except for the following buffers.

- Lysis buffer [20 m*M* Tris-Cl, pH 7.5, 10 m*M* 2-meraptoethanol (2-ME), and protease inhibitors]
- Wash buffer (20 m*M* Tris-Cl, pH 7.5, 10 m*M* 2-ME containing indicated concentration of imidazole)
- Elution buffer (20 m*M* Tris-Cl, pH 7.5, 10 m*M* 2-ME, 300 m*M* imidazole)
- Dialysis buffer (20 m*M* Tris-Cl, pH 7.5, 1 m*M* DTT)

## *In Vitro* Ubiquitination Assays

*In vitro* ubiquitination assays are powerful tools that can be used not only to show that particular substrates are ubiquitinated, but also to identify enzymes, namely E2s and E3s, which ubiquitinate specific substrates, or to identify new substrates of known E3s.

There are two ways to detect ubiquitination of substrates *in vitro*: (1) detection of high molecular weight signals by recognizing substrate after the *in vitro* reaction either by autoradiography when radioisotope-labeled substrates are used or by Western blotting with antibodies specific for substrates of interest or (2) detection of high molecular weight signals by detecting ubiquitin after immunoprecipitation of substrates. In general, the efficiency of *in vitro* ubiquitination of the substrates is the key factor in selecting the method. Usually, we use the former method when the efficiency of *in vitro* ubiquitination of the substrates is high. The following section introduces protocols for different *in vitro* ubiquitination assays that we used to detect IRP2 ubiquitination.

### *Protocol 1a.* In Vitro *Ubiquitination Assay Using Cell Lysates as an Enzyme Source*

This method was originally used to show that oxidation of IRP2 made the protein a better substrate for ubiquitination (Iwai *et al.*, 1998). Because it is very difficult to control oxidation levels of IRP2 *in vitro*, we first used the latter method described in the previous paragraph to detect ubiquitination of oxidized IRP2 in the *in vitro* ubiquitination assay. Because cell lysates contain deubiquitinating enzymes that hydrolyze ubiquitin conjugations generated during the *in vitro* reactions, ubiquitin aldehyde (Pickart and Rose, 1986) or *N*-ethylmaleimide (NEM) (Chen *et al.*, 1995) is added to the reaction mixture or stop buffer, respectively, to inactivate the deubiquitinating enzymes.

*Reagents*

- Oxidized IRP2
- Ubiquitin (Sigma, St. Louis, MO)
- Ubiquitin aldehyde, a synthetic inhibitor of deubiquitinating enzymes (BIOMOL, Exeter, UK)
- HeLa S100 lysate
- Fr. I of HeLa S100 lysate
- Fr. II of HeLa S100 lysate
- ATP and an ATP regeneration system (0.5 m*M* ATP, 10 m*M* creatine phosphate, 10 $\mu$g creatine phosphokinase)
- Stop buffer (1% NP-40, 0.5% deoxycholate, 50 m*M* Tris-Cl, pH 8.0, 150 m*M* NaCl, 10 m*M* NEM and 0.1% SDS): NEM, which inactivates most deubiquitinating enzymes by modifying active site cysteine residues irreversibly, is labile and should be prepared immediately before use
- Wash buffer (1% NP-40, 0.5% deoxycholate, 50 m*M* Tris-Cl, pH 8.0, 150 m*M* NaCl, and 0.1% SDS)
- 10% cysteine (Nacalai Tesque, Kyoto, Japan) prepared freshly in doubly distilled water
- Anti-*myc* antibody (9E10) (Santa Cruz Biotechnology, Santa Cruz, CA)
- Protein A Sepharose FF (Amersham Biosciences, Piscataway, NJ )
- Rabbit antiubiquitin antibody (gift from Dr. Aaron Ciechanover, Technion, Haifa, Israel)

*Procedure*

1. Prepare a 20-$\mu$l reaction mixture containing 1 $\mu$g oxidized IRP2; 20 m*M* Tris-Cl, pH 7.5; 5 m*M* $MgCl_2$; 1 m*M* DTT; 5 $\mu$g ubiquitin; 0.5 $\mu$g ubiquitin aldehyde; enzymes from HeLa S100 lysate (30 $\mu$g), Fr. I (5 $\mu$g), Fr. II (20 $\mu$g), or Fr. I (5 $\mu$g) plus Fr. II (20 $\mu$g); and ATP and an ATP regeneration system.
2. Incubate reaction mixtures at 37° for 60 min.
3. Stop the reactions by adding stop buffer and incubate on ice for 15 min.
4. Add cysteine (0.1% at a final concentration) to neutralize NEM.
5. Centrifuge samples at 15,000 rpm for 20 min at 4°.
6. Transfer supernatant to a new Eppendorf tube.
7. Add 5 $\mu$g of anti-*myc* antibody (9E10).
8. Incubate on ice for 2 h.
9. Add 10 $\mu$l of protein A Sepharose FF.
10. Rotate tubes for 45 min at 4°.
11. Wash immunoprecipitates five times with wash buffer.
12. Electrophorese samples on a 6% SDS–PAGE gel and Western blotting with a rabbit antiubiquitin antibody.

Figure 1 is an example of an *in vitro* ubiquitination assay using cell lysates as an enzyme source.

*Protocol 1b.* In Vitro *Ubiquitination Assay Using Purified E1 and E2*

Oxidized IRP2 was ubiquitinated in the presence of both Fr. I and Fr. II, but not in the presence of either fraction alone (Fig. 1). It has been suggested that Fr. II contains E1 and most E3s. Because both Fr. I and Fr. II are necessary for ubiquitination of oxidized IRP2, we suspect that Fr. I contains E2 for oxidized IRP2. About 30 E2s have been identified in the human genome by their domain structure. We can determine the E2 for IRP2 ubiquitination by adding purified recombinant E2s to the *in vitro* ubiquitination assay. Oxidized IRP2 was incubated with purified E1, Fr. II as an E3 source, and various recombinant E2s (purified by using bacterial expression system) instead of Fr. I to identify E2s specific for oxidized IRP2.

The reagents and procedures are identical to protocol 1a, with the exception of using purified E1 and E2s. The 20-$\mu$l reaction mixture contains 1 $\mu$g oxidized IRP2, 20 m*M* Tris-Cl, pH 7.5, 5 m*M* $MgCl_2$, 1 m*M* DTT, 5 $\mu$g ubiquitin, 0.5 $\mu$g ubiquitin aldehyde, 100 ng E1, 200 ng E2s, 20 $\mu$g Fr. II from the HeLa S100 lysate, and ATP and an ATP regeneration system.

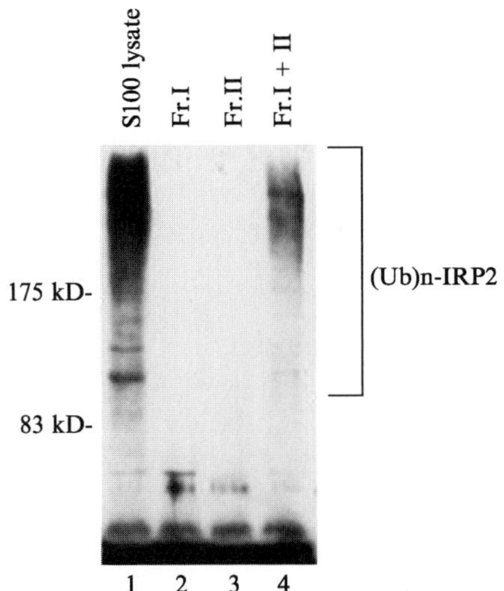


FIG. 1. Oxidized IRP2 was incubated with HeLa S100 lysate, fraction I (Fr. I), fraction II (Fr. II), or Fr. I and II (Fr. I + II) as indicated. IRP2 samples were precipitated with an anti-*myc* antibody and resolved on 6% SDS–PAGE, followed by blotting with anti-ubiquitin antibody to detect ubiquitinated IRP2.

Figure 2 shows that the UbcH5a, b, and c can function as E2s in the ubiquitination of oxidized IRP2.

*Protocol 1c.* In Vitro *Ubiquitination Assay to Identify E3s Specific for Oxidized IRP2*

Because we know that E3s exist in Fr. II, E3s for oxidized IRP2 could be identified using the *in vitro* ubiquitination assay containing purified E1 and E2 and adding Fr. II subfractions prepared by conventional column chromatographic techniques. However, previous observations indicate that the IDD domain is recognized by an E3 specific for IRP2, suggesting that we could copurify the E3 with the IDD domain from baculovirus-infected insect cells expressing the IDD domain. Purified IDD fractions were tested in the *in vitro* ubiquitination assay using purified E1 and E2. The IDD fraction from the insect cells did contain E3 activity.

The assay itself is identical to that used for isolation of a IRP2-specific E3 from chromatographically generated fractions of Fr. II. Reagents and procedures are identical to protocol 1a with the exception of using purified E1 and E2s in the reaction mixture. The 20-$\mu$l reaction mixture contains 1 $\mu$g oxidized IRP2; 20 m*M* Tris-Cl, pH 7.5; 5 m*M* $MgCl_2$; 1 m*M* DTT; 5 $\mu$g

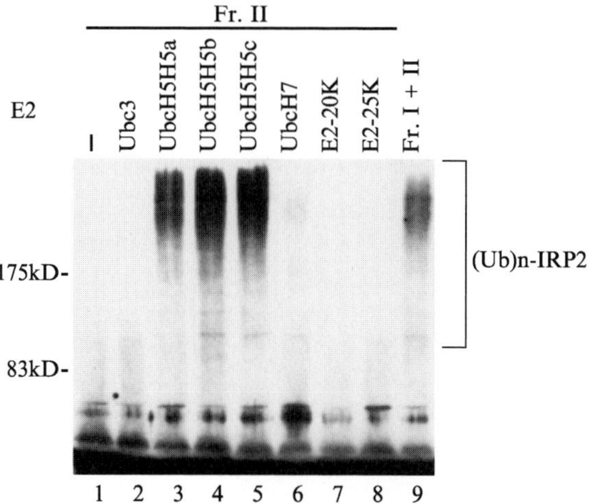


FIG. 2. Oxidized IRP2 was incubated with Fr. II in the absence (– ; lane 1) or presence of recombinant E2s (lanes 2–8) or Fr I (lane 9) as indicated. IRP2 samples were precipitated with an anti-*myc* antibody and resolved on 6% SDS–PAGE, followed by blotting with antiubiquitin to detect ubiquitinated IRP2.

ubiquitin; 0.5 $\mu$g ubiquitin aldehyde; 100 ng E1; 200 ng E2s; E3 sources (1$\mu$g of purified IDD domain from *E. coli* or baculovirus-infected insect cells); and ATP and an ATP regeneration system.

Oxidized IRP2 was efficiently ubiquitinated in the presence of the insect cell-derived IDD domain, but not in the presence of the bacterially derived IDD domain (Fig. 3). Because insect cells, but not bacteria, possess the ubiquitin system, these assays suggested that an E3 for oxidized IRP2 was copurified with the IDD domain from insect cell lysates.

### *Protocol 2.* In Vitro *Ubiquitin Conjugation Assay of Heme-Loaded IRP2*

It has been shown that the iron bound to IRP2 is contained in a heme moiety (Yamanaka *et al.*, 2003). Additionally, we isolated HOIL-1 ubiquitin ligase (see next section) and determined that HOIL-1 recognizes heme-loaded IRP2 using the *in vitro* ubiquitination assay with purified E1, E2 and HOIL-1 enzymes. Because we realized that heme-loaded IRP2 is a much better substrate for ubiquitination by HOIL-1 than *in vitro*-oxidized IRP2, we used substrate-specific antibodies to detect high molecular weight ubiquitinated IRP2 instead of anti-ubiquitin detection.

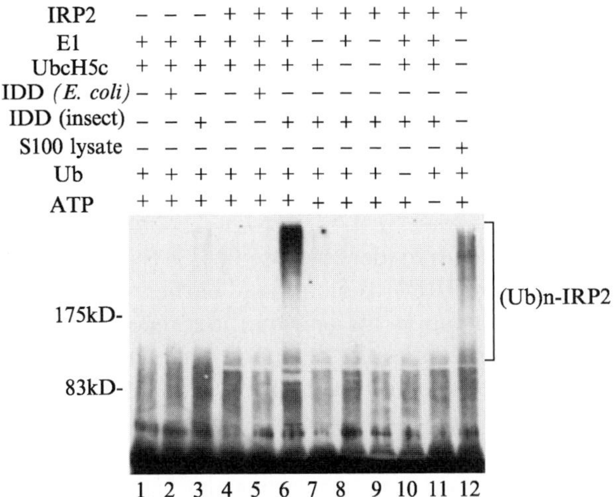


FIG. 3. Oxidized IRP2 (lanes 4–12) or buffer control (lanes 1–3) were incubated without (−) or with (+) ATP and an ATP-regenerating system (ATP), ubiquitin (Ub), E1, UbcH5c, and the IDD domain (IDD) purified from *E. coli*, or insect cells as indicated. IRP2 samples were precipitated with an anti-*myc* antibody and resolved on 6% SDS–PAGE, followed by blotting with anti-ubiquitin to detect ubiquitinated IRP2.

*Reagents*

Heme-loaded IRP2
GST-ubiquitin (BIOMOL, Exeter, UK)
Purified mouse E1
Purified UbcH5c
Purified HOIL-1
ATP and an ATP regeneration system (0.5 m*M* ATP, 10 m*M* creatine phosphate, 10 μg creatine phosphokinase)
4× SDS sample buffer (200 m*M* Tris-Cl, pH 6.8, 8% SDS, 400 m*M* DTT, 40% glycerol, 0.2 % bromphenol blue)
Anti-*myc* antibody (9E10) (Santa Cruz Biotechnology, Santa Cruz, CA)

*Procedure*

1. Prepare a 20-μl reaction mixture containing 1 μg heme-loaded IRP2; 20 m*M* Tris-Cl, pH 7.5; 5 m*M* $MgCl_2$; 1 m*M* DTT; 5 μg GST-ubiquitin; 100 ng E1; 50 ng UbcH5c; 1 μg HOIL-1; and ATP and an ATP regenerating system.

2. Incubate at 37° for 20 min.
3. Stop the reactions by the addition of 7 μl of 4× SDS sample buffer.
4. Separate samples on a 5% SDS–PAGE gel and Western blotting with anti-*myc* antibody.

## Method Used to Identify an E3 Ubiquitin Ligase Specific for Oxidized IRP2

In general, it is difficult to identify specific E3s for a particular substrate. However, in the case of IRP2, as mentioned earlier, we know that the IDD domain is the E3 recognition site and that it is recognized only when it is oxidized. This is consistent with the report that in low oxygen, IRP2 is stable even in the presence of iron (Hanson *et al.*, 1999).

Thus, we could formulate two approaches to identify the E3 specific for oxidized IRP2: (1) purify proteins that bind to the oxidized IDD domain or (2) use a yeast two-hybrid screening using the IDD domain as bait. It is difficult to control the oxidation state of IRP2 *in vitro* and mildly oxidized IRP2 is a good substrate for the E3, but strongly oxidized IRP2 is not (unpublished observations). This suggested that the first approach might not be suitable for E3 purification. Additionally, it is known that yeast cells can grow anaerobically in the presence of glucose as a carbon source if ergosterol and Tween 80 are added to the culture. We hypothesized that the IDD domain would not be oxidized in yeast cells cultured anaerobically, but would be oxidized in cells cultured aerobically. Therefore, we established a differential two-hybrid screening in which yeast cells were cultured in either aerobic or anaerobic conditions using the IDD domain as bait to identify E3s specific for oxidized IRP2.

### *Protocol for a Differential Yeast Two-Hybrid Screening*

#### *Reagents and Instruments*

AH109 cells

pGBKT7-IDD [pGBKT7 (BD Biosciences, Palo Alto, CA) subcloned with a cDNA fragment encoding the IDD domain]

A cDNA library of the human kidney cancer cell line UOK111 constructed in pGADT7 (prepared in our laboratory)

SD agar plates lacking histidine, tryptophan, and leucine, but containing 2.5 m*M* 3-aminotriazole

SD agar plates lacking histidine, tryptophan, and leucine, but containing 2.5 m*M* 3-aminotriazole, 40 μg/ml ergosterol, and 0.2% Tween 80

Anaerobic chamber (Coy Laboratory Products, Grass Lake, MI)

*Procedure*

1. Transform AH109 cells with pGBKT7-IDD using the lithium acetate method.
2. Sequentially transform a cDNA library of the human kidney cancer cell line UOK111 constructed in pGADT7 (prepared in our laboratory) into AH109 cells carrying pGBKT7-IDD.
3. Plate transformants onto SD agar plates lacking histidine, tryptophan, and leucine, but containing 2.5 m*M* 3-aminotriazole.
4. Incubate at 30° for 4 days aerobically.
5. Pick $His^+$, $Trp^+$, and $Leu^+$ colonies and plate them onto SD agar plates lacking histidine, tryptophan, and leucine, but containing 2.5 m*M* 3-aminotriazole.
6. Incubate at 30° for 3 days aerobically.
7. Transfer colonies to SD agar plate lacking histidine, tryptophan, and leucine, but containing 2.5 m*M* 3-aminotriazole, 40 $\mu$g/ml ergosterol, and 0.2% Tween 80 by replica plating.
8. Incubate at 30° for 4 days either aerobically or anaerobically in an anaerobic chamber.
9. Isolate colonies that can grow in the aerobic condition but not in the anaerobic condition.

One of the isolated clones, Clone H3, contains the entire coding region of a RING finger protein. Because the RING finger motif is one of the common structural features of ubiquitin ligases, we designate the clone as heme-oxidized IRP2 ubiquitin ligase–1 (HOIL-1) and are conducting further analysis of the protein. HOIL-1 has been cloned independently as a protein associated with the hepatitis B virus X protein (XAP3) (Cong *et al.*, 1997), protein kinase C (RBCK) (Tokunaga *et al.*, 1998), or UbcM4 (UIP28) (Martinez-Noel *et al.*, 1999).

## Concluding Remarks

The methods described here were used to identify HOIL-1 as an ubiquitin ligase specific for oxidation-induced ubiquitination of IRP2, but these methods should be broadly applicable to the isolation of substrate-specific E3s.

Isolation of the HOIL-1 ubiquitin ligase as an E3, which ubiquitinates physiologically oxidized protein, suggests that the ubiquitin system is involved in degrading oxidized proteins. Analysis of the function of HOIL-1 will shed more light on this issue.

## References

Benaroudj, N., Tarcsa, E., Cascio, P., and Goldberg, A. L. (2001). The unfolding of substrates and ubiquitin-independent protein degradation by proteasomes. *Biochimie* **83,** 311–318.

Chen, Z., Hagler, J., Palombella, V. J., Melandri, F., Scherer, D., Ballard, D., and Maniatis, T. (1995). Signal-induced site-specific phosphorylation targets I kappa B alpha to the ubiquitin-proteasome pathway. *Genes Dev.* **9,** 1586–1597.

Ciechanover, A., Heller, H., Elias, S., Haas, A. L., and Hershko, A. (1980). ATP-dependent conjugation of reticulocyte proteins with the polypeptide required for protein degradation. *Proc. Natl. Acad. Sci. USA* **77,** 1365–1368.

Cong, Y. S., Yao, Y. L., Yang, W. M., Kuzhandaivelu, N., and Seto, E. (1997). The hepatitis B virus X-associated protein, XAP3, is a protein kinase C-binding protein. *J. Biol. Chem.* **272,** 16482–16489.

Grune, T., Merker, K., Sandig, G., and Davies, K. J. (2003). Selective degradation of oxidatively modified protein substrates by the proteasome. *Biochem. Biophys. Res. Commun.* **305,** 709–718.

Hanson, E. S., Foot, L. M., and Leibold, E. A. (1999). Hypoxia post-translationally activates iron-regulatory protein 2. *J. Biol. Chem.* **274,** 5047–5052.

Hentze, M. W., Muckenthaler, M. U., and Andrews, N. C. (2004). Balancing acts: Molecular control of mammalian iron metabolism. *Cell* **117,** 285–297.

Hershko, A., Heller, H., Eytan, E., and Reiss, Y. (1986). The protein substrate binding site of the ubiquitin-protein ligase system. *J. Biol. Chem.* **261,** 11992–11999.

Iwai, K. (2003). An ubiquitin ligase recognizing a protein oxidized by iron: Implications for the turnover of oxidatively damaged proteins. *J. Biochem. (Tokyo)* **134,** 175–182.

Iwai, K., Drake, S. K., Wehr, N. B., Weissman, A. M., La Vaute, T., Minato, N., Klausner, R. D., Levine, R. L., and Rouault, T. A. (1998). Iron-dependent oxidation, ubiquitination, and degradation of iron regulatory protein 2: Implications for degradation of oxidized proteins. *Proc. Natl. Acad. Sci. USA* **95,** 4924–4928.

Iwai, K., Klausner, R. D., and Rouault, T. A. (1995). Requirements for iron-regulated degradation of the RNA binding protein, iron regulatory protein 2. *EMBO J.* **14,** 5350–5357.

Iwai, K., Yamanaka, K., Kamura, T., Minato, N., Conaway, R. C., Conaway, J. W., Klausner, R. D., and Pause, A. (1999). Identification of the von Hippel-Llindau tumor-suppressor protein as part of an active E3 ubiquitin ligase complex. *Proc. Natl. Acad. Sci. USA* **96,** 12436–12441.

Klausner, R. D., Rouault, T. A., and Harford, J. B. (1993). Regulating the fate of mRNA: The control of cellular iron metabolism. *Cell* **72,** 19–28.

La Vaute, T., Smith, S., Cooperman, S., Iwai, K., Land, W., Meyron-Holtz, E., Drake, S. K., Miller, G., Abu-Asab, M., Tsokos, M., Switzer, R. R., Grinberg, A., Love, P., Tresser, N., and Rouault, T. A. (2001). Targeted deletion of the gene encoding iron regulatory protein-2 causes misregulation of iron metabolism and neurodegenerative disease in mice. *Nature Genet.* **27,** 209–214.

Martinez-Noel, G., Niedenthal, R., Tamura, T., and Harbers, K. (1999). A family of structurally related RING finger proteins interacts specifically with the ubiquitin-conjugating enzyme UbcM4. *FEBS Lett.* **454,** 257–261.

Meyron-Holtz, E. G., Ghosh, M. C., Iwai, K., La Vaute, T., Brazzolotto, X., Berger, U. V., Land, W., Ollivierre-Wilson, H., Grinberg, A., Love, P., and Rouault, T. A. (2004). Genetic ablations of iron regulatory proteins 1 and 2 reveal why iron regulatory protein 2 dominates iron homeostasis. *EMBO J.* **23,** 386–395.

Pickart, C. M., and Rose, I. A. (1986). Mechanism of ubiquitin carboxyl-terminal hydrolase: Borohydride and hydroxylamine inactivate in the presence of ubiquitin. *J. Biol. Chem.* **261,** 10210–10217.

Shringarpure, R., Grune, T., Mehlhase, J., and Davies, K. J. (2003). Ubiquitin conjugation is not required for the degradation of oxidized proteins by proteasome. *J. Biol. Chem.* **278,** 311–318.

Tokunaga, C., Kuroda, S., Tatematsu, K., Nakagawa, N., Ono, Y., and Kikkawa, U. (1998). Molecular cloning and characterization of a novel protein kinase C-interacting protein with structural motifs related to RBCC family proteins. *Biochem. Biophys. Res. Commun.* **244,** 353–359.

Yamanaka, K., Ishikawa, H., Megumi, Y., Tokunaga, F., Kanie, M., Rouault, T. A., Morishima, I., Minato, N., Ishimori, K., and Iwai, K. (2003). Identification of the ubiquitin-protein ligase that recognizes oxidized IRP2. *Nature Cell Biol.* **5,** 336–340.

# [22] Purification and Assay of the Chaperone-Dependent Ubiquitin Ligase of the Carboxyl Terminus of Hsc70-Interacting Protein

*By* SHIGEO MURATA, MICHIKO MINAMI, and YASUFUMI MINAMI

## Abstract

It is notable that both chaperone and ubiquitin-proteasome systems are required for the removal of aberrant cellular proteins to ensure protein homeostasis in cells. However, the entity that links the two systems had remained elusive. The carboxyl terminus of Hsc70-interacting protein (CHIP), originally identified as a cochaperone of Hsc70, has both a TPR motif and a U-box domain. The TPR motif associates with Hsp70 and Hsp90, whereas the U-box domain executes ubiquitin ligase activity. Thus, CHIP is an ideal molecule, acting as a protein quality control ubiquitin ligase that selectively leads abnormal proteins recognized by molecular chaperones to degradation by the proteasome. This chapter describes methods of analyzing chaperone-dependent ubiquitin ligase activity of CHIP using firefly luciferase as a model substrate.

## Introduction

Over 30% of newly synthesized cellular proteins are discarded without being folded properly (Schubert *et al.*, 2000). Even when proteins are normally formed into tertiary structures, macromolecular crowding in the cytosol increases the spontaneous denaturation and consequently the likelihood of partially folded or unfolded proteins to undergo off-pathway

METHODS IN ENZYMOLOGY, VOL. 398

0076-6879/05 $35.00
DOI: 10.1016/S0076-6879(05)98022-1

reactions, such as aggregation. In addition, environmental stresses such as heat, oxidation (i.e., formation of free radicals), and ultraviolet could result in the production of impaired proteins. However, those proteins with nonnative or aberrant structures are rapidly removed inside the cells. The cellular apparatus monitoring the "normality" of protein architectures in the cell is usually referred to as "the protein quality control system" (Wickner *et al.*, 1999). This monitoring machinery is considered an integral cellular component involved in maintaining cell survival and homeostasis because it prevents the accumulation of abnormal proteins and the formation of toxic inclusion bodies followed by cell death, which is seen in various neurodegenerative diseases. Indeed, there are growing lines of evidence that failure of the protein quality control system leads to neurodegeneration (Sherman and Goldberg, 2001). Hence, it is important for our understanding of the molecular basis underlying neurodegenerative diseases to characterize the quality control mechanism of the cell. At present, the ubiquitin-proteasome system is considered to play a key role in protein homeostasis by catalyzing the immediate destruction of misfolded or impaired (i.e., abnormal) cellular proteins (Bercovich *et al.*, 1997; Jungmann *et al.*, 1993; Lee *et al.*, 1996). "Abnormal proteins" tend to expose hydrophobic regions, which are then recognized by molecular chaperones such as Hsp70 and Hsp90 (Terasawa *et al.*, 2005). Molecular chaperones prevent these abnormal proteins from irreversibly aggregating and assist in their reversion to a properly folded and functional state. However, when these chaperones fail to refold the abnormal proteins, the ubiquitin-proteasome system disposes of unfolded, nonfunctional proteins. Therefore, it is rational to predict that an E3 protein exists that is associated or cooperates with molecular chaperones. A 35-kDa protein called CHIP (carboxyl terminus of Hsc70-interacting protein) has been identified as a candidate for a ubiquitin ligase that plays a role in protein quality control (Connell *et al.*, 2001; Meacham *et al.*, 2001). CHIP possesses two characteristic domains: one is a TPR domain at its amino terminus and the other is a U-box domain at its carboxyl terminus (Ballinger *et al.*, 1999). TPR domains have been reported to interact with multiple proteins, including heat shock proteins, and are present in various molecules, such as phosphatase 5, cyclophilin 40, FKBP52, Hip (Hsc70-interacting protein), and Hop (Hsc70-Hsp90 organizing protein). CHIP is indeed associated with carboxyl termini of Hsp70, Hsc70, and Hsp90 through its TPR and adjacent charged domains. The important and unique feature of CHIP not found in other TPR-containing proteins is the U-box domain in the carboxyl-terminal region. The U-box domain has a tertiary structure highly similar to the RING finger domain and has shown to possess ubiquitin ligase activity. Therefore, there is a great interest in whether CHIP serves as a link between the chaperone system and the ubiquitin-proteasome system. In cells transfected with CHIP and glucocorticoid receptor (GR), one of the well-known client proteins of Hsp90, CHIP promoted ubiquitylation of GR (Connell *et al.*, 2001). Another study indicated that CHIP is also involved in the quality control of endoplasmic reticulum resident proteins. In cells, CHIP ubiquitylated the cystic fibrosis transmembrane conductance regulator (CFTR), a well-known Hsc70 substrate, in cooperation with Hsc70 in both U-box- and TPR motif-dependent manners (Meacham *et al.*, 2001). Finally, the ubiquitylation assay system, completely recapitulated *in vitro*, using firefly luciferase as a model substrate, directly demonstrated that CHIP is a chaperone-dependent E3 and selectively polyubiquitylates unfolded proteins (Murata *et al.*, 2001). Luciferase was ubiquitylated by CHIP, together with the E2 enzyme Ubc4 or 5, only when it was heat denatured in the presence of Hsp90 or Hsc70/Hsp40 complex, and CHIP did not ubiquitylate native luciferase or luciferase that was denatured in the absence of these molecular chaperones. Thus, CHIP is a "quality control E3" and recognizes the nonnative state of polypeptides with the assistance of molecular chaperones, thereby ubiquitylating a selected set of target proteins. This chapter describes methods of purification of chaperones and enzymes for ubiquitylation, as well as heat shock-dependent ubiquitylation of firefly luciferase by CHIP.

## Materials

### *Preparation of Hsp90*

We obtain both Hsp90 and Hsp70 from a single source of starting material in a considerable amount, as described by Minami *et al.* (2000).

Bovine or porcine brains (~400 g) are rinsed with cold water and buffer A (20 m*M* Tris-HCl, pH 7.5, 20 m*M* NaCl, 1 m*M* EDTA) and are then homogenized in 400 ml of buffer A containing 0.1% aprotinin and 0.5 m*M* phenylmethylsulfonyl fluoride (PMSF) with a blender. The homogenate is centrifuged at 9000*g* for 15 min at 4°, and the supernatant is ultracentrifuged at 90,000*g* for 30 min at 4°. The resulting crude lysate is mixed on ice for 1 h with 150 ml of DEAE-cellulose DE–52 (Whatman), which has been equilibrated with buffer A. The resin is washed with 1 liter of buffer A and packed into a column. Then it is washed further with 400 ml of a buffer solution (20 m*M* Tris-HCl, pH 7.5, 50 m*M* NaCl), followed by elution with a linear gradient of 50–500 m*M* NaCl in 20 m*M* Tris-HCl, pH 7.5. Because Hsp70 and Hsp90 are eluted from the column consecutively in this order, both peak fractions determined by SDS–PAGE are collected separately.

Fractions containing Hsp90 are applied directly to a hydroxyapatite HTP (Bio-Rad) column (25 ml) equilibrated with 20 m*M* potassium phosphate, pH 7.5. After washing with 50 ml of the same buffer, the proteins are eluted with a linear gradient of 20–300 m*M* potassium phosphate, pH 7.5. Peak fractions containing Hsp90 are concentrated by ultrafiltration with an Amicon PM–30 membrane and centrifuged briefly to remove aggregates after which the supernatant is subjected to gel filtration through a Sephacryl S–300 column (2.5 $\times$ 100 cm; Amersham Biosciences) equilibrated with buffer A. The obtained fractions containing Hsp90 are loaded onto a Q-Sepharose column (5 ml; Amersham Biosciences), which has been equilibrated with a buffer solution (50 m*M* Tris-HCl, pH 7.5, 10 m*M* NaCl), and elution with a linear gradient of 10 m*M*–1 *M* NaCl in 50 m*M* Tris-HCl, pH 7.5, is performed. Peak fractions are combined and condensed by ultrafiltration with an Amicon PM–30 membrane and dialyzed against a buffer solution (20 m*M* potassium phosphate, pH 7.5, 50 m*M* NaCl, 0.1 m*M* EDTA). Approximately 20 mg of Hsp90 is obtained consistently.

*Preparation of Hsp70*

Fractions containing Hsp70 eluted from a DE–52 column as mentioned earlier are applied to an HTP column (60 ml) equilibrated with 10 m*M* potassium phosphate, pH 7.5. After washing with 60 ml of the same buffer, the proteins are eluted with a linear gradient of 10–250 m*M* potassium phosphate, pH 7.5. Fractions containing Hsp70 determined by SDS–PAGE are supplemented with 3 m*M* $MgCl_2$ and loaded onto an ATP-agarose column (20 ml; Sigma C–8 linkage), which has been equilibrated with buffer B (20 m*M* Tris-HCl, pH 7.5, 20 m*M* NaCl, 3 m*M* $MgCl_2$). After washing successively with buffer B, buffer B containing 500 m*M* NaCl, and buffer B again, the column is eluted with buffer B containing 3 m*M* ATP. After peak fractions from an ATP-agarose column are applied to a DE–52 column (10 ml), the column is washed with a buffer solution (20 m*M* Tris-HCl, pH 7.5, 20 m*M* NaCl) and eluted with a linear gradient of 20–500 m*M* NaCl in 20 m*M* Tris-HCl, pH 7.5. Peak fractions containing Hsp70 are combined and condensed by ultrafiltration with an Amicon PM–30 membrane, followed by dialysis against a buffer solution (20 m*M* HEPES, pH 7.4, 25 m*M* KCl, 2 m*M* $MgCl_2$). About 5 mg of Hsp70 is obtained.

*Preparation of Hsp40*

Hsp40 is recombinantly expressed and purified. The expression plasmid pET/Hsp40 for human Hsp40 as described by Minami *et al.* (1996) is transformed into BL21(DE3) cells and incubated with 0.4 m*M* isopropyl–1-thio-$\beta$-D-galactoside (IPTG) for 2 h. Cells from the 1-liter culture are disrupted in a French pressure cell (SML Instruments, Inc.) in 25 ml of buffer A containing 1 m*M* PMSF. The lysate is mixed with 80 ml of DEAE-Sephacel (Amersham Biosciences), which has been equilibrated with buffer A on ice for 1 h. The unbound material is collected, and the resin is washed twice with 100 ml of buffer A. The flow through and wash fractions containing Hsp40 are combined and loaded onto an HTP column (10 ml) equilibrated with 10 m*M* potassium phosphate, pH 7.5. The column is washed with 30 ml of the same buffer, and Hsp40 is eluted with a linear gradient of 10–300 m*M* potassium phosphate, pH 7.5. Peak fractions are diluted with an equal volume of water and passed through a DEAE-Sephacel column (10 ml) followed by rechromatography on an HTP column (10 ml). Purified Hsp40 is concentrated by ultrafiltration with an Amicon PM–30 membrane and dialyzed against a buffer solution (20 m*M* HEPES, pH 7.4, 100 m*M* NaCl, 0.1 m*M* EDTA). Finally, 10–20 mg of Hsp40 is obtained.

*Preparation of Enzymes for Ubiquitylation Assay*

Recombinant His6-mouse E1 (Uba1) is produced from baculovirus-infected Sf9 insect cells. Cells are lysed with buffer containing 25 m*M* Tris-HCl (pH 8.0), 150 m*M* NaCl, and 0.5% Triton X–100 with complete (EDTA free) protease inhibitor cocktail (Roche) and clarified by centrifugation at 15,000*g* for 10 min at 4°. The supernatant is passed through a column packed with Ni-NTA agarose (Qiagen) five times repeatedly to allow binding to the resin. The resin is washed with buffer containing 20 m*M* sodium phosphate (pH 7.4), 500 m*M* NaCl, and 10 m*M* imidazole. Bound proteins are eluted with buffer containing 20 m*M* sodium phosphate (pH 7.4), 150 m*M* NaCl, and 300 m*M* imidazole. Eluted proteins are dialyzed against buffer containing 25 m*M* Tris-HCl (pH 7.5), 150 m*M* NaCl, 10 m*M* 2-ME, and 10% glycerol. About 2 mg of His-E1 is obtained from Sf9 cells cultured in 500 ml of Grace's insect cell culture medium supplemented with 5% of fetal bovine serum.

Recombinant His-UbcH5c, GST-Ub, and GST-mouse CHIP are produced in *Escherichia coli*. UbcH5c is subcloned into the pET28a vector (Novagen) to produce a recombinant enzyme with a His6 tag at the N terminus. Ubiquitin and CHIP are expressed as GST fusion protein using the pGEX6p–1 vector (Amersham). *E. coli* BL21(DE3)–Codonplus-RIL (Stratagene) are transformed with the plasmids. Transformed cells are grown at 37° in LB medium containing 50 $\mu$g/ml kanamycin for UbcH5c or 100 $\mu$g/ml ampicillin for Ub and CHIP. Protein expression is induced with 0.2 m*M* IPTG, and cells are incubated further for 3 h at 37°.

For purification of His-UbcH5c, cells are lysed with buffer containing 10 m*M* Tris-HCl (pH 8.0), 300 m*M* NaCl, 1% Triton X–100, 10 m*M* 2-ME, 1 mg/ml lysozyme, and 20 m*M* imidazole and are incubated on ice for 30 min. Lysates are sonicated to reduce the viscosity and centrifuged at 30,000*g* for 30 min to clarify the solution. The supernatant is passed through a column packed with Ni-NTA agarose five times to allow binding. The resin is washed with buffer containing 10 m*M* Tris-HCl (pH 8.0), 300 m*M* NaCl, 1% Triton X–100, 10 m*M* 2-ME, and 20 m*M* imidazole. Bound proteins are eluted with buffer containing 10 m*M* Tris-HCl (pH 8.0), 150 m*M* NaCl, and 300 m*M* imidazole. Eluted proteins are dialyzed against buffer containing 25 m*M* Tris-HCl (pH 7.5), 150 m*M* NaCl, 10 m*M* 2-ME, and 10% glycerol. About 20 mg of His-UbcH5c is obtained from cells cultured in 500 ml of LB media.

For purification of GST-Ub or GST-CHIP, cells are lysed with buffer containing 50 m*M* Tris-HCl (pH 8.0), 500 m*M* NaCl, 1% Triton X–100, 10 m*M* 2-ME, and 1 mg/ml lysozyme and are sonicated. The supernatant (centrifugation at 30,000*g* for 30 min) is passed through a column packed with glutathione-Sepharose resin (Amersham) five times. The resin is washed with buffer containing 50 m*M* Tris-HCl (pH 7.5), 500 m*M* NaCl, 1% Triton X–100, and 10 m*M* 2-ME. Bound proteins are eluted with buffer containing 20 m*M* glutathione and 100 m*M* Tris-HCl (pH 8.5). Eluted proteins are dialyzed against buffer containing 25 m*M* Tris-HCl (pH 8.0), 10 m*M* 2-ME, and 10% glycerol. About 25 mg of GST-CHIP and 50 mg of GST-Ub are obtained from cells cultured in 500 ml of LB media.

### *Other Materials*

Firefly luciferase (Roche) is dissolved in buffer containing 20 m*M* HEPES (pH 7.4), 50 m*M* NaCl, and 1 m*M* DTT at a concentration of 4 mg/ml, aliquoted, and stored at –80° until use. Repeated freeze–thaw should be avoided as firefly luciferase is labile. Use a fresh aliquot in every experiment. Bovine Ub and the antiluciferase antibody are from Sigma.

## Heat Denaturation of Firefly Luciferase

Heat treatment of firefly luciferase is carried out as described previously (Minami *et al.*, 1996, 2000). For assay of Hsp90-dependent ubiquitylation, 0.26 $\mu M$ of luciferase is heated at 45° for 5 min with or without 11 $\mu M$ of Hsp90 in 30 m*M* MOPS (pH 7.2) buffer containing 2 m*M* DTT. For assay of Hsp70-dependent ubiquitylation, 0.26 $\mu M$ of luciferase is heated at 43° for 10 min with 4.7 $\mu M$ of Hsp70 and/or 3.2 $\mu M$ of Hsp40 in the presence or absence of 4 m$M$ ATP in 50 m$M$ MOPS buffer (pH 7.2) containing 50 m$M$ KCl, 3 m$M$ $MgCl_2$, and 2 m$M$ DTT. After heating, they are chilled quickly in an ice bath.

## Ubiquitylation Assay

One microliter of the aforementioned heat-denatured or native luciferase is incubated for 2 h at 30° in a reaction volume of 50 $\mu$l containing 50 ng of recombinant mouse E1, 0.5 $\mu$g of UbcH5c, 2 $\mu$g of GST-CHIP, 5 $\mu$g of bovine Ub or 10 $\mu$g of GST-Ub, 50 m$M$ Tris-HCl (pH 7.5), 2m$M$ $MgCl_2$, 1 m$M$ DTT, and 4 m$M$ ATP. After terminating the reaction by the addition of 20 $\mu$l sample buffer for SDS–PAGE, the boiled supernatant is separated by 4–12% SDS–PAGE and visualized by Western blotting with antiluciferase antibody for detection of the ubiquitylating activity of CHIP (Fig. 1).

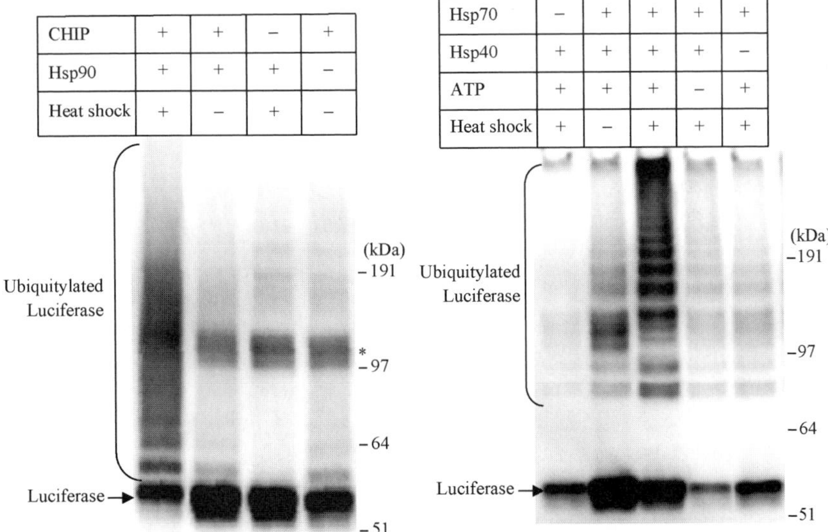


FIG. 1. Chaperone-dependent ubiquitylation of heat-denatured luciferase by CHIP. The effects of chaperones (Hsp90, Hsp70, Hsp40) and heat shock treatment on CHIP-dependent ubiquitylation of luciferase were assayed by the *in vitro* reconstituted system. (Left) Hsp90-dependent ubiquitylation of heat-denatured luciferase using bovine ubiquitin. (Right) Hsp70/Hsp40/ATP-dependent ubiquitylation of heat-denatured luciferase using GST-ubiquitin instead of bovine ubiquitin. Ubiquitylation of luciferase was detected by Western blotting with an antiluciferase antibody. Unmodified luciferase is shown by arrows. Ubiquitylated luciferase with a high molecular mass is shown by square brackets. An asterisk indicates nonspecific bands.

Comments

We used firefly luciferase as a model substrate for CHIP because the state of this protein, whether native or denatured and captured by chaperones, can be manipulated easily by established methods (Minami *et al.*, 1996, 2000). However, the requirement of heat treatment depends on substrates. While heat shock treatment is required for firefly luciferase to be captured by Hsp90 or Hsp70 and ubiquitylated by CHIP, CFTR undergoes ubiquitylation by CHIP in the presence of Hsp70 and Hsp40 (Hdj2) without heat treatment (Younger *et al.*, 2004). nNOS is also shown to be ubiquitylated with the help of Hsp70 and Hsp40 without heat treatment (Peng *et al.*, 2004). CFTR and nNOS are prone to misfolding in the normal intracellular environment and thus are natural substrates of Hsp70.

Although we used UbcH5c as E2 enzyme in this assay, CHIP can cooperate with all the Ubc4/5 family members (Ubc4, UbcH5a, b, c). Interestingly, CHIP is active as a ubiquitin ligase only when it forms a homodimer via coiled-coil domains located between the TPR and U-box domains (Nikolay *et al.*, 2004), although the mechanism for its activation is unclear. Finally, we illustrate the scheme of the mechanism of chaperone-dependent ubiquitylation of unfolded/misfolded substrates by CHIP in Fig. 2.

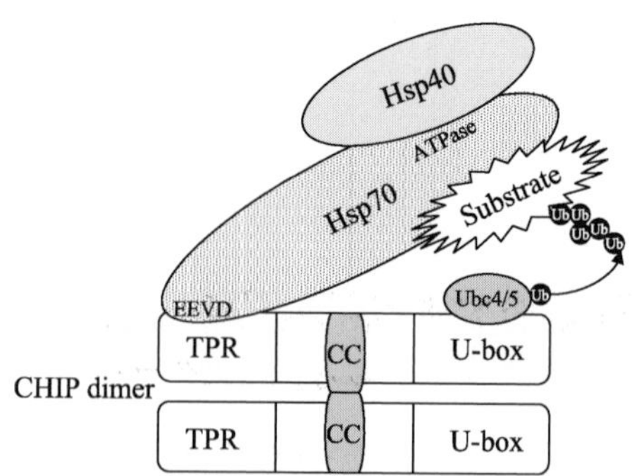


FIG. 2. A schematic representation of chaperone-dependent ubiquitylation of unfolded or misfolded proteins by CHIP. CHIP is active as a homodimer formed via coiled-coil domains. CHIP associates with the carboxyl-terminal EEVD sequence of Hsp70 through its TPR motif at the N terminus. Unfolded or misfolded proteins are captured by Hsp70 in a manner dependent on both Hsp40 and ATP. Specific E2 enzymes such as Ubc4 and Ubc5s are recruited on the U-box domain of CHIP and then transfer activated ubiquitins to the Hsp70-loaded substrate proteins.

## References

Ballinger, C. A., Connell, P., Wu, Y., Hu, Z., Thompson, L. J., Yin, L. Y., and Patterson, C. (1999). Identification of CHIP, a novel tetratricopeptide repeat-containing protein that interacts with heat shock proteins and negatively regulates chaperone functions. *Mol. Cell. Biol.* **19,** 4535–4545.

Bercovich, B., Stancovski, I., Mayer, A., Blumenfeld, N., Laszlo, A., Schwartz, A. L., and Ciechanover, A. (1997). Ubiquitin-dependent degradation of certain protein substrates *in vitro* requires the molecular chaperone Hsc70. *J. Biol. Chem.* **272,** 9002–9010.

Connell, P., Ballinger, C. A., Jiang, J., Wu, Y., Thompson, L. J., Hohfeld, J., and Patterson, C. (2001). The co-chaperone CHIP regulates protein triage decisions mediated by heat-shock proteins. *Nature Cell Biol.* **3,** 93–96.

Jungmann, J., Reins, H. A., Schobert, C., and Jentsch, S. (1993). Resistance to cadmium mediated by ubiquitin-dependent proteolysis. *Nature* **361,** 369–371.

Lee, D. H., Sherman, M. Y., and Goldberg, A. L. (1996). Involvement of the molecular chaperone Ydj1 in the ubiquitin-dependent degradation of short-lived and abnormal proteins in *Saccharomyces cerevisiae*. *Mol. Cell. Biol.* **16,** 4773–4781.

Meacham, G. C., Patterson, C., Zhang, W., Younger, J. M., and Cyr, D. M. (2001). The Hsc70 co-chaperone CHIP targets immature CFTR for proteasomal degradation. *Nature Cell Biol.* **3,** 100–105.

Minami, Y., Hohfeld, J., Ohtsuka, K., and Hartl, F. U. (1996). Regulation of the heat-shock protein 70 reaction cycle by the mammalian DnaJ homolog, Hsp40. *J. Biol. Chem.* **271,** 19617–19624.

Minami, Y., Kawasaki, H., Minami, M., Tanahashi, N., Tanaka, K., and Yahara, I. (2000). A critical role for the proteasome activator PA28 in the Hsp90-dependent protein refolding. *J. Biol. Chem.* **275,** 9055–9061.

Murata, S., Minami, Y., Minami, M., Chiba, T., and Tanaka, K. (2001). CHIP is a chaperone-dependent E3 ligase that ubiquitylates unfolded protein. *EMBO Rep.* **2,** 1133–1138.

Nikolay, R., Wiederkehr, T., Rist, W., Kramer, G., Mayer, M. P., and Bukau, B. (2004). Dimerization of the human E3 ligase CHIP via a coiled-coil domain is essential for its activity. *J. Biol. Chem.* **279,** 2673–2678.

Peng, H. M., Morishima, Y., Jenkins, G. J., Dunbar, A. Y., Lau, M., Patterson, C., Pratt, W. B., and Osawa, Y. (2004). Ubiquitylation of neuronal nitric-oxide synthase by CHIP, a chaperone-dependent E3 ligase. *J. Biol. Chem.* **279,** 52970–52977.

Schubert, U., Anton, L. C., Gibbs, J., Norbury, C. C., Yewdell, J. W., and Bennink, J. R. (2000). Rapid degradation of a large fraction of newly synthesized proteins by proteasomes. *Nature* **404,** 770–774.

Sherman, M. Y., and Goldberg, A. L. (2001). Cellular defenses against unfolded proteins: A cell biologist thinks about neurodegenerative diseases. *Neuron* **29,** 15–32.

Terasawa, K., Minami, M., and Minami, Y. (2005). Constantly updated knowledge of Hsp90. *J. Biochem.* **137,** 443–447.

Wickner, S., Maurizi, M. R., and Gottesman, S. (1999). Posttranslational quality control: Folding, refolding, and degrading proteins. *Science* **286,** 1888–1893.

Younger, J. M., Ren, H. Y., Chen, L., Fan, C. Y., Fields, A., Patterson, C., and Cyr, D. M. (2004). A foldable CFTR{Delta}F508 biogenic intermediate accumulates upon inhibition of the Hsc70-CHIP E3 ubiquitin ligase. *J. Cell Biol.* **167,** 1075–1085.

# [23] Methods for the Functional Genomic Analysis of Ubiquitin Ligases

*By* WEI LI, SUMIT K. CHANDA, IVANA MICIK, and CLAUDIO A. P. JOAZEIRO


## Abstract

Ubiquitin ligases (E3s) are critical components of the ubiquitin-proteasome system as they are the major determinants of specificity in ubiquitin conjugation. The number of predicted E3s in the mammalian genome is exceeding 400 and is represented by two major subfamilies: HECT domain-containing E3s and RING finger-type E3s. Given the size of this protein family and lack of knowledge on the functions of most of these 400 proteins, their functional annotation should benefit from modern genomic tools. This article presents a methodology consisting of the use of a cDNA expression library to identify suppressors of polyglutamine (polyQ)-mediated protein aggregate formation in cells, as an example of a genomic approach to assign functions to E3s. In this screen, we identified novel RING finger-type E3s exhibiting suppressor activity among >50% of all the potential E3s in the mouse and human genomes. This method could be adapted easily to identify E3s that function in other processes and signaling pathways.


## Introduction

The ubiquitin-proteasome system (UPS) is a posttranslational modification machinery that is conserved from lower eukaryotes to humans. There are three major components that drive the conjugation of ubiquitin to target proteins: the ubiquitin-activating enzyme (E1), ubiquitin-conjugating enzymes (E2), and ubiquitin ligases (E3). There is one ubiquitin E1 in the human genome, fewer than 60 E2s, and more than 400 potential E3s. Although it is known that E2s can confer a certain level of specificity toward the substrate, it is the huge variety of E3s that determines the specificity toward the large number of targets of ubiquitination in the cell.

Since the discovery of the HECT domain based on sequence homology to the ubiquitin ligase E6-AP (Huibregtse *et al.*, 1995), other families of proteins harboring distinct motifs have been described to be associated with ubiquitin ligase activity both *in vitro* and *in vivo*. Among them, the

METHODS IN ENZYMOLOGY, VOL. 398

0076-6879/05 $35.00
DOI: 10.1016/S0076-6879(05)98023-3

RING finger motif containing E3s has so far been the largest E3 family described (Joazeiro and Weissman, 2000). RING finger variants are also emerging as potential E3 motifs (Hatakeyama *et al.*, 2001; Lu *et al.*, 2002; Wertz *et al.*, 2004). Despite vast literature and intensive investigation on individual E3s, the majority of them remain poorly annotated and their functions remain unknown. Therefore, a genome-wide approach to understanding the functions of potential E3s is going to provide significant insight into the signaling pathways they are involved in and the mechanisms by which they operate.

The UPS is a major cellular protein degradation pathway and plays an important role in maintaining protein homeostasis. In many types of neurodegenerative diseases, protein homeostasis is disturbed due to the abnormal accumulation of misfolded proteins and the formation of intracellular protein aggregates. Although it is not clear whether aggregate formation is a causative factor for neurodegenerative disease, the phenomenon is tightly associated with degeneration of the neurons and most certainly contributes to disease progression. These aggregates are often found to stain immunopositive for ubiquitin as well as for proteasome subunits and other components of the UPS, such as ubiquitin-binding proteins. It has been proposed that such association with UPS components to aggregates results from a protective mechanism as cells attempt to degrade the misfolded proteins (Taylor *et al.*, 2003). However, aggregate formation is also likely to have deleterious consequences on the cell through sequestration and inhibition of the normal protein functions (Bence *et al.*, 2001; Chai *et al.*, 2002; Steffan *et al.*, 2000). This chapter describes the methodology of using an arrayed mammalian cDNA library to perform an imaging-based screen for the suppressors of poly Q-mediated protein aggregate formation in the neuro2a mouse neuroblastoma cell line. The screen led to the discovery of a novel suppressor E3 that could play a role in neurodegeneration. This cDNA library could be adapted to other types of imaging and reporter assays and could be used as a tool to unveil the different functions of ubiquitin ligases.

## Methods

### *Libraries and HTS Screening Methods*

Genome-scale collections of full-length cDNAs, transcribed short interfering RNAs (shRNAs), and chemically synthesized siRNAs have recently become available publicly. These sources include the Mammalian Genome Collection (MGC) (Strausberg *et al.*, 2002), the Origene TrueClone collection (Origene) (http://www.origene.com), the Cold Spring Harbor mouse short hairpin-transcribed siRNA collection (http://www.openbiosystems.com), the human druggable siRNA collection (http://www.qiagen.com), the Invitrogen Ultimate ORF collection (http://www.invitrogen.com), Riken Fantom Collection (http://www.gsc.riken.go.jp/e/FANTOM/), Genecopoeia's ORFExpress collection (http://www.genecopoeia.com/), and others. These reagent sets contain between 9000 and 25,000 individually addressable nucleic acid entities that encode, or are directed against, substantial portions of the human and mouse transcriptomes. In an effort to facilitate the focused analysis of gene family function, subsets of siRNAs and cDNAs representing restricted protein classes or pathways, such as GPCRs, kinases, ubiquitin-ligases, and apoptosis-related genes, are also accessible through vendors such as Dharmacon (http://www.dharmacon.com/), Guthrie Research Institute (http://www.cdna.org/), and Ambion (http://www.ambion.com/).

These nucleic acids collections are deployed for screening in two predominant formats: microarrays and multiwell plates. In the former, a standard microarray printing apparatus (i.e., PixSys5500 Robotic Arrayer) can be utilized to immobilize samples usually at densities of 9 spots/cm$^2$ (Ziauddin and Sabatini, 2001). Arrays are then desiccated and can be stored for periods of >3 months. Alternatively, tissue culture plates containing either 96 or 384 wells can be utilized for spotting nucleic acid collections employing any number of automated liquid handling devices (i.e., Packard Minitrack) (Chanda *et al.*, 2003). In multiwell plates, cDNAs cloned in expression vectors ready for transfection into mammalian cells are compartmentalized physically, allowing a broader range of phenotypic detection methods. Substantial cost benefits can be extracted by "collapsing" or "pooling" wells of multiwell plates (Berns *et al.*, 2004). However, this may result in a substantial reduction in assay sensitivity and the capacity for rare-event detection, as well as increased difficultly in deconvoluting the identities of putative phenotypic modulators.

Genes and short hairpin RNAs (siRNAs) are typically introduced into mammalian cells through transient transfection or viral transduction. Rapid and parallel transient introduction of nucleic acids is achieved most efficiently through "reverse" or "retro" transfection methodologies (Fig. 1). In this procedure, the transfection reagent (i.e., Fugene or Lipofectamine 2000) is added to cDNAs or siRNAs that have been prearrayed upon the collection matrix. After an appropriate incubation period, cells are added to the reagent/nucleic acid complex to complete the transfection reaction. Detailed methodologies can be found at http://function.gnf.org and http://jura.wi.mit.edu/sabatini_public/reverse_transfection.htm. Large collections of retroviruses and lentiviruses that harbor cDNAs or shRNAs can be manufactured in 96-well plates utilizing similar strategies (Berns *et al.*,

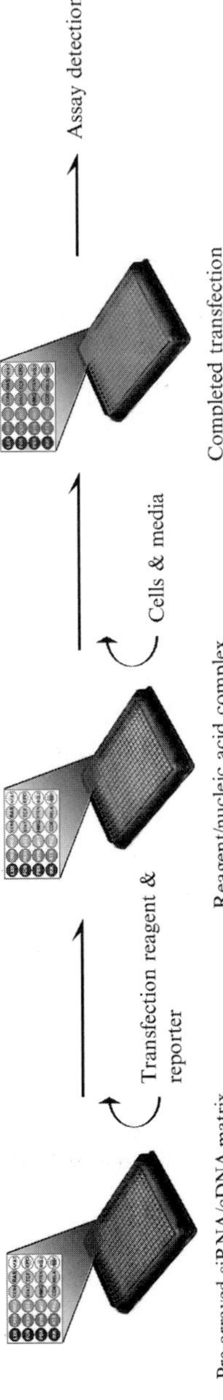


FIG. 1. Multiwell plate reverse transfection method. (See color insert.)

2004; Paddison *et al.*, 2004) and prearrayed in a manner comparable to siRNAs or cDNAs. The large-scale viral transduction of nucleic acids considerably expands the repertoire of cell types amenable to functional profiling analysis by enabling the transfection of cell lines and primary cells that are resistant to transfection by the calcium phosphate method or using liposome-based reagents.

Finally, a number of high-throughput assay detection methodologies are available to measure phenotypes of interest. Largely adapted from high-throughput small molecule screening systems, these instruments can rapidly extract whole-well luminescence, fluorescence intensity, fluorescence polarization, or absorbance measurements in 96-, 384-, and 1536-well plates (Hemmila and Hurskainen, 2002). This approach is ideally suited for analysis of surrogate transcriptional and posttranscriptional reporters such as GFP, luciferase, and $\beta$-galactosidase. More sophisticated technologies permit the investigation of biological events at the level of single cells and include automated microscopy and high-throughput FACS platforms (Carpenter and Sabatini, 2004; Kuckuck *et al.*, 2001). These tools enable the high-throughput measurement of discrete molecular events such as subcellular localization, cellular morphology, cell migration, and cytoskeletal rearrangements, as well as cell surface markers, DNA content, and many other phenotypes (Abraham *et al.*, 2004; Price *et al.*, 2002). Furthermore, this ability to measure single cell events allows for subpopulation analysis and may also result in significant increases in sensitivity.

The MGC and the Origene collections currently contain full-length cDNAs that represent more than 50% of all the potential ubiquitin ligases encoded by human and mouse genomes (S. Batalov and C. Joazeito, unpublished observations). Collectively, the expanding E3 cDNA and siRNA collections, the development of more efficient high-throughput delivery strategies, and evolving detection technologies will facilitate the execution of cellular assays in a miniaturized and highly parallel format, which will further enable annotation of E3 functions on a scale that cannot be achieved by conventional methods. The following sections present an imaging-based screen as an example of identifying E3s associated with a specific cellular process by genome-wide functional analysis.

### *Polyglutamine Screen as an Example for the Functional Genomic Analysis of E3s*

PolyQ disease is a type of neurodegenerative disorder that includes Huntington's disease, spinobulbar muscular atrophy, and several different forms of spinocerebellar ataxias (Zoghbi and Orr, 2000). This disease is caused by expansion of the CAG repeat sequences in the disease proteins, which code for polyQ tracts. To date there are nine known polyQ expansion-mediated neurondegenerative disorders. As in many other types of neurodegenerative diseases, intraneuronal aggregate formation is a hallmark of these diseases. Several different genetic screens in different biological systems have been performed to look for suppressors of polyQ-mediated aggregate formation. Ubiquitin, E1, and ubiquitin-conjugating enzymes are among the suppressors uncovered from such screens (Table I). However, no E3s have been found in any of the previous screens. We set out to perform a genome-wide screen to look for suppressors of polyQ aggregate formation for the first time in a mammalian system.

*Prescreen Optimization and Formatted Assay.* Ataxin-3 (MJD-1) is a polyQ-containing protein and the disease protein for spinocerebellar ataxia-3. The normal ataxin–3 protein has a polyQ sequence ranging from 12 to 41 amino acids. We used a GFP-Ataxin-3 fusion protein that contains a tract of 84 glutamines (hereafter referred to as GFP-Atx-Q84) as the reporter in a cell-based assay. The GFP-Atx-Q84 expression plasmid is transiently transfected into the Neuro2a cell line. This cell line was selected because of its high tendency to form aggregates upon overexpression of polyQ expanded proteins (Yoshizawa *et al.*, 2001). Twenty-four hours after transfection, GFP-Atx-Q84 forms visible perinuclear aggregates. In agreement with a previous report (Cummings *et al.*, 1998), aggregates are dispersed into a diffused signal throughout the cytoplasm and nucleus when the cells are cotransfected with molecular chaperones Hsp70 and Hsp40. Compared to the nonaggregated GFP-AtxQ84, the fluorescent intensity of the aggregate is much higher and is localized predominantly in the perinuclear region. The assay is optimized to the 384-well reverse transfection format. Each well is spotted with 62.5 ng of library cDNA. GFP-Atx-Q84 (31.25 ng/well) is mixed with 0.125 $\mu$l/well of Lipofectamine2000 in 10 $\mu$l of OPTI-MEM media, and the mixture is dispensed immediately after mixing into the wells using an automated dispensing apparatus. The plates are incubated at room temperature for 30–45 min before cells are dispensed 6000/well. The plates are then incubated at 37°, 5% $CO_2$ for 24 h. Ten microliters of Hoechst 33342 dye is added to each well to stain the nuclei before live images are taken.

*Image Acquisition.* The scanning and processing of images are performed using the Q3DM EIDAC 100 high-throughput microscopy system. The 384-well black optical bottom Greiner plates used in this assay are scanned under 10× magnification, and images are taken from each well in two different fluorescent channels: the UV channel, which reflects the nuclear staining, and the GFP channel for GFP-Atx-Q84. Four to six images are taken per well. Images are analyzed on a per-cell basis, and aggregate-forming cells are identified based on the aggregate formation

TABLE I
PREVIOUSLY IDENTIFIED MODIFERS OF THE POLYQ AGGREGATE FORMATION IN THE UPS

| PolyQ origin | Modification | Organism | Gene | Function | Ref. |
|---|---|---|---|---|---|
| Ataxin-1 | Enhancer | *Drosophila* | Ubi63E | Ubiquitin | *Nature* (2000) **408,** 101–106 |
| Ataxin-1 | Enhancer | *Drosophila* | UbcD1 | ubiquitin conjugating enzyme | *Nature* (2000) **408,** 101–106 |
| Ataxin-1 | Enhancer | *Drosophila* | dUbc-E2H | ubiquitin conjugating enzyme | *Nature* (2000) **408,** 101–106 |
| N/A | Suppressor | *C. elegans* | rpt-5 | ATPase subunit of the 19s proteasome | *Proc. Natl. Acad. Sci. USA* (2004) **101,** 6403–6408 |
| N/A | Suppressor | *C. elegans* | pas-4 | a4 20S proteasome subunit | *Proc. Natl. Acad. Sci. USA* (2004) **101,** 6403–6408 |
| N/A | Suppressor | *C. elegans* | phi-37 | ubiquitin | *Proc. Natl. Acad. Sci. USA* (2004) **101,** 6403–6408 |
| N/A | Suppressor | *C. elegans* | uba-1 | ubiquitin activating enzyme E1 | *Proc. Natl. Acad. Sci. USA* (2004) **101,** 6403–6408 |
| Huntington | Suppressor | *S. cerevisiae* | ubp13 | ubiquitin C-terminal hydrolase | *Science* (2003) **302,** 1769–1772 |

algorithm described in the following section. The total number of cells, GFP-positive cells, and aggregate-forming cells are counted for each well. The percentage of aggregate formation normalized by transfection efficiency, represented by the number of aggregate-forming cells over the number of GFP-positive cells in each well, is used as the final readout for the screen.

*An Algorithm to Measure Aggregate Formation.* Individual cells in a well are defined by the algorithm via their nuclear staining. Nuclear masks are generated by applying a nonlinear least-squares optimized image filter to create marked object-background contrast, followed by automatic histogram-based thresholding (Price *et al.*, 1996). Images are also corrected for shade distortion, and effects due to background fluorescence are corrected by estimating and subtracting the mean background image intensity, which is determined dynamically on a per-image basis.

Aggregates are determined by their high and localized fluorescence intensity. Aggregate masks are generated by the Aggregates algorithm by Beckman Coulter. Aggregates are assigned to individual cells based on the distance from the nuclear centroid.

Using the aggregate masks, three measurements are computed: the number of aggregates in each cell, the size of each aggregate, and the fractional localized intensity of the aggregates. The first two measurements are trivially computed from the aggregate masks. Fractional localized intensity captures the relative brightness of the aggregates. The measurement logic computes sums of intensities $I(x, y)$ for pixels belonging to aggregates and for the remaining pixels. The fractional localized intensity of the aggregates $F_a$ is the ratio of the total integrated intensity in the aggregates and the total integrated intensity in the cell:

$$F_a = \frac{\sum_{(x,y)\in a} I(x,y)}{\sum_{(x,y)\in cell} I(x,y)}$$

Figure 2 shows a gallery of cells where no aggregates were detected. Figures 3 and 4 show the gallery of cells where at least one aggregate was detected with and without the aggregate masks, respectively.

*Statistical Analysis.* A pilot screen of 1000 cDNA was performed in triplicate to assess the feasibility for a larger scale screen. cDNA clones from this 1K set, which suppressed GFP-Atx-Q84 mediated aggregate formation in Neuro2a cells, were subsequently used as positive controls in the genome-scale library screen. A total of 11,000 cDNAs derived from the MGC collection and all the E3s in the Origene collection were

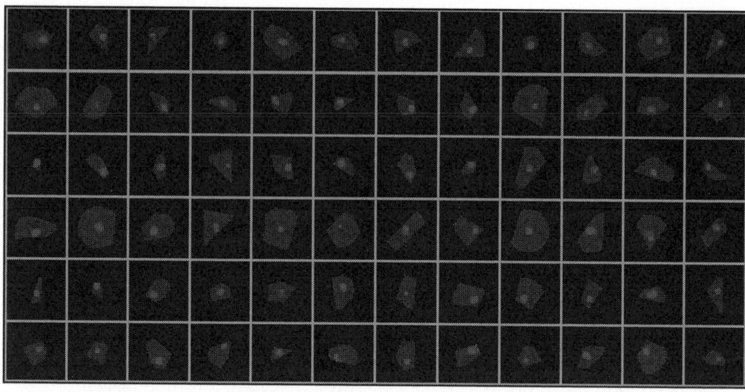

FIG. 2. Gallery of cells where no aggregates were detected. (See color insert.)

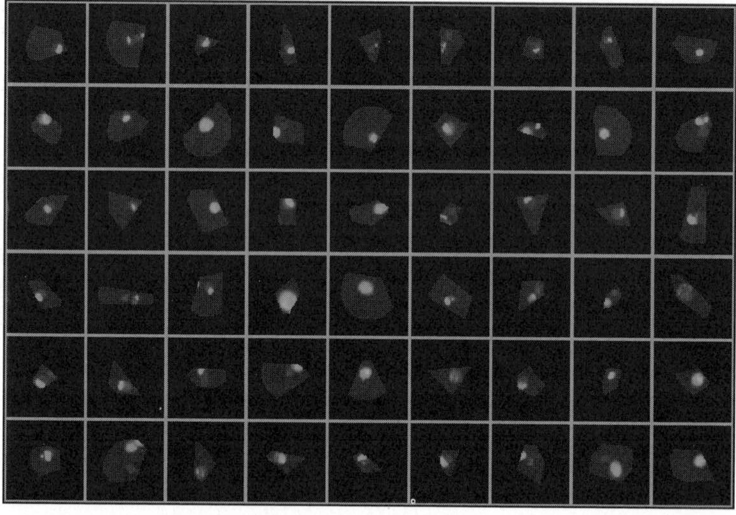

FIG. 3. Gallery of cells with at least one aggregate. (See color insert.)

interrogated in duplicate. All data were analyzed using the aggregate formation algorithm, and the fold reduction of percentage aggregate formation in each well was calculated against the plate mean. All cDNAs in the screen were ranked according to their fold reduction value, corrected by the replicate quality between the duplicates. cDNAs that reduced aggregate formation to the same extent or greater than the positive controls were considered to be hits. All the hits were repicked from the library and used in the same assay to confirm their suppressor activities against aggregate formation mediated by GFP-Atx3-Q84.

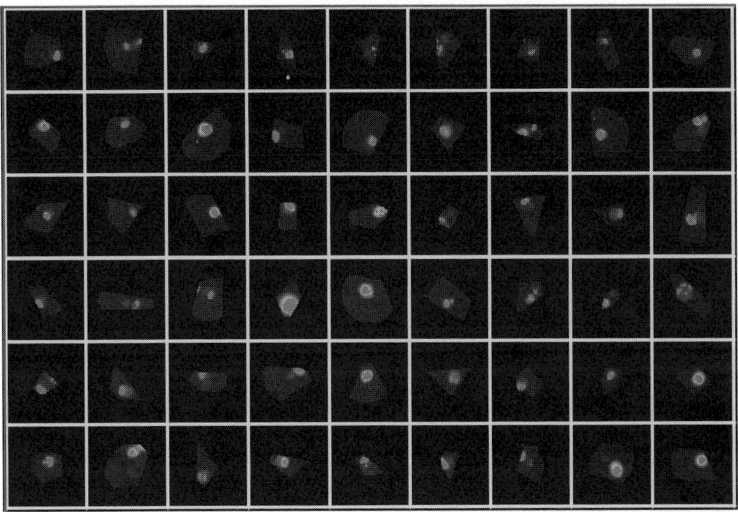

FIG. 4. Gallery of cells with at least one aggregate with the aggregate mask shown. (See color insert.)

*Secondary Assay and Future Perspectives.* The general toxicity of overexpressing the cDNAs in the neuro2a cell line was assessed by CellTiter-Glo luminescent cell viability assay (Promega) and cotransfection with a control GFP reporter. cDNAs that affected cell viability or GFP expression significantly when overexpressed were eliminated from the hits.

Semiquantitative RT-PCR was performed to ensure the similar level of transcription of the GFP-Atx-Q84 reporter when cotransfected with candidate suppressors of aggregate formation. This assay serves to filter out cDNAs that suppress aggregate formation indirectly due to effects on the GFP-Atx-Q84 mRNA levels.

A novel RING finger-type E3 exhibited suppressor activity in the primary screen and was chosen for follow-up studies. It will be determined whether the suppressor activity is RING finger dependent. This putative E3 will be expressed as a recombinant protein and tested for its ubiquitin ligase activity in an *in vitro* ubiquitination assay (Chen and Pickart, 1990). The function of this novel E3 will be explored further in animal models to understand whether this protein plays a role in neurodegenerative disease *in vivo.* Other potential substrates will be identified by yeast two-hybrid and immunoprecipitation followed by mass spectrometry analysis.

## Summary

Using a combination of arrayed cDNA expression library and an automated image acquisition and processing technique, we were able to identify a novel ubiquitin ligase with a biological function. This method can be expanded easily to mammalian siRNA libraries, as well as to other reporter systems, such as the luciferase coupled to promoter elements that respond to signaling pathways. Utilizing a comprehensive collection of all the potential E3s encoded by the mouse/human genome, it will be possible to systematically elucidate the roles of ubiquitin ligases in the modulation of cellular pathways and phenotypes.

## References

Abraham, V. C., Taylor, D. L., and Haskins, J. R. (2004). High content screening applied to large-scale cell biology. *Trends Biotechnol.* **22,** 15–22.

Bence, N. F., Sampat, R. M., and Kopito, R. R. (2001). Impairment of the ubiquitin-proteasome system by protein aggregation. *Science* **292,** 1552–1555.

Berns, K., Hijmans, E. M., Mullenders, J., Brummelkamp, T. R., Velds, A., Heimerikx, M., Kerkhoven, R. M., Madiredjo, M., Nijkamp, W., Weigelt, B., Agami, R., Ge, W., Cavet, G., Linsley, P. S., Beijersbergen, R. L., and Bernards, R. (2004). A large-scale RNAi screen in human cells identifies new components of the p53 pathway. *Nature* **428,** 431–437.

Carpenter, A. E., and Sabatini, D. M. (2004). Systematic genome-wide screens of gene function. *Nature Rev. Genet.* **5,** 11–22.

Chai, Y., Shao, J., Miller, V. M., Williams, A., and Paulson, H. L. (2002). Live-cell imaging reveals divergent intracellular dynamics of polyglutamine disease proteins and supports a sequestration model of pathogenesis. *Proc. Natl. Acad. Sci. USA* **99,** 9310–9315.

Chanda, S. K., White, S., Orth, A. P., Reisdorph, R., Miraglia, L., Thomas, R. S., De Jesus, P., Mason, D. E., Huang, Q., Vega, R., Yu, D. H., Nelson, C. G., Smith, B. M., Terry, R., Linford, A. S., Yu, Y., Chirn, G. W., Song, C., Labow, M. A., Cohen, D., King, F. J., Peters, E. C., Schultz, P. G., Vogt, P. K., Hogenesch, J. B., and Caldwell, J. S. (2003). Genome-scale functional profiling of the mammalian AP-1 signaling pathway. *Proc. Natl. Acad. Sci. USA* **100,** 12153–12158.

Chen, Z., and Pickart, C. M. (1990). A 25-kilodalton ubiquitin carrier protein (E2) catalyzes multi-ubiquitin chain synthesis via lysine 48 of ubiquitin. *J. Biol. Chem.* **265,** 21835–21842.

Cummings, C. J., Mancini, M. A., Antalffy, B., De Franco, D. B., Orr, H. T., and Zoghbi, H. Y. (1998). Chaperone suppression of aggregation and altered subcellular proteasome localization imply protein misfolding in SCA1. *Nature Genet.* **19,** 148–154.

Hatakeyama, S., Yada, M., Matsumoto, M., Ishida, N., and Nakayama, K. I. (2001). U box proteins as a new family of ubiquitin-protein ligases. *J. Biol. Chem.* **276,** 33111–33120.

Hemmila, I. A., and Hurskainen, P. (2002). Novel detection strategies for drug discovery. *Drug Discov. Today* **7,** S150–S156.

Huibregtse, J. M., Scheffner, M., Beaudenon, S., and Howley, P. M. (1995). A family of proteins structurally and functionally related to the E6-AP ubiquitin-protein ligase. *Proc. Natl. Acad. Sci. USA* **92,** 5249.

Joazeiro, C. A., and Weissman, A. M. (2000). RING finger proteins: Mediators of ubiquitin ligase activity. *Cell* **102,** 549–552.

Kuckuck, F. W., Edwards, B. S., and Sklar, L. A. (2001). High throughput flow cytometry. *Cytometry* **44,** 83–90.

Lu, Z., Xu, S., Joazeiro, C., Cobb, M. H., and Hunter, T. (2002). The PHD domain of MEKK1 acts as an E3 ubiquitin ligase and mediates ubiquitination and degradation of ERK1/2. *Mol. Cell* **9,** 945–956.

Paddison, P. J., Silva, J. M., Conklin, D. S., Schlabach, M., Li, M., Aruleba, S., Balija, V., O'Shaughnessy, A., Gnoj, L., Scobie, K., Chang, K., Westbrook, T., Cleary, M., Sachidanandam, R., McCombie, W. R., Elledge, S. J., and Hannon, G. J. (2004). A resource for large-scale RNA-interference-based screens in mammals. *Nature* **428,** 427–431.

Price, J. H., Goodacre, A., Hahn, K., Hodgson, L., Hunter, E. A., Krajewski, S., Murphy, R. F., Rabinovich, A., Reed, J. C., and Heynen, S. (2002). Advances in molecular labeling, high throughput imaging and machine intelligence portend powerful functional cellular biochemistry tools. *J. Cell Biochem.* **39**(Suppl.), 194–210.

Price, J. H., Hunter, E. A., and Gough, D. A. (1996). Accuracy of least squares designed spatial FIR filters for segmentation of images of fluorescence stained cell nuclei. *Cytometry* **25,** 303–316.

Steffan, J. S., Kazantsev, A., Spasic-Boskovic, O., Greenwald, M., Zhu, Y. Z., Gohler, H., Wanker, E. E., Bates, G. P., Housman, D. E., and Thompson, L. M. (2000). The Huntington's disease protein interacts with p53 and CREB-binding protein and represses transcription. *Proc. Natl. Acad. Sci. USA* **97,** 6763–6768.

Strausberg, R. L., Feingold, E. A., Grouse, L. H., Derge, J. G., Klausner, R. D., Collins, F. S., Wagner, L., Shenmen, C. M., Schuler, G. D., Altschul, S. F., Zeeberg, B., Buetow, K. H., Schaefer, C. F., Bhat, N. K., Hopkins, R. F., Jordan, H., Moore, T., Max, S. I., Wang, J., Hsieh, F., Diatchenko, L., Marusina, K., Farmer, A. A., Rubin, G. M., Hong, L., Stapleton, M., Soares, M. B., Bonaldo, M. F., Casavant, T. L., Scheetz, T. E., Brownstein, M. J., Usdin, T. B., Toshiyuki, S., Carninci, P., Prange, C., Raha, S. S., Loquellano, N. A., Peters, G. J., Abramson, R. D., Mullahy, S. J., Bosak, S. A., McEwan, P. J., McKernan, K. J., Malek, J. A., Gunaratne, P. H., Richards, S., Worley, K. C., Hale, S., Garcia, A. M., Gay, L. J., Hulyk, S. W., Villalon, D. K., Muzny, D. M., Sodergren, E. J., Lu, X., Gibbs, R. A., Fahey, J., Helton, E., Ketteman, M., Madan, A., Rodrigues, S., Sanchez, A., Whiting, M., Young, A. C., Shevchenko, Y., Bouffard, G. G., Blakesley, R. W., Touchman, J. W., Green, E. D., Dickson, M. C., Rodriguez, A. C., Grimwood, J., Schmutz, J., Myers, R. M., Butterfield, Y. S., Krzywinski, M. I., Skalska, U., Smailus, D. E., Schnerch, A., Schein, J. E., Jones, S. J., and Marra, M. A. (2002). Generation and initial analysis of more than 15,000 full-length human and mouse cDNA sequences. *Proc. Natl. Acad. Sci. USA* **99,** 16899–16903.

Taylor, J. P., Tanaka, F., Robitschek, J., Sandoval, C. M., Taye, A., Markovic-Plese, S., and Fischbeck, K. H. (2003). Aggresomes protect cells by enhancing the degradation of toxic polyglutamine-containing protein. *Hum. Mol. Genet.* **12,** 749–757.

Wertz, I. E., O'Rourke, K. M., Zhou, H., Eby, M., Aravind, L., Seshagiri, S., Wu, P., Wiesmann, C., Baker, R., Boone, D. L., Ma, A., Koonin, E. V., and Dixit, V. M. (2004). De-ubiquitination and ubiquitin ligase domains of A20 downregulate NF-kappaB signalling. *Nature* **430,** 694–699.

Yoshizawa, T., Yoshida, H., and Shoji, S. (2001). Differential susceptibility of cultured cell lines to aggregate formation and cell death produced by the truncated Machado-Joseph disease gene product with an expanded polyglutamine stretch. *Brain Res. Bull.* **56,** 349–352.

Ziauddin, J., and Sabatini, D. M. (2001). Microarrays of cells expressing defined cDNAs. *Nature* **411,** 107–110.

Zoghbi, H. Y., and Orr, H. T. (2000). Glutamine repeats and neurodegeneration. *Annu. Rev. Neurosci.* **23,** 217–247.

# Section IV

# Proteasome

# [24] Purification of PA700, the 19S Regulatory Complex of the 26S Proteasome

*By* GEORGE N. DEMARTINO

## Abstract

The 26S proteasome is a 2,400,000-Da protease complex that selectively degrades proteins modified by polyubiquitin chains. The 26S proteasome is composed of two 700,000-Da multisubunit complexes: the 20S proteasome, which serves as the proteolytic core of the complex, and PA700, an ATPase regulatory complex responsible for the binding, modification, and delivery of substrates to the proteolytic chamber. Thus, PA700 mediates multiple functions essential for ubiquitin-dependent proteolysis by the 26S proteasome. This chapter reviews briefly the structure and function of PA700, details the methodology for its large-scale purification from mammalian tissues, and describes a simple functional PA700 assay based on the stimulation of proteasome activity.

## Introduction

PA700 (proteasome activator of 700,000 Da, also known as 19S *r*egulatory *p*article), is the regulatory component of the 26S proteasome, an ATP-dependent molecular machine that degrades most cellular proteins modified covalently by a polyubiquitin chain (DeMartino and Slaughter, 1999; Voges *et al.*, 1999). The 26S proteasome is composed of one or two copies of PA700 bound to the 20S proteasome, the proteolytic core of the 26S proteasome (Fig. 1). The 20S proteasome contains multiple catalytic sites for the hydrolysis of substrate proteins. These catalytic sites, however, are located within a luminal chamber of the cylindrically shaped 20S proteasome and thus are physically sequestered from substrates (Baumeister *et al.*, 1998; Bochtler *et al.*, 1999; Groll *et al.*, 1997). To reach the catalytic sites, substrates must pass through a narrow, gated pore at either end of the 20S cylinder (Groll *et al.*, 2001). Each pore is formed by a ring of seven 20S proteasome subunits to which PA700 binds. Thus, a major function of PA700 is to regulate substrate entrance to the catalytic chamber of the proteasome (see later).

PA700 has a native molecular weight of approximately 700,000 and is composed of 18 different subunits, whose primary structures have been determined (Table I). Although a crystal structure of PA700 has not been

METHODS IN ENZYMOLOGY, VOL. 398

0076-6879/05 $35.00
DOI: 10.1016/S0076-6879(05)98024-5

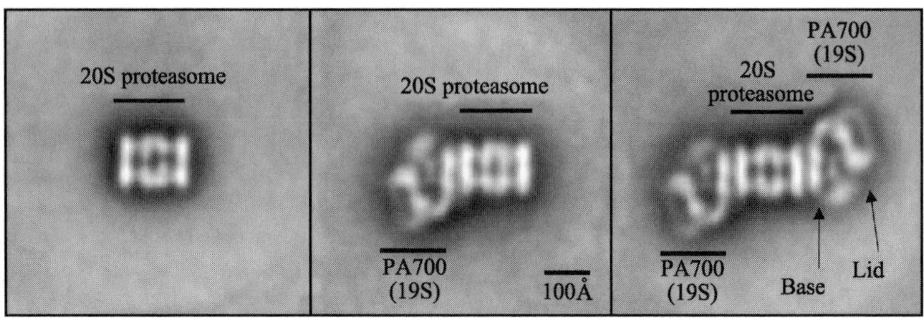


FIG. 1. Architecture of the proteasome and PA700. Averaged electron micrograph images of 20S proteasome (left) and 26S proteasomes with one copy of PA700 (center) and two copies of PA700 (right). 26S proteasomes were assembled from purified 20S proteasome and purified PA700.

solved (in contrast to the 20S proteasome), the general architecture of PA700 is established. For example, PA700 is composed of two subcomplexes termed "base" and "lid" (Glickman *et al.*, 1998). The base consists of six homologous members of the AAA ATPase family and two non-AAA subunits (Ogura and Wilkinson, 2001). The AAA subunits account for the ATPase activity of the 26S proteasome (isolated PA700 also displays ATPase activity) and appear to form a six-membered ring that binds directly to the ends of the proteasome. This topology suggests that to reach the catalytic chamber, substrates must pass through the centers of two attached rings formed by the 20S proteasome and the PA700 base, respectively. The two non-ATPase subunits of the base interact with several nonproteasomal proteins that contain ubiquitin-like (UBL) domains; because some proteins with UBL domains also bind polyubiquitin chains, they could mediate indirect targeting of multiubiquitinated substrates to PA700 (Hartmann-Petersen *et al.*, 2003). The lid consists of the remaining 10 PA700 subunits, most of whose functions are unclear. One lid subunit, Rpn11, catalyzes a deubiquitinating activity for the removal of multiubiquitin chains from substrates (Verma *et al.*, 2002; Yao and Cohen, 2002). Most eukaryotes, but not *Saccharomyces cerevisiae*, contain a second subunit with deubiquitinating activity, Uch37 (Lam *et al.*, 1997). The association of base and lid appears to be stabilized by Rpn10 (S5a), a subunit originally identified as a multiubiquitin chain-binding protein (Deveraux *et al.*, 1994; Glickman *et al.*, 1998).

A minimal model for the role of PA700 in the degradation of multiubiquitinated proteins by the 26S proteasome is illustrated in Fig. 2 and is described later. It is likely that some processes mediated directly by PA700 may be duplicated by certain non-PA700 proteins, thereby adding additional regulation and complexity to 26S proteasome function; this latter topic is beyond the scope of this chapter and is not discussed further.

1. *Proteasome activation*. PA700 binds to one or both ends of the 20S proteasome. This interaction activates the proteasome by opening the regulated gates at the end of the 20S cylinder (Groll *et al.*, 2001; Ma *et al.*, 1994). The molecular details of this mechanism are unknown for PA700, but are likely similar to those documented for PA28, a distinct protein activator of the 20S proteasome (Whitby *et al.*, 2001).

2. *Polyubiquitin chain binding*. PA700 binds polyubiquitin chains directly, thereby providing substrate specificity of the 26S proteasome for ubiquitin-modified as opposed to unmodified proteins. At least two PA700 subunits, Rpn10 (S5a) and Rpt 5 (S6′), display this property, but their relative roles and importance in substrate targeting remain unclear (Deveraux *et al.*, 1994; Lam *et al.*, 2002). The possible roles of other PA700 subunits in direct binding of polyubiquitin chains remain to be established. As noted earlier, some polyubiquitinated proteins may be delivered to PA700 indirectly via proteins that bind to both polyubiquitin and PA700.

3. *Substrate unfolding*. Many substrates of the 26S proteasome retain most or all of their three-dimensional structure after polyubiquitin modification. Because folded proteins are physically restricted from entering the catalytic chamber, such substrates must be unfolded prior to degradation. This process may be achieved by chaperone-like functions of PA700 (Braun *et al.*, 1999; Strickland *et al.*, 2000). Chaperone-like functions of PA700 may also be responsible for the nonproteolytic role of PA700 in the regulation of transcription (Gonzalez *et al.*, 2002).

4. *Substrate translocation*. With few known exceptions, most proteasome substrates are degraded completely once engaged by the proteasome, suggesting that they are degraded processively. Moreover, the catalytic chamber of the proteasome is physically distant from even the most proximal possible substrate-binding site on PA700. Thus, PA700 probably functions to translocate substrates from their initial binding sites to the catalytic chamber of the proteasome for degradation.

5. *Substrate deubiquitination*. Proteasomal degradation of polyubiquitinated proteins requires removal of the chain from the substrate, probably because the bulky chain would impede passage of the substrate polypeptide through the open pore. At least part of this function is accomplished by the Rpn11 subunit of the lid (Lundgren *et al.*, 2004; Verma *et al.*, 2002; Yao and Cohen, 2002).

6. *ATP hydrolysis*. The overall degradation of polyubiquitinated proteins by the 26S proteasome requires continuous ATP hydrolysis by PA700 (Voges

TABLE I
COMPONENT SUBUNITS OF PA700/19S REGULATORY COMPLEX[a]

| Rpn | S | Other common names | Subcomplex | Approximate molecular mass (Da) | Structural features | Function |
|---|---|---|---|---|---|---|
| Rpt1 | S7 | CIM5 | Base | 48,500 | AAA domain | ATPase |
| Rpt2 | S4 | YTA5, p56 | Base | 49,000 | AAA domain | ATPase; gate regulation |
| Rpt3 | S6b | TBP7, p48 | Base | 47,000 | AAA domain | ATPase |
| Rpt4 | S10b | SUG2, p42 | Base | 44,000 | AAA domain | ATPase |
| Rpt5 | S6a | TBP1, p50 | Base | 49,000 | AAA domain | ATPase; polyubiquitin binding |
| Rpt6 | S8 | SUG1, CIM3, p45 | Base | 45,500 | AAA domain | ATPase |
| Rpn1 | S2 | HRD2, NAS1, p97 | Base | 100,000 | LRR, KEKE motifs | UBL binding |
| Rpn2 | S1 | SEN3, p112 | Base | 106,000 | LRR, KEKE motifs | UBL binding |
| Rpn3 | S3 | SUN2, p58 | Lid | 61,000 | PCI domain | |

| | | | | | | |
|---|---|---|---|---|---|---|
| Rpn4 | | SON1 | | 60,000 | | Transcriptional regulation |
| | S5b | p50.5 | Lid | 55,000 | | |
| | | p42E | Lid | 42,000 | | |
| Rpn5 | - | p55 | Lid | 53,000 | PCI domain | |
| Rnp6 | S9 | p44.5 | Lid | 47,500 | PCI domain | |
| Rpn7 | S10a | p44 | Lid | 45,500 | PCI domain | |
| Rpn8 | S12 | Mov34, p40 | Lid | 37,000 | | |
| Rpn9 | S11 | Nas7, p40.5 | Lid | 43,000 | PCI domain | |
| Rpn10 | S5a | Mcb1, Mbp1, p54 | Base–lid interface | 41,000 | UIM motif | Polyubiquitin binding |
| Rpn11 | S13 | Poh1, Pad1 | Lid | 35,000 | JAMM/MPN domain | Deubiquitinating metalloprotease |
| Rpn12 | S14 | NIN1, p31 | Lid | 31,000 | | |
| Rpn13 | | Daq1 | | | | |
| | | UCH37, p37 | Lid | 37,000 | | Deubiquitinating |
| | | NAS6, p28, | | 28,000 | Ankyrin repeats | |
| | S15 | | | | PDZ domain | |

[a]Component subunits of PA700 are identified according to two commonly employed nomenclatures, "Rpn/Rpt" and "S". The table includes data for PA700 from all sources, including mammals and yeast.

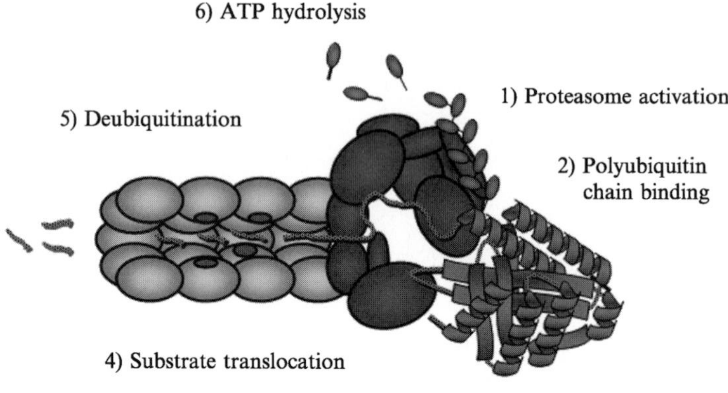


FIG. 2. Functions of PA700 in the degradation of ubiquitin-dependent protein degradation by the 26S proteasome. (See color insert.)

*et al.*, 1999). The exact mechanistic basis for this energy requirement remains obscure, but must be mediated by one or more of the six AAA subunits of the base. Several PA700-mediated features described earlier are regulated by ATP, including PA700 binding to and gating of the 20S proteasome (Köhler *et al.*, 2001; Ma *et al.*, 1994), polyubiquitin chain binding (Lam *et al.*, 2002), chaperone-like activity (Braun *et al.*, 1999), and substrate deubiquitination (Verma *et al.*, 2002; Yao and Cohen, 2002). A major challenge in understanding the molecular function of the 26S proteasome is to define how ATP hydrolysis is linked mechanistically to proteolysis.

## Purification of PA700

The purification described later results in PA700 free of both 20S proteasome and other nonproteasomal proteins reported to associate with 26S proteasome (Leggett *et al.*, 2002). PA700 appears to be partitioned between 20S proteasome-bound and 20S proteasome-free forms, although the exact distribution may vary considerably among different cells. PA700 purified by the following procedure likely is derived from both forms because the purification is optimized for conditions that promote dissociation of most of the 26S proteasome into 20S proteasome and PA700 components.

## Collection and Preparation of Bovine Blood

Bovine blood is obtained from a slaughterhouse in the presence of anticoagulant (500 ml of 150 m*M* sodium citrate, pH 7.6, 5 m*M* EDTA per gallon of blood) and transported immediately to the laboratory. The blood is centrifuged at 5000*g* for 1 h. The supernatant and white cell layer is removed by aspiration. The packed red blood cells are resuspended in phosphate-buffered saline and recentrifuged. This process is repeated four to five times or until the supernatant is free of protein. The washed red blood cells may be used immediately or may be frozen at –70° for future use. We have not detected differences in structural or functional properties of PA700 prepared from fresh as opposed to frozen red blood cells.

## Cell Lysis and Centrifugation

The following procedure describes the purification of PA700 from 1 liter of packed red blood cells. The size of the preparation may be scaled as required. All purification procedures are carried out at 4°.

Fresh or thawed cells are lysed hypotonically in 5 volumes of lysis buffer (20 m*M* Tris-HCl, pH 7.6, 1 m*M* EDTA, 5 m*M* $\beta$–mercaptoethanol), and centrifuged at 14,000*g* for 1 h. The supernatant is removed and saved; cell pellets are resuspended in lysis buffer and recentrifuged. The supernatant of the second centrifugation is combined with the first and is subjected to batch anion-exchange chromatography.

## Batch Anion-Exchange Chromatography

The lysate supernatant is added to DE52 anion exchanger (Whatman) equilibrated with lysis buffer (0.25 ml DE52 / ml supernatant) and stirred gently for 30 min. The resin is allowed to settle under gravity (about 60 min) and most of the supernatant is discarded directly. The DE52 resin is poured into a large Buchner funnel fitted with Whatman 1 filter paper and washed extensively with lysis buffer until the filtrate is colorless. The bound protein containing PA700 is eluted with about 1500 ml of lysis buffer containing 0.4 *M* NaCl.

## Ammonium Sulfate Precipitation

Solid ammonium sulfate is added slowly to the filtrate to 40% saturation (0.243 g/ml). After an additional 60 min, the precipitated material is collected by centrifugation (14,000*g* for 1 h), washed in lysis buffer containing 40% saturated ammonium sulfate, and recentrifuged. The pellet is resuspended in less than 30 ml of buffer X (20 m*M* Tris-HCl, pH 7.6, at 4°, 100 m*M* NaCl, 1 m*M* $MgCl_2$, 0.1 m*M* EDTA, 0.5 m*M* dithiothreitol, 20% glycerol) and dialyzed overnight against the same buffer.

### Gel Filtration Chromatography

The dialyzed material is centrifuged to remove undissolved proteins and remaining red cell ghosts. The clear supernatant is subjected to gel filtration chromatography on Sephacryl S-300 (140 × 5 cm) equilibrated with buffer X. The eluted fractions are assayed for PA700 activity as described later. The activity elutes as a homogeneous peak with an apparent molecular weight of approximately 700,000.

### DEAE Anion-Exchange Chromatography

Fractions with peak PA700 activity are pooled and applied directly to a DEAE ion-exchange column (20 × 2.5 cm) equilibrated in buffer X. We have successfully employed several types of DEAE resins but typically use DEAE Fractogel (EM Separations). The bound proteins are eluted with 1000 ml of a linear gradient of 100–450 m*M* NaCl in buffer X. Fractions are assayed for PA700 activity, which elutes as a single homogeneous peak at approximately 300 m*M* NaCl. Fractions with peak PA700 activity are pooled and dialyzed against 5 m*M* potassium phosphate buffer, pH 7.6, 5 m*M* $\beta$–mercaptoethanol, and 20% glycerol.

### Hydroxylapatite Chromatography

The dialyzed fractions from the DEAE Fractogel column are applied to a column (8 × 2.5 cm) of hydroxylapatite equilibrated in dialysis buffer. The bound proteins are eluted with 500 ml of a 5–200 m*M* linear gradient of potassium phosphate buffer, pH 7.6, containing 5 m*M* $\beta$–mercaptoethanol and 20% glycerol. Fractions are assayed for PA700 activity, which elutes as homogeneous peak at approximately 75 m*M* phosphate. In some preparations, a small second peak of activity elutes at higher phosphate concentrations. This peak contains small amounts of 26S proteasome and possibly a second form of PA700 whose exact structural relationship to that of the main PA700 peak is unclear. A pool of the main PA700 peak is made based on activity and on inspection of SDS–PAGE of eluted fractions. In most preparations, the PA700 at this stage is highly purified. In rare instances, the PA700 contains minor contaminants visible on SDS–PAGE after staining with Coomassie blue. In the latter event, the PA700 is subjected to a second round of gel filtration chromatography, as described earlier. Pooled fractions from the hydroxylapatite column (or from the second gel filtration column, if required) are dialyzed extensively against 20 m*M* Tris-HCl, pH 7.6, at 4°, 20 m*M* NaCl, 1.0 m*M* dithiothreitol, 1 m*M* EDTA, and 20% glycerol.

## Concentration, Storage, and Yield

The dialyzed PA700 is concentrated on an Amicon XM-300 membrane to at least 1 mg/ml, aliquoted, quick frozen in liquid nitrogen, and stored at –80°. We have found PA700 to be physically and functionally stable after 12 months of storage and can withstand several freeze–thaw cycles at high protein concentrations. Approximately 20–30 mg of PA700 is purified from 1 liter of bovine red blood cells.

## Assessment of Purity

PA700 purified as described earlier migrates as a single band on non-denaturing polyacrylamide gels and sediments as a homogeneous peak during glycerol density gradient centrifugation (Fig. 3). SDS–PAGE analysis of fractions from the glycerol gradient shows that all defined PA700 subunits migrate with a similar distribution and are coincident with the peak of PA700 activity.

## Assay for PA700 Activity

PA700 is assayed by its ability to stimulate the peptide-hydrolyzing activity of the 20S proteasome. Latent 20S proteasome, prepared from bovine red blood cells as described previously (McGuire *et al.*, 1989), displays a low capacity to hydrolyze 7-amino-4-methylcoumarin (AMC) from a model peptide substrate such as succinyl-leucyl-leucyl-valyl-tyrosyl-AMC. The increase in proteasome activity in this assay reflects the binding of PA700 to the latent proteasome and gating of the substrate entry pores (Adams *et al.*, 1998). PA700 stimulates proteasome activity by 20- to 50-fold in this assay (Ma *et al.*, 1994).

The standard assay contains 50 m$M$ Tris-HCl, pH 8.0 at 37°, 2 m$M$ dithiothreitol , 200 $\mu M$ ATP, 10 m$M$ $MgCl_2$, 10 n$M$ 20S proteasome, and PA700 (either 100 n$M$ purified protein or 5 $\mu$l of the column fractions described previously) in a final volume of 25 $\mu$l. After incubation for 45 min at 37°, 200 $\mu$l of 100 $\mu M$ succinyl-leucyl-leucyl-valyl-tyrosyl-AMC in 20 m$M$ Tris-HCl, pH 8.0, 1 m$M$ dithiotheitol is added to the reaction, and the rate of hydrolysis of AMC from the peptide is measured directly by the time-dependent increase of fluorescence at 360 excitation/480 emission. This assay is adapted easily to a 96-well fluorescent place reader. Controls assays include incubations lacking 20S proteasome, PA700, or both. Purified PA700 has no detectable peptide-hydrolyzing activity, but in early stages of the purification, PA700-containing fractions display low and variable amounts of endogenous 20S and 26S proteasome activity.

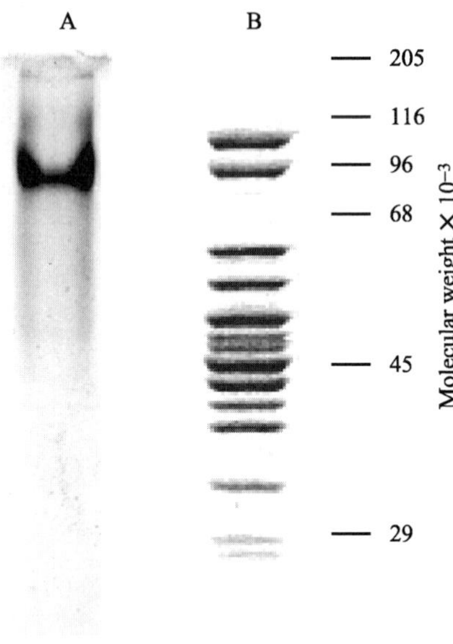


FIG. 3. Purification of PA700 from bovine red blood cells. Nondenaturing (A) and SDS-denaturing (B) polyacrylamide gel electrophoresis of PA700 purified according to the procedure described in the text. Molecular weight markers are indicated.

## Note on Generality of the Method

This purification procedure has been optimized for bovine red blood cells, the source with which we have the most experience. The procedure has been applied successfully to several other tissues sources, including human red blood cells, HeLa cells, and rabbit skeletal muscle. Other sources, including bovine heart and bovine spleen, have produced PA700 with much poorer yield and/or purity.

## References

Adams, G. M., Crotchett, B., Slaughter, C. A., DeMartino, G. N., and Gogol, E. P. (1998). Formation of proteasome-PA700 complexes directly correlates with activation of peptidase activity. *Biochemistery* **37,** 12927–12932.

Baumeister, W., Walz, J., Zühl, F., and Seemüller, E. (1998). The proteasome: Paradigm of a self-compartmentalizing protease. *Cell* **92,** 367–380.

Bochtler, M., Ditzel, L., Groll, M., Hartmann, C., and Huber, R. (1999). The proteasome. *Annu. Rev. Biophys. Biomol. Struct.* **28,** 295–317.

Braun, B. C., Glickman, M., Kraft, R., Dahlmann, B., Kloetzel, P.-M., Finley, D., and Schmidt, M. (1999). The base of the proteasome regulatory particle exhibits chaperone-like activity. *Nature Cell Biol.* **1,** 221–226.

DeMartino, G. N., and Slaughter, C. A. (1999). The proteasome, a novel protease regulated by multiple mechanisms. *J. Biol. Chem.* **274,** 22123–22126.

Deveraux, Q., Ustrell, V., Pickart, C., and Rechsteiner, M. (1994). A 26S protease subunit that binds ubiquitin conjugates. *J. Biol. Chem.* **269,** 7059–7061.

Glickman, M. H., Rubin, D. M., Coux, O., Wefes, I., Pfeifer, G., Cjeka, Z., Baumeister, W., Fried, V., and Finley, D. (1998). A subcomplex of the proteasome regulatory particle required for ubiquitin-conjugate degradation and related to the COP9-signalosome and eIF3. *Cell* **94,** 615–623.

Gonzalez, F., Delahodde, A., Kodadek, T., and Johnston, S. A. (2002). Recruitment of a 19S proteasome subcomplex to an activated promoter. *Science* **286,** 548–550.

Groll, M., Bajorek, M., Kohler, A., Moroder, L., Rubin, D. M., Huber, R., Glickman, M. N., and Finley, D. (2001). A gated channel into the proteasome core particle. *Nature Struct. Biol.* **11,** 1062–1067.

Groll, M., Ditzel, L., Lowe, J., Stock, D., Bochtler, M., Bartunik, H. D., and Huber, R. (1997). Structure of the 20S proteasome from yeast at 2.4A resolution. *Nature* **386,** 463–471.

Hartmann-Petersen, R., Seeger, M., and Gordon, C. (2003). Transfering substrates to the 26S proteasome. *Trends Biochem. Sci.* **28,** 26–31.

Köhler, A., Cascio, P., Leggett, D. S., Woo, K. M., Goldberg, A. L., and Finley, D. (2001). The axial channel of the proteasome core particle is gated by the Rpt2 ATPase and controls both substrate entry and product release. *Mol. Cell* **7,** 1143–1152.

Lam, Y. A., Lawson, T. G., Velayultham, M., Zweierm, J. L., and Pickart, C. M. A. (2002). Proteasomal ATPase subunit recognizes the polyubiquitin degradation signal. *Nature (London)* **416,** 763–767.

Lam, Y. A., Xu, W., DeMartino, G. N., and Cohen, R. E. (1997). Editing of ubiquitin conjugates by an isopeptidase in the 26S proteasome. *Nature* **385,** 737–740.

Leggett, D. S., Hanna, J., Borodovsky, A., Crosas, B., Schmidt, M., Baker, R. T., Waltz, T., Ploeugh, H., and Finley, D. (2002). Multiple associated proteins regulate proteasome structure and function. *Mol. Cell* **10,** 498–507.

Lundgren, J., Masson, P., Realini, C., and Young, P. (2004). Use of RNA interference and complementation to study the function of the *Drosophila* and human 26S proteasome subunits. *Mol. Cell. Biol.* **23,** 5320–5330.

Ma, C.-P., Vu, J. H., Proske, R. J., Slaughter, C. A., and DeMartino, G. N. (1994). Identification, purification, and characterization of a high-molecular weight, ATP-dependent activator (PA700) of the 20S proteasome. *J. Biol. Chem.* **269,** 3539–3547.

McGuire, M. J., McCullough, M. L., Croall, D. E., and DeMartino, G. N. (1989). The high molecular weight multicatalytic proteinase, macropain, exists in a latent form in human erythrocytes. *Biochim. Biophys. Acta* **995,** 181–186.

Ogura, T., and Wilkinson, A. J. (2001). AAA + superfamily ATPases: Common structure-diverse function. *Genes Cells* **6,** 575–597.

Strickland, E., Hakala, K., Thomas, P. J., and DeMartino, G. N. (2000). Recognition of misfolding proteins by PA700, the regulatory subcomplex of the 26S proteasome. *J. Biol. Chem.* **275,** 5565–5572.

Verma, R., Aravind, L., Oania, R., McDonald, W. H., Yates, J. R., Koonin, E. V., and Deshaies, R. J. (2002). Role of Rpn11 metalloprotease in deubiquitination and degradation by the 26S proteasome. *Science* **298,** 611–615.

Voges, D., Zwickl, P., and Baumeister, W. (1999). The 26S proteasome: A molecular machine designed for controlled proteolysis. *Annu. Rev. Biochem.* **68,** 1015–1068.

Whitby, F. G., Masters, E. I., Kramer, L., Knowlton, J. R., Yao, Y., Wang, C. C., and Hill, C. P. (2001). Structural basis for the activation of 20S proteasomes by 11S regulators. *Nature* **408,** 115–120.

Yao, T., and Cohen, R. E. (2002). A cryptic protease couples deubiquitination and degradation by the proteasome. *Nature* **419,** 403–407.

# [25] Purification and Analysis of Recombinant 11S Activators of the 20S Proteasome: *Trypanosoma brucei* PA26 and Human PA28$\alpha$, PA28$\beta$, and PA28$\gamma$

*By* Eugene I. Masters, Gregory Pratt, Andreas Förster, and Christopher P. Hill

## Abstract

Proteasomes perform the bulk of nonlysosomal degradation of aberrant, damaged, misfolded, and naturally short-lived regulatory proteins in eukaryotic cells. They are ~700-kDa assemblies whose hollow architecture sequesters the proteolytic sites inside a central chamber, thereby ensuring that the activity of isolated proteasomes is repressed. *In vivo*, proteasomes are activated by protein complexes, including the 11S activators (PA28 and PA26), which bind to one or both ends of the barrel-shaped structure. This chapter describes protocols for the purification of recombinant 11S regulators, characterization of their ability to stimulate proteasome activity, and crystallization of proteasome complexes.

## Introduction

The proteasome (also called 20S proteasome or core particle) performs the bulk of nonlysosomal protein degradation in eukaryotes. This activity is of central importance for a myriad of cellular processes, including housekeeping functions and regulatory pathways. Crystal structures of archaeal (Löwe *et al.*, 1995) and eukaryotic proteasome (Groll *et al.*, 1997; Unno *et al.*, 2002) revealed a barrel-shaped, ~700-kDa assembly that is composed of 28 subunits arranged in four heptameric rings, with the end rings composed of $\alpha$ subunits and the two central rings composed of $\beta$ subunits. Archaeal proteasomes generally consist of only one type of $\alpha$ and one type of $\beta$ subunit, each repeated seven times in their respective rings. In contrast, eukaryotic proteasomes contain seven different $\alpha$ subunits

METHODS IN ENZYMOLOGY, VOL. 398

0076-6879/05 $35.00
DOI: 10.1016/S0076-6879(05)98025-7

($\alpha$1–$\alpha$7) and seven different $\beta$ subunits ($\beta$1–$\beta$7), with each distinct subunit occupying a unique position in its ring.

The mechanism by which the proteasome avoids indiscriminate degradation of inappropriate substrates was revealed by the crystal structures, which showed that the proteolytic active sites are sequestered inside a central catalytic chamber that is formed by the double ring of $\beta$ subunits. Substrates enter the proteasome through the $\alpha$-annulus (Wenzel and Baumeister, 1995), which has a diameter of 17 Å (measured between atom centers), and products are believed to exit through this same opening also. This narrow opening prevents the entry of folded proteins. Access is further restricted by the N-terminal tails of $\alpha$ subunits, which in the case of eukaryotic proteasomes are ordered and form a precisely closed gate structure. For archaeal proteasomes, the $\alpha$ tails are generally disordered but still provide a barrier to entry of protein substrates (Benaroudj *et al.*, 2003; Förster *et al.*, 2003).

Proteasomes are activated *in vivo* by three different types of activator: 19S/PA700/RC (DeMartino and Slaughter, 1999; Voges *et al.*, 1999), PA200 (Ustrell *et al.*, 2002), and 11S (Dubiel *et al.*, 1992; Hill *et al.*, 2002; Ma *et al.*, 1992). This chapter concerns the 11S activators, which are heptamers of ~30-kDa subunits (Johnston *et al.*, 1997; Zhang *et al.*, 1999) and are broadly distributed in metazoans but appear to be absent in yeasts and plants. Jawed vertebrates encode three homologs known as PA28$\alpha$, PA28$\beta$, and PA28$\gamma$ (also known as REG$\alpha$, REG$\beta$, and REG$\gamma$) that typically share ~45% sequence identity between PA28$\alpha$ and PA28$\beta$, whereas the PA28$\gamma$ shares ~30% sequence identity with the $\alpha$ and $\beta$ homologs. Simpler species encode a single PA28 that is most closely related to PA28$\gamma$. The most distantly related 11S homolog is PA26 of *Trypanosoma brucei* (Yao *et al.*, 1999), which is an outlier within the family and shares 133 structurally equivalent residues, from a total of 231, with its closest relative, PA28$\alpha$. The RMSD for overlap of PA26 and PA28$\alpha$ on C$\alpha$ atoms is 1.7 Å, and only 9% of these residues are invariant between PA26 and all three human homologs (Förster *et al.*, 2005).

The biological roles of 11S activators are not entirely clear (Rechsteiner and Hill, 2005). Several observations suggest that PA28$\alpha$ and PA28$\beta$ function in the production of peptide ligands for MHC class I molecules. These data are not entirely consistent, however, and possible roles for PA28$\gamma$ and the single homolog of simpler species are not obvious. PA28 is known to form hybrid proteasomes in which 11S and 19S activators are bound to opposite ends of the same proteasome molecule, and given the relative abundance of 11S, 19S, and proteasome, it seems likely that 11S activators normally function as hybrid proteasomes (Hendil *et al.*, 1998; Tanahashi *et al.*, 2000). In this context, 11S activators may function to localize the 19S–proteasome complex to a specific intracellular site or substrate or they might enhance proteasome activity, for example, by facilitating product release. Regardless of their biological roles, biochemical and structural studies of 11S activators have revealed fundamental principles of the proteasome mechanism.

Here we provide details for the purification of recombinant 11S activators, biochemical assays of their binding to proteasome and stimulation of the peptidase activity of the proteasome, generation of mutants with altered activities, and protocols for crystallization of human PA28$\alpha$ and PA26 complexes with yeast and archaeal proteasomes.

## Activity Assays

PA28 was first identified and purified from bovine liver and heart (Ma *et al.*, 1992) and from human red blood cells (Dubiel *et al.*, 1992) on the basis of its ability to stimulate the hydrolysis of small fluorogenic peptide substrates by the proteasome. This assay uses model peptide substrates attached covalently by their C terminus to groups that fluoresce when the bond connecting peptide and fluorophor is cleaved. This assay is quick, convenient, and monitors the three distinct active sites of eukaryotic proteasomes. It is the method of choice for monitoring activator peaks during purification procedures. There is some concern, however, that it does not fairly report on the activity against larger, potentially more physiologically relevant substrates. Consequently, HPLC/MS assays are sometimes employed to monitor the degradation of longer peptide substrates (Dick *et al.*, 1996; Li *et al.*, 2001; Zhang *et al.*, 1998c). Another recently applied assay uses fluorescamine to quantify the number of amino groups liberated during proteolysis (Cascio *et al.*, 2002). It is important when interpreting the results of these various assays to remember that the connection between the biochemically observed stimulation of peptidase activity and the potential biological roles of 11S activators is currently unknown.

### *Fluorogenic Peptide Activity Assays*

Fluorogenic peptide substrates are usually modified at their N terminus with *t*-butyloxcarbonyl (Boc), succinyl (Suc), or benzyloxycarbonyl (Cbz). The C-terminal leaving groups are usually 7-amido-4-methylcoumarin (MCA) or $\beta$-naphthylamide ($\beta$NA). The substrates are commonly written according to the one-letter amino acid code. The identity of the C-terminal residue reports on each of the three distinct active sites of eukaryotic proteasomes, which are located at N termini of $\beta$1 (PGPH), $\beta$2 (T-like), and $\beta 5$ (C-like), and preferentially cleave following acid, basic, and hydrophobic residues, respectively (Bochtler *et al.*, 1999). Thus, commonly used substrates include Cbz-LLE-$\beta$NA for the PGPH site, Boc-FSR-MCA, PFR-MCA, and Boc-VLK-MCA for the T-like site, and Suc-LLVY-MCA, Suc-AAF-MCA, and Suc-LY-MCA for the C-like site. These reagents can be purchased from Sigma or Peptides International. Note that although preference for the P1 site of the short peptide substrates provides a useful tool, other determinants are also important for specificity in the hydrolysis of longer substrates (Bogyo *et al.*, 1998; Groll *et al.*, 2002; Harris *et al.*, 2001; Wang *et al.*, 2003).

Fluorogenic substrates, with the exception of Cbz-LLE-$\beta$NA, are prepared by dissolving the powdered solid in dimethyl sulfoxide (DMSO) to make a 10 m*M* stock, which is stored in aliquots at –80°. Immediately before use, a 10 m*M* substrate aliquot is thawed and brought to 200 $\mu M$, typically in TSD, pH 8.8 [10 m*M* Tris-HCl, 25 m*M* KCl, 10 m*M* NaCl, 1.1 m*M* $MgCl_2$, 0.1 m*M* EDTA, 1 m*M* dithiothreitol (DTT)]. Cbz-LLE-$\beta$NA is prepared by dissolving the powdered solid in DMSO to make a 10 m*M* stock and then storing at room temperature. Substrate solutions must be prepared fresh shortly before use or frozen aliquots must be used soon after thawing.

A typical assay incubates 170 ng proteasome with various amounts of activator (e.g., 100 nmol) for 10 min in 50 $\mu$l of buffer (e.g., 10 m*M* Tris, pH 7.5). The reaction is performed at a defined temperature, usually 37°. The reference reaction omits an activator. Enzymatic reactions are initiated by adding 50 $\mu$l of a 200 $\mu M$ solution of fluorogenic peptide substrate with mixing by pipeting up and down gently. After incubation for various time intervals, reactions are quenched by the addition of 200 $\mu$l of ice-cold ethanol with mixing. One hundred microliters of the quenched reaction mixture is transferred to a glass culture tube (6 × 50 mm, VWR), and the fluorescence signal is measured on a LS–5 fluorescence spectrophotometer (Beckman). For MCA substrates, the instrument is set to excitation at 380 nm and emission at 440 nm, with a slit width of 3. $\beta$NA substrates require settings of excitation at 335 nm, emission at 410 nm, and a slit width of 3.

Results are expressed as fold-stimulation, i.e., ratios of the rate of fluorescence generation in the presence of 11S compared to the reference reaction. For examples, see Realini *et al.* (1997). Higher concentrations are typically used for PA28$\beta$ (e.g., 440 nmol), which at lower concentrations shows significantly diminished activity, presumably because homomeric PA28$\beta$ has a much weaker heptamerization affinity (Realini *et al.*, 1997).

A simple version of this assay is used to identify chromatography fractions that contain an activator. Fifty-microliter reactions are made up with 10 $\mu$l of each fraction and 150 ng of purified proteasome in TSD, pH 8.8. A reference reaction with only TSD, pH 8.8, and 150 ng proteasome is also prepared. Reactions are incubated at 37° for 20 min, quenched, and fluorescence read as described earlier. Fractions corresponding to increased activity represent the activator peak.

### *HPLC Analysis of Longer Peptides*

An HPLC/MS method has been applied to assay proteasome stimulation by PA28/REG (Li *et al.*, 2001; Zhang *et al.*, 1998c). For example, see Zhang *et al.* (1998c), in which two substrates were used: P21, a 21 residue peptide (SADPELALALRVSMEEQRQRQ), and BBC1, a 49 residue peptide (MKKEKARVITEEEKNFKAFASLRMARANARLFGIRAK-RAKEAAEQDGSG). Substrates are incubated with proteasome and PA28, and samples are taken at various time points (e.g., 10 min, 30 min, 135 min, 5 h, and 12 h). Reaction products are separated on a C18 HPLC column and eluted with a gradient of 0–45% acetonitrile containing 0.1% trifluoroacetic acid. Product peptide masses are analyzed by submitting fractions to mass spectrometry. This assay has revealed more complex behavior than apparent from the fluorogenic peptide release assays. For example, PA28$\gamma$ decreases the rate of P21 hydrolysis but increases the rate of BBC1 hydrolysis, whereas PA28$\alpha$ accelerated degradation of both P21 and BBC1.

## Expression and Purification of Recombinant 11S Activators

The biochemical characterization of 11S activators has been facilitated greatly by the ability to express recombinant proteins in *Escherichia coli*, including homologs from rat (Song *et al.*, 1997). As described later, bacterial expression and purification procedures have been developed for recombinant human PA28$\alpha$, PA28$\beta$, and PA28$\gamma$, and we have utilized procedures for *T. brucei* PA26 that were initially developed elsewhere (Yao *et al.*, 1999).

### *Purification of PA26*

All of our studies with PA26 have utilized the recombinant protein expressed in *E. coli*. This protein includes a hexahistidine tag inserted after the initiator methionine. This tag has not been removed for the subsequent structural or biochemical studies. The expression plasmid, pBtpa, was created by Yao *et al.* (1999) and includes a threonine in place of the authentic serine residue at position 226. Our protocol is a minor modification of the published method. A 4-liter culture will normally yield ~25 mg of purified protein.

The following solutions and reagents are prepared for the purification of PA26:

1. Solutions for *E. coli* culture in low phosphate media:
   a. $10^6$ micronutrient stock. A 10-ml solution of deionized water, 37 mg of ammonium molybdate, 158 mg manganese chloride tetrahydrate, 247 mg boric acid, 84 mg cobalt chloride, 25 mg copper sulfate, and 18 mg zinc sulfate.
   b. 10× MOPS salts. A 1-liter solution is prepared with 0.4 *M* MOPS (3-[*N*-morpholino]propanesulfonic acid), pH 7.4, 40 m*M* tricine, 0.5 *M* sodium chloride, 95 m*M* ammonium chloride, 0.1 m*M* ferrous sulfate, 2.76 m*M* potassium sulfate, 0.005 m*M* calcium chloride. Finally, 10 $\mu$l of the $10^6$ micronutrient stock is added.
   c. 5× low phosphate media. A 1-liter solution is prepared with 500 ml 10× MOPS salts, 20 g glucose, 10 g casamino acids (Difco), 0.5 ml of 1 *M* potassium phosphate, pH 7.5, 5 ml of 1.5 m*M* thiamine, and 0.1 g adenine. The solution is sterile filtered with a 0.2-$\mu$m filter (Nalgene) into an autoclaved bottle and stored at 4°.
   d. Low phosphate media. Prepare shortly before use by adding 200 ml of sterile filtered 5× low phosphate media to 800 ml of sterile water. Ampicillin is added to 75 $\mu$g/ml and media swirled to mix.
2. Solutions for protein purification (all are 0.2 $\mu$m filtered and degassed for use on a chromatography workstation).
   a. TBS: 20 m*M* Tris-HCl, pH 7.9, 150 m*M* NaCl
   b. TBSI: TBS made up with 0.5 *M* imidazole
   c. $TS_{200}$: 10 m*M* Tris-HCl, pH 7.1, 190 m*M* KCl, 10 m*M* NaCl, 1.1 m*M* $MgCl_2$, 0.1 m*M* EDTA
3. Columns for purification: 20 ml Ni-NTA (Qiagen), HiPrep Superdex 200 26/60 HR (Amersham Biosciences).

*Escherichia coli* BL21(DE3) is transformed with pBtpa and grown overnight on a LB agar plate with 50 $\mu$g/ml ampicillin at 37° (Yao *et al.*, 1999). A 10-ml starter culture is initiated by inoculating a single colony into low phosphate media. The culture is grown overnight with constant shaking at 37° and centrifuged to pellet cells. The supernatant is decanted and cells are resuspended in 10 ml of fresh media. Four 2-liter culture flasks are each prepared with 1 liter of low phosphate media and 1 ml of the starter culture. Cells are grown at 37° with constant shaking (200 rpm) until reaching a density of $OD_{600} = 1.6$ (typically after 20–24 h) before harvesting by centrifugation at 8000*g* (4°, 15 min). Cell pellets are stored at –80°.

Cells are thawed on ice and resuspended in TBS (3–5 ml TBS/g of cell paste) to which a protease inhibitor cocktail tablet (complete, EDTA-free, protease inhibitor cocktail tablets, Roche) is added. Cells are lysed by a French pressure cell (two passes, minimum 5000 psi) followed by three rounds of 30 s of sonication interspersed with 30 s on ice. The lysate is clarified by centrifugation (39,000*g*, 30 min, 4°) and the supernatant is 0.2 $\mu$m filtered. All subsequent steps are performed at 4°.

The clarified supernatant is applied to a 20-ml Ni-NTA column equilibrated in TBS. The column is washed with 5 column volumes (CV) of TBS followed by 5 CV of 95% TBS:5% TBSI. Protein is eluted with 5 CV of 100% TBSI. Fractions (5 ml) are monitored by SDS–PAGE, and fractions corresponding to the main peak of PA26 are pooled and dialyzed against 2 liters of $TS_{200}$ + 4 m*M* EDTA for at least 2 h, followed by dialysis against 2 liters of $TS_{200}$ + 2 m*M* DTT for at least 2 h. The protein is concentrated by ultrafiltration (Amicon Stirred Cell with YM30 filter, Millipore) to not more than 20 mg/ml (Bradford method). Up to 1 ml of the concentrated sample is loaded onto the Superdex 200 column that has been equilibrated in $TS_{200}$ + 1 m*M* DTT, with multiple runs performed as needed for protein volumes in excess of 1 ml. Protein is eluted in $TS_{200}$ + 1 m*M* DTT at a flow rate of 0.5 ml/min; 2-ml fractions are collected. The protein elutes ~180 ml after injection, and the purest fractions (by SDS–PAGE) are pooled and the protein stored at 4°. For example chromatograms and gels, see Yao *et al.* (1999). We have always used the protein within a few days of completing the preparation and have not attempted longer term storage.

## Purification of Recombinant Human PA28$\alpha$, PA28$\beta$, and PA28$\gamma$

Our procedure for the purification of recombinant human PA28$\alpha$/REG$\alpha$, PA28$\beta$/REG$\beta$, and PA28$\gamma$/REG$\gamma$ is a minor adaptation of the published method (Realini *et al.*, 1997). These recombinant proteins do not have extraneous sequences, such as affinity tags. A 2-liter culture will typically yield ~10 mg of ~95% pure PA28.

The following solutions and reagents are prepared for the purification of PA28.

1. Solutions for protein purification (all are 0.2 $\mu$m filtered and degassed for use on a chromatography workstation).
   a. TSD, pH 8.8: 10 m*M* Tris-HCl, 25 m*M* KCl, 10 m*M* NaCl, 1.1 m*M* $MgCl_2$, 0.1 m*M* EDTA, and 1 m*M* DTT
   b. TSD400, pH 8.8: TSD, pH 8.8, + 365 m*M* KCl
   c. TSD pH 7.2: TSD, pH 8.8, with pH adjusted to pH 7.2
   d. TSD200, pH 7.2: TSD, pH 7.2, + 165 m*M* KCl

2. Columns for purification: DEAE (HR10/30 column packed with DEAE-Sepharose Fast Flow resin, Amersham Biosciences) and HiPrep Superdex 200 26/60 (Amersham Biosciences)

*Escherichia coli* BL21(DE3) is transformed with the appropriate plasmid (pAED4-REG$\alpha$, pAED4-REG$\beta$, pAED4-REG$\gamma$) (Realini *et al.*, 1997) and plated onto LB agar, ampicillin (50 $\mu$g/ml). Fifty milliliters of LB medium, supplemented with 100 $\mu$g/ml of ampicillin, is inoculated with a single colony. The culture is grown overnight (30°, 200 rpm) and is used to inoculate 2 liters of LB medium with ampicillin at 100 $\mu$g/ml. Following growth (30°, 200 rpm) to an $OD_{600}$ of 0.3, isopropyl-$\beta$-D-thiogalactopyranoside (IPTG) is added to 0.5 m*M*. Two hours postinduction, cells are harvested by centrifugation (4°, 15 min, 4000*g*) and resuspended in cold (4°) TSD, pH 8.8. Cells are harvested by centrifugation at 3700*g* for 15 min and stored at –80°.

Cells are thawed on ice and resuspended in TSD, pH 8.8, to a final volume of ~45 ml to which a protease inhibitor cocktail tablet (complete protease inhibitor cocktail tablets, Roche) is added. Lysis and clarification are performed as described earlier for PA26. All subsequent steps are performed at 4°.

The clarified, filtered lysate is applied to the DEAE column. The loaded column is washed with 1 CV of TSD, pH 8.8. A gradient is run over 26 CV from TSD pH 8.8, to TSD400, pH 8.8. Five-milliliter fractions are collected, and the purest fractions, as estimated by SDS–PAGE, are pooled and concentrated by ultrafiltration (Amicon Stirred Cell with YM30 filter, Millipore), without exceeding a protein concentration of 20 mg/ml (determined by the Bradford assay). For example gels and chromatograms for these preparations, see Realini *et al.* (1997). Overconcentration at this stage can lead to the formation of soluble aggregates, as revealed by dynamic light scattering. Up to 2 ml of the concentrated sample is loaded onto the Superdex 200 column equilibrated in TSD200, pH 7.2, with additional runs performed as necessary for volumes larger than 2 ml. The protein is eluted with the same solution at 0.5 ml/min. After 100 ml of elution, 5-ml fractions are collected until 300 ml, with protein elution monitored by $A_{280}$. In the case of PA28$\alpha$, PA28$\gamma$, and heteroligomeric PA28$\alpha/\beta$, it is common to see three distinguishable peaks: a small peak corresponding to soluble aggregates eluting in the void volume, a predominant peak corresponding to the heptameric state eluting at ~160 ml, and a small peak representing monomeric protein eluting after 200 ml. In the case of PA28$\beta$, the majority of protein elutes at retention volumes expected for a dimer or monomer state. Fractions are assayed for activity and SDS–PAGE is used to identify the purest fractions. Recombinant human

PA28 proteins run as single bands at $\sim$32 kDa on SDS–PAGE. Fractions showing the greatest purity are pooled and concentration is determined by the Bradford method. The protein is typically dialyzed against TSD, pH 7.2, and stored at 4°. The protein can also be dialyzed against TSD, pH 7.2, + 10% glycerol, flash frozen in liquid nitrogen, and stored at –80°.

### *Preparation of Heteroligomeric PA28 Complexes*

PA28$\alpha$ and PA28$\beta$ copurify from tissues (Mott *et al.*, 1994), and the recombinant proteins preferentially form a heteroligomer (Realini *et al.*, 1997). Preparation of heteroligomers by mixing the separately purified recombinant proteins has been used to investigate the roles of the separate subunits (Song *et al.*, 1997; Zhang *et al.*, 1998b) and the heteroligomer stoichiometry (Zhang *et al.*, 1999). This approach has also been used to investigate the potential for "dominant-negative" effects that might be achieved by incorporating one or a few defective subunits into an assembled heptamer (Zhang *et al.*, 1998a). Typically, equal amounts of purified recombinant PA28$\alpha$ and PA28$\beta$ are mixed and incubated overnight at 4°. The complex is purified on a Superdex 200 column as described earlier for homomeric complexes.

## Generation and Selection of PA28 Activity Mutants

An efficient error-prone polymerase chain reaction (PCR) screen has been developed to generate PA28 single residue mutants that show diminished activity or altered specificity. The major advantage of this approach is that it rapidly surveys the importance of essentially all 11S residues without reliance upon prior mechanistic assumptions. For example, the first use of this method (Zhang *et al.*, 1998a) identified an internal nine-residue loop (the "activation loop") as being especially important for the stimulation of proteasome activity without making dramatic contributions to binding affinity. This analysis was performed prior to determination of the PA28$\alpha$ crystal structure and contributed significantly to mechanistic interpretation of the structure (Knowlton *et al.*, 1997). An adaptation of this approach was also used to find a point mutant of PA28$\gamma$ that displayed altered specificity for the hydrolysis of model substrates (Li *et al.*, 2001). For an illustration of the method, see Zhang *et al.* (1998a).

Identification of the activation loop (Zhang *et al.*, 1998a) used pAED4-REG$\alpha$ as the template for PCR with the following components: 0.2 m*M* each of dGTP and dATP; 1 m*M* each of dCTP and dTTP; 10 m*M* Tris-HCl, pH 8.3, 50 m*M* KCl, 7 m*M* $MgCl_2$, 0.01% gelatin, and 2.5 units of *Taq* DNA polymerase (Perkin-Elmer) for 100-$\mu$l reactions. After an initial melting at

94° for 4 min, 30 cycles of denaturation (1 min at 94°), annealing (1 min at 50°), and extension (3 min at 72°) are performed. Subsequent analysis demonstrated that this procedure generates mostly (60% of transformants) single site point mutants.

For the identification of active and inactive constructs, PCR products are ligated into pET11a, transformed, and colonies selected to inoculate 50-$\mu$l cultures (H medium: 10 tryptone, 8 g NaCl per liter water) in a 96-well microtiter plate and grown overnight at 30°. Aliquots from each well are transferred to the corresponding wells of another 96-well microtiter plate and induced with 0.4 m*M* IPTG at 30° for 2 h. Each aliquot of the induced cells is lysed by the addition of 30 $\mu$l of 20 m*M* Tris-HCl, pH 7.5, 1% Triton X–100, and 0.6 mg/ml polymixin B sulfate. Ten microliters of 17 ng/$\mu$l proteasome and 50 $\mu$l of 200 m*M* *N*-succinyl-Leu-Leu-Val-Tyr-7-amido-4-methylcoumarin (Sigma) are added to each well. The tray is incubated at 37° for 30 min and is then visualized under near-UV illumination. Colonies producing active PA28$\alpha$ are highly fluorescent after about 30 min of incubation at 37°; those producing inactive PA28$\alpha$ remain dark.

To distinguish valid nonactive mutants from clones that fail to express PA28$\alpha$, reaction mixtures are transferred to a nitrocellulose membrane with a dot blot apparatus and anti-PA28$\alpha$ antibodies are used in a standard Western blot protocol (Harlow, 1988). To further validate the results, clones that are inactive in the peptide degradation assay yet positive for expression are rescreened using the same protocol. Clones passing the second round of screening are further characterized by sequencing their plasmids.

## 11S–Proteasome Binding Assays

The use of purified proteins in peptide hydrolysis assays indicates that the activator and the proteasome bind to each other. Other more direct binding assays have also proven useful. Sizing chromatography is not generally helpful because dissociation usually occurs on the column. Competition activity assays, in which nonactivating mutant PA28 proteins are tested for their ability to impair stimulation by wild-type PA28, suffer from the possibility of subunit mixing between a wild-type and a mutant activator. Our two preferred approaches are (1) an ELISA, which has the advantage that it can provide estimates of binding constants, but requires specific antibodies for proteasome and activator and (2) velocity ultracentrifugation, which can be used for any proteasome–activator pair, without need for antibodies, and is especially valuable for studies with archaeal proteasomes, which have high constitutive peptidase activity and therefore do not display large levels of further stimulation upon addition of activator in fluorogenic peptide assays.

*ELISA Binding Assay*

Briefly, proteasomes are tethered to the well of an ELISA tray and incubated with a wild type or mutant activator. Unbound proteins are washed away. Bound activator is eluted with high salt buffer and quantified by immunoblot. The assay relies on a monoclonal antibody, MCP20 (a kind gift from K. Hendil, University of Copenhagen), which binds native proteasome without interfering with activator binding. This antibody is now available commercially (Affiniti Research Products).

The ELISA tray is prepared by coating the microtiter well with 200 $\mu$l of goat antimouse IgG at 20 $\mu$g/ml in 0.05 *M* carbonate, pH 9.6. Coated wells are rinsed three times in Tris-buffered saline containing 0.1% Tween 20 (TBS-T) and blocked with 200 $\mu$l of 1.5% nonfat milk in TBS-T for 2 h. Wells are filled with 200 $\mu$l of a 1:2500 dilution of ascites fluid containing MCP20 and incubated overnight at 4°. After three washes with TBS-T, each well is filled with 200 $\mu$l of human red cell proteasomes at 30 $\mu$g/ml in TBS-T and incubated overnight. The wells are washed three times with TBS-T and incubated with activator in 10 m*M* Tris, pH 7.5, for 20 min at 37° followed by an additional 150 min at 4°. Each well is quickly rinsed twice with an excess of cold 10 m*M* Tris, pH 7.5, 0.1% Tween 20 and once with 10 m*M* Tris, pH 7.5. The bound activator proteins are eluted with 200 $\mu$l of 0.5 *M* NaCl in 20 m*M* Tris, pH 7.5, and the high salt eluate is dot blotted onto nitrocellulose for subsequent detection with PA28-specific antibodies. Specific monoclonal antibodies to PA28$\alpha$, $\beta$, and $\gamma$ are available commercially from a number of sources, including Affiniti Research Products (Exeter, United Kingdom).

This technique can be used to determine the relative binding affinities for different activators through a competitive binding assay (Realini *et al.*, 1997). For example, this technique shows that PA28$\gamma$ has a greater affinity for proteasome than PA28$\alpha$ because PA28$\gamma$ outcompetes PA28$\alpha$ when an equimolar mix is allowed to bind to the tethered proteasome. This approach has also been used to demonstrate that although PA28$\beta$ alone has a low affinity for proteasome, presumably because of its low heptamerization affinity, the heteroligomeric PA28$\alpha$/$\beta$ complex has a greater affinity for proteasome than PA28$\gamma$.

*Sedimentation Velocity Binding Assay*

We collect sedimentation velocity data on a Beckman Optima XL-I analytical ultracentrifuge. Immediately before the centrifugation run, samples are dialyzed extensively against 20 m*M* Tris, pH 7.5, 200 m*M* NaCl, and 2 m*M* DTT and the used dialysis buffer is retained as a blank for background correction. Samples (proteasome alone, activator alone, various proteasome:activator ratios; final protein concentration 0.9 mg/ml) are centrifuged at 20° at a rotor speed of 42,000 rpm, and 200 interference measurements are recorded at 30-s intervals. Interference data are averaged and corrected for background against the blank. The program dcdt+ (Philo, 2000) is used for g(s*) analysis to determine values for the sedimentation coefficients. Proteasomes and activators from different species display different s* values; typical values are activator alone 8.5 s*, proteasome alone 18 s*, single capped proteasome–activator complex 22 s*, and double-capped proteasome–activator complex 24 s*.

## Crystallization

We have made extensive efforts at crystallizing 11S activators and their complexes with proteasomes. Efforts at cocrystallization have been assisted by the remarkable lack of species specificity for proteasome–11S interactions; proteasome from almost any species will be stimulated by almost any activator homolog. This observation is explained by the high degree of structural and sequence conservation in proteasome residues that contact 11S activators (Förster *et al.*, 2005). We have therefore employed a combinatorial approach to cocrystallization that screens many of the available 11S activators with proteasomes from a variety of species. Thus far we have succeeded in crystallizing human PA28$\alpha$ (Knowlton *et al.*, 1997) and complexes of PA26 with yeast (*S. cerevisiae*) (Whitby *et al.*, 2000) and archaeal (*Thermoplasma acidophilum*) proteasomes (Förster *et al.*, 2005).

### *Crystallization of Recombinant Human PA28α*

PA28$\alpha$, expressed and purified as described earlier, is dialyzed into 10 m*M* MOPS, pH 7.2, 0.2 m*M* EDTA, and 1 m*M* DTT (precrystallization buffer) at 4°. Note that it is important to maintain the protein at less than 20 mg/ml throughout the purification, as higher concentrations seem to induce the formation of soluble aggregates that poison crystallization (monitored by dynamic light scattering). The protein is concentrated to 10 mg/ml using YM30 Centripep concentrators (Millipore), with concentration determined by Bradford assay. Crystals grow at 4° in sitting drops against a reservoir of 500 $\mu$l of 12% polyethylene glycol 6000 (PEG 6000), 0.1 *M* MOPS, pH 7.2, 2 *M* NaCl, and 0.9 m*M* $ZnCl_2$ . The drops are 2 $\mu$l of protein solution mixed with 2 $\mu$l of reservoir solution. Crystals normally grow in 5–7 days as thin plates. The highest quality crystals are grown from protein that has been freshly purified. Crystals are transferred to a cryoprotectant solution of 20% glycerol, 13.5% PEG 6000, 2 *M* NaCl, and 100 m*M* MOPS, pH 7.1, suspended in a rayon loop, and cooled for data collection by plunging into liquid nitrogen.

*Crystallization of PA26–Yeast Proteasome Complex*

Recombinant *T. brucei* PA26 is purified as described earlier, and the histidine-tagged yeast proteasome is prepared as described previously (Whitby *et al.*, 2000). PA26 and proteasome are combined in a molar ratio of 2.5:1, dialyzed against 10 m*M* Tris-HCl, pH 7.5, 1 m*M* EDTA, and 1 m*M* DTT (precrystal buffer), and concentrated to 10 mg/ml (determined by Bradford) with YM30 Centripep concentrators (Millipore) that have been washed in fresh precrystal buffer. It is important that PA26 is concentrated in the presence of proteasome, as it is prone to precipitate out of solution in isolation. Crystals are grown in sitting drops at 4° using a 500-$\mu$l reservoir of 0.1 *M* sodium HEPES, pH 7.5, 40% 2,4-methylpentanediol, and 0.2 *M* NaCl. The drop is composed of 4 $\mu$l of protein and 4 $\mu$l of reservoir solution. Crystals grow as rectangular blocks of 200 $\mu$m in the longest dimension and 50 $\mu$m in the smaller dimension after ~5–6 weeks of undisturbed growth. It is important not to move the trays during this time. Crystal growth is infrequent. Many preparations never yield crystals and even for those that do, X-ray grade crystals are generally only found in ~1 out of 24 drops. Crystals are prepared for data collection by suspending in a nylon loop directly from the drop and plunging into liquid nitrogen.

*Crystallization of PA26–Archaeal Proteasome Complexes*

PA26 and *T. acidophilum* proteasome are mixed at a molar ratio of 2.5:1 and concentrated to 10 mg/ml. Crystals grow at 21° in hanging drops. Optimization of conditions utilized the additive approach of Majeed *et al.* (2003). The reservoir solution is 90% (0.1 *M* Na citrate/phosphate, pH 4.2, 0.2 *M* $Li_2SO_4$, 15% PEG 1000) + 10% (1 *M* imidazole, pH 7.0). Sometimes, the initial crystals can be further improved by "feeding" (Bergfors, 2003), in which 2 $\mu$l of fresh protein solution is added to drops that contain a spray of tiny crystals. For data collection, crystals are transferred to 0.1 *M* Na citrate/phosphate, pH 5.7, 0.2 *M* $Li_2SO_4$, 20% PEG 1000, and 30% glycerol, suspended in a nylon loop, and plunged into liquid nitrogen.

## Acknowledgments

We thank Martin Rechsteiner for critical reading and comments on the manuscript and for the leading role he has played on proteasome activator research at Utah. We also thank C. C. Wang for many valuable discussions on PA26. Research on proteasome activators is supported in our laboratories by NIH Grants RO1 GM59135 (Hill) and RO1 GM60334 (Rechsteiner).

## References

Benaroudj, N., Zwickl, P., Seemüller, E., Baumeister, W., and Goldberg, A. L. (2003). ATP hydrolysis by the proteasome regulatory complex PAN serves multiple functions in protein degradation. *Mol. Cell* **11,** 69–78.

Bergfors, T. (2003). Seeds to crystals. *J. Struct. Biol.* **142,** 66–76.

Bochtler, M., Ditzel, L., Groll, M., Hartmann, C., and Huber, R. (1999). The proteasome. *Annu. Rev. Biophys. Struct.* **28,** 295–317.

Bogyo, M., Shin, S., McMaster, J. S., and Ploegh, H. L. (1998). Substrate binding and sequence preference of the proteasome revealed by active-site-directed affinity probes. *Chem. Biol.* **5,** 307–320.

Cascio, P., Call, M., Petre, B. M., Walz, T., and Goldberg, A. L. (2002). Properties of the hybrid form of the 26S proteasome containing both 19S and PA28 complexes. *EMBO J.* **21,** 2636–2645.

De Martino, G. N., and Slaughter, C. A. (1999). The proteasome, a novel protease regulated by multiple mechanisms. *J. Biol. Chem.* **274,** 22123–22126.

Dick, T. P., Ruppert, T., Groettrup, M., Kloetzel, P. M., Kuehn, L., Koszinowski, U. H., Stevanovic, S., Schild, H., and Rammensee, H. G. (1996). Coordinated dual cleavages induced by the proteasome regulator PA28 lead to dominant MHC ligands. *Cell* **86,** 253–262.

Dubiel, W., Pratt, G., Ferrell, K., and Rechsteiner, M. (1992). Purification of an 11S regulator of the multicatalytic protease. *J. Biol. Chem.* **267,** 22369–22377.

Förster, A., Whitby, F. G., Robinson, H., and Hill, C. P. (2005). The 1.9 Å structure of a proteasome 11 S activator complex and implication for proteasome-PAN/PA interactions. *Mol. Cell* **18,** 589–599.

Groll, M., Ditzel, L., Löwe, J., Stock, D., Bochtler, M., Bartunik, H. D., and Huber, R. (1997). Structure of 20S proteasome from yeast at 2.4Å resolution. *Nature* **386,** 463–471.

Groll, M., Nazif, T., Huber, R., and Bogyo, M. (2002). Probing structural determinants distal to the site of hydrolysis that control substrate specificity of the 20S proteasome. *Chem. Biol.* **9,** 655–662.

Harlow, E. L. D. (1988). *In* "Antibodies," pp. 493–506. Cold Spring Harbor Press, Cold Spring Harbor, NY.

Harris, J. L., Alper, P. B., Li, J., Rechsteiner, M., and Backes, B. J. (2001). Substrate specificity of the human proteasome. *Chem. Biol.* **8,** 1131–1141.

Hendil, K. B., Khan, S., and Tanaka, K. (1998). Simultaneous binding of PA28 and PA700 activators to 20S proteasomes. *Biochem. J.* **332,** 749–754.

Hill, C. P., Masters, E. I., and Whitby, F. G. (2002). The 11S regulators of 20S proteasome activity. *Curr. Top. Microbiol. Immunol.* **268,** 73–89.

Johnston, S. C., Whitby, F. G., Realini, C., Rechsteiner, M., and Hill, C. P. (1997). The proteasome 11S regulator subunit REG alpha (PA28 alpha) is a heptamer. *Protein Sci.* **6,** 2469–2473.

Knowlton, J. R., Johnston, S. C., Whitby, F. G., Realini, C., Zhang, Z., Rechsteiner, M., and Hill, C. P. (1997). Structure of the proteasome activator REGalpha (PA28alpha). *Nature* **390,** 639–643.

Li, J., Gao, X., Ortega, J., Nazif, T., Joss, L., Bogyo, M., Steven, A. C., and Rechsteiner, M. (2001). Lysine 188 substitutions convert the pattern of proteasome activation by REG$\gamma$ to that of REGs $\alpha$ and $\beta$. *EMBO J.* **20,** 3359–3369.

Löwe, J., Stock, D., Jap, B., Zwickl, P., Baumeister, W., and Huber, R. (1995). Crystal structure of the 20S proteasome from the archaeon T. acidophilum at 3.4 A resolution. *Science* **268,** 533–539.

Ma, C. P., Slaughter, C. A., and De Martino, G. N. (1992). Identification, purification, and characterization of a protein activator (PA28) of the 20 S proteasome (macropain). *J. Biol. Chem.* **267,** 10515–10523.

Majeed, S., Ofek, G., Belachew, A., Huang, C. C., Zhou, T., and Kwong, P. D. (2003). Enhancing protein crystallization through precipitant synergy. *Structure (Cambr.)* **11,** 1061–1070.

Mott, J. D., Pramanik, B. C., Moomaw, C. R., Afendis, S. J., De Martino, G. N., and Slaughter, C. A. (1994). PA28, an activator of the 20 S proteasome, is composed of two nonidentical but homologous subunits. *J. Biol. Chem.* **269,** 31466–31471.

Philo, J. S. (2000). A method for directly fitting the time derivative of sedimentation velocity data and an alternative algorithm for calculating sedimentation coefficient distribution functions. *Anal. Biochem.* **279,** 151–163.

Rechsteiner, M., and Hill, C. P. (2005). Mobilizing the proteolytic machine: Cell biological roles of proteasome activators and inhibitors. *Trends. Cell Biol.* **15,** 27–33.

Rechsteiner, M., Realini, C., and Ustrell, V. (2000). The proteasome activator 11S REG (PA28) and class I antigen presentation. *Biochem. J.* **345,** 1–15.

Realini, C., Jensen, C. C., Zhang, Z., Johnston, S. C., Knowlton, J. R., Hill, C. P., and Rechsteiner, M. (1997). Characterization of recombinant REGalpha, REGbeta, and REGgamma proteasome activators. *J. Biol. Chem.* **272,** 25483–25492.

Song, X., von Kampen, J., Slaughter, C. A., and De Martino, G. N. (1997). Relative functions of the $\alpha$ and $\beta$ subunits of the proteasome activator, PA28. *J. Biol. Chem.* **272,** 27994–28000.

Tanahashi, N., Murakami, Y., Minami, Y., Shimbara, N., Hendil, K. B., and Tanaka, K. (2000). Hybrid proteasomes. Induction by interferon-$\gamma$ and contribution to ATP-dependent proteolysis. *J. Biol. Chem.* **275,** 14336–14345.

Unno, M., Mizushima, T., Morimoto, Y., Tomisugi, Y., Tanaka, K., Yasuoka, N., and Tsukihara, T. (2002). The structure of the mammalian 20S proteasome at 2.75 Å resolution. *Structure* **10,** 609–618.

Ustrell, V., Hoffman, L., Pratt, G., and Rechsteiner, M. (2002). PA200, a nuclear proteasome activator involved in DNA repair. *EMBO J.* **21,** 3516–3525.

Voges, D., Zwickl, P., and Baumeister, W. (1999). The 26S proteasome: A molecular machine designed for controlled proteolysis. *Annu. Rev. Biochem.* **68,** 1015–1068.

Wang, C. C., Bozdech, Z., Liu, C. L., Shipway, A., Backes, B. J., Harris, J. L., and Bogyo, M. (2003). Biochemical analysis of the 20 S proteasome of *Trypanosoma brucei. J. Biol. Chem.* **278,** 15800–15808.

Wenzel, T., and Baumeister, W. (1995). Conformational constraints in protein degradation by the 20S proteasome. *Nature Struct. Biol.* **2,** 199–204.

Whitby, F. G., Masters, E. I., Kramer, L., Knowlton, J. R., Yao, Y., Wang, C. C., and Hill, C. P. (2000). Structural basis for the activation of 20S proteasomes by 11S regulators. *Nature* **408,** 115–120.

Yao, Y., Huang, L., Krutchinsky, A., Wong, M. L., Standing, K. G., Burlingame, A. L., and Wang, C. C. (1999). Structural and functional characterization of the proteasome-activating protein PA26 from *Trypanosoma brucei. J. Biol. Chem.* **274,** 33921–33930.

Zhang, Z., Clawson, A., Realini, C., Jensen, C. C., Knowlton, J. R., Hill, C. P., and Rechsteiner, M. (1998a). Identification of an activation region in the proteasome activator REGalpha. *Proc. Natl. Acad. Sci. USA* **95,** 2807–2811.

Zhang, Z., Clawson, A., and Rechsteiner, M. (1998b). The proteasome activator or PA28: Contribution by both $\alpha$ and $\beta$ subunits to proteasome activation. *J. Biol. Chem.* **273,** 30660–30668.

Zhang, Z., Kruchinsky, A., Endicott, S., Realini, C., Rechsteiner, M., and Standing, K. G. (1999). Proteasome activator 11S REG or PA28: Recombinant REG$\alpha$/REG$\beta$ hetero-oligomers are heptamers. *Biochemistry* **38,** 5651–5658.

Zhang, Z., Realini, C., Clawson, A., Endicott, S., and Rechsteiner, M. (1998c). Proteasome activation by REG molecules lacking homolog-specific inserts. *J. Biol. Chem,* **273,** 9501–9509.

# [26] Purification and Assay of Proteasome Activator PA200

*By* VICENÇA USTRELL, GREGORY PRATT, CARLOS GORBEA, and MARTIN RECHSTEINER

## Abstract

PA200, the most recently discovered activator of the 20S proteasome, is a nuclear protein thought to play a role in DNA repair. Homologs of PA200 have been found in rat, frog, birds, worms, and budding yeast, where it is called Blm3p (now known as Blm10p), but not in *Drosophila* or fission yeast. Western blots of SDS–PAGE transfers reveal 160 and 200K forms of mammalian PA200, and organ surveys demonstrate that the 200K species is highest in testis. PA200 purified from bovine testis binds the ends of the cylindrical 20S proteasome, forming volcano-shaped structures in negatively stained EM images. *In vitro* assays demonstrate that binding of PA200 activates peptide hydrolysis by the 20S proteasome. This chapter describes the purification and assay of bovine testis PA200.

## Introduction

The 20S proteasome is a major component of eukaryotic cells, accounting for as much as 1% of soluble cellular protein (Hendil and Hartmann-Petersen, 2004). The enzyme is found in the nucleus and cytosol, but not within membrane-enclosed organelles other than the nucleus (Gordon, 2002; Wojcik and DeMartino, 2003). Indiscriminate degradation of cellular proteins is prevented by sequestration of the catalytic sites of the proteasome within a central chamber of the cylindrical particle (Groll and Clausen, 2003; Heinemeyer *et al.*, 2004). The 20S proteasome becomes a functional protease when it binds other cellular components that open channels to the central chamber. One such protein is the 19S regulatory complex, a multimeric protein that associates with the 20S proteasome to form the 26S proteasome (Bajorek and Glickman, 2004; Ferrell *et al.*, 2000;

METHODS IN ENZYMOLOGY, VOL. 398

0076-6879/05 $35.00
DOI: 10.1016/S0076-6879(05)98026-9

Gorbea *et al.*, 1999). This larger enzyme degrades ubiquitylated proteins in the presence of ATP. Other activators are heptameric rings known as PA28. Unlike the 19S regulatory complex, PA28 activators do not increase the degradation of intact proteins. They do, however, greatly enhance the energy-independent degradation of peptides by the proteasome (Hill *et al.*, 2002; Rechsteiner *et al.*, 2000).

We have characterized a novel proteasome activator purified from bovine testis (Ustrell *et al.*, 2002). Because the protein is a monomer of approximately 200 kDa, it is called PA200 for proteasome activator with a molecular weight of 200 K. Peptide sequences obtained from PA200 map onto a hypothetical translation product of a cDNA (KIAA0077) obtained from human KG1 myeloid cells. Proteins with high homology to PA200 have been found in rat (AF_296169.1), chicken (XM_419293.1), and frog (BC_070702.1). Homologs are also present in the worm (NM_073752.1), in the mosquito (XM_312339.1), and in the budding yeast (NP_116648.2). Bioinformatic searches have identified HEAT-like motif repeats in the PA200 molecule (Kajava *et al.*, 2004). Similar structural motifs are also found in two subunits of the 19S regulatory complex, S1 and S2, and in the proteasome regulator Ecm29.

Immunofluorescence studies have shown that PA200 is located in the nucleus of HeLa cells. Like the heptameric PA28 proteasome activators, PA200 increases the rate of peptide cleavage by the 20S proteasome, but it does not promote the degradation of intact proteins. However, the pattern of activation by mammalian PA200 differs from that of PA28$\alpha\beta$. Whereas both activators enhance catalysis at all three proteasome active sites, PA200 binding results in greater activation of cleavage after acidic residues than after basic or hydrophobic amino acids.

The 20S proteasome is involved in a number of essential biological processes usually when complexed to either the 19S and/or to the 11S regulators (Pickart and Cohen, 2004). We proposed that PA200 has a role in the DNA damage response perhaps by recruiting proteasomes to sites of DNA damage. Support for this hypothesis includes (1) the presence of high levels of PA200 in testis, where meiosis generates numerous DNA double strand breaks; (2) localization of PA200 to nuclei of cultured mammalian cells; and (3) the fact that, like many DNA repair proteins, PA200 forms foci upon exposing cells to ionizing radiation. Also there was an early report that mutation of BLM3, the yeast homolog of PA200, rendered yeast hypersensitive to the radiomimetic agent bleomycin (Moore, 1991). However, this piece of evidence is now questionable, as deletion of BLM3 does not cause bleomycin hypersensitivity (Aouida *et al.*, 2004). Still the apparent association of Blm3p with Sir4p (Gavin *et al.*, 2002; Ho *et al.*, 2002), a protein that moves to DNA double strand breaks (Martin *et al.*,

1999; Mills *et al.*, 1999), can be interpreted in favor of a connection between PA200/Blm3p and DNA repair. Blm3p has been reported to be involved in proteasome maturation, and surprisingly it was not found to be an activator, but rather was found to inhibit the yeast 20S proteasome (Fehlker *et al.*, 2003). Clearly more studies are needed to elucidate the function(s) of PA200–proteasome complexes, as the protein is likely to play an important, although currently undefined, role in cellular metabolism.

## Purification of PA200 from Bovine Testis

Our approach for purifying PA200 involves two rounds of DEAE chromatography followed by gel filtration. During size-exclusion chromatography, PA200 molecules dissociate from the 20S proteasome and can be purified further. Chromatography on Mono Q and glycerol gradient centrifugation are employed to purify PA200 to homogeneity. Throughout purification the location of PA200 is monitored by Western blotting with antibodies to peptides within the PA200 sequence (Affinity Bioreagents, Inc; Golden, CO).

### *Materials*

Frozen bovine testes are from Pel-Freez Biologicals (Rogers, AR). TSK-DEAE-650 M is from Supelco (Bellefonte, PA). Q Sepharose, Superdex 200, and Mono Q are from Amersham-Pharmacia. Fluorogenic peptide substrates LLVY-MCA, LY-MCA, and LLE-$\beta$NA are from Sigma; LRR-MCA and IETD-MCA are from Peptides International, Inc. (Louisville, KY). LSTR-MCA is from Peninsula Laboratories, Inc. (Belmont, CA). The Complete protease inhibitor cocktail is from Roche Molecular Biochemicals (Mannheim, Germany).

### *Extract Preparation and First DEAE*

Frozen bovine testicles are shattered with a hammer and chisel, the outer connective tissue (the albuginea) is removed, and the remaining tissue is cut into small pieces. Diced tissue (200–600 g) is homogenized in a Waring blender in 2.4 volumes of 0.25% Triton X-100, 10 m*M* Tris, pH 7.0, and 1 m*M* dithiothreitol (DTT) to which 1 tablet of Complete protease inhibitor cocktail (Aniento *et al.,* 1996) is added per 50 ml. The homogenate is centrifuged at 100,000*g* for 60 min in a Beckman TY 50.2 rotor (or equivalent), and the supernate is batch adsorbed to 500 ml of DEAE 650-M resin equilibrated in 10 m*M* Tris, pH 7.0, 25 m*M* KCl, 10 m*M* NaCl, 5.5 m*M* $MgCl_2$, 0.1 m*M* EDTA, 1 m*M* DTT, and 10% glycerol (TSDG). The resin is then loaded into a Pharmacia XK50 column and washed with 1

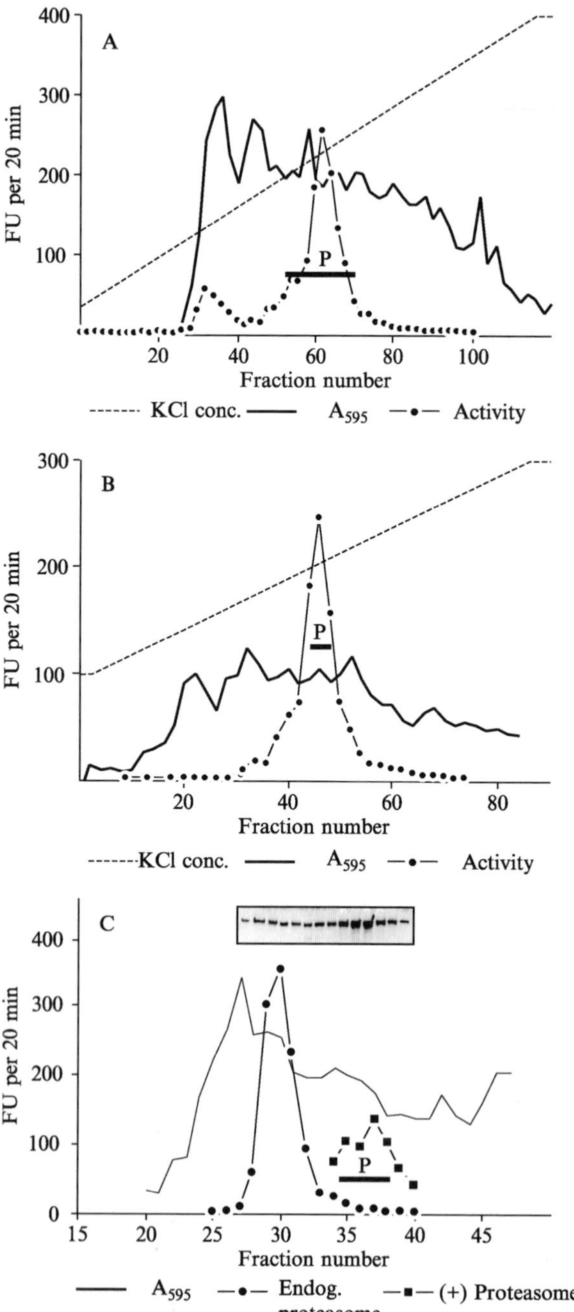
400
300
200
100
A
P
FU per 20 min
20
40
60
80
100
Fraction number
KCl conc.
$A_{595}$
Activity
300
200
100
B
P
FU per 20 min
20
40
60
80
Fraction number
KCl conc.
$A_{595}$
Activity
C
400
300
200
100
0
P
FU per 20 min
15
20
25
30
35
40
45
Fraction number
$A_{595}$
Endog. proteasome
(+) Proteasome

volume of TSDG. The column is developed at 1.5 ml/min with 5 column volumes of a linear 0 to 400 m*M* KCl gradient in TSDG (15-ml fractions are convenient). PA200 elutes from this column at about 225 m*M* Kcl, roughly coincident with 20S proteasome activity (see Fig. 1A).

*Second DEAE and Gel Filtration*

The entire region of 20S proteasome activity is pooled (solid bar in Fig. 1A), diluted with 2 volumes of TSDG, and applied to a 100-ml DEAE-65M column. The column is rinsed with 100 ml of TSDG and is developed with 1 liter of a linear gradient of 0–300 m*M* KCl in TSDG. Fractions of 5 ml are collected and assayed for 20S proteasome activity as well as for PA200 distribution by Western blotting. Once again the elution of PA200 closely mirrors 20S proteasome activity (see Fig. 1B). These fractions are pooled, diluted with an equal volume of TSDG, and concentrated by application to a 20-ml column of Pharmacia Q Sepharose Fast Flow resin in TSDG. The column is rinsed with 1 volume of TSDG, and proteins are eluted with 750 m*M* KCl in TSDG. Portions of the eluate (less than 15 mg/ml protein; less than 50 mg protein total) are applied to a Pharmacia Superdex 200 26/60 column equilibrated with TSDG containing 150 m*M* KCl. Apparently high salt elution from Q Sepharose and/or the slow gel filtration results in enough PA200-20S proteasome complexes dissociating that activation by free PA200 molecules can be assayed in fractions just behind the 20S proteasome (see Fig. 1C).

*Purification of Free PA200 on Mono Q and Glycerol Gradients*

Fractions from the Superdex 200 column containing free PA200 are applied to a 1-ml Mono Q column in TSDG containing 125 m*M* KCl, and the column is developed with a 20-ml linear gradient of 125 to 500 m*M* KCl in TSDG. Fractions of 0.25 ml are collected and PA200 is located by Western blotting. PA200-enriched fractions are pooled and sedimented on a 5 to 20% glycerol gradient for 19 h at 25,000 rpm in a SW28 rotor. PA200 is the deepest sedimenting protein and appears to be a mixture of

FIG. 1. Profiles obtained from the first three chromatographic steps in the purification of bovine testis PA200. Two DEAE columns and a gel filtration step constitute the initial steps in PA200 purification (see text for details). The distribution of PA200 closely mirrors proteasome activity in both DEAE steps (A and B) so it is safe to pool (P) based on 20S activity. During gel filtration, considerable amounts of PA200 dissociate from the proteasome and elute just behind the majority of 20S activity. The free PA200 can be detected by adding small amounts of 20S proteasome, which results in increased peptide hydrolysis due to PA200 (fractions 34 to 40 in C). (C, inset) A Western blot showing the distribution of PA200 in the various fractions.

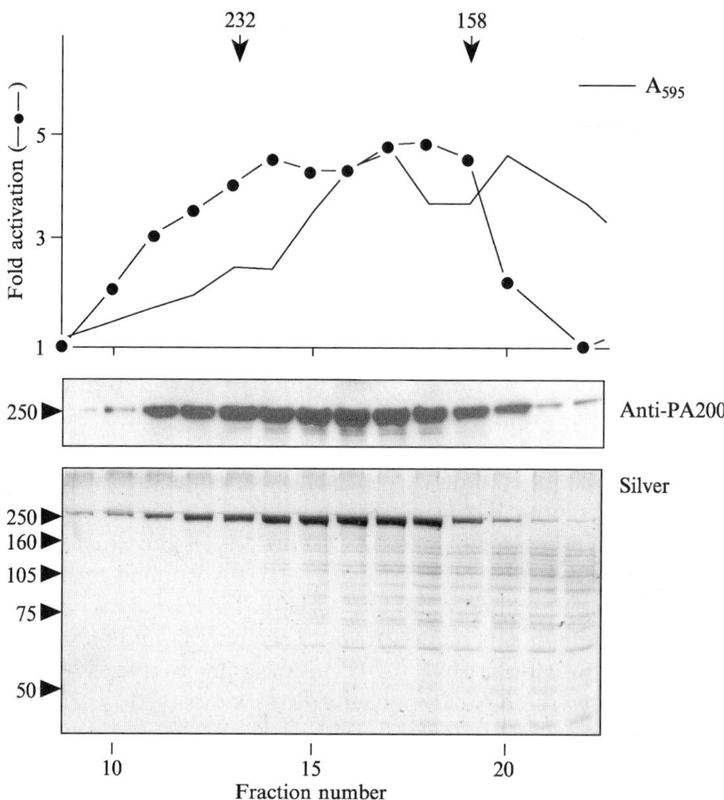


FIG. 2. Glycerol gradient sedimentation of bovine testis PA200. Profile obtained when the Mono Q pool was centrifuged on a 30-ml 5–20% glycerol gradient for 19 h at 85,500*g*. Activation and protein distribution (top), a Western blot (middle), and a silver-stained SDS–PAGE gel of the various gradient fractions (bottom).

monomers and dimers (Fig. 2). Several fractions of the glycerol gradient contain PA200 at greater than 90% purity.

### *Expected Yield and Purity*

Western blots and a protein-stained SDS–PAGE gel of various fractions from a typical purification are presented in Fig. 3. One can expect to recover about 600 $\mu$g of PA200 from 625 g of testis. The purified protein can be stored at –80° without significant loss of activity for at least a 6-month period.

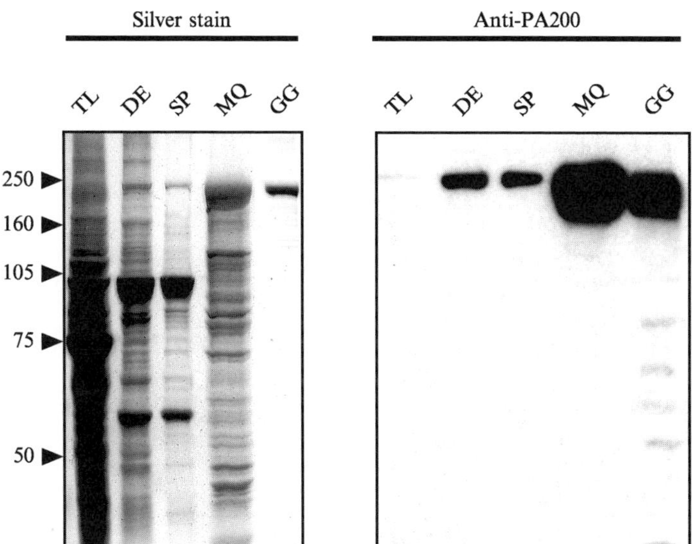


FIG. 3. SDS–PAGE analysis of pools obtained during PA200 purification. Samples are testis lysate (TL), second DEAE pool (DE), sizing pool (SP), Mono Q pool (MQ) and final glycerol gradient (GG).

## PA200 Activation Assay

Endogenous proteasome activity is measured by adding 5-$\mu$l aliquots from each column fraction to 100 $\mu$l of 100 $\mu M$ LLVY-MCA in 20 m$M$ Tris, pH 7.8, 5 m$M$ $MgCl_2$, 10 m$M$ KCl, and 1 m$M$ DTT. After incubating at 37° for 20 min, reactions are quenched by the addition of 200 $\mu$l ETOH, and fluorescence is measured using a spectrofluorimeter at an excitation wavelength of 380 nm and an emission wavelength of 440 nm with MCA peptide substrates and at an excitation of 335 nm and emission of 410 nm for LLE-$\beta$NA. Proteasome activation is assayed by adding 5 $\mu$l of the various fractions in a 100-$\mu$l reaction mix containing 200 ng of purified bovine red blood cell (RBC) proteasome and 100 $\mu M$ LLVY-MCA. After 20 min of incubation the reactions are quenched with 200 $\mu$l ETOH. Purified PA200 binding and activation of the proteasome are assayed by incubating increasing concentrations of PA200 (145 to 435 ng) with 200 ng of purified bovine RBC proteasome for 30 min at room temperature. An eightfold molar ratio of PA200 to 20S proteasome is saturating for activation.

## References

Aniento, F., Papavassiliou, A. G., Knecht, E., and Roche, E. (1996). Selective uptake and degradation of c-Fos and v-Fos by rat liver lysosomes. *FEBS Lett.* **390,** 47–52.

Aouida, M., Page, N., Leduc, A., Peter, M., and Ramotar, D. (2004). A genome-wide screen in *Saccharomyces cerevisiae* reveals altered transport as a mechanism of resistance to the anticancer drug bleomycin. *Cancer Res.* **64,** 1102–1109.

Bajorek, M., and Glickman, M. H. (2004). Keepers at the final gates: Regulatory complexes and gating of the proteasome channel. *Cell Mol. Life Sci.* **61,** 1579–1588.

Fehlker, M., Wendler, P., Lehmann, A., and Enenkel, C. (2003). Blm3 is part of nascent proteasomes and is involved in a late stage of nuclear proteasome assembly. *EMBO Rep.* **4,** 959–963.

Ferrell, K., Wilkinson, C. R., Dubiel, W., and Gordon, C. (2000). Regulatory subunit interactions of the 26S proteasome, a complex problem. *Trends Biochem. Sci.* **25,** 83–88.

Gavin, A. C., Bosche, M., Krause, R., Grandi, P., Marzioch, M., Bauer, A., Schultz, J., Rick, J. M., Michon, A. M., Cruciat, C. M., Remor, M., Hofert, C., Schelder, M., Brajenovic, M., Ruffner, H., Merino, A., Klein, K., Hudak, M., Dickson, D., Rudi, T., Gnau, V., Bauch, A., Bastuck, S., Huhse, B., Leutwein, C., Heurtier, M. A., Copley, R. R., Edelmann, A., Querfurth, E., Rybin, V., Drewes, G., Raida, M., Bouwmeester, T., Bork, P., Seraphin, B., Kuster, B., Neubauer, G., and Superti-Furga, G. (2002). Functional organization of the yeast proteome by systematic analysis of protein complexes. *Nature* **415,** 141–147.

Gorbea, C., Taillandier, D., and Rechsteiner, M. (1999). Assembly of the regulatory complex of the 26S proteasome. *Mol. Biol. Rep.* **26,** 15–19.

Gordon, C. (2002). The intracellular localization of the proteasome. *Curr. Top. Microbiol. Immunol.* **268,** 175–184.

Groll, M., and Clausen, T. (2003). Molecular shredders: How proteasomes fulfill their role. *Curr. Opin. Struct. Biol.* **13,** 665–673.

Heinemeyer, W., Ramos, P. C., and Dohmen, R. J. (2004). The ultimate nanoscale mincer: Assembly, structure and active sites of the 20S proteasome core. *Cell Mol. Life Sci.* **61,** 1562–1578.

Hendil, K. B., and Hartmann-Petersen, R. (2004). Proteasomes: A complex story. *Curr. Protein Pept. Sci.* **5,** 135–151.

Hill, C. P., Masters, E. I., and Whitby, F. G. (2002). The 11S regulators of 20S proteasome activity. *Curr. Top. Microbiol. Immunol.* **268,** 73–89.

Ho, Y., Gruhler, A., Heilbut, A., Bader, G. D., and Moore, L. (2002). Systematic identification of protein complexes in *Saccharomyces cerevisiae* by mass spectrometry. *Nature* **415,** 180–183.

Kajava, A. V., Gorbea, C., Ortega, J., Rechsteiner, M., and Steven, A. C. (2004). New HEAT-like repeat motifs in proteins regulating proteasome structure and function. *J. Struct. Biol.* **146,** 425–430.

Martin, S. G., Laroche, T., Suka, N., Grunstein, M., and Gasser, S. M. (1999). Relocalization of telomeric Ku and SIR proteins in response to DNA strand breaks in yeast. *Cell* **97,** 621–633.

Mills, K. D., Sinclair, D. A., and Guarente, L. (1999). MEC1-dependent redistribution of the Sir3 silencing protein from telomeres to DNA double-strand breaks. *Cell* **97,** 609–620.

Moore, C. W. (1991). Further characterizations of bleomycin-sensitive (blm) mutants of *Saccharomyces cerevisiae* with implications for a radiomimetic model. *J. Bacteriol.* **173,** 3605–3608.

Pickart, C. M., and Cohen, R. E. (2004). Proteasomes and their kin: Proteases in the machine age. *Nature Rev. Mol. Cell. Biol.* **5,** 177–187.

Rechsteiner, M., Realini, C., and Ustrell, V. (2000). The proteasome activator 11S REG (PA28) and class I antigen presentation. *Biochem. J.* **345,** 1–15.
Ustrell, V., Hoffman, L., Pratt, G., and Rechsteiner, M. (2002). PA200, a nuclear proteasome activator involved in DNA repair. *EMBO J.* **21,** 3403–3412.
Wojcik, C., and De Martino, G. N. (2003). Intracellular localization of proteasomes. *Int. J. Biochem. Cell Biol.* **35,** 579–589.

# [27] Purification, Crystallization, and X-Ray Analysis of the Yeast 20S Proteasome

*By* MICHAEL GROLL and ROBERT HUBER

## Abstract

Intracellular protein degradation is one of the most precisely regulated processes in living cells. The main component of the degradation machinery is the 20S proteasome present in eukaryotes as well as in prokaryotes. We have developed successful purification protocols for the 20S proteasome in its native state using an affinity tag strategy. This chapter describes in detail the purification protocols, proteolytic activity assays, crystallization, and structure determination for the yeast 20S proteasome. The crystal structure of the eukaryotic proteasome opens new possibilities for identifying, characterizing, and elucidating the mode of action for natural and synthetic inhibitors, which affect its function. Some of these compounds may find therapeutic applications in contemporary medicine.

## Introduction

At present, structural and functional characterization of eukaryotic heteromeric protein complexes remains a big challenge. Their elaborate architecture, instability, and transient association to other cellular proteins make purification and handling of such large protein complexes extremely difficult. Usually, conventional protein purification methods involve fractionation on ion-exchange resins and, consequently, exposure to high concentrations of salt, which may alter the composition and purity of protein complexes. Additionally, cellular proteases, which can only be removed stepwise during purification, can partially degrade surface-exposed flexible regions of complex subunits, causing inhomogeneities, which may lead to difficulties in crystal formation.

20S proteasomes show a compact fold due to tight interactions between their subunits and can be purified to homogeneity relatively easily, thus

METHODS IN ENZYMOLOGY, VOL. 398

0076-6879/05 $35.00
DOI: 10.1016/S0076-6879(05)98027-0

representing an exception (Groll *et al.*, 1997). Difficulties that arose in the purification of mammalian 20S proteasomes were due to the existence of various subpopulations, as mammalian proteasomes contain interferon-$\gamma$-inducible interchangeable $\beta$-type subunits (Haass and Kloetzel, 1989). This led to substoichiometries in the overall architecture of 20S particles and caused tremendous difficulties in crystallization (Morimoto *et al.*, 1995). In contrast, yeast 20S proteasomes lack inducible subunits and form a homogeneous protein population, which makes them well suited for crystallization (Heinemeyer *et al.*, 1991). We have developed a native purification protocol for the eukaryotic 20S proteasome from *Saccharomyces cerevisiae* (Groll *et al.*, 1997). At that time, affinity tagging strategies were not available for eukaryotes.

Fusion of different affinity tags (hexa-histidine, GST, or protein A) to the target proteins, which is widely used nowadays, significantly simplifies the purification process and provides an effective tool to obtain pure proteins in a relatively short time, as compared to conventional purification protocols (Schneider *et al.*, 1995; Wach *et al.*, 1997). However, the recombinant tagged proteins have a serious drawback in terms of protein crystallization: a hexa-histidine tag exposed on the surface of a molecule displays a flexible extension of about 21 Å, which, in most cases, disturbs crystal packing or may even prevent the formation of crystals. GST and protein A (ProA) form compact fusion domains, but may also interfere with crystallization because of the high flexibility in the hinge-spacer region. A new method of modifying eukaryotic proteins with cleavable affinity tags (ProA or GST) has been developed (Knop *et al.*, 1999). Proteolytic cleavage sites (TEV protease, PreScission protease) followed by the fusion protein and a dominant kanMX6-resistant marker are surrounded by flanking homologous sequences of the desired gene. This module is inserted into the chromosomal yeast DNA by homologous recombination and allows capture of the protein of interest from the crude cell extract by one-step affinity purification. Mild buffer conditions applied during purification make it possible to isolate native protein complexes without the problems caused by substoichiometric components and other inhomogeneities (Knop *et al.*, 1999). This method has already been used successfully for purification of the yeast proteasome (Leggett *et al.*, 2002; M. Groll, unpublished results).

This chapter describes a detailed purification protocol for the isolation of the yeast 20S proteasome in its native state from wild-type yeast cells and with a ProA affinity tag chromosomally introduced to the proteasomal subunit $\beta 2$ by homologous recombination. Furthermore, crystallization and structure determination of the yeast 20S proteasome are discussed.

## Native Purification of the Yeast 20S Proteasome

Eukaryotic heterooligomeric multifunctional complexes cannot be overexpressed in *Escherichia coli*, as they represent elaborate assemblies, consisting of different subunits. One way to obtain these proteins is to purify them in their native state. An efficient preparation of the 20S proteasome from yeast was established, giving reproducible yields of about 50 mg pure and crystallizable protein from 500 g of yeast cells. Yeast cells (*S. cerevisiae*) are from Bäko (München) (see Fig. 1A). The cells are washed twice with ice-cold water and resuspended in 20 m*M* Tris-HCl,

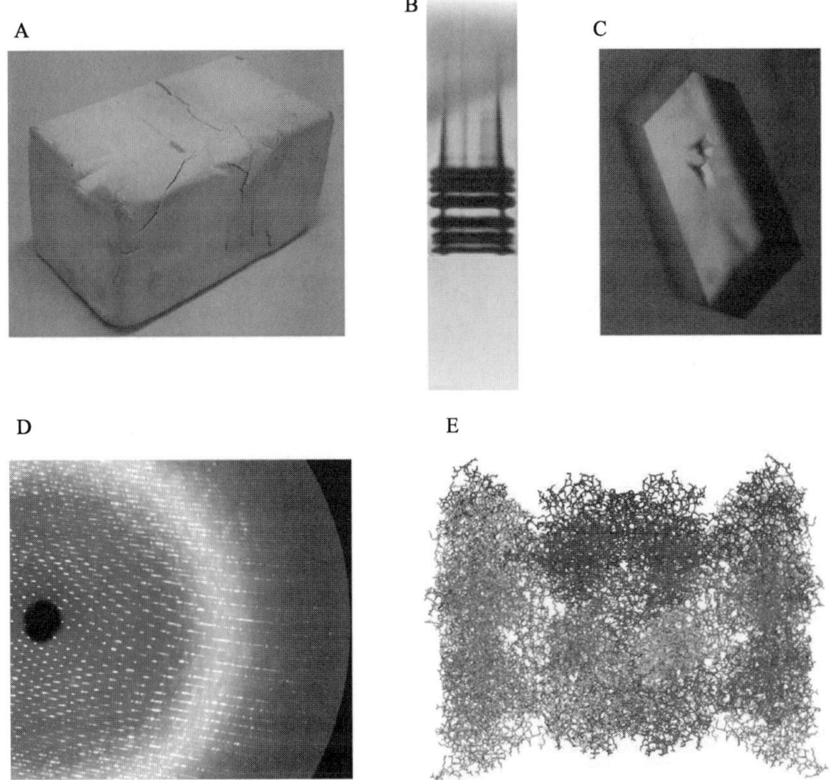


FIG. 1. Yeast 20S proteasome: From cells to the atomic resolution structure. (A) Block of 500 g yeast cells. (B) SDS–PAGE of the purified wild-type yeast 20S proteasome. (C) Crystal of the yeast 20S proteasome. (D) Typical diffraction image of proteasome crystals. (E) Crystal structure of the yeast 20S proteasome. Individual subunits are highlighted with specific shading.

pH 7.5, 1 m*M* EDTA, and 1 m*M* NaN3 (buffer A). The final suspension contains 60% buffer. Cells are broken in a French press using a pressure of 2200 psi. Under these conditions, 80% of the cells are broken, as confirmed by microscopic analysis. The main part of the lipids is removed after filtration through "glasswool" and the crude extract is centrifuged for 45 min at 10,000*g* in a Sorvall RC2B centrifuge to pellet the cell debris. In order to obtain a clear cell lysate, the supernatant is ultracentrifuged for 60 min at 130,000*g* in a Ti55.2 rotor (Beckman). The remaining lipids forming the surface layer of the supernatant are removed with a syringe, and the clarified lysate (total volume of approximately 700 ml, protein concentration of 30 mg/ml) is used for the following purification steps. Immediately after centrifugation the supernatant is applied to a Q-Sepharose column (Sigma, 1-liter volume), equilibrated with 280 m*M* NaCl in buffer A. The column is washed with 2 column volumes of 280 m*M* NaCl in buffer A at a flow rate of 10 ml/min. Bound proteins are eluted in 2 column volumes of buffer A using a gradient from 280 to 800 m*M* NaCl and 12-ml fractions are collected. The 20S proteasome elutes at 400–450 m*M* NaCl. Chymotrypsin-like (CL), peptidylglytamyl peptide hydrolase (PGPH), and trypsin-like (TL) activity (see later) is measured in all fractions. Fractions containing all three activities are combined, diluted threefold with buffer A, and applied to a ceramic hydroxy-apatite column (Biorad, 100 ml volume, flow rate 5 ml/min), equilibrated with 60 m*M* potassium phosphate, pH 7.5. The column is washed, and proteins are eluted with a linear gradient from 60 to 300 m*M* potassium phosphate. The proteasome elutes at 120 m*M* phosphate. CL, PGPH, and TL activity is measured in all fractions as mentioned earlier, and fractions containing proteasomal activity are combined and applied to a Mono Q column (Pharmacia, 8 ml volume, equilibrated in buffer A with 350 m*M* NaCl, flow rate 5 ml/min) using an ÄKTA-Prime system (Pharmacia). The column is washed with 10 column volumes of 350 m*M* NaCl in buffer A, and proteins are eluted with a linear gradient (20 column volumes) from 350 to 600 m*M* NaCl. The peaks are divided in 5-ml fractions, and purity control is performed by SDS–PAGE analysis. Fractions displaying the highest purity of proteasomes are combined (approximately 80 mg of protein, 40 ml volume), concentrated to a final volume of 4 ml by ultracentrifugation using an Amicon concentrator with an YM 100 membrane, and applied to a Superose 6 column (Pharmacia, volume 24 ml, flow rate 1 ml/min, equilibrated in buffer A). The 20S proteasome elutes at 11 ml. Fractions of 1 ml are collected and analyzed by SDS–PAGE, and fractions having the most pure protein are pooled (see Fig. 1B). All preparative steps are performed at 4°. 20S proteasomes obtained from one preparation can be stored at 4° for up to 4 weeks.

## Activity Assay

The yeast 20S proteasome possesses three major proteolytic activities against chromogenic substrates: CL, PGPH, and TL. The kinetic parameters of these activities are known to depend significantly on the assay conditions and enzyme preparation (Dahlmann *et al.*, 1985). The CL, TL, and PGPH activities of yeast proteasome are determined as follows: CL activity is monitored with 0.34 n*M* yeast proteasome in buffer A containing 0.025% SDS and 5% dimethyl sulfoxide (DMSO), with 7.5 $\mu M$ of the fluorogenic substrate peptide Suc-Leu-Leu-Val-Tyr-AMC (Bachem). Release of the fluorophore is detected at 360 nm excitation and 460 nm emission. TL activity is measured with 5.6 n*M* yeast proteasome in buffer A containing 100 m*M* NaCl, 0.01% Triton X–100, 5% DMSO, and 50 $\mu M$ Z-Ala-Arg-Arg-AMC (Bachem) as the fluorogenic substrate. Released fluorophore is measured as for CL. PGPH activity is monitored with 3.4 n*M* yeast proteasome in buffer A containing 100 m*M* NaCl, 0.001% SDS, 5% DMSO, 1 m*M* dithiothreitol (DTT), and 50 $\mu M$ Z-Leu-Leu-Glu-$\beta$-Na (Bachem) as the chromogenic substrate (345 nm excitation, 425 nm emission). The use of detergents for each activity has been optimized as described elsewhere (Arribas and Castano, 1990). The enzyme assays are carried out at 37° in a total volume of 500 $\mu$l on a spectrofluorimeter SM25 (Biotek-Kontron).

## Purification of Native 20S Proteasomes by Affinity Chromatography

The crystal structure of the yeast 20S proteasome allowed us to predict that a protease- cleavable affinity tag introduced into subunit $\beta 2$ would result in a surface-exposed accessible fusion complex, without disturbing the assembly. Therefore, subunit $\beta 2$ is chromosomally mutated at its C terminus by fusion of a TEV protease-cleavable protein A tag. The endogenous promoter is not altered so that protein synthesis is still at a physiological level. The strain is prepared by homologous recombination, and resulting cells show no conspicuous phenotype. A plasmid containing the Tev-ProA and geneticin modules is used as a template for the polymerase chain reaction (the plasmid library was kindly provided by Dr. Knop, Heidelberg). DNA is manipulated by standard techniques (Sambrook *et al.*, 1989). Genes of interest are amplified with *Taq* polymerase. Three to 5 $\mu$l of the purified constructs is used to transform competent yeast cells. The transformation protocol is based on the LiOAc method (Schiestl and Gietz, 1989). The transformed yeast cells are grown for at least 6 h or overnight to an optical density of 0.5–1.5 (600 nm) at 30° in YPAD medium (1% yeast extract, 2% peptone, and 2% glucose supplemented with 100 mg/ml adenine).

Twelve liters of YPD medium (1% yeast extract, 2% peptone, and 2% glucose, pH 5) is inoculated with the $\beta 2$-modified yeast strains, and cells are grown in Fernbach flasks to an OD between 2 and 4. Cells are harvested by centrifugation, resuspended in a twofold volume of buffer B (50 m*M* Tris-HCl, pH 8, 1m*M* EDTA), and lysed by French press at 2200 psi. The lysate is centrifuged at 45,000*g* for 45 min and is incubated with IgG resin (10 mg, ICN) for 1 h at 4°, and the resin is washed with 50 bed volumes of 500 m*M* NaCl in buffer B. At this step, SDS–PAGE analysis shows that the proteasome complex is already pure and the regulatory particles are completely removed. The IgG beads are washed with 3 volumes of TEV elution buffer (50 m*M* Tris-HCl, pH 7.5, 1 m*M* EDTA, 1 m*M* DTT), and the proteasome is cleaved from the ProA tag bound to the IgG resin by TEV protease under the following conditions: IgG beads are incubated with 1.5 volumes of buffer containing 150 U of 6His-TEV protease at 30° for 1 h. The cleaved proteasome is eluted with buffer B, and the TEV protease is removed by incubation of the elute with nickel-NTA resin (Qiagen, Hilden, Germany) at 4° for 15 min. The 20S proteasome is purified further using a Superose 6 (Pharmacia) size-exclusion chromatography column in buffer A. The yield is about 1–2 mg from 12 liters of cell culture. Crystals of the obtained proteasome complex diffract to a much higher resolution compared to those obtained by conventional purification methods (M. Groll, unpublished results).

## Crystallization of the Yeast 20S Proteasome and Structural Analysis

Crystals of the yeast 20S proteasome are grown by using the vapor diffusion method at 24°. The protein concentration used for crystallization is 40 mg/ml in 20 m*M* Tris-HCl (pH 7.5), 1 m*M* EDTA, and 1 m*M* $NaN_3$. Crystals are obtained in drops containing 4 $\mu$l of protein and 2 $\mu$l of a reservoir solution [28 m*M* magnesium acetate, 0.1 *M* morpholinoethanesulfonic acid (pH 6.5), and 12% 2,4-methylpentanediol]. Proteasome crystals in complex with inhibitors are prepared by cocrystallization (the concentration of the inhibitors is about 1 m*M*). Crystals appear within 1 week and grow to a final size of about $100 \times 50 \times 500$ nm$^3$ (see Fig. 1C). Proteasome crystals show high anisotropy (the resolution limit of measurable reflections is beyond 2.0 Å in the $b^*$ direction but only about 2.8 Å perpendicular to $b^*$) and increased mosaicity (see Fig. 1D). In general, more than 20 crystals have to be tested by X-ray diffraction before finding a proper candidate. The unit cell dimensions of the proteasome crystals are rather large (300 Å in the *b* axis), and synchrotron radiation is essential for obtaining adequate data sets at high resolution. Due to the instability of the crystals, they first have to be soaked in a cryoprotectant buffer (30% methylpentanediol, 24 m*M* magnesium acetate, 0.1 *M* morpholinoethane sulfonic acid) and then flash-frozen in a stream of cold nitrogen gas at 100K (Oxford Cryosystems). X-ray intensities are evaluated by commonly available software packages [DENZO, MOSFILM (Lesslie, 1994; Otwinowski and Minor, 1997)], as well as data reduction [usually CCP4 (Potterton *et al.*, 2003)]. Anisotropy of the reflections is corrected by comparing observed structure–factor amplitudes with those calculated from the model with isotropic temperature factors. Crystallographic refinement of the wild-type yeast 20S proteasome started from coordinates of the 20S proteasome from *Thermoplasma acidophilum* by rigid body refinement of the complete particle, and electron density map calculations were performed with CNS (Brünger *et al.*, 1998) using tight energy constants (Engh and Huber, 1991) and twofold noncrystallographic symmetry restraints, which will finally produce the refined model (see Fig. 1E). Determination of the crystal structure of proteasome:ligand complexes is now possible using the coordinates of the refined yeast 20S proteasome model.

## Acknowledgment

We thank Dr. Michael Knop, European Molecular Biology Laboratory, Heidelberg, Germany, for critical reading, advice, and discussions about the manuscript.

## References

Arribas, J., and Castano, J. G. (1990). Kinetic studies of the differential effect of detergents on the peptidase activities of the multicatalytic proteinase from rat liver. *J. Biol. Chem.* **265,** 13969–13973.

Brünger, A., Adams, P., Clore, G., De Lano, W., Gros, P., Grosse-Kunstleve, R., Jiang, J., Kuszewski, J., Nilges, M., Pannu, N., Read, R., Rice, L., Simonson, T., and Warren, G. (1998). Crystallography and NMR system: A new software suite for macromolecular structure determination. *Acta Crystallogr. D Biol. Crystallogr.* **1,** 905–921.

Dahlmann, B., Rutschmann, M., Kuehn, L., and Reinauer, H. (1985). Activation of the multicatalytic proteinase from rat skeletal muscle by fatty acids or sodium dodecyl sulphate. *Biochem. J.* **228,** 171–177.

Engh, R., and Huber, R. (1991). Accurate bond and angles parameters for X-ray protein structure refinement. *Acta Crystallogr.* **A47,** 392–400.

Groll, M., Ditzel, L., Löwe, J., Stock, D., Bochtler, M., Bartunik, H. D., and Huber, R. (1997). Structure of 20S proteasome from yeast at 2.4 A resolution. *Nature* **386,** 463–471.

Haass, C., and Kloetzel, P. M. (1989). The Drosophila proteasome undergoes changes in its subunit pattern during development. *Exp. Cell Res.* **180,** 243–252.

Heinemeyer, W., Kleinschmidt, J. A., Saidowsky, J., Escher, C., and Wolf, D. H. (1991). Proteinase yscE, the yeast proteasome/multicatalytic-multifunctional proteinase: Mutants unravel its function in stress induced proteolysis and uncover its necessity for cell survival. *EMBO J.* **10,** 555–562.

Knop, M., Siegers, K., Pereira, G., Zachariae, W., Winsor, B., Nasmyth, K., and Schiebel, E. (1999). Epitope tagging of yeast genes using a PCR-based strategy: More tags and improved practical routines. *Yeast* **15,** 963–972.

Leggett, D., Hanna, J., Borodovsky, A., Crosas, B., Schmidt, M., Baker, R., Walz, T., Ploegh, H., and Finley, D. (2002). Multiple associated proteins regulate proteasome structure and function. *Mol. Cell.* **10,** 495–507.

Lesslie, A. G. (1994). Mosfilm user guide, mosfilm version 5.2. MRC Laboratory of Molecular Biology, Cambrige, UK.

Morimoto, Y., Mizushima, T., Yagi, A., Tanahashi, N., Tanaka, K., Ichihara, A., and Tsukihara, T. (1995). Ordered structure of the crystallized bovine 20S proteasome. *J. Biochem. (Tokyo)* **117,** 471–474.

Otwinowski, Z., and Minor, W. (1997). Processing of X-ray diffraction data collected in oscillation mode. *Methods Enzymol.* **276,** 307–326.

Potterton, E., Briggs, P., Turkenburg, M., and Dodson, E. (2003). A graphical user interface to the CCP4 program suite. *Acta Crystallogr. D Biol. Crystallogr.* **59,** 1131–1137.

Sambrook, J., Fritsch, E., and Maniatis, T. (1989). "Molecular Cloning: A Laboratry Manual." Cold Spring Harbor Laboratory Press, Cold Spring Harbor, NY..

Schiestl, R., and Gietz, R. (1989). High efficiency transformation of intact yeast cells using single stranded nucleic acids as a carrier. *Curr. Genet.* **16,** 339–346.

Schneider, B., Seufert, W., Steiner, B., Yang, Q., and Futcher, A. (1995). Use of polymerase chain reaction epitope tagging for protein tagging in *Saccharomyces cerevisiae. Yeast* **11,** 1265–1274.

Wach, A., Brachat, A., Alberti-Segui, C., Rebischung, C., and Philippsen, P. (1997). Heterologous HIS3 marker and GFP reporter modules for PCR-targeting in *Saccharomyces cerevisiae. Yeast* **13,** 1065–1075.

# [28] Preparation of Hybrid (19S-20S-PA28) Proteasome Complexes and Analysis of Peptides Generated during Protein Degradation

*By* Paolo Cascio and Alfred L. Goldberg

## Abstract

PA28 (also named REG or 11S) is a ring-shaped (180-kDa) interferon-$\gamma$-induced complex that associates with the 20S proteasome and dramatically stimulates the breakdown of short peptides. Immunoprecipitation studies indicate that *in vivo* PA28 also exists in larger complexes that also contain the 19S particle, which is required for the ATP-ubiquitin-dependent degradation of proteins. However, because of its lability (e.g., it does not withstand exposure to high ionic strength buffers), this larger complex cannot be purified by standard biochemical protocols. Therefore, we developed a method to reconstitute *in vitro* such hybrid proteasomes (i.e., PA28-20S-19S) from highly purified components. This chapter

METHODS IN ENZYMOLOGY, VOL. 398

0076-6879/05 $35.00
DOI: 10.1016/S0076-6879(05)98028-2

describes conditions that allow the association of PA28 with "singly capped" 26S (i.e., 19S-20S) particles. In addition assays are described to measure absolute rates of degradation of several nonubiquitinated proteins by 26S and 20S proteasomes and methods to analyze the pattern and size distribution of peptides generated during the degradation of these proteins.

## Role of Proteasomes in MHC Class I Antigen Presentation

The immune system continually screens for viral infections and cancers by monitoring whether cells are synthesizing foreign or mutant proteins (Goldberg *et al.*, 2002; Rock and Goldberg, 1999). This surveillance process depends on the presence of MHC class I molecules that bind and display to cytotoxic lymphocytes (CTL) 8- to 10-residue peptides that are derived from the spectrum of proteins expressed in the cell. It is now firmly established that most antigenic peptides are generated during the degradation of intracellular proteins by the ubiquitin-proteasome pathway (Rock and Goldberg, 1999). These peptides are then translocated via the TAP transporter into the endoplasmic reticulum (ER), where they bind to MHC class I molecules. The resulting peptide-MHC class I complexes are then transported to the cell surface. The active form of the proteasome, which appears to degrade most cellular proteins, is the 26S proteasome (Baumeister *et al.*, 1998; Voges *et al.*, 1999). This 2.4-MDa complex is formed by the association of the 19S regulatory particle with one or both ends of the core 20S proteasome. The 26S proteasome degrades ubiquitinated and some nonubiquitinated proteins in an ATP-dependent manner (Coux *et al.*, 1996; Hershko and Ciechanover, 1998). Proteins are cleaved within the 20S (700 kDa) core proteasome, which is composed of four stacked rings. The two inner $\beta$ rings contain six proteolytic sites, which differ in substrate specificity: two have chymotrypsin-like, two trypsin-like, and two caspase-like activities (see Kisselev and Goldberg, 2005).

## The Proteasome Activator PA28

In higher eukaryotes, there is an additional proteasome regulatory complex, termed PA28 or 11S REG (Rechsteiner *et al.*, 2000), which appears to play an important role in MHC class I antigen presentation. However its precise function in this process is still unresolved, and its importance appears to vary depending on the antigen (Murata *et al.*, 2001). PA28 is a ring-shaped (180 kDa) multimeric complex that can bind to the two ends of the 20S proteasome and dramatically stimulate its capacity to hydrolyze small peptides (Dubiel *et al.*, 1992; Ma *et al.*, 1992). In mammals, PA28 is composed of two homologous subunits, PA28$\alpha$ and

PA28$\beta$, both of which are induced by interferon-$\gamma$. Consequently, PA28 is assumed to promote antigen presentation, and expression of PA28$\alpha$ alone (Groettrup *et al.*, 1996; Yamano *et al.*, 2002) or PA28$\alpha\beta$ (Schwarz *et al.*, 2000; Sun *et al.*, 2002; van Hall *et al.*, 2000; Yamano *et al.*, 2002) has been reported to enhance MHC class I presentation of some, but not all, antigens (Schwarz *et al.*, 2000; van Hall *et al.*, 2000). Furthermore, cells lacking this complex have a reduced ability to generate certain antigens (Murata *et al.*, 2001). Unlike components of the 26S proteasome, PA28 is not essential for growth or protein degradation. PA28 $\alpha$ and $\beta$ subunits have no homology to components of the 19S complex and must have arisen relatively recently in evolution (Tanahashi *et al.*, 1997). By itself, PA28$\alpha$ *in vitro* forms a heptameric ring (Knowlton *et al.*, 1997) that can stimulate peptide hydrolysis by 20S particles to the same extent as the heteromeric (3$\alpha$4$\beta$) complex (Song *et al.*, 1997) present *in vivo*. The PA28 subunits form a ring around a central opening through which substrates may pass into the 20S particle or products may exit (Knowlton *et al.*, 1997; Kohler *et al.*, 2001; Whitby *et al.*, 2000).

The precise roles of PA28$\alpha\beta$ and its homologs, PA28$\gamma$ in mammals (also called the Ki antigen) and PA26 in trypanosomes, remain unclear and controversial (Rechsteiner *et al.*, 2000; Rock and Goldberg, 1999). Mammalian PA28 was first isolated by DeMartino's (Ma *et al.*, 1992) and Rechsteiner's (Dubiel *et al.*, 1992) laboratories as an activator of hydrolysis of small peptide but not of proteins or ubiquitin-conjugated proteins by the 20S particle. A variety of biochemical actions have been proposed for this complex, including allosteric modification of active sites of the 20S proteasome (Li *et al.*, 2001; Ma *et al.*, 1992) or of noncatalytic modifier sites (Li *et al.*, 2001), stimulation of peptide entry into this particle (Kohler *et al.*, 2001; Ma *et al.*, 1992), stimulation of peptide exit (Whitby *et al.*, 2000), and facilitating the binding of proteasomes to chaperones or to components of the endoplasmic reticulum (Rechsteiner *et al.*, 2000).

X-ray diffraction of the complex of the trypanosome homolog PA26 with the yeast 20S proteasome has shown that binding of this complex opens the normally closed gate in the $\alpha$ ring of the proteasome (Forster *et al.*, 2003; Whitby *et al.*, 2000), thus facilitating substrate entry into the degradative particle (Groll *et al.*, 2000) and/or product exit (Kohler *et al.*, 2001). On this basis, PA28 was predicted to cause the release of peptide products of greater mean length. In fact, yeast proteasomes with a mutation that prevents closing of this gate do yield peptide products of greater mean size than normal particles (Kohler *et al.*, 2001). Because most (70%) proteasomal products are shorter than eight residues, the minimum length needed for binding to MHC molecules (Kisselev *et al.*, 1999), premature release of peptides from the proteasome would be expected to enhance the fraction of products capable of serving in antigen presentation, either directly or after trimming by aminopeptidases in the cytosol (Beninga *et al.*, 1998; Mo *et al.*, 1999; Stoltze *et al.*, 2000) or in the ER (Saric *et al.*, 2002; Serwold *et al.*, 2002; York *et al.*, 2002). However, as discussed later, we have not been able to confirm such a mechanism for the action of PA28 (Cascio *et al.*, 2002).

Although PA28 greatly stimulates proteasomal degradation of short peptides *in vitro*, it seems unlikely that PA28 functions *in vivo* to stimulate the hydrolysis of cytosolic oligopeptides within proteasomes because such peptides are hydrolyzed very rapidly by various other cytosolic peptidases (Saric *et al.*, 2001, 2004). Moreover, PA28-20S complexes cannot support the ATP-dependent hydrolysis of proteins or ubiquitin-conjugated proteins by the proteasome (Dubiel *et al.*, 1992; Kuehn and Dahlmann, 1996; Ma *et al.*, 1992). This process requires the 26S proteasome complex, as the 19S complex, unlike PA28, contains binding sites for ubiquitin chains (Thrower *et al.*, 2000) and ATPase subunits to catalyze protein unfolding and translocation into the 20S particle (Benaroudj and Goldberg, 2000; Benaroudj *et al.*, 2003; Braun *et al.*, 1999; Navon and Goldberg, 2001). Hendil *et al.* (1998) showed that immunoprecipitates of 19S particles from cells contain not only 20S proteasomes, but also the PA28 complex. These investigators further showed that interferon-$\gamma$ treatment of cells led to the appearance of these hybrid complexes (Tanahashi *et al.*, 2000). However, purification and characterization of these hybrid complexes were impossible due to their extreme lability (e.g., they do not withstand standard chromatographic methods or exposure to high ionic strength buffers).

## *In Vitro* Reconstitution of Hybrid 26S Proteasomes (PA28-20S-19S)

To avoid the problem, Cascio *et al.* (2002) and Kopp *et al.* (2001) developed a method to reconstitute *in vitro* these hybrid complexes from purified components. Using this approach and native gel electrophoresis and electron microscopy, it was possible to demonstrate that PA28 associates specifically with the asymmetric form of 26S proteasomes containing one 19S cap to generate a hybrid 19S-20S-PA28 complex. To produce sufficient amounts of these complexes for biochemical and structural studies, we normally purify recombinant mouse PA28$\alpha$ (Ahn *et al.*, 1995) expressed in *Escherichia coli* and 26S proteasomes from rabbit spleen, as described previously (Cascio *et al.*, 2001). The 26S proteasomes can thus be purified to apparent homogeneity, although as found by native PAGE, these preparations consist of approximately equal amounts of a mixture of singly capped complexes (19S-20S) and doubly capped symmetric complexes (19S-20S-19S) (Cascio *et al.*, 2002). Spleen was initially chosen because it contains immunoproteasomes exclusively (Cascio *et al.*, 2001; Eleuteri *et al.*, 1997; Noda *et al.*, 2000; Van Kaer *et al.*, 1994), the specialized forms that are induced together with PA28 by interferon-$\gamma$ (Realini *et al.*, 1994).

Recombinant mouse PA28$\alpha$ can be expressed in *E. coli* and purified as described (Song *et al.*, 1997) with minor modifications. After isolation of the inclusion bodies, the precipitated protein is denaturated and solubilized in 10 m$M$ Tris-HCl, pH 7.5, containing 8 $M$ urea and 0.1 $M$ dithiothreitol (DTT). The urea is then removed by gel filtration with a Sephadex G-25 column equilibrated with 25 m$M$ Tris-HCl, pH 7.5, 0.1 m$M$ EDTA, and 1 m$M$ DTT. At this stage, the protein is mostly soluble and able to stimulate the peptidase activities of the 20S proteasome severalfold. Purification to apparent homogeneity is then achieved using a Mono-Q HR column (Amersham Pharmacia) equilibrated with 25 m$M$ Tris-HCl, pH 7.5, 0.1 m$M$ EDTA, and 1 m$M$ DTT and is resolved with a linear gradient from 0 to 1 $M$ NaCl. Active fractions are pooled, and PA28$\alpha$ is purified further by gel filtration on a Superose 6 HR column (Amersham Pharmacia) equilibrated with the same buffer. The purified PA28$\alpha$ does not contain any exo- or endopeptidase activity as measured with several different fluorogenic substrates but dramatically stimulates all three peptidase activities of the 20S particles. This lack of contaminating peptidases is important to establish for rigorous analyses of peptides produced during protein degradation.

Furthermore, PA28$\alpha$ is also able to increase three- to fourfold the activity of the 26S proteasomes against Suc-LLVY-amc (Cascio *et al.*, 2002), a specific substrate of chymotrypsin-like active sites [see Kisselev and Goldberg (2005) for a detailed description of the continuous assay of several peptidase activities of the proteasome]. To perform this assay, 26S immunoproteasomes (10 p$M$) are preincubated in 500 $\mu$l of 20 m$M$ Tris-HCl, pH 7.5, 2 m$M$ ATP, 5 m$M$ $MgCl_2$, 1 m$M$ EDTA at 37° with and without PA28 (20 n$M$). Reactions are initiated by adding Suc-LLVY-amc at a final concentration of 100 $\mu M$, and the fluorescence of released amc is measured (excitation 380 nm, emission 460 nm) continuously in a recording spectrafluorometer. A similar three- to fourfold activation is also found with peptide substrates specific for the trypsin- and caspase-like sites of the proteasome. Our initial experiments utilized 26S immunoproteasomes isolated from spleen, which contain only interferon-$\gamma$-induced $\beta$ subunits, but very similar results are obtained upon addition of PA28$\alpha$ to 26S normal proteasomes isolated from rabbit muscle.

In similar experiments using native PA28$\alpha\beta$ (purchased from Boston Biochem, Cambridge, MA, or Affinity, Mamhead, UK) in place of recombinant PA28$\alpha$, a very similar three- to fourfold activation of peptide hydrolysis by 26S proteasomes is also seen. However, this stimulation of

26S proteasomes is obtained with significantly lower concentrations of the native molecule than of the recombinant PA28$\alpha$ lacking the $\beta$ subunits (i.e., $K_d$ values for this interaction with PA28$\alpha\beta$ are 2 n*M* and 10 n*M* for PA28$\alpha$). This stimulation of peptidase activity [also reported by Kopp *et al.* (2001)] is clearly not due to an activation of contaminating 20S particles, as no such particles can be detected in our 26S preparations by gel electrophoresis, even after several hours of incubation (Cascio *et al.*, 2002). In addition, upon native gel electrophoresis, no activity is seen in the region of the gel that corresponds to the PA28-20S-PA28 complexes studied extensively by others (Realini *et al.*, 1994).

To study further the biochemical and structural properties of hybrid proteasomes, such complexes can be reconstituted *in vitro* by incubating 26S particles (0.1 $\mu M$) with PA28$\alpha$ (1.5 $\mu M$) in 20 m*M* Tris-HCl, pH 7.5, 2 m*M* ATP, 5 m*M* $MgCl_2$, and 1 m*M* EDTA at 37° for 30 min. When such samples are then subjected to native PAGE, a clear change in the electrophoretic migration of the 26S particles can be detected (Cascio *et al.*, 2002). The band identified by Coomassie staining and peptidase activity as the singly capped form of 26S disappears and, simultaneously, there is a severalfold increase in the peptidase activity and Coomassie staining of a broad higher molecular weight band, which includes the doubly capped form. Thus, the greater peptidase activity found when 26S and PA28 are preincubated together is undoubtedly associated with PA28-induced modification and activation of all, or nearly all, the singly capped forms.

### *Electron Microscopic Analysis of Hybrid Particles*

To investigate further the structure of this new "hybrid" form of proteasomes, samples prepared for native PAGE electrophoresis can also be analyzed by negative stain electron microscopy. For this purpose, samples of the hybrid complexes are diluted 1:15 with incubation buffer immediately prior to adsorption onto glow-discharged carbon-coated copper grids. Grids are then washed with four drops of incubation buffer and stained with two drops of freshly prepared 0.75% uranyl formate. Specimens are inspected with an electron microscope usually operated at 120 kV, and images are taken at a nominal magnification of 52,000 using low-dose procedures. Images obtained are digitized with a scanner using a pixel size of 6.7 Å at the specimen level, and large number particles are selected for further computational processing (e.g., using the SPIDER image processing package). In an analysis that we performed based on 130 images and 4600 particles (Cascio *et al.*, 2002), the proteasomes on the grid could be grouped into six classes: (1) 20S proteasomes alone (comprising 3.5% of the total number of particles), (2) singly capped 26S proteasomes

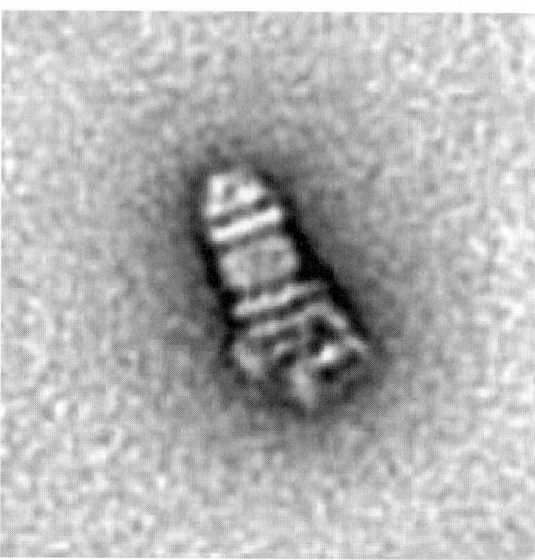

Fig. 1. Electron micrograph of negatively stained hybrid proteasome complexes (PA28-20S-19S): average image based on 266 particles.

(comprising 32%), (3) doubly capped 26S particles (31%), (4) the new hybrid proteasomes containing one PA28 ring and one 19S particle (10%), and (5) 20S proteasomes associated with one (2.5%) or (6) two (1.5%) PA28 rings. (Approximately 20% of the selected particles could not be classified unambiguously but would have belonged to classes 2, 3, or 4.) The particles of each class were then subjected to multireference alignment, and the average image of the hybrid proteasomes (class 4) is shown in Fig. 1. The relatively low amounts of the hybrid particles seen in electron micrographs must underestimate their actual abundance in solution, as electron microscopy involves relatively harsh procedures (e.g., staining samples) and a large dilution of the particles, which cause the dissociation of PA28$\alpha$. Consequently, the PAGE pattern, which indicated quantitative conversion of singly capped proteasomes into hybrid particles (Cascio *et al.*, 2002), is probably a much better measure of the frequency of hybrid proteasomes than EM.

### *Measurement of Rates of Protein Breakdown by Proteasomes with Fluorescamine*

Simple methods have been developed to measure the absolute rates of protein degradation by different forms of 26S particles (i.e., 26S proteasomes, 26S immunoproteasomes, 26S hybrid proteasomes) without the use of isotopes or ubiquitination. When these particles are incubated with various proteins that lack distinct tertiary structures (e.g., reduced and alkylated IGF-1, casein, oxidized ovalbumin), they are hydrolyzed without ubiquitination by 26S particles in an ATP-dependent fashion for many hours (Cascio *et al.*, 2001; Kisselev *et al.*, 1999). This approach offers several advantages compared to the use of chemically ubiquitinated protein substrates, which are very difficult to produce in sufficient amounts for product analysis and, in any case, are degraded by 26S proteasomes at very low rates for reasons that are not yet clear (Thrower *et al.*, 2000). In contrast, methods for the oxidation of ovalbumin (Kisselev *et al.*, 1999) and carboxymethylation of IGF-1 (Akopian *et al.*, 1997) to ensure denaturation and to prevent sulfhydryl bond formation are very easy and produce relatively good substrates for ATP-dependent degradation by 26S proteasomes (Cascio *et al.*, 2001). Casein in its native form has little tertiary structure and does not require prior chemical treatment for proteaosomal degradation. Furthermore, growing evidence shows that various proteins are also degraded by 26S proteasomes *in vivo* in an ATP-dependent process without ubiquitination (Benaroudj *et al.*, 2001; Tarcsa *et al.*, 2000; Verma and Deshaies, 2000), which suggests that degradation of these artificial substrates, aside from being convenient for analysis of the cleavage process, may also reflect an *in vivo* process. Peptide bond cleavage in proteins is then assayed by measuring the appearance of new amino groups with fluorescamine. To reduce the background in the fluorescamine assay the denatured ovalbumin, IGF-1, and casein are then reductively methylated to eliminate the free amino groups in the protein substrate (Akopian *et al.*, 1997) (see later). Degradation of reductively methylated IGF-1 (1.5 m*M*), casein (1 m*M*), and ovalbumin (10 $\mu M$) is typically carried out at 37° in 20 m*M* HEPES, pH 7.5, 2 m*M* ATP, 5 m*M* $MgCl_2$, and 1 m*M* EDTA with a concentration of the proteasomes between 10 and 30 n*M* depending on the substrate. The rates of protein breakdown can be followed easily during incubation of up to 10 h by removing aliquots at regular intervals, which are analyzed for the appearance of new amino groups (Akopian *et al.*, 1997). This rate of protein cleavage is tightly linked to and directly proportional to the disappearance of the polypeptide (Cascio *et al.*, 2001; Kisselev *et al.*, 1999) as expected, as degradation of these proteins is highly processive. Typically, a 2-$\mu$l aliquot for each time point is removed from the reaction mixture and is mixed immediately with 48 $\mu$l of 0.1 *M* sodium phosphate buffer (pH 6.8) and frozen. At the end of the incubation, samples are defrosted and 25 $\mu$l of an acetone solution of fluorescamine (0.3 mg/ml) is added. The mixture is vortexed for 1 min to allow fluorescamine to form a fluorescent adduct with N termini of peptides generated by proteasomal cleavage, and then water is added to a final volume of 0.5 ml. Fluorescence is measured immediately in a spectrafluorometer using an excitation wavelength of 370 nm and an emission of 480 nm. The amino groups generated are estimated using as a standard an equimolar mixture of peptides of known concentrations (e.g., EAA-$NH_2$, AEAA-$NH_2$, AAEAAG-$NH_2$, AAVAAG-$NH_2$, TTQRTRALV-$NH_2$). The instrument has to be calibrated each time before use, and once the linearity of the calibration curve has been established, it is possible to use just one concentration of the standard. For this purpose, we normally use 0.4 $\mu M$ of peptide solution in water and adjust the sensitivity of the instrument to measure a signal that is 10% of the full scale.

Using this method, we could show that even though the addition of PA28$\alpha$ causes a three- to fourfold increase in hydrolysis of several peptides by the 26S particles, the rate of breakdown of the proteins does not change or increases only slightly (Fig. 2). Importantly, under the incubation conditions (i.e., with a large excess of protein substrate), proteins are degraded, and different peptide products are each generated by 26S proteasomes

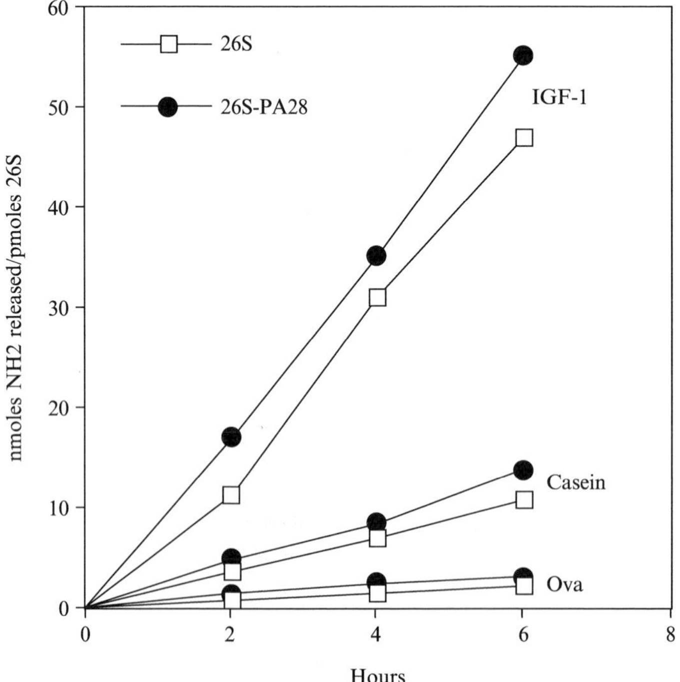


FIG. 2. Association of PA28 with the 26S proteasomes does not increase the rate of protein breakdown as measured by the generation of new peptides by fluorescamine.

at similar linear rates for many hours (Cascio *et al.*, 2001). Thus, the peptides, once generated, do not undergo significant further digestion by the proteasomes (Cascio *et al.*, 2001), and degradation of the proteins is carried out strictly by the 26S or the hybrid complexes, as even after 6 h, only these species (i.e., no 20S particles) are evident on native gels (Cascio *et al.*, 2002).

### *Methods Used to Analyze the Pattern of Peptides Released during Proteasomal Degradation of Proteins*

Analysis of the pattern of peptides produced during degradation of proteins by different forms of 26S proteasomes can be performed conveniently using the small polypeptide, insulin-like growth factor-1 (IGF-1) (8000 Da) as substrate. In fact, with larger proteins (e.g., casein or ovalbumin), the number of different peptides produced is too large to allow resolution of individual products by standard (RP)-HPLC methods. However with IGF-1, the number of products generated is much smaller, and the individual peaks can be resolved quite easily. This approach allows a rapid evaluation of differences in the cleavage specificities between different forms of proteasomes or under different degradation conditions. The recombinant human IGF-1 is first denatured by the reduction of disulfide bonds and carboxymethylation of the cysteines (Akopian *et al.*, 1997) and is then incubated with 26S proteasomes (10 n*M*) or hybrid complexes at 37° in 20 m*M* HEPES, pH 7.5, 2 m*M* ATP, 5 m*M* $MgCl_2$, and 1 m*M* EDTA for 5–6 h. To minimize the possibility of reentry and repeated cleavage of peptides already released by the proteasome, it is important to perform such a reaction in the presence of an initial large molar excess of the substrate (IGF-1) over the enzyme and to ensure that the fraction of substrate consumed never exceeds 10–15%, which can be estimated by the disappearance of the substrate peak (which has a retention time of 180 min and is not evident on the chromatogram shown). Peptides generated can be separated on a $C_8$ Vydac column (250 × 2.1 mm, 5 $\mu$m) equilibrated with 0.06% trifluoroacetic acid at a flow rate of 0.15 ml/min and developed with a gradient of acetonitrile from 0 to 8% in 20 min, to 28% in the next 100 min, to 36% in the subsequent 20 min, and to 44% in the last 10 min. In Fig. 3A the different patterns of peptides generated by 26S immunoproteasomes and 26S hybrid complexes are analyzed using this method; clearly, with PA28$\alpha$ added, there are some peptide peaks that are produced only upon formation of the hybrid complexes and some that are produced in much lower amounts or not at all.

Analysis of the patterns of products generated during the degradation of larger proteins is more difficult, as the complexity of the peptide

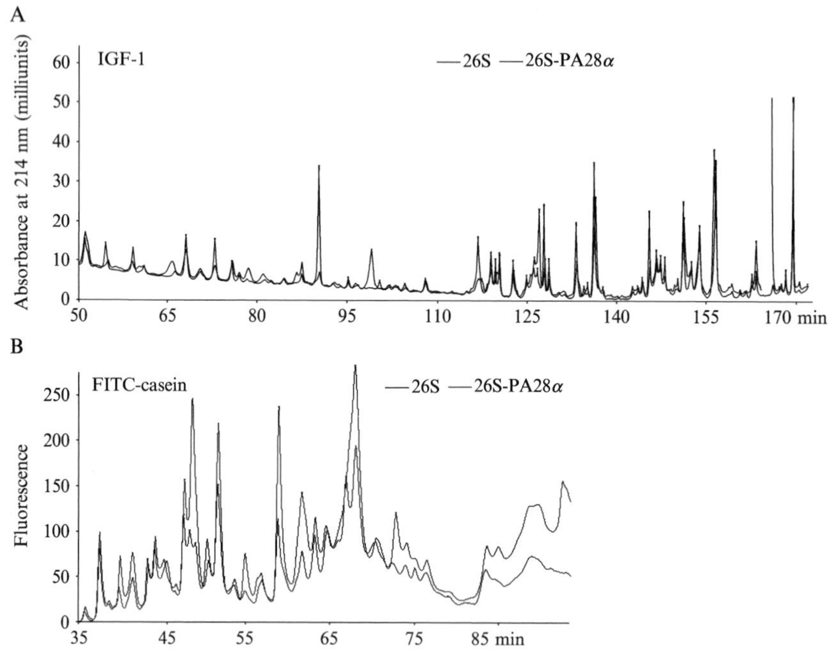


FIG. 3. The association of PA28$\alpha$ with 26S proteasomes modifies the patterns of peptides generated from IGF-1 (A) and FITC-casein (B).

mixtures increases greatly, while the amounts of each peptide decrease below the detection limit of usual UV detectors. One way to avoid these problems is to use fluorescein isothiocyanate (FITC)-modified casein as the proteasomal substrate (Akopian *et al.*, 1997), which increases the sensitivity of the detection of individual peptide products greatly and in parallel reduces the number of peptides that can be analyzed, as only a limited fraction of them contains the fluorescent group. FITC-casein (10 $\mu M$) can be degraded by 26S particles as described previously for other substrates (see earlier discussion). At the end of the reaction, an aliquot of the digestion reaction (normally 30–50 $\mu$l) is injected onto a $C_{18}$ Vydac column (250 $\times$ 2.1 mm, 5 $\mu$m) equilibrated with 10 m$M$ sodium phosphate (pH 6.8). Peptides are eluted by a linear gradient of acetonitrile from 0 to 50% in 100 min at a flow of 0.2 ml/min. Fluorescent peptides are detected at an excitation wavelength of 490 nm and an emission of 521 nm. Figure 3B shows such an analysis in which the patterns of peptides generated by 26S immunoproteasomes and 26S hybrid proteasomes differ greatly.

*Analysis of the Size Distribution of Peptides Produced by Proteasomes*

As part of the continuous turnover of cell proteins, the great majority of peptides generated by proteasomes must be degraded rapidly into amino acids by cytosolic endo- and exopeptidases (Saric *et al.*, 2004). In mammalian cells, some of the proteasomal products escape complete degradation and are presented to the immune system on the cell surface in complexes with MHC class I molecules (Goldberg *et al.*, 2002). Obviously, information on the sizes of the products of protein breakdown by mammalian proteasomes is essential for a full understanding of both MHC class I antigen presentation and postproteasomal steps in the complete degradation of proteins to amino acids. The size distribution cannot be analyzed simply by mass spectrometry of proteasomal products, as this method does not allow quantitative evaluation of peptides generated. Because of this limitation of mass spectrometry, we developed an alternative size-exclusion chromatographic method in which the total pool of peptides produced is derivatized with fluorescamine before injection into the HPLC, and the fluorescence that is eluted from the column is monitored continuously (Kohler *et al.*, 2001). This new method presents several advantages: it is fast, reliable, and the results obtained are highly reproducible. Furthermore, by derivatizing peptide mixtures prior to fractionation, it is possible to even resolve and quantitate amino acids and di- or tripeptides. Other size-exclusion methods fail to resolve such short peptides and the standard mass spectrometry methods systematically exclude them. In addition, the fluorescamine assay allows comparisons of the total amounts of peptides in mixed populations, which is not possible by other approaches (e.g., mass spectrometry).

Peptides generated during the degradation of IGF-1 are separated from the undigested protein on a $C_{18}$ Vydac column (Kisselev *et al.*, 1998) and lyophilized. After resuspension, this material is analyzed by size-exclusion chromatography with a polyhydroxy-ethyl aspartamide column (0.46 × 20 cm, Poly LC, Columbia, MD) equilibrated with 0.2 *M* sodium sulfate and 25% acetonitrile, pH 3.0, using a HPLC chromatographer and a fluorometer detector. Peptide products are resuspended in 0.1 *M* HEPES buffer, pH 6.8. For each analysis, 3 nmol of peptides in a total volume of 20 $\mu$l is added to 10 $\mu$l of fluorescamine (0.3 mg/ml acetone). The reaction is terminated after 1 min with 30 $\mu$l of $H_2O$, and the sample is immediately injected into the HPLC. To determine the apparent molecular mass of the peptides eluted, the column is calibrated each time before use with 11 standard amino acids and peptides in the 200- to 3500-Da range that had been derivatized with fluorescamine in the same way as the proteasome products. Prior control studies (Kisselev *et al.*, 1999; Kohler *et al.*, 2001) showed that retention times of these fluorescamine-derivatized peptides are highly reproducible, and a linear function of the logarithm of their molecular weights. This method has proven to be very useful in determining whether the association of PA28 with 26S proteasomes might cause a modification in the size of the peptide products. Whitby *et al.* (2000) suggested that PA28 promotes antigen presentation by allowing the premature release of peptides from the proteolytic chamber, which could result in the generation of longer peptides (Kisselev *et al.*, 1999; Kohler *et al.*, 2001). This proposal is attractive because normally most products of the proteasome are shorter than eight residues, which is too small to bind to MHC class I molecules (Kisselev *et al.*, 1999). Also, when ovalbumin is degraded by 26S proteasomes, over 90% of the time the immunodominant epitope SIINFEKL is cleaved internally (Cascio *et al.*, 2001). Therefore, the premature release of products should enhance the yield of longer peptides capable of serving as antigenic precursors. Accordingly, mutations that leave the gate in the proteasomal $\alpha$ ring in an open position cause a generation of products of greater mean size (Kohler *et al.*, 2001). However, despite the clear differences in the patterns of peptides generated from proteins upon formation of the hybrid complexes (Fig. 3), rigorous analysis by size-exclusion chromatography clearly shows that the size distributions of peptides released from 26S proteasomes are indistinguishable in the presence or absence of PA28 (Fig. 4). This result argues strongly against the conclusion that these ring complexes only facilitate gate opening, which would result in the release of longer products, and strongly suggests that PA28 allosterically enhances the activities of active sites of the proteasome.

## Conclusions

Although immunoprecipitation studies clearly had indicated the existence *in vivo* of a 19S-20S-PA28 complex (Hendil *et al.*, 1998) that is induced by interferon-$\gamma$ (Tanahashi *et al.*, 2000), various attempts to isolate this hybrid particle from cells or tissues have failed repeatedly because of its lability and sensitivity to salt (Tanahashi *et al.*, 2000). As an alternative approach to define its structure and properties, Cascio *et al.* (2002) and Kopp *et al.* (2001) have established conditions that allow its *in vitro* reconstitution. To convert the singly capped particles quantitatively into hybrid complexes, a large molar excess (301) of PA28$\alpha$ over 26S proteasomes is required. Thus, the affinity of the singly capped 26S for PA28$\alpha$, like that of 20S particles, is not high, and the complex dissociates readily upon dilution. *In vivo* additional factors probably facilitate the association between the core particle and PA28, and a role for phosphorylation in this process has been suggested (Dubiel *et al.*, 1992; Yang *et al.*, 1995).

To study biochemical and structural properties of these hybrid 26S complexes, we developed several methods that allow accurate measurements of

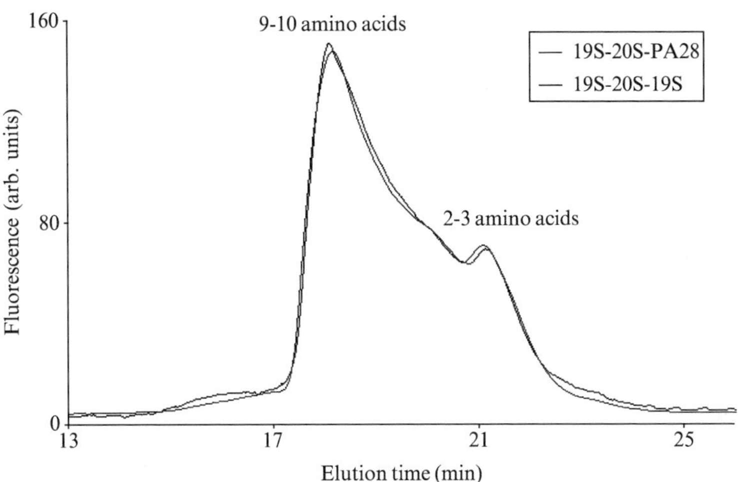


FIG. 4. 26S proteasomes and hybrid complexes generate peptides with the same size distribution as shown by size-exclusion chromatography of fluorescamine precolumn-derivatized peptides and online detection.

rates of proteins breakdown and qualitative and quantitative analysis of the peptides generated. Such methods are reliable and highly reproducible and are suitable for studies not only of different forms of proteasomes in different conditions, but can also be adapted easily for studies of other aspects of proteasome function or of other proteolytic enzymes.

## Acknowledgments

These studies have been supported by a grant to A. L. Goldberg from the National Institutes of General Medical Sciences (R01-GM46147). The authors are grateful to Mary Dethavong for her assistance in the preparation of this manuscript.

## References

Ahn, J. Y., Tanahashi, N., Akiyama, K., Hisamatsu, H., Noda, C., Tanaka, K., Chung, C. H., Shibmara, N., Willy, P. J., Mott, J. D., Slaughter, C. A., and DeMartino, G. N. (1995). Primary structures of two homologous subunits of PA28, a gamma-interferon-inducible protein activator of the 20S proteasome. *FEBS Lett.* **366,** 37–42.

Akopian, T. N., Kisselev, A. F., and Goldberg, A. L. (1997). Processive degradation of proteins and other catalytic properties of the proteasome from *Thermoplasma acidophilum*. *J. Biol. Chem.* **272,** 1791–1798.

Baumeister, W., Walz, J., Zuhl, F., and Seemuller, E. (1998). The proteasome: Paradigm of a self-compartmentalizing protease. *Cell* **92,** 367–380.

Benaroudj, N., and Goldberg, A. L. (2000). PAN, the proteasome-activating nucleotidase from archaebacteria, is a protein-unfolding molecular chaperone. *Nat. Cell Biol.* **2,** 833–839.

Benaroudj, N., Tarcsa, E., Cascio, P., and Goldberg, A. L. (2001). The unfolding of substrates and ubiquitin-independent protein degradation by proteasomes. *Biochimie* **83,** 311–318.

Benaroudj, N., Zwickl, P., Seemuller, E., Baumeister, W., and Goldberg, A. L. (2003). ATP hydrolysis by the proteasome regulatory complex PAN serves multiple functions in protein degradation. *Mol. Cell* **11,** 69–78.

Beninga, J., Rock, K. L., and Goldberg, A. L. (1998). Interferon-$\gamma$ can stimulate post proteasomal trimming of the N-terminus of antigenic peptides by inducing leucine aminopeptidase. *J. Biol. Chem.* **273,** 18734–18742.

Braun, B. C., Glickman, M., Kraft, R., Dahlmann, B., Kloetzel, P. M., Finley, D., and Schmidt, M. (1999). The base of the proteasome regulatory particle exhibits chaperone-like activity. *Nat. Cell Biol.* **1,** 221–226.

Cascio, P., Call, M., Petre, B. M., Walz, T., and Goldberg, A. L. (2002). Properties of the hybrid form of the 26S proteasome containing both 19S and PA28 complexes. *EMBO J.* **21,** 2636–2645.

Cascio, P., Hilton, C., Kisselev, A. F., Rock, K. L., and Goldberg, A. L. (2001). 26S proteasomes and immunoproteasomes produce mainly N-extended versions of an antigenic peptide. *EMBO J.* **20,** 2357–2366.

Coux, O., Tanaka, K., and Goldberg, A. (1996). Structure and functions of the 20S and 26S proteasomes. *Annu. Rev. Biochem.* **65,** 801–804.

Dubiel, W., Pratt, G., Ferrell, K., and Rechsteiner, M. (1992). Purification of an 11 S regulator of the multicatalytic protease. *J. Biol. Chem.* **267,** 22369–22377.

Eleuteri, A. M., Kohanski, R. A., Cardozo, C., and Orlowski, M. (1997). Bovine spleen multicatalytic proteinase complex (proteasome): Replacement of X, Y, and Z subunits by LMP7, LMP2, and MECL1 and changes in properties and specificity. *J. Biol. Chem.* **272,** 11824–11831.

Forster, A., Whitby, F. G., and Hill, C. P. (2003). The pore of activated 20S proteasomes has an ordered 7-fold symmetric conformation. *EMBO J.* **22,** 4356–4364.

Goldberg, A., Cascio, P., Saric, T., and Rock, K. (2002). The importance of the proteasome and subsequent proteolytic steps in the generation of antigenic peptides. *Mol. Immunol.* **39,** 147.

Groettrup, M., Soza, A., Eggers, M., Kuehn, L., Dick, T. P., Schild, H., Rammensee, H. G., Koszinowski, U. H., and Kloetzel, P. M. (1996). A role for the proteasome regulator PA28alpha in antigen presentation. *Nature* **381,** 166–168.

Groll, M., Bajorek, M., Kohler, A., Moroder, L., Rubin, D. M., Huber, R., Glickman, M. H., and Finley, D. (2000). A gated channel into the proteasome core particle. *Nature Struct. Biol.* **7,** 1062–1067.

Hendil, K. B., Khan, S., and Tanaka, K. (1998). Simultaneous binding of PA28 and PA700 activators to 20 S proteasomes. *Biochem. J.* **332,** 749–754.

Hershko, A., and Ciechanover, A. (1998). The ubiquitin system. *Annu. Rev. Biochem.* **67,** 425–479.

Kisselev, A. F., Akopian, T. N., and Goldberg, A. L. (1998). Range of sizes of peptide products generated during degradation of different proteins by archaeal proteasomes. *J. Biol. Chem.* **273,** 1982–1989.

Kisselev, A. F., Akopian, T. N., Woo, K. M., and Goldberg, A. L. (1999). The sizes of peptides generated from protein by mammalian 26 and 20S proteasomes: Implications for understanding the degradative mechanism and antigen presentation. *J. Biol. Chem.* **274,** 3363–3371.

Kisselev, A. F., and Goldberg, A. L. (2005). Monitoring activity and inhibition of 26S proteasomes with fluorogenic peptide substrates. *Methods Enzymol.* **398**(30), 2005 (this volume).

Knowlton, J. R., Johnston, S. C., Whitby, F. G., Realini, C., Zhang, Z., Rechsteiner, M., and Hill, C. P. (1997). Structure of the proteasome activator REGalpha (PA28alpha). *Nature* **390,** 639–643.

Kohler, A., Cascio, P., Leggett, D. S., Woo, K. M., Goldberg, A. L., and Finley, D. (2001). The axial channel of the proteasome core particle is gated by the Rpt2 ATPase and controls both substrate entry and product release. *Mol. Cell* **7,** 1143–1152.

Kopp, F., Dahlmann, B., and Kuehn, L. (2001). Reconstitution of hybrid proteasomes from purified PA700-20 S complexes and PA28alphabeta activator: Ultrastructure and peptidase activities. *J. Mol. Biol.* **313,** 465–471.

Kuehn, L., and Dahlmann, B. (1996). Reconstitution of proteasome activator PA28 from isolated subunits: Optimal activity is associated with an alpha,beta-heteromultimer. *FEBS Lett.* **394,** 183–186.

Li, J., Gao, X., Ortega, J., Nazif, T., Joss, L., Bogyo, M., Steven, A. C., and Rechsteiner, M. (2001). Lysine 188 substitutions convert the pattern of proteasome activation by REGgamma to that of REGs alpha and beta. *EMBO J.* **20,** 3359–3369.

Li, J., and Rechsteiner, M. (2001). Molecular dissection of the 11S REG (PA28) proteasome activators. *Biochimie* **83,** 373–383.

Ma, C. P., Slaughter, C. A., and De Martino, G. N. (1992). Identification, purification, and characterization of a protein activator (PA28) of the 20 S proteasome (macropain). *J. Biol. Chem.* **267,** 10515–10523.

Mo, X. Y., Cascio, P., Lemerise, K., Goldberg, A. L., and Rock, K. (1999). Distinct proteolytic processes generate the C and N termini of MHC class I-binding peptides. *J. Immunol.* **163,** 5851–5859.

Murata, S., Udono, H., Tanahashi, N., Hamada, N., Watanabe, K., Adachi, K., Yamano, T., Yui, K., Kobayashi, N., Kasahara, M., Tanaka, K., and Chiba, T. (2001). Immunoproteasome assembly and antigen presentation in mice lacking both PA28alpha and PA28beta. *EMBO J.* **20,** 5898–5907.

Navon, A., and Goldberg, A. L. (2001). Proteins are unfolded on the surface of the ATPase ring before transport into the proteasome. *Mol. Cell* **8,** 1339–1349.

Noda, C., Tanahashi, N., Shimbara, N., Hendil, K. B., and Tanaka, K. (2000). Tissue distribution of constitutive proteasomes, immunoproteasomes, and PA28 in rats. *Biochem. Biophys. Res. Commun.* **277,** 348–354.

Realini, C., Dubiel, W., Pratt, G., Ferrell, K., and Rechsteiner, M. (1994). Molecular cloning and expression of a gamma-interferon-inducible activator of the multicatalytic protease. *J. Biol. Chem.* **269,** 20727–20732.

Rechsteiner, M., Realini, C., and Ustrell, V. (2000). The proteasome activator 11 S REG (PA28) and class I antigen presentation. *Biochem. J.* **345,** 1–15.

Rock, K. L., and Goldberg, A. L. (1999). Degradation of cell proteins and the generation of MHC class I-presented peptides. *Annu. Rev. Immunol.* **17,** 739–779.

Saric, T., Beninga, J., Graef, C. I., Akopian, T. N., Rock, K. L., and Goldberg, A. L. (2001). Major histocompatibility complex class I-presented antigenic peptides are degraded in cytosolic extracts primarily by thimet oligopeptidase. *J. Biol. Chem.* **276,** 36474–36481.

Saric, T., Chang, S. C., Hattori, A., York, I. A., Markant, S., Rock, K. L., Tsujimoto, M., and Goldberg, A. L. (2002). An IFN-gamma-induced aminopeptidase in the ER, ERAP1, trims precursors to MHC class I-presented peptides. *Nat. Immunol.* **3,** 1169–1176.

Saric, T., Graef, C. I., and Goldberg, A. L. (2004). Pathway for degradation of peptides generated by proteasomes: A key role for thimet oligopeptidase and other metallopeptidases. *J. Biol. Chem.* **279,** 46723–46732.

Schwarz, K., Eggers, M., Soza, A., Koszinowski, U. H., Kloetzel, P. M., and Groettrup, M. (2000). The proteasome regulator PA28alpha/beta can enhance antigen presentation without affecting 20S proteasome subunit composition. *Eur. J. Immunol.* **30,** 3672–3679.

Serwold, T., Gonzalez, F., Kim, J., Jacob, R., and Shastri, N. (2002). ERAAP customizes peptides for MHC class I molecules in the endoplasmic reticulum. *Nature* **419,** 480–483.

Song, X., von Kampen, J., Slaughter, C. A., and De Martino, G. N. (1997). Relative functions of the alpha and beta subunits of the proteasome activator, PA28. *J. Biol. Chem.* **272,** 27994–28000.

Stoltze, L., Schirel, M., Schwarz, G., Schröter, C., Thompson, M., Hersh, L., Kalbacher, H., Stevaovic, S., Rammensee, H. G., and Schild, H. (2000). Two new proteases in the MHC class I processing pathway. *Nat. Immunol.* **1,** 413–418.

Sun, Y., Sijts, A. J., Song, M., Janek, K., Nussbaum, A. K., Kral, S., Schirle, M., Stevanovic, S., Paschen, A., Schild, H., *et al.* (2002). Expression of the proteasome activator PA28 rescues the presentation of a cytotoxic T lymphocyte epitope on melanoma cells. *Cancer Res.* **62,** 2875–2882.

Tanahashi, N., Murakami, Y., Minami, Y., Shimbara, N., Hendil, K. B., and Tanaka, K. (2000). Hybrid proteasomes. Induction by interferon-gamma and contribution to ATP-dependent proteolysis. *J. Biol. Chem.* **275,** 14336–14345.

Tanahashi, N., Yokota, K., Ahn, J. Y., Chung, C. H., Fujiwara, T., Takahashi, E., De Martino, G. N., Slaughter, C. A., Toyonaga, T., Yamamura, K., Shimbara, N., and Tanaka, K. (1997). Molecular properties of the proteasome activator PA28 family proteins and gamma-interferon regulation. *Genes Cells* **2,** 195–211.

Tarcsa, E., Szymanska, G., Lecker, S., O'Connor, C. M., and Goldberg, A. L. (2000). $Ca^{2+}$-free calmodulin and calmodulin damaged by *in vitro* aging are selectively degraded by 26S proteasomes without ubiquitination. *J. Biol. Chem.* **275,** 20295–20301.

Thrower, J. S., Hoffman, L., Rechsteiner, M., and Pickart, C. M. (2000). Recognition of the polyubiquitin proteolytic signal. *EMBO J.* **19,** 94–102.

van Hall, T., Sijts, A., Camps, M., Offringa, R., Melief, C., Kloetzel, P. M., and Ossendorp, F. (2000). Differential influence on cytotoxic T lymphocyte epitope presentation by controlled expression of either proteasome immunosubunits or PA28. *J. Exp. Med.* **192,** 483–494.

Van Kaer, L., Ashton-Rickardt, P. G., Eichelberger, M., Gaczynska, M., Nagashima, K., Rock, K. L., Goldberg, A. L., Doherty, P. C., and Tonegawa, S. (1994). Altered peptidase and antiviral activities in LMP2 mutant mice. *Immunity* **1,** 533–541.

Verma, R., and Deshaies, R. J. (2000). A proteasome whodunit: The case of the missing signal. *Cell* **101,** 341–344.

Voges, D., Zwickl, P., and Baumeister, W. (1999). The 26S proteasome: A molecular machine designed for controlled proteolysis. *Annu. Rev. Biochem.* **68,** 1015–1068.

Whitby, F. G., Masters, E. I., Kramer, L., Knowlton, J. R., Yao, Y., Wang, C. C., and Hill, C. P. (2000). Structural basis for the activation of 20S proteasomes by 11S regulators. *Nature* **408,** 115–120.

Yamano, T., Murata, S., Shimbara, N., Tanaka, N., Chiba, T., Tanaka, K., Yui, K., and Udono, H. (2002). Two distinct pathways mediated by PA28 and hsp90 in major histocompatibility complex class I antigen processing. *J. Exp. Med.* **196,** 185–196.

Yang, Y., Fruh, K., Ahn, K., and Peterson, P. A. (1995). *In vivo* assembly of the proteasomal complexes, implications for antigen processing. *J. Biol. Chem.* **270,** 27687–27694.

York, I. A., Chang, S. C., Saric, T., Keys, J. A., Favreau, J. M., Goldberg, A. L., and Rock, K. L. (2002). The ER aminopeptidase ERAP1 enhances or limits antigen presentation by trimming epitopes to 8-9 residues. *Nature Immunol.* **3,** 1177–1184.

# [29] Characterization of the Proteasome Using Native Gel Electrophoresis

*By* Suzanne Elsasser, Marion Schmidt, and Daniel Finley

## Abstract

Several features of the proteasome make it an excellent subject for analysis by native gel electrophoresis: its size, the multiplicity of variant complexes having proteasome activity, the ease of in-gel assays for proteasome activity, and even its relatively high cellular abundance. Accordingly, native gels have been used to analyze the composition, assembly, gating activity, and binding characteristics of the proteasome. This chapter describes methods for preparing, running, and developing native gels and the proteasome species that are routinely visualized. Additionally, the use of native gels to resolve proteasome complexes present in lysate and to characterize proteasome ligands are described. Following native gel electrophoresis, secondary analyses can be performed, such as activating the core particle, making specific activity assessments, Western blotting of the native gel, resolving native complexes with subsequent SDS–PAGE, and protein identification by mass spectrometry.

## Introduction

The proteasome has multiple forms, the most obvious distinction being between the 26S and the 20S proteasome. Rechsteiner and co-workers showed that the 20S proteasome is the proteolytic core particle of the 26S proteasome using native gels followed by fluorogenic peptide hydrolysis, excision of the active bands, and resolution of these complexes with SDS–PAGE (Hoffman *et al.,* 1992; Hough *et al.,* 1987). The 26S proteasome itself is resolved into two major forms on native gels, and second-dimension SDS–PAGE analysis showed that the lower mobility form has roughly twice the amount of non-20S polypeptides as compared with the higher mobility form (Glickman *et al.,* 1998a). Using this approach, a complex inactive toward suc-LLVY-AMC, but composed of proteins unique to the 26S proteasome, was also identified (Hoffman *et al.,* 1992). Furthermore, combining the 20S protease [now known also as the proteasome core particle (CP)] with this complex [now known either as PA700 or

METHODS IN ENZYMOLOGY, VOL. 398

0076-6879/05 $35.00
DOI: 10.1016/S0076-6879(05)98029-4

the 19S regulatory particle (RP)] reconstituted the 26S proteasome. These observations, combined with electron microscopy data (Peters *et al.,* 1991, 1993), formed the basis for the singly and doubly capped model of the proteasome (Hoffman *et al.,* 1992; Rechsteiner *et al.,* 1993). These studies also led to the identification of core particle variants. The CP purified from mammalian cells was resolved on native gels into three distinct complexes, with the most abundant and fastest-migrating complex being the least active among them. The activated CP bands were later determined to contain distinct activators now known as PA28 and PA200 (Ma *et al.,* 1992; Ustrell *et al.,* 2002).

Several further insights into the structure of the proteasome, e.g., that the regulatory particle is composed of base and lid subassemblies (Glickman *et al.,* 1998a), also involved a key role for native gel electrophoresis, but in these cases native gels were used to define the properties of mutant proteasomes from yeast. Glickman and co-workers (1998a) identified the base as proximal to the core particle and the lid as the more distal component. While investigating the role of the base-localized ATPase subunits, Rubin and co-workers (1998) found that mutations of one of the six ATPases affected the peptide hydrolysis activity, but not the assembly, of the proteasome, as determined by native gel analysis and subsequent Coomassie staining. The low specific activity of the mutant proteasome suggested that the mutant regulatory particle was defective in opening the axial gates of the core particle, which are sealed in the absence of regulators (Groll *et al.,* 2000). The hydrolysis defect of the ATPase mutant was suppressed by a core particle mutant possessing a constitutively open gate, an effect more clearly demonstrated using native gels (Kohler *et al.,* 2001).

After a mild affinity technique for purifying the proteasome was developed by Leggett and co-workers (2002), native gel electrophoresis revealed that the affinity-purified proteasome ran more slowly than the conventionally purified proteasome, indicating the presence of abundant components of the proteasome lost during conventional chromatography. Native gels have also been used to study the *in vivo* reassembly of the proteasome from the core particle and the regulatory particle as cells emerge from a deep stationary phase (Bajorek *et al.,* 2003) as well as core particle maturation (Frentzel *et al.,* 1994). The association of proteasome ligands, such as ubiquitin-like proteins Rad23 and Dsk2, has been demonstrated by following a mobility shift of proteasomes on native gels (Elsasser *et al.,* 2002). Similarly, a role for Rpn10 in the recognition of ubiquitin conjugates by the proteasome has been shown using this native gel mobility shift assay (Elsasser *et al.,* 2002, 2004).

## Preparing and Running Native Gels

### *Characteristics of the Gel*

Our standard native gels for characterization of the proteasome are prepared using the solutions described in Table I. A gel mixture is prepared in the resolving buffer to a final acrylamide concentration of 3.5% using a 40% stock solution of acrylamide and bisacrylamide in a 37.5:1 ratio (Bio-Rad, 161-0148), and the gel cocktail is polymerized with 0.1% TEMED and 0.1% APS. The gel mixture is poured into a minigel apparatus set with 1.5-mm spacers (we use Bio-Rad and Hoeffer) and should polymerize within 10 min. Due to the low acrylamide concentration of the native gels, they are susceptible to slow leaks. We have tried to circumvent leaks by pouring "plugs" to seal the bottom of plates, but have found that this leads to band distortion.

### *Preparing the Plates*

Plates should be absolutely clean. Native gels, because of their low acrylamide percentage, are particularly vulnerable to adhering to the plate. We use dish detergent (Ivory Soap) and a sponge to clean the plates, rinse generously with water, and follow with a rinse of nondenatured ethanol to enhance drying. Note that drying the plates with paper towels (as opposed to Kimwipes) may leave behind fibers that fluoresce in the UV or distort band morphology, which will become evident upon developing the gel.

TABLE I
RESOLVING BUFFER

| Final concentration | Stock |
|---|---|
| 90 m*M* Tris base, 90 m*M* boric acid | 450 m*M* each Tris base and boric acid, final pH 8.3 (no buffering required). Store at room temperature and prepare fresh every 2 weeks |
| 5 m*M* $MgCl_2$ | 1 *M* $MgCl_2$ |
| 0.5 m*M* EDTA | 0.5 *M* EDTA, pH 8.0 (titrated with HCl) |
| 1 m*M* ATP-$MgCl_2$ | 250 m*M* each ATP (disodium salt, Sigma, A3377) and $MgCl_2$, and 500 m*M* Tris base. Store in aliquots at –80° and thaw only once |

*Pouring the Gel*

Care must be taken in pouring the gel mixture into the plate sandwich and inserting the comb. Most workers have exhaustive experience preparing gels with stacks, for which the pouring procedure is not critical. After mixing the cocktail gently, the gel should be administered in a slow stream by dribbling into the plate sandwich along one of the spacers. This enhances even mixing and polymerization. Pouring the gel mixture down the center of the plate, for example, often results in irregular polymerization dynamics and consequent uneven band migration, such as a hump in the middle of the gel. The comb should be inserted such that there are no bubbles trapped below the comb teeth, as this will cause significant band distortion.

*Preparing and Loading the Sample*

Once polymerized, the native gel is clipped into its apparatus, and the reservoirs are filled with cold running buffer. It is worth testing for slow leaks in the apparatus while preparing the samples, as an interruption in electrophoresis caused by a depleted reservoir buffer leads to band diffusion and anomalous migration. Samples to be analyzed should be mixed thoroughly and carefully with 5× sample buffer [250 m*M* Tris-HCl (pH 7.4), 50% glycerol, 60 ng/ml xylene cyanol]. Loading the sample into wells using elongated pipette tips (Sorensen Bioscience Inc., 13790) or flat-bottomed Hamilton syringes facilitates depositing the sample at the bottom of the well and suppresses the mixing of sample with liquid in the well. To visualize proteasome holoenzyme using the fluorogenic overlay assay (see next section), 1–5 $\mu$g purified proteasome per millimeter of lane width is loaded.

*Running and Developing the Gel*

The gels are run from negative to positive, at 100–110 V (about 23–25 mA), for 3 h at 4° in a coldroom. Following electrophoresis, gels should be carefully dislodged from the plates and flipped into a dish containing developing buffer [50 m*M* Tris-HCl (pH 7.4), 5 m*M* $MgCl_2$, and 1 m*M* ATP]. After decanting the buffer, the gel is incubated with 50 $\mu M$ suc-LLVY-AMC in developing buffer without agitation for 10 to 30 min at 30°, depending on the proteasome amount loaded. suc-LLVY-AMC (Bachem, I-1395.0025) is a substrate of the chymotryptic active site of the proteasome, and cleavage results in the release of AMC, the fluorescing compound. During development, the gel should be submerged and absolutely flat, as the AMC can diffuse and leave strong ghost bands if the gel is folded back on itself. After incubation, the gel is exposed to UV light and photographed. We use the UV transilluminator (Model TL-33, UVP, Inc.), which has a peak of emission almost coincident with the excitation maximum for AMC (356 and 366 nm, respectively). Gels are conveniently transferred with a slotted, bevel-edged kitchen spatula. Because the quality of UV filters decreases with usage, giving a high background that can obscure signal, we reserve a filter for native gel photography only.

### *Variable Parameters*

*Well Width.* For native gels, wider lanes give a far superior band morphology and are best for distinguishing complexes of lower abundance. For the Bio-Rad minigel system, combs with 12-, 5-, and 3-mm teeth (5, 9, and 15 teeth per comb, respectively) are available. For all lane widths, trailing edges on the sides of the lanes can be seen (see Fig. 1). These trailing edges, which derive from sample diffusing and sometimes electrophoresing into the well dividers before electrophoresing to the bottom of the well, are less dominant in wider lanes, and greater resolution can be

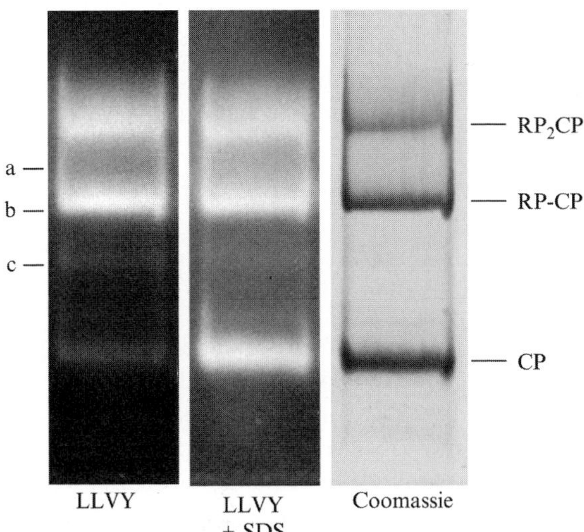


FIG. 1. Purified proteasome resolved on a native gel: 25 $\mu$g of affinity-purified proteasome and 20 $\mu$g of conventionally purified core particle were mixed, loaded into 12-mm wells, and resolved according to our standard conditions. The gel was developed first in the absence (left) and then in the presence (center) of 0.02% SDS. The gel was subsequently stained with Coomassie blue (right). The relative migration of other proteasomal subcomplexes is indicated: (a) base-CP-RP, (b) base$_2$-CP, and (c) base-CP (inferrered from Glickman *et al.*, 1998a).

obtained. These trailing edges can be minimized by keeping the sample volume small and commencing electrophoresis quickly after loading.

*Percentage Acrylamide.* Generally, we find 3.5% gels are optimal for most purposes. We have used final acrylamide concentrations ranging from 3 to 4% and have found that, while the mobility of the proteasome increases with decreasing gel concentration, the relative distance between the various proteasome species remains largely fixed under our standard buffer conditions. For mobility shift assays (see later), greater mobility is an advantage, but because the 3% gels are particularly difficult to handle, the 3.5% gels are a better choice for these assays. In comparison to 3.5% gels, 4% gels may give slightly sharper band morphology and are slightly easier to handle.

*Tris/Boric Acid Concentration.* For native gels, the Tris/boric acid concentration varies from 90 to 180 m*M*. Higher buffer conductivity may decrease proteasome mobility and result in greater heating of the gel, but at the same time may counter effects on mobility due to the presence of salt in the loaded sample and may also produce sharper bands.

*Temperature, Voltage, and Running Time.* While we typically run gels at 4° in a cold room, it is also possible to run native gels using a connected cryostat (Fisher Scientific Isotemp 1016S). Using a cryostat set at 10°, we have run native gels on the bench top, which appears to increase mobility without compromising band quality. At 4°, 2 h is a minimum for verifying the presence of the major proteasome species, and 3.5 h is useful for distinguishing complexes in a mobility shift assay. Decreasing the voltage in order to run overnight gives rise to more diffuse bands that can be difficult to analyze.

## Analysis

### *Proteasome Profile*

In yeast, three major species of proteasome are typically observed: core particle, core particle associated with one regulatory particle (RP-CP), and core particle associated with two regulatory particles ($RP_2CP$). The characteristic migration of these species in our standard native gels is shown in Fig. 1. The proteasome regulatory particle can be split into the base and lid, which are proximal and distal to the core particle, respectively. Loss of the lid from the holoenzyme has been seen in proteasomes prepared from cells lacking Rpn9, Rpn10, or the C terminus of Rpn11 (Glickman *et al.,* 1998a; Verma *et al.,* 2002). The lid also dissociates from the base in the presence of 1 *M* NaCl (Saeki *et al.,* 2000). The resulting complexes, base-CP-RP, $base_2CP$, and base-CP migrate above, coincident with, and below the RP-CP band, as indicated in Figs. 1a, 1b, and 1c, respectively.

### *Stimulation of the Core Particle*

In the absence of RP, the CP has low specific activity as compared with both forms of the holoenzyme (compare left and right panels, Fig. 1). This difference between CP and holoenzyme activity is a consequence of the axial gates of the proteasome being opened by the presence of the regulatory particle (Groll *et al.,* 2000). The activity of the CP toward suc-LLVY-AMC hydrolysis can be stimulated by adding 0.02% SDS to the developing buffer (compare left and center panels, Fig. 1). Developing the gel first in the absence and then in the presence of SDS reveals this characteristic feature of the CP and can be used to make a band assignment with reasonable confidence.

### *Coomassie and Silver Staining the Native Gel*

Protein staining the native gel subsequent to the in-gel activity assay is strongly advised (see Fig. 1). Comparison of the activity assay and the protein stain can reveal proteasome subcomplexes that have disassociated from the core particle following purification and also complexes containing both core particles and inhibitors. Additionally, this analysis can serve to evaluate, albeit qualitatively, the specific activity of mutant proteasome species and the interaction of activators with the proteasome (Schmidt *et al.*, 2005). Regulatory particle mutants that are defective in gating and core particle mutants that have a constitutively open gate have been characterized with this type of analysis (Kohler *et al.,* 2001; Rubin *et al.,* 1998).

### *Western Analysis*

The identities of proteins present in various bands can be determined by blotting the gel to a membrane and probing with antibodies specific to individual complexes. Before transfer, the entire gel is soaked in a solution of 25 m*M* Tris base, 192 m*M* glycine, and 1% SDS, followed by a 10-min incubation in transfer buffer (25 m*M* Tris base, 192 m*M* glycine), and transferred at 250 mA for 1.5 h (J. Roelofs, personal communication). Submerging the gel in transfer buffer during assembly into the transfer cassette limits the distortion of the gel. Once transfer is complete, all of the usual techniques for Western blotting apply.

### *SDS–PAGE Following the Native Gel*

Complexes present in the native gel can also be resolved by a subsequent second dimension of SDS–polyacrylamide gel electrophoresis. An entire lane from the native gel can be excised, inserted into a single long well of a denaturing gel, secured with 1% agarose prepared with 2× sample buffer, and resolved (see Fig. 2). Smaller regions of the gel,

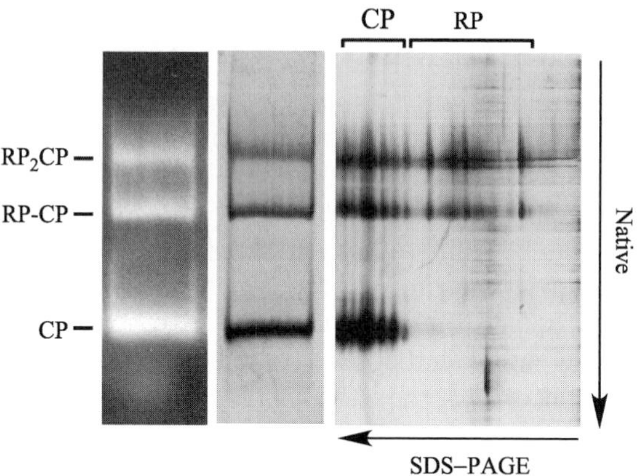


Fig. 2. Resolution of proteasome complexes by SDS–PAGE following native gel separation: 25 $\mu$g of affinity-purified proteasome were loaded into 12-mm wells and resolved according to our standard conditions. The gel was developed in the presence of 0.02% SDS (left). The central portion of the lane was excised, subjected to 10% SDS–PAGE, and silver stained (right). A sister lane was stained with Coomassie blue (center).

such as active bands, can be excised, boiled in an equal volume of 2× SDS–PAGE loading buffer, and the eluted proteins resolved. For each of these options, the best results will be achieved by excising the central portion of the lane so as to avoid the characteristic trailing that occurs at the side of each lane (see Fig. 1).

### *Mass Spectrometry*

The most general method for band assignment is mass spectrometry. This approach has several advantages, the greatest of which is sensitivity. Active bands may be excised directly following the fluorogenic assay (Schmidt *et al.,* 2005). Alternately, gels may be stained with Coomassie blue or silver before band excision, allowing greater precision in the complexes excised, as well as the identification of nonfluorogenic species. The key use of this method is to identify novel proteasome-associated proteins of low abundance and to precisely define components of a given proteasome subcomplex.

## Detecting Proteasome in Lysate

Native gels provide a surprisingly good means to characterize proteasomes from unfractionated lysates. Some proteasome-associated proteins are sensitive to the conditions used in conventional purifications (Glickman *et al.*, 1998b; Leggett *et al.*, 2002) and, by the same token, the association of some proteins with the proteasome may be weakened by the affinity tags currently in use. Resolving lysate on a native gel gives the closest picture of the *in vivo* state of the proteasome. Because purification is not required, proteasome species that have not been modified by affinity tagging can be characterized. suc-LLVY-AMC hydrolysis activity of resolved proteins will reveal active complexes, and Western blotting will reveal inactive complexes composed of proteasome components.

## Method

For analysis of lysate on native gels, we harvest cells and resuspend in lysis buffer [50 m*M* Tris-HCl (pH 8.0), 5 m*M* $MgCl_2$, 0.5 m*M* EDTA, and 1 m*M* ATP] at a ratio of 1.5 ml buffer per gram wet cell mass. Cells can be lysed by French press (one pass at 2000 psi) or by vortexing for 5 min at 4° with 0.5-mm acid-washed glass beads (Sigma, G8772; J. Roelofs, personal communication). The extract is cleared in a microfuge at 4° for 30 min (15,000*g*). The lysate will have a protein concentration of roughly 10 mg/ml. Between 50 and 300 $\mu$g total protein should be loaded onto the gel in order to see a signal in the fluorogenic assay. Native gels are quite sensitive to salt, and it is recommended to limit salt in the lysis buffer.

## Proteasome Binding Assays

Native gels can be used to evaluate ligand binding to the proteasome. This approach has been used to study the binding of ubiquitin conjugates as well as of ubiquitin-like proteins (Elsasser *et al.*, 2002, 2004). Some proteasome ligands will also impede the mobility of the proteasome on native gels, and the existence of a complex can be inferred from this observation. For ligands that do not alter proteasome mobility on native gels, complex formation may be analyzed by following the proteasome dependence of ligand mobility using Western blotting or radiolabeling combined with autoradiography.

In these binding assays, the time and temperature required for complex formation must be determined empirically. The mobility of proteasomes on native gels is sensitive to buffer conditions, and care must be taken that the buffer conditions are precisely the same in each sample. Additionally, because there is no stacking gel, loading uneven volumes will generate a higher trailing edge in samples with the greater volume. Because this can be mistaken easily for a subtle mobility shift, the volume loaded in each well of a binding assay should be exactly the same. In order to ascertain

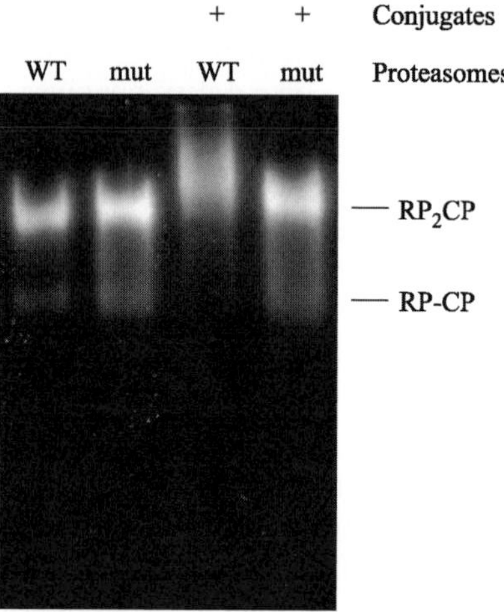


Fig. 3. A mobility shift assay. Five micrograms of proteasomes affinity purified from cells carrying a ProA-tagged Rpn11 (WT) or from cells additionally bearing a mutation in the ubiquitin chain-binding site of Rpn10 (mut) were incubated for 40 min at 30° with 8 pmol of ubiquitin conjugates or buffer only. Complexes were resolved by native PAGE and visualized by suc-LLVY-AMC hydrolysis using our standard conditions. Adapted from Elsasser *et al.* (2004) with permission from the American Society of Biochemistry and Molecular Biology.

that an observed mobility shift is not due to gel artifacts, the first and last lane of the gel should be loaded with identical samples in order to determine a baseline of migration (Fig. 3).

## References

Bajorek, M., Finley, D., and Glickman, M. H. (2003). Proteasome disassembly and downregulation is correlated with viability during stationary phase. *Curr. Biol.* **13,** 1140–1144.

Elsasser, S., Chandler-Militello, D., Muller, B., Hanna, J., and Finley, D. (2004). Rad23 and Rpn10 serve as alternative ubiquitin receptors for the proteasome. *J. Biol. Chem.* **279,** 26817–26822.

Elsasser, S., Gali, R. R., Schwickart, M., Larsen, C. N., Leggett, D. S., Muller, B., Feng, M. T., Tubing, F., Dittmar, G. A., and Finley, D. (2002). Proteasome subunit Rpn1 binds ubiquitin-like protein domains. *Nat. Cell Biol.* **4,** 725–730.

Frentzel, S., Pesold-Hurt, B., Seelig, A., and Kloetzel, P. M. (1994). 20S proteasomes are assembled via distinct precursor complexes: Processing of LMP2 and LMP7 proproteins takes place in 13–16S preproteasome complexes. *J. Mol. Biol.* **236,** 975–981.

Glickman, M. H., Rubin, D. M., Coux, O., Wefes, I., Pfeifer, G., Cjeka, Z., Baumeister, W., Fried, V. A., and Finley, D. (1998a). A subcomplex of the proteasome regulatory particle required for ubiquitin-conjugate degradation and related to the COP9-signalosome and eIF3. *Cell* **94,** 615–623.

Glickman, M. H., Rubin, D. M., Fried, V. A., and Finley, D. (1998b). The regulatory particle of the Saccharomyces cerevisiae proteasome. *Mol. Cell. Biol.* **18,** 3149–3162.

Groll, M., Bajorek, M., Kohler, A., Moroder, L., Rubin, D. M., Huber, R., Glickman, M. H., and Finley, D. (2000). A gated channel into the proteasome core particle. *Nat. Struct. Biol.* **7,** 1062–1067.

Hoffman, L., Pratt, G., and Rechsteiner, M. (1992). Multiple forms of the 20S multicatalytic and the 26 S ubiquitin/ATP-dependent proteases from rabbit reticulocyte lysate. *J. Biol. Chem.* **267,** 22362–22368.

Hough, R., Pratt, G., and Rechsteiner, M. (1987). Purification of two high molecular weight proteases from rabbit reticulocyte lysate. *J. Biol. Chem.* **262,** 8303–8313.

Kohler, A., Cascio, P., Leggett, D. S., Woo, K. M., Goldberg, A. L., and Finley, D. (2001). The axial channel of the proteasome core particle is gated by the Rpt2 ATPase and controls both substrate entry and product release. *Mol. Cell* **7,** 1143–1152.

Leggett, D. S., Hanna, J., Borodovsky, A., Crosas, B., Schmidt, M., Baker, R. T., Walz, T., Ploegh, H., and Finley, D. (2002). Multiple associated proteins regulate proteasome structure and function. *Mol. Cell* **10,** 495–507.

Ma, C. P., Slaughter, C. A., and De Martino, G. N. (1992). Identification, purification, and characterization of a protein activator (PA28) of the 20 S proteasome (macropain). *J. Biol. Chem.* **267,** 10515–10523.

Peters, J. M., Cejka, Z., Harris, J. R., Kleinschmidt, J. A., and Baumeister, W. (1993). Structural features of the 26 S proteasome complex. *J. Mol. Biol.* **234,** 932–937.

Peters, J. M., Harris, J. R., and Kleinschmidt, J. A. (1991). Ultrastructure of the approximately 26S complex containing the approximately 20S cylinder particle (multicatalytic proteinase/proteasome). *Eur. J. Cell Biol.* **56,** 422–432.

Rechsteiner, M., Hoffman, L., and Dubiel, W. (1993). The multicatalytic and 26S proteases. *J. Biol. Chem.* **268,** 6065–6068.

Rubin, D. M., Glickman, M. H., Larsen, C. N., Dhruvakumar, S., and Finley, D. (1998). Active site mutants in the six regulatory particle ATPases reveal multiple roles for ATP in the proteasome. *EMBO J.* **17,** 4909–4919.

Schmidt, M., Haas, W., Crosas, B., Santamaria, P. G., Gygi, S. P., Waltz, T., and Finley, D. (2005). The HEAT repeat protein Blm10 regulates the yeast proteasome by capping the core particle. *Nat. Struct. Mol. Biol.* **12,** 294–303.

Saeki, Y., Toh-e, A., and Yokosawa, H. (2000). Rapid isolation and characterization of the yeast proteasome regulatory complex. *Biochem. Biophys. Res. Commun.* **273,** 509–515.

Ustrell, V., Hoffman, L., Pratt, G., and Rechsteiner, M. (2002). PA200, a nuclear proteasome activator involved in DNA repair. *EMBO J.* **21,** 3516–3525.

Verma, R., Aravind, L., Oania, R., McDonald, W. H., Yates, J. R., 3rd, Koonin, E. V., and Deshaies, R. J. (2002). Role of Rpn11 metalloprotease in deubiquitination and degradation by the 26S proteasome. *Science* **298,** 611–615.

# [30] Monitoring Activity and Inhibition of 26S Proteasomes with Fluorogenic Peptide Substrates

*By* ALEXEI F. KISSELEV and ALFRED L. GOLDBERG

## Abstract

Eukaryotic proteasomes have three different types of active sites: two chymotrypsin-like, two trypsin-like, and two caspase-like (also termed PGPH) sites that differ in their specificity toward model fluorogenic peptide substrates. The chymotrypsin-like site is often considered the most important in protein breakdown, and the only one whose activity has to be assayed in order to assess the capacity of proteasomes to degrade proteins. However, recent results indicate that either trypsin-like or caspase-like sites also have to be inhibited in order to reduce breakdown of most proteins by 50%. Thus, the activities of all three types of active sites have to be assayed in order to evaluate the state of the proteasome inside cells or the potency of inhibitors. This chapter describes assays of purified 26S proteasomes with fluorogenic peptide substrates, including new substrates of the caspase- and trypsin-like sites. A novel assay of proteasome activity in crude cell extracts that allows rapid evaluation of the state of the proteasomes in cells treated with inhibitors is also described.

The 26S proteasome is an ATP-dependent proteolytic complex that consists of a 20S proteolytic core and one or two 19S regulatory complexes (Baumeister *et al.*, 1998; Coux *et al.*, 1996; Voges *et al.*, 1999). The 20S proteasome is a hollow cylindrical particle consisting of 28 polypeptides arranged in four stacked rings. Each of the inner $\beta$ rings contains three active sites, which are located in the proteolytic chamber inside the hollow cylinder (Groll *et al.*, 1997). Access of substrates and exit of degradation products are controlled by two narrow, gated axial channels each formed by seven subunits of the outer $\alpha$ rings (Groll *et al.*, 2000). The 20S core can also exist in a free form that is often referred to as the 20S proteasome and is actually more stable and easier to isolate than the 26S particles, which require the presence of ATP and glycerol in all buffers to maintain its integrity during purification (Coux *et al.*, 1996).

Although multiple forms of proteasomes exist within cells [e.g., 20S and 26S complexes (Yang *et al.*, 1995) and complexes containing PA28 (Tanahashi *et al.*, 2000)], only the 26S proteasome has been proven to play a significant role in intracellular protein degradation. Most proteins in mammalian cells are degraded by a process requiring ATP (Goldberg

METHODS IN ENZYMOLOGY, VOL. 398

0076-6879/05 $35.00
DOI: 10.1016/S0076-6879(05)98030-0

and St John, 1976) and most appear to require substrate ubiquitinylation (Hershko and Ciechanover, 1998). Unlike the 26S or related 19S–20S–PA28 complexes (Cascio *et al.*, 2002; Kopp *et al.*, 2001), the 20S proteasome functions independently of ATP and is incapable of degrading polyubiquitinylated proteins. Although a role for the 20S proteasome in the degradation of damaged or unfolded polypeptides has often been suggested (Grune *et al.*, 2003; Liu *et al.*, 2003), 20S particles in their most physiological form cannot degrade typical polypeptides and only acquire the ability to degrade after nonphysiological treatments such as the removal of stabilizing agents (e.g., glycerol), the addition of low concentrations of SDS (0.01–0.02%), or freeze–thaw cycles (Coux *et al.*, 1996). If extracts of mammalian tissues are prepared in a gentle manner to mimic intracellular conditions (presence of glycerol and intracellular concentrations of potassium ions), then the 20S particles are in a latent (inactive) form, unable to degrade denatured proteins or cleave peptides (Coux *et al.*, 1996; Köhler *et al.*, 2001). Therefore, free 20S proteasomes are unlikely to play a significant role in intracellular protein degradation.

Because the 20S proteasome contains the exact same peptidase sites as the 26S proteasomes, it is widely assumed that assaying the activity of the 20S particle is an accurate measure of the state of the 26S proteasome. This assumption has no experimental basis. When the activity of 26S particles is assayed in the presence of ATP, channels in the $\alpha$ rings of the 20S core are open by ATPases of the 19S regulatory particle (Köhler *et al.*, 2001), and peptide substrates and inhibitors have ready access to the 20S core. In contrast, the state of the channels in the free 20S proteasome depends on many nonphysiological factors that are difficult to control. The presence of detergents during purification and repeated freezing and thawing of samples cause gate opening and activation of the latent 20S particles (Coux *et al.*, 1996). Also, hydrophobic peptides, including commonly used fluorometric substrates (e.g., Suc LLVY-amc), trigger gate opening and thus stimulate the activity of the 20S particles (Kisselev *et al.*, 2002). In contrast, low concentrations of sodium and potassium ions retard spontaneous activation of the 20S (Köhler *et al.*, 2001). Thus, the activity of the 20S particle typically reflects not only the ability of the active sites to cleave substrates, but also the ability of substrates to traverse or even cause opening of the pore in the $\alpha$ ring (Kisselev *et al.*, 2002). These complications can give misleading results; in fact, on at least one occasion, we observed more than a 10-fold difference in the sensitivity of 20S and 26S proteasomes to inhibitors (Kisselev *et al.*, 2003). Even more strikingly, the HIV protease inhibitor ritonavir inhibits 20S proteasomes but activates 26S particles (Gaedicke *et al.*, 2002). Consequently, whenever possible, the peptidase activities of 26S proteasomes should be measured to evaluate the efficacy of inhibitors and the capacity of these particles to degrade proteins.

The absolute rate of cleavage of peptide bonds in protein substrates by purified 26S proteasomes can be assayed with the fluorescamine assay that reacts with $\alpha$ amino groups generated upon the cleavage of peptide bonds (Akopian *et al.*, 1997; Kisselev *et al.*, 1998; Udenfriend *et al.*, 1972). This assay, which is discussed in Cascio and Goldberg (2005), has proven to be a very useful tool in studies of the mechanisms of protein degradation and in the production of antigenic peptides (Cascio *et al.*, 2001; Kisselev *et al.*, 1998). However, assays of proteasome individual active sites with fluorogenic peptide substrates are more sensitive, accurate, easier to perform, and allow measurements of proteasome activity in crude lysates.

The six active sites of the proteasome differ in their specificity toward peptide bonds in proteins and model fluorogenic peptides. Two are "chymotrypsin-like" and cleave after hydrophobic residues, two are "trypsin-like" and cleave after basic residues, and two are "caspase-like" (also termed PGPH) sites and cut peptide bonds after acidic residues. Early genetic studies in yeast revealed distinct roles of these sites in protein degradation. Inactivation of the chymotrypsin-like sites by mutation caused significant inhibition of the degradation of three model proteins and significant impairment of cell growth (Chen and Hochstrasser, 1996; Heinemeyer *et al.*, 1997). Inactivation of the trypsin-like sites caused milder defects, and no changes were found in strains in which the caspase-like sites were mutated (Arendt and Hochstrasser, 1997; Heinemeyer *et al.*, 1997). These data led many investigators to believe that inhibition of the chymotrypsin-like sites adequately reflects inhibition of protein breakdown. However, studies indicate that inhibition of chymotrypsin-like sites alone causes little inhibition of protein breakdown, and that either the trypsin-like or the caspase-like sites have to be inhibited as well in order to inhibit proteasomal proteolysis by 50% (Kisselev *et al.*, 2005; Oberdorf *et al.*, 2001). Thus, in order to evaluate the potency of proteasome inhibitors (either inside cells or in a test tube), the activity of all active sites has to be measured.

## Fluorogenic Peptide Substrates Used to Assay Proteasome Activity and Inhibition

Ever since the 20S proteasome was discovered by Wilk and Orlowski (1983) more than two decades ago, fluorogenic peptide substrates have been used to measure proteasome activity. They provide a convenient and sensitive way to monitor the activity of proteasomes during purification and to measure the ability of new inhibitors to block different active sites of proteasomes. These substrates are three to four amino acid residue peptides with a fluorogenic reporter group at the C terminus. The proteasome cleaves an amido bond between an amino acid and the reporter group, resulting in release of a highly fluorescent product. Three fluorophores are used in proteasome substrates: 7-amino-4-methylcoumarin (AMC), 2-naphtylamine (NA), and 4-methoxy-2-naphtylamine (MNA). Of these three, AMC has the highest fluorescence, and substrates containing AMC are used most often. When these fluorophores are attached to a peptide, small letters are used to abbreviate them (amc, na, mna) in order to avoid confusion with a single letter amino acid code.

Several substrates are available to measure the activities of the three active sites of the proteasomes. Suc-LLVY-amc is used most often to assay chymotrypsin-like activity. Z-GGL-amc and Suc-AAF-amc can also be used for the same purpose, although they are cleaved at slower rates than Suc-LLVY-amc. Z-GGL-amc also has lower solubility and often precipitates at 100 $\mu M$.

Several commercially available substrates with R-amc scissile bond (Boc-LRR-amc, Z-ARR-amc, Bz-FVR-amc, Bz-VGR-amc, Boc-LSTR-amc) can be used for assaying trypsin-like activity. However, the $K_m$ of the 26S proteasome for these substrates is high (>0.5 m$M$), and the specific activity at low concentrations of substrates is low (Table I), sometimes requiring higher concentrations of expensive substrates. To overcome this problem, we used information about extended substrate specificity of proteasomes obtained from position scanning combinatorial libraries (Harris *et al.*, 2001; Nazif and Bogyo, 2001) to design the new substrate Ac-RLR-amc (synthesized by Enzyme System Products). It has a much lower $K_m$, higher specific activity (Table I), and is cleaved exclusively at

TABLE I
SUBSTRATES OF TRYPSIN-LIKE SITES[a]

| Substrate | $K_m$ ($\mu M$) | Concentration ($\mu M$) 100 | 600 | 1000 |
|---|---|---|---|---|
| | | | Specific activity (nmol/min*mg) | |
| Z-ARR-amc | | 7.7 | 19 | 22 |
| Bz-VGR-amc | | 13 | 57 | 83 |
| Boc-LRR-amc | $\gg$500 | 26 | 68 | 89 |
| Boc-LSTR-amc | $\gg$500 | 5.3 | 27 | 60 |
| Ac-RLR-amc | 78 | 45 | 91 | 110 |

[a] All values were determined with highly purified 26S proteasomes from rabbit muscles.

the R-amc bond (as tested by HPLC, data not shown). Therefore, it is our substrate of choice for assaying the trypsin-like activity of purified proteasomes.

Until 2003, caspase-like sites were the most difficult to assay. The only widely used substrate, Z-LLE-na, is cleaved with the release of the NA carcinogene, which is less fluorescent than AMC and fluoresces at a different wavelength. This may cause inconvenience to laboratories using filter-based instruments because NA requires a different set of filters. To eliminate this problem, Z-LLE-amc was synthesized. However, it is cleaved at a 600-fold lower rate than Z-LLE-na. The reason why replacement of the fluorogenic reporter group causes such a dramatic decrease in cleavage is unclear. Furthermore, Z-LLE-na is cleaved by proteasomes not only at the Glu-na but also at the Leu-Glu bond (Kisselev *et al.*, 1999). When used with extracts or partially purified preparations of proteasomes, the Glu-na generated by the latter cleavage may be cleaved further by aminopeptidases to produce free NA, generating a signal from nonproteasomal proteolysis. To alleviate this problem, we developed, using positional scanning libraries, two new substrates of the caspase-like sites: Ac-nLPnLD-amc and Ac-GPLD-amc (Kisselev *et al.*, 2003). These peptides are now available commercially (Bachem), and the better of these two, Ac-nLPnLD-amc, should be a substrate of choice for assays of caspase-like activity.

## Assays of Purified Proteasomes

Assays of proteasome activity with fluorogenic peptide substrates can be run with a continuous recording of the fluorescence of the product, AMC, by measuring its fluorescence in aliquots taken from the reaction mixtures at different times or simply by measuring fluorescence at the end of incubation. Fluorescence can be measured in a cuvette with a spectrofluorometer or in a plate with a fluorescent plate reader. Any instrument used for this measurement has to be calibrated with different concentrations of AMC in the reaction buffer. This procedure should result in a calibration curve that is linear over a wide range of AMC concentrations (0.1–10 $\mu M$). The instrument has to be calibrated on a daily basis, but after linearity of the calibration curve has been established, just one concentration of the standard can be used. We usually use 5 $\mu M$ AMC solution. If the concentration of a substrate in the assay is 100 $\mu M$, the signal from this standard will correspond to a signal from the product when 5% of the substrate is degraded. To ensure linearity of the assay, consumption of the substrate during the assay should not exceed 10%, i.e., if the signal generated during the assay is more than double the fluorescence of the 5 $\mu M$ AMC standard, the assay has run for too long or there was too much enzyme.

For routine assays of purified proteasomes, we use Suc-LLVY-amc as a substrate of the chymotrypsin-like sites, Ac-nLPnLD-amc as a substrate of the caspase-like sites, and Ac-RLR-amc as a substrate of the trypsin-like site, all at 100 $\mu M$. If Ac-RLR-amc is not available, 100 $\mu M$ Boc-LRR-amc can be used instead. All substrates are prepared as 100× stock solutions in dimethyl sulfoxide (DMSO). The final concentration of DMSO in the assay should not exceed 4%. All peptidase assays of the 26S proteasome use the same assay buffer, which consists of 50 m$M$ Tris-HCl, pH 7.5, 40 m$M$ KCl, 5 m$M$ $MgCl_2$, 0.5 m$M$ ATP, 1 m$M$ dithiothreitol (DTT). ATP has to be used to prevent dissociation of the 26S proteasome into its components, to ensure that it has maximal activity, and that indeed 26S but not 20S activity is measured. KCl is needed to further decrease contribution of the 20S proteasome to peptide cleavage by suppressing its spontaneous activation (Köhler *et al.*, 2001). When performing continuous assays in plastic cuvettes or multiwell plates, bovine serum albumin (BSA) or soybean trypsin inhibitor (SBTI) often has to be added to the buffer. BSA or SBTI is needed to prevent a decrease of 26S activity observed frequently during assays. We attribute this phenomenon to 26S adsorption to the plastic walls of the plate or cuvette. ATP, DTT, and SBTI or BSA should be added fresh at the day of the experiment from stock solutions stored at –20°.

### *Continuous Assay of Peptidase Activities on a Spectrafluorometer*

Because amc substrates are expensive, we recommend minimizing the consumption of the substrate by adjusting the height of the cuvette in the fluorometer so that the excitation beam goes as close to the bottom of the cuvette as possible. For example, if the height of a 1.5-ml semimicrocuvette is adjusted properly, 400 $\mu$l of substrate should be sufficient. For measuring AMC fluorescence (excitation: 380 nm, emission: 440–460 nm), quartz cuvettes can be replaced by disposable plastic cuvettes with clear sides.

Prepare solutions of substrates in the assay buffer containing 0.5 mg/ml BSA or 50 $\mu$g/ml SBTI. Transfer 400 $\mu$l of the substrate solution into a cuvette. If inhibitors are being tested, add 4 $\mu$l of a 100× stock and mix. Place the cuvettes in a thermostated turret. If using semimicrocuvettes, make sure that all cuvettes are inserted the same way, with the long side perpendicular to the excitation beam. The thermostat should be set to 37° for assaying mammalian proteasomes and to 30° for yeast proteasomes. If a thermostated cuvette holder is not available, eukaryotic 26S proteasomes can be assayed at room temperature. Preincubate the substrate at the assay temperature for at least 5 min. Add the enzyme (in a volume of 1–10 $\mu$l), mix well, and start recording the fluorescence (excitation: 380 nm, emission: 460 nm). The concentration and amount of enzyme should be determined experimentally so that the consumption of substrate at the end of the assay does not exceed 1/10 of its initial amount. A good starting point is 0.1 $\mu$g of pure 26S proteasomes per cuvette. If the assay is linear, incubation for 10–15 min is sufficient. For assays that are initially nonlinear, such as those observed in the presence of a slow-binding inhibitor, longer incubations (up to 1 h) may be needed to reach linearity. After the assay is completed, use the instrument software to determine the slopes of the linear portions of the reaction progress curves. They can then be used to determine activity remaining after treatment with an inhibitor (by dividing slopes in the presence of inhibitor by the slope in its absence). Specific activity of the enzyme can also be calculated according to the following equation:

Specific activity [nmol/(min*mg)] = [(slope(FU/min)*standard concentration ($\mu M$))]/[(concentration of enzyme stock (mg/ml)/enzyme dilution) *fluorescence of standard (FU)]

*Continuous Assay of Peptidase Activities on a Fluorescent Plate Reader*

This procedure describes an assay on a 96-well plate. The protocol of assays performed in a 384- or 1536-well format is the same except that the volumes of reagents should be decreased accordingly to reflect the smaller volume of the wells. The assays should be performed on a flat-bottom, black fluorescence plate. Proteasome adsorption to the well surface represents a serious problem, especially if polystyrene plates are used. After testing several different plates, we have chosen Greiner polypropylene plates because they seem to be least prone to adsorption. Even with these plates, 50 $\mu$g/ml BSA or SBTI has to be added to prevent proteasome adsorption and to maintain linearity of the assay.

We usually prepare an assay mixture by mixing 100 $\mu$l of 2× substrate, 50 $\mu$l of 4× inhibitor (or 50 $\mu$l of buffer), and 50 $\mu$l of 4× enzyme to give a final assay volume of 200 $\mu$l. We use the same substrates (Suc-LLVY-amc, Ac-nLPnLD-amc, Ac-RLR-amc) at the same final concentration (100 $\mu M$) as in fluorometer-based assays. The final concentration of the 26S protreasome is 0.3–1 $\mu$g/ml. Substrate and inhibitor solutions are prepared by diluting 100× stock solutions in the assay buffer, transferred into the plate with a multichannel pipetman, and mixed. The plate is covered and incubated at the assay temperature (see earlier discussion) for at least 5 min. During the incubation, a 4× solution of 26S proteasomes in assay buffer is prepared. The reaction is initiated by the transfer of 50 $\mu$l of the proteasome solution into each well with a multichannel pipetman and mixing. Most plate readers have a built-in automixing feature; set the program to mix for 5–10 s before the first read. When using a filter-based instrument, use 380-nm excitation and 460-nm emission filters. If these filters are not available, shorter wavelength filters may be used. The AMC emission maximum is at 440 nm; using a 440-nm emission filter will increase the AMC signal but will decrease the signal:background ratio because the uncleaved substrate fluoresces at 440 nm but not at 460 nm. One reading per minute is sufficient but we always use the shortest interval necessary to read all wells. Choose assay duration as described in the previous procedure. Assay each sample in duplicates (except when column fractions are screened); if there is a significant difference between them, reassay the sample.

The protocol can be modified with respect to how the reagents can be mixed. For example, when screening column fractions, we add 1–5 $\mu$l of a sample to 200 $\mu$l of 1× substrate that has been preincubated at the assay temperature on the plate. Importantly, the volume of the AMC standard used for calibration should be exactly the same as the final assay volume because fluorescence depends on the height of the liquid in a well.

### *Time Point(s) Assay of Peptidase Activities*

If equipment for performing continuous assays is not available or if large numbers of assays need to be performed (i.e., when screening small molecule libraries for inhibitors) and the reaction is linear, continuous assays can be replaced by end point measurements.

*Determining the Linearity of the Assay (if Equipment for Performing Continuous Assays Is Not Available).* Set up the assay as described earlier and incubate a plate or cuvettes with the assay mixtures at the assay temperature (30° or room temperature for yeast, 37° or room temperature for mammalian enzymes). Take fluorescence measurements at 0, 15, 30, and 60 min. After all aliquots are collected, measure the fluorescence and plot its values against time for each sample. If the assay is linear, subsequent experiments can be performed by measuring fluorescence at the end of the incubation.

*End Point Measurements.* Preincubate substrates in tubes (400 $\mu$l/tube) or on a plate (100 $\mu$l/well on a 96-well plate; 40 $\mu$l/well on a 384-well plate) for 5–10 min at the assay temperature (depending on the source of the proteasome). Add enzyme, leaving at least one tube or well per substrate as a control (i.e., no enzyme should be added) and incubate for 30 to 60 min (as determined in the previous experiment). Read the fluorescence immediately after completion of the reaction or freeze samples until the measurement. Proteasome activity is proportional to the difference in fluorescence between sample and control.

## Assaying Proteasome Activity in Crude Extracts: How to Test Whether the Proteasome Is Inhibited When Expected Stabilization of the Protein of Interest by Proteasome Inhibitors Does Not Occur

The ability to selectively measure the activity and inhibition of all three active sites of proteasomes in crude extracts can be useful for many experiments. For example, many investigators use one of the widely available proteasome inhibitors to test whether their protein of interest is a substrate of the proteasome. If these agents stabilize the protein, the conclusion is that it is indeed a substrate of the proteasome. Lack of inhibition may indicate involvement of a different proteolytic system or that not all active sites of proteasome were inhibited. In this case, a rapid test of activity and inhibition of all three active sites is needed. A similar test was used to measure inhibition of the 20S proteasome in extracts of blood cells of multiple myeloma patients treated with Velcade (Lightcap *et al.*, 2000). The ability to selectively measure proteasome activity in extracts will eliminate the need for laborious purification to measure $K_i$ and $k_{obs}$ of new inhibitors or to screen small molecule libraries for new inhibitors.

Here we describe an assay of proteasome activity in cytosolic extracts of certain cell lines (e.g., HeLa). It can be used to measure the inhibition of all three active sites in cells treated with irreversible inhibitors (vinyl sulfones, $\beta$-lactones, epoxyketones) or reversible but slowly dissociating agents such as Velcade or other boronates (Kisselev and Goldberg, 2001). Unlike rapidly reversible peptide aldehyde inhibitors (such as MG132 and PSI), these compounds will not dissociate during extract preparation and upon dilution of the extract into the assay mixture.

The major difference between the assay of proteasomes in extracts and the assay of purified proteasomes is that in extracts, proteasome substrates may be cleaved by other proteolytic enzymes. For example, Suc-LLVY-amc is also a substrate of calpains and mast cell chymases. Cleavage of Ac-nLPnLD-amc by caspases cannot be excluded, although amino acid residues in its P2-P4 positions have been optimized to match proteasome specificity that is different from those of caspases (Kisselev *et al.*, 2003). Numerous cellular proteases, such as furin and prohormone convertases (PCs), tryptase, and lysosomal cathepsins, cleave peptide bonds after basic residues, and thus can cleave substrates of the trypsin-like site. However, cathepsins, furin, and PCs are localized inside organelles, and their contribution to the cleavage of proteasome substrates should be lower if cytosolic extracts are used and if measurements are done at neutral pH. The contribution of different enzymes may also vary from cell line to cell line, from tissue to tissue (Rodgers and Dean, 2003), and perhaps from species to species. It may also depend on growth conditions (Fuertes *et al.*, 2003) and whether the cell is undergoing apoptosis (especially for caspase-like activity).

The contribution of nonproteasomal enzymes can be determined easily using a highly specific inhibitor of proteasomes such as epoxomicin. In 6 years since the discovery that this natural product is a proteasome inhibitor (Meng *et al.*, 1999), no other target of this agent has been reported. The most widely used proteasome inhibitor MG132 is not suitable for these purposes because it inhibits lysosomal enzymes (Fuertes *et al.*, 2003; Kisselev and Goldberg, 2001; Rodgers and Dean, 2003). At high concentrations (20 $\mu M$), epoxomicin blocks all three active sites completely and irreversibly. Brief treatment of extract with epoxomicin (30 min) followed by a peptidase assay allows us to determine the fraction of proteolytic activity that is not inhibited by epoxomicin and is therefore carried out by enzymes other than the proteasome.

Using this approach, we found that the contribution of nonproteasomal proteolysis to the cleavage of substrates of trypsin-like sites is always higher than to the cleavage of caspase- and chymotrypsin-like sites. In total extracts of many cell lines and tissues, most of the cleavages of basic substrates is not by proteasomes. In HeLa cells, this contribution decreases dramatically if cytosolic extracts are used, indicating that most proteases cleaving after basic residues are inside organelles. This contribution also depends on the substrate and varies from 25–35% for Ac-RLR-amc and Boc-LRR-amc to 10% for Boc-LSTR-amc (Table II). The contribution of other proteases to the cleavage of Suc-LLVY-amc and Ac-nLPnLD-amc is negligible in cytosolic HeLa cell extracts (Table II). Thus, the activity of all three active sites can be measured directly in cytosolic extracts of HeLa cells. However, the situation may be different in other cell lines, so further purification may be required.

TABLE II
CONTRIBUTION OF NONPROTEASOMAL PROTEASES TO CLEAVAGE OF PROTEASOME SUBSTRATES IN CYTOSOLIC HELA EXTRACTS

| Active site | Substrate ($\mu M$) | % nonproteasomal activity[a] |
|---|---|---|
| Chymotrypsin-like | Suc-LLVY-amc (100) | 5 ± 3 |
| Caspase-like | Ac-nLPnLD-amc (100) | 3 ± 3 |
| Trypsin-like | Boc-LRR-amc (1000) | 25 |
| | Ac-RLR-amc (100) | 34 ± 5 |
| | Boc-LSTR-amc (600) | 11 ± 3 |

[a] Defined as percentage activity remaining after treatment of extracts with epoxomicin. Numbers are mean ± range of two different extract preparations.

Special consideration has to be given to the preparation of cell extracts for proteasome activity measurements. Commonly used detergents such as Triton X-100 and Nonidet P-40 cannot be used because they cause significant inhibition of proteasome activity. In order to generate cytosolic extracts, we permeabilize cells with digitonin (Shamu *et al.*, 1999), a detergent that does not inhibit the proteasome, in a buffer that contains 0.25 *M* sucrose. Sucrose prevents the destruction of organelles and also helps preserve the integrity of 26S proteasomes. Cytosol is then "squeezed out" of the permeabilized cells by centrifugation.

### *Preparing Cell Extracts for Proteasome Activity Measurements*

This protocol was developed to measure proteasome activity in extracts of cells treated with irreversible inhibitors. The amount of cells needed for such experiment differs considerably from cell line to cell line and has to be determined experimentally in each case. For example, one 10-cm plate of 50–70% confluent HeLa cells is sufficient to perform this analysis, but at least two 15-cm plates are needed to measure proteasome activity in rat fibroblasts. After treating the cells with inhibitors, wash them three times with phosphate-buffered saline and harvest. At this stage, the cells may but do not have to be frozen. Resuspend the cells in 4 volumes of homogenization buffer (50 m*M* Tris-HCl, pH 7.5, 250 m*M* sucrose, 5 m*M* $MgCl_2$, 2 m*M* ATP, 1 m*M* DTT, 0.5 m*M* EDTA) containing 0.025% digitonin. Always add DTT, ATP, and digitonin fresh. Highly purified digitonin (E. Merck, Darmstadt, Germany) is added from a 5 m*M* stock solution in DMSO that is stored at –20°. Incubate resuspended cells on ice for 5 min to allow permeabilization by digitonin. "Squeeze out" the cytosol by centrifugation at 20,000*g* for 15 min at 4° and then transfer the supernatant to a separate tube. Determine the protein concentration in cytosolic extract by the Bradford assay (or any other commonly used method) and use it for the measurement of proteasome activity. If needed, the cell pellet can be resuspended in the homogenization buffer and then sonicated to extract the remaining components.

### *Assaying Proteasome Activity in Cell Extracts*

As discussed earlier, when assaying proteasome activity in crude extracts, the contribution of other proteases that may be present in the preparation to cleavage of proteasomal substrates has to be determined. For this purpose, incubate the extract at 37° in the presence or absence of 20 $\mu M$ epoxomicin for 30 min. The amount of cell extract needed depends on the cell line. For HeLa cell extracts, 1–6 $\mu$g of total protein (determined by Bradford)/200 $\mu$l substrate is sufficient, but more protein may be needed in cell lines where activity is lower. When comparing proteasome activity in different samples, make sure that equal amounts of total protein are added to all wells. Immediately after completion of the incubation with epoxomicin, measure proteasome activity using the same protocol as described for purified proteasomes except that 100 $\mu M$ Ac-RLR-amc is replaced by 600 $\mu M$ Boc-LSTR-amc for assay of the trypsin-like site. In our experience, nonproteasomal proteases contribute less to the cleavage of Boc-LSTR-amc than to the cleavage of Ac-RLR-amc (Table II). A high concentration of Boc-LSTR-amc (600 $\mu M$) is needed because the signal is too low for 100 $\mu M$ Boc-LSTR-amc (Table I). Substrates of the chymotryspsin-like and of the caspase-like sites are the same as with purified proteasomes: 100 $\mu M$ Suc-LLVY-amc and 100 $\mu M$ Ac-nLPnLD-amc. If activity against Ac-nLPnLD-amc is too low and the amount of extracts is limited, increase substrate in the concentration severalfold. The $K_m$ for this substrate is ~500 $\mu M$, and an increase in its concentration will result in a proportional increase of the signal. However, the $K_m$ of Suc-LLVY-amc is ~60 $\mu M$, and increasing its concentration above 100 $\mu M$ will not cause a significant increase in the signal. For each sample, calculate the proteasome activity by subtracting activity in the epoxomicin-treated extracts from the activity in the control extracts.

### *Partial Purification of Proteasomes by Ultracentrifugation*

If proteasomes contribute to less than 50% of substrate cleavage in crude extracts (most likely, it will happen with Boc-LSTR-amc and other substrates of the trypsin-like sites), then further purification by ultracentrifugation is needed. We centrifuge a sample for 2 h at 300,000*g* [instead of the traditional 5 h at 100,000*g* (Gaczynska *et al.*, 1993)]. The resulting proteasome-containing pellets are small, transparent, and hard to see. Immediately after removing a tube from the rotor, mark the location of the pellet with a black dot on the outside of the tube. Remove and discard the supernatant, and resuspend the amorphous, greasy pellet in the homogenization buffer by repeated pipetting. Leave the suspension on ice for 15–30 min to complete solubilization. Centrifuge samples for 10 min at 16,000*g* to remove material that is still insoluble. Measure protein concentration in the supernatant, incubate for 30 min in the presence or absence of 20 $\mu M$ epoxomicin, and determine proteasome activity as described in the previous paragraph. Amounts of protein per well may be reduced three- to fourfold compared with extracts because proteasomes are enriched in the pellet. It is likely that low molecular weight enzymes cleaving substrates of the trypsin-like sites will be removed by this procedure so that high concentrations of expensive Boc-LSTR-amc can be replaced by 100 $\mu M$ Ac-RLR-amc or by less expensive Boc-LRR-amc.

If initial tests reproducibly demonstrate more than 95% inhibition of cleavages of all substrates by 20 $\mu M$ epoxomicin and the amount of sample available for assay is limited, treatment of partially purified proteasomes with epoxomicin can be eliminated from subsequent experiments and all activity in extracts attributed to proteasomes.

*Buffers and Substrates Used for Measurements of Proteasome Activity*

26S proteasome assay buffer: 50 m*M* Tris-HCl, pH 7.5, 40 m*M* KCl, 5 m*M* $MgCl_2$, 0.5 m*M* ATP, 1 m*M* DTT, 0.05 (plate) or 0.5 (plastic cuvette) mg/ml BSA or SBTI. Always add DTT, ATP, and BSA or SBTI fresh.

Buffer for cytosolic extract preparation: 50 m*M* Tris-HCl, pH 7.5, 250 m*M* sucrose, 5 m*M* $MgCl_2$, 2 m*M* ATP, 1 m*M* DTT, 0.5 m*M* EDTA, and 0.025% digitonin. Always add DTT, ATP, and digitonin fresh.

Stock solutions (stored at –20°): 100 m*M* ATP, 1 *M* DTT, 50 mg/ml BSA, 50 mg/ml SBTI (all in water), and 5% digitonin (in DMSO).

Substrates: 100 $\mu M$ Suc-LLVY-amc (all assays), 100 $\mu M$ Ac-nLPnLD-amc (all assays), 100 $\mu M$ Ac-RLR-amc (purified or partially purified proteasomes), and 600 $\mu M$ Boc-LSTR-amc (extracts).

Stock solutions of substrates (in DMSO): 10 m*M* Suc-LLVY-amc, 10 m*M* Ac-nLPnLD-amc, 10 m*M* Ac-RLR-amc, 60 m*M* Boc-LSTR-amc.

Stock solution of epoxomicin: 10 m*M* in DMSO.

## Acknowledgments

We thank Alice Callard and Rebecca Evans for assistance in performing experiments. This work was supported by an NIGMS grant to Alfred Goldberg, a Special Fellowship of the Leukemia and Lymphoma Society, and start-up funds of the Norris Cotton Cancer Center at Dartmouth Medical School to Alexei Kisselev. We thank Omar Amir, Emlyn Samuel, Vicky Cowling, Jim DiRenzo, and Fedor Kisseljov for critical reading of the manuscript.

## References

Akopian, T. N., Kisselev, A. F., and Goldberg, A. L. (1997). Processive degradation of proteins and other catalytic properties of the proteasome from *Thermoplasma acidophilum*. *J. Biol. Chem.* **272,** 1791–1798.

Arendt, C. S., and Hochstrasser, M. (1997). Identification of the yeast 20S proteasome catalytic centers and subunit interactions required for active-site formation. *Proc. Natl. Acad. Sci. USA* **94,** 7156–7161.

Baumeister, W., Walz, J., Zühl, F., and Seemüller, E. (1998). The proteasome: Paradigm of a self-compartmentalizing protease. *Cell* **92,** 367–380.

Cascio, P., Call, M., Petre, B. M., Walz, T., and Goldberg, A. L. (2002). Properties of the hybrid form of the 26S proteasome containing both 19S and PA28 complexes. *EMBO J.* **21,** 2636–2645.

Cascio, P., and Goldberg, A. L. (2005). Preparation of hybrid (19S–20S–PA28) proteasome complexes and analysis of peptides generated during protein degradation. *Methods Enzymol.* **398**(30), 2005 (this volume).

Cascio, P., Hilton, C., Kisselev, A., Rock, K., and Goldberg, A. (2001). 26S proteasomes and immunoproteasomes produce mainly N-extended versions of an antigenic peptide. *EMBO J.* **20,** 2357–2366.

Chen, P., and Hochstrasser, M. (1996). Autocatalytic subunit processing couples active site formation in the 20S proteasome to completion of assembly. *Cell* **86,** 961–972.

Coux, O., Tanaka, K., and Goldberg, A. L. (1996). Structure and functions of the 20S and 26S proteasomes. *Annu. Rev. Biochem.* **65,** 801–847.

Fuertes, G., Martin D. E., Llano, J. J., Villarroya, A., Rivett, A. J., and Knecht, E. (2003). Changes in the proteolytic activities of proteasomes and lysosomes in human fibroblasts produced by serum withdrawal, amino-acid deprivation and confluent conditions. *Biochem. J.* **375,** 75–86.

Gaczynska, M., Rock, K. L., and Goldberg, A. L. (1993). Gamma-interferon and expression of MHC genes regulate peptide hydrolysis by proteasomes. *Nature* **365,** 264–267.

Gaedicke, S., Firat-Geier, E., Constantiniu, O., Lucchiari-Hartz, M., Freudenberg, M., Galanos, C., and Niedermann, G. (2002). Antitumor effect of the human immunodeficiency virus protease inhibitor ritonavir: Induction of tumor-cell apoptosis associated with perturbation of proteasomal proteolysis. *Cancer Res.* **62,** 6901–6908.

Goldberg, A. L., and St John, A. C. (1976). Intracellular protein degradation in mammalian and bacterial cells. *Annu. Rev. Biochem.* **45,** 747–803.

Groll, M., Bajorek, M., Kohler, A., Moroder, L., Rubin, D. M., Huber, R., Glickman, M. H., and Finley, D. (2000). A gated channel into the proteasome core particle. *Nature Struct. Biol.* **7,** 1062–1067.

Groll, M., Ditzel, L., Löwe, J., Stock, D., Bochtler, M., Bartunik, H., and Huber, R. (1997). Structure of 20S proteasome from yeast at 2.4 Å resolution. *Nature* **386,** 463–471.

Grune, T., Merker, K., Sandig, G., and Davies, K. J. (2003). Selective degradation of oxidatively modified protein substrates by the proteasome. *Biochem. Biophys. Res. Commun.* **305,** 709–718.

Harris, J. L., Alper, P. B., Li, J., Rechstainer, M., and Bakes, B. J. (2001). Substrate specificity of the human proteasome. *Chem. Biol.* **8,** 1131–1141.

Heinemeyer, W., Fischer, M., Krimmer, T., Stachon, U., and Wolf, D. H. (1997). The active sites of the eukaryotic 20S proteasome and their involvement in subunit precursor processing. *J. Biol. Chem.* **272,** 25200–25209.

Hershko, A., and Ciechanover, A. (1998). The ubiquitin system. *Annu. Rev. Biochem.* **67,** 425–479.

Kisselev, A. F., Akopian, T. N., and Goldberg, A. L. (1998). Range of sizes of peptide products generated during degradation of different proteins by archaeal proteasomes. *J. Biol. Chem.* **273,** 1982–1989.

Kisselev, A. F., Akopian, T. N., Woo, K. M., and Goldberg, A. L. (1999). The sizes of peptides generated from protein by mammalian 26S and 20S proteasomes: Implications for understanding the degradative mechanism and antigen presentation. *J. Biol. Chem.* **274,** 3363–3371.

Kisselev, A. F., Callard, A. V. E., and Goldberg, A. L. (2005). Importance of different active sites in protein breakdown by 26S proteasomes and efficacy of proteasome inhibition depends on the protein substrate. In preparation.

Kisselev, A. F., Garcia-Calvo, M., Overkleeft, H. S., Peterson, E., Pennington, M. W., Ploegh, H. L., Thornberry, N. A., and Goldberg, A. L. (2003). The caspase-like sites of proteasomes, their substrate specificity, new inhibitors and substrates, and allosteric interactions with the trypsin-like sites. *J. Biol. Chem.* **278,** 35869–35877.

Kisselev, A. F., and Goldberg, A. L. (2001). Proteasome inhibitors: From research tools to drug candidates. *Chem. Biol.* **8,** 739–758.

Kisselev, A. F., Kaganovich, D., and Goldberg, A. L. (2002). Binding of hydrophobic peptides to several non-catalytic sites promotes peptide hydrolysis by all active sites fo 20S proteasomes: Evidence for peptide-induced channel opening in the alpha-rings. *J. Biol. Chem.* **277,** 22260–22270.

Köhler, A., Cascio, P., Leggett, D. S., Woo, K. M., Goldberg, A. L., and Finley, D. (2001). The axial channel of the proteasome core particle is gated by the Rpt2 ATPase and controls both substrate entry and product release. *Mol. Cell* **7,** 1143–1152.

Kopp, F., Dahlmann, B., and Kuehn, L. (2001). Reconstitution of hybrid proteasomes from purified PA700–20S complexes and PA28 alpha beta activator: Ultrastructure and peptidase activities. *J. Mol. Biol.* **313,** 465–471.

Lightcap, E. S., McCormack, T. A., Pien, C. S., Chau, V., Adams, J., and Elliott, P. J. (2000). Proteasome inhibition measurements: Clinical application. *Clin. Chem.* **46,** 673–683.

Liu, C. W., Corboy, M. J., De Martino, G. N., and Thomas, P. J. (2003). Endoproteolytic activity of the proteasome. *Science* **299,** 408–411.

Meng, L., Mohan, R., Kwok, B. H., Elofsson, M., Sin, N., and Crews, C. M. (1999). Epoxomicin, a potent and selective proteasome inhibitor, exhibits *in vivo* antiinflammatory activity. *Proc. Natl. Acad. Sci. USA* **96,** 10403–10408.

Nazif, T., and Bogyo, M. (2001). Global analysis of proteasomal substrate specificity using positional-scanning libraries of covalent inhibitors. *Proc. Natl. Acad. Sci. USA* **98,** 2967–2972.

Oberdorf, J., Carlson, E. J., and Skach, W. R. (2001). Redundancy of mammalian proteasome beta subunit function during endoplasmic reticulum associated degradation. *Biochemistry* **40,** 13397–13405.

Rodgers, K. J., and Dean, R. T. (2003). Assessment of proteasome activity in cell lysates and tissue homogenates using peptide substrates. *Int. J. Biochem. Cell Biol.* **35,** 716–727.

Shamu, C. E., Story, C. M., Rapoport, T. A., and Ploegh, H. L. (1999). The pathway of US11-dependent degradation of MHC class I heavy chains involves a ubiquitin-conjugated intermediate. *J. Cell Biol.* **147,** 45–58.

Tanahashi, N., Murakami, Y., Minami, Y., Shimbara, N., Hendil, K. B., and Tanaka, K. (2000). Hybrid proteasomes: Induction by interferon-gamma and contribution to ATP-dependent proteolysis. *J. Biol. Chem.* **275,** 14336–14345.

Udenfriend, S., Stein, S., Bohlen, P., Dairman, W., Leimgruber, W., and Weigele, M. (1972). Fluorescamine: A reagent for assay of amino acids, peptides, proteins, and primary amines in the picomole range. *Science* **178,** 871–872.

Voges, D., Zwickl, P., and Baumeister, W. (1999). The 26S proteasome: A molecular machine designed for controlled proteolysis. *Annu. Rev. Biochem.* **68,** 1015–1068.

Wilk, S., and Orlowski, M. (1983). Evidence that pituitary cation-sensitive neutral endopeptidase is a multicatalytic protease complex. *J. Neurochem.* **40,** 842–849.

Yang, Y., Fruh, K., Ahn, K., and Peterson, P. A. (1995). *In vivo* assembly of the proteasomal complexes, implications for antigen processing. *J. Biol. Chem.* **270,** 27687–27694.

# [31] The Synthesis and Proteasomal Degradation of a Model Substrate $Ub_5DHFR$

*By* Y. Amy Lam, Jen-Wei Huang, and Oluwafemi Showole

## Abstract

The importance of substrate polyubiquitination in protein degradation has been established for many years. However, the many intricacies of substrate recognition and ubiquitination by E3 enzymes have greatly limited access to degradable substrate *in vitro*. Thus, detailed analysis of protein degradation using purified 26S proteasomes has been difficult. The ability to synthesize polyubiquitin chains in a test tube has provided a method to make large quantities of a specific polyubiquitinated substrate, $Ub_5DHFR$. This chapter focuses on the synthesis and degradation of this model substrate.

## Introduction

The importance of polyubiquitination *in vivo* was first illustrated by the preferential degradation of polyubiquitinated over monoubiquitinated substrates (Chau *et al.*, 1989). However, *in vitro* analysis of protein degradation by the 26S proteasome has been difficult due to the lack of a homogeneous polyubiquitinated substrate. This problem stems from the inherent complexity of substrate ubiquitination by E3 enzymes. While E3s recognize and bind their substrates with high affinity, the location and the extent of ubiquitin conjugation on the substrate are often unpredictable. Usually, multiple substrate lysines are competent to be sites of ubiquitination, and conjugation can result in the attachment of one to many ubiquitins at each site. This generates a complex mixture of polyubiquitinated substrates, whose individual degradation rates may vary (Petroski and Deshaies, 2003; Thrower *et al.*, 2000). While such products are generally acceptable for qualitative analysis of proteasome-mediated degradation, they are poorly suited for quantitative degradation assays (e.g., kinetic studies), where substrate homogeneity is essential. Additionally, for many substrates of the pathway, posttranslational modification is a prerequisite for E3 recognition. This requirement further limits the yield of polyubiquitinated substrate that can be produced via E3 catalysis.

The Pickart laboratory has developed several key reagents needed for the *in vitro* synthesis of a model substrate called $Ub_5DHFR$, which can be

METHODS IN ENZYMOLOGY, VOL. 398

0076-6879/05 $35.00
DOI: 10.1016/S0076-6879(05)98031-2

made without an E3 enzyme. That dihydrofolate reductase (DHFR) can be engineered to be degraded by the Ub pathway was first shown *in vivo* using the yeast *Saccharomyces cerevisiae* in the Varshavsky laboratory (Johnson *et al.*, 1995; Johnston *et al.*, 1995). These investigators found that when ubiquitin was fused to the N terminus of DHFR, it served to initiate the synthesis of a polyubiquitin chain, which then targeted DHFR for degradation by the 26S proteasome. To stabilize the UbDHFR fusion from enzymatic removal of the ubiquitin, the G76V mutation was introduced at the ubiquitin C terminus to prevent cleavage by endogenous deubiquitinating enzymes. This substrate and related ubiquitin fusion proteins were used in several genetic screens to identify conjugation factors in the so-called ubiquitin fusion degradation (UFD) pathway (Johnson *et al.*, 1995). Therefore, the *in vivo* degradation of polyubiquitinated–ubiquitin fusion proteins has been well established *in vivo*.

The *in vitro* synthesis of $Ub_5$DHFR relies on a ubiquitin-conjugating enzyme, E2-25K, that specifically links Ub into chains through Ub lysine 48. In a reaction containing a UbDHFR fusion protein (readily expressed and purified from bacteria) and a preformed Ub4 chain, E2-25K produces $Ub_5$DHFR as the sole product in the presence of E1 and ATP (Piotrowski *et al.*, 1997). Therefore, $Ub_5$DHFR generated in this reaction contains a single polyubiquitin chain of defined length, which is conjugated at just one site (the K48 residue of the fused ubiquitin, see Fig. 1). This substrate is efficiently recognized and degraded by the purified 26S proteasome (Thrower *et al.*, 2000). The use of this homogeneous substrate overcomes many of the difficulties described earlier. Additionally, the ability to synthesize milligram quantities of the substrate creates an ample supply of a proteasome substrate. All of these features facilitate detailed studies of protein degradation by the 26S proteasome. This chapter focuses on procedures to synthesize $Ub_5$DHFR and to assay its degradation by the 26S proteasome.

Original studies with $Ub_5$DHFR used *in vivo* labeling of UbDHFR with $^{35}$S-Met to introduce a signal to monitor degradation (degradation can be measured as the production of acid-soluble radioactivity) (Thrower *et al.*, 2000). While this method proved effective, it was necessary to use several millicuries of $^{35}$S-Met in each labeling to ensure adequate specific radioactivity. Subsequently, an alternative method was developed that uses minimal amount of radioactivity. A protein kinase A (PKA) site was engineered into the C terminus of DHFR (Raasi and Pickart, 2003). Once the protein substrate ($Ub_5$DHFR) has been synthesized and purified by conventional means, labeling can be achieved with a minimal amount of purified substrate, with commercially available protein kinase A and only 10 $\mu$Ci of radiolabeled ATP. Thus, labeling can be performed when needed, on a much smaller scale and with less radioactive material.

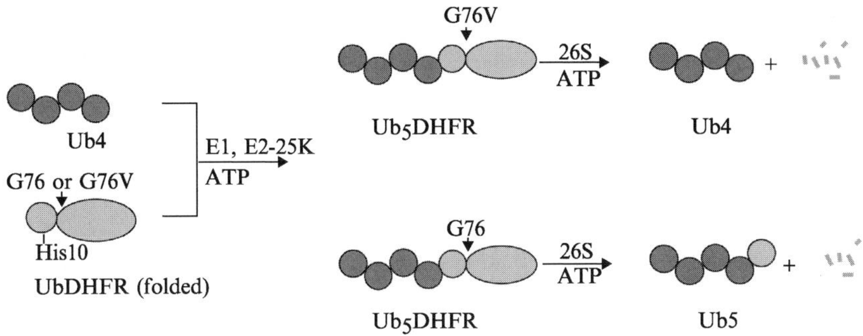


FIG. 1. Synthesis and degradation of $Ub_5DHFR$ (scheme). A His10 tag is located at the N terminus of the UbDHFR fusion. For ubiquitin conjugation, the key residues for chain elongation are K48 and G76 of ubiquitin. In order to control the chain synthesis reactions, these two residues are selectively blocked or deblocked for synthesis; in the final Ub4 product, the proximal chain terminus is blocked by the presence of an extra residue, D77. In order to activate Ub4 in the synthesis reaction, we add YUH1 to remove D77 and expose G76 on the proximal ubiquitin, while the distal ubiquitin remains blocked and unavailable for self-conjugation (Piotrowski *et al.*, 1997). For the UbDHFR fusion, the K48 of the fused ubiquitin is available for chain elongation, while the C terminus of this ubiquitin is blocked by the DHFR moiety. Thus, each reactant has a distinct reactive group; when they are combined, only the desired product is generated.

The normal C-terminal residue of ubiquitin is glycine 76 (G76). This residue was changed to valine 76 (V76) in the UbDHFR fusion protein used and described thus far (Johnson *et al.*, 1995; Thrower *et al.*, 2000). As discussed earlier, this change was originally introduced to prevent deubiquitinating enzymes from removing the DHFR moiety *in vivo*. Using $Ub_5DHFR$ that is based on this version of fusion protein, it was shown that the entire UbDHFR fusion protein is degraded by the 26S proteasome, while Ub4 is released. The fused ubiquitin in UbDHFR is degraded because the engineered V76 residue cannot be processed by the deubiquitinating activity of the 26S proteasome. In contrast, the remainder of the chain retains the normal G76 residue, which is cleavable by this activity. More recently, a UbG76DHFR fusion protein was produced and used to generate $Ub_5DHFR$. With this version, DHFR is degraded by the 26S proteasome, while a Ub5 chain is released (Fig. 1) (Yao and Cohen, 2002). The enzyme responsible for chain removal was later identified as the POH1/Rpn11 subunit in the 26S proteasome (Verma *et al.*, 2002; Yao and Cohen, 2002).

In this chapter, we have introduced the PKA site into the UbG76DHFR construct. The synthesis, purification, and degradation of the V76 and G76 versions of $Ub_5DHFR$ are described and compared.

Purification of UbV76DFHR and UbG76DHFR

Construction of a pET16b-based plasmid specifying UbV76DHFR with a C-terminal PKA site was described by Raasi and Pickart (2003). A UbG76DHFR version was created by site-directed mutagenesis of the valine moiety of UbV76DHFR using the Quickchange method (Stratagene).

The plasmid expressing UbV76DHFR or UbG76DHFR is transformed into the BL21(DE3)pLysS strain of *Escherichia coli*. Cells are grown overnight (250 rpm, 37°) in 20 ml of 2× YT medium with ampicillin (200 $\mu$g/ml) and chloramphenicol (30 $\mu$g/ml). The overnight cultures are used to inoculate 1 liter of 2× YT media (200 $\mu$g/ml ampicillin, 30 $\mu$g/ml chloramphenicol) in Fernbach flasks and grown with continuous shaking (250 rpm, 37°) to a cell density of $OD_{600} = 0.4–0.6$. Expression of the fusion protein is induced with 0.4 m*M* isopropyl-$\beta$-D-thiogalactopyranoside (IPTG, Research Products International). The cultures are grown with shaking for an additional 3 h. Cells are then pelleted by centrifugation at 6500 rpm (4°, 15 min) and the pellets are stored at –80°. A 1-liter culture produces approximately 4 g of cells.

Cell pellets are thawed on ice and resuspended in 8 ml of lysis buffer (50 m*M* HEPES, pH 7.0, 1 m*M* phenylmethylsulfonyl fluoride (PMSF), 100 $\mu$M *p*-tosyl-L-lysine chloromethyl ketone (TLCK), 100 $\mu$g/ml soybean trypsin inhibitor, and 100 $\mu$g/ml leupeptin; all from Sigma). Cell lysis is initiated with the addition of 0.4 mg/ml lysozyme, 0.02 mg/ml DNase I, and 10 m*M* $MgCl_2$. Mechanical homogenization of the lysozyme-treated suspension is performed using a disposable transfer pipette until the lysate is nonviscous and uniform in appearance. Cell lysis and subsequent purification by nickel affinity and ion-exchange chromatography are performed at 4° to minimize nonspecific proteolysis.

Cellular debris and insoluble proteins are removed from the cell lysate by means of centrifugation at 13,000 rpm (4°, 15 min) in a microcentrifuge. The fusion protein in the supernatant is purified by nickel affinity chromatography at 4°. The lysate is allowed to rotate with 2 ml of nickel-charged His-Bind resin (Novagen) preequilibrated with buffer (50 m*M* HEPES, pH 7.0) for 1 h at 4°. The His-Bind resin is prepared and maintained according to the manufacturer's recommendations. The bound UbDHFR is washed with four column volumes of each of (1) 60 m*M* and (2) 120 m*M* imidazole in buffer containing 50 m*M* HEPES, pH 7.0, and 0.1 m*M* folic acid. The protein is then eluted with two column volumes of 300 m*M* imidazole in the same buffer and dialyzed twice againist 2 liter of 50 m*M* HEPES, pH 7.0, for 4 h each time.

The dialyzed sample from nickel affinity chromatography is further subjected to subtractive anion-exchange chromatography to remove impurities from the fusion protein. The sample is loaded onto 0.25 ml of Q-Sepharose Fast Flow resin (Amersham-Pharmacia Biotech) preequilibrated with buffer (50 m*M* HEPES, pH 7.0). The unbound fraction contains the UbDHFR fusion protein. The resin is washed with two column volumes of buffer to improve recovery of UbDHFR. These pooled flow through and washes are concentrated in a 4-ml Amicon Ultra centrifugal concentrator (5000 MWCO, Millipore) until the protein concentration is greater than 5 mg/ml. Analysis of samples from each purification step by sodium dodecyl sulfate–polyacrylamide gel electrophoresis (SDS–PAGE) is recommended to confirm success of the procedure (see Fig. 2A). The purified UbDHFR is stored at –80°.

For the G76 version of UbDHFR protein, despite performing cell lysis and purification at 4°, some fusion protein is proteolysed, apparently generating free (His-tagged) Ub and DHFR during the purification. In order to separate the free Ub from the intact fusion protein, the product from the Q-Sepharose step is concentrated to 100 $\mu$l and is then purified further by gel filtration chromatography using a Superdex 75 GL column (30 mm $\times$ 100 cm, Amersham). The column is preequilibrated with buffer containing

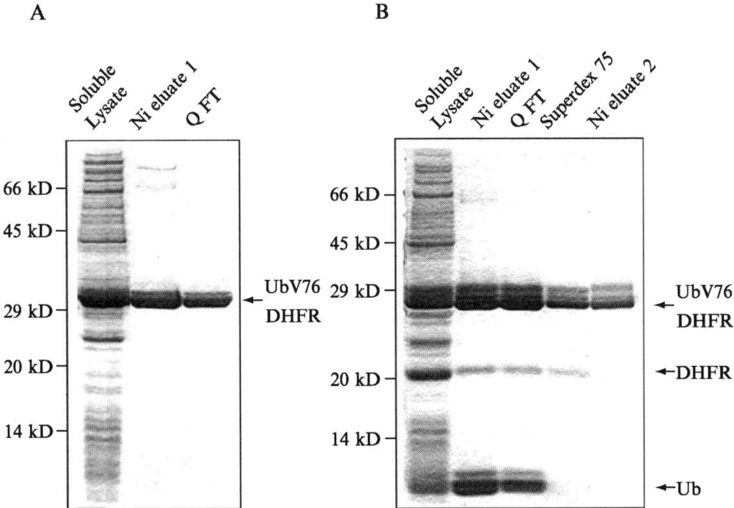


FIG. 2. Purification of UbV76DHFR (A) and UbG76DHFR (B). Both Ub fusion proteins were purified by nickel chelate chromatography, followed by subtractive anion exchange (passage through Q-Sepharose) to remove impurities. UbG76DHFR was purified further by gel filtration chromatography (to remove Ub-sized cleavage products) followed by a second nickel column (to remove DHFR-sized cleavage products). Protein samples were resolved by 13.5% SDS–PAGE and stained with Coomassie blue.

25 m$M$ Tris-HCl, pH 7.6, and 0.5 m$M$ EDTA. Peak fractions (as determined by UV absorbance at 280 nm) are analyzed further by SDS–PAGE. Fractions containing UbDHFR and DHFR are pooled. To remove DHFR (which lacks the His tag) from the pooled fractions, this mixture is reloaded onto nickel resin and purified as described earlier. The analysis of each purification step is shown in Fig. 2B. The eluted UbDHFR fusion protein is concentrated and diluted in a 4-ml Amicon Ultra device until the imidazole concentration is less than 10 m$M$ and the protein concentration reaches at least 5 mg/ml. With both fusion proteins, protein concentrations are estimated by comparison to standard quantities of bovine serum albumin (BSA) on an SDS–PAGE gel. The typical yield of each fusion protein from 1 liter of cell culture is 4–6 mg of purified protein.

### Synthesis of $Ub_5$DHFR

$Ub_4$ (90 $\mu M$) and UbDHFR (82 $\mu M$) are incubated at 37° for 3 h with 0.1 $\mu M$ E1, 20 $\mu M$ E2-25K, 0.6 $\mu M$ YUH1, 2 m$M$ ATP, 5 m$M$ $MgCl_2$, 50 m$M$ Tris, pH 8, and an ATP-regenerating system. (Note that E2-25K is only active with E1 from mammalian sources.) The $Ub_4$ is added in slight excess to ensure that most of the UbDHFR fusion is converted to product. (Note that the addition of YUH1, ubiquitin C-terminal hydrolase is used to remove the D77 residue in the proximal Ub of the $Ub_4$; this activates $Ub_4$ for conjugation.) $Ub_5$DHFR is purified on nickel resin, using 0.5 ml resin for a 1-ml reaction. The loaded column is first washed with buffer containing 50 m$M$ HEPES, pH 7.5, and 0.1 m$M$ folic acid (FA), followed by a wash with 60 m$M$ imidazole in the same buffer. $Ub_5$DHFR is eluted with 200 m$M$ imidazole in HEPES/FA buffer. The eluate is repeatedly concentrated and diluted with HEPES/FA buffer, as described previously, until the imidazole concentration is less than 10 m$M$ and the protein concentration reaches at least 2 mg/ml. Figure 3 shows results of a typical synthesis and purification of Ub5G76DHFR.

### Radiolabeling of $Ub_5$DHFR

For each reaction (20 $\mu$l), 25 $\mu$g of $Ub_5$DHFR and 75 ng of PKA are incubated at 37° for 15 min in 20 m$M$ MES, pH 6.5, 1 m$M$ dithiothreitol, 1 m$M$ $MgCl_2$, 10 mg/ml of ovalbumin as protein carrier, and 10 $\mu$Ci of [$\gamma$-$^{32}$P] ATP. We use recombinant PKA produced in our laboratory; however, one can also use commercial bovine heart PKA from Sigma, as described previously (Raasi and Pickart, 2003). Following incubation, the sample is loaded onto a micro Bio-Spin (Bio-Rad) chromatography column (prewashed with MES buffer containing 100 m$M$ NaCl and 10 mg/ml ovalbumin

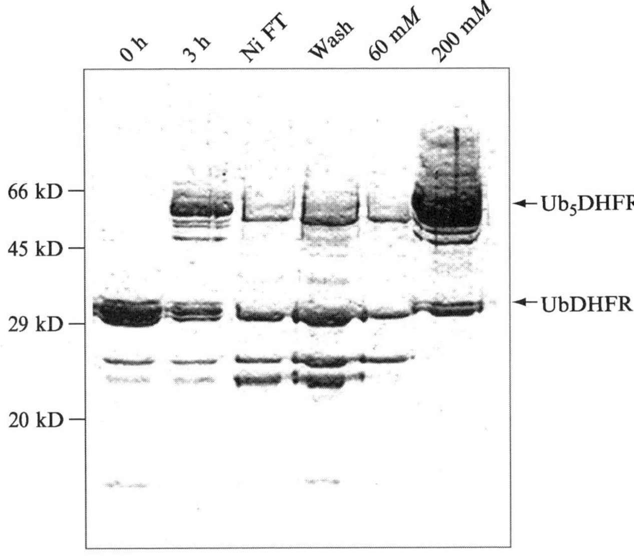


FIG. 3. Synthesis and purification of the V76 and G76 versions of $Ub_5$DHFR are identical (see text); only the G76 version is shown here. The conjugation reaction was sampled at the beginning and at the end of the reaction. Note that Ub4 (34 kDa) and UbDHFR (32 kDa) comigrate in a 13.5% gel. The $Ub_5$DHFR product is purified using Ni resin and eluted at 200 m*M* imidazole (see text). Because UbDHFR carries the His tag, any unreacted UbDHFR will copurify with $Ub_5$DHFR; therefore, Ub4 is used in slight excess during conjugation to minimize the level of unreacted UbDHFR in the final product. Protein samples were resolved on a 13.5% gel and stained with Coomassie blue.

before final spin) and microcentrifuged at 2000 rpm for 1 min to remove unincorporated ATP. Typical protein recovery is 25–40%. Protein concentration is determined by SDS–PAGE, using BSA as standard. Radiolabeled sample ($10^4$/pmol) is stored 4° and used within 2 weeks. The specific radioactivity of the G76 version is typically two-fold higher than V76 fusion protein.

## Degradation Studies

For each reaction (30 $\mu$l), 200 n*M* substrate and 20 n*M* purified bovine erythrocyte 26S proteasomes are incubated at 37° in 50 m*M* Tris, pH 7.6, 2 m*M* $MgCl_2$, 1 m*M* ATP, ATP regeneration system, 1 $\mu$*M* Ub-aldehyde (Boston Biochem.), and 1 mg/ml ubiquitin as protein carrier. Ubal serves to inhibit Uch37-deubiquitinating activity in the mammalian 26S proteasome (Lam *et al.*, 1997), and this inhibitor is preincubated with the proteasome for

5 min on ice. Substrates and proteasomes are preincubated separately at 37° for 2 min before combining them to initiate the degradation reaction. Two methods can be used to monitor the degradation of $Ub_5$DHFR: Western blotting (with anti-His tag antibody) provides a qualitative analysis of degradation, whereas measurement of trichloroacetic (TCA)-soluble radioactivity gives quantitative rates.

### TCA Precipitation

Degradation is measured by the radioactivity in the soluble fraction after precipitation of proteins using TCA. At designated times, 5 $\mu$l of the reaction is removed and added to 15 $\mu$l of cold 10 mg/ml BSA, vortexed to mix, and 17 $\mu$l of cold 40% TCA is added and vortexed immediately to achieve a final concentration of 10% TCA. Once precipitated, samples are kept on ice for 10 min or longer prior to further processing. Once samples from all times points in the time course have been collected and precipitated, TCA-treated samples are microcentrifuged at 13,000 rpm for 10 min at 4°. From each sample, 20 $\mu$l of the supernatant is removed, added to 5 ml scintillation fluid, and counted in a liquid scintillation counter.

### Western Blot Analysis

Because there is a polyHis tag at the N terminus of UbDHFR, the level of $Ub_5$DHFR can be monitored by Western blot analysis using anti-His10 antibodies (Santa Cruz). At specified times, 7.5 $\mu$l of the reaction is removed and added to SDS–PAGE sample buffer. Samples are run on a 13.5% SDS–PAGE gel and blotted onto a PVDF membrane for 1 h at 67 V at 4°. The blot is preincubated with 15 ml blocking buffer (10 mg/ml BSA in 20 m*M* Tris, pH 7.6, and 50 m*M* NaCl) for 5 min, and anti-His polyclonal antibodies are added at 1:1000 dilution to the blocking buffer and rocked at room temperature for 1 h. The blot is washed with buffer (50 m*M* Tris, pH 7.6, 150 m*M* NaCl, and 0.5% Tween 20) five times at 5-min intervals. For secondary antibodies, we use goat anti-rabbit antibodies conjugated with alkaline phosphatase at 1:1000 dilution (Santa Cruz) in blocking buffer; the blot is rocked at room temperature for 1 h. It is then washed with buffer (50 m*M* Tris, pH 7.6, 150 m*M* NaCl, and 0.5% Tween 20) five times at 5-min intervals. After the last wash buffer is removed, the blot is incubated with 15 ml of NBT/PCIP solution [(Bio-Rad) prewarmed to room temperature] and rocked at room temperature until a signal appears. The blot is then washed with water to stop further color development.

## Results

Protein degradation can be monitored qualitatively by Western blotting. As shown in Fig. 4, degradation of both versions of $Ub_5DHFR$ is strictly dependent on the addition of 26S proteasome, as the disappearance of both substrates occurs only when the 26S is added. Second, the loss of $Ub_5DHFR$ is not due to chain disassembly because the level of UbDHFR does not increase as $Ub_5DHFR$ disappears nor is there any detectable production of intermediates. Finally, degradation is chain dependent, such that only $Ub_5DHFR$ is degraded, while the background level of (contaminating) UbDHFR remains unchanged over time. Similarly, when UbDHFR is added as the only substrate to the 26S proteasome, it is stable. All data are in agreement with known properties of (poly)ubiquitin-dependent protein degradation by the 26S proteasome. As expected, degradation of the G76 version of $Ub_5DHFR$ is accompanied by the appearance of (His-tagged) $Ub_5$.

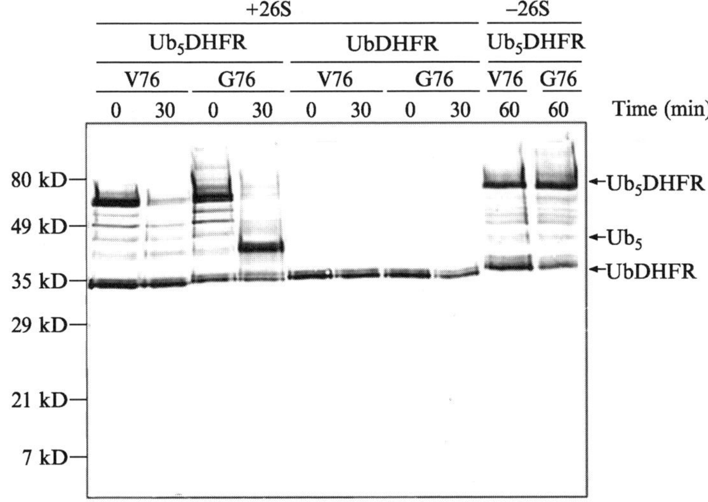


FIG. 4. Degradation of $Ub_5DHFR$ by 26S proteasomes (anti-His blot). Purified 26S proteasomes were incubated with $Ub_5DHFR$ or UbDHFR (V76 or G76, as indicated) for the indicated times. As controls, both forms of $Ub_5DHFR$ were incubated for 1 h at 37° without 26S proteasomes (last two lanes). With Ub5G76DHFR, the Ub5 product is detected because the fused Ub, which carries the His tag, is cleaved off with the chain. With the Ub5V76DHFR, this fused Ub and its His tag are degraded. The Ub4 product, although released (data not shown), is not detected by anti-His antibodies.

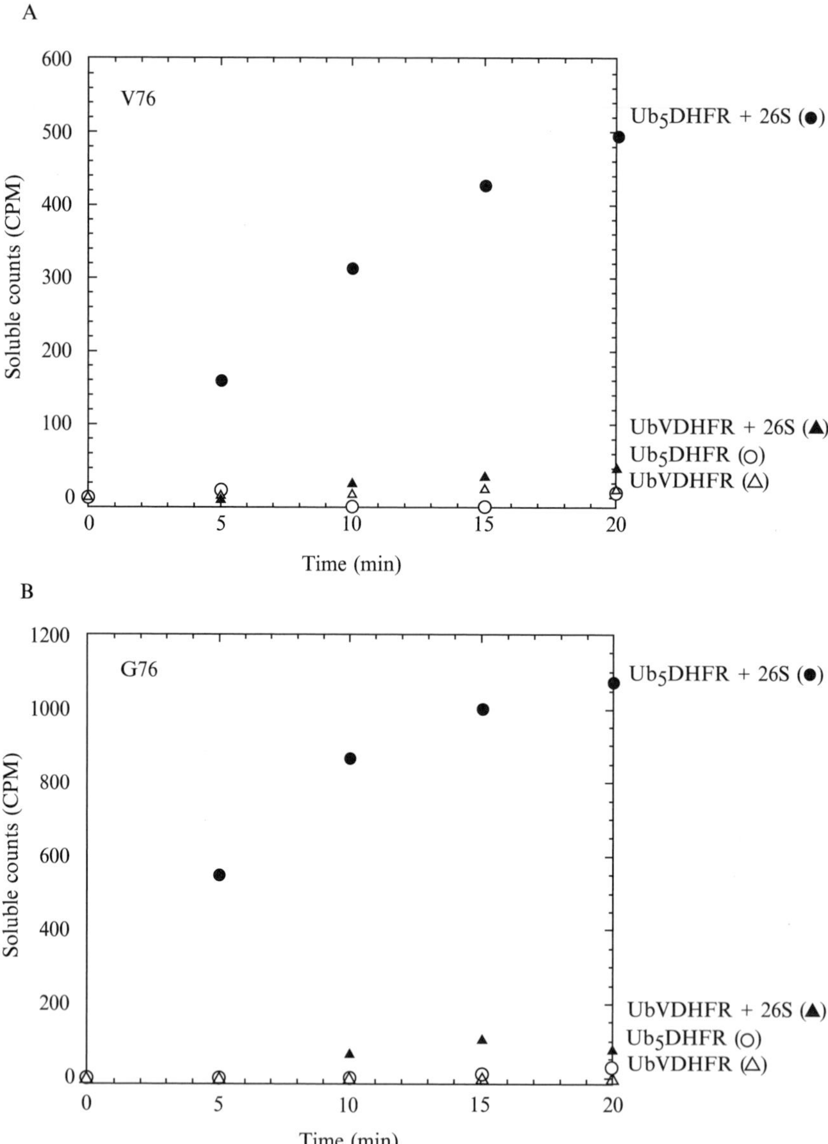


FIG. 5. Degradation of $Ub_5DHFR$ by 26S proteasomes (acid-soluble radioactivity). Purified 26S proteasomes were incubated with $^{32}P$-labeled $Ub_5DHFR$ or $^{32}P$-labeled UbDHFR. Both versions of the fusion protein, V76 (A) and G76 (B), were tested for degradation under identical conditions. As controls, degradation was also monitored in the absence of 26S proteasomes.

To determine the level of degradation in a more quantitative manner, we used radiolabeled substrates and measured the radioactivity in the soluble fraction after TCA precipitation. For both versions of $Ub_5DHFR$, there is a rapid production of soluble radioactivity only in the presence of 26S proteasomes. As expected, substrates that do not carry a chain are not degraded, as shown by the minimal release of radioactivity. These data reconfirm the importance of the polyubiquitin chain as a targeting signal. Thus, degradation of both versions of $Ub_5DHFR$ can be monitored by the rate of production of soluble radioactivity (Fig. 5).

The one suprising aspect of our results is that we found little difference in the rate of proteolysis of the V76 and G76 versions of the fusion protein. In earlier studies with $Ub_5DHFR$, degradation of the V76 version was shown to be 10-fold slower than the G76 version (Yao and Cohen, 2002). It was shown that when POH1/Rpn11 cannot cleave the linkage in the ubiquitin fusion (due to the presence of V76), the fused ubiquitin is degraded. One possible explanation is that the requirement to unfold ubiquitin, an extremely stable protein, slows the degradation process. This may provide an explanation for the essential role of Rpn11 and the need for ubiquitin recycling to prevent blockage of the proteasome. In all earlier versions of UbDHFR, including the ones used in Rpn11 studies, the C terminus of UbDHFR carried a hemagglutinin (HA) tag, whereas in our current version, this tag was replaced by a PKA recognition sequence. Thus, our data suggest that the sequence context at the C terminus of this substrate may also determine its rate of degradation. Other results are consistent with this model (M. Ajua-Alemanji *et al.*, manuscript in preparation).

In theory, one can use the ubiquitin fusion strategy to polyubiquitinate any substrate protein of interest and then study its degradation by the 26S proteasome. This method eliminates the necessity to discover and isolate cognate E3 enzymes or replicate posttranslational modifications required for substrate recognition. While more individual substrates need to be tested to confirm the broad generality of this method, several studies have shown that certain substrates, when either engineered to remove all lysines or are naturally lysine-less protein, can be polyubiquitinated within cells at the substrate N terminus and degraded in a 26S proteasome-dependent manner (Aviel *et al.*, 2000; Ben-Saadon *et al.*, 2004; Breitschopf *et al.*, 1998; Kuo *et al.*, 2004). Therefore, it is likely that the ubiquitin fusion technique provides a reliable and physiological relevant method to study proteasome-dependent degradation of a variety of substrates. Moreover, assays with $Ub_5DHFR$ will provide a convenient tool to discover and study activators and inhibitors of 26S proteasomes.

## References

Aviel, S., Winberg, G., Massucci, M., and Ciechanover, A. (2000). Degradation of the epstein-barr virus latent membrane protein 1 (LMP1) by the ubiquitin-proteasome pathway: Targeting via ubiquitination of the N-terminal residue. *J. Biol. Chem.* **275,** 23491–23499.

Ben-Saadon, R., Fajerman, I., Ziv, T., Hellman, U., Schwartz, A. L., and Ciechanover, A. (2004). The tumor suppressor protein p16(INK4a) and the human papillomavirus oncoprotein-58 E7 are naturally occurring lysine-less proteins that are degraded by the ubiquitin system: Direct evidence for ubiquitination at the N-terminal residue. *J. Biol. Chem.* **279,** 41414–41421.

Breitschopf, K., Bengal, E., Ziv, T., Admon, A., and Ciechanover, A. (1998). A novel site for ubiquitination: The N-terminal residue, and not internal lysines of MyoD, is essential for conjugation and degradation of the protein. *EMBO J.* **17,** 5964–5973.

Chau, V., Tobias, J. W., Bachmair, A., Marriott, D., Ecker, D. J., Gonda, D. K., and Varshavsky, A. (1989). A multiubiquitin chain is confined to specific lysine in a targeted short-lived protein. *Science* **243,** 1576–1583.

Johnson, E. S., Ma, P. C., Ota, I. M., and Varshavsky, A. (1995). A proteolytic pathway that recognizes ubiquitin as a degradation signal. *J. Biol. Chem.* **270,** 17442–17456.

Johnston, J. A., Johnson, E. S., Waller, P. R., and Varshavsky, A. (1995). Methotrexate inhibits proteolysis of dihydrofolate reductase by the N-end rule pathway. *J. Biol. Chem.* **270,** 8172–8178.

Kuo, M. L., den Besten, W., Bertwistle, D., Roussell, M. F., and Sherr, C. J. (2004). N-terminal polyubiquitination and degradation of the Arf tumor suppressor. *Genes Dev.* **18,** 1862–1874.

Lam, Y. A., Xu, W., De Martino, G. N., and Cohen, R. E. (1997). Editing of ubiquitin conjugates by an isopeptidase in the 26S proteasome. *Nature* **385,** 737–740.

Petroski, M. D., and Deshaies, R. J. (2003). Context of multiubiquitin chain attachment influences the rate of Sic1 degradation. *Mol. Cell* **11,** 1435–1444.

Piotrowski, J., Beal, R., Hoffman, L., Wilkinson, K. D., Cohen, R. E., and Pickart, C. M. (1997). Inhibition of the 26 S proteasome by polyubiquitin chains synthesized to have defined lengths. *J. Biol. Chem.* **272,** 23712–23721.

Raasi, S., and Pickart, C. M. (2003). Rad23 ubiquitin-associated domains (UBA) inhibit 26 S proteasome-catalyzed proteolysis by sequestering lysine 48-linked polyubiquitin chains. *J. Biol. Chem.* **278,** 8951–8959.

Thrower, J. S., Hoffman, L., Rechsteiner, M., and Pickart, C. M. (2000). Recognition of the polyubiquitin proteolytic signal. *EMBO J.* **19,** 94–102.

Verma, R., Aravind, L., Oania, R., McDonald, W. H., Yates, J. R., Koonin, E. V., and Deshaies, R. J. (2002). Role of Rpn11 metalloprotease in deubiquitination and degradation by the 26S proteasome. *Science* **298,** 611–615.

Yao, T., and Cohen, R. E. (2002). A cryptic protease couples deubiquitination and degradation by the proteasome. *Nature* **419,** 403–407.

# [32] Assaying Degradation and Deubiquitination of a Ubiquitinated Substrate by Purified 26S Proteasomes

*By* RATI VERMA and R. J. DESHAIES

## Abstract

The 26S proteasome is a multisubunit complex that catalyzes ATP-dependent proteolysis of cellular proteins. It eliminates misfolded proteins, as well as labile regulatory proteins, thereby serving a central role in maintaining cellular homeostasis. The bulk of the known substrates of the 26S proteasome are earmarked for proteolysis by covalent modification with a multiubiquitin chain, which is recognized by specific receptors. Once targeted, the substrate is deubiquitinated and degraded by the 26S proteasome. This chapter describes assays that monitor ATP- and ubiquitin-dependent proteolysis of the S-Cdk inhibitor Sic1.

## Introduction

The 26S proteasome is a 2-MDa complex that comprises a 20S proteolytic core that is sealed off from the cellular milieu at both ends by the 19S cap (Pickart and Cohen, 2004). The 19S cap contains (minimally) 20 proteins present in stoichiometric amounts and numerous other proteasome-interacting proteins (PIPs) present in substoichiometric amounts (Verma *et al.*, 2000). Known activities resident in the 19S cap are receptors such as Rpn10 that bind the multiubiquitin chain as a prelude to substrate degradation (Verma *et al.*, 2004), an isopeptidase activity (Rpn11) that removes the polyubiquitin chain (Verma *et al.*, 2002; Yao and Cohen, 2002), and six ATPases (Rpts 1–6) that contribute to the unfolding and translocation of the substrate into the 20S proteolytic core (Rubin *et al.*, 1998). The 20S protease itself is a cylinder composed of two seven-membered outer ($\alpha$) rings that are catalytically inactive and two seven-membered inner ($\beta$) rings. In each of the $\beta$ rings, three of the subunits possess a peptidase active site (Chen and Hochstrasser, 1996).

## Purification of 26S Proteasomes

The following method is a more detailed description of the protocol described previously (Verma *et al.,* 2000).

METHODS IN ENZYMOLOGY, VOL. 398

0076-6879/05 $35.00
DOI: 10.1016/S0076-6879(05)98032-4

*Reagents*

1. 5× lysis buffer (buffer A): 250 m$M$ Tris, pH 7.5, 750 m$M$ NaCl, 50% glycerol, 25 m$M$ $MgCl_2$, 5 m$M$ ATP
2. 10× ATP regenerating system: 10 mg/ml creatine phosphokinase, 100 m$M$ ATP, 200 m$M$ HEPES, pH 7.2, 200 m$M$ magnesium acetate, 1.5 $M$ creatine phosphate

*Individual Component Preparation*

Creatine phosphokinase (Sigma) stock is made up as 10 mg/ml in 50 m$M$ NaCl, 20 m$M$ HEPES, pH 7.2, 2 m$M$ $MgCl_2$, 1 m$M$ EDTA, 50% glycerol and stored at −20°

Creatine phosphate, Sigma, is made up in water and stored at −80°

ATP is prepared from ATP sodium salt (Calbiochem). Dissolve 5 g ATP in 90.7 ml water and pH to 7.5 with NaOH

3. 2× no-salt buffer (buffer B): 50 m$M$ Tris, pH 7.5, 20 m$M$ $MgCl_2$
4. Washed anti-FLAG M2 affinity gel: Anti-FLAG affinity resin (Product No. A2220, Sigma) is washed with 0.1 $M$ glycine, pH 3.5, and Tris-buffered saline as detailed in the manufacturer's protocol and resuspended in 1× buffer A as a 50% slurry
5. 20% Triton X-100
6. 26S elution buffer: 25 m$M$ Tris, pH 7.5, 10 m$M$ $MgCl_2$, 150 m$M$ NaCl, 15% glycerol
7. 100× Flag peptide: 10 mg/ml of NH2-ASP-TYR-LYS-ASP-ASP-ASP-ASP-LYS–COOH in water. Store in small aliquots at −20°

*Growth Medium Preparation*

10× synthetic medium (SD) containing 6.7% yeast nitrogen base minus amino acids, 5% casamino acids, 20% dextrose, 0.02% adenine sulfate, and 0.02% tryptophan

*Strain.* Purification of 26S proteasome complexes from yeast cells whose endogenous PRE1 locus (PRE1 encodes a 20S $\beta$ subunit) is modified to encode a FLAG-tagged polypeptide (RJD1144).

*Protocol*

1. Inoculate 25 ml of SD medium with a single colony of RJD1144 growing on a SD-Ura plate. Grow culture to saturation (~28 h) at 30°.

2. Dilute 1.5 ml of saturated culture into each of six 2.8-liter Fernbach flasks containing 1.5 liter of 1× SD and grow to an $OD_{600}$ of 3.0.

3. Harvest cells in six 1-liter Nalgene centrifuge bottles in a Sorvall RC 3B Plus centrifuge using the HBB6 rotor (6K rpm for 20 min). Collect

pellet into four 50-ml centrifuge tubes by washing cells in a total volume of 200 ml ice-cold sterile water.

4. Flash freeze in liquid nitrogen the cell pellet obtained by centrifuging the 50-ml tubes in the Sorvall H1000B tabletop rotor at 5000 rpm for 5 min and store at −80° for a minimum of 2 h such that a frozen cell pellet that can be ground is obtained.

5. Grind cell pellet manually using a mortar and pestle, with the mortar nestled inside an ice bucket filled with dry ice. The procedure typically takes 10–20 min (depending on the amount being ground) so keep the pellet cold by the periodic addition of liquid nitrogen every 2 min. Transfer the ground powder to a 50-ml centrifuge tube up to the 20-ml mark and flash freeze in liquid nitrogen. Store tubes at −80° until ready to use.

6. Add 10 ml of 1× lysis buffer per tube and supplement with ATP to 5 m*M* and ARS to 1×. Thaw cell powder on ice, while gently shaking occasionally to help bring material into suspension.

7. Transfer thawed lysate into a 30-ml screw-cap (i.e., Oak Ridge) centrifuge tube and centrifuge in a Sorvall SS34 rotor at 17,000 rpm for 20 min.

8. Collect pooled supernatants into a 50-ml tube and supplement again with ATP and $MgCl_2$ to 5 m*M* final concentrations. Add 800 $\mu$l of washed FLAG beads (50% slurry) to a clean 15-ml centrifuge tube.

9. Add 13 ml of supernatant (typically 130 mg protein) from step 8 and incubate at 4° on a rotating wheel for 1.5 h.

10. Centrifuge tubes at 3000 rpm in a Sorvall H1000B rotor for 5 min and aspirate all but 1 ml of supernatant.

11. Transfer bead suspension to a 2-ml microcentrifuge tube.

12. Centrifuge beads at 7000 rpm fmicrofuge for 30 s and wash beads three times with buffer A containing 0.2% Triton and an additional 2 m*M* ATP.

13. Wash twice with buffer B containing 2 m*M* ATP. Collect beads by centrifugation and use a 25-gauge needle to remove all traces of wash buffer.

14. Add a volume of 26S elution buffer that is three times the volume of the bead pellet. Add ATP to 2 m*M* (taking bead volume into account) and FLAG peptide to 1×. Elute at 4° for 3 h on a rotator.

15. Pellet beads in a microcentrifuge and collect 12 $\mu$l supernatant for SDS–PAGE analysis (12% gel). Bands should be visible by Coomassie blue staining if a typical yield is obtained (300–400 $\mu$g 26S/130 mg lysate). Flash freeze remaining supernatant and store at −80°. In addition to the aforementioned protocol, there are several other published procedures for purifying 26S proteasomes by one-step affinity chromatography. They are listed in Table I.

TABLE I
OTHER PROTOCOLS FOR ONE-STEP AFFINITY PURIFICATION OF 26S PROTEASOMES

| Nature of epitope tag | Chromatographic step | Assay | Comments | Ref. |
|---|---|---|---|---|
| 1. *PRE1-TEV-MYC9* | Anti-myc 9E10 antibody coupled to protein A-Sepharose | Degradation of UbSic1 | Elution using TEV protease, thus avoiding Flag peptide, which may interfere with peptidase assays | Petroski and Deshaies (2003) |
| 2. *RPN11-TEV-PROTEIN A* | IgG | — | Elution using TEV protease | Leggett *et al.* (2002) |
| 3. *RPN11-3XFLAG* | Anti-Flag M2 agarose | Degradation of T7Sic1PY | Elution with Flag peptide; good yield of doubly capped 26S proteasomes | Sone *et al.* (2004) |

## Monitoring Activity of Purified 26S Proteasome

### *Degradation*

*Substrate.* Mbp-Sic1-MycHis6 is purified from *Escherichia coli* (Verma *et al.*, 1997b) or the trimeric Sic1 complex comprising Gst-Cdc28,Clb5, and Sic1 is purified from Hi5 insect cells. Purified Sic1 is phosphorylated by the G1 Cdk complex and ubiquitinated by the SCF ubiquitin ligase complex in the presence of yeast E1, Cdc34 (E2), ubiquitin, 1× ARS, and magnesium acetate as described previously (Seol *et al.*, 1999; Verma *et al.*, 2001) [see Petroski and Deshaies (2005) for a detailed protocol for the preparation of ubiquitinated Sic1].

*Assay.* A typical degradation reaction contains 50 $\mu$l purified 26S proteasome (75–100 n$M$ final concentration), 6 $\mu$l 10× ARS, and 1 $\mu$l 200 m$M$ magnesium acetate and is assembled on ice. The reaction is initiated by the addition of 2–3 $\mu$l of ubiquitinated Sic1 (UbSic1; 250–300 n$M$ final concentration) and is incubated at 30° for 3 min (trimeric UbSic1) or 5 min (UbMbpSic1). The reaction is stopped by the addition of 5× Laemmli SDS–PAGE buffer. A 15-$\mu$l aliquot is resolved on an 8% (or 10%) polyacrylamide gel, and proteins are transferred to nitrocellulose, taking care to retain the stacker gel. The reaction is monitored by developing the nitrocellulose blot with a polyclonal antibody to Sic1. The ATP dependence of the degradation reaction can be assessed by pretreating 26S preparations and UbSic1 with apyrase (15 U/ml at 30° for 5 min) or glucose/hexokinase (5 U/ml hexokinase plus 30 m$M$ glucose) to deplete ATP. Ubiquitin dependence can be monitored by mixing in an equimolar amount of unmodified Sic1 (generated by a reaction equivalent to a ubiquitination reaction but lacking ubiquitin and E1) (Verma *et al.*, 2001).

### *Deubiquitination*

The deubiquitinating activity of the 26S proteasome can be unmasked when degradation is inhibited. Degradation can be blocked by using inhibitors of the 20S core peptidase activity (Meng *et al.*, 1999). Lactacystin and MG132 were ineffective at 200 $\mu M$ (not shown), whereas both epoxomicin and YU101 were effective in blocking degradation of UbSic1 at 100 $\mu M$ (Fig. 1). A time course experiment demonstrated that purified 26S proteasomes had to be preincubated with the inhibitor at 30° for a minimum of 20 min for maximal inhibition (Fig. 2). An epoxomicin dosage experiment demonstrated that the $IC_{50}$ was approximately 50 $\mu M$ (Fig. 3). This value is much higher than the reported $IC_{50}$ of epoxomicin (40–80 n$M$) for inhibiting chymotryptic activity of the proteasome (Meng *et al.*, 1999). However, it has been argued that inhibition of the chymotryptic site by

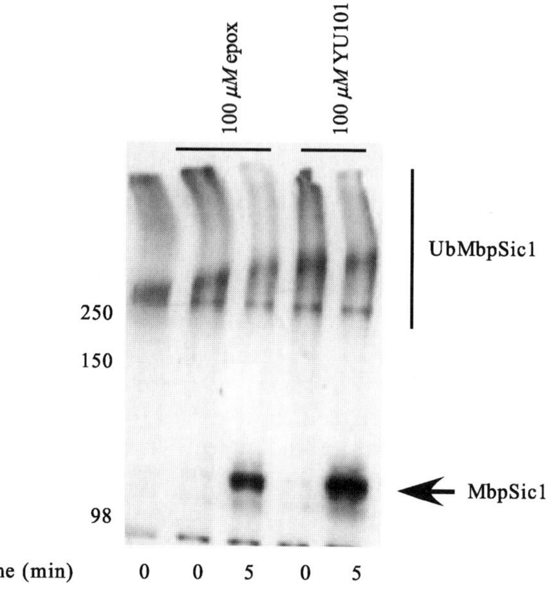


FIG. 1. UbSic1 is deubiquitinated but not degraded in the presence of the 20S peptidase inhibitors epoxomicin and YU101. 26S proteasomes were preincubated with either epoxomicin or YU101. The assay was initiated by the addition of UbSic1 and was terminated either immediately (lanes 2 and 4) or after a 5-min incubation at 30° (lanes 3 and 5). Lane 1 is time 0 with no added inhibitor.

itself causes little inhibition of protein breakdown and that either the tryptic or the caspase-like active site has to be inhibited in addition to the chymotryptic active site to achieve significant inhibition of protein breakdown (see Kisselev and Goldberg, 2005). At 100 $\mu M$ epoxomicin, all three active sites of the proteasome should be inhibited.

In our hands, the deubiquitination assay serves as a useful surrogate for measuring protein breakdown by the proteasome. Like proteolysis of UbSic1, its deubiquitination requires (i) intact 26S proteasome, (ii) ATP, (iii) multiubiquitin chain receptor function (provided by either Rpn10 or Rad23), and (iv) an intact Rpn11 active site. An advantage of assaying deubiquitination as opposed to degradation is that instead of monitoring the disappearance of high molecular weight UbSic1, one can instead monitor the appearance of a discrete, lower molecular weight species of a deubiquitinated substrate. This renders the assay much more sensitive, which makes it easier to see subtle defects in substrate processing (e.g.,

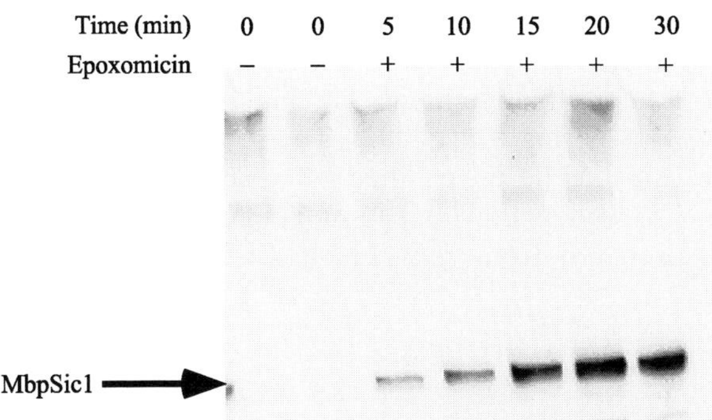


FIG. 2. Time course of deubiquitination. 26S proteasomes were pretreated with 100 $\mu M$ epoxomicin at 30° for different lengths of time as indicated. The reaction was then initiated by the addition of UbSic1 and was terminated after 5 min.

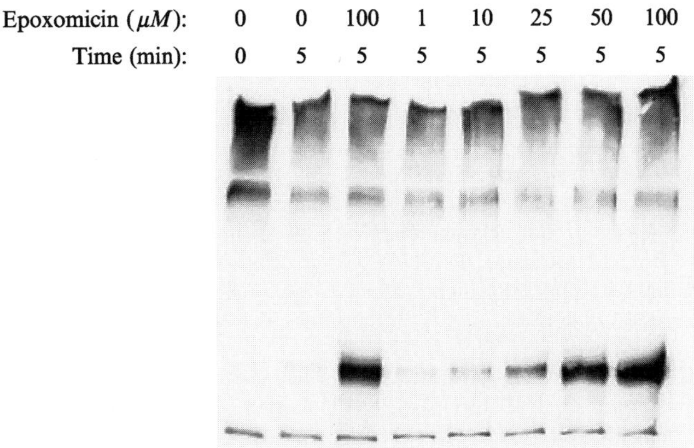


FIG. 3. Determination of $IC_{50}$ for epoxomicin. 26S proteasomes were pretreated with varying concentrations of epoxomicin for 45 min at 30° before addition of substrate.

such as that shown by 26S proteasomes isolated from *rad23Δ* cells; see Verma *et al.*, 2004).

*Assay.* A standard deubiquitination reaction is set up by incubating 50 $\mu$l of purified 26S proteasome with 100 $\mu M$ epoxomicin (0.5 $\mu$l of a 10 *mM* stock in dimethyl sulfoxide; store in aliquots at –20°) at 30° for 45 min. Ubiquitin aldehyde (UbAl) can also be added at a final concentration of 2.5 $\mu M$ (Chung and Baek, 1999) if contaminating cysteine isopeptidases are a problem. The preincubated 26S proteasomes are put on ice, and the reaction is assembled, initiated, and monitored as described earlier. The deubiquitination activity is dependent on the metalloisopeptidase Rpn11, a proteasomal subunit that contains the highly conserved JAMM motif ($EX_nHXHX_{10}D$) (Verma *et al.*, 2002). The specificity of the deubiquitination reaction can thus be determined by assaying in the presence of the metal chelator 1,10-phenanthroline (1 m*M*) or by assaying activity with 26S proteasomes prepared from *rpn11ts/rpn11AXA* mutant strains (Verma *et al.*, 2002).

## References

Chen, P., and Hochstrasser, M. (1996). Autocatalytic subunit processing couples active site formation in the 20S proteasome to completion of assembly. *Cell* **86,** 961–972.

Chung, C. H., and Baek, S. H. (1999). Deubiquitinating enzymes: Their diversity and emerging roles. *Biochem. Biophys. Res. Commun.* **266,** 633–640.

Kisselev, A. K., and Goldberg, A. F. (2005). Monitoring activity and inhibition of 26S proteasomes with fluorogenic peptide substrates. *Methods Enzymol.* **398**(30), 2005 (this volume).

Leggett, D. S., Hanna, J., Borodevsky, A., Crosas, B., Schmidt, M., Baker, R. T., Walz, T., Ploegh, H., and Finley, D. (2002). Multiple associated proteins regulate proteasome structure and function. *Mol. Cell* **10,** 495–507.

Meng, L., Mohan, R., Kwok, B. H., Elofsson, M., Sin, N., and Crews, C. M. (1999). Epoxomicin, a potent and selective proteasome inhibitor, exhibits *in vivo* antiinflammatory activity. *Proc. Natl. Acad. Sci. USA* **96,** 10403–10408.

Petroski, M. D., and Deshaies, R. J. (2003). Context of multiubiquitin chain attachment influences the rate of Sic1 degradation. *Mol. Cell* **11,** 1435–1444.

Petroski, M. D., and Deshaies, R. J. (2005). *In vitro* reconstitution of SCF substrate ubiquitination with purified proteins. *Methods Enzymol.* **398**(13), 2005 (this volume).

Pickart, C. M., and Cohen, R. E. (2004). Proteasomes and their kin: Proteases in the machine age. *Nature Rev. Mol. Cell. Biol.* **5,** 177–187.

Rubin, D. M., Glickman, M. H., Larsen, C. N., Dhruvakumar, S., and Finley, D. (1998). Active site mutants in the six regulatory particle ATPases reveal. *EMBO J.* **17,** 4909–4919.

Seol, J., Feldman, R., Zachariae, W., Shevchenko, A., Correll, C., Lyapina, S., Chi, Y., Galova, M., Claypool, J., Sandmeyer, S., *et al.* (1999). Cdc53/cullin and the essential Hrt1 RING-H2 subunit of SCF define a ubiquitin ligase module that activates the E2 enzyme Cdc34. *Genes Dev.* **13,** 1614–1626.

Sone, T., Saeki, Y., Toh-e, A., and Yokosawa, H. (2004). Sem1p is a novel subunit of the 26 S proteasom from *Saccharomyces cerevisiae*. *J. Biol. Chem.* **279,** 28807–29916.

Verma, R., Annan, R. S., Huddleston, M. J., Carr, S. A., Reynard, G., and Deshaies, R. J. (1997b). Phosphorylation of Sic1p by G1 Cdk required for its degradation and entry into S phase. *Science* **278,** 455–460.

Verma, R., Aravind, L., Oania, R., McDonald, W. H., Yates, J. R., 3rd, Koonin, E. V., and Deshaies, R. J. (2002). Role of Rpn11 metalloprotease in deubiquitination and degradation by the 26S proteasome. *Science* **298,** 611–615.

Verma, R., Chen, S., Feldman, R., Schieltz, D., Yates, J., Dohmen, R. J., and Deshaies, R. J. (2000). Proteasomal proteomics: Identification of nucleotide-sensitive proteasome-interacting proteins by mass spectrometric analysis of affinity-purified proteins. *Mol. Biol. Cell* **11,** 3425–3439.

Verma, R., McDonald, H., Yates, J. R., 3rd, and Deshaies, R. J. (2001). Selective degradation of ubiquitinated Sic1 by purified 26S proteasome yields active S phase cyclin-Cdk. *Mol. Cell* **8,** 439–448.
Verma, R., Oania, R., Graumann, J., and Deshaies, R. J. (2004). Multiubiquitin chain receptors define a layer of substrate selectivity in the ubiquitin-proteasome system. *Cell* **118,** 99–110.
Yao, T., and Cohen, R. E. (2002). A cryptic protease couples deubiquitination and degradation by the proteasome. *Nature* **419,** 403–407.

# [33] Probing the Ubiquitin/Proteasome System with Ornithine Decarboxylase, a Ubiquitin-Independent Substrate

*By* MARTIN A. HOYT, MINGSHENG ZHANG, and PHILIP COFFINO

## Abstract

Ornithine decarboxylase (ODC) is an unusual proteasome substrate—ubiquitin conjugation plays no part in its turnover. It can therefore be used as a probe to distinguish proteasome-mediated actions that do or do not depend on the activity of the ubiquitin system. A 37 residue region of ODC suffices for proteasome interactions, and within this sequence functionally critical residues have been identified. Because no posttranslational modifications are required for substrate preparation, ODC and derived constructs can be readily generated as substrates for either *in vitro* or *in vivo* studies. This chapter describes methodologies that allow the use of ODC as a reporter to examine the ubiquitin-proteasome system, both in reconstituted *in vitro* systems and in living cells.

## Introduction

Ornithine decarboxylase is degraded by the proteasome, but ubiquitin conjugation plays no role in this process. ODC is an enzyme required for polyamine biosynthesis. Proteasome degradation is an important element of its complex regulation. The protein has several characteristics that make it a favorable experimental substrate for studying proteasome action.

1. Active substrates are easily prepared without need for performing a complex series of enzymatic transfers of ubiquitin.
2. A small domain of ODC suffices as a degradation tag and functions autonomously upon transfer to other proteins.
3. The tag is conserved and recognized among diverse eukaryotes.

METHODS IN ENZYMOLOGY, VOL. 398

0076-6879/05 $35.00
DOI: 10.1016/S0076-6879(05)98033-6

4. Well-developed methods are available to ascertain specificity, as critical residues of the tag have been described.
5. Both *in vivo* and *in vitro* systems for studying ODC degradation have been described.
6. *In vitro* degradation can be performed with biochemically pure components.
7. Because ubiquitin conjugation is not involved, ODC turnover can be used to evaluate proteasome function within the more global ubiquitin/proteasome system.

The proteasome targeting signal of ODC consists of a conserved 37 amino acid C-terminal sequence (cODC) that is intrinsic to the primary sequence of ODC (Fig. 1). This degradation tag functions autonomously and can be appended to other proteins, making them labile. Although cODC has autonomous function, its affinity for the proteasome is enhanced about 10-fold when a second protein, antizyme (AZ), associates with ODC (Zhang *et al.*, 2003). Antizyme dissociates the ODC homodimer (the enzymatically active form) and forms an ODC:antizyme heterodimer (reviewed in Coffino, 2001). This rearrangement of ODC quaternary structure makes cODC more accessible. Antizyme is a physiologic regulator of ODC activity and abundance and is the effector element of a feedback system that controls cellular polyamines, the downstream products of ODC activity. Antizyme is encoded by an mRNA of unusual structure, consisting of a short ORF that overlaps a second ORF encoding the bulk of the protein; the second ORF lacks an initiation codon and is encoded by a +1 reading frame with respect to the first (Matsufuji *et al.*, 1995). The protein therefore requires translational frameshifting for its decoding. Superphysiologic levels of polyamines enhance the efficiency of frameshifting at a specific site of the mRNA, increasing production of the protein. Antizyme binds to ODC, increasing its affinity for the proteasome and thus accelerating its degradation. Because ODC is the initial and rate-limiting enzyme in

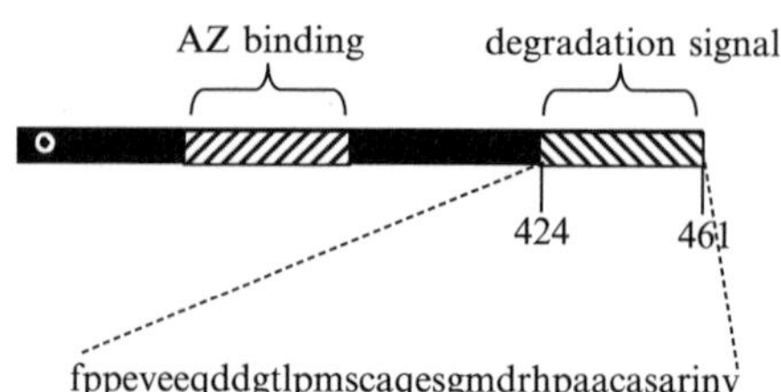


FIG. 1. Positions of regions within mouse ODC critical for degradation. Residues important for proteasome recognition in carboxy-terminal degradation domain are indicated by underlining.

the polyamine biosynthesis pathway, cellular polyamines are restored to normal by this means. The feedback system does not depend on allostery to control ODC enzyme but instead uses a combination of translational frameshifting and targeted degradation. This elaborate mechanism is conserved from fungi to mammals.

For both *in vivo* and *in vitro* studies, ODC provides a convenient means to distinguish between ubiquitin-dependent and -independent mechanisms of proteasome-mediated proteolysis. Notably, measurements of ODC turnover allow investigators to differentiate whether mutations, treatments, or biologic states impair proteasome function specifically or instead alter properties of the ubiquitin-proteasome pathway that are external to proteasomes.

## Structural Elements of Ornithine Decarboxylase and Antizyme Important for Degradation

The *cis*-acting residues of cODC critical for directing degradation have been identified. They function in native ODC and also in otherwise stable proteins that have been destabilized by fusion to cODC. By mutating these elements, one can test whether the ODC-specific degradation pathway is operating as intended in the experimental system under investigation. Varying the cODC sequence also offers opportunities for systematically manipulating substrate targeting.

Figure 2 shows an alignment of the C-terminal region of ODC proteins from diverse vertebrate species. Two subregions of cODC contain a high density of invariant residues. The first of these consists of residues 440–446 of the mouse enzyme. However, scanning mutagenesis has shown that among these, only Cys441 is intolerant to mutation (H. Chen and P. Coffino, unpublished data); even the conservative Cys441Ser mutation strongly stabilizes ODC (Hoyt, 2003; Zhang *et al.*, 2003). The second

```
                        * ****               ** *
421 QSHGFPPEVE EQDDGTLPMS CAQESGMDRH PAACASARIN V 461 Mouse
421 .......... .......... ........H. S......... . 461 Shrew mouse
421 .......... ...V...... .......... .......S.. . 461 Rat
415 .N........ ...V....I. .......... .......S.. . 455 Hamster
421 .NPD...... ...AS...V. ..W....K.. R......S.. . 461 Human
421 RTQD...G.. .P.V.P..V. ..W....K.. S.....T... . 461 Cow
411 KEQE.LA... ...VAS..L. ..C...IE-Y ..T....S.. . 450 Chicken
422 KE..IL...P --.LSA.HV. .......ELA ..V.TA.S.. . 460 Xenopus
```

Multi sequence alignment of ODC carboxy termini. Asterisks mark the position of invariant residues.

FIG. 2. Multisequence alignment of ODC carboxy termini. Asterisks mark the position of invariant residues.

subregion that shows invariance is at the very C-terminal end. Serial truncations that remove the last few amino acids progressively stabilize ODC, and removing the last five suffices for full stabilization (Ghoda *et al.*, 1992). These effects of Cys441 mutation and C-terminal truncation are hallmarks of the ubiquitin-independent ODC degradation pathway (Hoyt, 2003). The 37 terminal native residues of ODC provide an autonomous degradation domain when fused to the C terminus of other proteins. Those tested include green fluorescent protein (GFP) (Li *et al.*, 1998), luciferase (Leclerc *et al.*, 2000), Ura3 (M. Hoyt and P. Coffino, unpublished data), and dihydrofolate reductase (Loetscher *et al.*, 1991; Zhang *et al.*, 2003). Additionally, *Trypanosoma brucei* ODC is a stable protein about 70% identical to the mammalian enzyme in its core sequence, but lacking a region that corresponds to cODC. Grafting cODC to the end of *T. brucei* ODC labilizes the enzyme (Ghoda *et al.*, 1989, 1990).

Antizymes actually consist of a conserved gene family whose individual members have been independently conserved among eukaryotes (Ivanov *et al.*, 2000). Three distinct antizyme orthologs are found in mammals. Antizymes 1 and 2 are expressed ubiquitously; antizyme 3 is expressed only at a specific stage of sperm cell development. Although antizymes 1 and 2 both bind to ODC and inhibit its enzymatic activity, antizyme 1 accelerates ODC degradation by proteasomes much more effectively (Chen *et al.*, 2002; Zhu *et al.*, 1999). Antizyme 1 from *Rattus norvegicus*, which has 227 amino acids, has been dissected to reveal its working parts. Residues 120–227 are sufficient for ODC binding, but are insufficient to direct degradation (Li and Coffino, 1994). This additionally requires sequence information present in amino acids 70–119. Mixing and matching bits of antizymes 1 and 2 revealed that the capacity to strongly promote degradation *in vitro* maps to residues 131 and 145 of antizyme 1 (Chen *et al.*, 2002).

## Examples of the Experimental Use of ODC/AZ

### *GFP-cODC: A Labilized Reporter of Temporal Changes in Gene Expression*

The green fluorescent protein of *Aequorea victoria* is widely used as a reporter of transcription. Because GFP is a stable protein, its level does not track the rate of GFP synthesis, but instead provides a record of its accumulation over time. A labile reporter would more closely approximate synthetic rate and provide acute information about time-dependent changes in synthesis. Appending cODC to the GFP carboxy terminus (Li *et al.*, 1998) labilized the reporter, creating a system for tracking acute changes in gene expression.

### *Cyclin D1: A Target of AZ Degradation*

The observation that AZ expression inhibits the growth of some cells prompted an examination of AZ-associated changes in cyclin levels. AZ was shown to associate noncovalently with cyclin D1. Using purified components, AZ was found to direct cyclin D1 degradation by proteasomes without the involvement of ubiquitylation, an observation confirmed *in vivo* by showing that AZ induction accelerated the turnover of a mutant form of cyclin D1 that cannot be ubiquitylated (Newman *et al.*, 2004).

### *Ubistatin: Discerning Ubiquitin-Specific Proteasome Targeting*

A chemical library screen revealed a series of compounds that inhibited the proteolysis of cyclin B and ubiquitinated Sic1 by purified proteasomes. To determine the specificity of the inhibitors, their activity was investigated using ODC as a proteasome substrate. The compounds proved much less potent for the ubiquitin-independent substrate, supporting the conclusion that the inhibitory effect was directed at ubiquitin conjugates (Verma *et al.*, 2004).

### *ODC as a Platform for Investigating a cis-Acting Proteasome Inhibitory Sequence*

The Epstein–Barr virus EBNA1 protein contains a tract of 60–300 residues consisting entirely of glycines and alanines. When embedded in diverse proteins, it inhibits their degradation by the proteasome. The question of whether it impairs ubiquitin conjugation or acts instead on downstream events was resolved by embedding a glycine-alanine sequence within ODC (Zhang and Coffino, 2004). Its action was shown to be independent of ubiquitin conjugation; it instead functions in the proteasome by interrupting substrate insertion (Zhang and Coffino, 2004).

## Methods for *In Vivo* Degradation of ODC

### *Pulse–Chase or Immunoblot Analysis of Polyamine-Induced Turnover of ODC in Cultured Mammalian Cells*

One of the simplest means to assess the capacity for ODC turnover in cultured cells is to measure the change in ODC protein or enzymatic activity following polyamine administration. As discussed earlier, the addition of exogenous polyamines stimulates antizyme synthesis, resulting in accelerated ODC turnover. Polyamine administration has no apparent effect on the rate of ODC synthesis in cultured mammalian cells (van Daalen Wetters *et al.*, 1989).

The turnover of ODC in mammalian cells can be measured directly using pulse–chase analysis (Ghoda *et al.*, 1989; van Daalen Wetters *et al.*, 1989). For this purpose, either the endogenous ODC protein or the ectopically expressed ODC, introduced by transient or stable transfection, can be utilized. Endogenous ODC is normally expressed at very low levels in mammalian cells, sometimes making immunoblot analysis of ODC difficult. Ectopic ODC, expressed from plasmid vectors, confers the advantage of higher expression levels and the possibility of introducing epitope-tagged variants. For historical reasons, the mouse (*Mus musculus*) ODC cDNA (Gupta and Coffino, 1985) is used most often in such vectors, and studies of turnover utilized this ODC. Hereafter, ODC will designate mouse ODC specifically, unless otherwise stated. Due to the extensive homology of the primary peptide sequences, antisera prepared against recombinant mouse ODC react equally well with ODCs from other mammalian species.

For pulse–chase analysis, approximately $1.25 \times 10^5$ cells are transferred in 200 $\mu$l of Dulbecco's modified Eagle medium (DMEM) containing 10% fetal bovine serum to each well of a 24-well tissue culture plate (Falcon 353047). Putrescine is added to a final concentration of 500 $\mu M$, and the cells are cultured for 4–24 h in the presence of the polyamine. Exogenously added putrescine is converted rapidly to spermidine *in vivo* by the action of the biosynthetic enzyme spermidine synthase. Putrescine is added instead of spermidine because polyamine oxidases present in fetal bovine serum act on the latter polyamine, producing toxic products. If spermidine must be used, then horse serum, which lacks polyamine oxidases, is substituted for fetal bovine serum. The cells are washed three times with phosphate-buffered saline, followed by the addition of DMEM lacking methionine and containing 10% fetal bovine serum. Cells are incubated for 15 min in the methionine-free medium, followed by the addition of 50 $\mu$Ci of L-[$^{35}$S]methionine. The cells are labeled for 1 h at 37°, washed again three times with phosphate-buffered saline, and resuspended in complete medium containing 500 $\mu M$ putrescine and 5 m$M$ methionine to initiate the chase. After chase periods of 1.5, 3, and 6 h, cells are washed twice with phosphate-buffered saline, lysed on ice for 5 min in 400 $\mu$l of RIPA buffer (50 m$M$ Tris, pH 8.0, 150 m$M$ NaCl, 1.0% NP-40, 0.5% deoxycholate, 0.1% SDS), and frozen at –80°. After one freeze–thaw cycle, the lysates are clarified by centrifugation (14,000$g$), and trichloroacetic acid-insoluble radioactivity is determined. The amount of labeled ODC in each lysate is analyzed by immunoprecipitation followed by separation by SDS–polyacrylamide gel electrophoresis and autoradiography.

### *Example: Comparing the Effect on ODC of AZ1 and AZ2*

*In vitro* experiments have demonstrated that although AZ1 promotes ODC turnover, AZ2 does not. To compare the activities of the two paralogs *in vivo*, ODC, AZ1, and AZ2 ORFs are inserted into the mammalian expression vector pcDNA3.1(−) for ODC and pcDNA3.1(+) for the AZs (both vectors from Invitrogen). To facilitate Western blot analysis of expressed proteins, the Flag epitope tag (Asp-Tyr-Lys-Asp-Asp-Asp-Asp-Lys) is inserted immediately following the initial methionine of the native protein sequence in these constructs. CHO-K1 cells are transiently transfected with mammalian expression plasmids encoding FlagODC, FlagAZ1, FlagAZ2, or combinations of these using the Lipofectamine 2000 transfection reagent, as recommended by the manufacturer (Invitrogen). Control cells are transfected with empty vector, and empty vector DNA is added as needed to make total DNA concentrations identical for each transfection condition. After incubation for 24 h, cells are harvested, and the FlagODC, FlagAZ1, and FlagAZ2 are analyzed by Western blot using the anti-Flag M2 monoclonal (Sigma) as the primary antibody, horseradish peroxidase-conjugated secondary antibody, and ECL (Amersham Biosciences) for immunodetection.

CHO cells are transfected for expression of ODC, AZ1, or AZ2 and for coexpression of ODC with AZ1 or AZ2, as indicated (Fig. 3). Each of the three Flag-tagged proteins is detected following transfection. Adjacent lanes represent duplicate independently transfected cultures. For coexpression with ODC, each of the AZs is transfected using a series of 10-fold dilutions of DNA. Transfections for AZ2 are carried out with four times more DNA than corresponding transfections for AZ1 to partially compensate for the higher expression level reproducibly observed with the latter. AZ1 induces degradation of ODC, as does AZ2. Transfection with 0.1 $\mu$g FlagAZ1 DNA (lane 5) and 4.0 $\mu$g FlagAZ2 DNA (lane 7) resulted in similar levels of the FlagAZs and both reduced FlagODC to an undetectable level. At a further 10-fold reduction of transfecting DNA [0.01 $\mu$g FlagAZ1 DNA (lane 6) and 0.4 $\mu$g FlagAZ2 DNA (lane 8)] FlagAZ1 completely suppressed FlagODC but FlagAZ2 did so only partially. At further 10-fold dilution (lane 9), FlagAZ2 had no apparent effect on FlagODC. These results imply that AZ2 has the capacity to induce ODC turnover *in vivo*, but is less potent than AZ1.

### *Pulse–Chase Analysis of ODC in Yeast Cells*

The ubiquitin-independent degradation of mouse ODC in yeast cells has been well characterized (Hoyt *et al.*, 2003). All the structural requirements for ODC degradation are conserved in *Saccharomyces cerevisiae*,

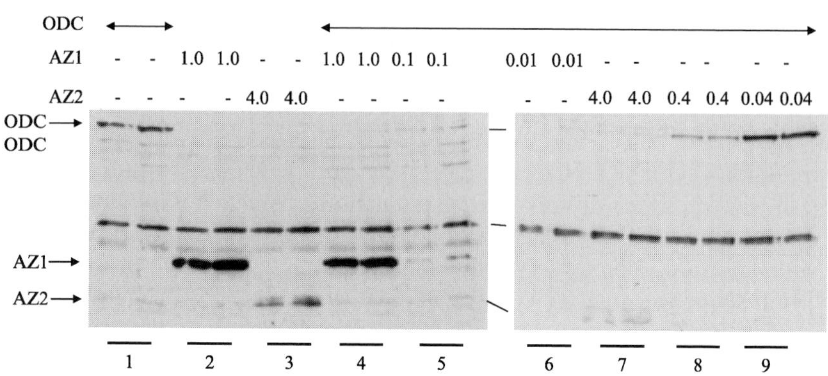


FIG. 3. Both AZ1 and AZ2 induce degradation of ODC in cultured cells, but the former is more potent. Both AZ1 and AZ2 cause degradation of ODC in cultured cells. CHO cells were transfected for expression of ODC, AZ1, or AZ2 and for coexpression of ODC with AZ1 or AZ2, as indicated. Adjacent lanes represent duplicate independently transfected cultures. For coexpression with ODC, each of the AZs was transfected using a series of 10-fold dilutions of DNA. Each of the transfected proteins was Flag tagged at the N terminus and expression was visualized by Western blotting with an anti-Flag antibody. Transfections for AZ2 were carried out with four times more DNA than corresponding transfections for AZ1 to partially compensate for the higher expression level reproducibly observed with the latter.

making mouse ODC an excellent probe of proteasome function in this fungus. Mouse ODC can be expressed using any number of existing yeast vectors designed for the heterologous expression of proteins in *S. cerevisiae.* We have found a set of vectors designed by Mumberg *et al.* (1995) ideal for this purpose. In particular, the ODC coding region expressed from vectors containing the *ADH1* promoter results in easily detectable levels of mature ODC protein with a half-life of 5–10 min in yeast cells. The following protocol, adapted from that of Suzuki and Varshavsky (1999), can be used to measure ODC turnover in yeast cells.

Yeast transformants expressing ODC are grown to mid-logarithmic growth phase ($OD_{600}$ ~0.5 to 1.0) in SD medium (Guthrie and Fink, 1991) lacking the appropriate components to maintain selection for the transforming plasmid. A 10-ml culture volume typically provides enough cells for the collection of three or four time points during the chase period.

Cells are harvested by centrifugation at 2000*g* for 3 min at room temperature, and the growth medium is discarded. The cells are washed twice by resuspension in 1 ml of SD medium lacking methionine (SD–Met) and transferred into a 1.5-ml microcentrifuge tube. The cells are incubated for 1 min at 30° in SD–Met and collected by brief centrifugation in a microcentrifuge.

For labeling, the cells are resuspended in 0.4 ml of SD–Met followed by the addition of 200 $\mu$Ci of L-[$^{35}$S]methionine. A L-[$^{35}$S]methionine and L-[$^{35}$S]cysteine mixture (Expre$^{35}$S$^{35}$S protein labeling mixture, Perkin Elmer Life Sciences) also works well for this purpose. Metabolic labeling is carried out for 5 min at 30°. Following the labeling period, the cells are collected by brief centrifugation in a microcentrifuge, and the labeling medium is removed. The chase period is initiated by resuspending the cells in 0.4 ml of SD medium containing 10 m*M* methionine, 1 m*M* cysteine (if the labeling mixture is used), and 0.5 mg/ml cycloheximide. For each time point during the chase period, a 0.1-ml sample of the cell suspension is collected and transferred to a 2-ml screw-cap tube containing 0.7 ml of ice-cold lysis buffer (50 m*M* HEPES, pH 7.5, 150 m*M* NaCl, 1 m*M* EDTA, 1% Triton X-100, 1 m*M* phenylmethylsulfonyl fluoride) and 0.5 ml of 0.5-mm glass beads. The cells are disrupted by four pulses of 30 s duration at the maximum speed setting in a Mini-beadbeater (Biospec Products), with cooling on ice between pulses. Alternatively, cells can be lysed using a bench-top vortexer for four pulses of 1 min duration with cooling on ice between pulses.

Following the chase period, the cell lysates are cleared by centrifugation in a microcentrifuge (13,000*g*, 10 min at 4°), and the amount of trichloroacetic acid-insoluble radioactivity is determined. The amount of labeled ODC in each lysate is analyzed by immunoprecipitation followed by separation by SDS–polyacrylamide gel electrophoresis and autoradiography. For lysates containing FLAG epitope-tagged ODC, we typically carry out immunoprecipitations using anti-FLAG M2 antibody-agarose (Sigma) using 10 $\mu$l of antibody-agarose conjugate in $\sim$0.5 ml of lysate (containing $\sim 2.0 \times 10^7$ cpm). Immunoprecipitations are done at 4° for 1–2 h followed by four washes with 1 ml of wash buffer (lysis buffer containing 0.1% SDS).

## Methods for *In Vitro* Degradation of ODC

AZ1 stimulated degradation of ornithine decarboxylase has been reconstituted *in vitro* in two forms, using either purified proteasomes or a crude rabbit reticulocyte lysate system as the source of proteasome activity. In the crude lysate, AZ1 usually can drive the degradation of ODC to near completion in 1 h; this degradation system is quite active, reliable, and accessible. In contrast, utilizing a degradation system containing fully defined components requires that AZ1, proteasomes, and labeled ODC each be prepared in a highly purified form, which is more technically demanding. Some special cases, such as quantitative determination of kinetic parameters, require the fully defined biochemical system.

### *ODC Degradation in Reticulocyte Lysate*

$^{35}$S-labeled ODC is produced by coupled *in vitro* transcription/translation in a rabbit reticulocyte lysate system (TNT, Promega) following the manufacturer's protocol. The plasmid used for transcription/translation of ODC has a T7 promoter positioned upstream of the ODC ORF. Incubation time for the coupled reaction is usually 90 min at 30°, and the reaction is stopped by chilling the samples on ice. The samples can be stored at –20° for several weeks. The degradation assay is performed in a 50-$\mu$l volume at 37° and contains 5 $\mu$l of lysate (which provides both labeled ODC and the proteasomes required for its degradation), 50 m*M* Tris-HCl, pH 7.5, 5 m*M* $MgCl_2$, an ATP regenerating buffer [2 m*M* dithiothreitol (DTT), 10 m*M* creatine phosphate, and 1.6 mg/ml creatine kinase], and 2 $\mu$g recombinant AZ1. AZ1 is produced as described under "ODC Degradation Using Purified Components." To document the progress of the degradation reaction, 10-$\mu$l aliquots are removed periodically (typically after 0, 15, 30, and 60 min) and mixed with an equal volume of 2× SDS–PAGE loading buffer. The fraction of ODC remaining undegraded at each time point is determined by SDS–PAGE and autoradiography. ODC mutated at a single residue (Cys441 → Ala or Ser) or with a C-terminal 37 amino acid truncation, each of which prevents proteasome-mediated ODC degradation, can be used as a negative control. Replacing AZ1 with its dilution buffer, or using proteasome-specific inhibitors, provides additional controls for specificity of the degradation reaction.

### *ODC Degradation Using Purified Components*

*Preparation of 26S Proteasomes.* Highly purified proteasomes from bovine red blood cells (a gift from C. Pickart ) or rat liver work equally well in the ODC degradation assay described here. Here a protocol for purification from rat liver, a modification (Zhang *et al.*, 2003) of a published procedure (Reidlinger *et al.*, 1997), is described. Sprague–Dawley rat livers are stored frozen prior to use. All steps should be performed at 0–4°. Rat livers (150 g wet weight) are homogenized in 500 ml buffer A [20 m*M* Tris-HCl, pH 7.5, containing 2 m*M* ATP, 5 m*M* $MgCl_2$, 1 m*M* DTT, 0.1 m*M* EDTA, 50 m*M* NaCl, and 20% (v/v) glycerol]. After removing cell debris by centrifugation ($RCF_{max}$ = 10,410*g*, 80 min) and filtration through cheesecloth, the sample is further cleared by centrifugation (30,000 rpm in a Beckman Type 35 rotor, $RCF_{max}$ = 105,000*g*, 1 h). The supernatant is filtered through Whatman filter paper, and the pH is adjusted to 7.5. The 26S proteasomes are sedimented by centrifugation (35,000 rpm in a Beckman Type 35 rotor, $RCF_{max}$ = 143,000*g*, 18 h), and the pellet is resuspended in 25 ml buffer A without glycerol. The pH is again adjusted to 7.5, if required. The sample is then layered onto a glycerol gradient (15–40% in buffer A, 32 ml/tube × 6) and centrifuged for 16 h (25,000 rpm in a Beckman SW27 rotor, $RCF_{max}$ = 110,000*g*). Fractions are assayed for peptidase activity (using LLVY-AMC or another fluorogenic peptide), and those containing peptidase activity are pooled and column fractionation is performed on an AKTA FPLC (Amersham Biosciences). Ten milligrams of protein per run is loaded onto a Mono Q HR 5/5 column (Amersham Biosciences) equilibrated with 20 m*M* Tris-HCl, pH 7.5, 1 m*M* DTT, 50 m*M* NaCl, and 10% (v/v) glycerol. Proteasomes are eluted using a linear salt gradient (50–800 m*M* NaCl, 40 column volumes) in the equilibration buffer. Active fractions are pooled, and 26S proteasomes are collected by centrifugation (31,000 rpm in a Beckman type 35 rotor, $RCF_{max}$ = 112,000*g*, 12 h). The purified 26S proteasomes (about 20 mg) are resuspended in 1.5 ml of the aforementioned equilibration buffer with 2 m*M* ATP and stored at –80°. The 26S proteasomes are analyzed by SDS–PAGE fractionation and should show a characteristic pattern of protein bands that closely resemble those reported previously for highly purified preparations. Purity and 26S mobility are further confirmed by nondenaturing gel electrophoresis and substrate overlays (Glickman *et al.*, 1998).

*Preparation of AZ1. Escherichia coli* strain M15[pREP4] (Stratagene) is transformed with the expression plasmid pQE30 (Qiagen) containing the full-length rat AZ1 ORF downstream of the 6x histidine tag. Endogenous AZ1 cDNA requires a +1 translational reading frameshift for expression; a modified ORF containing a single base deletion at the site of frameshifting is therefore used to artificially align the two reading frames. The cells are grown at 37° in 1.0 liter of LB supplemented with ampicillin (100 mg/liter) and kanamycin (50 mg/liter) to an $OD_{600}$ of 0.6 to 0.8. The culture is then induced with 1 m*M* isopropyl-$\beta$-D-thiogalactoside (IPTG) and allowed to grow for another 4 h. Cells are harvested by centrifugation at 4000*g* for 20 min. Cells are resuspended in 50 ml AZ lysis buffer (20 m*M* Tris, 300 m*M* NaCl, 5 m*M* 2-mercaptoethanol, and 10% glycerol, pH 8.2) with 0.75 mg/ml lysozyme and incubated at 4° for 30 min. The sample is subjected to two cycles of freezing at −80° and thawing at room temperature to ensure cell lysis. The lysate is cleared of cell debris via centrifugation at 12,000*g* for 30 min at 4°.

The clarified supernatant is incubated with 1.0 ml of Talon metal affinity resin (BD Bioscience) that has been equilibrated with AZ lysis buffer. Binding is allowed to proceed for 30 min at 4° with constant rocking. The resin is then batch washed by adding 20 volumes of AZ lysis buffer and centrifugation at 1000*g* for 3 min. This wash is repeated twice more, and the resin is then packed into a small volume column (~5 ml) and washed with 5 volumes of AZ lysis buffer supplemented with 10 m*M* imidazole. His6-AZ1 is eluted from the Talon resin with 5.0 ml AZ lysis buffer supplemented with 200 m*M* imidazole.

The eluate from the Talon resin is diluted with 2 volumes of AZ lysis buffer lacking NaCl to reduce the concentration of NaCl to 100 m*M* and is then supplemented with EDTA to 2.0 m*M* final concentration. Column fractionation is performed on an AKTA FPLC (Amersham Biosciences). The sample is loaded onto a 1.0-ml MonoQ anion-exchange column (Amersham Biosciences) preequilibrated in loading buffer (20 m*M* Tris, 100 m*M* NaCl, 5 m*M* 2-mercaptoethanol, 2 m*M* EDTA, and 10% glycerol, pH 8.2) and eluted via a 40 volume (40 ml) linear gradient of 100–500 m*M* NaCl. The AZ elutes as a sharp peak at ~280–310 m*M* NaCl. The yield is about 0.25–0.5 mg. Aliquots are frozen at –80°. Activity is stable for at least several months but may be lost on repeated freezing and thawing.

The activity of the AZ1 preparation can be tested in the aforementioned degradation assay using *in vitro*-translated labeled ODC as a substrate and reticulocyte lysate as a source of proteasomes.

*Metabolic Labeling of ODC. E. coli* strain M15[pREP4] (Stratagene) is transformed with the expression plasmid pQE30 (Qiagen) containing the full-length mouse ODC ORF downstream of the 6x histidine tag. The cells are grown in 3 ml of LB medium supplemented with ampicillin (100 mg/liter) and kanamycin (50 mg/liter) overnight at 37°. An 80-ml volume of the same prewarmed medium is inoculated with 1 ml of the overnight culture and is grown at 37° with vigorous shaking until an $OD_{600}$ of 0.4–0.6 is reached (2–3 h). The cells are then collected by centrifugation (1600*g*) at room temperature for 10 min and washed once with M9 medium, followed by resuspension in 40 ml of M9 medium containing 0.063% methionine assay medium (Fisher Scientific). After 30 min of incubation at 37°, protein expression is induced by the addition of IPTG (1 m*M*). Induction is carried out for 60 min, and [$^{35}$S]methionine (2.5 mCi) is added for 5 min, followed by unlabeled methionine (1 m*M*) for an additional 10 min. The cells are chilled on ice for 5 min and are then harvested by centrifugation at 1600*g* for 10 min at 4°. Cells are washed once by resuspension in 7.5 ml of cold 1× extraction/wash buffer (50 m*M* sodium phosphate, pH 7.0, 0.3 *M* NaCl). Lysozyme is added to a final concentration of 0.75 mg/ml, and the suspension is incubated at room temperature for 20 min. Cells are subsequently disrupted by sonication using 3× 10-s pulses with 30-s rests on ice between each burst. (*Note*: Great care should be taken during sonication steps to prevent dispersal of radioactive aerosols.) The cell extract is cleared by centrifugation at 10,000–12,000*g* at 4°. The resulting supernatant is transferred to 0.5 ml of Talon metal affinity resin (BD Bioscience) that has been equilibrated with 1× extraction/wash buffer. Binding is allowed to proceed for 30 min at 4° with constant rocking. The resin is then batch washed by adding 20 volumes of 1× extraction/wash buffer followed by centrifugation at 1000*g* for 3 min. This wash is repeated twice more. The resin is then packed into a small volume column (~5 ml) and washed with 5 volumes of 1× extraction/wash buffer. The labeled His6-mODC is then eluted from the Talon resin with 5.0 ml 1× extraction/wash buffer supplemented with 150 m*M* imidazole and 1 mg/ml bovine serum albumin (BSA), and 0.5 fractions are collected. The two fractions with highest radioactivity are pooled and concentrated to 100 μl during a buffer exchange to 50 m*M* Tris-HCl, pH 7.5, 10 m*M* KCl, 5 m*M* $MgCl_2$, 1 m*M* DTT, 1 m*M* ATP, and 10% glycerol using a column concentrator (Millipore, 30-kDa molecular mass cutoff). The specific radioactivity for $^{35}S$-labeled ODC is typically about $5 \times 10^4$ cpm/pmol. The labeled ODC can be stored at −80° for 1 month without loss of activity. The purity of the labeled proteins can be assessed by SDS–PAGE and radiography. Radiolabeled ODC should appear as a single band at 53 kDa. The quality of the labeled ODC can be tested by carrying out a degradation assay using rabbit reticulocyte lysate as a source of proteasomes before going on with the following purified system. Unlabeled ODC can be prepared similarly, omitting the labeling steps. The yield of unlabeled ODC is typically 10 mg/liter culture.

*Proteolysis Reactions.* Incubations are performed in a volume of 20 μl at 37° and contain 50 m*M* Tris-HCl, pH 7.5, 5 m*M* $MgCl_2$, 10 m*M* KCl, 10% glycerol, an ATP regenerating system (2 m*M* DTT, 10 m*M* creatine phosphate, 1.6 mg/ml creatine kinase), 2 mg/ml BSA, 50 n*M* rat proteasomes, about 2 μl $^{35}S$ ODC (about 10,000 cpm), and 2 μg AZ1 ($^{35}S$-labeled or cold). Reactions are preincubated for 10 min and initiated by the addition of proteasomes. Proteasome inhibitors, when present, are preincubated with all the other components of the reaction for 30 min at 37° before substrate is added. The incubation time for the degradation reaction is typically 30–40 min. The reaction is terminated by adding 140 μl of 20% ice-cold trichloroacetic acid. After incubation on ice for 10–30 min and microcentrifugation for 30 min at 14,000*g*, 150 μl of the supernatant is removed and subjected to scintillation counting to determine the amount of radiolabel converted from TCA-precipitable $^{35}S$-ODC to acid-soluble peptides. Total counts are obtained using water in place of trichloroacetic acid. The background of released counts obtained without proteasomes is usually about 0.5% of the total counts. The percentage degradation of labeled proteins is determined by percentage degradation = (released cpm-background cpm)/total cpm. By including unlabeled ODC in the reaction and using labeled ODC as a tracer, the initial velocity of degradation can be derived by v = [(percentage degradation) × (labeled + unlabeled substrate)]/incubation period. This method of measuring the peptide products of $^{35}$S-ODC is sensitive enough to accurately measure either AZ1-dependent or (the far less active) AZ1-independent degradation of ODC.

AZ1-stimulated degradation of labeled ODC can also be measured by SDS–PAGE and autoradiography. However, observing the disappearance of labeled substrate is much less sensitive than the production of radiolabeled acid-soluble peptides. Similarly, using this reconstituted system, the degradation of unlabeled ODC can be detected by Western blotting using His-tag or Flag-tag epitopes appended to the N terminus of ODC.

Usually in this reconstituted system, AZ1 will stimulate ODC degradation by at least fivefold. If this stimulation is not observed, then the activity of each component, i.e., AZ1, labeled ODC, and proteasome, should be checked as described earlier, using appropriate controls of specificity.

## Acknowledgment

This work was supported by USA National Institutes of Health Grant R01 GM–45335.

## References

Chen, H., Mac Donald, A., and Coffino, P. (2002). Structural elements of antizymes 1 and 2 required for proteasomal degradation of ornithine decarboxylase. *J. Biol. Chem.* **277,** 45957–45961.

Coffino, P. (2001). Regulation of cellular polyamines by antizyme. *Nature Rev. Mol. Cell. Biol.* **2,** 188–194.

Ghoda, L., Phillips, M. A., Bass, K. E., Wang, C. C., and Coffino, P. (1990). Trypanosome ornithine decarboxylase is stable because it lacks sequences found in the carboxyl terminus of the mouse enzyme which target the latter for intracellular degradation. *J. Biol. Chem.* **265,** 11823–11826.

Ghoda, L., Sidney, D., Macrae, M., and Coffino, P. (1992). Structural elements of ornithine decarboxylase required for intracellular degradation and polyamine-dependent regulation. *Mol. Cell. Biol.* **12,** 2178–2185.

Ghoda, L., van Daalen Wetters, T., Macrae, M., Ascherman, D., and Coffino, P. (1989). Prevention of rapid intracellular degradation of ODC by a carboxyl-terminal truncation. *Science* **243,** 1493–1495.

Glickman, M. H., Rubin, D. M., Fried, V. A., and Finley, D. (1998). The regulatory particle of the *Saccharomyces cerevisiae* proteasome. *Mol. Cell. Biol.* **18,** 3149–3162.

Gupta, M., and Coffino, P. (1985). Mouse ornithine decarboxylase: Complete amino acid sequence deduced from cDNA. *J.Biol. Chem.* **260,** 2941–2944.

Guthrie, C., and Fink, G. R. (1991). Guide to yeast genetics and molecular biology. *Methods Enzymol.* **194**.

Hoyt, M. A., Zhang, M., and Coffino, P. (2003). Ubiquitin-independent mechanisms of mouse ornithine decarboxylase degradation are conserved between mammalian and fungal cells. *J. Biol. Chem.* **278,** 12135–12143.

Ivanov, I. P., Matsufuji, S., Murakami, Y., Gesteland, R. F., and Atkins, J. F. (2000). Conservation of polyamine regulation by translational frameshifting from yeast to mammals. *EMBO J.* **19,** 1907–1917.

Leclerc, G. M., Boockfor, F. R., Faught, W. J., and Frawley, L. S. (2000). Development of a destabilized firefly luciferase enzyme for measurement of gene expression. *Biotechniques* **29,** 590–591.

Li, X., and Coffino, P. (1994). Distinct domains of antizyme required for binding and proteolysis of ornithine decarboxylase. *Mol. Cell. Biol.* **14,** 87–92.

Li, X., Zhao, X., Fang, Y., Jiang, X., Duong, T., Fan, C., Huang, C. C., and Kain, S. R. (1998). Generation of destabilized green fluorescent protein as a transcription reporter. *J. Biol. Chem.* **273,** 34970–34975.

Loetscher, P., Pratt, G., and Rechsteiner, M. (1991). The C terminus of mouse ornithine decarboxylase confers rapid degradation on dihydrofolate reductase. *J. Biol. Chem.* **266,** 11213–11220.

Matsufuji, S., Matsufuji, T., Miyazaki, Y., Murakami, Y., Atkins, J. F., Gesteland, R. F., and Hayashi, S. I. (1995). Autoregulatory frameshifting in decoding mammalian ornithine decarboxylase antizyme. *Cell* **80,** 51–60.

Mumberg, D., Muller, R., and Funk, M. (1995). Yeast vectors for the controlled expression of heterologous proteins in different genetic backgrounds. *Gene* **156,** 119–122.

Newman, R. M., Mobascher, A., Mangold, U., Koike, C., Diah, S., Schmidt, M., Finley, D., and Zetter, B. R. (2004). Antizyme targets cyclin D1 for degradation: A novel mechanism for cell growth repression. *J. Biol. Chem.* **279,** 41504–41511.

Reidlinger, J., Pike, A. M., Savory, P. J., Murray, R. Z., and Rivett, A. J. (1997). Catalytic properties of 26S and 20S proteasomes and radiolabeling of MB1, LMP7, and C7 subunits associated with trypsin-like and chymotrypsin-like activities. *J. Biol. Chem.* **272,** 24899–24905.

Suzuki, T., and Varshavsky, A. (1999). Degradation signals in the lysine-asparagine sequence space. *EMBO J.* **18,** 6017–6026.

van Daalen Wetters, T., Macrae, M., Brabant, M., Sittler, A., and Coffino, P. (1989). Polyamine-mediated regulation of mouse ornithine decarboxylase is posttranslational. *Mol. Cell. Biol.* **9,** 5484–5490.

Verma, R., Peters, N. R., D'Onofrio, M., Tochtrop, G. P., Sakamoto, K. M., Varadan, R., Zhang, M., Coffino, P., Fushman, D., Deshaies, R. J., and King, R. W. (2004). Ubistatins inhibit proteasome-dependent degradation by binding the ubiquitin chain. *Science* **306,** 117–120.

Zhang, M., and Coffino, P. (2004). Repeat sequence of Epstein–Barr virus EBNA1 protein interrupts proteasome substrate processing. *J. Biol. Chem.* **279,** 8635–8641.

Zhang, M., Pickart, C. M., and Coffino, P. (2003). Determinants of proteasome recognition of ornithine decarboxylase, a ubiquitin-independent substrate. *EMBO J.* **22,** 1488–1496.

Zhu, C., Lang, D. W., and Coffino, P. (1999). Antizyme2 is a negative regulator of ornithine decarboxylase and polyamine transport. *J. Biol. Chem.* **274,** 26425–26430.

# [34] Atomic Force Microscopy of the Proteasome

*By* PAWEL A. OSMULSKI and MARIA GACZYNSKA

## Abstract

The proteasome should be an ideal molecule for studies on large enzymatic complexes, given its multisubunit and modular structure, compartmentalized design, numerous activities, and its own means of regulation. Considering the recent increased interest in the ubiquitin-proteasome pathway, it is surprising that biophysical approaches to study this enzymatic assembly are applied with limited frequency. Methods including atomic force microscopy, fluorescence spectroscopy, surface plasmon resonance, and high-pressure procedures all have gained popularity in characterization of the proteasome. These methods provide significant and often unexpected insight regarding the structure and function of the enzyme. This chapter describes the use of atomic force microscopy for dynamic structural studies of the proteasome.

## Introduction

The size of proteasomal assemblies and their degree of complication make them desirable but often challenging objects for biophysical studies. One such challenge stems from the fact that not only is the proteasome built from modules, but every module is also a multisubunit protein complex (Zwickl *et al.*, 2001). The 19S cap or 11S activator attaches to the $\alpha$ face of the eukaryotic catalytic 20S core. A gating mechanism guarding access to the catalytic chamber of the core, missing in the archaebacterial 20S proteasome, is also formed by the $\alpha$ ring (Forster *et al.*, 2003; Groll and Huber, 2003). The crystal structures of eukaryotic and archaebacterial core particles of the human activator complex and a structure of a hybrid core-activator complex have been solved; however, the most physiologically relevant 26S assembly continues to elude successful X-ray analysis (Groll *et al.*, 1997; Lowe *et al.*, 1995; Whitby *et al.*, 2000; Unno *et al.*, 2002).

It is reasonable to predict that giant protein assemblies such as the proteasomes exhibit a significant degree of structural dynamics, essential for their biological activity. The rotational symmetry of the proteasome resembles that of membrane channels or chaperonins. It has been established both experimentally with cation channels and theoretically using Monte Carlo simulations that multiprotomeric proteins exhibiting rotational symmetry demonstrate rapid, synchronized transitions between

METHODS IN ENZYMOLOGY, VOL. 398

0076-6879/05 $35.00
DOI: 10.1016/S0076-6879(05)98034-8

conformations of all of their protomers (Duke *et al.*, 2001). Apparently, the proteasome also exhibits these stability-promoting transitions, although their amplitude is not as impressive as in the chaperonins (Horovitz *et al.*, 2001). Analysis of crystal structures, despite all of the fundamental knowledge it offers, often falls short of providing a wealth of dynamic data. Although the giant proteasome remains beyond the reach of the nuclear magnetic resonance (NMR) technique, it is a convenient object for atomic force microscopy because its size is actually an advantage and its global conformation changes can be monitored relatively easily. Fluorescence spectroscopy and high-pressure methods can also be used successfully with the proteasome. It is hoped that future research will see increased use of biophysical methods that are helpful to understanding the work of a biological nanodevice called the proteasome.

## Atomic Force Microscopy: An Overview of the Method

The atomic force microscope (AFM) was invented in 1986, the same year one of its inventors, Gerd Binnig, was awarded the Nobel Prize in Physics for his development of the scanning tunneling microscope (STM) in the early 1980s (Binnig, 1982, 1986). STM and AFM belong to a group of techniques currently growing in popularity, collectively known as scanning probe microscopy (SPM). Scanning probe microscopy is not strictly microscopy: it does not use lenses, and radiation is not transmitted through or reflected from the sample. Consequently, the wavelength of the radiation does not limit the resolution of the method. Instead, an SPM probe, which consists of a small and very sharp tip mounted on a cantilever, "feels" the surface of the sample. In the case of AFM, mechanical properties of the cantilever restrict the imaging accuracy. Deflection of a cantilever depends on the proximity of the sample and thus reflects the sample topography. Locking a sample/tip positioning system and a piezo element in a feedback loop facilitates constant corrections of the tip-sample distance in a $z$ (vertical) direction and allows for the creation of an image of the surface scanned in an $x$–$y$ plane (Fotiadis *et al.*, 2002; Yang *et al.*, 2003). The changes in tip position are monitored in modern AFMs by a laser beam deflection method (Fig. 1). What makes AFM so attractive for biological applications is that the sample can be immersed in liquid during scanning.

Atomic force microscopy has three major modes of operation: contact, oscillating, and force. In the contact, or DC mode ("deflection of cantilever" or "deflection change"), the tip and the atoms of the sample are in direct contact, causing the cantilever to deflect. The DC mode is rarely used with soft biological objects due to a potentially damaging shear force.

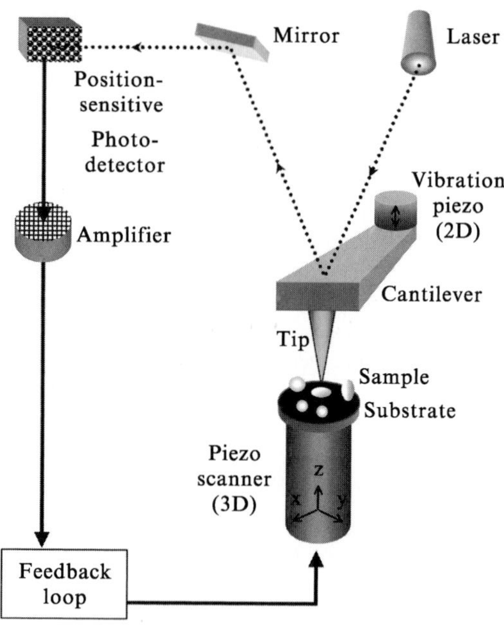


FIG. 1. Schematics of AFM in the oscillating mode with a laser beam deflection detection method.

In the oscillating mode (AC, "amplitude of cantilever" or "amplitude change"), the probe vibrates vertically while scanning. The proximity of the atoms of the sample causes changes in the amplitude of oscillations due to van der Waals and electrostatic forces. These vibrations are activated acoustically ("tapping mode") or magnetically. The oscillating mode is well suited for imaging biological objects because the tip touches the sample only very briefly and with a low force (Hansma *et al.*, 1994). Both contact and oscillating modes produce topography images with information about the height of the scanned objects, allowing the three-dimensional map of the objects to be reconstructed and analyzed. The third AFM mode of operation, the force mode, also called force spectroscopy, does not provide an image of the sample. In this mode, the strength of the interaction between the tip and the sample is measured in the order of piconewtons (Bustamante *et al.*, 1997; Santos and Castanho, 2004). For biological applications, the tip is usually modified with a receptor-specific ligand.

For successful imaging or force curve plotting, the sample must be immobilized on an atomically flat surface ("AFM substrate"). Muscovite mica is used as a substrate for most bioapplications, as a clean, flat, and negatively charged surface is obtained easily by peeling this layered mineral. Most proteins in physiological pH have enough surface positive charges to bind to mica by electrostatic forces, which are gentle enough to leave the protein in a native state and strong enough to keep it in place during oscillation mode scanning. Alternatively, the molecules can be attached to a modified surface of a substrate by cross-linking, His-tag binding, affinity binding, or embedding in a prepared lipid bilayer (see later).

Atomic force microscopy samples can be imaged either in air ("dry mode") or in liquid ("liquid" or "wet" mode). For the dry mode, which is often used with nucleic acids, a small droplet is deposited on the substrate, washed with water after a few minutes, and dried in a stream of clean air or nitrogen. For "wet" imaging, both the probe and the sample are immersed in a liquid during scanning and permitting the addition or removal of ligands. Although sharpness of the images is somewhat compromised in liquid, it is still possible to achieve the practical lateral resolution of 1–2 nm and vertical resolution of 0.1–0.2 nm. With such a resolution, proteins up to 50 kDa will generally be imaged as rods or granules without additional structural features recognizable. Larger proteins may show some structural features, but it takes multisubunit protein complexes to enjoy the full power of AFM imaging. The unique advantage of the liquid mode is studying biologically active, native molecule. One limitation of AFM imaging is that a standard tip of 10 nm diameter at the very end will obviously not penetrate deep channels in protein assemblies, imaging holes as craters or shallow depressions. The temporal resolution of AFM reaches milliseconds for force measurements. When scanning, a tip needs milliseconds to run across a molecule (one-dimensional imaging). A two-dimensional image of a single molecule is created in a time scale of seconds to minutes. Examples of some studies on native proteins include GroEL/GroES chaperone, ion pumps and membrane channels, antibodies, ATP synthases, toxins, and photosystem proteins, in addition to others (Czajkowsky *et al.*, 2000; Ehrenhofer *et al.*, 1997; Fotiadis *et al.*, 2002; Viani *et al.*, 2000; Vie *et al.*, 2001; Yang *et al.*, 2003).

## Imaging of Proteasomes by Tapping Mode AFM in Liquid

Proteasomal assemblies, with their large sizes and easily recognizable shapes, are perfect subjects for AFM imaging. The major advantage of AFM is the possibility to observe biologically active molecules in real time with lateral resolution close to electron microscopy and with excellent vertical resolution. The limitations of AFM technique include (i) lateral resolution, poorer than in X-ray crystallography or NMR spectroscopy;

(ii) temporal resolution of standard topography imaging in the range of seconds and minutes; fortunately, the force mode does not suffer from this limitation; and (iii) inability to explore the molecule beyond its surface. Nevertheless, the relative ease of imaging and the availability of a dynamic aspect make the AFM a very valuable technique for studies on the proteasome. The method to image randomly dispersed, electrostatically attached to mica 20S particles is described later. Such particles are perfect subjects for studies of conformational dynamics.

*Equipment and Supplies*

We use a Nanoscope IIIa microscope (Digital Instruments) with an E scanner and a glass cantilever holder for scanning in fluids ("liquid cell," "wet chamber"). The Nanoscope IIIa uses acoustically activated vibrations. Molecular Imaging Corp. offers similar microscopes with magnetically activated vibrations. Muscovite mica glued with a two-component epoxy resin (available from hardware or craft stores) to a steel disk serves as the AFM substrate. The steel disks and the mica in sheets or ready-to-use disks are available from electron microscopy suppliers. If the mica is purchased as inexpensive sheets, it can be cut with scissors into squares of roughly the size of steel disks, or mica disks can be punched out from the sheet with a large-diameter hole puncher. The steel disks with mica should be prepared in advance. It is convenient to glue several dozens of mica pieces at a time, wait for at least 24 h for the resin to cure, and store the ready-to-use substrates in a dust-free container. The fresh surface of a mica should be exposed just before the experiment, no more than about a minute before deposition of a protein solution. Removal of an external layer of mica can be achieved by peeling it out with a sticky tape (Scotch-type, pressure-sensitive). If the exposed surface is cracked, the next layer should be peeled out. The substrate can be reused for as long as a fresh surface of mica can be exposed. When the mica is used up, the steel disk should be cleaned with acetone to remove remaining glue before attaching another piece of mica.

For imaging proteasomes in the tapping mode in liquid, we use oxide-sharpened silicon probes (NP-ST or NP-STT) on cantilevers with a nominal spring constant of 0.32 N/m (Digital Instruments). Other probes of similar parameters include OTR8 and ORC8 (Olympus) and CSC (MicroMash). Not all tips on probes are of equal quality, which is apparent when imaging an AFM standard grid or any sample of known features. It is advisable to discard blunt tips and to use only the sharpest for the best results. Standard grids, or calibration gratings (several kinds are available from MicroMash), are usually made from silicon and represent arrays of steps, tips, or pillars of well-defined, micrometer-scale dimensions. They are used for tip evaluation and for calibration of piezoelectric scanners.

*Sample Preparation and Scanning*

For AFM imaging, purified proteasomes are diluted in 5 m*M* Tris-HCl buffer, pH 7.0 (imaging buffer), to give a nanomolar concentration. A 50 m*M* buffer can be used with similar results. It is advisable to filter all buffers with a 0.2-$\mu$m filter and to perform all manipulations in a clean air hood (PCR hood) to avoid contamination with dust particles. The protein preparation has to be as pure as possible, as any contaminant protein will be imaged as well as the proteasome. A 2-$\mu$l droplet of the diluted proteasomes is deposited on a freshly cleaved mica surface. After a few minutes of binding, but before drying of the droplet, the sample should be overlaid with about 30 $\mu$l of the imaging buffer and mounted in the head of the microscope. Additional buffer, enough to fill a cell, is needed when a liquid cell is used with O rings to seal off the liquid and to enable an easy exchange of buffers. Resonant frequency of the tip is manually tuned to 9–10 kHz, with the best results obtained for frequency slightly below the actual resonant peak. The amplitude of the tip ranges from 200 to 500 mV and set point ranges from 1.4 to 2 V. The higher the set point, the lower the force of tapping. Integral gain should be set at the highest value not causing distortion of the image ("ringing"), usually 0.2–0.4. Fields of 1 $\mu m^2$ or smaller are scanned at rates of 2–3 Hz with the highest available resolution of 512 $\times$ 512 pixels. It is advisable to collect both trace and retrace images (the probe moving left to right or right to left while scanning), as sometimes objects blurred in the image created in one direction will be acceptable in the other image. Ideally, the proteasome particles should be abundantly present in a scanned field, but should not touch each other. If desired, a monomolecular layer of proteasomes can be obtained by increasing the concentration of sample deposited on a mica. The proteasomes are stable during subsequent scans; they do not change orientation and location. The lateral drift reaches only about 10 nm per scan and it is possible to image the same particle for more than an hour. Ligands (substrates, inhibitors) or extra buffer can be added to the sample with a pipette through openings in the cantilever holder, even if O rings are not used. The ligands are not detected by the probe unless they are large enough (at least in the range of a few thousands daltons) and immobilized on the surface of a studied particle or, rather unlikely, on the surface of mica (Fig. 2B).

After prolonged scanning, the quality of the image may decline. This will be most likely the effect of a tip losing its sharpness due to wearing down or contamination. A worn-down tip should be discarded. A contaminated tip

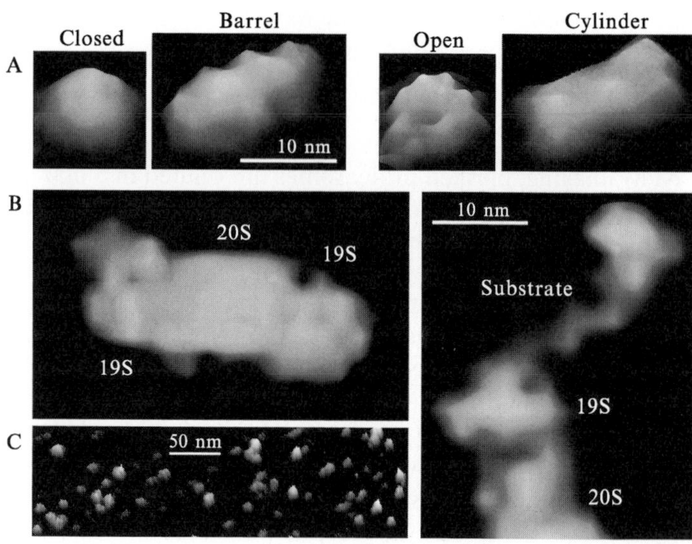


FIG. 2. Examples of AFM images of proteasomes. (A) Side plots (tilted topography images) of 20S proteasomes from fission yeast (*Schizosaccharomyces pombe*). (Left to right) Top-view particle in the closed conformation, side-view particle in the barrel conformation, open particle in the top view, and side-view cylinder. (B) Images of the human 26S proteasomes. (Left) A typical particle with two caps and the 20S core recognizable. (Right) A particle touching, presumably bound to, the 19S cap is ubiquitinated lysozyme, which was added to the mica-attached 26S proteasomes. The ubiquitinated protein appears unusually large, most likely because it was not attached stably to the mica and shifted during scanning. (C) A side plot of a fragment of a field of 20S proteasomes from *Saccharomyces cerevisiae*. The particles were distributed randomly on a mica substrate. The lightest shades of gray correspond to the highest-positioned areas.

can be cleaned by rinsing in ultrapure water or by irradiation with ultraviolet light for a few hours; however, the cleaning is not always successful. An obvious sign of tip contamination is a "double image": all topographical features are repeated in regular intervals. This is the result of a contaminating particle attached to the AFM tip and acting as a second tip. This "second tip" is often removable by "shaking off" the contaminant during several seconds of a fast scanning (at least 4 Hz) without engagement with the sample.

Buffers with high ionic strength are sometimes used to improve imaging due to a reduction of electrostatic forces between a tip and a sample. Addition of NaCl up to 200 m*M* during scanning, however, does not influence the results for proteasomes (Osmulski and Gaczynska, 2000).

*Image Processing*

Each company producing STM microscopes also supplies specialized software capable of collecting and, to varying degrees, manipulating and analyzing data. There are also numerous stand-alone software packages for reading data in several formats, processing the images, performing feature analysis (e.g., contour recognition, particle counting, profiling). and preparing images for publication. None of them is perfect in all of the aforementioned aspects, and the reader is strongly encouraged to test several sources to find the software that fits the particular needs. Although the existence of multiple data formats does not help this endeavor, most provide free trials and there are several good packages in the public domain. The most comprehensive collection of the imaging software with short descriptions of their capabilities is provided by the following Web pages:

http://www.microscopy.info (MC Services)
http://www.weizmann.ac.il/Chemical Research Support/surflab/peter/software/index.html (Weizmann Institute of Science, Israel)
http://web.mit.edu/cortiz/www/Software.html (Ortiz Laboratory at MIT, Cambridge, MA)
http://www.uksaf.org/software.html (UK Surface Analysis Forum)

These sites also list ample academic and commercial links to groups and companies working with STM images. From our perspective, two packages are the best for both the beginner and the advanced user: Image SXM by Steve Barrett (http://reg.ssci.liv.ac.uk) and ImageJ (http://rsb.info.nih.gov/ij/). They are available in most platforms, simple to use, comprehensive in their capabilities, provide an open source code, and are freely available.

The topographical images are presented in false color scale, where a particular color or shade corresponds to a height of the sample. The amber and gray-scale color palettes are the most popular. Tilted images (side plots; see Fig. 2) are often presented to better show particular features of objects. The collected images often require processing, as they show a certain level of tilting of a background surface and electronic and thermal noise. The most pronounced but easy to remove artifacts arise from peculiar properties of a piezo element and shape of the tip. For this purpose, a standard image processing consisting of flattening and plain fit provided by the Nanoscope software is applied. The lateral dimensions of particles are measured with the help of a cross-section option (Nanoscope, others). Because the tip cannot reproduce steep banks of small particles well, it is necessary to consider that the dimensions will be enlarged due to a tip-broadening effect. This effect can be removed conveniently and additional processing performed with software such as Image SXM. Section analysis in Nanoscope software, SXM Image, and others used for measuring particles.

## The Proteasome under AFM: Observations

Proteasome molecules imaged by the aforementioned method revealed a high level of dynamics when imaged with the practical resolution of 1–2 nm (Fig. 2). After analysis of AFM images performed under different conditions, it was possible to correlate the dynamic behavior of the proteasomes with their catalytic actions, to monitor a ligand binding, and to develop a model of allosteric signaling between the active centers and the gate (Gaczynska *et al.*, 2003; Osmulski and Gaczynska, 2000, 2002). The $\alpha$ rings of particles in the top-view orientation were either cone shaped or had a crater-like cavity in the middle, positioned where the entrance to the central channel was expected to be guarded by the gate. Therefore, we named the conformants "closed" and "open," respectively, hypothesizing that they represent the closed and open gate to the proteasomal central channel (Fig. 2A; Osmulski and Gaczynska, 2000). This hypothesis was supported by the observation that images of top-view proteasomes from *Thermoplasma acidophilum*, which do not have a gate in the $\alpha$ ring, showed exclusively open particles (P. A. Osmulski *et al.*, submitted; proteasomes kindly provided by A. L. Goldberg). In contrast, when eukaryotic proteasomes were analyzed, both the closed and the open conformants were present in all scans, and all of the particles were able to switch conformations between subsequent scans. Addition of peptide or protein substrates changed the partition between closed and open forms (Osmulski and Gaczynska, 2000). The conformational changes were observed in side-view proteasomes as well, with a partition between barrel-shaped and cylinder-shaped molecules closely following the partition between closed and open conformants (Osmulski and Gaczynska, 2002). These results prompted us to hypothesize that a "closed barrel" and an "open cylinder" represent allosteric forms analogous to R (relaxed) and T (tense) forms in a classical two-state model of allosteric transitions. We have found that gate opening observed under AFM is coupled with a tetrahedral transition state in any of the active centers. Interestingly, kinetic parameters calculated on the basis of the open/closed partition as a function of substrate concentration were in perfect agreement with the parameters obtained in the traditional "macroscopic" way, with the increase in a product concentration measured instead of the abundance of conformers (P. A. Osmulski *et al.*, submitted). Core particles isolated from human and mouse cultured cells, rabbit muscle, *Saccharomyces cerevisiae*, or *Schizosaccharomyces pombe* were undistinguishable in AFM images and revealed the same dynamic behavior (Gaczynska *et al.*, 2003; P. A. Osmulski and M. Gaczynska, unpublished observations). Analysis of AFM images is very useful for the characterization of ligand binding to the proteasomes. This was shown on the example of proline- and arginine-rich peptides (PR peptides; see Gaczynska and Osmulski, 2005). These peptides bind to the 20S core particles, presumably to the $\alpha$ ring. AFM images revealed that the 20S proteasome treated with the peptides was unable to switch between open and closed conformers and remained in a quasi-open form permanently (Gaczynska *et al.*, 2003).

The 20S core particles are not the only proteasomal assemblies imaged under AFM. Studies of the interactions of human 26S complexes with PR peptides showed dramatic conformational changes in the inhibitor-treated proteasomes (Gaczynska *et al.*, 2003). The yeast 26S assemblies (Fig. 2B) and human proteasomes complexed with the 11S activator or the PI31 inhibitor (kindly provided by G. DeMartino) were visualized as well (P. A. Osmulski and M. Gaczynska, unpublished observations). The power of AFM, however, extends well beyond simple topography imaging, and more applications are waiting to be implemented in proteasomal studies. The force measurements, for example, will be very useful to study binding of the ubiquitinated substrates to 19S cap, their defolding, and translocation to the core particle.

In addition to the aforementioned method of imaging electrostatically attached particles, several studies have explored the immobilized and organized 20S complexes under AFM. For example, proteasomes from archaeon *T. acidophilum* with histidine tags introduced at the C terminus of the $\alpha$ subunit were specifically attached to a mica-supported, ultraflat chelator lipid membrane (Dorn *et al.*, 1999). As a result, images were obtained of a dense layer of the 20S particles oriented in a side-view position. Interactions of the core particles with different supported lipid bilayers were explored for bovine proteasomes (Furuike *et al.*, 2003). In another study, contact mode AFM was used to successfully image the shear-resistant two-dimensional crystals of *T. acidophilum* proteasomes (Thess *et al.*, 2002). The 20S proteasomes $His_6$-tagged at the N termini of $\alpha$ subunits crystallized readily on a nickel-chelating lipid bilayer prepared on mica and optimized their packing in a crystal by interlocking both laterally and vertically (Thess *et al.*, 2002). The immobilized and organized proteasomes constitute very useful research objects: however, the dynamic information may be, in most parts, lost.

## Acknowledgments

The work was funded by R01 GM069819 (M.G.) and San Antonio Cancer Institute grants (P. A. O., M. G.).

## References

Binnig, G., Quate, C. F., and Gerber, C. H. (1986). Atomic force microscope. *Phys. Rev. Lett.* **56,** 930–933.

Binnig, G., and Rohrer, H. (1982). Scanning tunneling microscopy. *Helv. Phys. Acta* **55,** 726–735.

Bustamante, C., Rivetti, C., and Keller, D. J. (1997). Scanning force microscopy under aqueous solutions. *Curr. Opinion Struct. Biol.* **7,** 709–716.

Czajkowsky, D. M., Iwamoto, H., and Shao, Z. (2000). Atomic force microscopy in structural biology: From the subcellular to the submolecular. *J. Electr. Micr.* **49,** 395–406.

Dorn, I. T., Eschrich, R., Seemuller, E., Guckenberger, R., and Tampe, R. (1999). High-resolution AFM-imaging and mechanistic analysis of the 20 S proteasome. *J. Mol. Biol.* **288,** 1027–1036.

Duke, T. A., Novere, N. L., and Bray, D. (2001). Conformational spread in a ring of proteins: A stochastic approach to allostery. *J. Mol. Biol.* **308,** 541–553.

Ehrenhofer, U., Rakowska, A., Schneider, S. W., Schwab, A., and Oberleithner, H. (1997). The atomic force microscope detects ATP-sensitive protein clusters in the plasma membrane of transformed MDCK cells. *Cell Biol. Int.* **21,** 737–746.

Forster, A., Whitby, F. G., and Hill, C. P. (2003). The pore of activated 20S proteasomes has an ordered 7-fold symmetric conformation. *EMBO J.* **22,** 4356–4364.

Fotiadis, D., Scheuring, S., Muller, S. A., Engel, A., and Muller, D. J. (2002). Imaging and manipulation of biological structures with the AFM. *Micron* **33,** 385–397.

Furuike, S., Hirokawa, J., Yamada, S., and Yamazaki, M. (2003). Atomic force microscopy studies of interaction of the 20S proteasome with supported lipid bilayers. *Biochim. Biophys. Acta* **1615,** 1–6.

Gaczynska, M., and Osmulski, P. A. (2005). Characterization of noncompetitive regulators of proteasome activity. *Methods Enzymol.* **398**[35] 2005 (this volume).

Gaczynska, M., Osmulski, P. A., Gao, Y., Post, M. J., and Simons, M. (2003). Proline- and arginine-rich peptides constitute a novel class of allosteric inhibitors of proteasome activity. *Biochemistry* **42,** 8663–8670.

Groll, M., and Huber, R. (2003). Substrate access and processing by the 20S proteasome core particle. *Int. J. Bioch. Cell Biol.* **35,** 606–616.

Groll, M., Ditzel, L., Lowe, J., Stock, D., Bochtler, M., Bartunik, H. D., and Huber, R. (1997). Structure of 20S proteasome from yeast at 2.4 A resolution. *Nature* **386,** 463–471.

Hansma, P. K., Cleveland, J. P., Radmacher, M., Walters, D. A., Hillner, P. E., Bezanilla, M., Fritz, M., Vie, D., Hansma, H. G., Prater, C. B., Massie, J., Fukunaga, L., Gurley, J., and Elings, V. (1994). Tapping mode atomic force microscopy in liquids. *Appl. Phys. Lett.* **64,** 1738–1740.

Horovitz, A., Fridmann, Y., Kafri, G., and Yifrach, O. (2001). Review: Allostery in chaperonins. *J. Struct. Biol.* **135,** 104–114.

Lowe, J., Stock, D., Jap, B., Zwickl, P., Baumeister, W., and Huber, R. (1995). Crystal structure of the 20S proteasome from the archaeon *T. acidophilum* at 3.4 A resolution. *Science* **268,** 533–539.

Osmulski, P. A., and Gaczynska, M. (2000). Atomic force microscopy reveals two conformations of the 20S proteasome from fission yeast. *J. Biol. Chem.* **275,** 13171–13174.

Osmulski, P. A., and Gaczynska, M. (2002). Nanoenzymology of the 20S proteasome: Proteasomal actions are controlled by the allosteric transition. *Biochemistry* **41,** 7047–7053.

Santos, N. C., and Castanho, M. A. (2004). An overview of the biophysical applications of atomic force microscopy. *Biophys. Chem.* **107,** 133–149.

Thess, A., Hutschenreiter, S., Hofmann, M., Tampe, R., Baumeister, W., and Guckenberger, R. (2002). Specific orientation and two-dimensional crystallization of the proteasome at metal-chelating lipid interfaces. *J. Biol. Chem.* **277,** 36321–36328.

Unno, M., Mizushima, T., Morimoto, Y., Tomisugi, Y., Tanaka, K., Yasuoka, N., and Tsukihara, T. (2002). Structure determination of the constitutive 20S proteasome from bovine liver at 2.75 A resolution. *J. Biochem.* **131,** 171–173.

Viani, M. B., Pietrasanta, L. I., Thompson, J. B., Chand, A., Gebeshuber, I. C., Kindt, J. H., Richter, M., Hansma, H. G., and Hansma, P. K. (2000). Probing protein-protein interactions in real time. *Nature Struct. Biol.* **7,** 644–647.

Vie, V., Van Mau, N., Pomarede, P., Dance, C., Schwartz, J. L., Laprade, R., Frutos, R., Rang, C., Masson, L., Heitz, F., and Le Grimellec, C. (2001). Lipid-induced pore formation of the Bacillus thuringiensis Cry1Aa insecticidal toxin. *J. Membr. Biol.* **180,** 195–203.

Whitby, F. G., Masters, E. I., Kramer, L., Knowlton, J. R., Yao, Y., Wang, C. C., and Hill, C. P. (2000). Structural basis for the activation of 20S proteasomes by 11S regulators. *Nature* **408,** 115–120.

Yang, Y., Wang, H., and Erie, D. A. (2003). Quantitative characterization of biomolecular assemblies and interactions using atomic force microscopy. *Methods* **29,** 175–187.

Zwickl, P., Seemuller, E., Kapelari, B., and Baumeister, W. (2001). The proteasome: A supramolecular assembly designed for controlled proteolysis. *Adv. Prot. Chem.* **59,** 187–222.

# [35] Characterization of Noncompetitive Regulators of Proteasome Activity

*By* Maria Gaczynska and Pawel A. Osmulski

## Abstract

The success of bortezomib, a competitive proteasome inhibitor and a drug approved to treat multiple myeloma, spurred interest in compounds targeting catalytic sites of the enzyme. The aim of this chapter, however, is to focus attention on the small molecule, natural or synthetic compounds binding far away from the catalytic centers, yet modifying the performance of the proteasome. Defining allostery broadly as any kind of ligand-induced, long-distance transfer of conformational signals within a molecule, most such compounds are allosteric effectors capable of regulating the proteasome *in vitro* and *in vivo* in a manner more diverse and precise than competitive inhibitors. Proline- and arginine-rich peptides (PR peptides) are examples of such compounds and are currently being considered as

METHODS IN ENZYMOLOGY, VOL. 398

0076-6879/05 $35.00
DOI: 10.1016/S0076-6879(05)98035-X

potential drugs with anti-inflammatory and proangiogenic activities. This chapter describes a set of methods useful for characterizing the effects of such inhibitors on the proteasome.

## Introduction

Small molecule inhibitors may affect the activity of the proteasome in two basic ways: by competing with substrates for binding to one or more catalytic sites or by binding to other regulatory sites in the protein. In the latter case the compounds are known collectively as noncompetitive inhibitors. The competitive inhibitors have been studied extensively, have been utilized broadly, and are fairly well understood. For example, bortezomib (Velcade, PS-341), a boronic peptide derivative approved for the treatment of multiple myeloma, is a competitive inhibitor (Adams, 2004). MG-232 and epoxomycin are examples of competitive inhibitors commonly used to study the effects of proteasomal inhibition *in vitro* or *in vivo* (for a review, see Gaczynska and Osmulski, 2002). In contrast, noncompetitive inhibitors are utilized much less frequently and are not well understood. The antiproteasomal activities of all such currently known compounds were discovered by accident, as a surprising addition to actions targeting other proteins, usually bacterial or viral (Gaczynska and Osmulski, 2002). Despite their serendipitous discovery, for reasons outlined later, noncompetitive inhibitors are definitely worthy of attention, especially those that interact with the giant protease in an allosteric manner. The proteasome, with its complex structure and multiple activities, invites questions about the extent and role of allosteric interactions, defined here as any type of long-range, ligand-induced conformational signals, which modulate the functional properties of the protein. Briefly, the proteasome is constructed from multisubunit modules: 20S core, 19S caps, and 11S activator (Zwickl *et al.*, 2001). The tube-shaped proteolytic core module contains three pairs of active centers in the catalytic chamber, formed by two $\beta$ rings (Groll *et al.*, 1997). The centers are named after their specificities: chymotrypsin-like (ChT-L), trypsin-like (T-L), or peptidylglutamyl peptide hydrolyzing (PGPH, postacidic, caspase-like). The entrance to the chamber is guarded by two external $\alpha$ rings with gates and attachment sites for additional modules. Currently, allosteric interactions are suspected between the following elements of proteasomal complexes:

- Activator complex (11S) and catalytic centers of the 20S core (Li and Rechsteiner, 2001)
- Postulated regulatory sites in the 20S proteasome and catalytic centers, mediated by substrates or mixed action inhibitor ritonavir

(Kisselev *et al.*, 2002, 2003; Myung *et al.*, 2001; Schmidtke *et al.*, 2000)

Distant sites on the $\alpha$ ring and the gate, mediated by PR peptides (Gaczynska *et al.*, 2003)

Catalytic centers and the gate, mediated by substrates (Osmulski and Gaczynska, 2002)

The mechanism for and role of these transitions are far from understood. Nevertheless, the diversity and the precision of allosteric effects are worth noting: regulation of a particular endopeptidase activity, changes in product size and pattern, promoting the uptake of substrates, and blocking attachment of the modules. Moreover, in contrast to all of the competitive inhibitors that influence the 20S, 26S (19S–20S–19S), or activated proteasomes by just blocking the active centers, allosteric regulators may differentiate between distinct forms of the giant protease. All of these effects ultimately facilitate more thorough control of proteasome performance, both *in vitro* and *in vivo*, compared to the competitive blocking of one or more active centers. Although the notion of harnessing the power of proteasomal allostery may be perceived as far-fetched, it is worth remembering that not long ago the proteasome was considered too deeply involved in all aspects of cell life to allow for any kind of specific therapeutic intervention. While not minimizing the significance of specific competitive inhibitors such as PS-341, it is hoped that noncompetitive allosteric effectors will join them in the mainstream of proteasome-targeting compounds.

Methods useful for characterizing noncompetitive inhibitors of the proteasome are outlined, with special emphasis on approaches aimed at allosteric regulators.

## Materials

For *in vitro* testing of the potential regulator, pure proteasomes of defined assembly status (20S, 26S or activated) are needed. Yeast (*Saccharomyces cerevisiae*) or human cultured cells and sometimes mammalian tissues (rabbit muscle, bovine blood, and others) are the most popular sources. There are several advantages to using purified yeast 20S proteasomes instead of, or in addition to, mammalian enzyme. First, preparation of the enzyme from yeast culture is inexpensive. Second, yeast does not have immunosubunits and activators so it is easier to obtain a homogeneous proteasome preparation from yeast than from mammalian sources. Third, if the compound of interest affects both human and yeast enzymes in similar ways, the compound can be considered a universal modulator of proteasomal activities. Whatever the source of the proteasome, differential centrifugation combined with gel-filtration and anion-exchange chromatographies are usually employed with good results to obtain pure 20S and 26S proteasomes. Affinity chromatography or native polyacrylamide gel electrophoresis can be used to separate distinct proteasomal assemblies from each other (see other chapters in this volume). The 26S proteasome should always be handled in the presence of ATP, $Mg^{2+}$, and dithiothreitol, typically used in concentrations of a few millimoles to stabilize the assembly. When the 20S proteasome is purified, care should be taken to obtain a latent enzyme, as putative allosteric transitions can be abolished or compromised in the already activated core particle. The addition of glycerol to 20% of the final concentration in the storage buffer (e.g., 50 m*M* Tris-HCl, pH 7) efficiently prevents activation of the latent core particle. Even after taking these precautions, storage of the purified proteasomes is strongly discouraged, especially 26S complexes, for longer than 3 days at −20°.

## Measurements of Activity of the Proteasome

The enzymatic activity of proteasomal assemblies can be measured *in vitro* with catalytic site-specific model fluorogenic peptide substrates (appropriate for all assembly types), with polypeptides and denatured or unorganized proteins (appropriate for all assembly types), or with ubiquitinated proteins (appropriate for 26S and 19S/activator mixed complexes only). Clearly, the short fluorogenic model peptides are not physiological substrates. However, they allow for assessment of the performance of a single type of active center without the bias introduced by a specific primary structure of a long polypeptide. Moreover, model peptides equipped with different fluorescent tags can be used in mixtures to study allosteric interactions between active centers and regulatory sites (Kisselev *et al.*, 2002, 2003; Myung *et al.*, 2001). When mixing fluorogenic peptides, however, care should be taken to avoid coprecipitation of the substrates in the presence of each other, which hampered earlier attempts to apply this method (Kisselev *et al.*, 1999). In addition, compatibility of the substrates with ATP, Mg salts, and reducing agents used to help stabilize the 26S proteasome should be tested. The use of multiple peptide substrates offers the advantage of mimicking the "perfectly flexible polypeptide" with multiple, equally accessible cleavage sites. Stock solutions of the model peptide substrates, available from Bachem, Calbiochem, BioMol, Sigma, and other suppliers, are usually prepared in millimolar concentrations in dimethyl sulfoxide (DMSO; use a high-quality, molecular biology grade reagent) and stored at –20 or −80°. For consistency, it is recommended to always add the substrates up to 1% of the volume of the reaction mixture containing the proteasome and a reaction buffer. A 50 m*M* Tris-HCl, pH 8.0 (stored at 4°), buffer constitutes a convenient solvent for the reaction; however, phosphate buffers of similar concentration and pH can be used, obtaining comparable results. Up to 3% of DMSO in the reaction mixture will not affect the activity of the proteasome. The content of glycerol added to aid the stability of the purified proteasome should be limited to less than 10%, as proteasome as a hydrolase uses water molecules to complete its catalytic cycle and glycerol competes with water for binding sites in an active center of the enzyme. Bovine serum albumin (BSA) is sometimes added to highly purified but diluted samples of proteasome to increase the stability of the complex. We have found that proteasome is fully functional for at least 1 h and structurally stable for at least 2 h when assessed with atomic force microscopy, even when studied at very low concentrations. Because BSA may change a mode of their interactions by binding nonspecifically to both proteasome and/or tested compounds, its addition in this case is strongly discouraged.

The most commonly employed model peptide substrates include succinyl-LeuLeuValTyr-7-amido-4-methylcoumarin (SucLLVY-MCA; specific for the ChT-L activity), butoxycarbonyl-LeuArgArg-MCA (BocLRR-MCA; T-L activity), and carbobenzoxy-LeuLeuGlu-MCA (Z-LLE-MCA; post-acidic activity). Other three or four residue peptides with a blocked C terminus and a fluorescent tag (MCA or other) can also be used (Kisselev *et al.*, 1999). A free fluorophore AMC (7-amino-4-methylcoumarin) is released from model substrates as a product of the proteolytic action. The abbreviation MCA (4-methylcoumaryl-7-amide) indicates the peptide-bound tag with its fluorescence effectively quenched. Optimal excitation and emission wavelengths of AMC are about 380 and 460 nm, respectively. When using a fluorometer with filters, the pair of excitation/emission filters used for 4′-6-diamidino-2-phenylindole (a common DNA stain) is appropriate for AMC as well.

It is convenient to set up the activity assay in a black flat-bottom 96-well plate; however, an appropriately scaled up format in a standard fluorometric cuvette (four transparent walls) may be adequate for a small number of samples. The quality of plates available on the market varies, and care should be taken not to use plates that are visibly uneven, as this will compromise the results, especially for low fluorescence readings. It is prudent to test the flatness of a particular brand of plates by pipetting the same volume of AMC solution to all 96 wells. The difference between the measurements in any two wells will depend on the quality of the instrument used, but it should be less than 2% without any detectable trends throughout the rows and columns in a plate. Brand-name plates such as Falcon or Costar meet the flatness criteria. Most plates bind only a negligible amount of peptides and AMC, but this should be verified rigorously. Any unexpected decrease of fluorescence of the AMC standard or blank (no enzyme added) during the course of kinetic measurements will indicate that the plastic of the plate exhibits substantial affinity to these compounds.

The example reaction mixture in a single well has a total volume of 100 $\mu$l, with 1 $\mu$l of a substrate stock solution, with up to 10 $\mu$l of the proteasome preparation, and 1–2 $\mu$l of an inhibitor. One well with a free fluorophore of a known concentration should be included in every plate as a standard of fluorescence to enable calculation of the actual amount of the product released. One nanomole per well (10 $\mu M$) works well in the case of AMC. If a fluorometer has the "kinetic measurement" option, it is convenient to incubate the plate at 37° and measure fluorescence every 2–5 min for up to 1 h. Calculation of the rate of reaction (v) as an average increase in the amount of product generated per unit of time provides reliable and accurate results. If end point measurements are preferred, it may be convenient to quench the reaction after the chosen time, e.g., after 30 min or 1 h. Fluorescence in the quenched samples remains stable for at least several hours, eliminating the necessity of immediate measurements. The commonly used quenchers are 70% ethanol or 1% sodium dodecyl sulfate (SDS) added in at least the equal volume to the samples. It is advisable to test first if the studied inhibitor does not precipitate with the quencher. SDS, however, is not practical to use in plates due to its foaming.

Regardless of the convenience of model peptide substrate, polypeptides or protein substrates are necessary for studies on the processivity of the proteasome. Incubation times for protein substrates with the proteasome are usually longer than for model peptide substrates; however, incubations longer than 3 h at 37° should be avoided because there is danger of promoting nonspecific cleavages. Degradation of protein substrates can be measured as a decrease in the amount of a substrate or as an increase in the amount of products, as in the amount of unquenched fluorophore on peptide products, for example. Of note, monitoring for an increase in product formation is considered more accurate and reliable. It is worthwhile to analyze the pattern of peptide products to determine whether specificity and processivity of the proteasome are preserved upon its treatment with an inhibitor. The pattern can be assessed by reverse-phased chromatography (RP-HPLC), mass spectrometry (ES-MS), and, for large products, SDS–PAGE combined with mass spectrometry or peptide sequencing. To test specificity, comparison of one time point (e.g., 1 h at 37°) product profiles for the control and inhibited samples and the blank (no proteasome) should be sufficient. To assess processivity, at least three time points should be compared, e.g., at 30 min, 1 h, and 3 h. If the time-dependent changes in product profiles are quantitative only, then the cleavages are considered processive, as is the case with wild-type 20S and 26S complexes. If the changes are both qualitative and quantitative, such that early products are further cleaved, then the processivity has not been upheld.

## Specific *In Vitro* Assays for Noncompetitive Inhibitors and Allosteric Regulators

The testing of a novel compound in a context of its usefulness as a potential inhibitor is typically performed in the following steps: (a) identification of types of proteasomal complexes and their activities affected by the compound; (b) if the compound is a peptide, verifying that it is not a proteasomal substrate; (c) determination of the inhibition reversibility; and (d) finding the type of inhibition. Identification of the compound as a noncompetitive inhibitor usually triggers (e) further characterization of its properties. This section outlines briefly the methods used in the steps just described and discusses approaches applied in step e. For a more thorough description of this basic biochemical characterization, see Gaczynska and Osmulski (2005).

### *Basic Biochemical Characterization of a Novel Inhibitor*

To simplify the preliminary biochemical analysis of an inhibitor, the compound is first tested on the purified 20S proteasome using changes in degradation rates of the model peptides as probes. Because ChT-L activity is the most robust, it is examined most often. However, to identify the type of inhibition, it is necessary to inspect the response of all active centers for the presence of the compound. Initially, inhibitors are screened in a broad range of the working concentration reaching up to 100 $\mu M$ and consisting of at least five data points, followed by a more refined series of concentrations if more precise numbers are needed. The proteasomes are used in concentrations ranging from 1 to 5 n*M*. It is on this basis that the inhibitor concentration, which causes 50% inhibition of a particular proteasomal activity ($IC_{50}$), is determined. Activity of the proteasome that is unexposed to the inhibitor ("control") is set as 100% activity or 0% inhibition. Of note, however, complete inhibition is often not achievable at any inhibitor concentration for the thermodynamic (binding constant) or physical (solubility) limitations.

Peptide and peptide derivative-based inhibitors or, in general, compounds containing a cleavable amide bond constitute a rich pool of potential proteasomal substrates. Therefore, it is mandatory to determine whether they have been hydrolyzed by the enzyme. Revealing the products of their digest immediately discredits these compounds as inhibitors. A good method to evaluate the compound status is after prolonged incubation (up to 3 h) at 37° in the presence of the purified 20S proteasome. After desalting on a ZipTip C18 pipette tip (Millipore), the reaction mixture is analyzed with the combination of RP-HPLC and ES-MS to detect the formation of cleavage products. This is the preferred approach, as it provides both qualitative and quantitative data, albeit with a longer time commitment. At bare minimum, sacrificing some sensitivity, the sample should be analyzed on a reputable analytical C18 column with detection at 223 nm. An exclusive decrease of the initial content of inhibitor is not acceptable evidence for the enzymatic processing of the compound.

Results of reversibility determination of the inhibitor–proteasome interactions shape the downstream research strategy on the compound. For example, identification of a binding domain and participating residues is relatively straightforward for nonreversible inhibitors. Moreover, the mode of the interactions may affect future pharmacological approaches. Reversibility is tested on samples treated with the inhibitor following their few fold dilution. Subsequently, the activity of the diluted samples is compared to identically diluted control proteasomes. Instead of the dilution method, removal of the inhibitor with centrifugal membrane filter washes using at least 100 volumes of a reaction buffer is often employed, if it has been established in advance that a particular membrane does not bind proteasome or the inhibitor appreciably (Gaczynska and Osmulski, 2005; Gaczynska *et al.*, 2003). If the inhibitor is reversible, activities of the diluted control and treated proteasomes should be similar. If such is the case, the inhibition constant ($K_i$) can be determined from kinetic plots. In general, $K_i = k_{-1}/k_{+1}$, where $k_{-1}$ is a reaction rate of dissociation of an enzyme–inhibitor complex and $k_{+1}$ is a reaction rate of enzyme and inhibitor binding (Gaczynska *et al.*, 2003). If the inhibitor is nonreversible, the association constant ($k_{assoc}$) can be determined (Bogyo *et al.*, 1997).

The complete procedure just outlined is also applicable to other precisely defined proteasomal complexes. Their response to a particular compound may be drastically different from that observed for 20S particles because of structural changes imposed on the core complex by the regulatory particles. It is also possible that the inhibitor interacts specifically only with the additional complexes of proteasome or that the compound competes with them for the binding sites on the 20S proteasome (see later).

An order of addition of the substrate and noncompetitive inhibitor may influence the response of the proteasome to the inhibitor. The source of this phenomenon is not completely understood; however, it may be speculated that the inhibitor exhibits distinct affinities for the latent and already "working" proteasomes (Osmulski and Gaczynska, 2000). Thus, the inhibitor with a high affinity for the latent proteasome should be preincubated with the enzyme before addition of the substrate and *vice versa*. The activated 20S proteasome generated with 0.01– 0.03% SDS displays severalfold higher ChT-L activity. The response of such proteasomes to inhibitors should be treated with caution, as the detergent most likely drastically changes surface charge and allosteric capabilities of the proteasome molecule, in addition to the presumed gate opening (P. A. Osmulski and M. Gaczynska, unpublished observations).

*Competitive versus Noncompetitive Inhibitors*

Competitive inhibitors contend with substrates for binding to catalytic centers. In the case of proteasomes, the inhibitors may affect all or only selected types of active centers. Even if they bind to all of the active centers, the strength of their interactions usually differs manyfold between the individual sites. The reversible competitive inhibitors can be overpowered by supplying a few fold excess of the substrate. Noncompetitive inhibitors bind to noncatalytic sites independent of occupation of an active center by a substrate. Generally, their effects are not sensitive to the presence of the substrate. Uncompetitive inhibitors, very rare and often included in the noncompetitive group, also bind to noncatalytic sites, but only when an enzyme is complexed with the substrate. To date, no uncompetitive inhibitors of the proteasome have been identified. Some inhibitors are classified as a mixed type because they bind to both an active center(s) and noncatalytic sites.

Identification of a type of inhibition is usually based on the analysis of the Michaelis–Menten kinetic parameters of the purified control (not inhibited) and inhibited proteasome using model fluorogenic peptides or any other convenient substrate. Most often these parameters are calculated from Lineweaver–Burk plots as described, e.g., by Gaczynska *et al.* (2003) for proline- and arginine (PR)-rich peptides. Because of its algebraic transformation, Lineweaver–Burk plots often suffer from an overdependency on velocity values collected at low concentrations of the substrates. As such, alternative transformations of the classical Michaelis–Menten equation can be used. A convenient protocol using model peptides begins with the determination of initial velocities of degradation ($v$) of several different concentrations of substrate ($S$). The reactions are performed by control proteasomes incubated only with the identical volume of solvent used to dissolve the inhibitor and by proteasomes incubated with the compound of interest at a concentration close to its $IC_{50}$. Stock solutions of substrates are prepared in ranges covering concentrations below and above expected $K_M$ values, usually from 5 to 500 $\mu M$. To ensure accuracy, no less than 6, but an optimum of 9 or 12 different substrate concentrations should be used. All experimental points should be collected in duplicate or triplicate. Plotting values of the $1/v$ versus $1/S$ allows the determination of the maximal velocity ($V_{max}$) and Michaelis constant ($K_M$), as the plot crosses the $y$ axis at $1/V_{max}$ and the $x$ axis at $-1/K_M$. Proteasomes treated with noncompetitive inhibitors maintain their Michaelis constants [$K_{M(i)}$] equal to control (uninhibited) proteasome ($K_M$), but their maximal velocity [$V_{max(i)}$] is lower than that of the control ($V_{max}$). In contrast, with competitive inhibitors, $V_{max(i)}$ equals $V_{max}$, and $K_{M(i)}$ is bigger than $K_M$. The effect of mixed competitive/noncompetitive inhibitors is characterized by different $V_{max}$ and different $K_M$ values, with $K_{M(i)}$ of the inhibited sample larger than $K_M$ of the control. The type of inhibition is immediately recognizable by the relation of Lineweaver–Burk plots collected in the absence and presence of inhibitors (Gaczynska and Osmulski, 2005; Gaczynska *et al.*, 2003): the plot for a competitive inhibitor crosses the plot for the controlled proteasome at the $y$ axis; for a noncompetitive inhibitor at the negative site of the $x$ axis; and for a uncompetitive inhibitor the plots are parallel to each other (they never cross).

Generally, small molecule activators of the proteasome follow the noncompetitive type of kinetics, with their corresponding Michaelis constants equal and $V_{max}$ for the activated enzyme larger than $V_{max}$ for the control. Analysis may be further complicated if a compound selectively inhibits one type of peptidase and activates another. Moreover, classification of a particular compound as a competitive inhibitor does not necessarily rule out its participation in allosteric transitions. For example, the anti-AIDS drug ritonavir, a peptide analog and an inhibitor of the essential aspartyl protease of the HIV virus, inhibits ChT-L and PGPH peptidase activities, but activates the T-L peptidase of the 20S core *in vitro* (Schmidtke *et al.*, 1999). Detailed examination of proteasomal peptidase kinetics in the presence of ritonavir allowed Schmidtke *et al.* (2000) to develop the "two-site modifier" model of interactions between proteasomal active and regulatory sites. Finally, allostery induced by the specific binding of an inhibitor or even a substrate to a particular active center may affect all or just select catalytic centers, only in the same ring or also in its twin (Myung *et al.*, 2001).

### *Noncompetitive Ligands versus Allosteric Regulators*

There is growing support for the hypothesis that allostery is one of the fundamental properties of all proteins (Gunasekaran, 2004). A broad definition of allostery implies that most of the noncompetitive ligands should be considered allosteric. Indeed, if the binding of a regulator apart from active centers changes the catalytic performance of the proteasome, then its mechanism of action, in most cases, will include allosteric transitions. In turn, proteasomal regulators, which act as "plugs" or "filters," are deemed nonallosteric ligands. We now present strategies used to distinguish among a plug, a filter, and an allosteric effector using an N-terminal part of $\alpha 3$ subunit, gliotoxin, and PR peptides (proline- and arginine-rich peptides) as examples.

The "plug" bound to the gate area or inside the central channel, in the simplest case, obstructs the traffic of substrates and products, presumably without sending structural signals to a distance larger than the area of its own productive binding. Binding of the plug clogs the passage for all substrates; thus, all of the active centers are affected to the same extent, leading to a proportional decrease of degradation rates of all substrates specific for the centers. A straightforward comparison of degradation rates of the model substrates between control and ligand–treated proteasome is sufficient to conclude that the proteasome is capped over its central channel.

A peptide plugging the permanently open proteasome is likely the best example of a compound exploiting this type of inhibition. The gate of the proteasome is built with N-terminal parts of $\alpha$ subunits, among which the $\alpha 3$ subunit is the major contributor to the structural occlusion of the gate. The mutant proteasome with that part of the $\alpha 3$ subunit deleted remains open, as shown in its crystal structure and atomic force microscopy (AFM) images (Groll *et al.*, 2000; M. Gaczynska and P. A. Osmulski, unpublished observations). Addition of the N-terminal part of the $\alpha 3$ subunit in *trans* to the mutated 20S proteasome seals the entrance to the central channel, effectively substituting the missing element of the gate. As a result of gate clinching, all three peptidase activities are evenly shut down (Groll *et al.*, 2000). Although the plug may send allosteric signals of equal strength to all of the catalytic sites, this seems a less plausible explanation for the aforementioned observations.

The assumption of less stringent conditions of gate plugging led to the development of a hypothesis for the existence of a regulatory filter. Such a compound acts as a nonallosteric inhibitor by blocking traffic of selected substrates, by charge or size exclusion, for example. Alternatively, similar effects, although restricted to only one type of peptidase activity, are observed if the inhibitor binds inside of the proteasome in proximity to the targeted catalytic center, constituting a selective barrier. Thus, it is suspected that a filter binds either near the gate in the $\alpha$ rings or inside the 20S barrel. A cross-linking of the inhibitor to its target and a yeast two-hybrid screen are typically employed to identify such binding sites.

An unexpectedly complex example of an inhibitor acting like a filter is a secondary fungal natural product, gliotoxin. It inhibits noncompetitively all three proteasomal peptidases with a substantially higher affinity toward ChT-L (Kroll *et al.*, 1999; Paugam *et al.*, 2002). An essential group for the antiproteasomal activity of gliotoxin is a disulfide bridge in its heterobicyclic core, which introduces a reversible covalent modification of disulfide bonds at or near the ChT-L catalytic center. Therefore, it is postulated that gliotoxin directly obscures access to this catalytic center but simultaneously imposes a steric hindrance, leading to the inhibition of the other two proteasomal peptidases (Kroll *et al.*, 1999). On this basis, gliotoxin is also classified as an inhibitor of mixed characteristics involving both a plug and a filter mechanism.

In contrast, PR peptides provide examples of allosteric ligands. The peptides are derived from PR39, a 39 residues long, natural antimicrobial agent from the family of cathelicidins, taking part of the innate immunity in mammals (Zanetti, 2004). PR39 has several intracellular protein targets, among them the proteasome (Gao *et al.*, 2000). A yeast two-hybrid screen implicates an acidic C-terminal sequence of the $\alpha 7$ subunit, positioned on the outer rim of the $\alpha$ ring far away from the gate as an interaction site for the strongly basic peptide (Gao *et al.*, 2000). PR39 and its N-terminal-derived peptides inhibit ChT-L and postacidic activities of the purified human 20S proteasomes. The elimination of a strongly basic character of the peptides by substituting N-terminal arginines with alanines abrogates inhibition of the proteasome (Gaczynska *et al.*, 2003). Thus, for PR peptides, two conditions supporting allostery are fulfilled: the nonuniform pattern of noncompetitive inhibition and a postulated distinctive binding site. A more direct proof of the long-distance transfer of structural signals, however, is served by AFM imaging of the purified proteasomes in the presence of the peptides (see Osmulski and Gaczynska, 2005). They induce a surprisingly dramatic structural effect involving freezing of the gate in a quasi-open state. Interestingly, the 26S proteasome is also inhibited by PR peptides, which generate distinctive swelling of the 19S caps in the AFM images. Moreover, the peptides apparently compete with the 19S cap and 11S activator for binding sites on the $\alpha$ ring, confirming the previous localization of these sites on the $\alpha 7$ subunit (Gaczynska *et al.*, 2003; M. Gaczynska and P. A. Osmulski, unpublished observations). Results of the competition experiments, which attempt to reconstruct the 26S or activated complexes in the presence of inhibitors, if performed as described (Gaczynska *et al.*, 2003), provide valuable clues about possible binding sites of the compounds and their possible mode of interactions. Certainly, this type of experiment should be included in the plan of a novel inhibitor characterization.

Unfortunately, the *in vivo* effects for many ligands binding to the $\alpha$ ring are obscured by the superposition of a direct modulation of the 20S core activity and a lack of the effects on the "outcompeted" natural module. Such mixed actions are difficult to recognize *in vivo*, but their detection is easier *in vitro*, as all of the components of the reaction—proteasome, the ligand of interest, and the potential competitor—are available in a pure form and can be tested with substrates of choice.

Activity tests or the characterization of binding sites may be criticized for providing merely a circumstantial proof for allostery. In addition to those mentioned in this chapter, AFM and EM images and the analysis of crystal structures or fluorescence emission spectra offer direct proof for the involvement of ligand-induced structural transitions (see also Osmulski and Gaczynska, 2005).

## Acknowledgments

The work was funded by R01 GM069819 and DAMD17-03-1-0636 grants (M. G.) and San Antonio Cancer Institute Pilot Grant (P. A. O). The authors acknowledge the support from the San Antonio Cancer Institute Cancer Center Support Grant (P30 CA54174).

## References

Adams, J. (2004). The development of proteasome inhibitors as anticancer drugs. *Cancer Cell* **5,** 417–421.

Bogyo, M., McMaster, J. S., Gaczynska, M., Tortorella, D., Goldberg, A. L., and Ploegh, H. (1997). Covalent modification of the active site threonine of proteasomal beta subunits and the *Escherichia coli* homolog HslV by a new class of inhibitors. *Proc. Natl. Acad. Sci. USA* **94,** 6629–6634.

Gaczynska, M., and Osmulski, P. A. (2002). Inhibitor at the gates, inhibitor in the chamber: Allosteric and competitive inhibitors of the proteasome as prospective drugs. *Curr. Med. Chem. Imunol. Endoc. Metab. Agents* **2,** 279–301.

Gaczynska, M., and Osmulski, P. A. (2005). Small-molecule inhibitors of the proteasome activity. *In* "ubiquitin-proteasome protocols." *Methods Mol. Biol.* (C. Patterson and D. M. Cyr, eds.), pp. 3–22. Humana Press.

Gaczynska, M., Osmulski, P. A., Gao, Y., Post, M. J., and Simons, M. (2003). Proline- and arginine-rich peptides constitute a novel class of allosteric inhibitors of proteasome activity. *Biochemistry* **42,** 8663–8670.

Gao, Y. H., Lecker, S., Post, M. J., Hietaranta, A. J., Li, J., Volk, R., Li, M., Sato, K., Saluja, A. K., Steer, M. L., Goldberg, A. L., and Simons, M. (2000). Inhibition of ubiquitin-proteasome pathway-mediated I kappa B alpha degradation by a naturally occurring antibacterial peptide. *J. Clin. Invest.* **106,** 439–448.

Groll, M., Ditzel, L., Lowe, J., Stock, D., Bochtler, M., Bartunik, H. D., and Huber, R. (1997). Structure of 20S proteasome from yeast at 2.4 A resolution. *Nature* **386,** 463–471.

Groll, M., Bajorek, M., Kohler, A., Moroder, L., Rubin, D. M., Huber, R., Glickman, M. H., and Finley, D. (2000). A gated channel into the proteasome core particle. *Nature Struct. Biol.* **7,** 1062–1067.

Gunasekaran, K., Ma, B., and Nussinov, R. (2004). Is allostery an intrinsic property of all dynamic proteins? *Proteins Struct. Funct. Bioinform.* **57,** 433–443.

Kisselev, A. F., Akopian, T. N., Castillo, V., and Goldberg, A. L. (1999). Proteasome active sites allosterically regulate each other, suggesting a cyclical bite-chew mechanism for protein breakdown. *Mol. Cell* **4,** 395–402.

Kisselev, A. F., Garcia-Calvo, M., Overkleeft, H. S., Peterson, E., Pennington, M. W., Ploegh, H. L., Thornberry, N. A., and Goldberg, A. L. (2003). The caspase-like sites of proteasomes, their substrate specificity, new inhibitors and substrates, and allosteric interactions with the trypsin-like sites. *J. Biol. Chem.* **278,** 35869–35877.

Kisselev, A. F., Kaganovich, D., and Goldberg, A. L. (2002). Binding of hydrophobic peptides to several non-catalytic sites promotes peptide hydrolysis by all active sites of 20 S proteasomes: Evidence for peptide-induced channel opening in the alpha-rings. *J. Biol. Chem.* **277,** 22260–22270.

Kroll, M., Arenzana-Seisdedos, F., Bachelerie, F., Thomas, D., Friguet, B., and Conconi, M. (1999). The secondary fungal metabolite gliotoxin targets proteolytic activities of the proteasome. *Chem. Biol.* **6,** 689–698.

Li, J., and Rechsteiner, M. (2001). Molecular dissection of the 11S REG (PA28) proteasome activators. *Biochimie* **83,** 373–383.

Myung, J., Kim, K. B., Lindsten, K., Dantuma, N. P., and Crews, C. M. (2001). Lack of proteasome active site allostery as revealed by subunit-specific inhibitors. *Mol. Cell* **7,** 411–420.

Osmulski, P. A., and Gaczynska, M. (2002). Nanoenzymology of the 20S proteasome: Proteasomal actions are controlled by the allosteric transition. *Biochemistry* **41,** 7047–7053.

Osmulski, P. A., and Gaczynska, M. (2005). Atomic force microscopy of the proteasome. *Methods Enzymol.* **398**[34] 2005 (this volume).

Paugam, A., Creuzet, C., Dupouy-Camet, J., and Roisin, P. (2002). *In vitro* effects of gliotoxin, a natural proteasome inhibitor, on the infectivity and proteolytic activity of *Toxoplasma gondii*. *Parasitol. Res.* **88,** 785–787.

Schmidtke, G., Emch, S., Groettrup, M., and Holzhutter, H. G. (2000). Evidence for the existence of a non-catalytic modifier site of peptide hydrolysis by the 20 S proteasome. *J. Biol. Chem.* **275,** 22056–22063.

Schmidtke, G., Holzhutter, H. G., Bogyo, M., Kairies, N., Groll, M., de Giuli, R., Emch, S., and Groettrup, M. (1999). How an inhibitor of the HIV-I protease modulates proteasome activity. *J. Biol. Chem.* **274,** 35734–35740.

Zanetti, M. (2004). Cathelicidins, multifunctional peptides of the innate immunity. *J. Leukocyte Biol.* **75,** 39–48.

Zwickl, P., Seemuller, E., Kapelari, B., and Baumeister, W. (2001). The proteasome: A supramolecular assembly designed for controlled proteolysis. *Adv. Prot. Chem.* **59,** 187–222.

# [36] Development and Use of Antiproteasome Monoclonal Antibodies

*By* KLAVS B. HENDIL

## Abstract

This chapter describes the production of hybridomas and screening of monoclonal antibodies for use in enzyme-linked immunosorbent assay (ELISA), immunoblotting, and immunoprecipitation of components of the ubiquitin-proteasome pathway. Purification of antibodies and their use in affinity chromatography are also described.

## Introduction

With their extreme specificity, monoclonal antibodies have been valuable tools for the analysis of proteasome complexes. Their use in immunoblotting and immunoprecipitation of proteasomes or proteasome precursors is widespread. Monoclonal antibodies have also been used as conformational probes (Conconi *et al.*, 1999; Weitman and Etlinger, 1992), for identification of binding sites of cofactors (Kania *et al.*, 1996), for localization of subunits within proteasomes (e.g., Kopp and Kuehn, 2003; Kurucz *et al.*, 2002), in affinity chromatography (Hendil and Uerkvitz, 1991), for localization of proteasomes in cells (e.g., Brooks *et al.*, 2000), and for quantitative analysis of proteasomes (Dutaud *et al.*, 2002). Potentially, antibodies can also be microinjected into cells in order to inactivate their antigen (Lamb and Fernandez, 1997). This chapter provides a short guide for production and use of monoclonal antibodies to proteasomes but the procedures can of course be used equally well for other interesting proteins. More thorough descriptions are given in several handbooks (Goding, 1996; Harlow and Lane, 1999; Shepherd and Dean, 2000).

The major part of the work involved in making monoclonal antibodies is not in the production of hybridomas, but in the screening and characterization of the antibodies and subsequent cloning of the most promising hybridomas. In the procedures described here, hybridomas are given primary screenings before being cloned in order to avoid futile clonings of hybridomas, which later prove useless for the intended experiments.

METHODS IN ENZYMOLOGY, VOL. 398

0076-6879/05 $35.00
DOI: 10.1016/S0076-6879(05)98036-1

## Hybridoma Production

### *Antigen Preparation*

Antigens for immunization and screening are typically conventionally purified protein complexes or individual, tagged subunits expressed in yeasts or *Escherichia coli* and purified by affinity chromatography. The antigen need not be pure, in contrast to the situation when conventional, polyclonal antibodies are produced. Proteasomes and many other components in the ubiquitin/proteasome system are phylogenetically preserved and such proteins, when isolated from mammals, are therefore not good antigens in mice or rabbits. The immunogenicity may be improved by chemical modification, e.g., dinitrophenylation (Hudson and Hay, 1989).

Some recombinant proteins are poorly soluble and are typically purified from inclusion bodies dissolved in 8 *M* urea. Such proteins often precipitate when urea is removed by dialysis but suspensions of protein are still useful for the production of antibodies.

Peptides are sometimes used as antigens but need coupling to a protein, e.g., keyhole limpet hemocyanin, in order to elicit an antibody response. Antibodies against peptides are usually inferior to those produced by immunization with larger antigens but provide a shortcut around sometimes tedious cloning or purification procedures. The peptides are usually around 15 residues long. They are selected for being low in hydrophobic amino acids in order to have good solubility in water and to increase the chance of getting antibodies, reacting to the protein surface and therefore useful for immunoprecipitation and not only for immunoblotting. Because N and C termini of proteins are often exposed at the surface, use of peptides from these ends theoretically increases the chance of getting an antibody useful for immunoprecipitation and immunocytochemistry. Using a basic local alignment search tool (BLAST), search, ensure that the amino acid sequence, you have selected, is not found in other proteins. Peptides are coupled to carrier proteins before immunization. We prefer to extend the peptide chain by a cysteine residue at the N or C terminus and couple it to keyhole limpet hemocyanin via *m*-maleimidobenzoyl-*N*-hydroxysuccinimide ester (MBS), as the alternative coupling via amino groups modifies internal lysine residues. Pierce (http://www.piercenet.com) provides MBS and instructions for peptide coupling.

### *Immunization*

The antigen should preferably be at a concentration of 0.1–5 mg/ml in a nontoxic buffer, e.g., phosphate-buffered saline (PBS; 137 m*M* NaCl, 2.7 m*M* KCl, 4.1 m*M* $Na_2HPO_4$, and 0.73 m*M* $KH_2PO_4$, pH 7.4). Mix antigen solution or suspension with an equal volume of Freund's incomplete adjuvant and emulsify in a probe sonicator. Afterward, keep a beaker under the probe and wash it with a squeeze bottle with distilled water. Check that drops of emulsion, collected in the beaker, do not disperse but remain as white droplets on the wash water, indicating that a water-in-oil emulsion has formed. Alternatively, emulsification can be attained by connecting two syringes via a connector (e.g., from Sigma-Aldrich) and repeatedly forcefully injecting the antigen/adjuvant mixture from one barrel to the other. Immunize Balb/c mice by injecting 10–50 $\mu$g of antigen in 100–200 $\mu$l of emulsion into the peritoneal cavity of each mouse on day 1. Repeat the injection on days 14 and 28. With difficult antigens it may be advisable to immunize several mice. If so, anesthetize the mice lightly with ethyl ether on day 37, make a cut in the vein running at the ventral side of the tail, and collect a drop of blood from each mouse. Make a serial dilution of serum in buffer A (150 m*M* NaCl, 1.9 m*M* $NaH_2PO_4$, 8.1 m*M* $Na_2HPO_4$, 0.1 m*M* EDTA, 0.05% Tween 20, pH 7.4) with 1% newborn calf serum. Test the dilution series for antibodies by ELISA in a procedure like that used for the initial hybridoma screening, described later. The ELISA titer is defined as the reciprocal of the dilution, giving half-maximal color development. Boost the mouse with the highest ELISA titer by injecting 20–50 $\mu$g of antigen in saline (no adjuvant) into the peritoneal cavity on days 38, 41, and 42. Proceed with cell fusion on day 45.

*Solutions for Fusion, Cloning, and Hybridoma Propagation*

To make PEG solution, put 10 g PEG (polyethylene glycol 4000 for gas chromatography, E. Merck, Germany) into a sterile 100-ml bottle. Cap, but do not close tightly: air should be able to escape during sterilization or the flask may explode. Place in a pot with water and cover with aluminium foil. Boil for 30 min. Allow to cool so that the bottle can be handled and add slowly 10 ml of Dulbecco's modified Eagle's medium (DMEM) without serum while the bottle is turned by hand so that the contents mix well before the PEG solidifies. Distribute into portions of 2 ml and freeze. HT stock $\times$ 50: dissolve 68 mg of hypoxanthine and 19 mg of thymidine into 100 ml of hot water. Sterilize by filtration. Store in a refrigerator. Aminopterin stock $\times$ 100: dissolve 1.9 mg of aminopterin in 100 ml of 10 m*M* NaOH. Sterilize by filtration. Store frozen. Mercaptoethanol stock $\times$ 100: 18 $\mu$l of $\beta$-mercaptoethanol in 100 ml of water. Sterilize by filtration and store frozen.

The HAT medium is 50 ml supernatant from Sp2/0 cultures ("conditioned" medium, obtained as described later), 125 ml DMEM, 75 ml RPMI 1640 medium, 25 ml fetal calf serum, 5 ml Nutridoma CS (http://www.roche-applied-science.com), 5 ml HT stock $\times$ 50, 2.5 ml aminopterin stock × 100, and 2.5 ml mercaptoethanol stock × 100. Nutridoma CS is an additive used for growing cells at low culture densities, as in cloning. Other such additives are available commercially from several sources and render feeder cells superfluous.

HT medium used later in expansions of hybridoma cultures is a 2:1 mix of DMEM and RPMI 1640, supplied with fetal calf serum, HT, and mercaptoethanol as for HAT medium. Penicillin G, 50 U/ml, and streptomycin, 150 $\mu$g/ml, may be added to all media.

### *Fusion*

Start a culture of Sp2/0-Ag14 cells (available from http://www.atcc.org or http://www.ecacc.org.uk) a week before the fusion. Grow the cells in DMEM/10% fetal calf serum at a density below $10^6$ cells/ml. The day before fusion, seed 2–3 × $10^5$ cells/ml in each of three 175-cm$^2$ tissue culture flasks with 30 ml of medium.

On the day of fusion, warm all media in a 37° bath. Kill an immunized mouse by cervical dislocation. Dip it into 70% ethanol, place it on a slab of polystyrene foam, and open the skin on the left side of the abdomen with sterile forceps and scissors to expose the peritoneum. Keep the skin away from the peritoneum by fixing it to the polystyrene slab with needles. Move the mouse to a sterile hood and swab the peritoneum with 70% ethanol. Resterilize forceps and scissors (dip into 96% ethanol and ignite). The spleen is a brownish red, oblong organ at the left side of the mouse. It should be visibly enlarged compared to spleens from nonimmunized mice. Dissect the spleen free from connective tissue and move it to a large petri dish.

Add 2 drops of growth medium and section the spleen in the middle with scalpels. Squeeze the spleen fragments to empty them—like a tube of toothpaste. Cut remaining spleen fragments into pieces of less than 1 mm. Take your time. Add 10 ml of DMEM and transfer cell suspension and small tissue fragments with a wide-bore pipette to a 10-ml centrifuge tube. Let it stand for a few minutes to allow tissue fragments to settle. Move supernatant to a 50-ml centrifuge tube. Add another 2 ml of DMEM to the 10-ml tube with tissue fragments, and vortex for a few minutes. Fill up with DMEM without serum and let it stand for a few minutes for debris to settle before the supernatant is combined with the first preparation of spleen cells in the 50-ml tube. Spin cells (1500 rpm, 5 min), aspirate supernatant, and suspend cells into 10 ml of fresh DMEM without serum.

Spin the Sp2/0 cells (1–5 × $10^7$ cells) in 50-ml centrifuge tubes at 1500 rpm for 5 min. Aspirate and save the supernatant (= "conditioned medium") for preparation of medium for the hybridomas. Wash cells in DMEM without serum by centrifugations at 1500 rpm for 5 min. Combine spleen cells and washed Sp2/0 cells in a 50-ml tube and wash once in DMEM without serum. Aspirate all supernatant carefully. Break up the pellet by tapping the tube on the table. Add slowly, over 30–60 s, 1.5 ml of the PEG solution while the tube is rotated by hand in a 2-liter beaker with 40° water. Slowly rotate the tube in the 40° bath for another 30 s and then add 1 ml of warm (37°) DMEM without serum in the course of 30–60 s while the tube is rotated in order to mix the fluids inside. Leave for 1 min in the beaker with warm water. Add another 3 ml of warm DMEM without serum over 1 min and then about 20 ml of warm DMEM without serum over another minute. Spin for 5 min at 1500*g* (about 3000 rpm). Leave for 5 min in a 37° water bath without suspending the cell pellet and then aspirate the supernatant.

Resuspend cells very gently into 10 ml of HAT medium and break up large clumps by pipetting up and down a few times. Dilute into 250 ml of warm HAT medium and distribute into 96-well sterile flat-bottomed microtiter plates with 2 drops or 150 $\mu$l per well. There should be enough suspension for about 14 plates.

Pack plates in pairs in thin polyethylene bags and grow cells in an incubator with 10% $CO_2$. Polyethylene is permeable to $CO_2$ but not to water vapor. Check periodically and feed as needed with 1 or 2 drops of HT medium per well. Screen for positive clones when colonies reach a size that they are easily visible to the naked eye, usually at days 8–10. A good fusion should give at least 1000 colonies. Sometimes more than 5000 are obtained so that most wells contain several hybridomas.

## Screening and Cloning

### *Initial Screening for Antibodies (ELISA)*

Often, the same monoclonal antibody is not well suited for all potential applications, such as immunoblotting, immunoprecipitation, and immunocytochemistry. The antibodies should therefore ideally be screened by the same method for which they are intended to be employed. However, ELISA is used commonly in an initial screening because it is well suited for handling the hundreds of hybridomas.

Perform ELISA screening in the following way. Dilute antigen to about 2 $\mu$g/ml in PBS. Insoluble antigens may first be dissolved by adding crystals of urea to a small volume of suspension and then rapidly diluted into the final volume of buffer. Peptides may be adsorbed directly. If they adsorb poorly, they may be coupled to a protein different from the one used in immunizations. For instance, if the peptide was coupled to keyhole limpet hemocyanin for immunizations, it may be coupled to bovine serum albumin for adsorption to microtiter plates.

Add 1 drop (or 50 $\mu$l) of antigen solution per well in flat-bottomed microtiter plates and leave them overnight in the refrigerator. At this stage, plates may also be stored in the freezer, but prolonged storage may cause structural changes (denaturation) of the adsorbed protein and therefore loss of epitopes (antibody-binding sites).

Flood plates with 5% fat-free dry milk in PBS (may be preserved with 10 m$M$ $NaN_3$). Leave for 30–60 min. Wash three times in buffer A and flick the plates empty. Use a multichannel pipette to move 150 $\mu$l of culture medium from each well with hybridomas to corresponding wells in the microtiter plates with adsorbed antigen. Incubate >2 h at room temperature or overnight in a cold room. Wash three times in buffer A. Add 100 $\mu$l of diluted peroxidase-conjugated rabbit antibody to mouse immunoglobulins to each well [e.g., 10 $\mu$l Dako antibody (http://www.dakocytomation.com) +10 ml 1% calf serum in buffer A]. A Finnpipette Multistepper (http://www.thermo.com) or similar is convenient for repeated pipettings. Leave for 1–2 h. Wash four times in buffer A. Make substrate solution (dissolve 30 mg *o*-phenylenediamine in 100 ml of 25 m$M$ citric acid, 51 m$M$ $Na_2HPO_4$, pH 5.0, and add 30 $\mu$l 30% $H_2O_2$ immediately before use) and add 150 $\mu$l per well. Wells with antibodies will turn brown. Stop color development when appropriate (often 5–20 min) by adding 50 $\mu$l of 12.5% $H_2SO_4$ per well. Read plate at 490 nm in a microplate reader for a permanent record.

### *Freezing Hybridomas*

Use sterile glass Pasteur pipettes to move hybridomas from positive wells to wells in 24-cluster dishes with 1 ml of HT medium with Nutridoma CS. The growth medium should be the same as the hybridoma medium described earlier, but without culture supernatant from Sp2/0 cells and without aminopterin (i.e., HT medium with Nutridoma CS). Keep the cluster dishes in a 37° humidified incubator with 5–10% $CO_2$ and in polyethylene bags to reduce evaporation. Move cultures to 25-cm$^2$ tissue culture flasks with 1–2 ml of HT medium when they become sufficiently dense (around $10^6$ cells/ml) and let the flasks stand on end to reduce the surface. After a few days shake the culture lightly, dilute with another 3–4 ml of HT medium, and lay the flask down to normal position. When the cell density has reached about 5–10 $\times$ $10^5$ cells/ml, shake the culture lightly to detach cells, which may adhere to the surface. Centrifuge the cell suspension (1500 rpm, 5 min), move the supernatant to another tube, and preserve it by adding 1 $M$ stock solution of $NaN_3$ to 10 m$M$. Suspend cell pellet into 1–2 ml of 7% dimethyl sulfoxide in fetal calf serum and move to cryotubes. Wrap tubes in tissue paper (to reduce the cooling rate), collect into plastic boxes, and place in an –80° freezer, where the cells survive for several weeks while the corresponding supernatants are screened further. Cloned hybridomas are frozen similarly but are moved to liquid nitrogen the next day for long-term storage.

*Secondary Screenings*

False positives from the initial ELISA screening may be identified in a screen with ELISA plates with adsorbed antigen and parallel plates without antigen, with both sets of plates blocked with milk protein. Remaining supernatants may then be screened by immunoblotting, immunoprecipitation, or other methods for which the antibodies are intended to be used. If a tagged, recombinant protein was used for immunization, it is also important to check that the antibodies are not directed to the tag.

*Screening by Immunoblotting*

Immunoblotting (Western blotting) also provides an easy screen for species specificity. Some proteasome subunits are structurally related so that they share motifs and therefore potential epitopes. Occasionally, antibodies may therefore react with several subunits from diverse species.

For immunoblotting, run total cell protein on mini SDS–PAGE gels ($7 \times 8$ cm) along with molecular mass markers. Samples with cell proteins are made by suspending 1 g (wet weight) of cells with 7 ml of SDS sample buffer. Use 10–15 $\mu$l/lane. Blot proteins onto nitrocellulose sheets and stain reversibly with Ponceau S (0.5% Ponceau S in 5% acetic acid for 10 min, destain in water) so that the positions of the molecular mass standards can be marked up with a ballpoint pen. Use a shaker platform for the incubations and washings described later. Cut out individual lanes and incubate in Octaline trays (http://www.pateof.com), first for 60 min with 5% fat free dry milk in PBS and then overnight with culture supernatants diluted 1:10 in buffer B (50 m*M* Tris-HCl, pH 7.4, 150 m*M* NaCl, 5 m*M* $NaN_3$, 0.01% Tween 20). Wash blots $3 \times 10$ min in buffer B, incubate with peroxidase-conjugated antibody to murine immunoglobulin (e.g., Dako Rabbit anti-body to mouse immunoglobulin diluted 1:1000 in buffer B) for 1 h, wash $4 \times 10$ min in buffer B, and develop with TMB/$H_2O_2$ [stock solutions: 50 m*M* Na-acetate buffer, pH 5.0; dioctyl sodium sulfosuccinate, 0.4 % (w/v) in 96% ethanol; tetramethyl benzidine (TMB), 50 mg/ml acetone, keep in refrigerator. Immediately before use mix 20 ml of acetate buffer with 5 ml of dioctyl sodium sulfosuccinate stock and, with swirling, 0.2 ml of TMB stock and 10 $\mu$l 30% $H_2O_2$]. Good antibodies give green bands at the appropriate positions in a few minutes. Wash blots for 20 min in poststain (50 ml dioctyl sodium sulfosuccinate stock brought to 200 ml with water) before drying. Scan or photocopy the blots, which may fade with time. Enhanced chemiluminescence (ECL) is more sensitive than TMB but is less convenient when many strips have to be handled simultaneously.

*Screening for Immunoprecipitation*

Many antibodies, which react well in ELISA or in immunoblotting, do not bind the native antigen and are therefore not suitable for immunoprecipitation. A screen for immunoprecipitation after an initial ELISA screen may therefore be necessary.

For 20 samples, wash 250 $\mu$l protein G-agarose beads in 2 $\times$ 10 ml of PBS and suspend into 250 $\mu$l of PBS. Use a pipette with a wide-bore tip (cutoff tip) to aliquot the suspension to twenty 15-ml centrifuge tubes. Spin hybridoma culture supernatants at 5000 rpm for 5 min. Tumble 1 ml of each supernatant for 1 h with the protein-G suspension in a 15-ml tube in order to bind the antibody. Wash the beads with 2 $\times$ 15 mL of PBS with 0.1% Triton. Prepare a 10% homogenate of suitable cells in PBS with 0.1% Triton X-100. Centrifuge at 10,000*g* for 20 min and save the supernatant. Transfer 300 $\mu$l (or more, for less abundant proteins) of supernatant from cell extract to each tube with protein G-beads and tumble for 2 h at room temperature or overnight in a cold room. Wash beads in 4 $\times$ 10 ml of PBS with 0.1% Triton X-100 by centrifugations (1000 rpm, 3 min). Aspirate all fluid and suspend beads into 20 $\mu$l of SDS sample buffer. Run PAGE gels. If the precipitated antigen is abundant (like proteasomes) and has a mass different from those of heavy and light chains of immunoglobulin (around 25,000 and 50,000, respectively), it may be seen on a Coomassie-stained gel. Alternatively, the cell extract must be prepared from cells, labeled with [$^{35}$S]methionin, and the antigen detected by autoradiography of a nitrocellulose filter with proteins, blotted from the gel. Finally, one of the antibodies suitable for immunoblotting may be biotinylated (Shepherd and Dean, 2000) and used together with peroxidase-conjugated streptavidin to detect blotted antigens. The latter method may, however, be uncomfortably sensitive, as some antibodies precipitate traces of denatured antigens but not the native one. An antibody may therefore score positive without being good for immunoprecipitation.

Some antibodies react with free subunits but not with intact complexes (Jorgensen and Hendil, 1999). This can be useful for studies of assembly of macromolecular complexes (Nandi *et al.*, 1997). Antibodies may also react with denatured antigen but not with native ones. For instance, antibodies that do not precipitate the native antigen may precipitate an SDS-denatured antigen from a solution with Triton X-100 (Hartmann-Petersen *et al.*, 2001).

### *Isotyping*

Murine antibodies come in different isotypes with different physiological functions. As laboratory reagents, IgG1, IgG2a, and IgG2b are the most useful because they bind well to protein G and may therefore be suitable for immunoprecipitation. IgM antibodies often give unspecific binding. Several kits for isotyping of antibodies from cloned hybridomas are available commercially. However, we often make a preliminary isotyping before cloning in order not to waste time on less desirable isotypes. An ELISA-based kit from Zymed (http://www.invitrogen.com) provides an economic method for screening a large number of antibodies. With a modified protocol, the isotype of antibodies from uncloned cells can be determined: use microtiter plates with adsorbed antigen, as for hybridoma screening. Block by flooding with 5% fat-free dry milk in buffer A and incubate for >1 h. Wash three times in buffer A. Pipette 100 $\mu$l of culture supernatant from the same hybridoma into each well in column 1. Fill wells in column 2 with the next culture supernatant and so on. Incubate overnight in the refrigerator. Wash three times with buffer A. Dilute isotype-specific rabbit antibodies to mouse immunoglobulins as recommended by the supplier. Pipette 50 $\mu$l of anti-IgG1 into all wells in row A, anti-IgG2a in row B, anti-IgG2b in row C, anti-IgG3 in row D, anti-IgM in row E, anti-IgA in row F, anti-$\kappa$ light chain in row G, and anti-$\lambda$ light chain in row H. Incubate for 1–2 h. Wash three times in buffer A. Pipette 100 $\mu$l of diluted, peroxidase-conjugated antibody to rabbit immunoglobulin into all wells and incubate for 1 h. Wash four times in buffer A. Add 150 $\mu$l of *o*-phenylenediamine/$H_2O_2$ substrate into each well and stop color development with 50 $\mu$l of 12.5% $H_2SO_4$ when a suitable color intensity has been reached. Read the isotype by visual inspection or in a microplate reader.

### *Cloning of Hybridomas*

When the antibodies have been screened for usefulness as described earlier, they need to be cloned. Newly thawed cultures in HT medium should first be allowed to reach exponential growth. Then dilute cells to 50 cells/ml and 10 cells/ml in HT medium with Nutridoma CS. Make 5 ml of each suspension. Fill the two halves of a sterile microtiter plate with 1 drop of each of the suspensions. Add 1 drop/well of the same medium without cells. Incubate, feed, and screen by ELISA as after the fusion. Pick a positive culture, preferably from the half of the plate with the lowest density of positive wells. Expand as described for primary hybridoma cultures and clone once more. The number of cells per well is Poisson distributed: cells that have been cloned twice at a density where only about a fourth of the wells have cell colonies in them are 98% likely to be true clones. A further cloning improves this to 99.7%.

Some hybridomas are unstable so that a proportion of cells lose expression of antibody. Such hybridomas must be recloned from time to time or they will be lost, as nonproducers grow slightly faster.

## Antibody Production and Purification

Hybridomas usually grow best when kept at between $2 \times 10^5$ and $10^6$ cells/ml. Culture supernatants from dense cultures often contain about 20 $\mu$g of immunoglobulin per milliliter or more, which is sufficient for many applications. Larger amounts of antibodies were formerly produced in syngenic mice, as described elsewhere (Goding, 1996). This practice is, however, legally restricted in many countries and alternatives exist. We have good experience with CELLine devices (http://www.bdbiosciences.com), where the cells are grown inside a dialysis bag so that the antibodies are not diluted into the much larger nutrient compartment. Follow the instructions from the supplier. Stable hybridomas can be grown for several weeks and may yield antibody concentrations of several hundred micrograms per milliliter, comparable to concentrations obtained in peritoneal fluids. However, some hybridomas are unstable so that antibody production falls in the course of a few weeks, making this method uneconomical. The medium in the cell compartment is usually supplied with fetal calf serum at up to 30%. For some purposes, such as immunoprecipitation, the dilution of monoclonal antibodies by bovine immunoglobulin may be undesirable. If so, use Ultra-Low IgG fetal calf serum [e.g., from Invitrogen (http://www.invitrogen.com)].

IgG immunoglobulins may be purified on small columns of immobilized protein A or protein G, available from Amersham, Bio-Rad, and others, and used according to the manufacturer's instructions. Note that most murine antibodies only bind well to protein A at high pH and ionic strength. Some monoclonal antibodies lose titer when purified in that way, probably due to the harsh pH used for elution. Worse, the columns are difficult to strip for antibody, even with denaturants, so that traces of an antibody may elute together with the next antibody, purified on the same column. This causes problems with specificity. It may be costly to have a column for each antibody, so we favor purification by salt fractionation and ion-exchange chromatography: filter peritoneal fluid or culture supernatant through a small, wetted paper filter. Dilute the peritoneal fluid with 1 volume of 50 m*M* Tris/HCl, pH 7.5. Culture supernatants should not be diluted. Add ammonium sulfate to 40% saturation (240 g/liter), slowly and under vigorous stirring. Tumble or stir for >2 h. Spin at 7000 rpm for 10 min. Wash pellet by centrifugation in 40% saturated ammonium sulfate using a Dounce homogenizer to suspend the pellet.

Suspend the final pellet into 10 m*M* Tris-HCl, pH 8.0, and dialyze overnight against 1 liter of the same buffer. Filter through a 0.45-$\mu$m filter (spin first, if necessary). Apply to a DEAE or Q ion-exchange column. For instance, with a MonoQ HR 5/5 column (volume 1 ml, Amersham), equilibrate the column with 20 m*M* Tris-HCl, pH 8.0, and apply sample at 0.5 ml/min. Wash with 5 ml of the same buffer and elute with a gradient of 0–0.4 *M* NaCl in 20 m*M* Tris-HCl, pH 8.0, over 35 min at 500 $\mu$l/min. Collect fractions of 0.5 ml.

The first major peak is probably IgG. Assay the fractions by SDS–PAGE (IgG gives bands at around 25 and 50 kDa). Alternatively, dilute 1 $\mu$l from each fraction in 1 ml of buffer A with 1% calf serum and assay by ELISA, as for screening of hybridomas. Before the next use, the ion-exchange column should be purged with 0.1 *M* NaOH to denature and remove traces of adsorbed antibody. Pool appropriate fractions and dialyze against 150 m*M* NaCl, 22.1 m*M* $K_2HPO_4$, 2.9 m*M* $KH_2PO_4$, pH 7.6. Add 5 m*M* $NaN_3$ for storage. Alternatively, antibodies may be dialyzed against 0.1 *M* $NaHCO_3$ if the antibody is going to be coupled to a support such as NHS- or CNBr-activated matrices for use in immunoprecipitation or affinity chromatography. Activated supports are available from Amersham, Bio-Rad, or Pierce. Follow the procedures provided by the supplier.

## Affinity Purification of Antigen

Antibodies, which bind to the native antigen, are potentially useful for affinity purification. The antibody is purified (as described earlier) and coupled to NHS- or CNBr-activated matrices, which are available from several sources and used as recommended by the manufacturer. The coupling density is typically 2–5 mg of immunoglobulin per milliliter gel. The sample can be a supernatant of a raw cell extract centrifuged at 50,000*g* for 20 min, but a preliminary purification by ammonium sulfate precipitation extends column life considerably. Pump the sample into the column at a rate of 2 cm/min, follow with $A_{280}$, and wash with buffer until the baseline is low and stable. Invert the flow and elute with elution buffer at 2 cm/min. With nondenaturing buffers, it may be an advantage to decrease the flow rate. Glycine/HCl buffer, pH 2.5, is often used for elution but denatures many proteins, including proteasomes. If the sample is eluted with buffers at extreme pH, the fractions must therefore be neutralized immediately.

With some antibodies the antigen can be eluted by far milder conditions, e.g., even in distilled water or high concentrations of salt. ELISA can be used to screen for useful elution buffers (Bonde *et al.*, 1991): antigen is adsorbed to a microtiter plate, as described for the ELISA procedure, used to screen hybridomas (see earlier discussion). The plate is blocked with milk protein, washed, and incubated with monoclonal antibody, e.g., in the form of a supernatant from a hybridoma culture. After washing the plate, as described earlier, a large number of different potential elution buffers can be pipetted into individual wells. Wash the plate after 5–10 min and incubate with the peroxidase-coupled antibody to mouse immunoglobulin, as for hybridoma screenings. Wash the plates and incubate with the peroxidase substrate. Elution buffers, which release more than 20% of the monoclonal antibody, are worthwhile trying for elution of antigen bound in a column with antibody. Try the following elution buffers: 0.1 *M* glycine-HCl, pH 2.5; 0.1 *M* citrate/phosphate, pH 4; 0.1 *M* acetate buffer, pH 5; 0.1 *M* glycine/NaOH, pH 11; 100 m*M* Tris-HCl, pH 7, with 3 *M* NaCl; 100 m*M* Tris-HCl, pH 7, with 3 *M* KSCN; 100 m*M* Tris-HCl, pH 7, with 2 *M* NaI and a trace of sodium thiosulfate; 100 m*M* Tris-HCl, pH 7, with 4 *M* $MgCl_2$; 6 *M* guanidine-HCl, pH 3; 100 m*M* Tris-HCl, pH 7, with 3 *M* urea; 100 m*M* Tris-HCl, pH 7, in 50% ethylene glycol; distilled water. Depending on the results, buffers of other pH values may be tested, and denaturants may be diluted further. Next, check if the mildest potential elution conditions, found in the ELISA, irreversibly affect the antigen to be purified. For instance, columns with immobilized antibody MCP21 proved useful for the purification of human 20S proteasomes, which can be eluted with 2 *M* NaCl in a native state (Hendil and Uerkvitz, 1991) [the hybridoma is available from http://www.ecacc.org.uk and ready made MCP21-agarose from BioMol (http://www.biomol.com)].

## Examples: Use of Subunit-Specific Monoclonal Antibodies to Proteasomes

Immunoblotting is done as described for the screening of hybridomas. The same procedures can be followed when analyzing samples for the presence of proteasome subunits. The optimal dilution of the monoclonal antibody must be worked out for the particular experiment but is often more than 1000-fold for peritoneal fluids. Enhanced chemiluminescence detection of peroxidase-conjugated secondary antibodies gives sensitivities in the nanogram range and provides permanent records on X-ray film. Follow the procedures recommended by the supplier, e.g., Amersham or Pierce. The commercially available antibody MCP231 (http://www.biomol.com) reacts with several proteasome $\alpha$-type subunits from all eukaryotic species tested, namely human, mouse, rat, lobster, moth, potato, fission yeast, and budding yeast (Hendil *et al.*, 1995).

Immunoprecipitation of 26S proteasomes is often done with buffers containing ATP and glycerol and with low ionic strength, which stabilize 26S proteasomes. Work with chilled buffers. For mammalian proteasomes, homogenize the cell or tissue sample in 4 volumes of buffer C (25 m*M* Tris-HCl, pH 7.6, 1 m*M* dithiothreitol, 1 m*M* ATP, 2 m*M* EDTA, 10% glycerol, 0.05% Triton X-100), e.g., by sonication for a few seconds (cell cultures) or with a Potter–Elvehjem homogenizer or similar (for tissues). Protease inhibitors (e.g., Complete, Roche) may be added to the buffer. Check homogenization by microscopy. Centrifuge at 8000 rpm for 20 min and move the supernatant to a new tube with antibody and protein G-agarose. Use about 20 $\mu$g of antibody, typically corresponding to 10 $\mu$l of peritoneal fluid, and 10 $\mu$l of protein G-agarose for each milliliter of cell extract. Tumble the supernatant for >1 h in a cold room and wash the agarose beads in 7 × 10 ml of cold buffer C by centrifugations (1000 rpm for 5 min). Instead of using protein G, the antibody can be purified and coupled to CNBr-activated or NHS-activated beads (see earlier discussion). Antibodies, which are useful in immunoprecipitations, are available commercially (http://www.biomol.com). For instance, antibody S5a-18 reacts with subunit S5a/Rpn10 in the regulatory complex (PA700) and antibody MCP34 reacts with subunit XAPC7/$\alpha$4 of the 20S core proteasome. Most proteasome cofactors seem to bind to the regulatory complex or to the ends of the 20S proteasomes, i.e., to $\alpha$-type subunits. An antibody reacting with a $\beta$-type subunit (e.g., antibody MCP444; Kopp *et al.*, 1995) should therefore have the theoretical advantage of not interfering sterically with the binding of such cofactors.

Methods for the localization of antigens by immunocytochemistry are described in detail in Harlow and Lane (1999). As an example, tissue culture cells can be grown directly on sterilized coverslips in petri dishes. Take six coverslips with attached cells, wash them in serum-free medium, and fix for 10 min in freshly prepared 4% formaldehyde [in a hood, stir 4 g of paraformaldehyde in 100 ml of PBS (see earlier discussion) with a few drops of 1 *M* NaOH, heating to no more than 60°. Cool to room temperature and titrate back to pH 7.4 when the paraformaldehyde has dissolved]. After fixation, wash the coverslips in PBS and incubate for a few minutes in PBS with 0.1% Triton X-100. Wash twice in PBS. Place the coverslips, cells upward, on a wetted filter paper in a petri dish. Dilute the antibody, e.g., MCP20 (http://www.biomol.com), in 2% calf serum in PBS. Try dilutions of 1:10; 1:30, and 1:90. Add 1 drop of each antibody solution to each of two coverslips. Incubate for 1–2 h at room temperature. Wash the coverslips in three changes of 0.1% Triton X-100 in PBS and place them in a new dish with moist filter paper. Dilute fluorescence-labeled secondary antibody [e.g., Alexa Fluor 546-conjugated goat antimouse IgG, Molecular Probes (http://www.invitrogen.com)]. The supplier often recommends a dilution range. If not, try 1:200 and 1:1000 in 2% calf serum in PBS. Add 1 drop to each coverslip and incubate for 1 h at room temperature. Wash the coverslips in three changes of 0.1% Triton X-100 in PBS. Wipe the cell-free side of the coverslip with tissue paper and place the coverslip with cells downward on a drop of commercial mounting medium or glycerol on a microscope slide. Observe the slides using a fluorescence microscope with appropriate filters. Fluorescence will fade in the intense illumination, sometimes within minutes, so take photographs. Stained cells are often pretty. However, before rejoicing, see the next section on controls for specificity.

## Conclusions

After the hybridomas have been cloned, the properties of the antibodies should again be checked by blotting, precipitation, and so on, as in successful fusions there is a risk that the original well contained more than one hybridoma clone, reacting with the same antigen.

More detailed descriptions of immunostaining of cells and tissues can be found elsewhere (Harlow and Lane, 1999). Shepherd and Dean (2000) can be consulted on ELISA methods for the quantification of antigens.

Antibodies are wonderful reagents. The snag is in the controls. Preimmune serum, which provides a good control reagent with conventional antibodies, does not exist for monoclonal antibodies. Instead, one may use another monoclonal antibody of irrelevant specificity and of the same isotype. If the antigen is available in sufficient amounts, one may also block the antibody with a surplus of antigen, although this is not always a fail-safe proof of specificity (Kartner and Riordan, 1998). Where possible, identification of the antigen on a blot from a two-dimensional gel or of a precipitated antigen by mass spectrometry is preferred.

## Acknowledgment

I thank Dr. Rasmus Hartmann-Petersen for suggestions on the manuscript.

## References

Bonde, M., Frøkier, H., and Pepper, D. S. (1991). Selection of monoclonal antibodies for immunoaffinity chromatography: Model studies with antibodies against soy bean trypsin inhibitor. *J. Biochem. Biophys. Methods* **23,** 73–82.

Brooks, P., Murray, R. Z., Mason, G. G., Hendil, K. B., and Rivett, A. J. (2000). Association of immunoproteasomes with the endoplasmic reticulum. *Biochem. J.* **352,** 611–615.

Conconi, M., Djavadiohaniance, L., Uerkvitz, W., Hendil, K. B., and Friguet, B. (1999). Conformational changes in the 20S proteasome upon macromolecular ligand binding analyzed with monoclonal antibodies. *Arch. Biochem. Biophys.* **362,** 325–328.

Dutaud, D., Aubry, L., Henry, L., Levieux, D., Hendil, K. B., Kuehn, L., Bureau, J. P., and Ouali, A. (2002). Development and evaluation of a sandwich ELISA for quantification of the 20S proteasome in human plasma. *J. Immunol. Methods* **260,** 183–193.

Goding, J. W. (1996). "Monoclonal Antibodies: Principles and Practice." Academic Press, San Diego.

Harlow, E., and Lane, D. (1999). "Using Antibodies: A laboratory Manual." Cold Spring Habor Laboratory Press, Cold Spring Harbor, NY.

Hartmann-Petersen, R., Tanaka, K., and Hendil, K. B. (2001). Quaternary structure of the ATPase complex of human 26S proteasomes determined by chemical cross-linking. *Arch. Biochem. Biophys.* **386,** 89–94.

Hendil, K. B., Kristensen, P., and Uerkvitz, W. (1995). Human proteasomes analysed with monoclonal antibodies. *Biochem. J.* **305,** 245–252.

Hendil, K. B., and Uerkvitz, W. (1991). The human multicatalytic proteinase: Affinity purification using a monoclonal antibody. *J. Biochem. Biophys. Method* **22,** 159–165.

Hudson, L., and Hay, F. C. (1989). "Practical Immunology." Blackwell, Oxford.

Jorgensen, L., and Hendil, K. B. (1999). Proteasome subunit zeta, a putative ribonuclease, is also found as a free monomer. *Mol. Biol. Rep.* **26,** 119–123.

Kania, M. A., DeMartino, G. N., Baumeister, W., and Goldberg, A. L. (1996). The proteasome subunit, C2, contains an important site for binding of the PA28 (11S) activator. *Eur. J. Biochem.* **236,** 510–516.

Kartner, N., and Riordan, J. R. (1998). Characterization of polyclonal and monoclonal antibodies to cystic fibrosis transmembrane conductance regulator. *Methods Enzymol.* **292,** 629–652.

Kopp, F., Kristensen, P., Hendil, K. B., Johnsen, A., Sobek, A., and Dahlmann, B. (1995). The human proteasome subunit HsN3 is located in the inner rings of the complex dimer. *J. Mol. Biol.* **248,** 264–272.

Kopp, F., and Kuehn, L. (2003). Orientation of the 19S regulator relative to the 20S core proteasome: An immunoelectron microscopic study. *J. Mol. Biol.* **329,** 9–14.

Kurucz, E., Ando, I., Sumegi, M., Holzl, H., Kapelari, B., Baumeister, W., and Udvardy, A. (2002). Assembly of the *Drosophila* 26S proteasome is accompanied by extensive subunit rearrangements. *Biochem J.* **365,** 527–536.

Lamb, N. J. C., and Fernandez, A. (1997). Microinjection of antibodies into mammalian cells. *Methods Enzymol.* **283,** 72–83.

Nandi, D., Woodward, E., Ginsburg, D. B., and Monaco, J. J. (1997). Intermediates in the formation of mouse 20S proteasomes: Implications for the assembly of precursor beta subunits. *EMBO J.* **16,** 5363–5375.

Shepherd, P., and Dean, C. (2000). "Monoclonal Antibodies: A Practical Approach." Oxford University Press, Oxford.

Weitman, D., and Etlinger, J. D. (1992). A monoclonal antibody that distinguishes latent and active forms of the proteasome (multicatalytic proteinase complex). *J. Biol. Chem.* **267,** 6977–6982.

# Section V

# Isopeptidases

# [37] Preparation and Characterization of Yeast and Human Desumoylating Enzymes

*By* SHYR-JIANN LI, WILLIAM HANKEY, and MARK HOCHSTRASSER

## Introduction

As is true for ubiquitin, most ubiquitin-like proteins (Ubls) are processed from C-terminally extended precursor forms and are reversibly linked to substrate proteins. Ubl-specific proteases (ULPs) are proteases that cleave off the C-terminal peptides from Ubl precursors or cleave Ubls from molecules to which they had been added posttranslationally (or both). The first ULPs identified were a family of cysteine proteases that are specific for the Ubl called SUMO (small ubiquitin-like modifier) (Li and Hochstrasser, 1999, 2000). These proteases were shown to specifically cleave the peptide ($\alpha$-amino) or isopeptide ($\varepsilon$-amino) linkages after the C terminus of SUMO. The remainder of this chapter focuses almost exclusively on SUMO-specific ULPs.

SUMO-specific ULPs are responsible both for the processing of SUMO precursors to the active form and for selectively deconjugating SUMO from proteins (and presumably from other molecules, such as abundant cellular nucleophiles) (Hochstrasser, 2000; Johnson, 2004; Melchior *et al.*, 2003). Analogous to deubiquitinating enzymes (DUBs), the ability of ULPs to cleave isopeptide linkages between the C-terminal glycine of SUMO and the lysine side chains of modified substrate proteins allows the SUMO modification of proteins to be highly dynamic. For example, protein sumoylation profiles fluctuate dramatically as a function of cell cycle phase or growth condition (Li and Hochstrasser, 1999).

Despite the similarities shared by the ubiquitin and SUMO pathways, both in their biochemistry and in the sequences of many key enzymes (Schwartz and Hochstrasser, 2003), SUMO-specific ULPs and DUBS exhibit little obvious sequence similarity to one another. These ULPs are characterized by a ~200 residue segment called the ULP domain (UD); the core of this region bears distant similarity to the active site sequences of certain viral and bacterial cysteine proteases, including the adenovirus protease Avp, whose three-dimensional structure had been known (Li and Hochstrasser, 1999). A cocrystal structure of the C-terminal domain—the Ulp—of the yeast *Saccharomyces cerevisiae* Ulp1 protein linked covalently through its active site cysteine to yeast SUMO (called Smt3) has been determined (Mossessova and Lima, 2000). It revealed that the predicted

METHODS IN ENZYMOLOGY, VOL. 398

0076-6879/05 $35.00
DOI: 10.1016/S0076-6879(05)98037-3

His, Cys, and Asp catalytic triad residues were in fact arranged in the classical cysteine protease arrangement, with a conserved Gln residue serving as the oxyanion hole. The overall fold of the UD was similar to that of Avp.

*Saccharomyces cerevisiae* have two SUMO proteases, Ulp1 and Ulp2, which have distinct substrate specificities and participate in different cellular processes (Li and Hochstrasser, 1999, 2000). Ulp1 accounts for most of the SUMO precursor processing *in vivo* and desumoylates a subset of isopeptide-linked SUMO–protein conjugates. Deletion of *ULP1* is lethal. Conditional *ulp1-ts* mutants arrest as large-budded cells with short spindles and undivided nuclei at the nonpermissive temperature, indicating that Ulp1 function is required in G2/M prior to the metaphase-to-anaphase transition.

Ulp2 is less active than Ulp1 against the SUMO precursor *in vivo*, and *in vitro* Ulp2 has shown much lower activity against both recombinant SUMO substrates bearing peptide bond-linked C-terminal peptides and isopeptide-linked conjugates isolated from yeast (Li and Hochstrasser, 2000). The *ulp2Δ* mutant is viable, but exhibits a pleiotropic phenotype that includes slow growth, a severe sporulation defect, and hypersensitivity to UV radiation, elevated temperature, and chemicals that damage the mitotic spindle or DNA. Some *ulp2Δ* defects are apparently caused by the accumulation of isopeptide-linked SUMO homopolymers, suggesting that Ulp2 also functions in limiting the levels of free SUMO chains *in vivo* (Bylebyl *et al.*, 2003). Ulp1- and Ulp2-deficient strains accumulate different sumoylated proteins, based on anti-SUMO Western immunoblot analysis (Li and Hochstrasser, 2000). This observation and the distinctive phenotypes of these mutants indicate that Ulp1 and Ulp2 act on distinct substrates in the cell. This is at least partially attributable to their different subcellular localization. Ulp1 concentrates at nuclear pore complexes, and Ulp2 localizes to the nucleus (Li and Hochstrasser, 2000; Panse *et al.*, 2003; Schwienhorst *et al.*, 2000). The N-terminal domain of Ulp1 binds to specific karyopherins, tethering the enzyme to the NPC (Panse *et al.*, 2003), which appears to limit the access of Ulp1 to Ulp2 substrates inside the nucleus (Li and Hochstrasser, 2003).

Sequence database searches for potential orthologs of the yeast ULPs yielded seven putative hULPs (or SENPs, for sentrin-specific proteases; sentrin is an alternative name for SUMO), based on the presence of the UD domain (Li and Hochstrasser, 1999; Yeh *et al.*, 2000). Most of these have been shown to have desumoylating activity. Interestingly, one of them, hULP8/SENP8/Den1/NEDP1, was shown to process a distinct Ubl, called NEDD8 or Rub1, rather than SUMO (Mendoza *et al.*, 2003; Wu *et al.*, 2003). The human ULPs and their splice variants are localized to different cellular compartments and are anticipated to have different *in vivo* substrate specificities as well (Seeler and Dejean, 2003).

In addition to the large number of different SUMO targets in humans (over 50 substrates have been identified thus far), SUMO is actually encoded by a small gene family (*SUMO1* through *SUMO4*) (Bohren *et al.*, 2004; Guo *et al.*, 2004). Thus, human ULPs, and those from many other organisms, may distinguish not only between different SUMO-linked substrates, but also between variant SUMO precursors and SUMO variant-specific protein conjugates. The availability of purified ULPs and assays for their activity provide a starting point from which to approach questions of ULP substrate specificity and cellular function.

## Preparation of Yeast Desumoylating Enzymes Ulp1 and Ulp2

One of the advantages to using a bacterial system of expression for ULP purification is that *Escherichia coli* contains no endogenous desumoylating activity (Li and Hochstrasser, 1999). Active, full-length glutathione *S*-transferase (GST) fusions of yeast Ulp1 and Ulp2 are purified readily by one-step glutathione affinity chromatography. One or more subsequent column chromatography steps can be used in conjunction with this initial purification to yield preparations of higher purity. Plasmids for the expression of full-length recombinant GST-Ulp1 and GST-Ulp2 have been constructed using the pGEX-KG vector (Li and Hochstrasser, 1999, 2000). *E. coli* JM101 transformed with pGEX-Ulp1 or pGEX-Ulp2 are maintained on an M9 minimal plate containing 50 $\mu$g/ml ampicillin. Expression levels have been observed to decrease if cultures are inoculated from old plates rather than freshly transformed cells.

An overnight culture of JM101 carrying either pGEX-Ulp1 or pGEX-Ulp2 is diluted 1:100 into 4 liters of LB containing 100 $\mu$g/ml ampicillin. GST-Ulp expression is then induced with 0.4 m*M* isopropyl-$\beta$-D-thiogalactoside (IPTG), which is added during midlog phase ($OD_{600} = 0.8$) for 3 h at 30°. The cells are pelleted (5000*g* for 10 min), washed with ice-cold TBS (50 m*M* Tris-HCl, pH 7.5, 150 m*M* NaCl), and resuspended in 2–5 cell pellet volumes of lysis buffer [50 m*M* Tris-HCl, pH 7.5, 150 m*M* NaCl, 2 m*M* dithiothreitol (DTT), 5 m*M* EDTA, 2 m*M* phenylmethylsulfonyl fluoride (PMSF), and 20 $\mu$g/ml each of the protease inhibitors antipain, aprotinin, chymostatin, leupeptin, and pepstatin]. It is important to include 2 m*M* DTT in the buffer to preserve cysteine protease activity. The cell suspension is supplemented with lysozyme to a final concentration of 100 $\mu$g/ml and is kept on ice for 30 min. Cells are lysed with 0.1% Triton X-100 by gentle mixing and sonication with a Misonix ultrasonic processor (setting 4; 1-s pulse cycle for 5 min) to improve cell lysis and to reduce lysate viscosity.

After clearing of cell debris by centrifugation at 10,000*g* in a Beckman JA-20 rotor for 30 min at 4°, the supernatant is ready for fractionation by glutathione affinity chromatography. For GST-Ulp1, ~60% of the protein is found in the soluble fraction.

Ten milliliters (settled volume) of glutathione-Sepharose resin (Amersham) preequilibrated in lysis buffer is bound to the GST-Ulp fusion protein in the bacterial lysate during 60 min of gentle mixing on a rotating platform at 4°. The resin is then packed into a disposable column (Bio-Rad) by gravity and is washed with 20 column volumes of ice-cold lysis buffer with 0.1% Triton X-100 until the $OD_{280}$ is less than 0.1. The GST-Ulp fusion protein is eluted with 2 column volumes of lysis buffer containing 20 m*M* reduced glutathione (Sigma, St. Louis, MO). Most of the fusion protein is eluted in the breakthrough fraction. The presence and purity of the GST-Ulp fusion can be tested by an activity assay (see later) and by SDS–PAGE. Fractions containing active protein are then pooled and dialyzed overnight against 1000 volumes of dialysis buffer (50 m*M* Tris-HCl, 150 m*M* NaCl, 2m*M* DTT, 1 m*M* PMSF, and 20% glycerol) at 4°. Both GST-Ulp1 and GST-Ulp2 eluates from the glutathione-Sepharose column contain proteolytic fragments derived from the GST-fusion proteins. Roughly 25% of GST-Ulp1 is full-length, but proteolysis is more severe for the ~140-kDa GST-Ulp2 fusion protein. The glutathione-Sepharose eluates can be fractionated further on an S-200 gel filtration column (16 × 60 cm, Amersham). The column is equilibrated in 50 m*M* Tris-HCl, 150 m*M* NaCl, and 2 m*M* DTT and is developed in the same buffer. This produces satisfactory purifications of the full-length GST-Ulps: ~85% pure in the case of GST-Ulp1 and ~50% pure for GST-Ulp2, which are suitable for many applications. Additional chromatography procedures can be used if higher purity is required. The yield of purified GST-Ulp1 is approximately 50 μg per liter of culture, whereas that of GST-Ulp2 is roughly 10 μg per liter.

As an alternative to using full-length GST-Ulp1, several N-terminal truncations of Ulp1 that retain SUMO protease activity have been generated and characterized *in vivo* and *in vitro* (Li and Hochstrasser, 2003; Mossessova and Lima, 2000). GST-Ulp1C275 contains only the C-terminal 275 amino acids of Ulp1 and is both expressed at a higher level and subject to less degradation in *E. coli* than full-length GST-Ulp1. The affinity purification scheme for GST-Ulp1C275 is identical to that of the full-length protein. Because GST-Ulp1C275 has a basic isoelectric point, a CM-Sephadex (Amersham) purification step can be used to purify the affinity-purified enzyme further. The pooled GST-Ulp1C275 fractions collected from the glutathione-Sepharose column are dialyzed into a low-salt buffer containing 50 m*M* Tris-HCl, pH 7.4, and 2 m*M* DTT and are then loaded onto a preequilibrated 10-ml Hi-Trap CM-Sephadex cartridge (Amersham). The column is washed with 2 volumes of loading buffer and eluted with an NaCl gradient (from 0 to 0.5 *M*). The GST-Ulp1C275 protein elutes at 0.25 *M* NaCl, separate from the degradation products, which come off in lower salt fractions. The approximate yield of purified GST-Ulp1C275 is 600 $\mu$g per liter of culture.

Attempts to use a GST fusion of the catalytic domain of Ulp2 alone (GST-UD2) have been less successful. Although the protein is produced at much higher levels in *E. coli* than the full-length protein and is purified readily by affinity chromatography, very little activity against model SUMO fusion substrates has been detected. As an alternative, we have also attempted to express full-length Ulp2 in *E. coli* with a different purification tag, specifically an N-terminal $His_6$ peptide (pQE30-Ulp2). This is not significantly better than GST-Ulp2 with respect to expression level, yield, or activity (S.-J. Li and M. Hochstrasser, unpublished result).

## Preparation of Human Desumoylating Enzymes

We have also generated *E. coli* plasmids for the expression of GST fused to different human ULPs (hULP or SENP) (Table I) (Li and Hochstrasser, 1999; Yeh *et al.*, 2000). We have found that full-length GST-hULP3 (SENP3) and GST-hULP8 (SENP8) are readily expressed in and purified from bacteria. However, deletion of N-terminal sequences from hULP1, hULP2, and hULP5 is necessary to facilitate their expression in *E. coli*. After a series of trials, we have found that deletion of the first 300 amino acids from hULP1, the first 100 amino acids from hULP2, and the first 60 amino acids from hULP5 are sufficient to enable reasonable

TABLE I
HUMAN ULP PROTEINS[a]

| Name | SENP | Synonyms/orthologs |
|---|---|---|
| hULP1 | SENP1 | SuPr-2 (mouse) |
| hULP2 | SENP2 | SSP3, Smt3ip2 (splice variant), SuPr-1 (splice variant), Axam (rat) |
| hULP3 | SENP3 | SMT3IP1, SUSP3, SENP4 (splice variant), SuPr-3 (mouse) |
| hULP5 | SENP5 | |
| hULP6 | SENP6 | SUSP1 |
| hULP7 | SENP7 | SUSP2 |
| hULP8 | SENP8 | DEN1, NEDP1 |

[a] We use the original ULP nomenclature here (Li and Hochstrasser, 1999) but have carried over the numbering from SENP names (Yeh *et al.*, 2000) to avoid future confusion.

expression and facilitate purification of these proteins. We have so far been unable to express either full-length or truncated forms of hULP6 and hULP7 as GST fusions in *E. coli.* The conditions for the growth and induction of bacterial cultures and for affinity purification of the human GST-ULP fusions are essentially the same as those described for yeast GST-Ulp1 and GST-Ulp2. We have purified GST-hULP1($\Delta$N) and GST-hULP2($\Delta$N) and have shown peptidase and isopeptidase activity against human SUMO substrates for both proteins (S.-J. Li, unpublished results).

## Desumoylating Enzyme Activity Assays

This section describes enzyme assays that use a chimeric recombinant SUMO fusion protein as substrate and also describes a method of generating crude preparations of isopeptide-linked sumoylated substrates that can be used for *in vitro* analysis of ULP activity by anti-SUMO immunoblotting. Intein-based peptide ligation has been used to synthesize fluorogenic SUMO-AMC (7-amido-4-methyl coumarin) substrates (Gan-Erdene *et al.*, 2003). The availability of fluorogenic substrates simplifies kinetic analyses of desumoylating enzymes, although the use of SUMO-AMC by itself cannot address questions of ULP specificity for different SUMO-protein conjugates. We have also tried, so far unsuccessfully, to develop an AMC-labeled peptide substrate for yeast Ulp1 based on the C-terminal sequence of SUMO. This approach may have failed because a larger contact area between Ulp1 and SUMO might be needed for stable binding (Mossessova and Lima, 2000).

### *Peptidase Activity Assays*

*Recombinant Model Substrate Protein.* A chimeric substrate protein containing an N-terminal $His_6$-tag fused to yeast SUMO (Smt3) with a C-terminal HA tag has been used successfully to characterize yeast Ulp activities (Li and Hochstrasser, 1999, 2000). Cleavage by Ulp1 of the nine-amino acid HA peptide that follows the SUMO C terminus produces a size shift detected readily by SDS–PAGE. This cleavage reaction can be monitored either by immunoblotting with antibodies against $His_6$ or Smt3 or, if a radiolabeled substrate is used, by phosphorimager measurement or autoradiography. The use of radiolabeled substrate facilitates quantitation.

A plasmid expressing the chimeric $His_6$-SUMO-HA substrate construct from a *lac* promoter is transformed into an *E. coli* lacI$^Q$ strain such as JM101 (Li and Hochstrasser, 1999). Genes encoding yeast Smt3, human SUMO1, or human SUMO2 have been cloned into pQE30 (Qiagen) for this purpose. For purifying the radiolabeled SUMO chimeras, expression of the recombinant substrate is induced with 1 m$M$ IPTG in a 50-ml LB + ampicillin culture of log phase JM101 cells and incubated for 1 h at 37°. The cells are then pelleted, washed twice in M9 minimal medium (1 m$M$ $MgSO_4$, 0.2% glucose, 10 $\mu$g/ml thiamine, 1 m$M$ $CaCl_2$), and resuspended in 25 ml of M9 minimal medium with 25 $\mu$g/ml ampicillin and 1 m$M$ IPTG. $^{35}$S-Translabel (1 mCi; Amersham) is then added, and the culture is placed on a shaker platform for an additional hour at 37°. Cells are harvested and washed once with M9 minimal medium and are then resuspended in 0.5 ml spheroplasting buffer (25% sucrose, 50 m$M$ Tris-HCl, pH 8.0). Lysozyme (100 $\mu$g) is added at this point, and the suspension is left on ice for 5 min. The cell suspension is diluted to 1 ml with 50 m$M$ Tris-HCl, pH 8.0, buffer, and 10 $\mu$l of 10% Triton X-100 is added to lyse the cells. The resulting lysate is sonicated briefly if necessary and is centrifuged at 14,000$g$ in a microcentrifuge at 4° for 10 min. The NaCl concentration in the resulting supernatant is adjusted to 150 m$M$ with 5 $M$ NaCl. These Triton-based lysis conditions do not require intensive sonication (especially if used in conjunction with DNase I treatment to reduce lysate viscosity); prolonged sonication should be conducted with caution to minimize the risk of producing radioactive aerosol.

One milliliter of cobalt-affinity matrix (BD Talon-Clontech) preequilibrated in TBST (150 m$M$ NaCl, Tris-HCl, pH 7.5, 0.05% Triton X-100) is incubated with the cell lysate using gentle mixing for 1 h on a rotator at 4° to bind the $His_6$-tagged protein to the resin. The beads are poured into a disposable column (Bio-Rad, Polyprep) and washed with 50 ml TBST. The bound proteins are eluted with 100 m$M$ imidazole in TBS and dialyzed overnight into TBS at 4° against 1000 volumes of dialysis buffer. The purity of the substrate is checked by SDS–PAGE on a 12.5% gel. The expected yield of the purified substrate is approximately 1 mg/liter. Specific radioactivity can be measured by radioactive counting of an aliquot of the purified protein and comparing it to the protein concentration. The specific activity of GST-Ulp1 is 120 nmol of substrate cleaved per milligram of enzyme per minute, whereas the activity of the purified GST-Ulp2 is too low to provide a meaningful measurement of specific activity.

Substrate cleavage assays are performed at 30° with varying concentrations of substrate and GST-Ulp1 or GST-Ulp2 in a reaction buffer containing 150 m$M$ NaCl, 1 m$M$ DTT, 50 m$M$ Tris-HCl, pH 7.5, and 0.2% Triton X-100 (the latter for GST-Ulp2 only). Typical substrate concentrations are 1–10 $\mu M$, and enzyme concentrations are usually in the range of 0.3–3 n$M$. At each time point, an aliquot of the reaction mix is added to an equal volume of 2× SDS gel-loading buffer and heated to 95° for 5 min to stop the reaction. For assaying GST-Ulp1 activity, 5-min time points are usually taken over a 30-min period, but for GST-Ulp2, the reaction times need to be extended to 3 h and longer. Samples from each time point can then be separated on a 12.5% SDS–polyacrylamide gel and analyzed by phosphorimaging. The percentage of cleavage at each time point is calculated from the ratio of pixels in the cleavage product (the fast-migrating His6-SUMO band) to the combined pixels of the cleavage product plus remaining substrate (the slow-migrating His6-SUMO-HA band) in the same lane. (The HA tag does not contain any methionine or cysteine, so no radioactivity is lost due to its cleavage.) When we substitute human SUMO1 or SUMO2 for the yeast Smt3 sequence in the chimeric substrate, they are readily cleaved in assays of human ULPs.

## SUMO-AMC Substrate

To obtain a model substrate for SUMO-specific proteases that can be monitored by fluorescence, we generated a fusion of human SUMO1 (Gly96) and intein-CBD (*Bacillus circulans* chitin-binding domain) in a pTYB vector backbone (New England Biolab). The fusion protein product can be expressed in *E. coli* and purified using a chitin affinity column as described by the manufacturer (Impact system, New England Biolab). SUMO1-MESNa ($\beta$-mercaptoethanesulfonic acid) is generated by cleavage of the fusion protein with MESNa. SUMO-MESNa is then converted to SUMO-AMC (7-amido-4-methyl coumarin) by a chemical ligation reaction performed with a large excess of glycine-AMC. Details of the fusion protein purification and chemical ligation reaction have been published (Gan-Erdene *et al.*, 2003). In addition, human SUMO1-AMC is now available from a commercial source (Boston Biochem).

Hydrolysis of the SUMO-AMC substrate is determined spectrofluorometrically in a final volume of 100 $\mu$l. Typical assays contain 10 to 100 p*M* concentrations of enzyme, and the substrate is present in $10^3$ to $10^4$ molar excess over enzyme in the reaction buffer (50 m*M* Tris HCl, pH 7.5, at 25°). The estimated $K_m$ of human GST-hUlp1 for this substrate is 300 n*M*, measured at 10 p*M* concentration of the purified enzyme (Li and Craig Hill, unpublished result).

## Isopeptidase Activity Assays

### In Vitro-*Translated RanGAP1*

Sumoylated RanGAP1 is one of the most abundant and best-characterized SUMO1-protein conjugates in mammalian cells. RanGAP1 contains a single SUMO modification site at Lys526 (Sampson *et al.*, 2001), and approximately 20–30% of RanGAP1 translated in a rabbit reticulocyte lysate system is conjugated to SUMO present in the reticulocyte lysate, presumably by the conjugating enzymes also contained in the lysate. A typical translation reaction consists of an aliquot of transcription/translation coupling mix (STP3 system, Novagen), [$^{35}$S]methionine/cysteine Translabel (Amersham), and purified supercoiled plasmid DNA (the vector pAlter-Max A contains a MYC-tagged mouse RanGAP1 cDNA controlled by a T7 promoter; kindly provided by Dr. M. Matunis of John Hopkins University). The translation reaction mix is incubated for 1 h at 30°, and the reaction is terminated by the addition of RNase A (10 $\mu$g/ml, final concentration). *N*-Ethylmaleimide (NEM) is then added to the mixture (1 m*M*, final concentration) and incubated for 15 min at 23° to inactivate both sumoylating and desumoylating activities in the reticulocyte lysate. $\beta$-Mercaptoethanol and L-cysteine (2 m*M* each, final concentration) are added to neutralize excess NEM, and the reaction mixture can be used directly as substrate in the isopeptidase reaction.

We have shown that Ulp1 from *S. cerevisiae* cleaves Lys526-linked mammalian SUMO1-RanGAP1 in a time-dependent manner (Li and Hochstrasser, 1999). The reaction mixture contains 5 $\mu$l of *in vitro*-translated RanGAP1, 1 n*M* purified GST-Ulp1, 20 m*M* Tris-HCl, pH 8.0, 150 m*M* NaCl, and 1 m*M* DTT in a final volume of 10 $\mu$l. The progress of the isopeptidase reaction is monitored by the disappearance of the slower-migrating $^{35}$S-labeled sumoylated RanGAP1 band on a 12.5% SDS–PAGE gel and is quantified by phosphorimaging. SUMO1-RanGAP1 is processed with a $t_{1/2}$ of roughly 7 min in the presence of ~1 n*M* GST-Ulp1 (Li and Hochstrasser, 1999).

### *NEM-Treated Cell Lysate*

While sumoylated RanGAP1 is a well-defined isopeptide-linked substrate, some ULPs may have restricted activity toward this particular substrate (Kim *et al.*, 2000). We have therefore incubated either GST-Ulp1 or GST-Ulp2 with NEM-treated yeast cell lysate to demonstrate their abilities to cleave endogenous sumoylated proteins (Li and Hochstrasser, 1999, 2000). Wild-type or *ulp* mutant yeast cells (~$10^9$ cells) are harvested from log-phase cultures in YPD, washed once with buffer A (1.2 *M* sorbitol, 50 m*M* Tris-HCl, pH 7.5), and incubated in 1 ml of buffer A containing 0.5 mg of zymolyase 100T/ml for 30 min at 30°. After cell wall digestion, the cells are washed once in cold buffer A and lysed by sonication on ice in 0.5 ml buffer B (50 m*M* Tris-HCl, pH 8.0, 5 m*M* EDTA, 150 m*M* NaCl, 0.2% Triton X-100, 2 m*M* NEM, 2 m*M* PMSF, and 20 $\mu$g/ml each of leupeptin, pepstatin, and antipain). The resulting lysate is centrifuged at 14,000*g* for 10 min to clear cell debris, and soluble protein concentrations are determined by the bicinchoninic acid protein assay (Pierce). Prior to initiation of the cleavage reactions, L-cysteine and $\beta$-mercaptoethanol are added to 2 m$M$ each and incubated at 23° for 15 min to consume any remaining unreacted NEM. Reactions are initiated by the addition of 50 ng of purified GST-Ulp to a 20-$\mu$l reaction mixture containing 25 $\mu$g of soluble yeast protein in 50 m$M$ Tris-HCl, pH 8.0, 150 m$M$ NaCl, 1 m$M$ DTT, and 0.1% Triton X-100 and are terminated by boiling in SDS sample buffer. Reactions are monitored by anti-SUMO immunoblot analysis. Ulp activity in yeast cell lysates has been observed to be unaffected by various protease inhibitors.

In addition to the methods described earlier, we have developed other methods to demonstrate ULP enzyme activity. For example, we immobilized the chimeric substrate $His_6$-ubiquitin-Smt3-HA to a 96-well plate (His sorb plate, Qiagen) and monitored the cleavage reaction with an HRP-linked anti-HA antibody in an ELISA-based colorimetric reaction (using hydrogen peroxide and the substrate 3,3′,5,5′-tetramethyl benzidine). This method gives rapid results and is particularly useful for following ULP activity during enzyme purification (S.-J. Li, unpublished results). Purified SUMO E1 and E2 proteins (UBA2-AOS1 and UBC9) are now available from commercial sources (Alexis Biochemicals, San Diego, CA, or Boston Biochem, Boston, MA) so it is also possible, albeit expensive, to generate isopeptide-linked SUMO homopolymers as additional substrates for the detection of isopeptidase activity.

## References

Bohren, K. M., Nadkarni, V., Song, J. H., Gabbay, K. H., and Owerbach, D. (2004). A M55V polymorphism in a novel SUMO gene (SUMO-4) differentially activates heat shock transcription factors and is associated with susceptibility to type I diabetes mellitus. *J. Biol. Chem.* **279,** 27233–27238.

Bylebyl, G. R., Belichenko, I., and Johnson, E. S. (2003). The SUMO isopeptidase Ulp2 prevents accumulation of SUMO chains in yeast. *J. Biol. Chem.* **278,** 44113–44120.

Gan-Erdene, T., Nagamalleswari, K., Yin, L., Wu, K., Pan, Z. Q., and Wilkinson, K. D. (2003). Indentification and characterization of DEN1, a deneddylase of the ULP family. *J. Biol. Chem.* **278,** 28882–28891.

Guo, D., Li, M., Zhang, Y., Yang, P., Eckenrode, S., Hopkins, D., Zheng, W., Purohit, S., Podolsky, R. H., Muir, A., Wang, J., Dong, Z., Brusko, T., Atkinson, M., Pozzilli, P., Zeidler, A., Raffel, L. J., Jacob, C. O., Park, Y., Serrano-Rios, M., Larrad, M. T., Zhang, Z., Garchon, H. J., Bach, J. F., Rotter, J. I., She, J. X., and Wang, C. Y. (2004). A functional variant of SUMO4, a new I kappa B alpha modifier, is associated with type 1 diabetes. *Nature Genet.* **36,** 837–841.

Hochstrasser, M. (2000). Evolution and function of ubiquitin-like protein-conjugation systems. *Nature Cell Biol.* **2,** E153–E157.

Johnson, E. S. (2004). Protein modification by SUMO. *Annu. Rev. Biochem.* **73,** 355–382.

Kim, K. I., Baek, S. H., Jeon, Y. J., Nishimori, S., Suzuki, T., Uchida, S., Shimbara, N., Saitoh, H., Tanaka, K., and Chung, C. H. (2000). A new SUMO-1-specific protease, SUSP1, that is highly expressed in reproductive organs. *J. Biol. Chem.* **275,** 14102–14106.

Li, S.-J., and Hochstrasser, M. (1999). A new protease required for cell-cycle progression in yeast. *Nature* **398,** 246–251.

Li, S. J., and Hochstrasser, M. (2000). The yeast ULP2 (SMT4) gene encodes a novel protease specific for the ubiquitin-like Smt3 protein. *Mol. Cell. Biol.* **20,** 2367–2377.

Li, S.-J., and Hochstrasser, M. (2003). The Ulp1 SUMO isopeptidase: Distinct domains required for viability, nuclear envelope localization, and substrate specificity. *J. Cell Biol.* **160,** 1069–1081.

Melchior, F., Schergaut, M., and Pichler, A. (2003). SUMO: Ligases, isopeptidases and nuclear pores. *Trends Biochem. Sci.* **28,** 612–618.

Mendoza, H. M., Shen, L. N., Botting, C., Lewis, A., Chen, J., Ink, B., and Hay, R. T. (2003). NEDP1, a highly conserved cysteine protease that deNEDDylates Cullins. *J. Biol. Chem.* **278,** 25637–25643.

Mossessova, E., and Lima, C. D. (2000). Ulp1-SUMO crystal structure and genetic analysis reveal conserved interactions and a regulatory element essential for cell growth in yeast. *Mol. Cell* **5,** 865–876.

Panse, V. G., Kuster, B., Gerstberger, T., and Hurt, E. (2003). Unconventional tethering of Ulp1 to the transport channel of the nuclear pore complex by karyopherins. *Nature Cell Biol.* **5,** 21–27.

Sampson, D. A., Wang, M., and Matunis, M. J. (2001). The small ubiquitin-like modifier-1 (SUMO-1) consensus sequence mediates Ubc9 binding and is essential for SUMO-1 modification. *J. Biol. Chem.* **276,** 21664–21669.

Schwartz, D. C., and Hochstrasser, M. (2003). A superfamily of protein tags: Ubiquitin, SUMO and related modifiers. *Trends Biochem. Sci.* **28,** 321–328.

Schwienhorst, I., Johnson, E. S., and Dohmen, R. J. (2000). SUMO conjugation and deconjugation. *Mol. Gen. Genet.* **263,** 771–786.

Seeler, J. S., and Dejean, A. (2003). Nuclear and unclear functions of SUMO. *Nature Rev. Mol. Cell. Biol.* **4,** 690–699.

Wu, K., Yamoah, K., Dolios, G., Gan-Erdene, T., Tan, P., Chen, A., Lee, C. G., Wei, N., Wilkinson, K. D., Wang, R., and Pan, Z. Q. (2003). DEN1 is a dual function protease capable of processing the C terminus of Nedd8 and deconjugating hyper-neddylated CUL1. *J. Biol. Chem.* **278,** 28882–28891.

Yeh, E. T., Gong, L., and Kamitani, T. (2000). Ubiquitin-like proteins: New wines in new bottles. *Gene* **248,** 1–14.

# [38] Purification of the COP9 Signalosome from Porcine Spleen, Human Cell Lines, and *Arabidopsis thaliana* Plants

*By* Suchithra Menon, Vicente Rubio, Xiping Wang, Xing-Wang Deng, and Ning Wei

## Abstract

COP9 signalosome (CSN) is an evolutionarily conserved multisubunit protein complex involved in diverse cellular and developmental processes in eukaryotes. CSN functions in the cell as proteases that deconjugate Nedd8/Rub1 from cullin family proteins and depolymerize ubiquitin chains. As such, CSN represents an important regulator of multiple cullin-based E3 ubiquitin ligases. CSN has also been shown to associate with protein kinase activities. This chapter describes purification of the CSN complex by classical chromatography from porcine spleen and by immunoaffinity purification procedures from cultured human cells and transgenic *Arabidopsis* plants expressing epitope-tagged CSN subunits. It also describes *in vitro* deneddylation assays using the HeLa cell extract or the *Arabidopsis* cell extract, which we have used to test and compare the activity of purified CSN complexes.

## Introduction

COP9 signalosome is a conserved protein complex (~500 kDa) that regulates cullin-based E3 ubiquitin ligases. Thus far, three distinct biochemical activities have been ascribed to this complex: deconjugation of Nedd8/Rub1 from cullins (deneddylation), deubiquitinating (DUB) activity, and associated protein kinase activities (Lyapina *et al.*, 2001; Schwechheimer *et al.*, 2001; Seeger *et al.*, 1998; Zhou *et al.*, 2003). Among them, the cullin deneddylation activity of CSN is most characterized both genetically and biochemically. We have used the deneddylation assay routinely to characterize the activity of the purified CSN complex.

Cullins usually serve as scaffold components of SCF-like E3 ubiquitin ligases. The SCF complex consists of cullin1 (Cul1), a RING-finger protein Rbx1/Roc1/Hrt1, Skp1, and an F-box protein that acts as substrate receptor (reviewed in Deshaies, 1999). Cullin family members are modified by ubiquitin-like protein Nedd8 (human) or Rub1 (yeast and *Arabidopsis*) (Hochstrasser, 2000), which is shown to promote the polyubiquitination of

METHODS IN ENZYMOLOGY, VOL. 398

0076-6879/05 $35.00
DOI: 10.1016/S0076-6879(05)98038-5

SCF substrates (Pan *et al.*, 2004). CSN interacts with Cul1 and Rbx1/Roc1/Hrt1 and cleaves Nedd8/Rub1 from Cul1 through a metalloprotease activity centered in the JAMM motif of CSN5 (Cope *et al.*, 2002; Lyapina *et al.*, 2001; Schwechheimer *et al.*, 2001).

CSN, or the COP9 complex, was first identified genetically as a repressor of photomorphogenesis in *Arabidopsis* (Wei *et al.*, 1994). The first biochemical purification of the CSN complex was from cauliflower head due to ample availability of the source (Chamovitz *et al.*, 1996). The chromatographic procedure was later combined with an immunoaffinity step using specific anti-CSN antibodies to purify and identify all of the plant CSN subunits (Serino *et al.*, 1999). The mammalian CSN complex was purified to homogeneity from porcine (pig) spleen (Wei and Deng, 1998; Wei *et al.*, 1998) and from human red blood cells (Seeger *et al.*, 1998). These efforts have identified the complete subunit composition of CSN (CSN1 to CSN8) from plants and animals, revealing the evolutionary conservation and its remarkable homology to the proteasome lid subcomplex (reviewed in Wei and Deng, 2003). The chromatographically purified mammalian CSN complex has been used in many activity assays, such as *in vitro* cullin deneddylation and *in vitro* ubiquitination assays (Cope *et al.*, 2002; Lyapina *et al.*, 2001; Yang *et al.*, 2002), DUB assays (Zhou *et al.*, 2003), and cell microinjection experiments (Yang *et al.*, 2002).

More recently, purification of the CSN complex has been accomplished by a one-step immunopurification method using epitope-tagged CSN subunits in humans (Wu *et al.*, 2003; this article), *Schizosaccharomyces pombe* (Liu *et al.*, 2003), and *Arabidopsis thaliana* (this article). The following section describes in detail the procedures used to purify CSN from porcine spleen by chromatography, from human 293 cells by an immunoaffinity purification method, and from *Arabidopsis* seedlings by tandem affinity purification (TAP) and an immunopurification procedure. *In vitro* cullin deneddylation assays using human or plant cell extracts, which we have used to compare the activity of the CSN complex purified by various procedures, are also described.

## Biochemical Purification of CSN Complex from Porcine Spleen

The initial survey of CSN levels among various mouse organ types by an anti-CSN8 blot indicated that CSN is highly enriched in thymus, spleen, placenta, and brain (Wei and Deng, 1998). We therefore obtained frozen porcine spleen from Pel-Freez Biologicals (AR, www.pelfreez-bio.com) as the starting source for CSN biochemical purification. All of the chromatographic procedures are performed with the aid of an automatic FPLC system (Amersham-Pharmacia) in a cold room. Except for those specified,

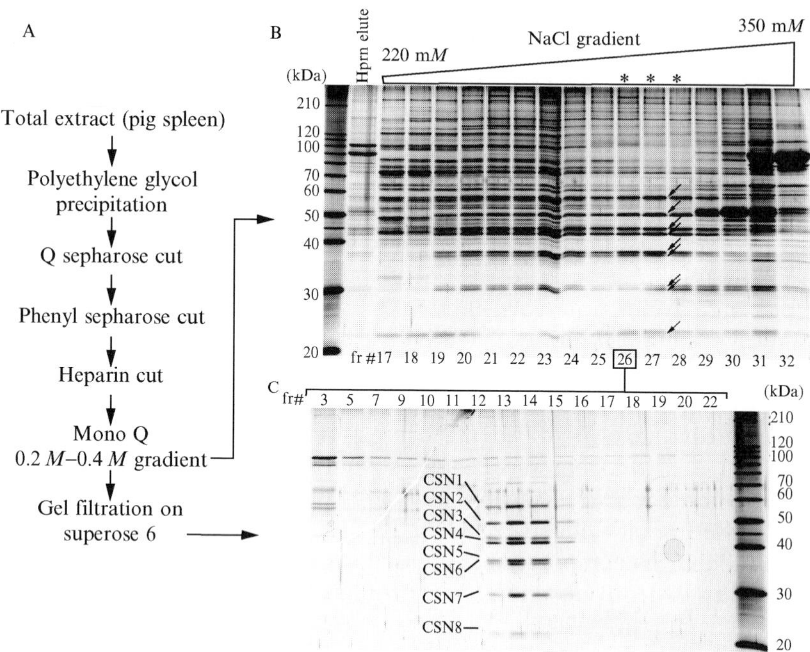


FIG. 1. Purification of CSN complex from porcine spleen. (A) Diagram depicting the chromatographic steps of CSN purification. (B) Examination of Mono-Q eluate. Two microliters from each of the 0.5-ml fractions of the Mono-Q gradient eluate was mixed with SDS sample buffer and loaded onto a 12% SDS–PAGE. Proteins were visualized by silver staining. CSN was eluted over a wide salt concentration range, but it was the predominant species in fractions #26 to #28. (C) The sample from Mono-Q fraction #26 (200 $\mu$l) was separated further onto a Superose 6 gel filtration column. Fractions were examined by silver staining. The CSN complex was found in fractions #12 to #14, corresponding to a molecular size of 450–500 kDa.

all buffers are cooled to 4° before use. We have used up to 300 g of fresh weight starting material for large-scale purification, but a smaller amount of 100 g is recommended for better yield, higher purity, and easier operation. An overview of the purification scheme is diagrammed in Fig. 1A.

*Homogenization*

• Use a blender to homogenize the tissue in 2.5–3× volumes of extraction buffer I [50 m$M$ Tris-HCl, pH 7.0, 1.5 m$M$ $MgCl_2$, 10 m$M$ KCl, 0.2 m$M$ EDTA, 5% glycerol, and freshly added 4 m$M$ dithiothreitol

(DTT), 2 m*M* phenylmethylsulfonyl fluoride (PMSF), and the 1× protease inhibitor cocktail (Roche)].

- Filter the mixture through four layers of cheesecloth.
- Centrifuge at 8000*g* for 10 min and then again at 12,000*g* for 15 min. Discard the pellet. The supernatant, or the total soluble fraction, is designated as fraction A. We usually obtain approximately 3% total proteins over fresh weight.

*PEG Cut*

- Make a 60% polyethylene glycol (PEG, MW 3350) stock in extraction buffer I. Slowly mix 1/4 volume of PEG stock with fraction A to obtain a final concentration of 12% on ice for 20 min.
- Centrifuge the sample at 10,000*g* for 10 min. CSN should be in the pellet.
- Dissolve the pellet in approximately 1× starting volume of extraction buffer II [50 m*M* Bis-Tris, pH 6.4, 1.5 m*M* $MgCl_2$, 10 m*M* KCl, 0.2 m*M* EDTA, 0.01% NP-40, 100 m*M* NaCl, 10% glycerol, and freshly added 4 m*M* DTT, 2 m*M* PMSF, and the 1× protein inhibitor cocktail (Roche)]. Manually break up the pellet in the buffer as finely as possible to facilitate the dissolving process. Leave the mixture on a rotator and allow the protein pellet to go into solution, which takes about 1–4 h depending on the size of the pellet.
- Centrifuge the mixture at 15,000*g* for 30 min and discard the pellet. Clarify the supernatant further by filtering through a 0.8-$\mu$m filter.

*Ion-Exchange Column Cut*

- Equilibrate a 300-ml Q Sepharose Fast Flow column (Amersham-Pharmacia) with 20 m*M* Bis-Tris, pH 6.4, 10% glycerol. Load the supernatant from the PEG step onto the column. Wash the column with equilibration buffer that contains 200 m*M* NaCl and elute CSN from the column with 400 m*M* NaCl in the equilibration buffer.

*Hydrophobic Column Cut*

- Add 1/3 volume of 600 m*M* $Na_2SO_4$ to the Q-column eluate so that the salt concentration of the solution reaches 300 m*M* NaCl and 150 m*M* $Na_2SO_4$.
- Equilibrate a 40-ml phenyl Sepharose high-performance column (Amersham-Pharmacia) with 10 m*M* Tris-propane, pH 7.0, 10% glycerol, 300 m*M* NaCl, and 150 m*M* $Na_2SO_4$. Load the CSN sample, wash with the column equilibration buffer, and elute with 10 m*M* Bis-Tris propane, pH 7.0, 10% glycerol. Immediately after elution, add 1/2 volume of extraction buffer II to the eluate. In this step a small amount of CSN may be lost, but

it will remove some contaminants that are otherwise difficult to remove at the end of the procedure.

*Heparin Column Cut*

- Equilibrate a 10-ml heparin column with 10 m*M* Bis-Tris propane, pH 7.0, 10% glycerol, 100 m*M* NaCl. Wash the column using the same buffer after loading the CSN-containing solution from the last step.
- Elute with 100 m*M* sodium phosphate, pH 7.2, 200 m*M* NaCl. The elution buffer should be made and kept at room temperature to avoid precipitation.

*Mono-Q Gradient*

- Equilibrate a 1-ml prepacked Mono Q HR 5/5 column (Amersham-Pharmacia) with 20 m*M* Bis-Tris propane, pH 7.0, 10% glycerol, and 200 m*M* NaCl.
- Dilute the CSN-containing heparin eluate with 3 volumes of 20 m*M* Bis-Tris propane, pH 7.0, 10% glycerol, and load onto the Mono-Q column. Wash the column with the Mono-Q equilibration buffer.
- Program a gradient elution from 200 to 400 m*M* NaCl in 20 m*M* Bis-Tris propane, pH 7.0, 10% glycerol over a 12-ml volume. Collect 0.5-ml fractions throughout the gradient elution.
- CSN-containing fractions can be identified by running a SDS–PAGE using 1 to 5 $\mu$l from each of the 0.5-ml fractions for silver staining (Fig. 1B). The Mono-Q fraction contains highly concentrated CSN that is suitable for many CSN-related biochemical assays such as cullin deneddylation and in combination with *in vitro* ubiquitin E3 activity assays (Yang *et al.*, 2002). For the deneddylation assay, the Mono-Q fraction of CSN can be used directly as in Figs. 2B and 3C. The sample may also be dialyzed into desired buffer if necessary for other applications.

*Gel Filtration*

- To purify the complex further, a final gel filtration chromatography is found to be highly effective. Equilibrate a Superose 6 HR 10/30 gel filtration column (Amersham-Pharmacia) with 20 m*M* Tris, pH 7.2, 150 m*M* NaCl, 10% glycerol, 2 m*M* $MgCl_2$. Load the Mono-Q sample, and elute in the same buffer. CSN usually elutes in fractions corresponding to approximately 500 kDa (Fig. 1C). The proteins may be concentrated further before the activity assay. The purified CSN from this fraction was shown to be active in cullin deneddylation and is suitable for cell microinjection experiments (Yang *et al.*, 2002).

When the columns are set up and equilibrated, we usually complete the entire procedure within 2 days without freezing the samples in the middle of the procedure. As an example of a purification experiment, from the 100*g* of fresh weight starting material (or 3.8*g* of crude porcine spleen soluble proteins), we have obtained approximately 0.75 mg of CSN complex after the Mono-Q step (counting Mono-Q fractions #26, #27, and #28 only, Fig. 2B).

## Purification of CSN from Cultured Cells Using Epitope-Tagged CSN Subunit

This method of purification utilizes cultured cells stably expressing a Flag epitope-tagged CSN subunit and involves one-step immunoaffinity purification. We have generated human 293 cell lines stably expressing Flag-CSN1 (Tsuge *et al.*, 2001) or Flag-CSN2 (Yang *et al.*, 2002) in which the Flag tag was fused to N termini of the respective proteins. We also developed cell lines expressing CSN2-MH and CSN3-MH, in which the subunits are fused to a myc tag followed by a 6xHis tag at C termini. However, because the latter two lines expressed significantly less fusion proteins relative to their endogenous counterparts, we reasoned that these fusion proteins are probably less efficient when used as a tag to pull down the entire CSN complex. Therefore, only N-terminal Flag-tagged lines were used, and the procedure we use to purify the Flag-CSN1 complex is described.

### *Protocol*

- Flag-CSN1 expressing cells are cultured in high glucose DMEM with 10% fetal bovine serum until the cells are confluent at midlate log phase. Collect the cells (25 × 15-cm plates) in phosphate-buffered saline and wash quickly with hypotonic buffer (20 m*M* HEPES, pH 7.2, 1.5 m*M* $MgCl_2$, 5 m*M* KCl, 0.5 m*M* DTT). Resuspend the cells in one packed cell volume of cold hypotonic buffer containing freshly added 1 m*M* PMSF and 1× protease inhibitor cocktail (Roche). Keep the samples on ice whenever possible.
- Lyse the cells using a sonicator (Branson Sonifier 250, 50% duty cycle, output control 5, 2 × 50 strokes). To the lysate, add 0.2% NP-40 and high salt extraction buffer (15 m*M* Tris-HCl, pH 7.5, 1 m*M* EDTA, 0.4 M NaCl, 0.1 m*M* DTT, 10% glycerol) to obtain a final salt concentration of 300 m*M*.

• Leave the lysate on ice for 15 min and then clear by centrifugation at 4000*g* for 10 min followed by 14,000*g* for 20 min at 4°. Clarify the supernatant further by filtering through a 0.8-$\mu$m filter. The rest of the procedure is carried out with the aid of an automatic FPLC system (Amersham-Pharmacia) at 4°.

• Equilibrate a 1-ml anti-Flag (M2) affinity column (Sigma) with 5 column volumes of buffer A (15 m*M* Tris-HCl, pH 7.5, 275 m*M* NaCl, 8% glycerol). Load the supernatant onto the column. In this experiment, a total cell extract of 50 ml containing 100 mg of protein is loaded at a rate of 0.1 ml/min.

• Wash the column with 15 ml of buffer A and then equilibrate with 5 ml of buffer B (15 m*M* Tris-HCl, pH 7.5, 50 m*M* NaCl, 10% glycerol) at 0.5 ml/min. Elute the Flag-CSN1 complex with 1 ml of 1 mg/ml Flag peptide (Sigma) in buffer B at 0.1 ml/min.

• We regenerate the anti-Flag M2 column by washing with 1 ml of 100 m*M* glycine, pH 2.7, followed by 15 ml of 20 m*M* Tris, pH 7.5. The column can be reused many times.

$CSN^{Flag\text{-}CSN1}$ complex-containing fractions are determined by SDS–PAGE and silver staining as shown in Fig. 2A. The complex can be concentrated further using microcon YM-3 filter devices (Millipore). We have obtained approximately 76 $\mu$g of the CSN complex from 100 mg of starting protein extract. The $CSN^{Flag\text{-}CSN1}$ complex was found to be active in the cullin deneddylation assay as described later (Fig. 2B). The $CSN^{Flag\text{-}CSN2}$ complex was isolated similarly. The cullin deneddylation activities of the CSN complexes purified by different procedures were compared.

## Affinity Purification of the CSN Complex from *Arabidopsis* Plants

Tandem affinity purification has emerged as an efficient and rapid means to purify protein complexes from plants (Rohila *et al.*, 2004). We have developed an alternative tandem affinity purification (TAPa) system (Fig. 3A) in which, as in the original TAP strategy, the first affinity step involves IgG binding. For the second affinity step, the original calmodulin-binding domain has been replaced with a tag containing 6xHis and a 9 repeats of myc epitope so that metal affinity and/or anti-myc immunoaffinity chromatography may be performed. Between the two affinity tags, we have inserted the cleavage site for the lower temperature active 3C protease, which allows the purification procedure to be performed at 4° throughout.

We obtained *Arabidopsis* plants harboring a *CSN3-TAPa* transgene (Fig. 3A) that can functionally rescue the *csn3* mutant (V. Rubio and

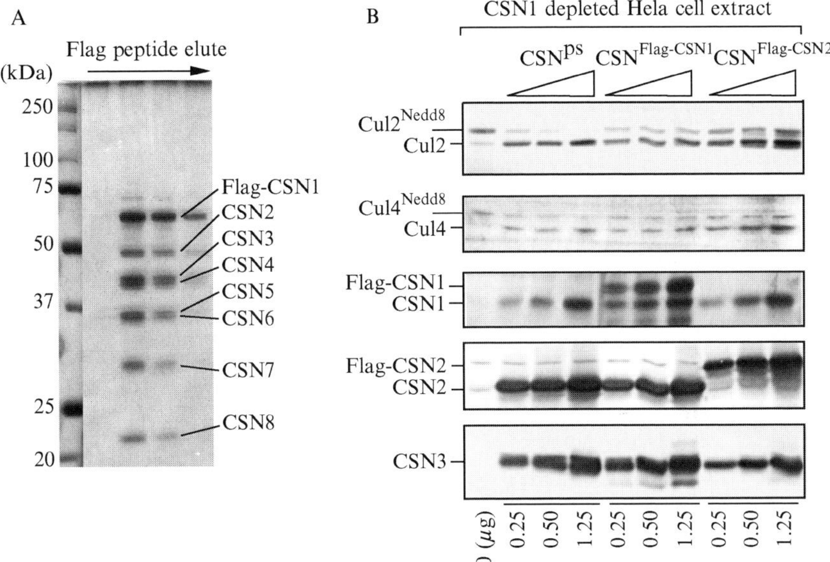


FIG. 2. (A) Silver staining of the Flag-CSN1 complex (CSN$^{Flag\text{-}CSN1}$). The cell extract from the Flag-CSN1 stable expressing cell line was bound to a column with anti-Flag (M2) resin. The complex was eluted from the column with Flag peptide (1 mg/ml). Eluate fractions were loaded onto a 7–15% gradient SDS–PAGE and visualized by silver staining. (B) Deneddylation activity of the purified CSN complex. The CSN1-depleted HeLa cytosolic extract of 20 $\mu$g was incubated in hypotonic buffer alone or with 0.25, 0.5, or 1.25 $\mu$g of porcine spleen CSN complex Mono-Q fraction #26 (CSN$^{ps}$), Flag-CSN1 complex (CSN$^{Flag\text{-}CSN1}$), or Flag-CSN2 complex (CSN$^{Flag\text{-}CSN2}$). Extracts were then subjected to immunoblot analysis using anti-Cul2 and anti-Cul4 to detect neddylation levels and to anti-CSN1, anti-CSN2, and anti-CSN3 antibodies to confirm the increasing amount of CSN complex.

X. W. Deng, unpublished data). Another *Arabidopsis* transgenic line used here is *fus6/Flag-CSN1-3-4*, which expresses AtCSN1 with an N-terminally fused 3x Flag epitope tag in the csn1 mutant background (Wang *et al.*, 2002, 2003). This transgenic allele completely rescues the mutant phenotype and expresses Flag-CSN1 at the same level as the endogenous CSN1 in wild-type plants (Wang *et al.*, 2003). This section describes the protocols used in CSN complex purification from both transgenic lines. It is important to note that these transgenes were crossed into their respective mutant backgrounds so that the corresponding endogenous gene products were absent. Finally, the activities of the purified CSN$^{Flag\text{-}CSN1}$ and CSN$^{CSN3\text{-}TAPa}$ complexes were evaluated by analyzing RUB1 deconjugation (derubylation) from *Arabidopsis* cullin1 (CUL1) (Fig. 3E).

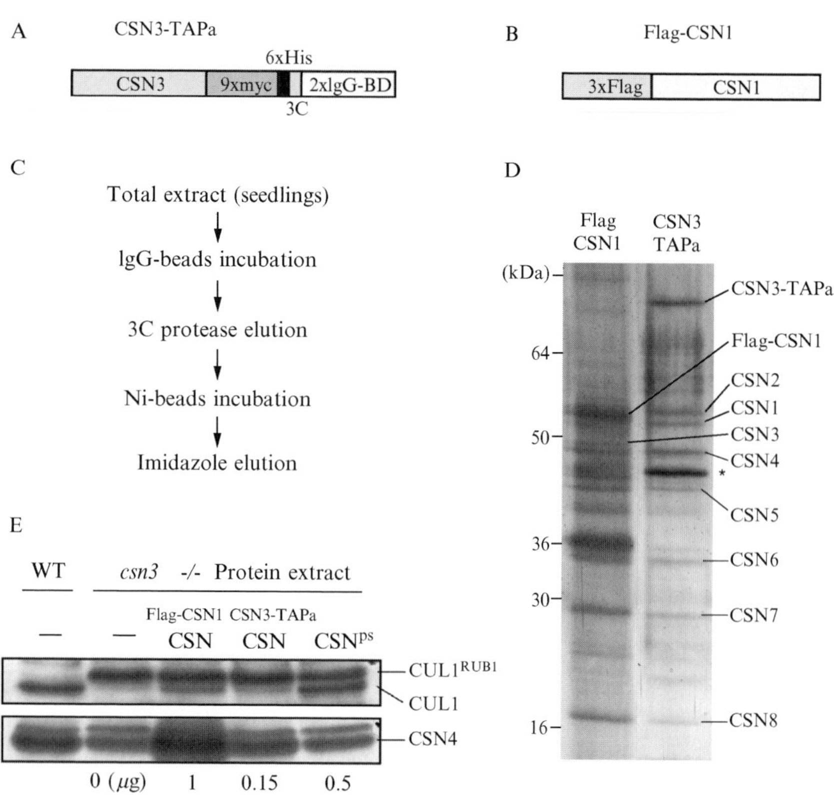


FIG. 3. Purification and activity assay for *Arabidopsis* CSN complexes. (A and B) Diagrams of the CSN3-TAPa and Flag-CSN1 constructs. (C) Summary of TAPa purification scheme. (D) Visualization of the CSN$^{Flag\text{-}CSN1}$ and CSN$^{CSN3\text{-}TAPa}$ components. The purified Flag-CSN1 complex and CSN3-TAPa complex were resolved in a 7.5–15% gradient SDS–PAGE and visualized by silver staining. The position for each CSN component is indicated. An asterisk indicates the position of the 3C protease band. The CSN3 band is absent in the CSN3-TAPa complex but appears as the larger CSN3-TAPa fusion protein. The Flag-CSN1 fusion protein runs at a similar position as the endogenous CSN2. (E) *Arabidopsis* cullin1 derubylation assay. Protein extracts (12 μg) corresponding to wild-type (WT) and homozygous *csn3* (*csn3* -/-) mutant seedlings were incubated in reaction buffer alone or in the presence of CSN$^{Flag\text{-}CSN1}$(1 μg), CSN$^{CSN3\text{-}TAPa}$ (0.15 μg), or porcine spleen CSN complex (CSN$^{ps}$, 0.5 μg) for 20 min at room temperature. Immunoblots of anti-Cul1 and anti-CSN4 antibodies are shown. Rubylated and unrubylated CUL1 can be distinguished clearly.

## *Protocol for Immunopurification of* Arabidopsis Flag-CSN1 *Complex*

• Surface sterilize 300 mg of *fus6/Flag-CSN1-3-4* seeds using 70% bleach treatment for 15 min with gentle rotation. Wash the seeds three times with sterile water, place them on MS plates (Gibco) containing 0.3% sucrose, and cold treat for 3 days. Later, transfer the plates to continuous white light conditions (110 $\mu$mol • $m^{-2}$ • $s^{-1}$).

• After 9 days, harvest the seedlings and grind them in liquid nitrogen using a mortar and pestle. Thaw the homogenate in 1 volume of extraction buffer [20 m*M* Tris-HCl, pH 7.5, 10 m*M* $MgCl_2$, 150 m*M* NaCl, 10% glycerol, 0.01% NP-40, 0.5 m*M* PMSF, and 1× protease inhibitor cocktail EDTA-free (Roche)], and centrifuge at 13,000*g* for 25 min at 4°. Save the supernatant and determine protein concentration by the Bradford assay (Bio-Rad).

• Wash 350 $\mu$l anti-Flag (M2) affinity beads (Sigma) once with 10 ml 100 m*M* glycine, pH 2.5, and three times with 10 ml extraction buffer each. Incubate the total extract with the M2 beads for 3 h at 4° with gentle rotation. Wash the beads four times with 10 ml of extraction buffer. We perform all centrifugations in a benchtop centrifuge at 150*g* for 3 min at 4°.

• Elute the Flag-CSN1 complex by incubating with 600 $\mu$l of 1 mg/ml Flag peptide (Sigma) for 12 h at 4° with gentle rotation.

• Concentrate the protein complex using microcon YM–3 filter devices (Millipore). Examine 4 $\mu$l of the sample by silver staining of SDS–PAGE (Fig. 3D). Approximately 16 $\mu$g of the $CSN^{Flag\text{-}CSN1}$ complex is obtained from 5 g fresh weight (26 mg total protein) Flag-CSN1 expressing seedlings described earlier.

*Protocol for* TAPa *Purification of* Arabidopsis CSN3-TAPa *Complex*

• Sterilize and plant 1 g of *CSN3-TAPa* seeds on MS plates (Gibco) containing 0.3% sucrose as described earlier.

• After 18 days, harvest the plant tissue and grind in liquid nitrogen using a mortar and a pestle. Thaw the homogenate in 2 volumes of extraction buffer [50 m*M* Tris-HCl, pH 7.5, 150 m*M* NaCl, 10% glycerol, 0.1% NP-40, 1 m*M* PMSF, and 1× protease inhibitor cocktail (Roche)], filter it through four layers of cheesecloth, and centrifuge at 12,000*g* for 15 min at 4°. The supernatant is the total extract (Fig. 3C).

• Incubate the total protein extract for 2 h at 4° with 500 $\mu$l IgG beads (Amersham Biosciences) that have been prewashed three times with 10 ml extraction buffer. Wash the beads three times with 10 ml of washing buffer (50 m*M* Tris-HCl, pH 7.5, 150 m*M* NaCl, 10% glycerol, 0.1% NP-40) and once with 10 ml of cleavage buffer (50 m*M* Tris-HCl, pH 7.5, 150 m*M* NaCl, 10% glycerol, 0.1% NP-40, 1 m*M* DTT).

• Elute proteins from the IgG beads by incubation with 50 $\mu$l (100 units) of 3C protease (Prescission protease, Amersham Biosciences) for 2 h in 5 ml of cleavage buffer at 4° with gentle rotation. Perform another elution and pool with the first eluate.

- Load the entire 10-ml eluate onto a 1-ml Ni-NTA resin column (Qiagen). Reload the flow through onto the column once. Wash with 30 ml of washing buffer.
- Elute the proteins from the Ni-NTA resin using 1 ml of imidazole elution buffer (50 m*M* Tris-HCl, pH 7.5, 150 m*M* NaCl, 10% glycerol, 0.1% NP-40, 0.05 *M* imidazole). Examine the purified complex by silver staining (Fig. 3D).
- Prior to the derubylation activity assay, remove imidazole and change into reaction buffer using desalting spin columns (Pierce). Concentrate the sample using microcon YM-3 filter devices (Millipore). Using this procedure, we obtain approximately 1.5 $\mu$g of CSN complex from 15 g fresh weight (33 mg total protein) of *CSN3-TAPa* seedlings grown as described earlier. The yield is considerably less compared to the Flag-CSN1 complex purification procedure.

## *In Vitro* Nedd8/Rub1 Deconjugation Assay Using Extracts from HeLa Cells or *Arabidopsis* CSN Mutants

The HeLa cytosolic extract is prepared according to Yang *et al.* (2002). The extract is depleted of CSN by passing through (twice) a 0.5-ml column with immobilized anti-CSN1 antibody. More than 90% of the CSN complex is removed based on immunoblots of anti-CSN3 (Fig. 2B) and anti-CSN8 (not shown). Because of the reduction in the level of CSN, Cul2, Cul4 (Fig. 2B), and Cul1 (not shown) preferentially accumulated in Nedd8-modified forms, and therefore were chosen as substrates to test CSN deneddylation activity. The total HeLa cell extract has also been used to detect CSN deneddylation activity (Yang *et al.*, 2002), although it is less sensitive than the CSN depleted extract.

The CSN-depleted HeLa extract (20 $\mu$g) is incubated in hypotonic buffer containing different amounts of CSN complex purified by various procedures or in hypotonic buffer alone in a total volume of 20 $\mu$l. The reaction is carried out at room temperature for 20 min and is then stopped by adding SDS–PAGE loading buffer for immunoblots. Neddylated cullin levels are detected by immunoblot analysis with anti-Cul2 (Zymed) and anti-Cul4 antibodies (Fig. 2B). The CSN complex purified by all three methods is active in this assay, with the CSN complex purified from porcine spleen ($CSN^{ps}$) being the most efficient compared to epitope tag-purified CSN complexes (Fig. 2B).

To test the RUB1-deconjugation activity of the *Arabidopsis* CSN complex, we used *csn* mutant extracts that contain hyperrubylated cullin 1 (CUL1) (Schwechheimer *et al.*, 2001; Wang *et al.*, 2002), an ideal substrate for *in vitro* derubylation assay. A similar strategy using the fission yeast *csn* mutant extract has been used to test the deconjugation activity of the mammalian CSN complex (Lyapina *et al.*, 2001). In this experiment (Fig. 3E), we made a protein extract from the *csn3* mutant (Salk_000593, Alonso *et al.*, 2003) in the reaction buffer (50 m*M* Tris-HCl, pH 7.5, 50 m*M* NaCl, 10% glycerol, 1 m*M* $MgCl_2$). Twelve microgram of *csn3* extract is incubated in a final volume of 14 $\mu$l containing purified $CSN^{Flag\text{-}CSN1}$ complex (1 $\mu$g), $CSN^{CSN3\text{-}TAPa}$ complex (0.15 $\mu$g), mammalian CSN complex purified from porcine spleen ($CSN^{ps}$; 0.5 $\mu$g), or the reaction buffer alone as a control. Reactions are carried out at room temperature for 20 min and are then stopped by adding SDS–PAGE sample buffer. The presence of rubylated and derubylated CUL1 is visualized by immunoblot analysis using the anti-CUL1 antibody. As shown in Fig. 3E, 1 $\mu$g of $CSN^{Flag\text{-}CSN1}$ complex was sufficient to perform derubylation of *Arabidopsis* CUL1, whereas 0.15 $\mu$g of $CSN^{CSN3\text{-}TAPa}$ complex was not. The CSN complex purified from porcine spleen ($CSN^{ps}$) was demonstrated to be more efficient than any of the affinity-purified CSN complexes tested in deconjugating RUB1 from plant CUL1 (Fig. 3E) or deconjugating Nedd8 from human Cul2 and Cul4 (Fig. 2B).

## Acknowledgments

Our research is supported by a grant from the National Health Institute (R01 GM61812) to N. W. and by grants from National Science Foundation (MCB-0077217 and NSF 2010 MCB-0115870) to X. W. D. V. R. is a long-term postdoctoral fellow of the Human Frontier Science Program.

## References

Alonso, J. M., Stepanova, A. N., Leisse, T. J., Kim, C. J., Chen, H., Shinn, P., Stevenson, D. K., Zimmerman, J., Barajas, P., Cheuk, R., Gadrinab, C., Heller, C., Jeske, A., Koesema, E., Meyers, C. C., Parker, H., Prednis, L., Ansari, Y., Choy, N., Deen, H., Geralt, M., Hazari, N., Hom, E., Karnes, M., Mulholland, C., Ndubaku, R., Schmidt, I., Guzman, P., Aguilar-Henonin, L., Schmid, M., Weigel, D., Carter, D. E., Marchand, T., Risseeuw, E., Brogden, D., Zeko, A., Crosby, W. L., Berry, C. C., and Ecker, J. R. (2003). Genome-wide insertional mutagenesis of *Arabidopsis thaliana*. *Science* **301,** 653–657.

Chamovitz, D. A., Wei, N., Osterlund, M. T., von Arnim, A. G., Staub, J. M., Matsui, M., and Deng, X. W. (1996). The COP9 complex, a novel multi-subunit nuclear regulator involved in light control of a plant developmental switch. *Cell* **86,** 115–121.

Cope, G. A., Suh, G. S., Aravind, L., Schwarz, S. E., Zipursky, S. L., Koonin, E. V., and Deshaies, R. J. (2002). Role of predicted metalloprotease motif of Jab1/Csn5 in cleavage of Nedd8 from Cul1. *Science* **298,** 608–611.

Deshaies, R. J. (1999). SCF and cullin/ring H2-based ubiquitin ligases. *Annu. Rev. Cell Dev. Biol.* **15,** 435–467.

Hochstrasser, M. (2000). Evolution and function of ubiquitin-like protein-conjugating systems. *Nature Cell Biol.* **2,** E153–E157.

Liu, C., Powell, K. A., Mundt, K., Wu, L., Carr, A. M., and Caspari, T. (2003). Cop9/signalosome subunits and Pcu4 regulate ribonucleotide reductase by both checkpoint-dependent and –independent mechanisms. *Genes Dev.* **17,** 1130–1140.

Lyapina, S., Cope, G., Shevchenko, A., Serino, G., Tsuge, T., Zhou, C., Wolf, D. A., Wei, N., Shevchenko, A., and Deshaies, R. J. (2001). Promotion of NEDD8-CUL1 conjugate cleavage by COP9 signalosome. *Science* **292,** 1382–1385.

Pan, Z. Q., Kentsis, A., Dias, D. C., Yamoah, K., and Wu, K. (2004). Nedd8 on cullin: Building an expressway to protein destruction. *Oncogene* **23,** 1985–1997.

Rohila, J. S., Chen, M., Cerny, R., and Fromm, M. E. (2004). Improved tandem affinity purification tag and methods for isolation of protein heterocomplexes from plants. *Plant J.* **38,** 172–181.

Schwechheimer, C., Serino, G., Callis, J., Crosby, W. L., Lyapina, S., Deshaies, R. J., Gray, W. M., Estelle, M., and Deng, X. W. (2001). Interactions of the COP9 signalosome with the E3 ubiquitin ligase $SCF^{TIRI}$ in mediating auxin response. *Science* **292,** 1379–1382.

Seeger, M., Kraft, R., Ferrell, K., Bech-Otschir, D., Dumdey, R., Schade, R., Gordon, C., Naumann, M., and Dubiel, W. (1998). A novel protein complex involved in signal transduction possessing similarities to 26S proteasome subunits. *FASEB J.* **12,** 469–478.

Serino, G., Tsuge, T., Kwok, S., Matsui, M., Wei, N., and Deng, X. W. (1999). Arabidopsis cop8 and fus4 mutations define the same gene that encodes subunit 4 of the COP9 signalosome. *Plant Cell* **11,** 1967–1980.

Tsuge, T., Matsui, M., and Wei, N. (2001). The subunit 1 of COP9 signalosome suppresses gene expression through the N-terminal domain and incorporates into the complex through the PCI domain. *J. Mol. Biol.* **305,** 1–9.

Wang, X., Feng, S., Nakayama, N., Crosby, W. L., Irish, V., Deng, X. W., and Wei, N. (2003). The COP9 signalosome interacts with $SCF^{UFO}$ and participates in *Arabidopsis* flower development. *Plant Cell* **15,** 1071–1082.

Wang, X., Kang., D., Feng, S., Serino., G., Schwechheimer, C., and Wei, N. (2002). CSN1 N-terminal-dependent activity is required for Arabidopsis development but not for Rub1/Nedd8 deconjugation of cullins: A structure-function study of CSN1 subunit of COP9 signalosome. *Mol. Biol. Cell* **13,** 646–655.

Wei, N., Chamovitz, D. A., and Deng, X. W. (1994). *Arabidopsis* COP9 is a component of a novel signaling complex mediating light control of development. *Cell* **78,** 117–124.

Wei, N., and Deng, X. W. (1998). Characterization and purification of the mammalian COP9 complex, a conserved nuclear regulator initially identified as a repressor of photomorphogenesis in higher plants. *Photochem. Photobiol.* **68,** 237–241.

Wei, N., and Deng, X. W. (2003). The COP9 signalosome. *Annu. Rev. Cell Dev. Biol.* **19,** 261–286.

Wei, N., Tsuge, T., Serino, G., Dohmae, N., Takio, K., Matsui, M., and Deng, X.-W. (1998). The COP9 complex is conserved between plants and mammals and is related to the 26 S proteasome regulatory complex. *Curr. Biol.* **8,** 919–922.

Wu, K., Yamoah, K., Dolios, G., Gan-Erdene, T., Tan, P., Chen, A., Lee, C. G., Wei, N., Wilkinson, K. D., Wang, R., and Pan, Z. Q. (2003). DEN1 is a dual function protease capable of processing the C terminus of Nedd8 and deconjugating hyper-neddylated CUL1. *J. Biol. Chem.* **278,** 28882–28891.

Yang, X., Menon, S., Lykke-Andersen, K., Tsuge, T., Xiao, D., Wang, X., Rodriguez-Suarez, R., Zhang, H., and Wei, N. (2002). The COP9 signalosome inhibits p27kip1 degradation and impedes G1-S phase progression via deneddylation of SCF Cul1. *Curr. Biol.* **12,** 667–672.

Zhou, C., Wee, S., Rhee, E., Naumann, M., Dubiel, W., and Wolf, D. A. (2003). Fission yeast COP9/Signalosome suppresses cullin activity through recruitment of the deubiquitylating enzyme Ubp12. *Mol. Cell.* **11,** 927–938.

# [39] Purification Method of the COP9 Signalosome from Human Erythrocytes

*By* BETTINA K. J. HETFELD, DAWADSCHARGAL BECH-OTSCHIR, and WOLFGANG DUBIEL

## Abstract

The COP9 signalosome (CSN) is a multimeric protein complex that occurs in all eukaryotic cells. Originally described in plants as a regulator of photomorphogenesis, its purification and characterization from mammalian cells revealed significant sequence homologies to subunits of the 26S proteasome lid complex, as well as of the eukaryotic translation initiation factor 3. Recent studies disclosed its participation in processes such as DNA repair, cell cycle regulation, development, and angiogenesis. At the moment, the pleiotropic effects of the CSN point to a regulatory role in the ubiquitin/26S proteasome system, but its exact function still remains to be clarified. This chapter describes the method to purify human CSN from red blood cells. Two outdated erythrocyte concentrates are sufficient to prepare approximately 0.5 mg of CSN. Washed cells are first lysed and then proteins are separated by a DEAE anion-exchange column. The CSN-containing fractions are pooled and subjected to an ammonium sulfate precipitation followed by dialysis. The concentrated proteins are then loaded onto a glycerol density gradient and ultracentrifugation is performed. The purification procedure is continued using two succeeding anion-exchange columns, resulting in a sufficiently pure CSN complex. Optionally, an additional density gradient centrifugation can be attached. The purified CSN complex possesses kinase, deneddylase, and deubiquitinase activities and can be stored for at least 2 months on ice at 4°.

METHODS IN ENZYMOLOGY, VOL. 398

0076-6879/05 $35.00
DOI: 10.1016/S0076-6879(05)98039-7

## Introduction

The COP9 signalosome was originally described as a negative regulator of photomorphogenesis in *Arabidopsis* (Wei *et al.*, 1994). In 1998 its ortholog was identified in mammals (Seeger *et al.*, 1998; Wei and Deng, 1998). Although there is accumulating evidence on the CSN as a component of the ubiquitin (Ub) system, its exact functions remain to be elucidated. In our group the complex has been identified during preparations of the 26S proteasome from red blood cells (Seeger *et al.*, 1998). In the course of the characterization of 26S isoforms, a new protein was cloned and sequenced that was originally believed to be part of the proteasome. Later it was identified as a subunit of the CSN. Therefore, a modification of the 26S proteasome preparation scheme led to isolation of the whole new complex (Seeger *et al.*, 1998). Using CSN purified by this procedure, a first insight into the architecture of the complex could be obtained by electron microscopy, revealing a striking resemblance to the 26S proteasome lid complex. These studies demonstrated that both the CSN and the lid lack any symmetry in subunit arrangement and exhibit a central groove (Kapelari *et al.*, 2000). Based on electron microscopy studies and on subunit–subunit interaction analyses, a first model of the subunit interactions was proposed (see Fig. 1). The model has been confirmed by yeast two-hybrid studies (Fu *et al.*, 2001). The identification of Jun activating binding protein (JAB1 = CSN5) and thyroid hormone receptor-interacting protein 15 (TRIP15 = CSN2) as components hinted to an involvement of the new complex in signal transduction. Kinase activity was found associated with the CSN, which led to phosphorylation of the transcription factors p53 (Bech-Otschir *et al.*, 2001) and c-Jun (Naumann *et al.*, 1999; Seeger *et al.*, 1998), as well as the CSN2 and CSN7 subunits (Kapelari *et al.*, 2000). Inhibition of CSN-associated kinase activity could be achieved by the anti-inflammatory and -carcinogenic agent curcumin (Henke *et al.*, 1999). Hitherto the inositol 1,3,4-trisphosphate 5/6-kinase (Wilson *et al.*, 2001), as well as casein kinase 2 (CK2) and protein kinase D (PKD) (Uhle *et al.*, 2003), was identified associated with the CSN. The CK2 and the PKD are present in our purified CSN from red blood cells.

A metalloprotease motif within the CSN5 subunit (JAMM–Jab1/MPN domain metalloenzyme motif) has been identified that is responsible for deubiquitination (Groisman *et al.*, 2003) as well as deneddylation (Lyapina *et al.*, 2001; Zhou *et al.*, 2001), supporting the fact that the CSN is a component of the Ub/26S proteasome system. In addition, a deubiquitinating enzyme could be identified associated with the CSN that is also present in our final CSN preparation (Zhou *et al.*, 2003).

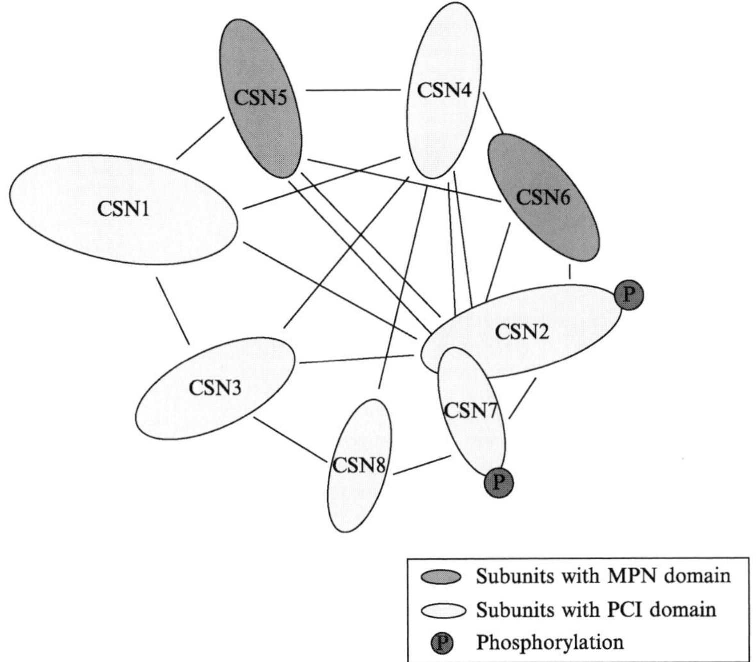


FIG. 1. Subunit interaction model of the COP9 signalosome. On the basis of mass distribution and subunit–subunit interactions, a model of the CSN complex was created (Kapelari *et al.*, 2000). Known phosphorylation sites and subunits with specific domains are indicated. (See color insert.)

## Purification Procedure of the Human COP9 Signalosome

### *Purification Strategy*

To purify the human CSN from outdated erythrocyte concentrate conserves a preparation scheme was developed (see Fig. 2), which is based on 26S proteasome purification, and derived from earlier CSN isolation (Seeger *et al.*, 1998). First, lysates of red blood cells are separated by a linear salt gradient on a DEAE Sepharose column. Following a 45% ammonium sulfate precipitation and subsequent dialysis the relevant fractions are subjected to a density gradient centrifugation (10–40% glycerol gradient). The density gradient is followed by two successive anion-exchange chromatographies performed with the FPLC system. The first uses a Resource Q and the second uses a Mono Q column, which are both eluted with a linear salt gradient. Finally, a second density gradient

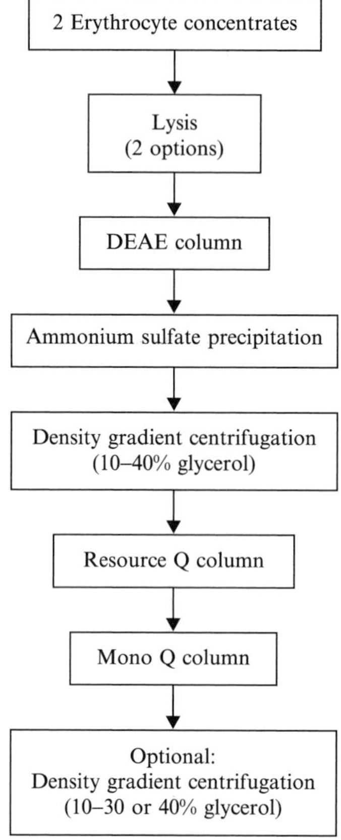


FIG. 2. Overview of the purification procedure of human CSN isolated from outdated erythrocyte conserves.

centrifugation (10–30 or 40% glycerol) can be carried out to obtain an even more purified CSN.

### *Cell Lysis*

The entire purification procedure, including the production of buffers and the use of instruments, is carried out at 4°. For purification of the CSN described here, two units of outdated erythrocyte concentrates are optimal. First, blood cells are washed three times with the same volume of phosphate-buffered saline (PBS; 137 m*M* NaCl, 100 m*M* potassium phosphate, pH 7.2) and are each time centrifuged at 4000*g* for 10 min at 4°. The light red supernatant plus the white fatty remnants in the interphase between supernatant and blood cells have to be removed carefully with a pipette (easier when connected with a vacuum pump). The cell pellet has to be resuspended each time with a glass stick after adding fresh PBS to allow sufficient washing. The final pellet amounts to approximately 400–500 ml of packed cells.

Next, for cell lysis, two alternative buffer conditions are available depending on the requested end products. If purification of the CSN alone is required, the cells are lysed by adding twice the volume of lysis buffer [40 m*M* Tris, 0.2% NP-40, 2 m*M* $\beta$-mercaptoethanol ($\beta$-Me)]. As an alternative, a 1:3 dilution adding just deionized water to the cells is recommended if a simultaneous purification of intact 26S proteasome is desired. In both cases, cells are lysed by osmotic shock. The addition of NP-40 seems to remove large amounts of the CSN associated with the erythrocyte membrane. The mixture should be stirred for at least 1 h at 4° for complete cell lysis, as can be seen by significant darkening of hemoglobin.

The cell debris is pelleted by centrifugation at 16,000*g* for 1 h at 4°. The dark red supernatant of low viscosity is recovered by decanting carefully, avoiding shaking. It is important not to mix cell debris into the supernatant and therefore an additional centrifugation for 30 min is advisable.

### *DEAE Anion-Exchange Chromatography Column*

During the washing procedure, 65 g of DEAE (DEAE cellulose anion exchanger, Sigma) is equilibrated three times in buffer A (20 m*M* Tris, pH 7.2; 50 m*M* KCl, 10% glycerol, 1 m*M* $\beta$-Me). The lysate (approximately 1.5 liter) is batched with the DEAE slurry by rotating at 200 rpm on a rotary shaker for at least 2 h. Shaking overnight is recommended. Next the supernatant of the batch is removed after sedimentation by gravity and then the batch is washed twice with the same volume of buffer A. Each time the DEAE beads are recovered by gravity (approximately 1 h). As the first purification step, DEAE ion-exchange chromatography is performed by tightly packing 200 ml of DEAE slurry into a suitable column (i.e., column SR25/45, Amersham Biosciences Europe, Freiburg, Germany) and washing it with at least 3 volumes of buffer A at a flow rate of 2 ml/min (approximately 5 h). During washing the red color of the DEAE beads dilutes significantly while removing hemoglobin. The bound proteins are eluted with a linear gradient of 50–400 m*M* KCl using 240 ml of buffer A and B1 (20 m*M* Tris, pH 7.2; 400 m*M* KCl, 10% glycerol, 1 m*M* $\beta$-Me) at a flow rate of 1 ml/min overnight. Approximately 65 fractions are collected and tested for the CSN by Western blotting (Fig. 4A). The CSN elutes between 180 and 230 m*M* KCl, corresponding to approximately 10 fractions with a pooled volume of 60–70 ml.

*Ammonium Sulfate Precipitation*

Proteins of the pooled fractions are precipitated with ammonium sulfate (45%). The salt is added slowly for 30 min while stirring. Subsequently, the obtained slurry is mixed for at least another 30 min. Sedimentation of the proteins is achieved by centrifugation at 13,000*g* for 15 min at 4°. The pellet is dissolved in a suitable volume and is dialyzed overnight against buffer C (20 m*M* Tris, pH 7.2; 50 m*M* KCl, 5% glycerol, 1 m*M* $\beta$-Me) using an appropriate dialysis membrane (regenerated cellulose dialysis membrane, MWCO 6-8000, Spectrum Europe, Breda, The Netherlands). In anticipation of the following glycerol gradient, a volume of 1–3 ml buffer C is recommended for dissolving the pellet, taking into account expansion of the suspension while desalting. Thus ammonium sulfate precipitation is an important step, providing both additional purification and concentration of the proteins.

*Density Gradient Centrifugation*

The proteins are then separated by density gradient centrifugation (10–40% glycerol) in a Beckman SW28 rotor. Linear gradients of buffer A and buffer B2 (20 m*M* Tris, pH 7.2; 50 m*M* KCl, 40% glycerol, 1 m*M* $\beta$-Me) are generated using 17.5 ml of each buffer (25 $\times$ 89-mm tubes, Beranek, Weinheim, Germany) and precooled for at least 1 h prior to loading of the proteins. A maximum of 1 ml containing up to 40–50 mg of proteins can be loaded onto this type of gradient. The protein amount obtained from two blood conserves generally requires two to four gradients. Centrifugation is performed at 27,000 rpm (96,500*g*) for 22 h at 4°. After centrifugation, gradients are fractionated using a fraction collector (RediFrac Fraction Collector, Amersham Biosciences). Twenty fractions per gradient are collected at a maximum speed of 1 ml/min, and 10-$\mu$l aliquots of the fractions are screened by Western blotting (Fig. 4B). Under these conditions, the CSN normally sediments into fractions 10–14, possessing a volume of approximately 1.8 ml each.

*Resource Q Anion-Exchange Chromatography*

Pooled CSN fractions from the glycerol gradient are purified further by Resource Q anion-exchange chromatography (Amersham Biosciences). The large volume after pooling is best applied to the column by a superloop device. Using buffers A and B3 (20 m*M* Tris, pH 7.2; 1 *M* KCl, 10% glycerol, 1 m*M* $\beta$-Me), a linear gradient is generated from 145 to 430 m*M* KCl (corresponds to 10–40% buffer B3), and the proteins are obtained at a flow rate of 1 ml/min. The CSN complex elutes in two main peaks between 280 and 350 m*M* KCl (Fig. 3A) as determined by Western blotting (Fig. 4C) and Coomassie staining (Fig. 4D). Clearly these two CSN peaks differ in their charge and in associated proteins. The kinases CK2 and PKD are mainly coeluted with the CSN peak at about 330 m*M* KCl, whereas the inositol 1,3,4-trisphosphate 5/6-kinase is found at lower salt concentrations (Uhle *et al.*, 2003). The nature of the different charges of the two CSN fractions is not known. It could be caused by a different phosphorylation status of CSN subunits.

*Mono Q Anion-Exchange Chromatography*

Further purification of the pooled fractions requires desalting. To concentrate the fractions and to reconstitute the retentate to buffer A, a suitable centrifugal filter device is used (Amicon Ultra–15, 10K NMWL, Millipore, Schwalbach, Germany). It is recommended to centrifuge at a speed of 1000*g* to obtain the best yield. A final volume of 10–15 ml is most convenient for subsequent loading on a Mono Q anion-exchange column (5/50 GL, Amersham Biosciences) by the superloop device. The CSN is released from the column using a linear gradient from 145 to 430 m*M* KCl of buffers A and B3 and a flow rate of 1 ml/min (Fig. 3B). Although the same conditions are used as in the preceding FPLC with the Resource Q column, the different anion-exchange material provides further purification from contaminating proteins. The CSN is eluted at a salt concentration of 300–350 m*M* KCl as demonstrated by Western blotting and Coomassie staining (Fig. 4E), as well as mass spectrometry. Usually there are two to three fractions containing pure CSN after this step. The complex is quite stable in a high salt concentration and can be stored on ice for approximately 2 months without disassembling and without losing kinase, deneddylase, and deubiquitinase activities.

*Density Gradient Centrifugation*

Optionally, as a last purification step, additional glycerol gradient centrifugation can be performed if the received CSN after Mono Q is not satisfying or further experiments require lower salt concentrations and prolonged storage. The relevant fractions are pooled and concentrated to a volume of 0.5–1.0 ml in a suitable filter device rotating at a maximum speed of 1000*g* (Amicon Ultra-15, 10K NMWL, Millipore). Using 6 ml

A

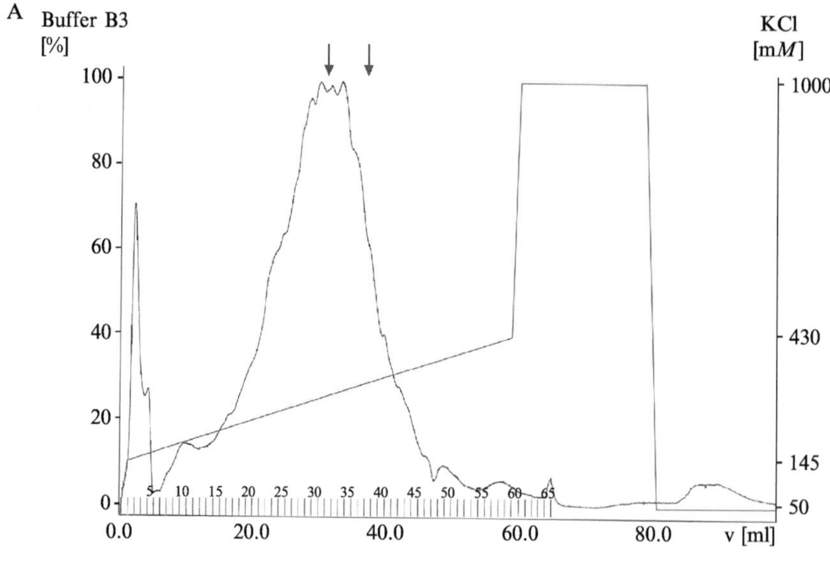


B

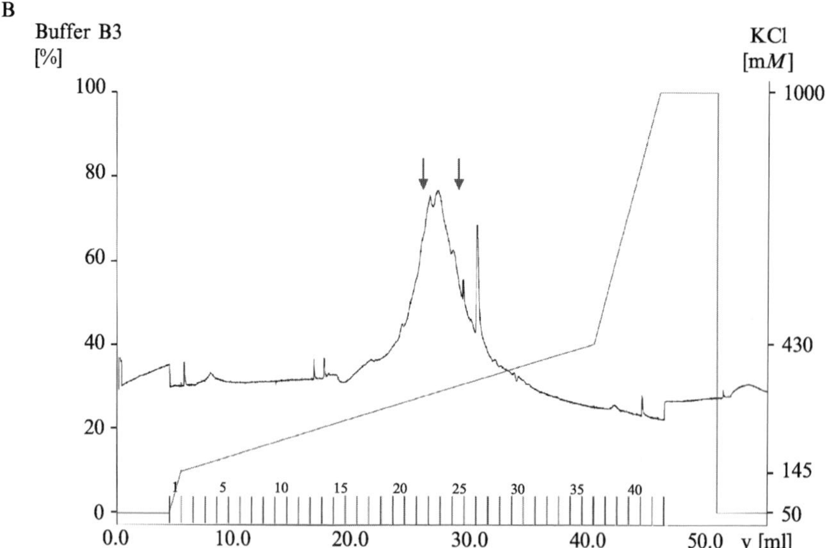


FIG. 3. Elution profile from FPLC columns. Resource Q column (A) and Mono Q column (B) with typical eluted protein profile and applied gradient. The red line shows the applied gradient from 10 to 40% buffer B3 (145–430 m*M* KCl) with subsequent regeneration of the column at 100% buffer B3 followed by buffer A. The black line indicates eluted protein concentration detected by an inline UV monitor (Monitor UV-M II, Amersham Biosciences). Numbers of collected 1-ml fractions are indicated above the *x* axis. Red arrows indicate fractions pooled for the next preparation step. (See color insert.)

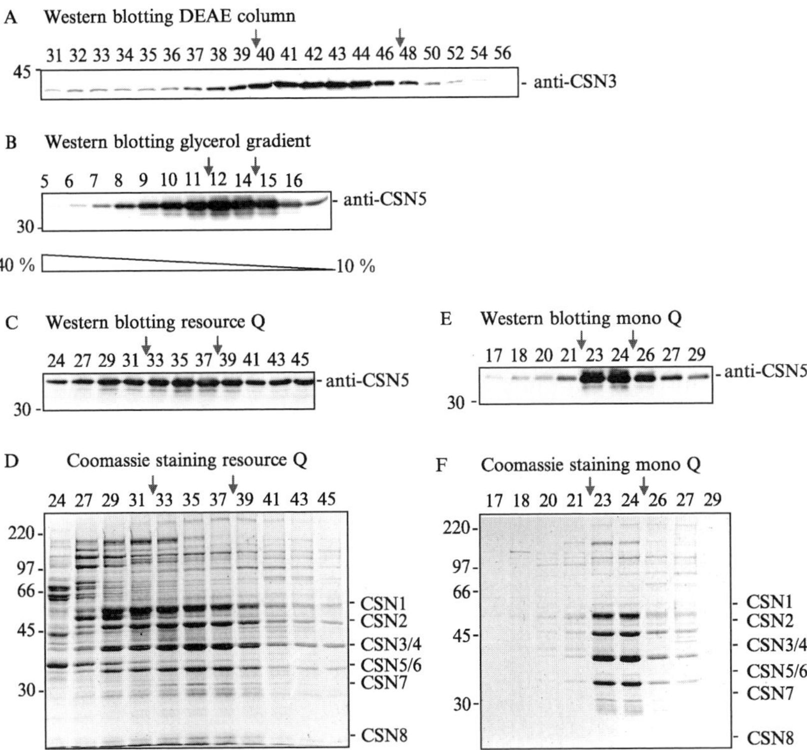


FIG. 4. Western blottings and Coomassie stainings at different purification steps. Red arrows indicate fractions pooled for next preparation step. (A) Western blotting of fractions 31–56 after DEAE column detected with anti-CSN3 antibody. (B) Western blotting of fractions 5–16 after density gradient centrifugation (10–40% glycerol) detected with anti-CSN5 antibody. (C and D) Western blotting and Coomassie staining of fractions 24–45 after Resource Q column. The CSN was detected with the anti-CSN5 antibody. (E and F) Western blotting and Coomassie staining of fractions 17–29 after Mono Q column. The CSN was detected with the anti-CSN5 antibody. (See color insert.)

each of buffer A and buffer B2 (40% glycerol) or buffer B4 (20 m*M* Tris, pH 7.2; 50 m*M* KCl, 30% glycerol, 1 m*M* $\beta$-Me), one linear gradient is cast and after applying the concentrated CSN, centrifugation for 22 h at 27,000 rpm at 4° is performed in a Beckman rotor SW40. A gradient from 10 to 30% is recommended if the CSN concentration is comparatively high and the Mono Q fractions reasonably pure. The 10–30% gradient results in more diluted but more purified CSN. If necessary, the salt concentration can be varied from 50 to 150 m*M* KCl in both buffers.

## Characteristics of the Purified CSN

After the purification procedure, each CSN subunit, including CSN7b, can be clearly identified by Coomassie staining, mass spectrometry, and Western blotting. Depending on the preparation, further proteins can be found, including subunits of the 26S proteasome. The purified CSN complex possesses kinase, deneddylase, and deubiquitinase activities even after 2 months of storage. During the purification procedure, a 18,000-fold purification of specific curcumin-sensitive kinase activity is achieved (Uhle *et al.*, 2003). Prolonged storage at 4° in a high salt concentration leads to generation of a high molecular weight complex, presumably a CSN dimer, that can be seen in nondenaturing gel electrophoresis with subsequent Western blotting. For longer storage, it is recommended to keep the purified CSN at −80° and thaw the complex only once. After thawing, CSN activities are still detectable. Following these purification procedures, approximately 0.5 mg of purified CSN can be obtained.

## Acknowledgment

This study was funded by a grant from the German Israel Foundation for Scientific Research and Development to W. D.

## References

Bech-Otschir, D., Kraft, R., Huang, X., Henklein, P., Kapelari, B., Pollmann, C., and Dubiel, W. (2001). COP9 signalosome-specific phosphorylation targets p53 to degradation by the ubiquitin system. *EMBO J.* **20,** 1630–1639.

Fu, H., Reis, N., Lee, Y., Glickman, M. H., and Vierstra, R. D. (2001). Subunit interaction maps for the regulatory particle of the 26S proteasome and the COP9 signalosome. *EMBO J.* **20,** 7096–7107.

Groisman, R., Polanowska, J., Kuraoka, I., Sawada, J., Saijo, M., Drapkin, R., Kisselev, A. F., Tanaka, K., and Nakatani, Y. (2003). The ubiquitin ligase activity in the DDB2 and CSA complexes is differentially regulated by the COP9 signalosome in response to DNA damage. *Cell* **113,** 357–367.

Henke, W., Ferrell, K., Bech-Otschir, D., Seeger, M., Schade, R., Jungblut, P., Naumann, M., and Dubiel, W. (1999). Comparison of human COP9 signalsome and 26S proteasome lid. *Mol. Biol. Rep.* **26,** 29–34.

Kapelari, B., Bech-Otschir, D., Hegerl, R., Schade, R., Dumdey, R., and Dubiel, W. (2000). Electron microscopy and subunit-subunit interaction studies reveal a first architecture of COP9 signalosome. *J. Mol. Biol.* **300,** 1169–1178.

Lyapina, S., Cope, G., Shevchenko, A., Serino, G., Tsuge, T., Zhou, C., Wolf, D. A., Wei, N., and Deshaies, R. J. (2001). Promotion of NEDD-CUL1 conjugate cleavage by COP9 signalosome. *Science* **292,** 1382–1385.

Naumann, M., Bech-Otschir, D., Huang, X., Ferrell, K., and Dubiel, W. (1999). COP9 signalosome-directed c-Jun activation/stabilization is independent of JNK. *J. Biol. Chem.* **274,** 35297–35300.

Seeger, M., Kraft, R., Ferrell, K., Bech-Otschir, D., Dumdey, R., Schade, R., Gordon, C., Naumann, M., and Dubiel, W. (1998). A novel protein complex involved in signal transduction possessing similarities to 26S proteasome subunits. *FASEB J.* **12,** 469–478.

Uhle, S., Medalia, O., Waldron, R., Dumdey, R., Henklein, P., Bech-Otschir, D., Huang, X., Berse, M., Sperling, J., Schade, R., and Dubiel, W. (2003). Protein kinase CK2 and protein kinase D are associated with the COP9 signalosome. *EMBO J.* **22,** 1302–1312.

Wei, N., Chamovitz, D. A., and Deng, X. W. (1994). Arabidopsis COP9 is a component of a novel signaling complex mediating light control of development. *Cell* **78,** 117–124.

Wei, N., and Deng, X. W. (1998). Characterization and purification of the mammalian COP9 complex, a conserved nuclear regulator initially identified as a repressor of photomorphogenesis in higher plants. *Photochem. Photobiol.* **68,** 237–241.

Wilson, M. P., Sun, Y., Cao, L., and Majerus, P. W. (2001). Inositol 1,3,4-trisphosphate 5/6-kinase is a protein kinase that phosphorylates the transcription factors c-Jun and ATF–2. *J. Biol. Chem.* **276,** 40998–41004.

Zhou, C., Seibert, V., Geyer, R., Rhee, E., Lyapina, S., Cope, G., Deshaies, R. J., and Wolf, D. A. (2001). The fission yeast COP9/signalosome is involved in cullin modification by ubiquitin-related Ned8p. *BMC Biochem.* **2,** 7.

Zhou, C., Wee, S., Rhee, E., Naumann, M., Dubiel, W., and Wolf, D. A. (2003). Fission yeast COP9/signalosome suppresses cullin activity through recruitment of the deubiquitylating enzyme Ubp12p. *Mol. Cell* **11,** 927–938.

# [40] UBP43, an ISG15-Specific Deconjugating Enzyme: Expression, Purification, and Enzymatic Assays

*By* Keun Il Kim and Dong-Er Zhang

## Abstract

UBP43 is the only deconjugating enzyme for the protein ISGylation system thus far identified. UBP43 activity is not critical for precursor processing, as UBP43-deficent cells can generate ISG15 conjugates upon type I interferon treatment. However, UBP43 deficiency caused a defect in the negative regulation of type I interferon signaling, resulting in enhanced and prolonged activation of Jak/Stat upon type I interferon treatment. This chapter describes the expression, purification, and enzymatic assays for UBP43.

## Introduction

The protein ISG15 conjugation process (ISGylation) follows an analogous mechanism to other ubiquitin-like protein (ubl) modifications, based on the sequence similarity of ISG15 to ubiquitin and the involvement of a similar set of enzymes for the conjugation and deconjugation processes

METHODS IN ENZYMOLOGY, VOL. 398

0076-6879/05 $35.00
DOI: 10.1016/S0076-6879(05)98040-3

(Kim and Zhang, 2003; Ritchie and Zhang, 2004). However, the ISGylation system is unique among ubl systems: expression of ISG15 and currently known enzymes for ISGylation and deISGylation are highly induced by type I interferon (IFN). As a result, protein ISGylation occurs when cells or bodies are exposed to viral or bacterial infection (Kim and Zhang, 2003). ISG15 is expressed as a precursor protein with a C-terminal extension in human and rodent cells (Ritchie and Zhang, 2004) and is processed prior to conjugation in order to expose the Gly-Gly residues for conjugation (Loeb and Haas, 1992). Protease activity for precursor processing seems to be independent of IFN signaling, although the exact nature of the protease has not been identified (Potter *et al.*, 1999).

Currently, three enzymes, UBE1L (E1), Ubc8 (E2), and UBP43 (USP18), have been identified for the ISGylation and deISGylation process (Kim *et al.*, 2004; Malakhov *et al.*, 2002; Yuan and Krug, 2001; Zhao *et al.*, 2004). Among them, UBP43 is the enzyme relatively well studied. UBP43 was identified as a putative ubiquitin isopeptidase (Kang *et al.*, 2001; Li *et al.*, 2000; Liu *et al.*, 1999; Schwer *et al.*, 2000; Zhang *et al.*, 1999). UBP43 shares high homology with other ubiquitin-specific proteases, especially in the area of catalytic sites that contain well-conserved Cys and His domains (Chung and Baek, 1999; D'Andrea and Pellman, 1998). However, elaborate biochemical characterization revealed an enzymatic specificity of this enzyme for ISG15 (Malakhov *et al.*, 2002). UBP43 can cleave linear fusions between ISG15 and other proteins or peptides as well as isopeptide bonds between target proteins and ISG15 *in vitro* (Malakhov *et al.*, 2002). However, UBP43 is not required for precursor processing *in vivo,* as UBP43-deficient cells can generate ISGylated proteins upon IFN treatment (Malakhova *et al.*, 2003). This chapter describes how we express and purify UBP43 and what the enzymatic assays for UBP43 are to distinguish the specificity of ISG15 from other ubls.

## Expression and Purification Of UBP43 in Insect Cells

Expression of mouse UBP43 (mUBP43) in *Escherichia coli* as either GST or 6His fusions produced less than 30% of full-length protein. Human UBP43 was even harder to express in *E. coli.* Therefore we used a baculoviral expression system to express mUBP43 or its catalytically inactive mutant form, mUBP43C61S, in insect cells (Sf9).

### *Construction of Baculoviral Expression Plasmid*

We introduced a 6His tag on the N terminus of mUBP43 to facilitate protein purification. The mouse *UBP43* gene is amplified by polymerase chain reaction (PCR) to add *Eco*RI and *Kpn*I restriction sites before the

ATG start codon and after the stop codon of the *UBP43* gene, respectively (the following primers were used—forward: 5′-TTTTGAATTCATGG-CAAGGGGTTTGGG-3′, reverse: 5′-TTTTGGTACCTCAGGATC-CAGTCTTCGTG-3′). The PCR product is then subcloned into the pAcHLT vector, which provides the 6His tag, from BD Biosciences, Pharmingen (San Diego, CA), resulting in pAcHLT-mUBP43.

### *Generation of Baculovirus Expressing 6His-mUBP43*

Recombinant baculovirus expressing 6His-mUBP43 is generated by cotransfecting the pAcHLT-mUBP43 construct with linearized-baculovirus DNA (BD Biosciences, Pharmingen) into Sf9 cells. Two micrograms of pAcHLT-mUBP43 construct and 0.2 $\mu$g of linearized baculovirus DNA are cotransfected into Sf9 cells ($5 \times 10^6$ cells/6-cm dish) using the Cellfectin transfection reagent (Invitrogen, Carlsbad, CA). These cells are grown for 5 days at 27° and culture supernatant containing viruses (6 ml/6 cm dish) is collected. This is the original viral stock.

### *Test of the Expression of 6His-mUBP43*

After generating recombinant baculovirus, we first test whether this virus expresses 6His-mUBP43 properly. Sf9 cells ($2 \times 10^6$ cells) are plated on 6-cm dishes and infected with 100 $\mu$l of baculovirus. Sf9 cells are cultured for 60 h, harvested, and then disrupted in buffer A (50 m*M* Tris-HCl, pH 7.6, 5 m*M* $MgCl_2$, 100 m*M* NaCl, 1% NP-40, and protease inhibitor cocktail). Twenty micrograms of cell extracts from uninfected or virus-infected Sf9 cells is separated on 10% SDS–PAGE and subjected to immunoblotting using an anti-mUBP43 antibody. A clear single band is detected from virus-infected cells, but not from Sf9 control, with a molecular size around 46 kDa, corresponding to 6His-mUBP43 (Fig. 1A). After confirmation of the expression of 6His-mUBP43, the original viral stock is amplified by infecting 100 $\mu$l of virus to Sf9 cells grown in 6 cm dish for 3 days. Five hundred $\mu$l of amplified viruses were then used for large-scale amplification by infecting Sf9 cells cultured in 150 mm dish (30ml). Viruses were harvested 3 days after infection.

### *Titration of Virus to Get the Maximum Expression of 6His-mUBP43*

The proper amount of virus for the maximum expression of 6His-mUBP43 in insect cells is measured with adherent cultures of Sf9 cells as follows. Sf9 cells ($2 \times 10^6$ cells) plated in 6-cm dishes are infected with 10, 25, 50, 100, and 200 $\mu$l of baculovirus and cultured for 60 h at 27°. Cells are harvested, washed with phosphate-buffered saline (PBS) once, and disrupted in buffer A. Twenty micrograms of protein extracts is subjected to

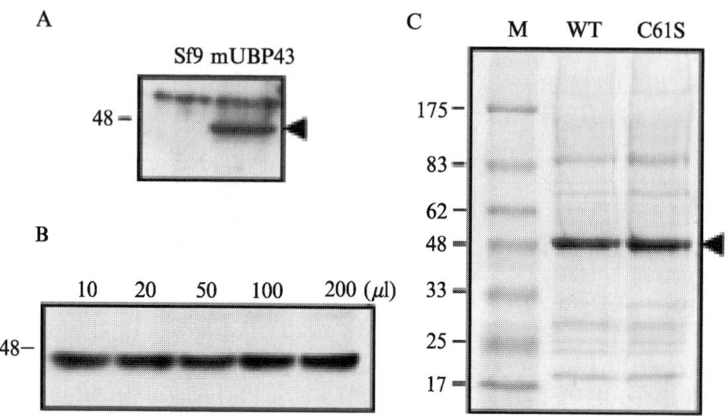


FIG. 1. Expression and purification of mUBP43. (A) Sf9 cells either uninfected (Sf9) or infected with baculovirus expressing 6His-mUBP43 (mUBP43) were cultured for 2.5 days at 27°, and cells extracts were subjected to SDS–PAGE followed by immunoblotting against the anti-mUBP43 antibody. (B) Sf9 cells ($2 \times 10^6$ cells) were infected with increasing amounts of baculovirus expressing 6His-mUBP43. The expression of 6His-mUBP43 was checked as in A. (C) 6His-mUBP43 and 6His-mUBP43C61S were purified using Ni-NTA-agarose affinity chromatography. Around 5 $\mu$g of proteins was separated by SDS–PAGE and visualized by Coomassie blue staining.

SDS–PAGE followed by Western blotting for mUBP43. In this titration assay, 100 $\mu$l of baculovirus is enough to infect $2 \times 10^6$ cells and to express the protein to a high level (Fig. 1B). The titer of virus decided in this step is applied for the following large-scale infection of Sf9 cells in suspension culture.

*Large-Scale Expression and Purification of 6His-mUBP43*

For the large-scale expression of 6His-mUBP43, Sf9 cells are cultured in 500 ml of Sf-900 II media (Invitrogen) until the cell density reaches $2 \times 10^6$ cells/ml and infects with the proper amount of baculovirus expressing 6His-mUBP43, titered as described earlier in the 6 cm dish. The virus-infected Sf9 cells are cultured for an additional 60 h at 27° in a shaking incubator. Cells are harvested by spinning at 3000 rpm (Sorval GS-3 rotor) for 10 min and washed with PBS once. Cells are disrupted in buffer A containing 10 m*M* imidazole, and the lysate is centrifuged at 15,000 rpm (Sorval SS-34 rotor) for 30 min. The cleared lysate is loaded onto 500 $\mu$l of Ni-NTA-agarose and washed with 20 column volumes of buffer A containing 20 m*M* imidazole. After washing, bound proteins are eluted with buffer A containing 200 m*M* imidazole (Fig. 1C). Fractions containing high concentrations of protein are pooled and dialyzed against buffer A containing

0.5 m*M* dithiothreitol (DTT) and 20% glycerol. Purified 6His-mUBP43 is kept at −80° until use. We generally get 500 μg–1 mg of purified 6His-mUBP43 from 1 liter of Sf9 culture.

## Enzymatic Assay for UBP43

### *ISG15-PESTc Cleavage Assay* In Vitro

The plasmid construct for GST-ISG15-gsPESTc (pGEX-ISG15-gsPESTc) is generated by replacing the *Nedd8* part of pGEX-Nedd8-gsPESTc (from K. Tanaka, The Tokyo Metropolitan Institute of Medical Science, Japan) with *ISG15* using *Bam*HI digestion. The plasmid is transformed into BL21 (DE3) *E. coli,* and expression is induced with 0.2 m*M* isopropyl-1-thio-β-D-galactopyranoside (IPTG) for 3 h at 37°. The GST-ISG15-gsPESTc fusion protein is purified on GSH-agarose (Sigma, St. Louis, MO). We originally used $^{125}$I-labeled ISG15-gsPESTc to show ISG15-specific activity of UBP43. We found that the cleavage product GST-ISG15 can be distinguished from GST-ISG15-gsPESTc on SDS–PAGE so we could avoid the use of $^{125}$I. The accuracy of the cleavage reaction with this substrate by mUBP43 has been confirmed by mass spectrometry analysis of the cleavage product gsPESTc (Malakhov *et al.*, 2002).

The proteolytic activity of UBP43 is assayed as follows. Two micrograms of GST-ISG15-PESTc is incubated with 0.1 or 1 μg of either 6His-mUBP43-WT or 6His-mUBP43-C61S in 50 m*M* Tris buffer (pH 7.6) containing 5 m*M* $MgCl_2$ and 0.5 m*M* DTT in a 20-μl reaction volume for 2 h at 37°. Reactions are stopped by adding 20 μl of 2× SDS sample buffer. Ten microliters of reactions out of a total 40 μl is subjected to SDS–PAGE followed by Coomassie blue staining. We detected a band shift of GST-ISG15-PESTc to a lower molecular weight form (GST-ISG15) in reactions containing wild-type UBP43 (6His-mUBP43-WT), whereas no changes were detected with its active site mutant form (6His-mUBP43-C61S) (Fig. 2A). We used Coomassie blue staining to detect proteins on the gel in this experiment. However, immunoblotting against the GST protein can also be used as an alternative to visualize the band shift so that we can get higher sensitivity of the assay using even less enzyme and substrate.

### *ISG15-β-Galactosidase (ISG15-β-Gal) Cleavage Assay in* Escherichia coli

The plasmid construct expressing ISG15-β-gal protein (pACISG15-β-gal) is generated by replacing the *ubiquitin (Ub)* part of the pACUb-β-gal construct (Kim *et al.*, 2000) with the *ISG15* gene using *Spe*I/*Bam*HI digestion.

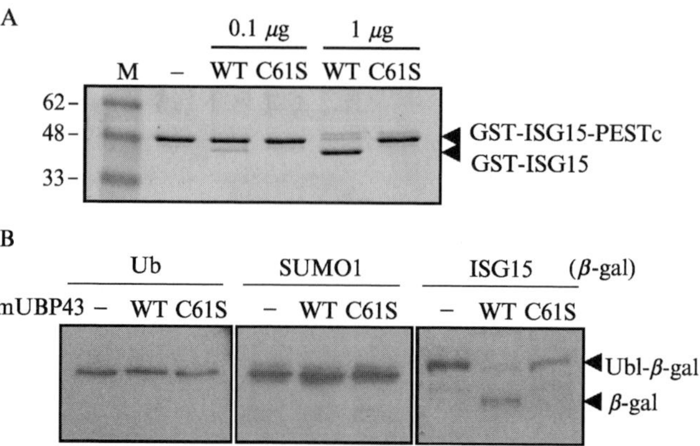


FIG. 2. Activity assay for mUBP43 *in vitro* and in *E. coli.* (A) Two micrograms of GST-ISG15-PESTc was incubated with 0.1 or 1 μg of either 6His-mUBP43-WT or 6His-mUBP43-C61S. The control reaction (–) contains no enzyme. One-fourth of the reaction (10 μl out of a total 40 μl) was subjected to SDS–PAGE followed by Coomassie blue staining. (B) Ubl-β-gal (Ub-β-gal, SUMO1-β-gal, and ISG15-β-gal) was coexpressed with either GST-mUBP43-WT or GST-mUBP43-C61S in *E. coli.* The expression of GST-mUBP43s was induced by treating cells with 0.2 m*M* IPTG for 3 h. *E. coli* cells were harvested, boiled in SDS–PAGE sample buffer, and subjected to SDS–PAGE on an 8% gel followed by immunoblotting against β-gal.

This construct is transformed into the MC1061 *E. coli* strain, and the expression of the ISG15-β-gal protein is confirmed by immunoblotting using both anti-β-gal and anti-ISG15 antibodies. We then generate a MC1061 strain expressing both ubl-β-gal and mUBP43 in the same cell (pACUb-β-gal/pGEX-mUBP43-WT, pACUb-β-gal/pGEX-mUBP43-C61S, pACSUMO1-β-gal/pGEX-mUBP43-WT, pACSUMO1-β-gal/pGEX-mUBP43-C61S, pACISG15-β-gal/pGEX-mUBP43-WT, pACISG15-β-gal/pGEX-mUBP43-C61S). In order to test ISG15-specific cleavage of UBP43 among ubl-β-gals, *E. coli* cells are cultured in LB media until the cell density reaches $A_{600} = 0.8$ and are induced with 0.2 m*M* IPTG for 3 h. *E. coli* cells are harvested, disrupted in SDS–PAGE sample buffer by boiling, and subjected to SDS–PAGE followed by immunoblotting using an anti-β-gal antibody (Cappel ICN, Irvine, CA). As shown in Fig. 2B, we detected a band shift from ubl-β-gal to β-gal only in the *E. coli* expressing ISG15-β-gal and the active form of mUBP43. The expression level of ISG15-β-gal was low compared to other ubl-β-gal in the same amount of protein loading. Thus, the amount of protein extract loaded on the gel was adjusted to get similar band intensities by Western blotting. One thing to keep in mind in this experiment is to use the proper amount of anti-β-gal antibody to get a clear background. Using too much antibody causes a high background and nonspecific bands close to the $\beta$-gal position. For the $\beta$-gal antibody used in the experiment, 5000- to 10,000-fold dilutions were used in our Western blotting.

### *ISG15-UBP43 Self-Cleavage Assay in* Escherichia coli

We developed another *E. coli*-based activity assay for UBP43. In this assay system, we assessed the ability of ISG15-mUBP43 fusions to undergo self-hydrolysis. *E. coli* were transformed with pET-Ub-mUBP43-His and pET-ISG15-mUBP43-His carrying wild-type or inactive versions of mUBP43. Production of mUBP43 in *E. coli* grown in rich media, such as LB, resulted in a high yield but a massive (more than 70%) degradation of mUBP43. However, growth of *E. coli* in nutrient-poor medium M9 and induction of expression at low temperature resulted in the production of small quantities of full-length mUBP43. Therefore, in this experiment, *E. coli* cultures are grown in 5 ml of M9 minimal medium to $A_{600} = 0.8$ and induced with IPTG for 4 h at 21°. Cells are harvested and lysed by sonication, and after removal of insoluble material at 18,000*g* for 5 min, supernatants are resolved on 10% SDS–PAGE and then subjected to immunoblotting against mUBP43. In this experiment we detected a band shift from ISG15-mUBP43 to mUBP43 but no band shift from Ub-mUBP43 (Malakhov *et al.*, 2002).

### *DeISGylating Activity Assay for UBP43*

The three assay methods described earlier mainly address the proteolytic activity of UBP43 to linear fusions between ISG15 and peptide or proteins. This section describes method used to detect the ISG15-deconjugating activity of UBP43. We used cytoskeleton-enriched fractions of *UBP43*$^{-/-}$ mouse lungs as a source of ISG15 conjugates because it has been reported that a substantial portion of cytoskeletal proteins are ISGylated (Malakhov *et al.*, 2002). For the preparation of cytoskeleton-enriched fraction, lungs are homogenized in 5 ml of 20 m*M* Tris buffer (pH 7.5) containing 2 m*M* EGTA, 100 m*M* NaCl, 2 m*M* $MgCl_2$, 2 m*M* phenylmethylsulfonyl fluoride, and 10 $\mu$g/ml leupeptin. The homogenate is centrifuged for 10 min at 16,000*g*. The pellet is washed twice with 7 ml of the same buffer and is then dissolved in 1 ml of 10 m*M* Tris buffer (pH 8.0) containing 5 m*M* EDTA, 100 m*M* NaCl, 5 m*M* $\beta$-mercaptoethanol, 2 m*M* $CaCl_2$, and 0.5% *n*-octyl $\beta$-D-glucopyranoside. The suspension is sonicated briefly and any insoluble material is removed by spinning for 10 min at 15,000*g*. The cytoskeleton-enriched fraction can be stored at −80° until use.

Proteins of cytoskeleton-enriched fraction are incubated with GST-mUBP43–6His or GST-mUBP43C61S–6His in 50 m*M* Tris buffer (pH

7.6) containing 5 m*M* $MgCl_2$ and 0.5 m*M* DTT. Reaction products are separated on 8–18% SDS–PAGE and subjected to immunoblotting against ISG15. Incubation of purified GST-mUBP43-6His with the cytoskeleton-enriched fraction leads to an obvious decrease of ISG15 conjugates, which results in an increase of free ISG15 (Malakhov *et al.*, 2002).

## Concluding Remarks

We described the expression, purification, and enzymatic assay of ISG15-specific deconjugating protease UBP43. UBP43 is the only deISGylating enzyme thus far identified. Interestingly, this enzyme is not involved in the precursor-processing step of ISG15, as UBP43-deficent cells can generate ISG15 conjugates (Ritchie *et al.*, 2002). This finding suggests that there is more than one ISG15 protease in the system. The IFN-independent enzymatic activity that might be responsible for the precursor processing of ISG15 has been identified, although the identity of the enzyme is unknown (Potter *et al.*, 1999). Isopeptidase T/USP5 was shown to react with ISG15-VS, suggesting a possible role of this enzyme in the precursor-processing step of ISG15 (Hemelaar *et al.*, 2004). Our assay method to characterize the ISG15-specific protease activity of UBP43 will be helpful in identifying or characterizing other putative ISG15 proteases as they become known.

## Acknowledgments

We thank members of the DEZ laboratory for valuable discussions and critical reading of the manuscript. This work is supported by National Institutes of Health Grant CA079849 (D.E.Z). The Stein Endowment Fund has partially supported the departmental molecular biology service laboratory for DNA sequencing and oligonucleotide synthesis. This is manuscript 16910-MEM from The Scripps Research Institute.

## References

Chung, C. H., and Baek, S. H. (1999). Deubiquitinating enzymes: Their diversity and emerging roles. *Biochem. Biophys. Res. Commun.* **266,** 633–640.

D'Andrea, A., and Pellman, D. (1998). Deubiquitinating enzymes: A new class of biological regulators. *Crit. Rev. Biochem. Mol. Biol.* **33,** 337–352.

Hemelaar, J., Borodovsky, A., Kessler, B. M., Reverter, D., Cook, J., Kolli, N., Gan-Erdene, T., Wilkinson, K. D., Gill, G., Lima, C. D., Ploegh, H. L., and Ovaa, H. (2004). Specific and covalent targeting of conjugating and deconjugating enzymes of ubiquitin-like proteins. *Mol. Cell. Biol.* **24,** 84–95.

Kang, D., Jiang, H., Wu, Q., Pestka, S., and Fisher, P. B. (2001). Cloning and characterization of human ubiquitin-processing protease-43 from terminally differentiated human melanoma cells using a rapid subtraction hybridization protocol RaSH. *Gene* **267,** 233–242.

Kim, K. I., Baek, S. H., Jeon, Y. J., Nishimori, S., Suzuki, T., Uchida, S., Shimbara, N., Saitoh, H., Tanaka, K., and Chung, C. H. (2000). A new SUMO-1-specific protease, SUSP1, that is highly expressed in reproductive organs. *J. Biol. Chem.* **275,** 14102–14106.

Kim, K. I., Giannakopoulos, N. V., Virgin, H. W., and Zhang, D. E. (2004). Interferon-inducible ubiquitin E2, Ubc8, is a conjugating enzyme for protein ISGylation. *Mol. Cell. Biol.* **24,** 9592–9600.

Kim, K. I., and Zhang, D. E. (2003). ISG15, not just another ubiquitin-like protein. *Biochem. Biophys. Res. Commun.* **307,** 431–434.

Li, X. L., Blackford, J. A., Judge, C. S., Liu, M., Xiao, W., Kalvakolanu, D. V., and Hassel, B. A. (2000). RNase-L-dependent destabilization of interferon-induced mRNAs: A role for the 2-5A system in attenuation of the interferon response. *J. Biol. Chem.* **275,** 8880–8888.

Liu, L. Q., Ilaria, R., Jr., Kingsley, P. D., Iwama, A., van Etten, R. A., Palis, J., and Zhang, D. E. (1999). A novel ubiquitin-specific protease, UBP43, cloned from leukemia fusion protein AML1-ETO-expressing mice, functions in hematopoietic cell differentiation. *Mol. Cell. Biol.* **19,** 3029–3038.

Loeb, K. R., and Haas, A. L. (1992). The interferon-inducible 15-kDa ubiquitin homolog conjugates to intracellular proteins. *J. Biol. Chem.* **267,** 7806–7813.

Malakhov, M. P., Malakhova, O. A., Kim, K. I., Ritchie, K. J., and Zhang, D. E. (2002). UBP43 (USP18) specifically removes ISG15 from conjugated proteins. *J. Biol. Chem.* **277,** 9976–9981.

Malakhova, O. A., Yan, M., Malakhov, M. P., Yuan, Y., Ritchie, K. J., Kim, K. I., Peterson, L. F., Shuai, K., and Zhang, D. E. (2003). Protein ISGylation modulates the JAK-STAT signaling pathway. *Genes Dev.* **17,** 455–460.

Potter, J. L., Narasimhan, J., Mende-Mueller, L., and Haas, A. L. (1999). Precursor processing of pro-ISG15/UCRP, an interferon-beta-induced ubiquitin-like protein. *J. Biol. Chem.* **274,** 25061–25068.

Ritchie, K. J., Malakhov, M. P., Hetherington, C. J., Zhou, L., Little, M. T., Malakhova, O. A., Sipe, J. C., Orkin, S. H., and Zhang, D. E. (2002). Dysregulation of protein modification by ISG15 results in brain cell injury. *Genes Dev.* **16,** 2207–2212.

Ritchie, K. J., and Zhang, D. E. (2004). ISG15: The immunological kin of ubiquitin. *Semin. Cell Dev. Biol.* **15,** 237–246.

Schwer, H., Liu, L. Q., Zhou, L., Little, M. T., Pan, Z., Hetherington, C. J., and Zhang, D. E. (2000). Cloning and characterization of a novel human ubiquitin-specific protease, a homologue of murine UBP43 (Usp18). *Genomics* **65,** 44–52.

Yuan, W., and Krug, R. M. (2001). Influenza B virus NS1 protein inhibits conjugation of the interferon (IFN)-induced ubiquitin-like ISG15 protein. *EMBO J.* **20,** 362–371.

Zhang, X., Shin, J., Molitor, T. W., Schook, L. B., and Rutherford, M. S. (1999). Molecular responses of macrophages to porcine reproductive and respiratory syndrome virus infection. *Virology* **262,** 152–162.

Zhao, C., Beaudenon, S. L., Kelley, M. L., Waddell, M. B., Yuan, W., Schulman, B. A., Huibregtse, J. M., and Krug, R. M. (2004). The UbcH8 ubiquitin E2 enzyme is also the E2 enzyme for ISG15, an IFN-alpha/beta-induced ubiquitin-like protein. *Proc. Natl. Acad. Sci. USA* **101,** 7578–7582.

# [41] Strategies for Assaying Deubiquitinating Enzymes

*By* Sung Hwan Kang, Jung Jun Park, Sung Soo Chung, Ok Sun Bang, and Chin Ha Chung

## Abstract

A general method for assaying deubiquitinating enzymes (DUBs) has been developed. This new method employs an indirect enzyme assay for determining the activity of DUBs using a linear fusion of polyHis-glutathione-*S*-transferase-ubiquitin-ecotin (His-GST-Ub-ecotin) as a substrate. Because ecotin, a trypsin inhibitor protein from *Escherichia coli*, is heat stable, the activity of DUBs can be assayed indirectly by determining the ability of ecotin to inhibit trypsin after incubation of any DUB with His-GST-Ub-ecotin followed by heating at 100°. In the substrate construction, His-GST fusion to Ub was used for facilitation of the substrate purification as well as for assisting the heat precipitation of His-GST-Ub and uncleaved His-GST-Ub-ecotin, as Ub itself is also heat stable. This method can also be used for assaying the proteases that process Ub-like proteins (Ubls) using the substrates, in which Ub is replaced by Ubls.

## Introduction

Many cellular processes are controlled by the Ub posttranslational modification to target proteins, including protein degradation by the 26S proteasome (Hershko and Ciechanover, 1998; Laney and Hochstrasser, 1999; Pickart, 2004). Ubs are ligated to target proteins by a multiple enzyme system, called E1, E2, and E3, through isopeptide bonds (Pickart, 2001). Reversal of ubiquitination catalyzed by DUBs also plays important roles in the regulation of numerous biological pathways, such as by the stabilization of critical regulatory proteins (D'Andrea and Pellman, 1998; Kim *et al.*, 2003; Wilkinson, 1997). Because Ub is encoded as precursor molecules, including tandem repeats of Ubs and fusions of Ub to ribosomal proteins consisting of 52 or 76–80 amino acids (Lund *et al.*, 1985; Ozkanynak *et al.*, 1984), DUBs play an additional important role in the generation of free Ub as well as of certain ribosomal proteins.

DUBs comprise four groups: the Ub C-terminal hydrolase (UCH) family, the Ub-specific processing protease (UBP) family, and two new groups, the Ub-specific JAMM motif containing metalloproteases (the JAMM family) and cysteine proteases containing the OTU domain (otubain family) (Kim *et al.*, 2003). The presence of increasing numbers of DUBs in all

METHODS IN ENZYMOLOGY, VOL. 398

0076-6879/05 $35.00
DOI: 10.1016/S0076-6879(05)98041-5

eukaryotes strongly suggests that distinct enzymes have specialized functions to specific substrates.

Numerous methods for assaying DUBs *in vitro* have been developed. So far, the most widely used method is SDS–PAGE, resolving the cleavage products after the incubation of DUBs with Ub fusions (e.g., Ub-$\beta$-galactosidase) or branched Ub polymers (e.g., tetra-Ub that are linked by either Lys-48 or Lys-63 isopeptide bonds). After staining the gels with Coomassie blue R-250 or with appropriate antibodies, DUB activity can be detected by the increase in the mobility of cleavage products or by the accumulation of Ub monomers. Although this method is very simple, it is laborious and requires a relatively long time to obtain results.

An easier method for assaying DUBs was described using $^{125}$I-labeled Ub-$\alpha$NH-MHISPPEPESEEEEEHYC (referred to as Ub-PESTc) as a substrate (Lee *et al.*, 1998; Woo *et al.*, 1995). Because the Tyr residue in the extension peptide was almost exclusively radioiodinated under a mild labeling condition, such as using IODO-BEADS (Pierce), DUBs could be assayed by simple measurement of the radioactivity released into acid-soluble products. In addition, the recombinant Ub-PESTc protein can be purified easily by heating the *E. coli* extracts at 85° because it is heat stable, like the Ub molecule. Using this assay protocol, a number of DUBs have been purified from various sources, such as yeast and rat and chick muscle (Baek *et al.*, 1998; Park *et al.*, 1997, 2002). However, this method involves the use of radioisotopes.

Stein *et al.* (1995) synthesized a fluorogenic peptide, *N*-benzyloxycarbonyl-Leu-Arg-Gly-Gly-7-amido-4-methylcoumarin (Z-LRGG-AMC), as a substrate of DUBs, as Ub ends with the amino acid sequence of LRGG. They also synthesized Ub-AMC by introduction of the leaving group (AMC) directly to the C-terminal end of Ub (Dang *et al.*, 1998). Therefore, DUBs can be assayed by simple measurement of the fluorescence of released AMC. Using the method, the rate constants ($k_c/K_m$) for the hydrolysis of Ub-AMC were shown to be $10^4$- and $10^7$-fold over those for cleavage of Z-LRGG-AMC for isopeptidase T and UCH-L3, respectively. Thus, the use of Ub-AMC appears to be the most time-saving and sensitive way for assaying DUBs. Here we introduce an additional DUB assay method that is relatively less expensive and can also be applied for the assay of Ubl-specific processing proteases (Ulps).

## Principle of the Method

Ecotin is a periplasmic protein in *E. coli* that is capable of inhibiting trypsin and other pancreatic serine proteases, including chymotrypsin and elastase (Chung *et al.*, 1983). This inhibitor can be assayed by measuring its

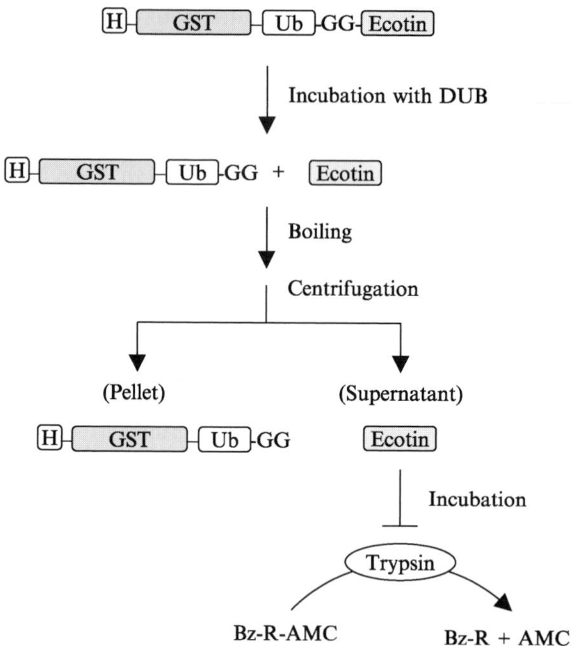


FIG. 1. Schematic outline of a procedure for assaying DUBs. H, 6× histidines; GST, glutathione-*S*-transferase; Ub, ubiquitin; Bz-R-AMC, *N*-benzoyl-Arg-7-amido-4-methylcoumarin.

ability to inhibit the activity of trypsin toward fluorogenic substrates, such as *N*-benzoyl-Arg-7-amido-4-methylcoumarin (Bz-R-AMC). Thus, the activity of DUBs can be measured indirectly by assaying ecotin after incubation of any DUB with a new substrate, His-GST-Ub-ecotin. In this construct, His-GST fusion facilitates substrate purification upon two-step affinity chromatography. Ecotin is unusually stable to heating at 100° at least for 30 min and to exposure to pH 1. Thus, boiling of the incubation mixtures followed by centrifugation can remove His-GST-Ub as well as uncleaved His-GST-Ub-ecotin as the pellets, while leaving ecotin in the supernatants. Because Ub itself is also heat stable, His-GST should assist in the heat precipitation of both His-GST-Ub and His-GST-Ub-ecotin. The overall procedure for assaying DUBs is outlined in Fig. 1.

## Assay of Deubiquitinating Enzymes

### *Buffers and Other Materials*

Phosphate-buffered saline (PBS): 10 m*M* $Na_2HPO_4$, 1.76 m*M* $KH_2PO_4$, 137 m*M* NaCl, 2.69 m*M* KCl (pH 7.4)
Buffer A: 50 m*M* Tris-HCl, 10 m*M* glutathione (pH 8.0)

Buffer B: 50 m*M* $NaH_2PO_4$, 300 m*M* NaCl (pH 8.0)
Buffer C: 500 m*M* Tris-HCl, 5 m*M* DTT, 5 m*M* EDTA, 25% (v/v) glycerol (pH 7.8)
Buffer D: 200 m*M* Tris-HCl, 40 m*M* $CaCl_2$ (pH 8.0)
M15[pREP4] (Qiagen)
BL21-CodonPlus(DE3)-RIL (Stratagene)
Glutathione-Sepharose 4B (Amersham Pharmacia)
$Ni^{2+}$-nitrilotriacetate (NTA) agarose (Qiagen)
Bovine serum albumin (BSA): 0.1% (w/v) in distilled $H_2O$ (Sigma)
Trypsin: 1 $\mu$g/ml in 10 m*M* glycine-HCl (pH 3.5) (Sigma)
Bz-R-AMC: 10 m*M* in 100% dimethyl sulfoxide (DMSO) (Sigma)
UBP69: purified as GST-fused form as described by Park *et al.* (2002)
Fluorometer: FLUOSTAR optima (BMG Labtechnologies)

### *Vector Construction and* E. coli *Strains*

To produce His-GST-Ub-ecotin, cDNAs for GST, Ub, and mature ecotin are inserted into pQE30 serially (Qiagen). The resulting plasmid is referred to as pQE/HGUb-ecotin. To validate the correct cleavage at the C-terminal Gly76 residue of Ub by DUBs, the Gly residue is substituted with Ala by site-directed mutagenesis using pQE/HGUb-ecotin as the template. These constructs are transformed into M15[pREP4]. A plasmid expressing GST-UBP69 is constructed as described previously (Park *et al.*, 2002) and is transformed to BL21-CodonPlus(DE3)-RIL.

### *Purification of His-GST-Ub-Ecotin*

M15[pREP4] cells carrying pQE/HGUb-ecotin are grown at 37° in Luria broth (1 liter) containing ampicillin (100 $\mu$g/ml) and kanamycin (50 $\mu$g/ml) until the optical density at 600 nm reaches 0.4. The cells are treated with isopropyl-$\beta$-D-thiogalactoside to a final concentration of 1 m*M* and are cultured further for the next 2 h. They are then harvested by centrifugation at 7,700*g* for 10 min, resuspended in 50 ml of ice-cold PBS, and disrupted by a French press at 14,000 psi. The cell lysates are centrifuged at 100,000*g* for 2 h. The resulting supernatants are loaded onto a glutathione-Sepharose 4B column (2 ml) equilibrated with PBS. After washing the column extensively with PBS, proteins bound to the column are eluted with buffer A by collecting 2-ml fractions. SDS–PAGE of each fraction reveals that the purity of His-GST-Ub-ecotin is over 90%, which is sufficient for assaying DUBs. However, further purification could be achieved by affinity chromatography using an NTA-agarose column. Fractions containing His-GST-Ub-ecotin are pooled, dialyzed in buffer B containing 10 m*M* imidazole, and loaded onto the affinity column (2 ml) equilibrated with the same buffer. After washing the column with buffer B containing 20 m*M* imidazole, the bound proteins are eluted with the same buffer containing 250 m*M* imidazole by

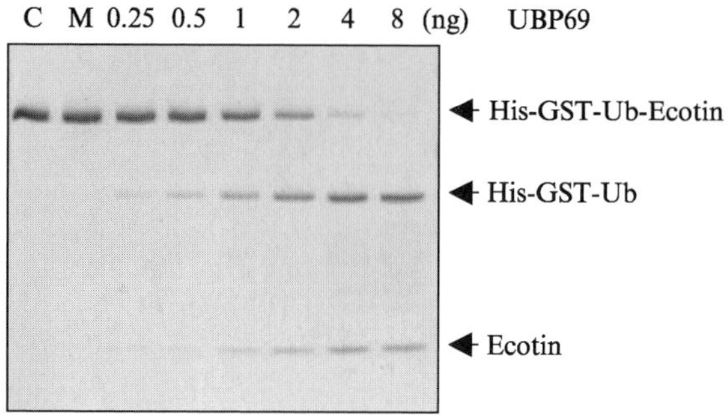


FIG. 2. Hydrolysis of His-GST-Ub-ecotin by UBP69. Reaction mixtures (50 $\mu$l) containing 1 $\mu$g of His-GST-Ub-ecotin and 10 $\mu$l of buffer C were incubated at 37° for 1 h in the absence (lane C) or presence (0.25–8 ng) of UBP69. The mixtures were then subjected to SDS–PAGE using 12% gels followed by staining with Coomassie blue R-250. Lane M indicates the sample incubated with 8 ng of UBP69 and 1 $\mu$g of His-GST-Ub(GA)-ecotin.

collecting 1-ml fractions. SDS–PAGE of the fractions showed that His-GST-Ub-ecotin was purified to apparent homogeneity (Fig. 2, lane C). From 1 liter culture of the cells, 20 mg of purified His-GST-Ub-ecotin is obtained with a yield of about 50%. A mutant form of the substrate, in which the C-terminal Gly76 was replaced by Ala [His-GST-Ub(GA)-ecotin], was also purified to apparent homogeneity as described earlier (Fig. 2, lane M).

*Assay Procedure*

1. Prepare on ice the reaction mixtures (final volume: 50 $\mu$l) containing 5 $\mu$l of 0.2 mg/ml His-GST-Ub-ecotin, 10 $\mu$l of buffer C, and appropriate amounts of any DUB sample.
2. Incubate the reaction mixtures at 37° for 1 h.
3. Stop the reaction by adding 150 $\mu$l of ice-cold 0.1% BSA followed by heating in a boiling water bath for 10 min.
4. Centrifuge at 20,000*g* for 10 min at 4°.
5. Take 10 $\mu$l of the supernatants, mix with 2 $\mu$l of 1 $\mu$g/ml trypsin, 28 $\mu$l of distilled $H_2O$, and 50 $\mu$l of buffer D, and incubate for 10 min at room temperature.
6. Add 10 $\mu$l of 1 m*M* Bz-R-AMC to the incubation mixtures on ice. Note that the stock solution of the peptide substrate (10 m*M* in 100% DMSO) is diluted 10-fold with distilled $H_2O$ just prior to use.
7. Monitor continuously the release of AMC from Bz-R-AMC by incubation of the samples at 37° in a fluorometer (FLUOSTAR optima).

Excitation and emission wavelengths are set at 355 and 460 nm, respectively. Alternatively, to measure individually the fluorescence of released AMC in each sample, incubate the samples in a 37° water bath for appropriate periods and stop the reaction by adding 0.1 ml of ice-cold $H_2O$ followed by heating in a boiling water bath for 10 min.

*Verification of Assay Method*

UBP69 is a DUB in rat muscle, which promotes the differentiation of embryonic myoblasts (Park *et al.*, 2002). To verify the assay method, reaction mixtures containing 0–8 ng of UBP69 are prepared and incubated as described in steps 1 and 2. The samples are then subjected to SDS–PAGE followed by staining with Coomassie blue R-250. Figure 2 shows that upon incubation of His-GST-Ub-ecotin with increasing amounts of UBP69, the band intensity of His-GST-Ub and ecotin increases with a concomitant decrease in that of His-GST-Ub-ecotin. However, His-GST-Ub(GA)-ecotin that has Ala in place of Gly76 in Ub was not cleaved by UBP69 (lane M). These results indicate that UBP69 processes the carboxyl side of the C-terminal Gly76 of Ub correctly, thus generating both ecotin and His-GST-Ub from His-GST-Ub-ecotin.

To verify the remaining procedure, samples obtained from step 6 are incubated in a fluorometer, and the release of AMC from Bz-R-AMC by trypsin is monitored continuously as described in step 7. Figure 3 shows that the extent of trypsin inhibition increases in parallel with the increase in the amount of UBP69 used for the incubation with His-GST-Ub-ecotin, confirming that ecotin generated by the UBP enzyme is responsible for the inhibitory effect. Using the same assay protocol, we could assay other UBP enzymes successfully, such as UBP46, UBP52, and UBP66 from chick muscle (Baek *et al.*, 1998), UBP45 from rat muscle (Park *et al.*, 2002), and HAUSP from mouse and human (Li *et al.*, 2002; Lim *et al.*, 2004) (data not shown). These results indicate that the indirect enzyme assay developed in this study can be used as a general method for assaying DUBs.

Of note was the finding that YUH1, a member of the UCH family from *Saccharomyces cerevisiae* (Miller *et al.,* 1989), could not release ecotin from His-GST-Ub-ecotin (data not shown). Thus, the assay method using His-GST-Ub-ecotin as a substrate may not be suitable for detecting the activity of UCH family members, as UCHs in general are incapable of cleaving Ub fusions with a size of the fusions exceeding 20 kDa (Wilkinson, 1997). It is also possible that other family members of DUBs, which are capable of releasing Ub monomers from poly-Ub chains linked through isopeptide bonds, may not hydrolyze His-GST-Ub-ecotin. However, a number of DUBs that belong to the UBP, otubain, and JAMM families have been shown to generate free Ubs from Ub fusions as well as from proteins

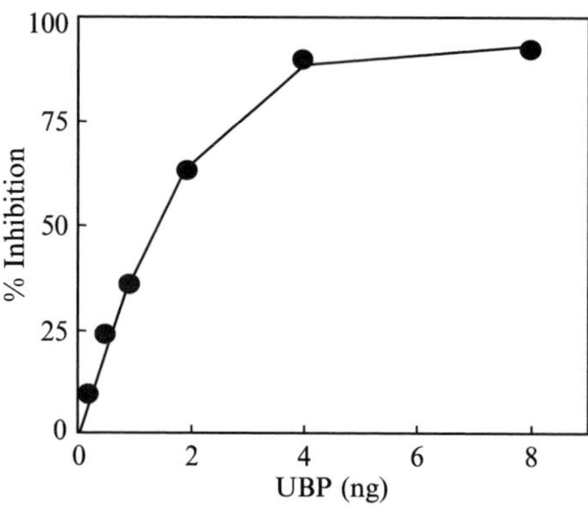


FIG. 3. Inhibition of trypsin by ecotin generated from His-GST-Ub-ecotin. Reaction mixtures were prepared and incubated as in Fig. 1. The reaction was stopped by adding 150 $\mu$l of ice-cold 0.1% BSA followed by heating in a boiling water bath for 10 min. Precipitates were removed by centrifugation at 20,000*g* for 10 min at 4°. Aliquots (10 $\mu$l) of the supernatants were mixed with 2 $\mu$l of 1 $\mu$g/ml trypsin, 28 $\mu$l of distilled $H_2O$, and 50 $\mu$l of buffer D and were incubated for 10 min at room temperature. After incubation, samples were added with 10 $\mu$l of 1 m*M* Bz-R-AMC on ice and were incubated further at 37° in a fluorometer (FLUOSTAR optima). The release of AMC from Bz-R-AMC was then monitored continuously at the excitation and emission wavelengths of 355 and 460 nm, respectively.

conjugated with branched poly-Ub chains or from synthesized branched Ub chains (Baek *et al.*, 1998; Balakirev *et al.*, 2003; Evans *et al.*, 2003; Verma *et al.*, 2002; Yao and Cohen, 2002). Thus, His-GST-Ub-ecotin can be used as a general substrate for assaying other DUB family members.

## Assay of Ubl-Specific Processing Proteases

A number of Ubls have been identified that share sequence and/or structural similarity to Ub (Jentsch and Pyrowolakis, 2000; Yeh *et al.*, 2000). These small proteins are conjugated to a variety of target proteins in a similar manner with Ub. These Ubls include SUMO, Nedd8, ISG15, and Ufm1 (Komatsu *et al.*, 2004), all of which are synthesized as precursors with one or more amino acids following the C-terminal Gly residue of the mature Ubls. Therefore, the tail amino acids or peptides of Ubl precursors need to be removed by Ulps. The method developed for assaying DUBs in the present study can readily be used for assaying various Ulps. The procedure for assaying Ulp should be identical to that for assaying DUBs, except for the use of His-GST-Ubl-ecotin, which can be prepared by the simple replacement of Ub in His-GST-Ub-ecotin by any Ubl, such as Ufm1 or ISG15 (e.g., His-GST-Ufm1-ecotin). We are presently using this method for the isolation of Ulps that are specific to Ufm1 and ISG15 precursors.

## Comments

In the construction of the substrate for DUBs, GST in His-GST-Ub-ecotin can be replaced by any protein, such as MBP, as long as it can be used for affinity purification and is large enough to facilitate precipitation during heating. Because ecotin can inhibit a number of other Ser proteases, chymotrypsin or elastase can be used with its specific fluorogenic or colorimetric peptide substrates in place of trypsin as a target protease for ecotin.

## Acknowledgments

This work was supported by grants from the Korea Research Foundation (DS0041) and Korea Science and Engineering Foundation (through RCPM). S. H. Kang was a recipient of the fellowship from the BK21 Program.

## References

Baek, S. H., Park, K. C., Lee, J. I., Kim, K. I., Yoo, Y. J., Tanaka, K., Baker, R. T., and Chung, C. H. (1998). A novel family of ubiquitin-specific proteases in chick skeletal muscle with distinct N- and C-terminal extensions. *Biochem. J.* **334,** 677–684.

Balakirev, M. Y., Tcherniuk, S. O., Jaquinod, M., and Chroboczek, J. (2003). Otubains: A new family of cysteine proteases in the ubiquitin pathway. *EMBO Rep.* **4,** 517–522.

Chung, C. H., Ives, H. E., Almeda, S., and Goldberg, A. L. (1983). Purification from *Escherichia coli* of a periplasmic protein that is a potent inhibitor of pancreatic proteases. *J. Biol. Chem.* **258,** 11032–11038.

D'Andrea, A., and Pellman, D. (1998). Deubiquitinating enzymes: A new class of biological regulators. *Crit. Rev. Biochem. Mol. Biol.* **33,** 337–352.

Dang, L. C., Melandri, F. D., and Stein, R. L. (1998). Kinetic and mechanistic studies on the hydrolysis of ubiquitin C-terminal 7-amido-4-methylcoumarin by deubiquitinating enzymes. *Biochemistry* **37,** 1868–1879.

Evans, P. C., Smith, T. S., Lai, M. J., Williams, M. G., Burke, D. F., Heyninck, K., Kreike, M. M., Beyaert, R., Blundell, T. L., and Kilshaw, P. J. (2003). A novel type of deubiquitinating enzyme. *J. Biol. Chem.* **278,** 23180–23186.

Hershko, A., and Ciechanover, A. (1998). The ubiquitin system. *Annu. Rev. Biochem.* **67,** 425–479.

Jentsch, S., and Pyrowolakis, G. (2000). Ubiquitin and its kin: How close are the family ties? *Trends Cell Biol.* **10,** 335–342.

Kim, J. H., Park, K. C., Chung, S. S., Bang, O., and Chung, C. H. (2003). Deubiquitinating enzymes as cellular regulators. *J. Biochem.* **134,** 9–18.

Komatsu, M., Chiba, T., Tatsumi, K., Iemura, S., Tanida, I., Okazaki, N., Ueno, T., Kominami, E., Natsume, T., and Tanaka, K. (2004). A novel protein-conjugating system for Ufm1, a ubiquitin-fold modifier. *EMBO J.* **23,** 1977–1986.

Laney, J. D., and Hochstrasser, M. (1999). Substrate targeting in the ubiquitin system. *Cell* **97,** 427–430.

Lee, J. I., Woo, S. K., Kim, K. I., Park, K. C., Baek, S. H., Yoo, Y. J., and Chung, C. H. (1998). A method for assaying deubiquitinating enzymes. *Biol. Proced. Online* **1,** 92–99.

Li, M., Chen, D., Shiloh, A., Luo, J., Nikolaev, A. Y., Qin, J., and Gu, W. (2002). Deubiquitination of p53 by HAUSP is an important pathway for p53 stabilization. *Nature* **416,** 648–653.

Lim, S. K., Shin, J. M., Kim, Y. S., and Baek, K. H. (2004). Identification and characterization of murine mHAUSP encoding a deubiquitinating enzyme that regulates the status of p53 ubiquitination. *Int. J. Oncol.* **24,** 357–364.

Lund, P. K., Moats-Staats, B. M., Simmons, J. G., Hoyt, E., D' Ercode, A. J., Martin, F., and Van Wyk, J. J. (1985). Nucleotide sequence analysis of a cDNA encoding human ubiquitin reveals that ubiquitin is synthesized as a precursor. *J. Biol. Chem.* **260,** 7609–7613.

Miller, H. I., Henzel, W. J., Ridgay, J. B., Kuang, W.-J., Chisholm, V., and Liu, C.-C. (1989). Cloning and expression of a yeast ubiquitin-protein cleaving activity in *Escherichia coli. Biotechnology* **7,** 698–704.

Ozkaynak, E., Finely, D., and Varshavsky, A. (1984). The yeast ubiquitin gene: Head-to-tail repeats encoding a polyubiquitin precursor protein. *Nature* **312,** 663–666.

Park, K. C., Kim, J. H., Choi, E. J., Min, S. W., Rhee, S., Baek, S. H., Chung, S. S., Bang, O., Park, D., Chiba, T., Tanaka, K., and Chung, C. H. (2002). Antagonistic regulation of myogenesis by two deubiquitinating enzymes, UBP45 and UBP69. *Proc. Natl. Acad. Sci. USA* **99,** 9733–9738.

Park, K. C., Woo, S. K., Yoo, Y. J., Wyndham, A. M., Baker, R. T., and Chung, C. H. (1997). Purification and characterization of UBP6, a new ubiquitin-specific protease in *Saccharomyces cerevisiae. Arch. Biochem. Biophys.* **347,** 78–84.

Pickart, C. M. (2001). Mechanisms underlying ubiquitination. *Annu. Rev. Biochem.* **70,** 503–533.

Pickart, C. M. (2004). Back to the future with ubiquitin. *Cell* **116,** 181–190.

Stein, R. L., Chen, Z., and Melandri, F. (1995). Kinetic studies of isopeptidase T: Modulation of peptidase activity by ubiquitin. *Biochemistry* **34,** 12616–12623.

Verma, R., Aravind, L., Oania, R., McDonald, W. H., Yates, J. R., Koonin, E. V., and Deshaies, R. J. (2002). Role of Rpn11 metalloprotease in deubiquitination and degradation by the 26S proteasome. *Science* **298,** 611–615.

Wilkinson, K. D. (1997). Regulation of ubiquitin-dependent processes by deubiquitinating enzymes. *FASEB J.* **11,** 1245–1256.

Woo, S. K., Lee, J. I., Park, I. K., Yoo, Y. J., Cho, C. M., Kang, M. S., Ha, D. B., Tanaka, K., and Chung, C. H. (1995). Multiple ubiquitin C-terminal hydrolases from chick skeletal muscle. *J. Biol. Chem.* **270,** 18766–18773.

Yao, T., and Cohen, R. E. (2002). A cryptic protease couples deubiquitination and degradation by the proteasome. *Nature* **419,** 403–407.

Yeh, E. T., Gong, L., and Kamitani, T. (2000). Ubiquitin-like proteins: New wines in new bottles. *Gene* **248,** 1–14.

# [42] *In Vitro* Cleavage of Nedd8 from Cullin 1 by COP9 Signalosome and Deneddylase 1

*By* KOSJ YAMOAH, KENNETH WU, and ZHEN-QIANG PAN

## Abstract

Enzymatic cleavage of Nedd8 from its cullin conjugates plays a critical role in ubiquitin-dependent proteolysis by regulating the activity of the cullin-based RING-H2 E3 ubiquitin ligases. This chapter provides methods for the preparation of two Nedd8 isopeptidases: the COP9 signalosome and human deneddylase 1. It also describes the development of cell-free systems for cleavage of the Nedd8-cullin 1 isopeptide bond formed *in vitro* or *in vivo*.

## Introduction

Conjugation of Nedd8, a small ubiquitin-like protein, to cullin family proteins plays a critical role in ubiquitin-dependent proteolysis by enhancing the capacity of the cullin-ROC1/Rbx1/Hrt1-based E3 ubiquitin ligases to promote the ubiquitination of substrate proteins (reviewed by Pan *et al.*, 2004). Nedd8 is covalently conjugated to a cullin protein through a process termed neddylation, characterized by formation of an isopeptide bond linking the carboxyl end of Nedd8 Gly-76 to the $\varepsilon$ amino group of a conserved cullin lysine residue (reviewed by Hochstrasser, 1998).

Neddylation can be reversed by the action of a Nedd8 isopeptidase (deneddylase). One well-documented deneddylase is the COP9 signalosome (CSN), which is an eight-subunit complex, originally isolated as a suppressor of plant photomorphogenesis (reviewed by Wei and Deng, 2003). As demonstrated by accumulating *in vivo* and *in vitro* evidence, CSN is an evolutionarily conserved, JAMM-catalytic domain-containing metalloprotease, efficiently hydrolyzing the cullin-Nedd8 conjugate (reviewed by Cope and Deshaies, 2003; also see Wu *et al.*, 2003). Another deneddylating enzyme is deneddylase 1 (DEN1; Gan-Erdene *et al.*, 2003; Wu *et al.*, 2003) or NEDP1 (Mendoza *et al.*, 2003), which was previously annotated as SENP8 due to its homology with the Ulp1/SENP cysteinyl SUMO-deconjugating enzyme family. DEN1/NEDP1 distinguishes itself from CSN in its capacity to bind to Nedd8 selectively and to hydrolyze C-terminal derivatives of Nedd8 efficiently and specifically (Gan-Erdene *et al.*, 2003; Mendoza *et al.*, 2003; Wu *et al.*, 2003), suggesting a role for this protease in processing the C terminus of the Nedd8

METHODS IN ENZYMOLOGY, VOL. 398

0076-6879/05 $35.00
DOI: 10.1016/S0076-6879(05)98042-7

precursor (-$G^{75}G^{76}$GGLRQ), thereby generating the functional form of Nedd8 (-$G^{75}G^{76}$) for conjugation to cullins. DEN1/NEDP1 contains an isopeptidase activity capable of removing Nedd8 from its cullin targets (Mendoza *et al.*, 2003; Wu *et al.*, 2003), albeit with much reduced efficiency compared with CSN (Wu *et al.*, 2003).

Cellular deneddylase activities including CSN and possibly DEN1 likely play a critical role in governing the Nedd8 modification status of a cullin, which may vary in response to developmental and environmental cues to regulate proteolytic events that are promoted by the cullin-based E3 ubiquitin ligases. Characterization of deneddylation by CSN and DEN1 using *in vitro* reconstitution systems is of fundamental importance to understanding the mechanistic role of these enzymes. To this end, we describe in detail the methods that we employ to prepare CSN and DEN1 and to analyze their proteolytic activities in hydrolyzing the Nedd8 conjugates with cullin 1 (CUL1).

## Methods for Isolation of Human Deneddylating Enzymes CSN and DEN1

### *Generation of a Stable Cell Line (293F23V5) for Affinity Purification of Human CSN*

In order to facilitate analysis of the deneddylating activity by human CSN, we developed a procedure for affinity purification of the CSN complex assembled in a stable HEK293 cell line (293F23V5) that was engineered to constitutively express two CSN subunits, CSN-2 and CSN-3, with Flag and V5 tags inserted at their N and C terminus, respectively. The 293F23V5 cell line was generated using a protocol as described previously (Dias *et al.*, 2002). HEK293 cells, maintained in growth medium [Dulbecco's modified Eagles medium supplemented with 10% bovine serum and 100 $\mu$g/ml of penicillin–streptomycin (GIBCO/BRL)], are plated at an approximate density of $2 \times 10^6$ cells per plate (10 cm) 20–24 h prior to transfection with pcDNA-Flag-CSN2 (7 $\mu$g) and pcDNA-CSN3-V5-zeo (7 $\mu$g), which carries the zeocin resistance gene (Invitrogen), using the standard calcium phosphate precipitation protocol. The pcDNA-Flag-CSN2 plasmid is constructed by polymerase chain reaction (PCR) using a human fetal kidney cDNA library (Clontech) with primers designed based on the CSN2 sequence from the GenBank database (accession no. AF084260). (A detailed description of the primer sequences and cloning can be given upon request.) The pcDNA-CSN3-V5-zeo plasmid is from Resgen Genestorm (clone ID RG002289). At 40 h posttransfection, cells are trypsinized, diluted at a ratio of 1:5, 1:50, or 1:500, and subsequently plated onto 15-cm plates with the addition of selection medium containing the growth medium (25 ml) and zeocin (1.25 $\mu$g/ml). The selection medium is changed daily to remove dead cells in the subsequent 3 days and then every third day for another 2 weeks. Individual colonies, becoming visible 10–15 days after the initial zeocin treatment, are picked by sterile cloning disks (Scienceware) presoaked in trypsin-EDTA (GIBCO/BRL) and are placed onto 24-well plates containing the selection medium. The cloned cells are released from the disks via gentle agitation and grown. Positive clones are identified by screening expression of both Flag-CSN2 and CSN3-V5 via immunoblot analysis.

### *Affinity Purification of Human CSN from 293F23V5 Cells*

To affinity purify human CSN, fifty 15-cm plates worth of 293F23V5 cells are harvested, and extracts are prepared using a procedure as described previously (Dias *et al.*, 2002). Extracts (15 ml; 5 mg of protein per milliliter), maintained in buffer A [15 m*M* Tris-HCl, pH 7.4, 0.5 *M* NaCl, 0.35% NP-40, 5 m*M* EDTA, 5 m*M* EGTA, 1 m*M* phenylmethylsulfonyl fluoride (PMSF), 2 $\mu$g/ml of antipain, and 2 $\mu$g/ml of leupeptin], are adsorbed to M2 agarose beads (0.5 ml; Sigma) by rotating the mixture at 4° for 14–16 h. The resulting beads are packed into a column, which is washed sequentially with 80 ml of buffer A and then 10 ml of buffer B [25 m*M* Tris-HCl, pH 7.5, 1 m*M* EDTA, 0.01% NP-40, 10% glycerol, 0.1 m*M* PMSF, 1 m*M* dithiothreitol (DTT), 0.2 $\mu$g/ml of antipain, and 0.2 $\mu$g/ml of leupeptin] plus 0.1 *M* NaCl. Bound proteins are eluted with Flag-peptide (1 mg/ml) in buffer B plus 0.1 *M* NaCl for three consecutive times, 0.7 ml per time. The pooled elutes are concentrated to 0.5 ml with a 5K NMWL Ultrafree-15 centrifugal filter device (Millipore), yielding CSN at a concentration of 0.5 pmol/$\mu$l. The enzyme preparation is stored at −80°. No significant loss of proteolytic activity is observed with repeated freeze–thaw for 10 times in a period of 1 year.

### *Sizing Chromatographic Analysis of Affinity-Purified Human CSN*

To determine the molecular weight of the affinity-purified human CSN complex, we carried out sizing chromatographic analysis. CSN is gel filtrated through the Superose 6 10/300 GL column (Amersham Biosciences) with buffer B plus 150 m*M* NaCl. Silver staining analysis revealed that all eight CSN subunits comigrated, peaking at fraction 25 (Fig. 1A, lane 4), which corresponds to a Stokes radius of 75 Å (Fig. 1D). In parallel, the purified CSN complex is analyzed by sedimentation through a glycerol gradient (15–35%, 5 ml) that contains buffer C (25 m*M* Tris-HCl, pH 7.5, 1 m*M* EDTA, 0.01% NP-40, 0.1 m*M* PMSF, 1 m*M* DTT, 0.2 $\mu$g/ml of

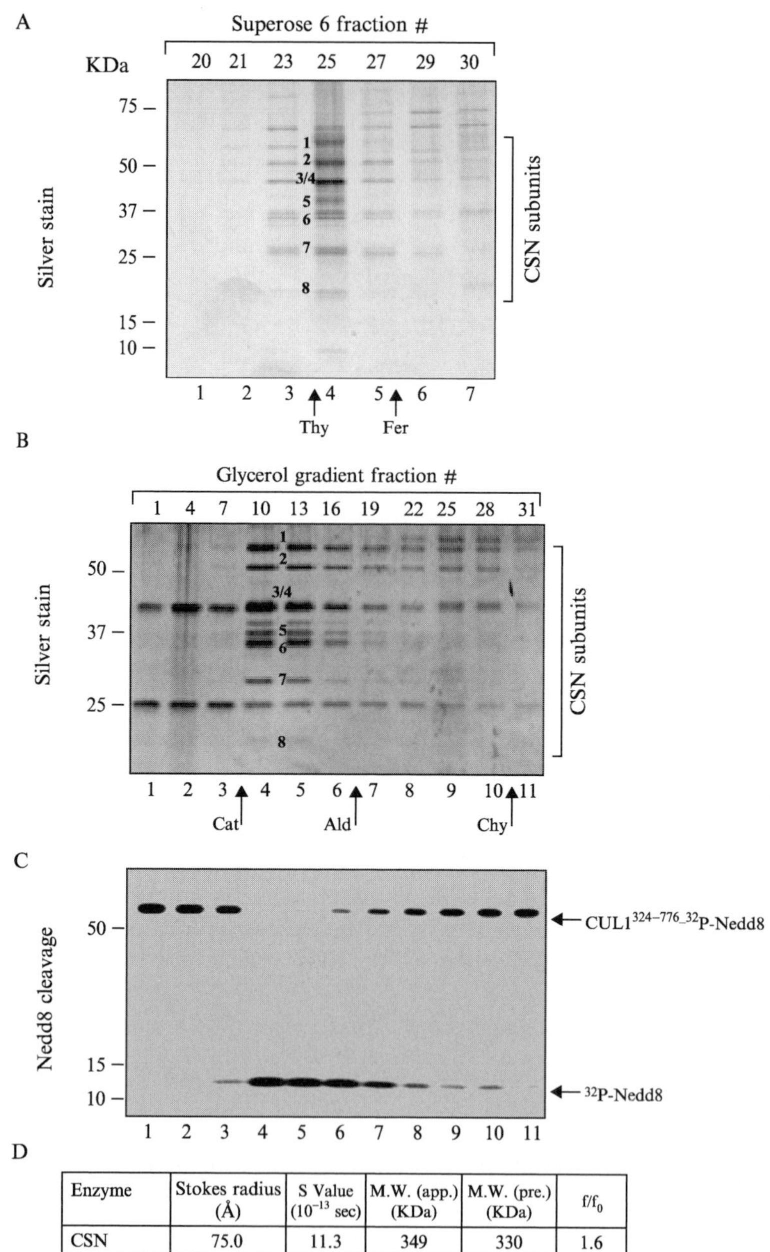


| Enzyme | Stokes radius (Å) | S Value ($10^{-13}$ sec) | M.W. (app.) (KDa) | M.W. (pre.) (KDa) | $f/f_0$ |
|---|---|---|---|---|---|
| CSN | 75.0 | 11.3 | 349 | 330 | 1.6 |

FIG. 1. Sizing chromatographic analysis of the affinity-purified human CSN complex. (A) Gel filtration analysis of the affinity-purified human CSN complex. Following gel filtration of the affinity-purified human CSN complex (200 pmol) through the Superose 6 column, an

antipain, and 0.2 $\mu$g/ml of leupeptin) plus 0.25 *M* NaCl. After centrifugation at 45,000 rpm (Beckman, SW50.1) for 18 h at 4°, fractions are collected and their aliquots are electrophoresed followed by staining with silver. The result showed comigration of all eight CSN subunits, peaking at fraction 10 (Fig. 1B, lane 4), corresponding to an S value of 11.3 (Fig. 1D).

Taken together, these results demonstrate that the M2 matrix-based single-step affinity purification yields an eight-subunit CSN complex. When calculated based on the formula by Siegel and Monty (1966), the apparent molecular mass of CSN in our preparation is 349 kDa, which is very close to the predicted molecular mass of 330 kDa, assuming that all eight subunits are present in single copies. Thus, the isolated human CSN contains eight subunits that appear to be stoichiometric. In addition, these analyses revealed that CSN possesses a frictional coefficient ($f/f_o$) of 1.6 (Fig. 1D), suggesting that the complex is moderately elongated.

### *Purification of Recombinant DEN1*

Large amounts of soluble DEN1 can be obtained easily using the *E. coli* expression system. Construction of the DEN1 expression plasmid for this purpose (pDEST-17-DEN1) has been described previously (Wu *et al.*,

aliquot of the indicated fractions (15 $\mu$l) was electrophoresed in 4–20% gradient denaturing gels and analyzed by silver staining. Numbers in the middle indicate positions of individual CSN subunits. Arrows at the bottom mark the migration positions of size markers thyroglobulin (669 kDa) and ferritin (440 kDa). (B) Glycerol gradient analysis of CSN. Purified CSN (100 pmol) was analyzed by glycerol gradient fractionation, and aliquots (15 $\mu$l) of the indicated fractions were separated on 4–20% SDS–PAGE followed by silver staining. Numbers in the middle mark CSN subunits. Arrows at the bottom indicate the migration positions of size markers catalase (240 kDa), aldolase (160 kDa), and chymotrypsinogen A (25 kDa). (C) CSN contains an intrinsic Nedd8 isopeptidase activity. Aliquots (3 $\mu$l) of the indicated glycerol gradient fractions were analyzed by the $^{32}$P-Nedd8-cleavage assay. An autoradiograph is shown. Positions of the CUL1$^{324-776}$-$^{32}$P-Nedd8 substrate and the cleavage product, $^{32}$P-Nedd8, are indicated. (D) Molecular parameters of the affinity-purified CSN complex are shown. The Stokes radius and sedimentation coefficient of CSN were obtained from standard curves generated by gel filtration and glycerol gradient analyses of molecular size markers. The apparent molecular mass of CSN was calculated by employing the Siegel and Monty formula (Siegel and Monty, 1966): $M_a = 6\pi\eta Nas/(1\text{-}v\rho)$, where a is Stokes radius, s is sedimentation coefficient, $\eta$ is viscosity of medium (0.01 poise), v is partial specific volume (0.725 cm$^3$/g), $\rho$ is density of medium (1), and N is Avogadro's number ($6.023 \times 10^{23}$). The frictional coefficient ($f/f_o$) of CSN was calculated based on $f/f_0 = a/(3vM_p/4\pi N)^{1/3}$, where $M_p$ is predicted molecular weight. $M_p$ was calculated assuming that CSN subunits exist as single copies. The accession numbers and respective molecular masses of CSN subunits are as follows: CSN1 (U20285; 56 kDa), CSN2 (AF084260; 53 kDa with a Flag epitope), CSN3 (AF031647; 48 kDa with a V5 epitope), CSN4 (BC009292; 46 kDa), CSN5/Jab1 (U65928; 38 kDa), CSN6 (BC002520; 36 kDa), CSN7 (BC011789, 30 kDa), and CSN8 (CR456994; 23 kDa).

2003). Purification is facilitated by the inclusion of an N-terminal glutathione *S*-transferase (GST) tag. A thrombin protease site is also present for removal of the GST tag during the purification process. Coomassie staining of SDS–PAGE gels is performed to monitor the purity after each step.

GST-DEN1 is highly expressed in BL21(DE3) cells. A 40-ml starter culture grown in LB media with 20 $\mu$g/ml ampicillin is incubated overnight at 37° in a shaking incubator. Four liters of LB media with 0.4% glucose and 20 $\mu$g/ml ampicillin is then inoculated with this starter culture and incubated until the optical density ($OD_{600}$) reaches 0.5 (approximately 3 h). Recombinant GST-DEN1 expression is induced by the addition of isopropyl-$\beta$-D-thiogalactoside (IPTG) at a final concentration of 0.8 m*M*. The culture is incubated for another 3 h and then cells are pelleted by centrifugation at 5000 rpm for 15 min in a Sorvall GS-3 rotor at 4°.

Following resuspension of cell pellet in buffer D (50 m*M* Tris-HCl, pH 8.0, 1% Triton X-100, 0.5 *M* NaCl, 10 m*M* EDTA, 10 m*M* EGTA, 10% glycerol, 2 m*M* PMSF, 5 m*M* DTT) and sonication (4× 20-s pulses, Misonix Sonicator 3000), the resulting extract is then clarified by centrifugation at 17,000 rpm for 30 min in a Sorvall SS-34 rotor at 4°. The majority of the GST-DEN1 is found in the soluble supernatant fraction. To purify GST-DEN1, the clarified extract supernatant is passed through a 30-ml glutathione-Sepharose column preequilibrated with buffer D. The column is then washed with 10 column volumes of buffer D followed by 10 column volumes of buffer E (25 m*M* Tris-HCl, pH 7.5, 1 m*M* EDTA, 0.01% NP-40, 10% glycerol, 1 m*M* DTT) plus 50 m*M* NaCl. GST-DEN1 is then eluted with 90 ml of 20 m*M* glutathione in buffer E plus 50 m*M* NaCl. Peak fractions are pooled, yielding approximately 700 mg of protein.

Because the GST tag is relatively large, its removal is desirable to prevent any possibility of interference with normal DEN1 activity. The pooled GST-DEN1 fractions are dialyzed against buffer F (20 m*M* Tris-HCl, pH 8.4, 150 m*M* NaCl, 5% glycerol, and 2.5 m*M* $CaCl_2$). Biotinylated thrombin (Novagen; 350 units) is then added to the dialyzed pool and incubated at room temperature overnight. A Millex GP syringe filter (Millipore) is used to remove any precipitates formed during this incubation. The filtrated material is passed through a preequilibrated strepavidin Sepharose high-performance column (Amersham Biosciences; 2 ml) to capture the biotinylated thrombin. The flow through is then put back onto the regenerated, preequilibrated glutathione-Sepharose column (30 ml) to capture the cleaved GST tag. The resulting flow-through fractions are then pooled, yielding approximately 200 mg of DEN1.

Any contaminating proteins remaining in the preparation is removed via gel filtration chromatography. By buffer exchange, the aforementioned DEN1 fraction is adjusted to contain buffer E plus 150 m*M* NaCl and is concentrated to approximately 50 mg/ml. It is then loaded onto a preequilibrated HiLoad 16/60 Superdex 75-pg column (SD75; Amersham Biosciences) in two batches. Figure 2A shows a representative elution profile for DEN1, peaking at fraction 9, that corresponds to chymotrypsinogen A of 25 kDa (lane 7). This result, combined with glycerol gradient analysis (data not shown), demonstrates that the bacterially expressed DEN1 exists as a monomer. The peak fractions from the two runs are pooled and concentrated again, yielding approximately 150 mg of pure DEN1.

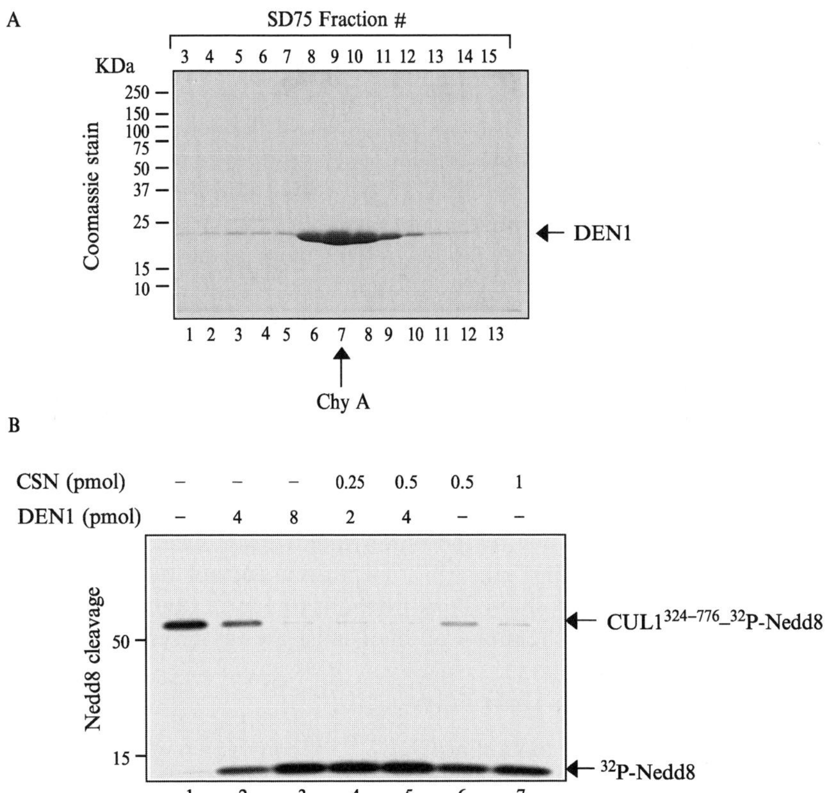


FIG. 2. Sizing chromatographic analysis of DEN1. (A) Gel filtration analysis of DEN1. Proteins derived from SD75 gel filtration fractions (10 $\mu$l) were separated by 4–20% gradient denaturing gels and stained by Coomassie. Arrow denotes fraction 9, where chymotrypsinogen A (25 kDa) migrates. (B) Comparison of *in vitro* deneddylase activity by DEN1 and CSN. DEN1 and/or CSN, in amounts indicated, was analyzed by the $^{32}$P-Nedd8 cleavage assay. The reaction products were analyzed as described for Fig. 1C.

## Preparation of Substrates

### *Preparation of the ROC1-CUL1$^{324–776}$ Complex*

In order to analyze the deneddylation activity by CSN and DEN1 *in vitro*, we generated a substrate containing $^{32}$P-labeled Nedd8 that was conjugated to a CUL1 moiety. For this purpose, we prepared a heterodimeric complex containing ROC1 and a CUL1 C-terminal fragment, CUL1$^{324–776}$. (Preparation of CUL1 alone has been unsuccessful due to its insolubility.) The complex, designated as ROC1-CUL1$^{324–776}$, was obtained by simultaneously overexpressing GST-HA-ROC1 and His-Flag-CUL1$^{324–776}$ in a bacteria strain that was transformed with a previously constructed plasmid called pGEX-4T3/pET-15b-(GST-HA-ROC1)/His-FLAG-CUL1$^{324–776}$ (Wu *et al.*, 2000). Following induction with IPTG (0.2 m*M*) at 25° for 14 h, the induced cell culture (2 liters) is harvested by centrifugation, and the resulting pellet is resuspended (in 1/25 culture volume) with buffer G (50 m*M* Tris-HCl, pH 8.0, 1% Triton X-100, 0.5 *M* NaCl, 10 m*M* EDTA, 10 m*M* EGTA, 10% glycerol, 2 m*M* PMSF, 0.4 $\mu$g/ml of antipain, 0.2 $\mu$g/ml of leupeptin, and 5 m*M* DTT). After sonication, the extracts are cleared by centrifugation, yielding 60 ml of supernatant, half of which is then bound to glutathione-Sepharose (2 ml) by rotating the mixture for 4 h at 4°, and the resulting beads are washed sequentially with buffer G (100 ml) and buffer B plus 50 m*M* NaCl (100 ml).

To obtain the GST-free ROC1-CUL1$^{324–776}$ complex, the aforementioned GST-ROC1-CUL1$^{324–776}$-immobilized glutathione-Sepharose are incubated with the biotinylated thrombin (20 units in 2 ml of buffer F) for 12–14 h at 14°, resulting in greater than 95% cleavage of GST-ROC1-CUL1$^{324–776}$ and release of the ROC1-CUL1$^{324–776}$ complex to the unbound fraction. Following removal of the biotinylated thrombin by strepavidin-Sepharose (50 $\mu$l), ROC1-CUL1$^{324–776}$ is purified further by gel filtration through the Superose 6 column. Peak fractions (0.5 ml; 1 mg/ml of protein) are pooled and used for the preparation of substrates as described later.

### In Vitro *Conjugation of $^{32}$P-Nedd8 to CUL1$^{324–776}$*

To prepare substrates containing CUL1$^{324–776}$ conjugated with $^{32}$P-Nedd8, a two-step reaction is carried out. Purified PK-Nedd8 (15 $\mu$g), constructed previously to contain an N-terminally located cAMP kinase phosphorylation site (Wu *et al.*, 2003), is incubated in a reaction mixture (200 $\mu$l) that contains 40 m*M* Tris-HCl, pH 7.4, 12 m*M* $MgCl_2$, 2 m*M* NaF, 50 m*M* NaCl, 25 $\mu M$ ATP, 50 $\mu$Ci of [$\gamma$-$^{32}$P]ATP, 0.1 mg/ml bovine serum albumin, and 20 units of cAMP kinase (Sigma). The reaction is incubated at 37° for 30 min, yielding $^{32}$P-Nedd8. A second reaction mixture is then added to include 4 m*M* ATP, APP-BP1/Uba3 (20 ng), 10 $\mu$g of Ubc12, and 80 pmol of ROC1-CUL1$^{324-776}$. The reaction is incubated further at 37° for 60 min, producing ROC1-CUL1$^{324-776}$_$^{32}$P-Nedd8, which is then purified by adsorbing the reaction mixture to the agarose matrix cross-linked with the anti-HA antibody (Sigma) at 4° for 2 h. (This affinity purification is based on the presence of a HA epitope at the N terminus of ROC1.) Following washing with buffer A and then B plus 50 m*M* NaCl, bound proteins are eluted with the HA peptide (2 mg/ml) in buffer B plus 50 m*M* NaCl, yielding ROC1-CUL1$^{324-776}$_$^{32}$P-Nedd8 with an approximate concentration of 0.26 $\mu M$.

### *Isolation of Flag-Nedd8-Conjugated Substrates Assembled* In Vivo

To prepare Nedd8 substrates that were conjugated *in vivo*, we created a Flag-Nedd8 expressing cell line, 293FN8, which allows affinity purification of cellular proteins covalently conjugated with Nedd8. The method used for the generation of 293FN8 was similar to that for 293F23V5 as described earlier with the following exceptions. HEK293 cells are cotransfected with pcDNA3.1-Flag-Nedd8 (7 $\mu$g) and Puro2-vector (2 $\mu$g) that carries the puromycin resistance gene (Clontech). Hence, colonies are grown under the selection with puromycin (Sigma; 2.5 $\mu$g/ml) in place of zeocin. The pcDNA3.1-Flag-Nedd8 plasmid is constructed by PCR using pET3a-Nedd8 (kindly provided by C. Pickart, John Hopkins University) as a template. (A detailed description of the primer sequences and cloning can be given upon request.) To isolate Flag-Nedd8 conjugates, two hundred and fifty 15-cm plates worth of 293FN8 cells are harvested, and the resulting extracts (75 ml) are used for M2 matrix-based affinity purification as described earlier for the CSN, yielding a final fraction (0.5 ml; 0.2 $\mu$g/$\mu$l of protein).

To detect the presence of cullin conjugates in the aforementioned Flag-Nedd8 immunoprecipitates, we performed immunoblot assays as shown in Fig. 3A. Consistent with previous *in vivo* findings (Hori *et al.*, 1999), Flag-Nedd8 was found to conjugate endogenous cullins 1–5 (Fig. 3A, lanes 1–5). In addition, when probed with antibodies recognizing Nedd8 or Flag, it was revealed that Flag-Nedd8 immunoprecipitates contained Nedd8 antibody-reactive bands with molecular masses unmatched with mononeddylated cullins (Fig. 3A, lanes 6 and 7; marked by arrows). This finding suggests that Flag-Nedd8 may conjugate with noncullin cellular proteins or form conjugates that contain cullins attached with multiple Nedd8 moieties. It was evident, however, that the Nedd8–cullin 1–5 conjugates are the predominant neddylated species, as reflected by their strong reactivity to the Flag antibody in comparison to other polypeptides (Fig. 3A, lane 7).

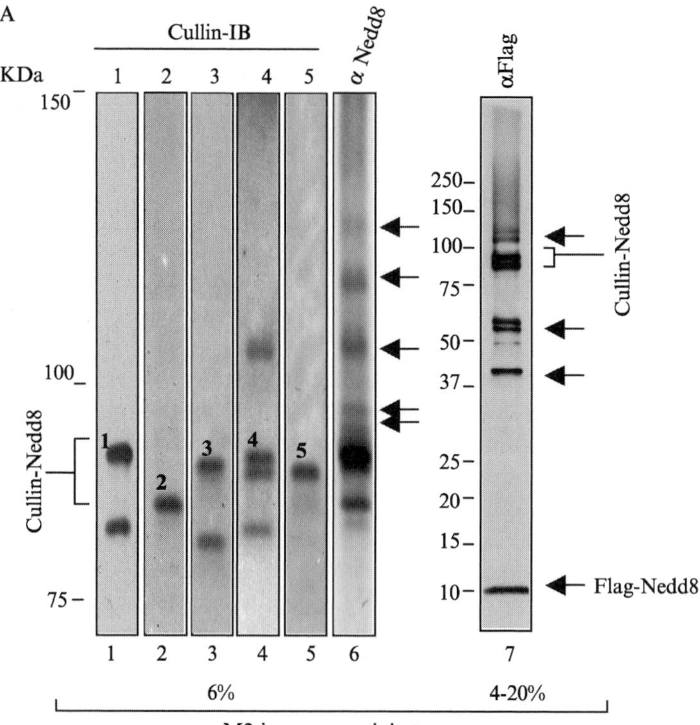


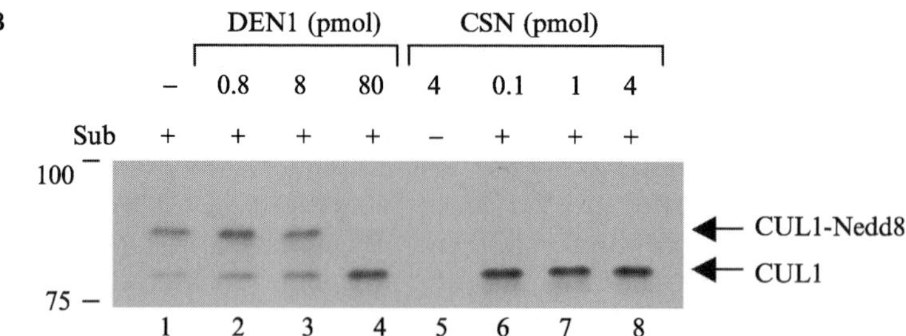


FIG. 3. Analysis of *in vivo*-assembled Flag-Nedd8 conjugates. (A) Immunoblot analysis of Flag-Nedd8 immunoprecipitates. An aliquot of the concentrated Flag-peptide eluant (7 $\mu$l) was subjected to 6% (lanes 1–6) or 4–20% (lane 7) SDS–PAGE followed by immunoblotting with specific antibodies against cullins 1–5 (lanes 1–5), Nedd8 (lane 6), or Flag (lane 7). Positions of neddylated cullins are indicated. Arrows mark polypeptides recognized by the Nedd8 or Flag antibody with apparent molecular masses unmatched with mononeddylated cullins 1–5. (B). Cleavage of *in vivo*-conjugated CUL1-Flag-Nedd8 isopeptide bond by DEN1 and CSN. An aliquot of the concentrated Flag-peptide eluant (4 $\mu$l) was incubated with indicated amounts of DEN1 (lanes 2–4) or CSN (lanes 6–8). The reaction mixture was analyzed by immunoblot with the anti-CUL1 antibody.

## *In Vitro* Cleavage of Nedd8 from CUL1 by CSN and DEN1

We developed a sensitive *in vitro* assay to measure the proteolytic activity of CSN and DEN1 in deconjugation of Nedd8 from CUL1. This autographic assay monitors CSN- or DEN1-dependent cleavage of Nedd8 from a radiolabeled ROC1-CUL1$^{324-776}$-$^{32}$P-Nedd8 substrate, which is characterized by the reduced levels of the CUL1$^{324-776}$-$^{32}$P-Nedd8 covalent conjugates and concomitant accumulation of $^{32}$P-Nedd8. The cleavage reaction is carried out in a mixture (10 $\mu$l) that contains 40 m*M* Tris-HCl, pH 7.4, 0.6 m*M* DTT, 2 m*M* NaF, 10 m*M* okadaic acid, 0.2 mg/ml BSA, 26 n*M* ROC1-CUL1$^{324-776}$-$^{32}$P-Nedd8, and deneddylating enzyme. Following incubation at 37° for 30 min, the reaction products are separated by 4–20% SDS–polyacrylamide gel electrophoresis and visualized by autoradiograms. Using this assay, we determined whether CSN contains an intrinsic deneddylase activity. For this purpose, aliquots of the glycerol gradient fractions derived from the affinity-purified human CSN complex (as shown in Fig. 1B) were tested for their ability to hydrolyze the radioactive substrate. The results of this experiment revealed a single peak of proteolytic activity, characterized by the disappearance of CUL1$^{324-776}$-$^{32}$P-Nedd8 and the accumulation of $^{32}$P-Nedd8, which was located in fractions 10–16 (Fig. 1C, lanes 4–6), coinciding with the migration of the CSN complex (Fig. 1B, lanes 4–6). Thus, CSN contains an intrinsic isopeptidase activity, capable of hydrolyzing the CUL1-Nedd8 linkage.

As shown in Fig. 2B, purified recombinant DEN1 was able to cleave Nedd8 from CUL1$^{324-776}$ as well, albeit much higher levels of DEN1 were required compared to CSN (lanes 2, 3, 6, and 7; also see Wu *et al.*, 2003). No *in vitro* synergistic effect by these two enzymes was observed as their combination yielded a mere additive cleavage (Fig. 2B, compare lanes 2, 5, and 6). Future experiments are required to determine whether these two enzymes function cooperatively *in vivo*.

To determine whether CSN and/or DEN1 was able to cleave the CUL1-Nedd8 isopeptide bond formed *in vivo*, we carried out the *in vitro* cleavage reaction using the aforementioned Flag-Nedd8 immunoprecipitates as substrate. The reaction was carried out identically to that with the radioactive substrate with the exception that both NaF and okadaic acid were omitted. In this case, the conversion from neddylated CUL1 to an unmodified species, as a result of incubation with DEN1 (Fig. 3B, lanes 2–4) or CSN (lanes 6–8), was monitored by immunoblot analysis using the anti-CUL1 antibody. As shown, while both enzymes were able to cleave the CUL1-Nedd8 conjugates, CSN appears to be much more active than DEN1 (by nearly three orders of magnitude) in this cleavage reaction (compare Fig. 3B, lanes 4 and 6). This result is in agreement with our findings with the *in vitro*-conjugated radioactive substrate (Fig. 2B; see Wu *et al.*, 2003), suggesting a prominent role for CSN in hydrolyzing the isopeptide bond formed between Nedd8 and CUL1 Lys720 residue.

Given that CSN does not interact with Nedd8 (data not shown), we reasoned that it must recognize ROC1-CUL1$^{324-776}$ for proteolysis. To test this, we performed *in vitro*-binding experiments in which glutathione-Sepharose-immobilized GST-HA-ROC1 or GST-HA-ROC1-CUL1$^{324-776}$ is incubated with CSN in a mixture (20 $\mu$l) containing buffer B plus 50 m$M$ NaCl for 20 min at 30°. After removing unbound proteins by washing, the bound proteins are analyzed by immunoblot with the anti-CSN5 (Jab1) antibody. As shown in Fig. 4, Jab1 was bound to both GST-HA-ROC1-CUL1$^{324-776}$ (lanes 2–4) and GST-HA-ROC1 (lanes 5–8). Removal of

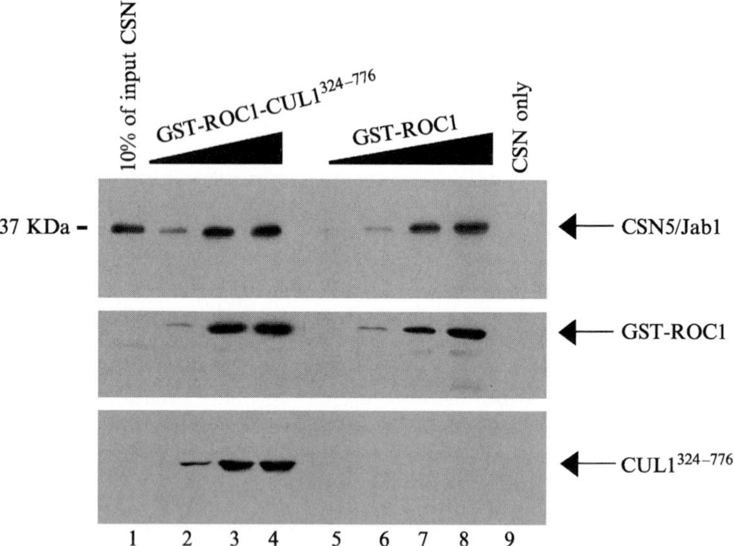


FIG. 4. CSN binds directly to the ROC1-CUL1 complex or ROC1 alone. (A) Purified CSN (2.5 pmol) was incubated alone (lane 9) or with increasing amounts of GST-HA-ROC1-CUL1$^{324-776}$ (lanes 2–4) or GST-HA-ROC1 (lanes 5–8), immobilized onto glutathione-Sepharose (20 $\mu$l). The amounts of bound GST-HA-ROC1-CUL1$^{324-776}$ corresponded to 4, 40, and 100 pmol (lanes 2–4), respectively. In reactions shown in lanes 5–8, 0.4, 4, 40, and 100 pmol of GST-ROC1 were coupled to the beads. Immunoblot analysis was carried out with antibodies as specified. Lane 1 shows immunoblot analysis of 10% of input CSN. GST-HA-ROC1 was prepared by overexpressing the fusion protein in bacteria transformed with the pGEX-4T3/pET-15b-(GST-HA-ROC1) plasmid (Wu *et al.*, 2000) using the procedures for GST-HA-ROC1-CUL1$^{324-776}$ as described earlier.

GST-HA-ROC1-CUL1$^{324–776}$ or GST-HA-ROC1 (lane 9), or substitution with GST (data not shown), abolished the interaction, indicating a specificity of the interaction. These results, consistent with previous findings using yeast two-hybrid assays (Lyapina *et al.*, 2001; Pintard *et al.*, 2003; Schwechheimer *et al.*, 2001), suggest a direct interaction between CSN and ROC1-CUL1 and imply a primary role for ROC1 in recognizing CSN.

## Acknowledgment

This study was supported by Public Health Service Grants GM61051 and CA095634.

## References

Cope, G. A., and Deshaies, R. J. (2003). COP9 signalosome: A multifunctional regulator of SCF and other cullin-based ubiquitin ligases. *Cell* **114**(6), 663–671.

Dias, D. C., Dolios, G., Wang, R., and Pan, Z. Q. (2002). CUL7: A DOC domain-containing cullin selectively binds Skp1·Fbx29 to form a novel SCF-like complex. *Proc. Natl. Acad. Sci. USA* **99**(26), 16601–16606.

Gan-Erdene, T., Kolli, N., Yin, L., Wu, K., Pan, Z. Q., and Wilkinson, K. D. (2003). Identification and characterization of DEN1, a deneddylase of the ULP family. *J. Biol. Chem.* **278**(31), 28892–28900.

Hochstrasser, M. (1998). There's the rub: A novel ubiquitin-like modification linked to cell cycle regulation. *Genes Dev.* **12**(7), 901–907.

Hori, T., Osaka, F., Chiba, T., Miyamoto, C., Okabayashi, K., Shimbara, N., Kato, S., and Tanaka, K. (1999). Covalent modification of all members of human cullin family proteins by NEDD8. *Oncogene* **18**(48), 6829–6834.

Lyapina, S., Cope, G., Shevchenko, A., Serino, G., Tsuge, T., Zhou, C., Wolf, D. A., Wei, N., Shevchenko, A., and Deshaies, R. J. (2001). Promotion of NEDD-CUL1 conjugate cleavage by COP9 signalosome. *Science* **292**(5520), 1382–1385.

Mendoza, H. M., Shen, L. N., Botting, C., Lewis, A., Chen, J., Ink, B., and Hay, R. T. (2003). NEDP1, a highly conserved cysteine protease that deNEDDylates cullins. *J. Biol. Chem.* **278**(28), 25637–25643.

Pan, Z. Q., Kentsis, A, Dias, D. C., Yamoah, K., and Wu, K. (2004). Nedd8 on cullin: Building an expressway to protein destruction. *Oncogene* **23,** 1985–1997.

Pintard, L., Kurz, T., Glaser, S., Willis, J. H., Peter, M., and Bowerman, B. (2003). Neddylation and deneddylation of CUL-3 is required to target MEI-1/Katanin for degradation at the meiosis-to-mitosis transition in *C. elegans*. *Curr. Biol.* **13**(11), 911–921.

Schwechheimer, C., Serino, G., Callis, J., Crosby, W. L., Lyapina, S., Deshaies, R. J., Gray, W. M., Estelle, M., and Deng, X. W. (2001). Interactions of the COP9 signalosome with the E3 ubiquitin ligase SCFTIRI in mediating auxin response. *Science* **292**(5520), 1379–1382.

Siegel, L. M., and Monty, K. J. (1966). Determination of molecular weights and frictional ratios of proteins in impure systems by use of gel filtration and density gradient centrifugation: Application to crude preparations of sulfite and hydroxylamine reductases. *Biochim. Biophys. Acta* **112**(2), 346–362.

Wei, N., and Deng, X. W. (2003). COP9 signalosome. *Annu. Rev. Cell. Dev. Biol.* **19,** 261–286.

Wu, K., Chen, A., and Pan, Z. Q. (2000). Conjugation of Nedd8 to CUL1 enhances the ability of the ROC1-CUL1 complex to promote ubiquitin polymerization. *J. Biol. Chem.* **275,** 32317–32324.

Wu, K., Yamoah, K., Dolios, G., Gan-Erdene, T., Tan, P., Chen, A., Lee, C. G., Wei, N., Wilkinson, K. D., Wang, R., and Pan, Z. Q. (2003). DEN1 is a dual function protease capable of processing the C-terminus of Nedd8 deconjugating hyper-neddylated CUL1. *J. Biol. Chem.* **278**(31), 28882–28891.

# [43] Ubiquitin-Ovomucoid Fusion Proteins as Model Substrates for Monitoring Degradation and Deubiquitination by Proteasomes

*By* TINGTING YAO and ROBERT E. COHEN

## Abstract

Protein degradation by 26S proteasomes requires the coordinated action of multiple binding and catalytic activities to process ubiquitinated protein substrates. For the purpose of studying conjugate degradation independently of substrate targeting and unfolding steps, we have developed substrates based on an N-terminal fusion of ubiquitin to an irreversibly unfolded protein, the 83 amino acid HA epitope-tagged first domain of chicken ovomucoid. Fluorescent labeling of the six cysteines in the ovomucoid moiety (OM) with Lucifer Yellow iodoacetamide yields $UbOM^{LY}$; the ubiquitin in the fusion protein can be extended by the addition of a K48-linked polyubiquitin chain to form $Ub_nOM^{LY}$. $UbOM^{LY}$ derivatives provide versatile substrates to monitor both protein degradation and deubiquitination by 26S proteasomes *in vitro*. Comparisons of polyubiquitin conjugates of unfolded $OM^{LY}$ with folded dihydrofolate reductase (DHFR) in degradation assays can help resolve and identify the rate-limiting steps in proteasome degradation.

## Introduction

Degradation of ubiquitin (Ub)-protein conjugates by the 26S proteasome involves multiple steps: substrate targeting to the proteasome, unfolding and translocation of the substrate to the interior of the catalytic (20S) core, proteolysis, release of peptide products, and recycling of the polyubiquitin targeting signal (Pickart and Cohen, 2004). To understand how these different functions are accomplished and coordinated will require a variety of specific *in vitro* biochemical assays. Typically, proteasome assays

METHODS IN ENZYMOLOGY, VOL. 398

0076-6879/05 $35.00
DOI: 10.1016/S0076-6879(05)98043-9

have relied upon the use of small chromogenic or fluorogenic peptide substrates (e.g., Suc-LLVY-AMC; Stein *et al.*, 1996). These substrates afford convenient and sensitive measures of the peptidase activities, but they report only on accessibility and efficiency of the internal catalytic sites of the proteasome. In order to understand protein degradation more fully, systems using purified proteasomes and defined protein substrates are needed; in particular, analyses of the processing of the (poly)Ub targeting signal require Ub-protein conjugates. To date, few such substrates have been described.

Pioneering work from the Pickart laboratory employed Ub(V76)-DHFR, a fusion of Ub to murine dihydrofolate reductase, as the basis of an *in vitro* substrate (Thrower *et al.*, 2000). Earlier work showed that Ub-(V76)DHFR expressed in yeast is ubiquitinated and degraded rapidly (Johnson *et al.*, 1995). The G76V mutation in the Ub moiety prevents the N-terminal Ub from being removed by deubiquitinating enzymes. *In vitro*, a preassembled tetraUb chain can be transferred to K48 in the Ub moiety of the fusion protein, resulting in a K48-linked pentaUb chain attached to the N terminus of DHFR. In subsequent incubations with proteasomes, the chain was protected from disassembly by inclusion of the potent deubiquitinating enzyme (DUB) inhibitor ubiquitin-aldehyde. $Ub_5$(V76)DHFR is targeted to the proteasome very efficiently ($K_M = 35$ n$M$), and inhibition studies with polyUb chains of different lengths unambiguously demonstrated that, for $Ub_n$-protein conjugates, a tetraUb chain is the minimal efficient targeting signal. However, the degradation of $Ub_5$(V76)DHFR was surprisingly slow ($k_{cat} = 0.05$ min$^{-1}$), and it was concluded that unfolding could be the rate-limiting step of degradation based on two observations. First, the proximal Ub(V76) was degraded along with the DHFR moiety. Ub is an exceedingly stable molecule and its unfolding is expected to be difficult. Second, folic acid, a small ligand of DHFR, inhibited degradation uncompetitively. Presumably, ligand binding stabilized the DHFR moiety, thereby making unfolding and translocation into the proteasome more difficult.

An *in vitro* substrate developed by Deshaies' group (Petroski and Deshaies, 2003; Verma *et al.*, 2001) employs a natural proteasome substrate, Sic1, which is an S-Cdk inhibitor that undergoes cell cycle-dependent phosphorylation, ubiquitination, and degradation in yeast. Recombinant E1, E2, and E3 enzymes that ubiquitinate Sic1 were expressed and purified from insect cells and used to form Sic1 conjugates *in vitro*. Single-turnover experiments with the $Ub_n$-Sic1 products revealed degradation rates ($t_{1/2} \sim 1$ min) similar to those measured *in vivo*. The advantage of the physiological relevance of $Ub_n$-Sic1 conjugates is offset somewhat by the difficulty in preparing large amounts of the substrate and, due to the processive nature of the conjugating enzymes, the heterogeneity of the $Ub_n$-Sic1 products. It was estimated that there are up to six to eight polyUb chains of various lengths attached to each wild-type Sic1 molecule. Despite the potential to restrict the ubiquitination site by the use of "single-lysine" Sic1 substrates (Petroski and Deshaies, 2003), the heterogeneity in the polyUb targeting signal, together with the unknown structure of Sic1, complicates interpretations of the degradation kinetics. In particular, the rate-limiting step(s) for $Ub_n$-Sic1 degradation is unclear, and precisely how the attached polyUb chains are processed is difficult to evaluate.

We now know that three DUBs, UCH37 (Lam *et al.,* 1997), USP14/Ubp6 (Borodovsky *et al.*, 2001; Leggett *et al.*, 2002), and POH1/Rpn11 (Verma *et al.,* 2002; Yao and Cohen, 2002), associate tightly with 26S proteasomes. Whereas the studies cited earlier focused on the targeting and unfolding steps of proteasomal processing, our aim was to isolate the degradation and deubiquitination steps for investigation. For this purpose, we developed a series of substrates based on an N-terminal Ub fusion to an irreversibly unfolded protein, the 69 amino acid first domain of chicken ovomucoid (Fig. 1); the UbOM fusion protein, like UbDHFR, can be extended by the addition of polyUb. The substrate was designed to meet the following criteria: (1) it is homogeneous and structurally well defined; (2) it is easy to prepare in large quantities; (3) it is degraded by 26S proteasomes in an ATP-dependent manner; and (4) degradation and deubiquitination can be monitored quantitatively and with high sensitivity.

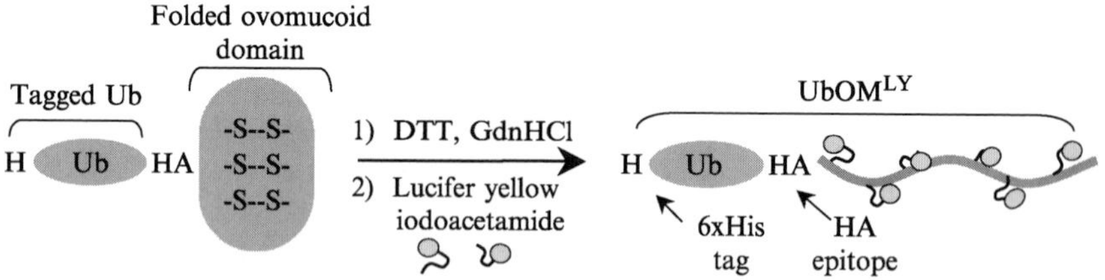


OM (ovomucoid domain):
MAEVDCSRFPNATDKEGKDVLVCNKDLRPICGTDGVTYTNDCLLCAYSIEFGTNISKEHDGECKETVPM

FIG. 1. Design and synthesis of UbOM$^{LY}$. Recombinant UbOM contains an N-terminal Ub moiety from which polyUb chains can be extended and a C-terminal OM moiety that contains three disulfide bonds, reduction of which unfolds this domain. The six cysteines are distributed along the entire length of the OM domain. Alkylation of the cysteines with Lucifer Yellow iodoacetamide not only makes unfolding irreversible, but also introduces fluorescent labels that can be traced when the Lucifer Yellow-labeled domain OM$^{LY}$ is degraded. An N-terminal 6xHis tag facilitates purification, and an HA epitope placed between the two domains is available for detection in Western blots. Adapted from Yao and Cohen (2002).

This chapter focuses on the preparation of fluorescently labeled UbOM substrates and their use to study deubiquitination as well as degradation by 26S proteasomes.

## Procedures

### *Materials*

Creatine phosphate, creatine phosphokinase, yeast inorganic pyrophosphatase, apyrase, and protease inhibitors are from Sigma (St. Louis, MO). The 1× protease inhibitor cocktail contains 16 μg/ml benzamidine HCl, 1.6 μg/ml aprotinin, 30 μg/ml bestatin, 40 μg/ml antipain, 0.2 μg/ml pepstatin A, and 140 μg/ml phenylmethylsulfonyl fluoride; a 50× stock can be stored frozen. $Ni^{2+}$-NTA resin is from Qiagen (Valencia, CA), and Superdex 75 HR10/30 gel filtration and HiTrap desalting columns are from Amersham-Pharmacia (Piscataway, NJ). Lucifer Yellow iodoacetamide and tris(2-carboxyethyl)phosphine-HCl (TCEP) are from Molecular Probes (Eugene, OR); succinyl-LLVY-AMC is from Bachem (King of Prussia, PA). Bovine 26S proteasomes are purified essentially as described by Hoffman *et al.* (1992), and purified bovine PA700 (i.e., proteasome 19S regulatory complex) and 20S proteasomes are gifts from G. DeMartino (UT Southwestern Medical Center, Dallas, TX). OM domain protein, a gift from A. Robertson (University of Iowa, Iowa City, IA), is made as described (DeKoster and Robertson, 1997). The preparation of ubiquitin aldehyde (Ubal) is by periodate oxidation of ubiquitin-diol (Dunten and Cohen, 1989; Hu *et al.*, 2002).

### *DNA Constructs*

The bacterial expression plasmid pRSUbOM encodes the UbOM fusion protein. The N-terminal 6xHis-Ub moiety is derived from pHISUB (Beers and Callis, 1993), and the C-terminal OM moiety is from pOMCH1-C (DeKoster and Robertson, 1997). To construct pRSUbOM, a 6xHis-Ub fragment is amplified by polymerase chain reaction with primers that include the sequence for an HA epitope in the 3′ primer, and the fragment is cloned into the pRSet vector between *Nde*I and *Hind*III sites. An OM fragment flanked by *Hind*III sites (Johnston *et al.*, 1999) is then inserted into C-terminal to the 6xHis-Ub. The resulting fusion protein contains 6×His-Ub and OM with the linker GYPYDVPDYAGESF in between. The plasmid that encodes Ub(V76)DHFR is from C. Pickart (Johns Hopkins University); note that it was engineered to contain a 10xHis-Ub at the N terminus and an HA epitope tag at the C terminus (Thrower *et al.*,

2000). Quikchange (Stratagene) site-directed mutagenesis is used to generate the G76 → V76 mutation in Ub(G76)OM and V76 → G76 mutation in Ub(V76)DHFR.

*Expression and Purification of Ub Fusion Proteins Containing the OM domain*

UbOM and mutants derived from it are expressed in *Escherichia coli* BL21(DE3)pLysS cells. Briefly, a fresh transformant is inoculated into a 50 ml of standard Luria broth supplemented with ampicillin (50 $\mu$g/ml) and chloramphenicol (30 $\mu$g/ml). After 12 h growth with shaking at 37°, the overnight culture is used to inoculate 2 liters of the same medium. When the large culture reaches midlog phase ($A_{600\text{ nm}} \sim 0.5$), 0.4 m*M* isopropyl-$\beta$-D-thiogalactopyranoside (IPTG) is added to induce expression. Growth is continued for 3 h, and then the cells are harvested by centrifugation, quick-frozen in dry ice, and stored at −20°. The OM moiety is unable to fold in the reducing environment within *E. coli* and also seems to be toxic to the host cell. In the case of pRSUbOM, transformants tend to die when growth enters the stationary phase. Therefore, it is important to prepare fresh transformants and to take precautions not to let the cultures overgrow. Prolonged growth usually leads to poor expression or severe proteolysis of the OM moiety.

Purifications of His-tagged UbOM fusion proteins with $Ni^{2+}$-NTA agarose resin (Qiagen) use denaturing conditions at room temperature and follow the manufacturer's general procedure. Briefly, harvested cells are thawed in 4 volumes of freshly prepared denaturing buffer A (8 *M* urea, 0.1 *M* NaPi, 0.01 *M* Tris-Cl, pH 8) and lysed by stirring for 1 h at room temperature. The lysate is then centrifuged at 17,000*g* for 20 min, and the supernatant fraction is filtered through a 0.45-$\mu$m membrane. $Ni^{2+}$-NTA agarose preequilibrated with buffer A is added to the filtrate (~1 ml of 50% resin slurry is used for every 4 ml of filtrate) and mixed gently for 1 h before it is loaded into a column. Unbound material is allowed to flow through and is discarded, and the resin is then washed with 10 column bed volumes of denaturing buffer B (8 *M* urea, 0.1 *M* NaPi, 0.01 *M* Tris-Cl, pH 6.3). To elute the His-tagged proteins, 250 m*M* imidazole is added to denaturing buffer B, and fractions are collected; EDTA (1 m*M* final concentration) is added to the fractions to scavenge excess $Ni^{2+}$ that might have dissociated from the resin during elution. UbOM-containing fractions are identified by SDS–PAGE and pooled, and the protein concentration is estimated from the absorbance at 280 nm based on a predicted extinction coefficient (Gill and von Hippel, 1989) of 0.41 ml/mg. Yields are typically 5–10 mg per liter of culture, and purity is ~80% based on SDS–PAGE and

Coomassie blue staining (not shown). Fusion proteins containing the OM moiety often appear as light brown after purification from the lysate. This color is likely due to metal ions bound to the cysteine residues in OM and can be removed by the addition of 10 m*M* EDTA.

*Fluorescent Labeling of the OM Domain*

Before further purification, the $Ni^{2+}$–NTA-purified Ub(G76)OM and Ub(V76)OM are modified covalently with Lucifer Yellow. Dithiothreitol (DTT; 1 m*M*) is added to the pooled eluate from the $Ni^{2+}$ column to reduce all the cysteine residues. A 5-ml Sephadex G25 HiTrap desalting column is then used to exchange the proteins into labeling buffer, which contains 50 m*M* Tris-Cl, pH 7.8, 1 m*M* tris(2-carboxyethyl)phosphine-HCl, 0.5 m*M* EDTA, and 6 *M* guanidine HCl. Lucifer Yellow iodoacetamide is added as a powder in an amount calculated to give a ninefold molar excess over total protein cysteines. After 5 h in the dark at room temperature, the reaction is stopped by the addition of 20 m*M* $\beta$-mercaptoethanol. Desalting through Sephadex G25 or dialysis (SpectraPor membrane; 3500 MWCO) is performed to remove free dye (Fig. 2A); this also serves to remove some of the small proteolytic products that are common contaminants after $Ni^{2+}$ affinity purification. After labeling, the fluorescent derivatives are purified further on a Superdex 75 (1 $\times$ 30 cm) column eluted with 100 m*M* NaCl in 50 m*M* Tris, pH 7.5 (Fig. 2B). For the OM domain alone (DeKoster and Robertson, 1997), lyophilized protein is directly dissolved in labeling buffer, and the subsequent labeling and gel filtration procedures are as described earlier for UbOM. The amounts of protein and extent of labeling are determined by using extinction coefficients of 15 $\mu g^{-1}$ $cm^{-1}$ at 215 nm for protein (Scopes, 1993) and 11,000 $M^{-1}$ $cm^{-1}$ at 426 nm for Lucifer Yellow. Labeling efficiencies are typically greater than 80% (i.e., $\geq$5 Lucifer Yellow moieties incorporated into the OM domain).

*Synthesis of polyUb-Protein Conjugates*

$Ub_nOM^{LY}$ conjugates are made by transfer of a presynthesized K48-linked polyUb chain to the purified $UbOM^{LY}$. TetraUb chain synthesis follows the published procedure (Piotrowski *et al.*, 1997; Raasi and Pickart, 2005), and chain transfer reactions are done essentially as described by Thrower *et al.* (2000). In the case of $Ub_5OM^{LY}$, 36 $\mu M$ tetraUb and 20 $\mu M$ $UbOM^{LY}$ chain acceptor are incubated with the enzymes as described, and >95% $UbOM^{LY}$ is converted to $Ub_5OM^{LY}$ after 4 h incubation at 37°. The 1$\times$ protease inhibitor cocktail is included in all the chain transfer reactions. $Ub_5OM^{LY}$ conjugates are isolated by $Ni^{2+}$ affinity purification in the same way as the monoUb fusion proteins. Unavoidably, the eluate from the $Ni^{2+}$

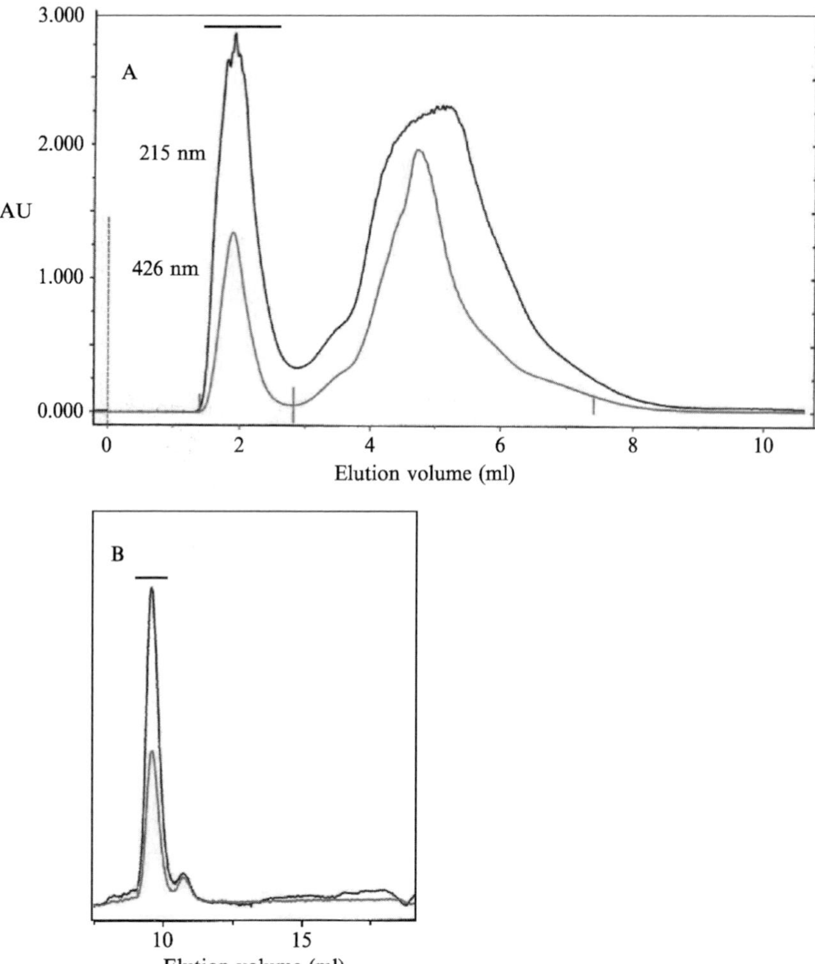


FIG. 2. Purification of fluorescently labeled UbOM$^{LY}$. (A) After labeling with Lucifer Yellow iodoacetamide, the reaction mixture (~1 ml) was desalted through a 5-ml HiTrap Sephadex G25 cartridge eluted with 100 m*M* NaCl in 50 m*M* Tris, pH 7.5. The eluate was monitored by absorbance at 215 nm (top trace) and 426 nm (bottom trace). Fractions were pooled (bar) and concentrated by ultrafiltration (e.g., Amicon Ultra-4 centrifugal filter). (B) A portion of the pooled protein from (A) was purified further by gel filtration through a Superdex 75 column (1 × 30 cm) eluted with the same buffer.

resin contains a small amount of unreacted $UbOM^{LY}$; because the $Ub_5$ chain increases targeting efficiency greatly (see later), the effect of the contamination in degradation assays is negligible.

Radiolabeled UbDHFR is synthesized by *in vitro* translation using a bacterial cell-free system (Rapid Translation System, Roche). Typically, a 300-$\mu$l reaction contains 50 $\mu M$ unlabeled methionine, 0.36 mCi [$^{35}$S] methionine, 2 $\mu$g plasmid, 1× protease inhibitor cocktail, and other components as specified by the manufacturer. The mixture is incubated at 30° for 4 h on a rotator. Purification with $Ni^{2+}$-NTA resin followed immediately. In general, the purification procedure was the same as when UbDHFR is expressed in *E. coli* (Thrower *et al.*, 2000); 2 m*M* folic acid and 10% glycerol are supplemented in all the buffers to stabilize the protein. The eluate from the $Ni^{2+}$ column is purified further on Superdex 75 (1 × 30 cm) eluted with 50 m*M* NaPi, pH 8, 150 m*M* NaCl, 1 m*M* DTT, 0.1 m*M* EDTA, and 0.05 mg/ml ovalbumin. Transfer of K48-linked tetraUb to $^{35}$S-labeled Ub(G76)-DHFR and Ub(V76)DHFR is less efficient than with the UbOM acceptors; therefore, a 15-fold excess of tetraUb is used with 4 $\mu M$ UbDHFR and the conjugation reactions are allowed to proceed overnight. The pentaUb conjugates are purified with $Ni^{2+}$-NTA resin.

### *Proteasome Degradation Assays*

*In vitro* degradation assays are essentially as described (Thrower *et al.*, 2000). A typical reaction contains 50 m*M* Tris Cl, pH 8, 10 m*M* $MgCl_2$, 1 m*M* DTT, 0.1 mg/ml ovalbumin, 10% glycerol, 1 m*M* ATP, 0.075 mg/ml creatine phosphokinase, 14 m*M* creatine phosphate, 2 unit/ml inorganic pyrophosphatase, 1× protease inhibitor cocktail (these inhibitors do not block any of the proteolytic activities of the proteasome), 42 n*M* of 26S proteasome, and 0–75 $\mu M$ substrate. To prevent deubiquitination by intrinsic UCH37 DUB activity (Lam *et al.*, 1997), USP14 (Borodovsky *et al.*, 2001), and possible trace amounts of other contaminating DUBs, the proteasome is preincubated with 2 $\mu M$ Ubal. Note that, unlike UCH37, very little USP14 remains associated with conventionally purified 26S proteasomes (Yao and Cohen, 2002). All incubations are at 37°. Depletion of ATP (e.g., as a control to assess 26S proteasome-independent degradation) is achieved by omission of the 1 m*M* ATP and inclusion of 10 units/ml apyrase during preincubation.

For the Lucifer Yellow-derivatized substrates, degradation can be quantified as trichloroacetic acid (TCA)-soluble fluorescence. For a typical reaction, 3.5-$\mu$l aliquots are removed at various times and mixed immediately with 100 $\mu$l ice-cold 20% TCA and 10 $\mu$l of 50 mg/ml bovine serum albumin (added as a coprecipitant). After incubation on ice for 1 h, the

TCA-treated samples are centrifuged at 20,000$g$ for 10 min at 4°. A portion (90 $\mu$l) of the supernatant liquid is removed and neutralized with 410 $\mu$l of 1 $M$ Tris-Cl, pH 8 (see note later) and then fluorescence is measured (we use a JOBIN YVON Spex Fluorolog–3 instrument) using wavelengths of 426 nm (excitation) and 531 nm (emission) with 6-nm bandwidth slits. Samples of completely digested $UbOM^{LY}$ substrate are made by a 1-h incubation with 10 $\mu M$ trypsin followed by the TCA precipitation/neutralization protocol. The ratio of fluorescence obtained from the proteasome reaction to that of the completely digested substrate is used to determine the fraction of $UbOM^{LY}$ degraded. Note that some batches of Tris base contain contaminants that give high background fluorescence. These can be eliminated by filtering (0.45-$\mu$m membrane) the Tris stock solution after treatment with activated charcoal powder (e.g., Norit SG; ~ 2 g per 500 ml).

### *Deubiquitination Assays*

Reaction conditions are essentially as for the degradation assays described earlier. Ubal is omitted in order to monitor UCH37 activity, whereas preincubation with Ubal revealed specific deubiquitination by POH1/Rpn11. Preincubation of 26S proteasomes or the 19S (PA700) complex with a $Zn^{2+}$ chelator [e.g., 0.5 m$M$ N,N,N′,N′-tetrakis(2-pyridylmethyl)ethylenediamine (TPEN), or *o*-phenanthroline] can be used to suppress POH1/Rpn11 activity (Verma *et al.*, 2002; Yao and Cohen, 2002). Deubiquitination can be assayed as the release of intact (poly)Ub or, in the absence of substrate degradation (e.g., when incubation is with the 19S complex rather than 26S proteasomes), release of intact HA-tagged $OM^{LY}$ from $Ub_nOM^{LY}$. Detection is by Western blotting (see later) or direct visualization of Lucifer Yellow fluorescence in gels; for the latter, we use a UVP BioImaging system equipped with a CCD camera.

### *Western Blotting*

Proteins resolved by SDS–PAGE are transferred to nitrocellulose (0.2-$\mu$m pore size) at 100 V for 1 h using a Mini Trans-Blot cell (Bio-Rad). For blocking, the membrane is incubated with 5% nonfat dry milk in blotting buffer (phosphate-buffered saline supplemented with 0.1% Tween 20) for 10 min. It is then incubated with immunopurified primary antibody overnight. After three washes with the blotting buffer, the membrane is incubated with the secondary antibody for 1 h. Detection is accomplished with either the SuperSignal West Pico or the West Dura chemiluminescent substrate from Pierce Chemical Co. The affinity-purified rabbit anti-Ub antibody is made as described (Haas and Bright, 1985) and used at either a 1:2000 to 1:1000 dilution; alternatively, many commercial sources of anti-Ub antibodies are available. The monoclonal anti-HA antibody is from Covance (clone 16B12) and is used at 1:1000 dilution. Horseradish peroxidase-conjugated secondary antibodies are from Pierce and are used at 1:10,000 dilution.

### *Kinetic Fitting*

Initial velocities for the degradation of $OM^{LY}$-based substrates are determined by linear fitting of the portions of the progress curves that correspond to <5% substrate consumed. $K_M$ and $k_{cat}$ values are determined by nonlinear least-squares fits of initial velocity data to the Michaelis–Menten equation using the program Ultrafit (Biosoft, Cambridge, UK). Simulations of $Ub_5$DHFR binding and degradation kinetics used KINSIM (Barshop *et al.*, 1983).

## Results and Discussion

### *An Unfolded Proteasome Substrate*

Figure 1 illustrates the design and synthesis of an unfolded proteasome substrate, $UbOM^{LY}$. UbOM is well expressed in *E. coli* and is purified easily via its N-terminal 6xHis tag. The N-terminal Ub moiety serves as an acceptor for polyUb chain extension. The OM domain (chicken ovomucoid first domain) is a 69 residue polypeptide that contains a triple-stranded $\beta$ sheet, a helix, and a small loop. Three disulfide bonds in the domain are essential to maintain the native state. Reduction of the disulfides and alkylation of the cysteines irreversibly unfold the OM moiety. This was confirmed by comparison of the circular dichroism spectra of thermally denatured OM and Lucifer Yellow derivatized OM ($OM^{LY}$), which indicated that $OM^{LY}$ lacks any appreciable secondary or tertiary structure (not shown). Note that many other fluorescent alkylating agents are available that may be used to alkylate the OM cysteines; we chose Lucifer Yellow iodoacetamide principally because of its high water solubility and because its large Stokes shift markedly simplifies detection in gels or solution.

When incubated with the 26S proteasome and ATP, $UbOM^{LY}$ is degraded rapidly. This is demonstrated both by disappearance of the substrate in a Western blot (Fig. 3A) or by direct visualization of fluorescent bands after SDS–PAGE (not shown) and by increasing TCA-soluble fluorescence over time (Fig. 3B). Importantly, degradation of $UbOM^{LY}$ is ATP dependent (Fig. 3B); this property distinguishes 26S from 20S proteasome-mediated degradation. The low level of degradation seen when ATP was depleted probably originated from proteolysis by 20S proteasomes,

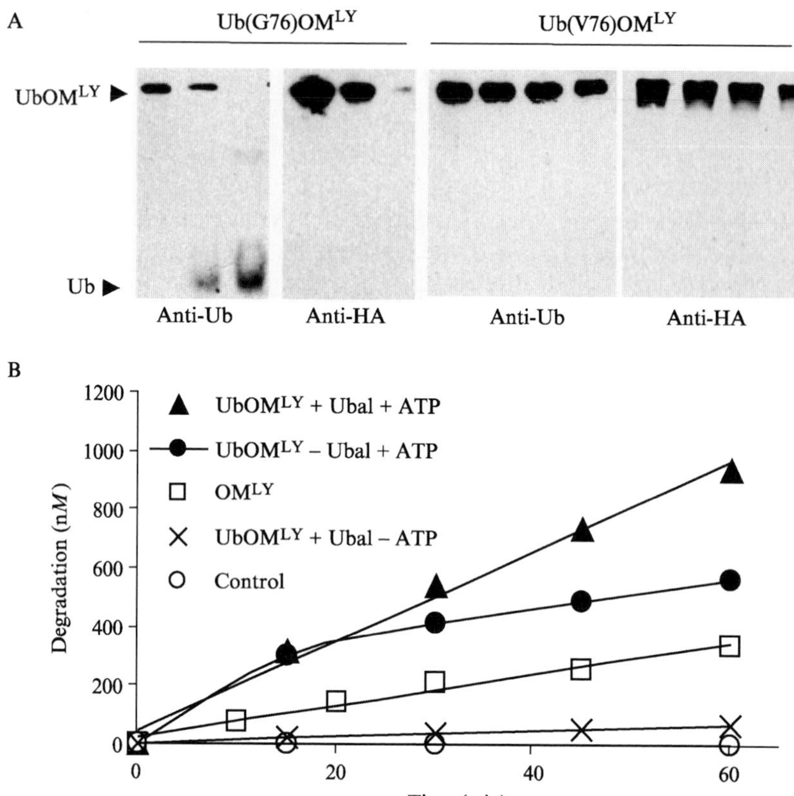


FIG. 3. $UbOM^{LY}$ degradation by the 26S proteasome is ATP dependent and is accompanied by deubiquitination. (A) $Ub(G76)OM^{LY}$ or $Ub(V76)OM^{LY}$ (0.5 $\mu M$) was incubated with 42 n$M$ bovine 26S proteasomes in the presence of 2 $\mu M$ Ubal. Disappearance of the substrate is accompanied by the appearance of Ub. Note that the Ub size product is not recognized by the anti-HA antibody; this is evident even when the film is overexposed. (B) Quantitation of $UbOM^{LY}$ degradation. $Ub(G76)OM^{LY}$ or free $OM^{LY}$ (16 $\mu M$) was incubated with 42 n$M$ 26S proteasome without Ubal. At various times, aliquots were removed and TCA-soluble fluorescence was measured. Fluorescence units were converted to amounts of products by comparison with samples of completely digested $UbOM^{LY}$ (see text). In the absence of enzyme (control) or ATP, degradation of $UbOM^{LY}$ was negligible. From Yao and Cohen (2002).

which formed because of dissociation of the 26S complexes in the absence of ATP.

Western blot detection with anti-Ub and anti-HA antibodies suggests that the N-terminal Ub moiety is released while the $(HA)OM^{LY}$ moiety is degraded. Previous studies have shown that proteasome degradation is processive and generally produces peptides with sizes ranging from 3 to 22 amino acids (Kisselev *et al.*, 1999). Typically, such small peptides remain in the supernatant when subject to TCA precipitation. The $OM^{LY}$ domain contains six cysteines distributed throughout the whole domain, five of which were labeled by Lucifer Yellow on average. Thus, TCA-soluble fluorescence is a good measure of complete degradation of the $OM^{LY}$ domain.

Interestingly, the presence of the DUB inhibitor Ubal also plays a significant role in $UbOM^{LY}$ degradation. Without Ubal, $UbOM^{LY}$ is deubiquitinated rapidly (thereby generating $OM^{LY}$) and then degrades at the same rate as $OM^{LY}$ alone. In the presence of Ubal, $UbOM^{LY}$ degradation is faster, although, as shown in the Western blots (Fig. 3A), a Ub size product is still released. This is due to the Ubal-insensitive POH1/Rpn11 deubiquitinating activity that, in addition to the Ubal-sensitive UCH37 DUB, is associated with the 26S proteasomes (Yao and Cohen, 2002). As an approach complementary to use of DUB inhibitors such as Ubal, substrates such as $Ub(V76)OM^{LY}$ that resist deubiquitination can be used in these assays. As illustrated in Fig. 3, comparisons of $Ub(G76)OM^{LY}$, $Ub(V76)$-$OM^{LY}$, and $OM^{LY}$ can be used to address the role of deubiquitination during proteasomal processing.

## Targeting of $OM^{LY}$-Based Substrates

*In vivo*, mono- and polyubiquitination have different biological consequences, which suggests that polyUb chains have unique features that are absent in monoUb. This was confirmed by *in vitro* kinetic studies of $Ub_5(V76)DHFR$ degradation and its inhibition by polyUb chains, which indicated that for K48-linked chains, tetraUb is the minimal targeting signal for efficient degradation (Thrower *et al.*, 2000). To investigate the role of the Ub moiety in $UbOM^{LY}$ degradation, degradations of $UbOM^{LY}$ and $OM^{LY}$ alone were compared (Table I). The two substrates have exactly the same $K_M$, suggesting that the $OM^{LY}$ moiety alone is responsible for targeting. Moreover, adding mono-Ub in a 50-fold excess does not inhibit degradation of $UbOM^{LY}$ (data not shown), which further indicates a negligible binding contribution from the Ub moiety. This is consistent with previous findings that monoUb does not bind to the proteasome appreciably. It is known that ATPases in the 19S regulatory particle have chaperone-like activities (Braun *et al.*, 1999), and it is likely that an unfolded polypeptide such as $OM^{LY}$ can bind to one or more of the ATPases. The $K_M$ values for $UbOM^{LY}$ and $OM^{LY}$ (16 $\mu M$) are much higher than that for $Ub_5(V76)DHFR$ (35 n$M$), indicating that targeting through an unfolded domain is not efficient.

TABLE I
DEGRADATION KINETICS OF MODEL SUBSTRATES[a]

| Substrate | $K_M$ ($\mu M$) | $k_{cat}$ ($min^{-1}$) |
|---|---|---|
| Ub(G76)$OM^{LY}$ | 16.4 | 1.17 |
| Ub(V76)$OM^{LY}$ | 16.5 | 0.12 |
| $OM^{LY}$ | 16.5 | 0.54 |
| $Ub_5$(G76)DHFR | 0.018[b] | 0.71 |
| $Ub_5$(V76)DHFR | 0.035[c] | 0.05[c] |

[a] Adapted from Yao and Cohen (2002).
[b] This is a $K_d$ value.
[c] FromThrower *et al.* (2000).

## *A Presteady State Burst of $Ub_5OM^{LY}$ Degradation*

When the Ub moiety in $UbOM^{LY}$ is extended with a tetraUb chain, the resulting protein contains a K48-linked pentaUb and an unfolded $OM^{LY}$ moiety. Based on previous studies with $UbOM^{LY}$ and $Ub_5$(V76)DHFR, we expected that $Ub_5OM^{LY}$ would be an excellent substrate for the 26S proteasome. Figure 4A shows that degradation of $Ub_5OM^{LY}$ was accompanied by the release of pentaUb. Surprisingly, progress curves of $Ub_5OM^{LY}$ degradation reveal an initial phase of rapid proteolysis followed by a much slower steady state (Fig. 4B). When the steady-state rates obtained at a range of substrate concentrations were fit to a Michaelis–Menten equation, the $K_M$ value was 0.7 $\mu M$ and the $k_{cat}$ value was 0.05 $min^{-1}$ (Fig. 4C). Thus, compared with $UbOM^{LY}$ ($K_M = 16.9$ $\mu M$, $k_{cat} = 1.17$ $min^{-1}$), $Ub_5OM^{LY}$ is targeted to the proteasome much more efficiently, yet is degraded 20-fold more slowly.

More rigorous kinetic analysis is needed to test whether the presteady state burst represents the first round of catalysis. It is likely that the products generated during the presteady state inhibited future turnovers. A complication is that targeting of $Ub_5OM^{LY}$ can be mediated by both the pentaUb chain and the unfolded $OM^{LY}$ moiety. To avoid this complication, we employed a folded substrate, $Ub_5$DHFR (Fig. 4C and Table I). Previously, Thrower *et al.* (2000) studied the steady-state kinetics of a related substrate, $Ub_5$(V76)DHFR. Degradation of $Ub_5$(V76)DHFR is solely dependent on the pentaUb chain. It is targeted to the proteasome efficiently with a $K_M$ of 35 n$M$, yet degraded slowly ($k_{cat} = 0.05$ $min^{-1}$). Upon restoring G76 in the proximal Ub, we found that $Ub_5$DHFR is degraded 15-fold faster ($k_{cat} = 0.71$ $min^{-1}$). Unlike $Ub_5$(V76)DHFR, degradation of $Ub_5$(G76)DHFR is accompanied by pentaUb release (Yao and Cohen, 2002). Notably, strong product inhibition was observed with $Ub_5$(G76)DHFR degradation. The fitted $K_i$ value for pentaUb is similar to the $K_M$ for this substrate (Fig. 4 and Table I). These results indicate that efficient removal of polyUb chains is required in proteasome-mediated degradation. The *in vitro* system in this study may lack the components needed for polyUb removal. Possibly, because highly purified 26S proteasomes were used and Ubal was present in these reactions, a deubiquitinating enzyme that normally is responsible for polyUb disassembly was inhibited.

*Degradation Rates: What Is Limiting?*

Ub(G76)$OM^{LY}$ is degraded 20-fold faster than $Ub_5$(V76)DHFR (Table I). However, when the proximal Ub(V76) is reverted to wild type, $Ub_5$DHFR is degraded almost as fast as Ub$OM^{LY}$. Unlike Ub$OM^{LY}$, $Ub_5$DHFR is a folded substrate. In the case of $Ub_5$(V76)DHFR, because of the G76V mutation, the proximal Ub is also degraded (Thrower *et al.*, 2000; Yao and Cohen, 2002). Thus, a single turnover of Ub(G76)$OM^{LY}$ involves degradation of an unfolded (HA)$OM^{LY}$ moiety of 83 residues, whereas with $Ub_5$(G76)DHFR a single turnover involves degradation of a folded DHFR moiety of 233 residues. In contrast, turnover of $Ub_5$(V76) DHFR involves degradation of the 310 residue Ub(V76)DHFR, which contains two folded domains.

A few possibilities can account for the rate differences. One is that unfolding is rate limiting. The free energies of unfolding for Ub and DHFR are 7.2 and 4.8 kcal $mol^{-1}$, respectively (Clark and Frieden, 1999; Khorasanizadeh *et al.*, 1996), and therefore it is likely that, unassisted, unfolding will be very slow. Furthermore, folic acid, a small ligand of DHFR, inhibits $Ub_5$(V76)DHFR degradation (Thrower *et al.*, 2000), which suggests that DHFR unfolding can become rate limiting. However, because Ub(G76)$OM^{LY}$ and $Ub_5$(G76)DHFR are degraded with similar rates, DHFR unfolding appears not to be rate limiting in these reactions. Little data are available for proteasome degradation rates *in vivo*. Many physiological substrates have extremely short half-lives, but without any information about the enzyme:substrate ratio, it is impossible to estimate the true rate of proteasomal degradation. The *in vitro* system developed by Deshaies's group uses a substrate that closely resembles a physiological substrate (Verma *et al.*, 2001). When incubated with purified proteasome, $Ub_n$-Sic1 has a degradation rate of 1 to 2 $min^{-1}$. This is similar to the rates for Ub(G76)$OM^{LY}$ and $Ub_5$(G76)DHFR. Degradation of these substrates is likely to reflect the true rate at which the proteasome can operate.

What then is rate limiting? Targeting is most likely very rapid. Based on data from $Ub_5$DHFR degradation (Fig. 4C), the substrate-binding kinetics were simulated using KINSIM (Barshop *et al.*, 1983). Although an accurate

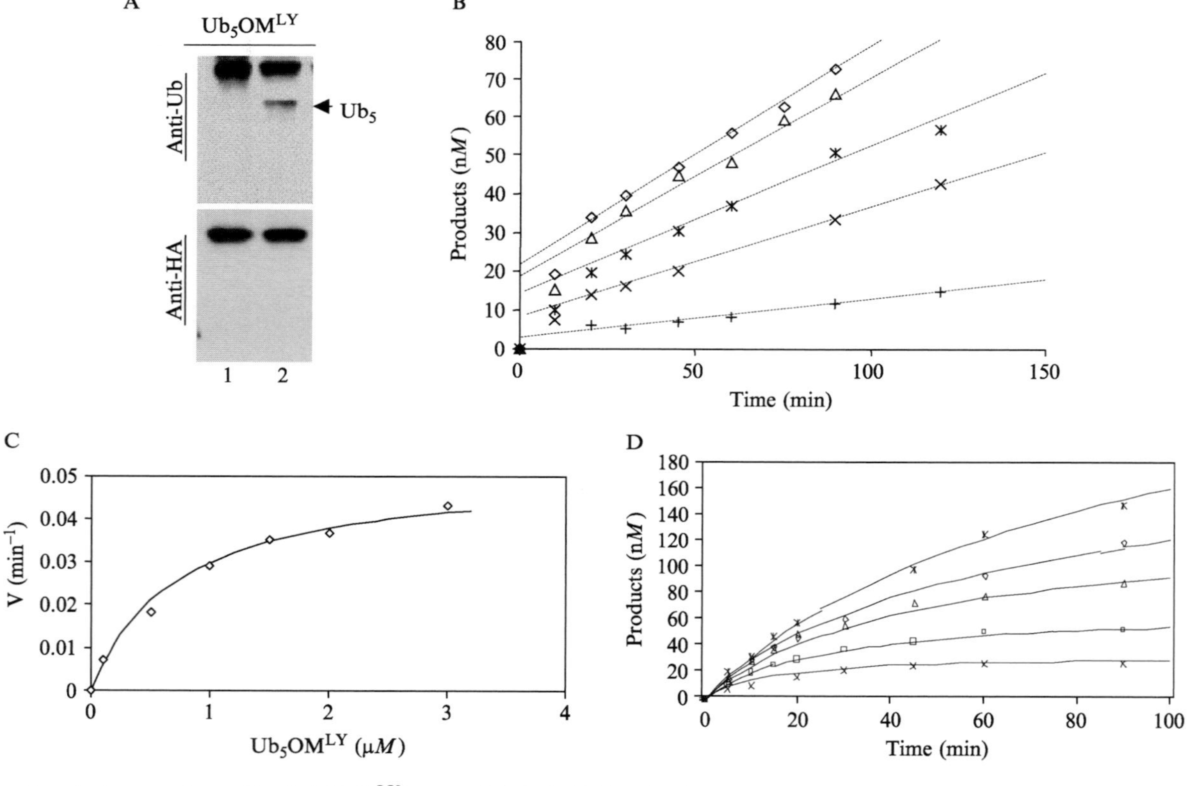


FIG. 4. Degradation of $Ub_5(G76)OM^{LY}$ but not $Ub_5(G76)DHFR$ by the 26S proteasome shows a presteady state burst. (A) Degradation of $Ub_5(G76)OM^{LY}$ monitored by Western blot. Lanes 1 and 2 correspond to 0- and 1-h incubation with 42 n*M* 26S

value cannot be obtained from steady-state data, a range for $k_{on}$ of between $10^6$ and $10^7$ $M^{-1}$ $s^{-1}$ can be approximated. Given that $K_d$ is 18 n$M$, the off rate must be extremely fast as well. This also explains why cosedimentation of polyUb chains and the proteasome is not observed. Second, unfolding should not be rate limiting in the case of $UbOM^{LY}$ and $Ub_5$DHFR degradation (see earlier discussion). Third, proteolysis is fast. The fluorogenic peptide Suc-LLVY-AMC is cleaved by the 26S proteasome with a $k_{obs}$ of 98 $min^{-1}$ (Thrower *et al.*, 2000), and this reflects only two of the six active sites in the proteasome, as this substrate is only susceptible to chymotryptic activity. That the 83 residue (HA)OM and 233 residue DHFR are degraded with similar rates argues against proteolysis being rate limiting. Fourth, deubiquitination at the proximal fused Ub is not rate limiting because $Ub_5$(G76)DHFR and $UbOM^{LY}$ are degraded with similar rates, and both are degraded faster than $OM^{LY}$ ($k_{cat}$ values, Table I). Having excluded the aforementioned four steps, remaining possible rate limitations include translocation, product release, and transitions between any of two steps. In the future, single turnover assays using the $UbOM^{LY}$ substrates may help to address this issue.

*Deubiquitination*

Figure 5 illustrates how deubiquitination by the UCH37 and POH1/Rpn11 proteasome-associated DUBs can be resolved and detected with $UbOM^{LY}$ substrates. UCH37 is a cysteine protease and is strongly inhibited by Ubal (Lam *et al.,* 1997), whereas the POH1/Rpn11 metalloprotease DUB activity is insensitive to Ubal but inactivated by chelation of its active site $Zn^{2+}$ by TPEN (Yao and Cohen, 2002). Moreover, deubiquitination by POH1/Rpn11 within the 26S proteasome is coupled to substrate degradation (or translocation) and, as such, depends on ATP hydrolysis. Note that Ub release by the Ubal-insensitive DUB (top, Fig. 5A) not only depends on ATP, but also corresponds to the generation of $OM^{LY}$ degradation products (fluorescent "peptide" bands in Fig. 5A, bottom). UCH37 activity shows no ATP dependence. Note also that because Ub-(G76)$OM^{LY}$

proteasome plus 2 $\mu M$ Ubal (see text). The substrate decrease is accompanied by the release of pentaUb. (B) Quantitative analysis of $Ub_5$(G76)$OM^{LY}$ degradation at different substrate concentrations; dotted lines (bottom to top) correspond to 0.1, 0.5, 1.0, 1.5, and 2.0 $\mu M$ $Ub_5$(G76)$OM^{LY}$. (C) Steady-state rates from b were fit by Michaelis–Menten kinetics, and $K_M$ and $k_{cat}$ values were obtained (see text). (D) Degradation of [$^{35}$S]$Ub_5$(G76)DHFR by the 26S proteasome in the presence of 2 $\mu M$ Ubal; data correspond (bottom to top) to 25, 50, 100, 150, and 250 n$M$ substrate. Lines are fitted results using KINSIM (Barshop *et al.*, 1983). Values obtained for the substrate $K_d$ (for the substrate) and $k_{cat}$ are shown in Table I; fitting for inhibition by the released $Ub_5$ product gave a $K_i$ of 8 n$M$.

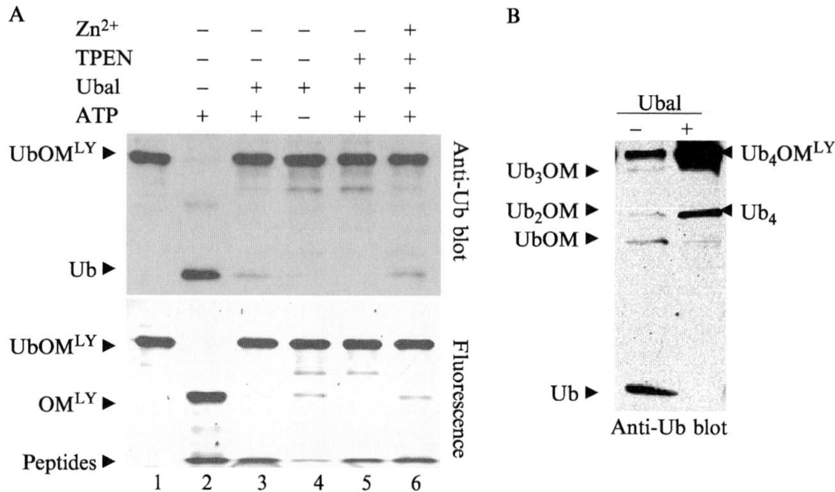


FIG. 5. Deubiquitination of $Ub_nOM^{LY}$ by activities in 26S proteasomes and the 19S regulatory complex. (A) Ub(G76)$OM^{LY}$ (3 $\mu M$) was incubated 1 h at 37° with 42 n$M$ bovine 26S proteasomes (lanes 2–6) with or without 1 m$M$ ATP (and ATP-regenerating system), 2 $\mu M$ Ubal, 0.5 m$M$ TPEN, or 0.5 m$M$ TPEN plus 0.75 m$M$ $ZnCl_2$ as indicated; proteasomes were omitted from the sample in lane 1. After separation by SDS–PAGE, products were visualized by Western blot and detection with anti-Ub antibody (top) or directly via Lucifer Yellow fluorescence (bottom). (B) $Ub_4$(G76)$OM^{LY}$ (3 $\mu M$) was incubated 1 h at 37° with 40 n$M$ bovine PA700 (19S complex) with or without 2 $\mu M$ Ubal. Ub-containing species were detected by SDS–PAGE and Western blotting with anti-Ub antibody. Without Ubal, the major products were free Ub and $Ub_{1-3}OM^{LY}$, which is consistent with the distal end-specific UCH37 activity being dominant. In the presence of Ubal, only $Ub_4$ was released, indicative of cleavage at the proximal Ub of the chain by POH1/Rpn11.

deubiquitination by UCH37 is much more efficient than by POH1/Rpn11, inhibition with Ubal is essential to unmask the metalloprotease activity.

Interestingly, the ATP dependence observed for POH1/Rpn11 in 26S proteasomes is absent from the activity in the isolated 19S (PA700) regulatory complex (Fig. 5B). By use of the $Ub_4$(G76)$OM^{LY}$ substrate, another fundamental difference between UCH37 and POH1/Rpn11 is apparent: whereas the UCH37 activity specifically cleaves Ub monomers from the distal end of the polyUb chain, POH1/Rpn11 cleaves only at the proximal Ub, which, in this example, releases an intact tetraUb chain. This specificity is consistent with the function of POH1/Rpn11 to remove polyUb chains from conjugates in order to facilitate complete translocation of the substrate polypeptide into the proteolytic chamber of the proteasome.

## Acknowledgments

We thank Sue Travis for assistance with the figures. T. Y. is a Fellow of The Leukemia and Lymphoma Society. This work was funded by NIH Grant R01 GM37666 to R. E. C.

## References

Barshop, B. A., Wrenn, R. F., and Frieden, C. (1983). Analysis of numerical methods for computer simulation of kinetic processes: Development of KINSIM—a flexible, portable system. *Anal. Biochem.* **130,** 134–145.

Beers, E. P., and Callis, J. (1993). Utility of polyhistidine-tagged ubiquitin in the purification of ubiquitin-protein conjugates and as an affinity ligand for the purification of ubiquitin-specific hydrolases. *J. Biol. Chem.* **268,** 21645–21649.

Borodovsky, A., Kessler, B. M., Casagrande, R., Overkleeft, H. S., Wilkinson, K. D., and Ploegh, H. L. (2001). A novel active site-directed probe specific for deubiquitylating enzymes reveals proteasome association of USP14. *EMBO J.* **20,** 5187–5196.

Braun, B. C., Glickman, M., Kraft, R., Dahlmann, B., Kloetzel, P. M., Finley, D., and Schmidt, M. (1999). The base of the proteasome regulatory particle exhibits chaperone-like activity. *Nat. Cell Biol.* **1,** 221–226.

Clark, A. C., and Frieden, C. (1999). Native *Escherichia coli* and murine dihydrofolate reductases contain late-folding non-native structures. *J. Mol. Biol.* **285,** 1765–1776.

DeKoster, G. T., and Robertson, A. D. (1997). Thermodynamics of unfolding for Kazal-type serine protease inhibitors: Entropic stabilization of ovomucoid first domain by glycosylation. *Biochemistry* **36,** 2323–2331.

Dunten, R. L., and Cohen, R. E. (1989). Recognition of modified forms of ribonuclease A by the ubiquitin system. *J. Biol. Chem.* **264,** 16739–16747.

Gill, S. C., and von Hippel, P. H. (1989). Calculation of protein extinction coefficients from amino acid sequence data. *Anal. Biochem.* **182,** 319–326.

Haas, A. L., and Bright, P. M. (1985). The immunochemical detection and quantitation of intracellular ubiquitin-protein conjugates. *J. Biol. Chem.* **260,** 12464–12473.

Hoffman, L., Pratt, G., and Rechsteiner, M. (1992). Multiple forms of the 20 S multicatalytic and the 26 S ubiquitin/ATP-dependent proteases from rabbit reticulocyte lysate. *J. Biol. Chem.* **267,** 22362–22368.

Hu, M., Li, P., Li, M., Li, W., Yao, T., Wu, J.-W., Gu, W., Cohen, R. E., and Shi, Y. (2002). Crystal structure of a UBP-family deubiquitinating enzyme in isolation and in complex with ubiquitin aldehyde. *Cell* **111,** 1141–1154.

Johnson, E. S., Ma, P. C., Ota, I. M., and Varshavsky, A. (1995). A proteolytic pathway that recognizes ubiquitin as a degradation signal. *J. Biol. Chem.* **270,** 17442–17456.

Johnston, S. C., Riddle, S. M., Cohen, R. E., and Hill, C. P. (1999). Structural basis for the specificity of ubiquitin C-terminal hydrolases. *EMBO J.* **18,** 3877–3387.

Khorasanizadeh, S., Peters, I. D., and Roder, H. (1996). Evidence for a three-state model of protein folding from kinetic analysis of ubiquitin variants with altered core residues. *Nat. Struct. Biol.* **3,** 193–205.

Kisselev, A. F., Akopian, T. N., Woo, K. M., and Goldberg, A. L. (1999). The sizes of peptides generated from protein by mammalian 26 and 20S proteasomes: Implications for understanding the degradative mechanism and antigen presentation. *J. Biol. Chem.* **274,** 3363–3371.

Lam, Y. A., Xu, W., DeMartino, G. N., and Cohen, R. E. (1997). Editing of ubiquitin conjugates by an isopeptidase in the 26S proteasome. *Nature* **385,** 737–740.

Leggett, D. S., Hanna, J., Borodovsky, A., Crosas, B., Schmidt, M., Baker, R. T., Walz, T., Ploegh, H., and Finley, D. (2002). Multiple associated proteins regulate proteasome structure and function. *Mol. Cell* **10,** 495–507.

Petroski, M. D., and Deshaies, R. J. (2003). Context of multiubiquitin chain attachment influences the rate of Sic1 degradation. *Mol. Cell* **11,** 1435–1444.

Pickart, C. M., and Cohen, R. E. (2004). Proteasomes and their kin: Proteases in the machine age. *Nat. Rev. Mol. Cell. Biol.* **5,** 177–187.

Piotrowski, J., Beal, R., Hoffman, L., Wilkinson, K. D., Cohen, R. E., and Pickart, C. M. (1997). Inhibition of the 26S proteasome by polyubiquitin chains synthesized to have defined lengths. *J. Biol. Chem.* **272,** 23712–23721.

Raasi, S., and Pickart, C. M. (2005). Synthesis of ubiquitin chains of defined length and chain linkage. *Methods Enzymol.* **399,** 21–36.

Scopes, R. K. (1993). "Protein Purification Principles and Practice," 3rd Ed., pp. 46–48. Springer-Verlag, New York.

Stein, R. L., Melandri, F., and Dick, L. (1996). Kinetic characterisation of the chymotryptic activity of the 20S proteasome. *Biochemistry* **35,** 3899–3908.

Thrower, J. S., Hoffman, L., Rechsteiner, M., and Pickart, C. M. (2000). Recognition of the polyubiquitin proteolytic signal. *EMBO J.* **19,** 94–102.

Verma, R., Aravind, L., Oania, R., McDonald, W. H., Yates, J. R., 3rd, Koonin, E. V., and Deshaies, R. J. (2002). Role of Rpn11 metalloprotease in deubiquitination and degradation by the 26S proteasome. *Science* **298,** 611–615.

Verma, R., McDonald, W. H., Yates, J. R., 3rd, and Deshaies, R. J. (2001). Selective degradation of ubiquitinated Sic1 by purified 26S proteasome yields active S phase cyclin-Cdk. *Mol. Cell* **8,** 439–448.

Yao, T., and Cohen, R. E. (2002). A cryptic protease couples deubiquitination and degradation by the proteasome. *Nature* **419,** 403–407.

# [44] Using Deubiquitylating Enzymes as Research Tools

*By* Rohan T. Baker, Ann-Maree Catanzariti, Yamuna Karunasekara, Tatiana A. Soboleva, Robert Sharwood, Spencer Whitney, and Philip G. Board

## Abstract

Ubiquitin is synthesized in eukaryotes as a linear fusion with a normal peptide bond either to itself or to one of two ribosomal proteins and, in the latter case, enhances the yield of these ribosomal proteins and/or their incorporation into the ribosome. Such fusions are cleaved rapidly by a variety of deubiquitylating enzymes. Expression of heterologous proteins as linear ubiquitin fusions has been found to significantly increase the yield of unstable or poorly expressed proteins in either bacterial or eukaryotic hosts. If expressed in bacterial cells, the fusion is not cleaved due to the absence of deubiquitylating activity and can be purified intact.

METHODS IN ENZYMOLOGY, VOL. 398

0076-6879/05 $35.00
DOI: 10.1016/S0076-6879(05)98044-0

We have developed an efficient expression system, utilizing the ubiquitin fusion technique and a robust deubiquitylating enzyme, which allows convenient high yield and easy purification of authentic proteins. An affinity purification tag on both the ubiquitin fusion and the deubiquitylating enzyme allows their easy purification and the easy removal of unwanted components after cleavage, leaving the desired protein as the only soluble product.

Ubiquitin is also conjugated to ε amino groups in lysine side chains of target proteins to form a so-called isopeptide linkage. Either a single ubiquitin can be conjugated or other lysines within ubiquitin can be acceptors for further conjugation, leading to formation of a branched, isopeptide-linked ubiquitin chain. Removal of these ubiquitin moieties or chains *in vitro* would be a valuable tool in the ubiquitinologists tool kit to simplify downstream studies on ubiquitylated targets. The robust deubiquitylating enzyme described earlier is also very useful for this task.

## Introduction

The expression of a cloned gene to isolate large quantities of its protein product demands a highly efficient expression system where protein can be purified to homogeneity, especially for crystallographic and therapeutic purposes. This is often difficult to achieve. Several systems have emerged that involve fusing the gene of interest downstream of a second gene to produce a fusion protein (Uhlen and Moks, 1990). This strategy generally gives reliably high protein yields and can also allow simple purification methods due to the affinity of certain fusion partners for a particular ligand. A major drawback of this approach is the covalent linkage of the two proteins, where the presence of the fusion partner may prevent or interfere with subsequent uses. A protease recognition site can be engineered between the two fused proteins; however, this usually involves altering the amino terminus of the desired product, resulting in the expression of an unauthentic protein (Butt *et al.*, 1989). Furthermore, cleavage of the fusion protein is rarely complete, causing a reduction in protein yield, and may also occur nonspecifically within the fused protein (Baker, 1996).

A fusion partner that has been used for some years is ubiquitin (Ub). This small eukaryotic protein provides two benefits. First, like other fusion partners, it offers a natural yield enhancement in both eukaryotic and prokaryotic hosts. Second, and uniquely, the Ub moiety can be removed by highly specific proteases known as deubiquitylating enzymes (DUBs) that do not cleave nonspecific sequences and do not leave additional amino acids at the amino terminus of the protein of interest (Baker, 1996; Hondred *et al.*, 1999). This cleavage occurs precisely after the final glycine residue at the carboxyl-terminal of Ub irrespective of the amino acid immediately following, with the sole exception of proline, which is cleaved inefficiently in yeast (Bachmair *et al.*, 1986), but can be cleaved by some mammalian DUBs (Angelats *et al.*, 2003; Baker *et al.*, 1999). To date, the main drawbacks of the Ub fusion technique have been no simple affinity purification for Ub and no readily available deubiquitylating enzyme. Most DUBs that have been isolated from various species have been relatively large enzymes and difficulties have been encountered with expressing and purifying large quantities, along with problems in finding a stable DUB with general activity against a range of fusion proteins (Varshavsky, 2000).

We have developed an efficient *Escherichia coli*-based expression system where the protein of interest is expressed as a fusion to poly-histidine-tagged Ub, enabling a simple one-step purification of the fusion protein by immobilized metal affinity chromatography (IMAC) (Catanzariti *et al.*, 2004). We have also engineered a mouse DUB, Usp2, to provide a minimal catalytically active DUB domain and expressed and purified this as a poly-histidine-tagged protein. The tagged protease allows *in vitro* cleavage of Ub from the desired protein as well as its selective removal from the cleavage reaction, along with the cleaved Ub, any uncleaved fusion protein, and any copurified contaminants, leaving the desired protein as the only soluble product. This system was found to be very effective and applicable to the expression of a broad range of proteins and peptides and should be useful for high-throughput applications (Catanzariti *et al.*, 2004). The relatively small size of Ub also makes this system attractive for metabolic labeling of proteins for applications such as nuclear magnetic resonance. It should also be noted that Wang *et al.* (2003) have described a similar expression system that uses biotin-tagged Ub and a biotin-tagged chicken DUB.

We have also found that the same engineered DUB domain was capable of removing isopeptide-linked Ub moieties from Ub conjugates *in vitro*, both mono-Ub conjugates and branched multi-Ub chains. This ability is of benefit in the study of ubiquitylated target proteins, enabling the demonstration that the conjugate is actually Ub, and/or removing the Ub moieties to allow simpler downstream proteomic analysis.

## Constructing Ubiquitin Fusion Protein Vectors

### *The Expression Vector pHUE*

The His-tagged ubiquitin fusion protein expression vector pHUE was modified from pET15b (Novagen). The *Eco*RI, *Cla*I, and *Hind*III sites are removed from the vector backbone by *Eco*RI–*Hind*III digestion, end filling, and self-ligation. A double-stranded oligonucleotide is then inserted into the *Bam*HI site to create an extended polylinker (Fig. 1). A polymerase chain reaction (PCR)-amplified human Ub open reading frame (Baker and Board, 1991; GenBank X56998) is then inserted between the *Nde*I and the *Bam*HI sites (Catanzariti *et al.*, 2004). pHUE thus contains the inducible T7 RNA polymerase promoter, a histidine tag at the 5′ end of a Ub open reading frame, and an extended polylinker (Fig. 1). The vector confers ampicillin resistance and also expresses the lac repressor lacI$^q$. The pHUE sequence is deposited in GenBank with accession number AY751539.

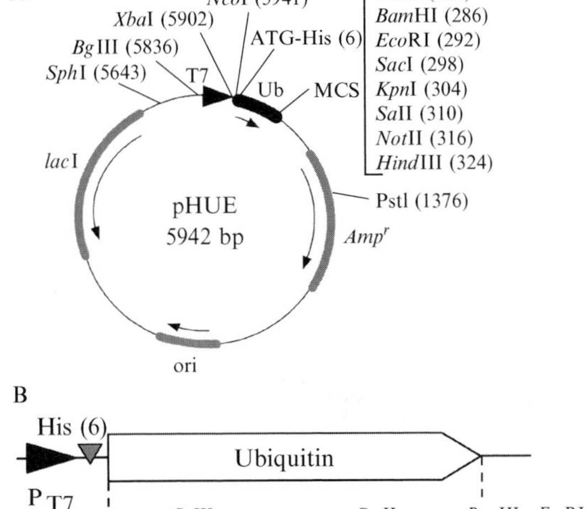


Fig. 1. The histidine-tagged ubiquitin expression vector, pHUE. (A) Plasmid map of pHUE showing the ubiquitin (Ub) coding region (black box), the T7 polymerase promoter (black triangle), and other regions (shaded boxes). Arrows indicate the direction of transcription. Restriction enzyme recognition sites within the multiple cloning site (MCS) are listed and other useful recognition sites in the vector backbone are also shown (unique, except *Bgl*II); locations are given relative to the start codon upstream of the his-tag, ATG = 1. *(His)6*, poly histidine tag; *Amp*$^r$, $\beta$-lactamase gene; *ori*, colE1 origin of replication; *lacI*, lacI repressor gene. (B) DNA and encoded protein sequence of the 5′ and 3′ end of the Ub-coding region showing the engineered *Sac*II site (underlined) within codons Leu73, Arg74, and Gly75 and the 3′ polylinker. Restriction sites and protein translation are given above and under the DNA sequence, respectively. Reproduced with permission from Catanzariti *et al.* (2004).

*General Considerations of Designing and Constructing Ubiquitin Fusions*

DUBs cleave precisely after the final glycine residue at the carboxyl-terminal of Ub irrespective of the amino acid immediately following, with the sole exception of proline, which is cleaved inefficiently in yeast (Bachmair *et al.*, 1986). The Ub-Pro bond can be cleaved by some mammalian DUBs (Angelats *et al.*, 2003; Baker *et al.*, 1999), but this has never been studied in detail *in vitro* and the catalytic efficiency of this reaction remains to be determined. Thus, it is best to avoid proline immediately after Gly76 of Ub wherever possible. If proline is essential, it would be best to engineer a Gly76-Met-Pro linkage and remove the Met with methionine aminopeptidase *in vitro*.

To engineer a precise Ub fusion, such that the desired residue is at the N terminus of the cleaved product, it is necessary to use a restriction enzyme site within the 3′ end of the Ub open reading frame and recreate the C terminus of Ub with the PCR primer. In some cases, one could use a restriction site spanning the Gly76 codon of Ub, but the N-terminal residue of the desired protein would also be encoded by the restriction site. For example, in the pHUE vector, Gly76 is encoded within a *Bam*HI site, GGA-TCC. Thus open reading frames can be ligated at this *Bam*HI site, but the resulting cleaved protein would always have an N-terminal serine residue. Several restriction enzyme variations can be envisaged that include a Gly-76 codon, but these will all constrain the resulting N-terminal residue. Thus, for maximum versatility, a *Sac*II site has been engineered within the Ub codons for Leu73, Arg74, and Gly75. The ligated DNA fragment must encode the Gly75-Gly76 residues of Ub, which are essential for cleavage by DUBs (Ecker *et al.*, 1989). This can be achieved by PCR amplifying the gene of interest using a primer with the 5′ extension dCT<u>C-CGC-GG</u>T-GGT-$(NNN)_6$, encoding Leu73, Arg74, Gly75, and Gly76 and containing the *Sac*II site (underlined; Baker, 1996; Baker *et al.*, 1994; Catanzariti *et al.*, 2004) followed by six codons from the 5′ end of the gene of interest. The reverse PCR primer should be complementary to the final five codons of the gene, the stop codon, and a unique restriction site chosen from the pHUE polylinker (Fig. 1B). A proofreading thermostable DNA polymerase should be used, and the product digested with *Sac*II and the 3′ polylinker enzyme and ligated into pHUE digested with the same enzymes. Recombinant vectors should be constructed in general laboratory host strains lacking T7 RNA polymerase (e.g., DH5$\alpha$, XL1-blue) and sequenced to confirm fidelity. Sequence primers that we use include a forward primer near the 3′ end of Ub (5′ ACTCTCTCAGACTACAACATCC 3′) and a pET-reverse primer (5′ TAGTTATTGCTCAGCGGTGGC 3′). The only downfall of this approach is when a *Sac*II site is present within the gene of interest; however, this should be uncommon, as *Sac*II is a rare cutter. Should it occur, then it could either be mutated prior to amplifying the gene of interest or a two-step cloning strategy can be employed. The Ub-fusion approach is also not suited to a recombination cloning technique, such as the Gateway technique (Invitrogen), as this would introduce extra amino acids at the N terminus of the cleaved protein. Also, if extra N-terminal residues can be tolerated, there are several restriction sites in the polylinker in pHUE downstream of Ub (Fig. 1B). The Ub-fusion approach can also be used to synthesize peptides of any length, encoded either by synthetic double-stranded oligonucleotides or by DNA fragments amplified from larger open reading frames (Catanzariti *et al.*, 2004).

## Expressing and Purifying Ubiquitin Fusion Proteins

The fusion protein vector constructed as described earlier in pHUE should then be transformed into a strain capable of expressing T7 RNA polymerase [e.g., BL21(DE3)]. Transformed cells are then grown and induced under the standard conditions for this T7-inducible system (Catanzariti *et al.*, 2004; Studier and Moffat, 1986). Our standard expression conditions are as follows.

1. Grow a 5-ml overnight culture at 37° in BL21(DE3) in Luria-Bertani (LB) broth supplemented with 100 $\mu$g/ml ampicillin.
2. Subculture this 1:100 into prewarmed 400 ml LB/ampicillin in a baffled 2-liter flask and shake at 37° at 250 rpm for 2 h (to mid/late exponential phase). Lower temperatures can also be used to promote protein solubility; we have used 16, 22, and 30° depending on the individual protein.
3. Add isopropyl-$\beta$-D-thiogalactoside (IPTG) to 0.4 m*M* and continue shaking for 4–6 h. The IPTG concentration and length of induction can also be varied to optimize the yield of the soluble protein. Although we have not yet used autoinduction media with this system, there is no reason that this should not also work.
4. Harvest cells and resuspend in 20 ml buffer A [50 m*M* $Na_2HPO_4$/$NaH_2PO_4$, pH 7.4, 300 m*M* NaCl, 10 m*M* imidazole, 10 m*M* 2-mercaptoethanol (2-ME), 30% glycerol] containing 1 m*M* 4-(2-aminoethyl) benzenesulfonyl fluoride (AEBSF). Freeze overnight at –70°. The glycerol and 2-ME help prevent bacterial proteins copurifying but could be omitted on a per-protein basis. We have also used Tris-based buffers (50 m*M* Tris, pH 8.0) and varied the NaCl concentration between 150 and 300 m*M*.
5. Thaw cells completely and lyse by sonication (3× 1-min bursts on ice with a microtip) or in a prechilled French press at 140 MPa. Recover the soluble fraction after centrifugation for 20 min at 20,000*g* at 4°.

6. Purify the his-tagged protein by a standard IMAC protocol. We have used Ni-NTA and Cobalt agaroses from commercial sources (Qiagen; Clontech) or Ni-iminodiacetate agarose made according to Porath and Olin (1983). We have used either batch mode for binding, washing, and elution or loaded into a disposable column for washing and elution. Proteins are generally eluted with either a Tris- or a phosphate-based buffer containing 200 m*M* imidazole, 150–300 m*M* NaCl, 1 m*M* 2-ME, and glycerol if required.

7. After SDS–PAGE analysis to judge yield and purity, dialyze the eluted protein against dialysis buffer (50 m*M* Tris- or phosphate-based buffer, pH 8.0, containing 150–300 m*M* NaCl and 1 m*M* 2-ME, and glycerol as necessary) to remove imidazole. This is necessary for subsequent IMAC to remove unwanted his-tagged proteins.

## Expressing, Purifying, and Optimizing the Usp2 Catalytic Core

The minimal Usp2 catalytic core (cc) open reading frame was obtained by PCR amplification of a mouse Usp2–45 cDNA from IMAGE clone 1922050 (AY255637; Gousseva and Baker, 2003). The PCR primers used are 5′-CGTGGATCCTCTGCTCACCAAAGCCAAGAATTC-3′ and 5′-TCCGGATCCTTACATACGGGAGGGTGGACTG–3′. The PCR product is digested with *Bam*HI and ligated into the *Bam*HI site of pET15b in the correct orientation, resulting in pHUsp2-cc (Catanzariti *et al.*, 2004). The pHUsp2-cc sequence is deposited in GenBank with accession number AY751540. The resulting protein (termed $H_6$-Usp2-cc) is expressed and purified exactly as described previously for his-tagged Ub fusions, with the note that the inclusion of 30% glycerol in all lysis, wash, and elution buffers is essential to prevent protein precipitation, and the inclusion of 20 m*M* 2-ME in the lysis and wash buffers greatly reduces copurification of *E. coli* proteins. Protease inhibitors can be used when preparing $H_6$-Usp2-cc, but cysteine protease inhibitors (e.g., antipain, E-64, leupeptin) should NEVER be used, given that Usp2 is a cysteine protease. Beware of protease inhibitor cocktails! Removal of imidazole from $H_6$-Usp2-cc is not essential but can be performed by dialysis as described earlier, but with 30% glycerol in the dialysis buffer. $H_6$-Usp2-cc is a very soluble ~42-kDa protein and is expressed at high levels and can be stored at –20 or –70° in the 30% glycerol elution or dialysis buffers. It is stable for at least 6 months at –20°.

The pH optimum of $H_6$-Usp2-cc was determined by cleavage of a $H_6$-Ub-M-GSTP1 fusion protein constructed in pHUE (Catanzariti *et al.*, 2004) at a 1:100 enzyme:substrate ratio (Fig. 2A). Under these conditions, the enzyme showed the highest activity around pH 8.5 and was least active

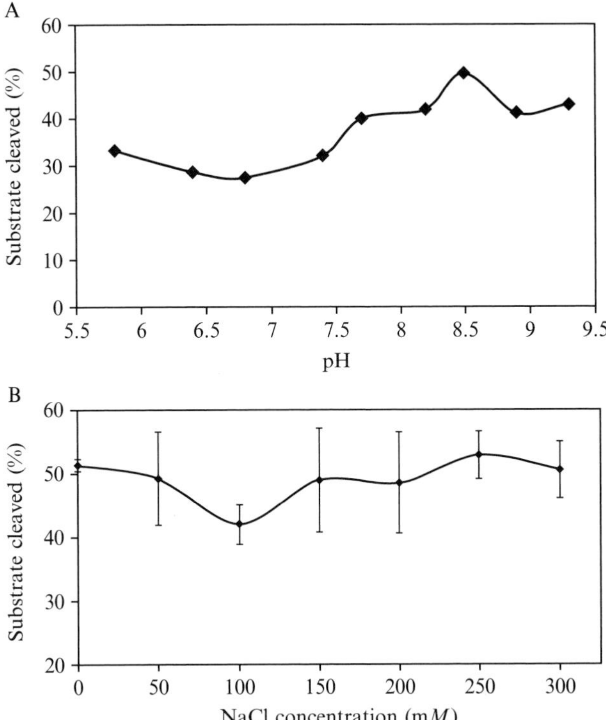


FIG. 2. pH and NaCl optimization of H6-Usp2-cc. (A) pH Optimum. $H_6$-Usp2-cc was combined with $H_6$-Ub-M-GSTP1 at a 1:100 enzyme:substrate ratio at 37° for 20 min in the presence of 137 m*M* NaCl with either 50 m*M* $Na_2HPO_4/NaH_2PO_4$ for pH 5.5–8 or 50 m*M* Tris-HCl for pH 8–9.5. Products were resolved by SDS–PAGE, Coomassie blue staining, and densitometry scanning. From the latter the percentage of substrate cleaved was calculated and then plotted against pH. (B) NaCl optimum. As for A except that a 50 m*M* $Na_2HPO_4/NaH_2PO_4$ pH 8.0 buffer was used, with varying NaCl concentration. Assays were repeated four times and plotted as percentages of substrate cleaved vs [NaCl]. Error bars represent standard deviations.

around pH 6.5–7.0. The enzyme retained >80% activity between pH 7.5 and 9. As many proteins are more stable around physiological pH, we generally use pH 8 as a good compromise. Usp2-cc activity was also investigated at various NaCl concentrations between 0 and 300 m*M* NaCl. In the four repetitions of this assay, activity was relatively independent of NaCl concentration (Fig. 2B). Thus the enzyme is quite versatile and cleavage conditions can be chosen to suit individual substrates.

## Cleaving Ubiquitin Fusions

The efficiency of Usp2-cc against several Ub fusions was initially assayed by incubating different Ub-fusion substrates with $H_6$-Usp2-cc at a 1:100 enzyme to substrate molar ratio for 60 min at 37° (Catanzariti *et al.*, 2004). The variables when cleaving a Ub fusion protein with $H_6$-Usp2-cc are buffer, temperature, and time. As described earlier, the enzyme is relatively insensitive to NaCl concentration, which can be chosen to suit the requirements of the fusion protein. A pH ~8 is preferred, and the presence of a reducing agent is critical for this cysteine protease; we usually supplement the digest with 1 m*M* dithiothreitol. A high glycerol concentration (>5%, v/v) will inhibit the enzyme, and if glycerol is present in the Ub fusion protein buffer, remove it by dialysis or buffer exchange. Being a mouse enzyme, we assume it is most active at 37°, but this has not been tested. It still retains substantial activity at room temperature and even at 16°, and we have often had better success doing overnight cleavage reactions at 16°. We occasionally observe apparent nonspecific cleavage at 37°, presumably due to copurifying *E. coli* proteases; this is greatly reduced or abolished at 16°. It may also be prevented with inhibitors of noncysteine proteases, although we have not investigated this to date. Cleavage is monitored by SDS–PAGE and Coomassie blue staining. If necessary, the enzyme concentration can be increased to a 1:10 molar ratio or even more; Usp2-cc has very high specificity for Ub and will not cleave nonspecifically. However, Usp2-cc was not able to cleave a Ub-P bond even at a 1:1 molar ratio (Catanzariti *et al.*, 2004), consistent with known limitations of DUBs against the Ub-Pro bond (Gilchrist *et al.*, 1997; but see later). Notably, other proteases such as enterokinase and factor Xa are also inhibited when their recognition sequence is followed by proline (Stevens, 2000).

## Recovering the Cleaved Product

The strategy of this expression system includes a final purification step using IMAC to isolate the cleaved product from the cleavage reaction. Following the complete cleavage described earlier, Ni-NTA agarose (20–50 $\mu$l bed volume) is added to the solution to bind the $H_6$-Usp2-cc, $H_6$-tagged Ub, any uncleaved $H_6$-tagged fusion protein, and any copurified contaminants. After binding for 30 min at 4°, the mixture is centrifuged for 1–2 min and the supernatant is collected for analysis by SDS–PAGE. Alternatively, the mixture can be centrifuged through any spin column that does not bind protein. Several examples are given in Catanzariti *et al.* (2004), and examples of three different plant proteins are shown in Fig. 3, including steps in the initial purification, cleavage, and final purification.

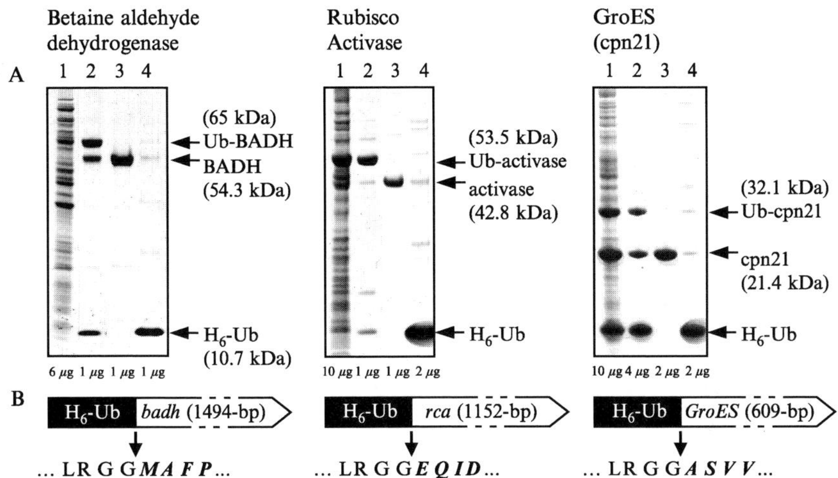


FIG. 3. Purification of recombinant plant proteins expressed in *E. coli* by ubiquitin fusion. (A) Coomassie-stained SDS–PAGE analysis of protein fractions during purification. Proteins are identified above the gels. Lane 1, total *E. coli* cellular soluble protein; lane 2, dialyzed Ni-NTA purified Ub-fused (Ub-) recombinant protein; lane 3, purified protein post $H_6$-Usp2-cc proteolysis that does not bind to Ni-NTA; and lane 4, products that bind to Ni-NTA post $H_6$-Usp2-cc proteolysis, where 6xHis-tagged Ub ($H_6$-Ub) is the major species. The amount ($\mu$g) of protein loaded per lane is shown at the bottom of the gel. (B) Schematic representation of the cloned plant gene showing the DUB recognition sequence and the four N-terminal amino acids residues of the mature plant protein (italics) downstream of the cleavage site (arrow).

This procedure was found to work extremely well, recovering almost all of the cleaved protein, while successfully removing all other proteins present in the digest (Fig. 3 and data not shown). Example growth conditions, cleavage conditions, and yields that can be obtained through this system are given in Table I. The use of Ub fusions to enhance yield has been well documented (Baker, 1996), and another example is betaine aldehyde dehydrogenase (BADH). When expressed from pHUE, we obtained BADH at 2.5% of total cellular soluble protein (CSP; Table I), whereas the same gene expressed with a 6xHis tag in pET28a, grown under the same conditions, resulted in less than 0.2% of CSP. Another protein, CBBX, did not give detectable expression from pET28a, but gave 5–10% of CSP from pHUE (data not shown).

N-terminal sequencing of several cleaved and purified proteins reveals that $H_6$-Usp2-cc cleaves precisely where expected (Catanzariti *et al.*, 2004). In addition, enzymes or peptides produced using this system retain their full activity/biological properties (Board *et al.*, 2004; Catanzariti *et al.*, 2004; Robinson *et al.*, 2004). We have also modified the procedure so that no

TABLE I
SUMMARY OF RECOMBINANT PLANT PROTEIN EXPRESSION AND PURIFICATION FROM *E. COLI* USING UBIQUITIN FUSION STRATEGIES

| | | Induction with 0.5 m*M* IPTG | | Ub-fused recombinant protein (Ub-P) | | H6-Usp2-cc digestion | | | | | Pure recombinant protein | |
|---|---|---|---|---|---|---|---|---|---|---|---|---|
| Plasmid name | Growth temp-erature[a] | Culture-density | Dura-tion (h) | Size (kDa) | Yield (% of CSP)[b] | Amt of Usp2-cc per mg Ub-P | Temper-ature | Dura-tion (h) | Any protein precipitate formed? | % digested | Size (kDa) | Yield (% of CSP)[b] |
| pHueBADH | 28° | $A_{600} = 1.0$ | 5 | 65 | 3.5 | 10 μg | 37° | 1 | No | 100 | 54.3 | 2.5 |
| pHueAct | 30° | $A_{600} = 1.0$ | 6 | 53.5 | 25 | 25 μg | 37° | 1 | Yes | 70 | 42.8 | 10 |
| | | | | | | 25 μg | 16° | 16 | No | 100 | | 15 |
| pHueCPN21 | 26° | $A_{600} = 1.5$ | 4 | 32.1 | 15 | 25 μg | 37° | 1 | No | 100 | 21.4 | 10 |
| | | | | | | 25 μg | 16° | 16 | No | 100 | | 10 |
| pHueCBBX | 26° | $A_{600} = 1.0$ | 4 | 45.8 | 8 | 10 μg | 37° | 1 | Yes | 80 | 35.1 | 5 |
| | | | | | | 10 μg | 16° | 16 | No | 100 | | 6 |

[a] All cultures were grown in BL21(DE3) in LB medium (pH 7) containing 100 μg/ml ampicillin.
[b] Yield of recombinant protein expressed as a percentage of the total *E. coli* cellular-soluble protein (CSP).

elution steps are performed; the purified $H_6$-tagged Ub-fusion protein bound to beads can be mixed with $H_6$-Usp2-cc bound to beads, and providing the beads are kept in suspension, cleavage occurs, resulting in the release of the desired protein as the only soluble molecule (data not shown).

## Cleavage of Branched Ubiquitin Conjugates

We have also tested $H_6$-Usp2-cc for its ability to cleave a branched Ub chain. The multiubiquitylated protein we have studied the most is the Lys-48-linked multi-Ub chain on the artificial Ub pathway substrate Ub-proline-$\beta$-galactosidase (Ub-P-$\beta$-gal) (Chau *et al.*, 1989). Ub-P-$\beta$-gal bearing a multi-Ub chain was immunoprecipitated from yeast lysates and incubated at 37° with $H_6$-Usp2-cc. This caused a time-dependent removal of the higher molecular weight multiubiquitylated forms of Ub-P-$\beta$-gal, resulting in predominantly monoubiquitylated Ub-P-$\beta$-gal ($Ub_2$-P-$\beta$-gal) after 60 min (Fig. 4A). More extended incubation with $H_6$-Usp2-cc resulted in complete removal of the multi-Ub chain, and also generated a product the size of P-$\beta$-gal, suggesting that $H_6$-Usp2-cc can cleave the Ub-Pro bond at a low efficiency, possibly only after the branched Ub chain has been removed from Ub-Pro-$\beta$-gal (Fig. 4B). Notably, mutation of the active site cysteine in Usp2-cc prevented its cleavage activity (Fig. 4B, lane 3). We have also used $H_6$-Usp2-cc to remove a Ub chain from the immunoprecipitated TNF receptor (Wertz *et al.*, 2004) and from a substrate that can be both mono- and multiubiquitylated (N. Gousseva and R. T. Baker, manuscript in preparation). While we have not yet investigated its activity against different Lys-linked Ub chains, $H_6$-Usp2-cc should be a versatile enzyme for the *in vitro* removal of Ub from Ub conjugates.

## Troubleshooting

This system is fairly robust given the established yield-enhancing properties of Ub fusions; high specificity of DUBs for Ub; and the ability to tolerate contaminating proteins that copurify by IMAC, given that these are removed by the final IMAC step after $H_6$-Usp2-cc digestion. In some instances we observe copurification of *E. coli* chaperones that appear to bind to the large amounts of fusion protein, and these can appear in the final product. These can be removed either by an ion-exchange chromatography step on the final product or, alternatively, the first IMAC-bound protein(s) can be washed with denaturing buffer (6 *M* guanidine, 50 m*M* Tris-Cl, pH 8) to remove associated proteins and then allowing refolding to occur in wash buffer, before eluting with imidazole.

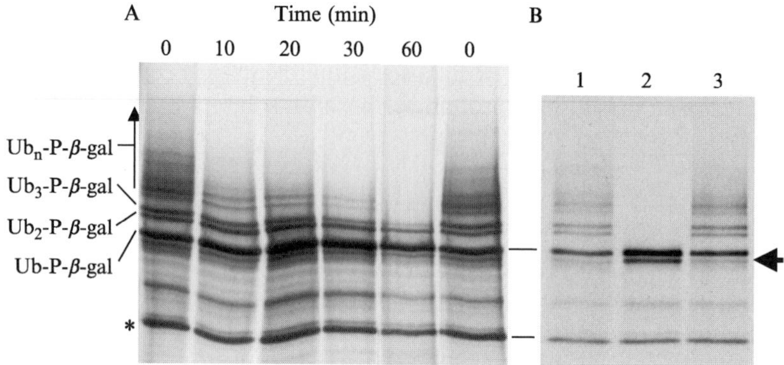


FIG. 4. Removal of branched Ub chains using Usp2-cc. (A) Time course of branched Ub cleavage. An artificial substrate of the N-end rule pathway in yeast, Ub-proline-β-galactosidase (Ub-P-β-gal), was expressed in yeast, labeled metabolically with [$^{35}$S] methionine, and immunoprecipitated from yeast lysates with a monoclonal antibody against β-gal (Sigma) exactly as described by Baker and Varshavsky (1991). The β-gal/antibody/protein A-Sepharose bead slurry was incubated at 37° with $H_6$-Usp2-cc for the time indicated above the lanes (minutes), resolved by 6% SDS–PAGE, treated with En$^3$Hance (NEN/DuPont), and exposed to X-ray film. Ub-β-gal species are labeled on the left, with multiubiquitylated forms indicated by the vertical arrow, and di- and triubiquitylated forms labeled $Ub_2$-P-β-gal and $Ub_3$-P-β-gal, respectively. An asterisk indicates a stable cleavage product of β-gal observed in yeast (Baker and Varshavsky, 1991). (B) Complete cleavage of a branched Ub chain and partial cleavage of the Ub-Pro bond. Lane 1, immunoprecipitated Ub-Pro-β-gal bearing a branched Ub chain. Lane 2, same as lane 1 but incubated for 60 min at 37° with $H_6$-Usp2-cc. Lane 3, same as lane 2, but the active-site Cys residue of Usp2-cc was mutated to serine. An arrowhead points to a protein of the size of Pro-β-gal (see text).

In some cases we observe apparent cleavage of the $H_6$-tagged Ub-fusion protein in *E. coli*, which is surprising given the absence of DUBs from *E. coli*. This is quite apparent in the GroES/cpn21 example in Fig. 3, where a protein of the expected mature cpn21 size (21 kDa) is observed in crude lysates, as well as in the intact $H_6$-Ub-cpn21 (32 kDa) and free $H_6$-Ub (10 kDa). The free cpn21 copurifies by IMAC, presumably because GroES is a multimeric protein, and only one 6Xhis tag would be required to purify the whole complex. While we have not investigated the cpn21 case, in other cases we have sequenced the N terminus of the apparent *E. coli*-cleaved fusion partners to find that they have the expected N terminus if Ub had been cleaved precisely (Catanzariti *et al.*, 2004), implying a cryptic protease activity in *E. coli*. We are presently trying to characterize this activity, and it is possible that it only occurs in certain *E. coli* strains. However, given that the cleavage occurs at the desired position, the worst case is a slight reduction in yield.

## References

Angelats, C., Wang, X.-W., Jermiin, L. S., Copeland, N., Jenkins, N., and Baker, R. T. (2003). Characterization of the mouse Usp15 ubiquitin specific protease. *Mammalian Genome* **14,** 31–46.

Bachmair, A., Finley, D., and Varshavsky, A. (1986). *In vivo* half-life of a protein is a function of its amino-terminal residue. *Science* **234,** 179–186.

Baker, R. T. (1996). Protein expression using ubiquitin fusion and cleavage. *Curr. Opin. Biotechnol.* **7,** 541–546.

Baker, R. T., and Board, P. G. (1991). The human Ub-52 amino acid fusion protein gene shares several structural features with mammalian ribosomal protein genes. *Nucleic Acids Res.* **19,** 1035–1040.

Baker, R. T., Smith, S. A., Marano, R., McKee, J., and Board, P. G. (1994). Protein expression using cotranslational fusion and cleavage of ubiquitin: Mutagenesis of the glutathione-binding site of human Pi class glutathione S-transferase. *J. Biol. Chem.* **269,** 25381–25386.

Baker, R. T., and Varshavsky, A. (1991). Inhibition of the N-end rule pathway in living cells. *Proc. Natl. Acad. Sci. USA* **88,** 1090–1094.

Baker, R. T., Wang, X.-W., Woollatt, E., White, J., and Sutherland, G. R. (1999). Identification, functional characterization and chromosomal localisation of USP15, a novel human ubiquitin-specific protease related to the Unp/USP4 oncoprotein, and a systematic nomenclature proposal for human ubiquitin-specific proteases. *Genomics* **59,** 264–274.

Board, P. G., Coggan, M., Watson, S., Gage, P. W., and Dulhunty, A. F. (2004). Clic-2 modulates cardiac ryanodine receptor $Ca^{2+}$ release channels. *Int. J. Biochem. Cell Biol.* **36,** 1599–1612.

Butt, T. R., Jonnalagadda, S., Monia, B. P., Sternberg, E. J., Marsh, J. A., Stadel, J. M., Ecker, D. J., and Crooke, S. T. (1989). Ubiquitin fusion augments the yield of cloned gene products in *Escherichia coli*. *Proc. Natl. Acad. Sci. USA* **86,** 2540–2544.

Catanzariti, A.-M., Soboleva, T. A., Jans, D. A., Board, P. G., and Baker, R. T. (2004). An efficient system for high-level expression and easy purification of authentic recombinant proteins. *Protein Sci.* **13,** 1331–1339.

Chau, V., Tobias, J. W., Bachmair, A., Marriott, D., Ecker, D. J., Gonda, D. K., and Varshavsky, A. (1989). A multiubiquitin chain is confined to specific lysine in a targeted short-lived protein. *Science* **243,** 1576–1583.

Ecker, D. J., Stadel, J. M., Butt, T. R., Marsh, J. A., Monia, B. P., Powers, D. A., Gorman, J. A., Clark, P. E., Warren, F., Shatzman, A., and Crooke, S. T. (1989). Increasing gene expression in yeast by fusion to ubiquitin. *J. Biol. Chem.* **264,** 7715–7719.

Gilchrist, C. A., Gray, D. A., and Baker, R. T. (1997). A ubiquitin-specific protease that efficiently cleaves the ubiquitin-proline bond. *J. Biol. Chem.* **272,** 32280–32285.

Gousseva, N., and Baker, R. T. (2003). Gene structure, alternate splicing, tissue distribution, cellular localization, and developmental expression pattern of mouse deubiquitinating enzyme isoforms Usp2-45 and Usp2-69. *Gene Expr.* **11,** 163–179.

Hondred, D., Walker, J. M., Mathews, D. E., and Vierstra, R. D. (1999). Use of ubiquitin fusions to augment protein expression in transgenic plants. *Plant Physiol.* **119,** 713–724.

Porath, J., and Olin, B. (1983). Immobilized metal ion affinity adsorption and immobilized metal ion affinity chromatography of biomaterials: Serum protein affinities for gel-immobilized iron and nickel ions. *Biochemistry* **22,** 1621–1630.

Robinson, A., Huttley, G. A., Booth, H. S., and Board, P. G. (2004). Modelling and bioinformatics studies of the human Kappa class glutathione transferase predict a novel third glutathione transferase family with homology to prokaryotic 2-hydroxychromene-2-carboxylate (HCCA) isomerases. *Biochem. J.* **379,** 541–552.

Stevens, R. C. (2000). Design of high-throughput methods of protein production for structural biology. *Struct. Fold. Des.* **8,** R177–R185.

Studier, F. W., and Moffatt, B. A. (1986). Use of bacteriophage T7 RNA polymerase to direct selective high-level expression of cloned genes. *J. Mol. Biol.* **189,** 113–130.

Uhlen, M., and Moks, T. (1990). Gene fusions for purpose of expression: An introduction. *Methods Enzymol.* **185,** 129–143.

Varshavsky, A. (2000). Ubiquitin fusion technique and its descendants. *Methods Enzymol.* **327,** 578–593.

Wang, T., Evdokimov, E., Yiadom, K., Yan, Z., Chock, P. B., and Yang, D. C. (2003). Biotin-ubiquitin tagging of mammalian proteins in *Escherichia coli*. *Protein Expr. Purif.* **30,** 140–149.

Wertz, I. E., O'Rourke, K. M., Zhou, H., Eby, M., Aravind, L., Seshagiri, S., Wu, P., Wiesmann, C., Baker, R., Boone, D. L., Ma, A., Koonin, E. V., and Dixit, V. M. (2004). De-ubiquitination and ubiquitin ligase domains of A20 downregulate NF-kappaB signalling. *Nature* **430,** 694–699.

# [45] Functional Annotation of Deubiquitinating Enzymes Using RNA Interference

*By* Annette M. G. Dirac, Sebastian M. B. Nijman, Thijn R. Brummelkamp, and René Bernards

## Abstract

Protein ubiquitination is a dynamic process, depending on a tightly regulated balance between the activity of ubiquitin ligases and their antagonists, the ubiquitin-specific proteases or deubiquitinating enzymes. The family of ubiquitin ligases has been studied intensively and it is well established that their deregulation contributes to diverse disease processes, including cancer. Much less is known about the function and regulation of the large group of deubiquitinating enzymes. This chapter describes how RNA interference against deubiquitinating enzymes can be used to elucidate their function. The application of this technology will greatly improve the functional annotation of this family of proteases.

## Deubiquitinating Enzymes

Modification of proteins by ubiquitin conjugation plays a major part in many biological processes. Ubiquitination not only targets proteins for destruction by the proteasome, but also triggers a number of nondestructive processes. These include membrane receptor endocytosis, vesicular

METHODS IN ENZYMOLOGY, VOL. 398

0076-6879/05 $35.00
DOI: 10.1016/S0076-6879(05)98045-2

trafficking, and chromatin structure and transcription, as well as viral budding (Aguilar and Wendland, 2003; Bach and Ostendorff, 2003; Gill, 2004; Zhang, 2003). While the enzyme families that mediate ubiquitin conjugation have been studied intensively, much less is known about the ubiquitin-specific proteases [also known as deubiquitinating enzymes (DUBs)], which mediate the removal of ubiquitin from cellular substrates. DUBs are cysteine proteases that specifically cleave ubiquitin monomers from ubiquitin-conjugated substrates. This enzyme family is composed of four subfamilies, the two best studied ones being ubiquitin-specific proteases (UBPs) and ubiquitin C-terminal hydrolases (UCHs), which remove ubiquitin from proteins and smaller peptides, respectively (Chung and Baek, 1999; D'Andrea and Pellman, 1998; Wilkinson, 2000). The common theme for UBPs and UCHs is the presence of three separate catalytic domains, Cys, His, and Asp, which each contribute one conserved residue to the DUB Cys-His-Asp catalytic triad. Apart from their catalytic domain, there is virtually no sequence homology among DUBs. It is believed that this variability among DUBs determines their substrate specificity and thereby their biological function.

## RNA Interference

The cellular response to double-stranded RNA molecules, first identified in worms and plants, constitutes an effective tool to inhibit gene expression in almost all higher organisms. A cellular RNase named DICER can rapidly process large double-stranded RNA molecules into short (21–23 bp) duplex RNAs. These RNAs then serve as a signal to program a ribonuclease complex, the RNA-induced (RNAi) silencing complex (RISC), for mRNA destruction. Thus, introduction of double-stranded RNAs that match a cellular transcript will cause degradation of the cognate mRNA molecules by RISC-directed mRNA cleavage (Fire, 1999; Hannon, 2002; Sharp, 2001). However, until recently the use of RNAi in mammalian cells was hampered by the fact that the introduction of long double-stranded RNA triggers a potent interferon response in most somatic cells, leading to a general shut off of protein synthesis. This problem can be circumvented by the use of 21- to 23-bp double-stranded RNAs, named siRNAs (for short interfering RNAs). These RNAs are short enough to evade the interferon response, but are sufficiently long to recruit the mammalian RISC complex to specific target mRNAs and induce their degradation (Elbashir *et al.*, 2001). Although transfection of human cells with *in vitro*-synthesized siRNAs is effective to silence a gene of interest, siRNAs can also be expressed using siRNA-encoding plasmid vectors (Brummelkamp *et al.*, 2002; Miyagishi and Taira, 2002; Paddison *et al.*, 2002). These vectors often contain RNA polymerase III promoters to direct the synthesis of short hairpin RNA (shRNA) molecules. Such shRNA transcripts contain a perfectly double-stranded stem of 19–29 bp, which is identical in sequence to the mRNA targeted for suppression by RNA interference, connected by a loop of 6–9 bases. This loop is efficiently removed *in vivo*, generating a short duplex RNA that functions as a siRNA (Brummelkamp *et al.*, 2002).

To functionally annotate the DUB gene family, we have designed a collection of RNA interference vectors to suppress 50 of the human DUB enzymes (Brummelkamp *et al.*, 2003). The DUB library was constructed using the pSUPER shRNA expression vector (Brummelkamp *et al.*, 2002). Apart from reduced cost compared to synthesis of synthetic siRNAs, the advantage of such a vector-based strategy is a persistent knockdown of target genes. This makes it possible to study phenotypes that develop over prolonged periods of time. The details of the construction of this library and its application in functional annotation of the DUB gene family are discussed here.

## Construction of a shRNA Library Targeting the Deubiquitinating Enzyme Family

### *RNA Interference Target Site Selection and Oligonucleotide Design*

The ENSEMBL and NIH databases were searched for genes encoding DUBs based on the presence of conserved DUB catalytic domains. This yielded sequences corresponding to 50 known and putative DUB enzymes from the UBP and UCH subfamilies. The sequences were retrieved, and for each DUB gene, four different RNA interference target sequences were selected. Multiple sequences were selected for each transcript, as not all shRNAs are equally effective in silencing the intended target. By pooling four different shRNAs that target the same transcript, we estimate that we obtain on average 80% suppression of gene expression for approximately 80% of the DUB transcripts targeted. As described previously (Brummelkamp *et al.*, 2002), we selected 19 nucleotides from the DUB transcript for each shRNA vector. These 19-mer sequences were selected according to the following criteria:

1. The 19-mer sequences should preferably be located within the open reading frame of the DUB transcript, approximately 50–100 nucleotides downstream of the start codon. This was done in order to decrease the likelihood of selecting target sequences that are protected by RNA-binding proteins.

2. We searched for the sequence motif AA(G/CN18)TT (i.e., first residue a G or C, followed by any 18 nucleotides, flanked by 5′ AA and 3′ TT), in which G/CN18 corresponds to the siRNA sense strand and has a G/C content of 30–70%. If such sequences could not be found, the search was extended to include AA(G/CN18)GT or AA (G/CN18)CT sequences.
3. In addition, sequences containing more than three consecutive T residues were avoided, as a stretch of five thymidines functions as a termination signal for RNA polymerase III-driven transcription that is used in short hairpin RNA (shRNA) expression vectors.
4. The selected target sequences should share minimal sequence identity with other genes to avoid "off target" effects of the shRNA vectors.

The 200 selected target sequences for all DUBs in the library are listed elsewhere (Brummelkamp *et al.*, 2003). Since the construction of this DUB library was completed, an important new design rule for siRNA choice has been developed (Reynolds *et al.*, 2004; Schwarz *et al.*, 2003). This rule is based on the finding that the two strands of a siRNA duplex are not equally eligible for recruitment into the RISC complex. More specifically, the stability of the base pairs at the 5′ ends of the two siRNA strands determine to which extent each strand participates in the RNAi pathway. The strand that has the lowest thermostability at its 5′ end will be chosen as the guide RNA for the RISC complex. A number of siRNA target site selection search engines, both of academic and of commercial origin, are publicly available. Some of these incorporate the selection criteria mentioned in this article, whereas others use algorithms that have been developed upon analysis of large numbers of active siRNAs. It should, however, be noted that most of these selection criteria are based on experiments with synthetic siRNAs rather than vector-based RNAi.

Examples of siRNA design Web sites are

http://bioweb.pasteur.fr/seqanal/interfaces/sirna.html
http://jura.wi.mit.edu/siRNAext/, http://sfold.wadsworth.org/index.pl
http://rnaidesigner.invitrogen.com/sirna/
http://bioweb.pasteur.fr/seqanal/interfaces/sirna.html
http://www.promega.com/siRNADesigner/program/
http://www.dharmacon.com/
http://sonnhammer.cgb.ki.se/siSearch/siSearch_1.6.html

Since increasing data collections of active siRNAs are being generated, algorithms for target selection are under continuous development. As a

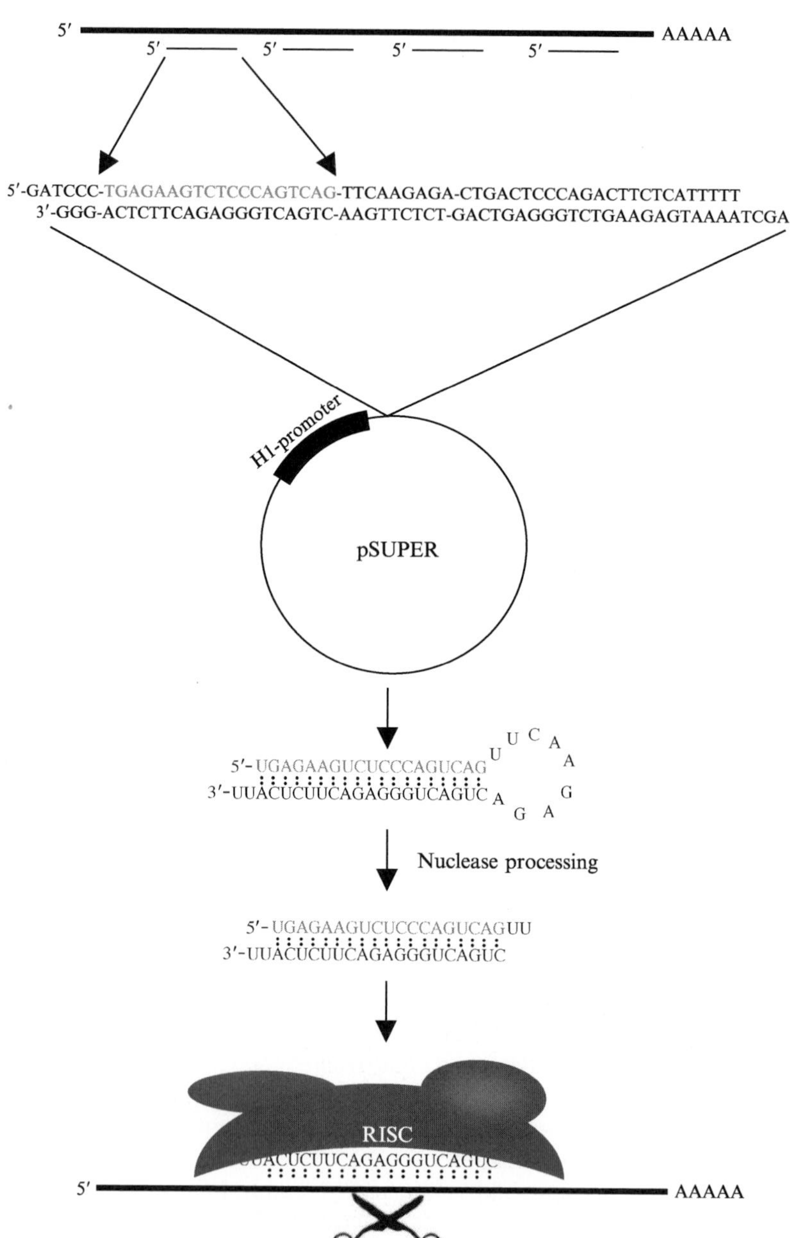


FIG. 1. Vector-based RNA interference. Four 19-mer targeting sequences are selected from each DUB transcript and two complementary 64-mer oligonucleotides are designed as

result, the efficiency of gene suppression by siRNAs will likely continue to improve.

For each siRNA target sequence, a 64-mer oligonucleotide set was designed, as described in more detail by Brummelkamp *et al.* (2002) (see also Fig. 1). The target sequence is separated from its reverse complement by a nine nucleotide spacer, which will form the loop of shRNAs upon transcription. Furthermore, the reverse complement sequence is followed by a TTTTT stretch, providing the RNA polymerase III termination signal. Finally, 64-mers are designed in such a way that 5′ *Bgl*II and 3′ *Hind*III overhangs are formed upon annealing of complementary oligonucleotides (Fig. 1). A total of 400 64-mer oligonucleotides were ordered from Sigma-Genosys. It should be noted that although the coupling efficiency per nucleotide during oligonucleotide synthesis is approximately 99%, it is likely that a significant amount of 64-mers will contain mutations ($0.99^{63} \sim 53\%$ full length), rendering a fraction of the cloned knockdown vectors inactive due to mutations. We have previously compared the knockdown efficiency of sequence-verified shRNA vectors with that of their polyclonal counterparts (that contain a mixture of correct and mutant RNA interference target sequences) and found comparable knockdown efficiencies in transient transfection assays. We therefore chose to clone the DUB library in a polyclonal format as described later (see Fig. 2).

### *DUB Library Cloning*

Oligonucleotides are dissolved in 50 $\mu$l demineralized $H_2O$ at an approximate concentration of 0.6 $\mu M$. The subsequent cloning steps are performed in a 96-well format. In order to anneal complementary 64-mer oligonucleotide sets, 96-well plates are filled with 48 $\mu$l annealing buffer (100 m*M* KAc, 30 m*M* HEPES, pH 7.4, 2 m*M* MgAc) per well. To each of 200 wells, 1 $\mu$l of corresponding sense- and antisense strand 64-mer oligonucleotide is added. The 96-well plates are heated to 92° for 10 min, followed by slow cooling to room temperature for efficient annealing.

Annealed 64-mers are phosphorylated according to a standard protocol by transferring 1 $\mu$l annealing mix to new 96 wells, each containing 5 units polynucleotide kinase (Roche) in 9 $\mu$l kinase buffer supplemented with

indicated and cloned into pSUPER. Polymerase type III-driven transcription of the insert sequence generates a transcript having a stem–loop structure in which the 19-mer target sequence forms the stem. After cleavage by an endogenous nuclease, siRNAs are generated. The antisense sequence incorporated into the RISC interacts with the target mRNA, leading to cleavage of this target. (See color insert.)

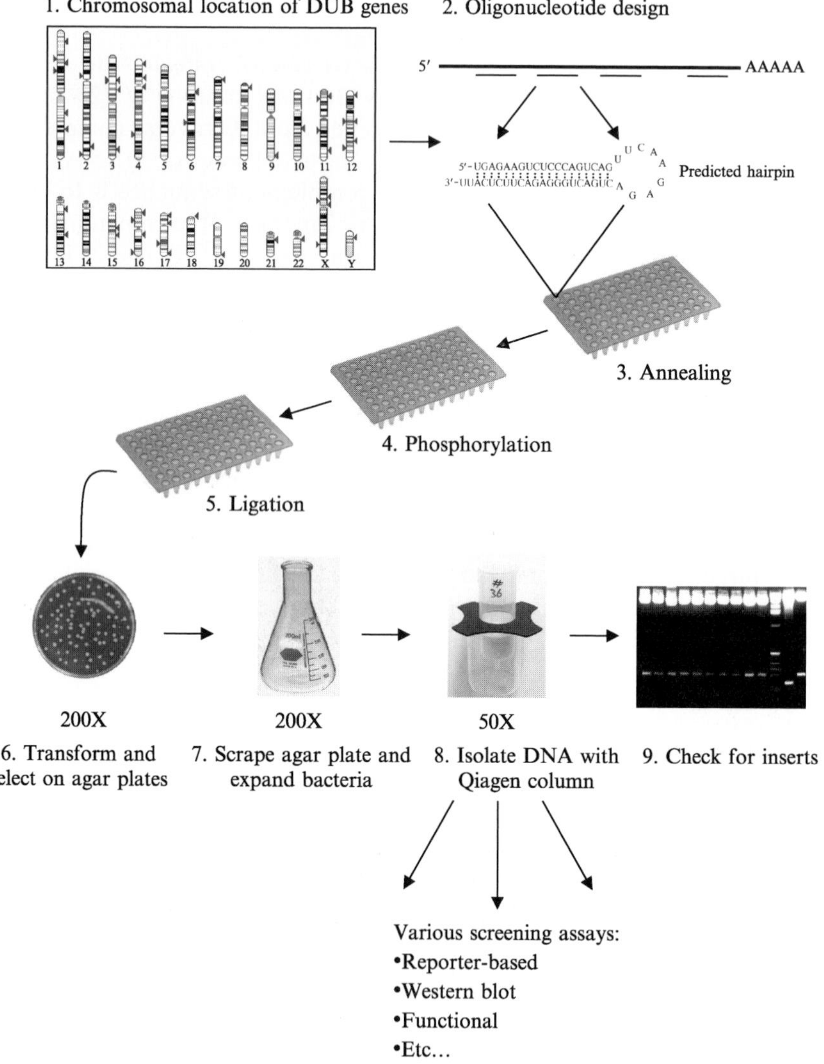


FIG. 2. Schematic overview of DUB library cloning. DUB sequences were selected, and four sets of oligonucleotides per transcript were designed. The 64-mers were annealed, phosphorylated, and ligated into pSUPER in a 96-well format. Ligated pSUPER plasmids were transformed into bacteria and selected on LB-ampicillin plates. Plates were scraped and bacteria were propagated in 50-ml TB cultures. Bacterial cultures were then pooled (four vectors targeting the same DUB per pool) and plasmid DNA was isolated using Qiagen columns. After checking the obtained DNA for inserts, various screening assays were performed. (See color insert.)

1 m*M* ATP. Phosphorylation reactions are incubated at 37° for 30 min followed by a 10-min heat inactivation at 70°.

Ten micrograms of pSUPER vector is digested overnight with 20 units of *Bgl*II and *Hind*III (Roche), followed by a 1-h treatment with 1 unit calf intestine alkaline phosphatase (Roche). The digested vector is gel purified using a QIA-EX II gel-extraction kit.

For ligations, 50 ng of the digested pSUPER vector is incubated in 96 wells with 2 $\mu$l of the annealed, phosphorylated 64-mer oligonucleotide in the presence of 400 units of T4 DNA ligase (New England Biolabs) in a volume of 10 $\mu$l for 1 h at room temperature. Electrocompetent DH10$\beta$ *Escherichia coli* (Invitrogen) are used for transformations. Bacteria are plated on LB-ampicillin plates and grown overnight at 37°. In order to make the library in a polyclonal format, all colonies from each plate are scraped and pooled into individual 50-ml cultures of Terrific broth, which are then grown overnight at 37°. The resulting 200 bacterial cultures are pooled in sets of four, such that each pool contains four pSUPER siRNA constructs targeting the same DUB. Finally, 50 DNA plasmid preparations are made from these pooled cultures (Fig. 2).

### *Library Verification*

In order to assess the knockdown efficiency conferred by sets of four pSUPER knockdown vectors targeting one DUB, we selected four different DUB enzymes and fused their open reading frames to that of green fluorescent protein, giving rise to four GFP-DUB fusion proteins. These chimeric constructs are then transfected in the absence and presence of their corresponding pSUPER-DUB knockdown vector sets, and GFP-Western blots are performed (Brummelkamp *et al.*, 2003). A significant reduction in GFP-DUB protein levels was observed in all cases. We conclude that four knockdown vectors per targeted DUB are sufficient to obtain a significant decrease in protein expression in the majority of cases.

## Assay Systems for Screening of a DUB RNAi Library

A variety of screening assays can be used to obtain insight into the function of individual members of the family of deubiquitinating enzymes. Because the number of known deubiquitinating enzymes is rather small, it is manageable to analyze the effect of a given shRNA directed against a particular DUB enzyme in detail. This section describes some examples that are widely applicable.

### *Microscopy-Based Screens*

Protein localization and degradation are two consequences of ubiquitination that can be studied easily by microscopy. In certain cases, ubiquitination only affects the activity and localization of a target protein, not its proteasomal degradation. It is well established that monoubiquitination can cause relocalization of a target protein to certain cellular compartments or substructures, whereas this modification does not lead to an increase in protein turnover (Aguilar and Wendland, 2003; Garcia-Higuera *et al.*, 2001). If such ubiquitin-dependent protein relocalization within cells can be detected by microscopy, a screen can be performed in living cells by constructing a GFP fusion of the target protein and cotransfecting it with the DUB shRNA library. By high-throughput microscopy it would then be possible to identify DUB enzymes that affect the subcellular localization of the GFP-target fusion protein. Obviously, one could study the influence of DUB enzymes on target protein turnover as well as target protein localization.

### *Phenotypic Screens*

Phenotypic analysis of cells that express shRNA molecules directed against one of the deubiquitinating enzymes can provide insight into the function(s) of individual DUB family members. One could study the effect of DUB enzymes on a number of cellular characteristics, such as morphology or cell division, or on possible cell fates, such as apoptosis, differentiation, oncogenic transformation, or senescence. These assays can be performed in different cell types to obtain an overview of individual DUB loss-of-function phenotypes.

### *Protein (Modification) Assay*

The relatively small size of the DUB enzyme family makes it feasible to study the effect of *in vivo* inhibition of a certain DUB enzyme on a target protein directly. After transfection of cells with individual sets of DUB knockdown vectors of the shRNA library, protein lysates can be isolated and potential DUB target proteins in these lysates can be analysed, e.g., by Western blot or immunoprecipitation-Western blot analysis. In this way one can directly measure the effect of DUB inhibition on the ubiquitination state of a specific target. We have employed this strategy to identify a DUB that reverses monoubiquitination of the Fanconi anemia protein FANCD2 (Nijman *et al.*, 2005). Monoubiquitination of FANCD2 leads to formation of a slower migrating form of the protein, which is detected easily using Western blotting. As a consequence, changes in FANCD2

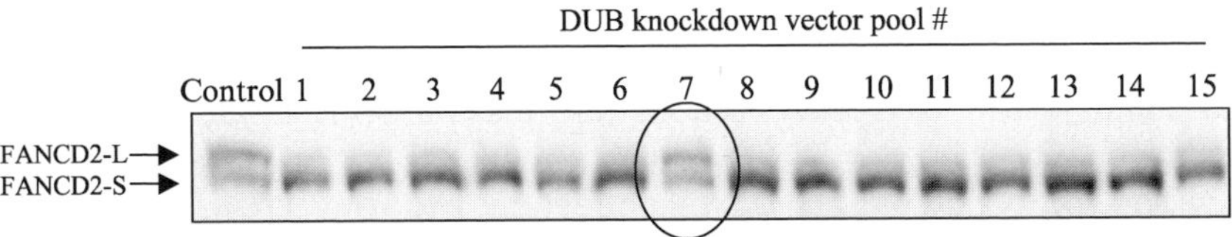


Fig. 3. Western blot analysis of FANCD2 monoubiquitination in cells transfected with various DUB knockdown vector pools. The unmodified (FANCD2-S) and monoubiquitinated (FANCD2-L) forms are indicated.

ubiquitination levels that are induced by knockdown of a specific DUB can also be seen directly on FANCD2 Western blots (Fig. 3).

*Reporter-Based Screens*

Components of a signaling pathway may be either activated or inactivated by covalent modification of one or more ubiquitin moieties. The activity of such signal transduction pathways can often be measured using reporter genes. One system that is widely used is that of fire fly luciferase reporter constructs, whose activity can be readily measured in transfected cells. Cotransfection of the DUB knockdown library can then be used in combination with the measurement of reporter gene activity to identify DUB enzymes that affect the activity of the studied signal transduction pathway under various conditions. A major advantage of this technique is that it is quantitative in nature and therefore can reveal subtle changes in activity. As a consequence it is highly suited in combination with RNAi, as RNAi may only lead to a partial reduction in gene expression that may not result in the induction of severe cellular phenotypes. A typical example of a luciferase reporter-based screen with DUB knockdown vectors is discussed next.

*Example of a Reporter-Based DUB Library Screen*

To illustrate the "pathway-reporter screen" approach, we describe here a screen for DUBs that modulates the cellular response to hypoxia. The Von Hippel–Lindau tumor suppressor protein (pVHL) regulates hypoxia-inducible factor 1-$\alpha$ (HIF1-$\alpha$) protein levels by polyubiquitination. Under normoxic conditions, the affinity of pVHL for HIF1-$\alpha$ is high, resulting in efficient ubiquitination and subsequent degradation by the proteasome. In contrast, under low oxygen conditions, affinity is low, resulting in the rapid stabilization of HIF1-$\alpha$ (Kaelin, 2002a,b). Under these conditions, HIF1-$\alpha$ can bind to HIF-responsive elements (HREs) present in genes such as vascular endothelial growth factor (VEGF) and erythropoietin, thereby activating their transcription. As a consequence, HIF1-$\alpha$ enables the cell to adapt to low oxygen conditions and plays a critical role in (tumor) angiogenesis.

To identify potential DUBs that affect this hypoxia response, we used a reporter construct containing three HREs driving the expression of the fire fly luciferase gene (Fig. 4). We cotransfected 0.5 $\mu$g of this reporter,

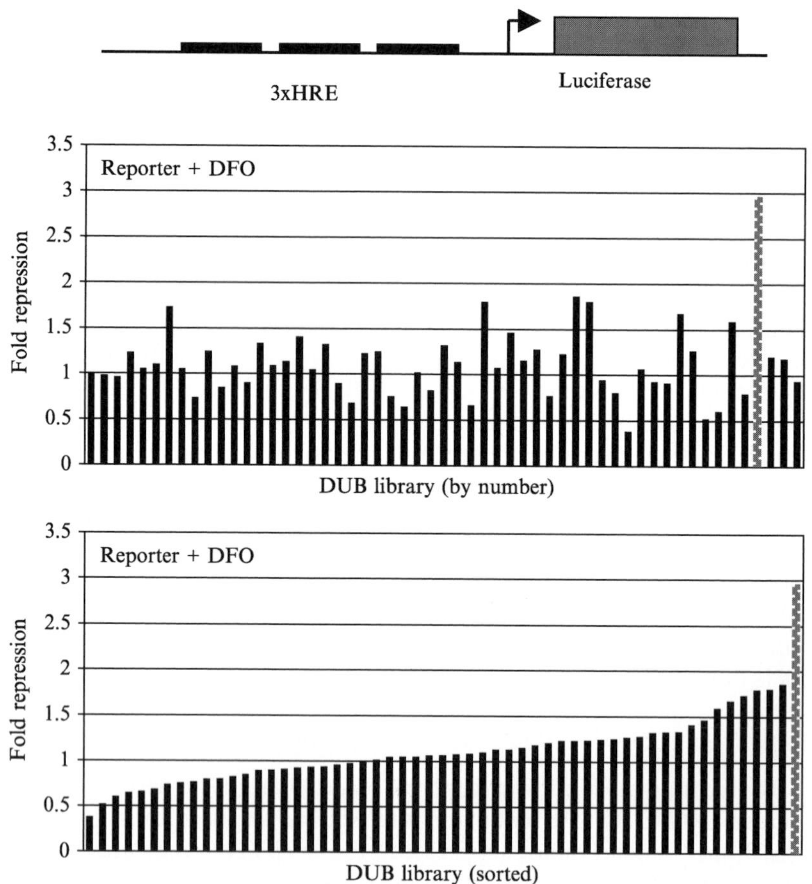


Fig. 4. Suppression of VDU1 inhibits a HIF1-$\alpha$ reporter construct. U2-OS cells were cotransfected with a hypoxia responsive element (HRE) reporter and the individual pools of the DUB knockdown library. Forty-eight hours later, cells were stimulated overnight with desferrioxamine (DFO) to mimic hypoxia, and luciferase activity was measured. A CMV-driven Renilla luciferase served as an internal control. Plotted is the fold repression per pool in random order (top) and sorted by ratio (bottom).

together with 3.0 μg of each of the pools of DUB knockdown library, in U2-OS osteosarcoma cells using the calcium–phosphate method. We also cotransfected 0.01 μg of SV40 promoter-driven renilla luciferase as an internal transfection control. Next we stimulated the cells overnight with 0.1 m*M* Fe-chelator desferrioxamine (DFO) to simulate hypoxia. We assayed firefly/renilla luciferase levels using standard procedures as described in the manufacturer's protocol (Promega) and plotted the luciferase ratios in a graph (Fig. 4). Only cotransfection of DUB knockdown pool #52 significantly affected the hypoxia reporter gene, repressing the reporter approximately twofold. This observation was confirmed when repeating the screen with a selected subset of DUB library pools (data not shown). Pool #52 contains four hairpin vectors aimed to suppress the mRNA of the deubiquitinating enzyme VDU1. Interestingly, VDU1 was found previously in a yeast two-hybrid screen using pVHL as bait (Li *et al.*, 2002). Furthermore, VDU1 stability appears to be influenced by pVHL, suggesting that, like HIF1-α, VDU1 is a pVHL target. Although the relevance of this latter finding remains unclear, the physical interaction, together with the newly identified genetic interaction, suggests that VDU1 is a player in HIF1-α signaling.

*Screen Follow-Up*

Upon performing a functional shRNA screen, a number of steps need to be taken in order to further validate candidate hits and to characterize these hits. First, part of the screen should be repeated to ask if the activity conferred by a specific pool of siRNA vectors is reproducible. If a candidate hit reproduces, one needs to identify the active shRNA vector(s) that causes the observed effect in the screen. In some cases, several of the four independent shRNA vectors present in each pool are active in the screen, but often only one of the four vectors is active.

Although the expression of hairpin RNA molecules usually results in gene-specific RNA interference, care should be taken, as these hairpin RNA molecules may also cause effects by influencing other (unintended) cellular factors (so-called off-target effects). To make sure that the observed effect is indeed caused by inhibition of the intended DUB enzyme, it is important to generate several independent shRNA vectors that inhibit the candidate "hit" from the initial screen. These vectors should then have a comparable effect as the original vector that was obtained in the screen. Furthermore, it is important to show that the shRNA vectors that resulted in a "hit" in a screen also inhibit the expression of the designated DUB enzyme. Another important question that should be addressed is whether the observed biological effect is dependent on the inhibition of deubiquitinating enzyme activity or whether other activities of the targeted DUB are responsible. Generating an expression vector for the candidate DUB enzyme in which the catalytic activity of the protease is eliminated by point mutation can help address this (Papa and Hochstrasser, 1993). Upon introduction in cells, such a "DUB-dead" version may act as a dominant negative, thereby resulting in a phenocopy of the shRNA-treated cells. If that is indeed the case, one can conclude that the observed effects are likely caused by the inhibition of deubiquitinating enzyme activity. Finally, depending on the experimental context of the system, one may wish to study the binding of the DUB to candidate DUB substrates, e.g., by performing coimmunoprecipitation experiments. The influence of a particular DUB on the ubiquitination status of such candidates may also be investigated.

## Concluding Remarks

The availability of the human genome sequence has allowed us to select DUB family members based on homology of the catalytic domain. By combining this information with RNAi technology and a diverse set of assays, we have started to functionally annotate DUBs in a relatively high-throughput manner. The development of better reagents and assays to study DUB function will make it possible to investigate DUB enzyme–substrate relationships further. The observation that inhibition of any particular DUB only affects a limited number of pathways in a highly specific fashion indicates that DUBs are relatively target specific. This, in combination with their involvement in many disease-relevant pathways (including cancer), makes DUBs promising drug targets. Insights into DUB enzyme function, combined with the development of specific DUB inhibitors, may yield a new class of protease inhibitors that can be applied in the clinic in the future.

## References

Aguilar, R. C., and Wendland, B. (2003). Ubiquitin: Not just for proteasomes anymore. *Curr. Opin. Cell Biol.* **15,** 184–190.

Bach, I., and Ostendorff, H. P. (2003). Orchestrating nuclear functions: Ubiquitin sets the rhythm. *Trends Biochem. Sci.* **28,** 189–195.

Brummelkamp, T. R., Bernards, R., and Agami, R. (2002). A system for stable expression of short interfering rnas in mammalian cells. *Science* **296,** 550–553.

Brummelkamp, T. R., Nijman, S. M., Dirac, A. M., and Bernards, R. (2003). Loss of the cylindromatosis tumour suppressor inhibits apoptosis by activating NF-kappaB. *Nature* **424,** 797–801.

Chung, C. H., and Baek, S. H. (1999). Deubiquitinating enzymes: Their diversity and emerging roles. *Biochem. Biophys. Res. Commun.* **266,** 633–640.

D'Andrea, A., and Pellman, D. (1998). Deubiquitinating enzymes: A new class of biological regulators. *Crit. Rev. Biochem. Mol. Biol.* **33,** 337–352.

Elbashir, S. M., Harborth, J., Lendeckel, W., Yalcin, A., Weber, K., and Tuschl, T. (2001). Duplexes of 21-nucleotide RNAs mediate RNA interference in cultured mammalian cells. *Nature* **411,** 494–498.

Fire, A. (1999). RNA-triggered gene silencing. *Trends Genet.* **15,** 358–363.

Garcia-Higuera, I., Taniguchi, T., Ganesan, S., Meyn, M. S., Timmers, C., Hejna, J., Grompe, M., and D'Andrea, A. D. (2001). Interaction of the Fanconi anemia proteins and BRCA1 in a common pathway. *Mol. Cell* **7,** 249–262.

Gill, G. (2004). SUMO and ubiquitin in the nucleus: Different functions, similar mechanisms? *Genes Dev.* **18,** 2046–2059.

Hannon, G. J. (2002). RNA interference. *Nature* **418,** 244–251.

Kaelin, W. G., Jr. (2002a). How oxygen makes its presence felt. *Genes Dev.* **16,** 1441–1445.

Kaelin, W. G., Jr. (2002b). Molecular basis of the VHL hereditary cancer syndrome. *Nat. Rev. Cancer* **2,** 673–682.

Li, Z., Na, X., Wang, D., Schoen, S. R., Messing, E. M., and Wu, G. (2002). Ubiquitination of a novel deubiquitinating enzyme requires direct binding to von Hippel-Lindau tumor suppressor protein. *J. Biol. Chem.* **277,** 4656–4662.

Miyagishi, M., and Taira, K. (2002). U6 promoter-driven siRNAs with four uridine 3′ overhangs efficiently suppress targeted gene expression in mammalian cells. *Nat. Biotechnol.* **20,** 497–500.

Nijman, S. M. B., Huang, T., Dirac, A. M. G., Brummelkamp, T. R., Kerkhoven, R. M., D'Andrea, A. D., and Bernards, R. (2005). The de-ubiquitinating enzyme USP1 regulates the Fanconi anemia pathway. *Mol. Cell.* **17**(3), 331–339.

Paddison, P. J., Caudy, A. A., Bernstein, E., Hannon, G. J., and Conklin, D. S. (2002). Short hairpin RNAs (shRNAs) induce sequence-specific silencing in mammalian cells. *Genes Dev.* **16,** 948–958.

Papa, F. R., and Hochstrasser, M. (1993). The yeast DOA4 gene encodes a deubiquitinating enzyme related to a product of the human tre-2 oncogene. *Nature* **366,** 313–319.

Reynolds, A., Leake, D., Boese, Q., Scaringe, S., Marshall, W. S., and Khvorova, A. (2004). Rational siRNA design for RNA interference. *Nat. Biotechnol.* **22,** 326–330.

Schwarz, D. S., Hutvagner, G., Du, T., Xu, Z., Aronin, N., Zamore, P. D., Ding, H., Keene, A., Affar el, B., Fenton, L., Xia, X., Shi, Y., Haley, B., and Tang, G. (2003). Asymmetry in the assembly of the RNAi enzyme complex. *Cell* **115,** 199–208.

Sharp, P. A. (2001). RNA interference—2001. *Genes Dev.* **15,** 485–490.

Wilkinson, K. D. (2000). Ubiquitination and deubiquitination: Targeting of proteins for degradation by the proteasome. *Semin. Cell Dev. Biol.* **11,** 141–148.

Zhang, Y. (2003). Transcriptional regulation by histone ubiquitination and deubiquitination. *Genes Dev.* **17,** 2733–2740.

# Author Index

## A



## B

## C

## D

## E



## F

## G

## H

## K

## L

## M

## N

## O



## P

## Q

## R

## S

## T

## U



## V



## W

## X



## Y

## Z

# Subject Index

## A

## D

## L



## M



## N

## R

## S

## T



## U

## W



## X



## Y

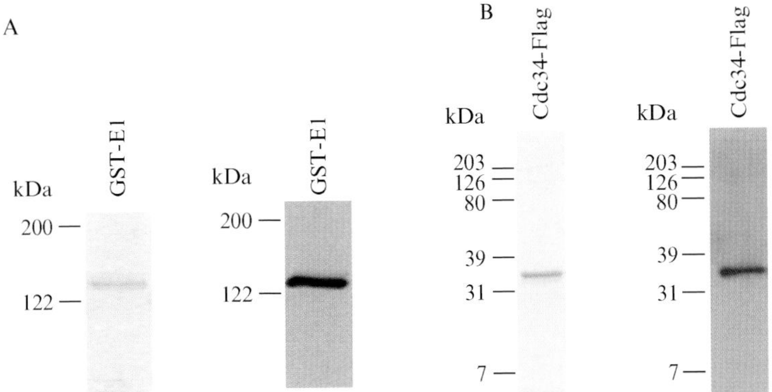


Hauser and Hofmann, Chapter 9, Fig. 1. Purified recombinant GST-E1 and Cdc34-Flag. (A) Affinity-purified GST-E1 (2 $\mu$g) was analyzed by SDS–PAGE and Coomassie blue staining. The identity of the purified protein (200 ng) was confirmed by immunostaining of a corresponding Western blot with an anti-E1 (isoform a) antibody (Calbiochem, No. 662102). (B) Affinity-purified and factor Xa-cleaved Cdc34-Flag (2 $\mu$g) was analyzed by SDS–PAGE and GelCode staining. The identity of the purified protein (0.5 ng) was confirmed by immunostaining of a corresponding Western blot with a monoclonal anti-Cdc34 antibody (136.1.12; Butz *et al.*, 2005).

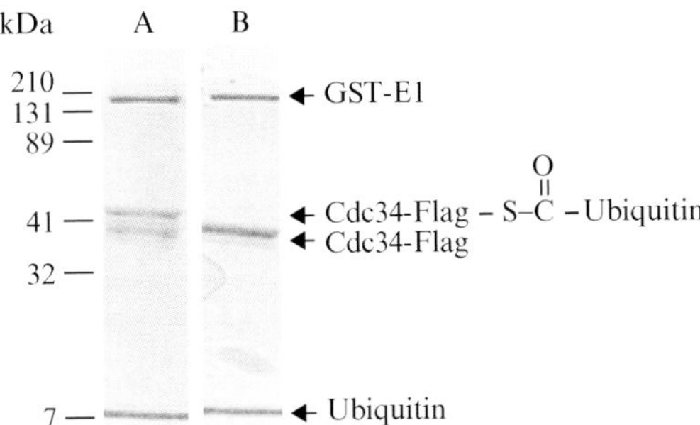


Hauser and Hofmann, Chapter 9, Fig. 2. Activity test. Activity of the purified enzymes was confirmed by assessing formation of a thioester intermediate between Cdc34 and ubiquitin, which can be monitored as a SDS–PAGE mobility shift under nonreducing conditions (A). Upon treatment with DTT, the thioester bond between the active site cysteine of Cdc34 and the carboxy terminus of ubiquitin is cleaved and the mobility shift is abolished (B).

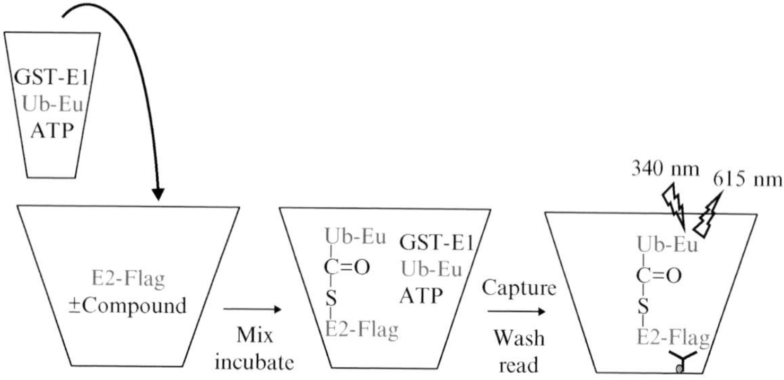


HAUSER AND HOFMANN, CHAPTER 9, FIG. 3. E2-ubiquitin thioester assay. A 96-well-based assay to monitor formation of the thioester intermediate between a ubiquitin-conjugating enzyme and ubiquitin was developed. Ubiquitin labeled with Europium was activated by E1 in the presence of ATP and added to a flag-tagged E2 in the presence or absence of a small molecular weight compound. Separation of the E2 from the remaining reagents was achieved by capturing the flag-tagged E2 on streptavidin-coated plates via a biotinylated anti-Flag antibody. Formation of the E2-ubiquitin thioester intermediate was then monitored by measuring Europium time-resolved fluorescence (excitation 340 nm; emission 615 nm).

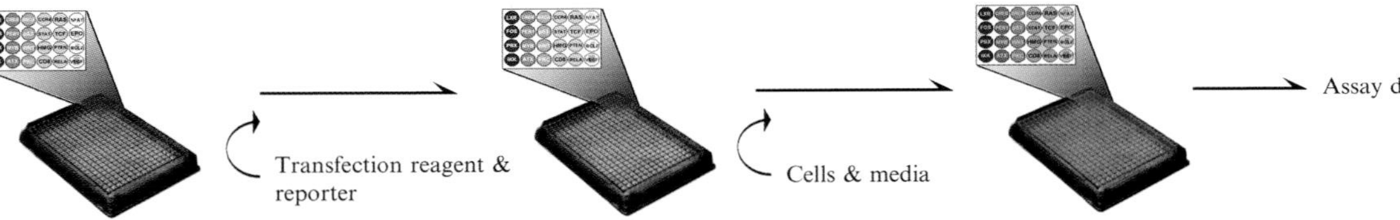


LI *ET AL.*, CHAPTER 23, FIG. 1. Multiwell plate reverse transfection method.

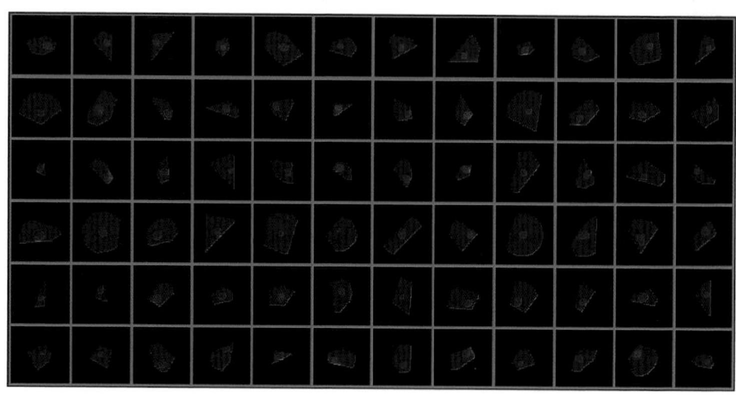

LI *ET AL.*, CHAPTER 23, FIG. 2. Gallery of cells where no aggregates were detected.

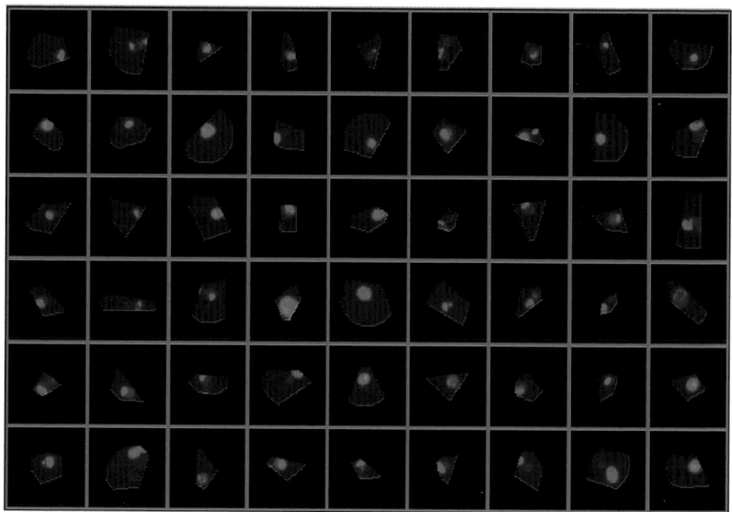

LI *ET AL.*, CHAPTER 23, FIG. 3. Gallery of cells with at least one aggregate.

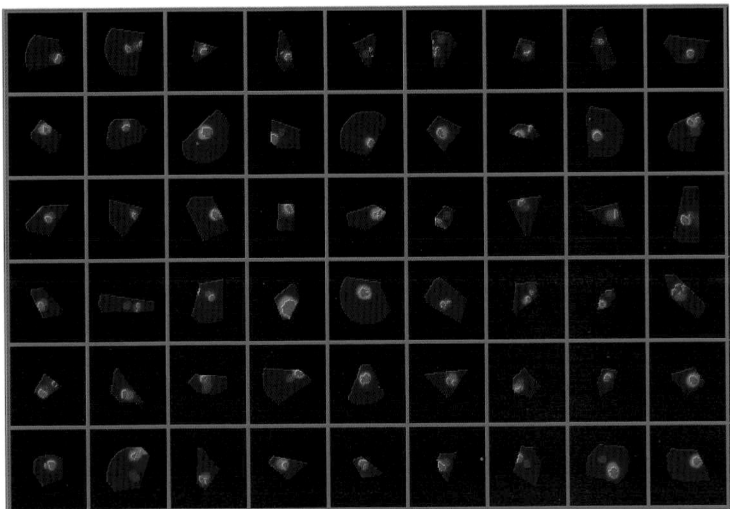

LI *ET AL.*, CHAPTER 23, FIG. 4. Gallery of cells with at least one aggregate with the aggregate mask shown.

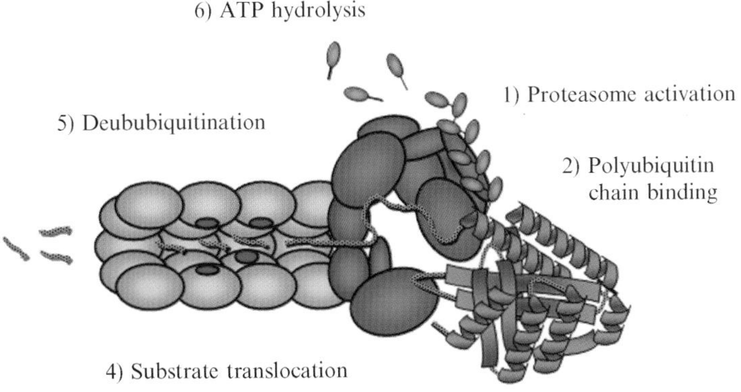


DEMARTINO, CHAPTER 24, FIG. 2. Functions of PA700 in the degradation of ubiquitin-dependent protein degradation by the 26S proteasome.

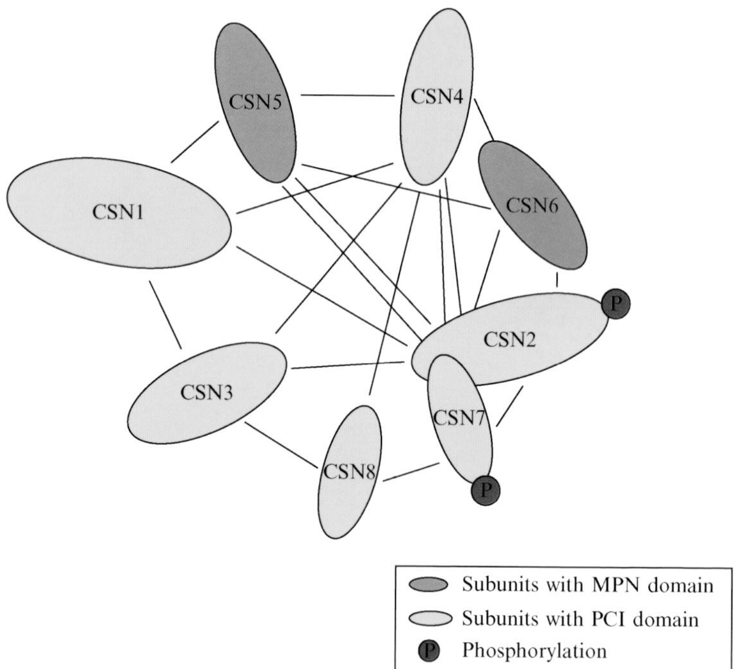


HETFELD *ET AL.*, CHAPTER 39, FIG. 1. Subunit interaction model of the COP9 signalosome. On the basis of mass distribution and subunit–subunit interactions, a model of the CSN complex was created (Kapelari *et al.*, 2000). Known phosphorylation sites and subunits with specific domains are indicated.

A

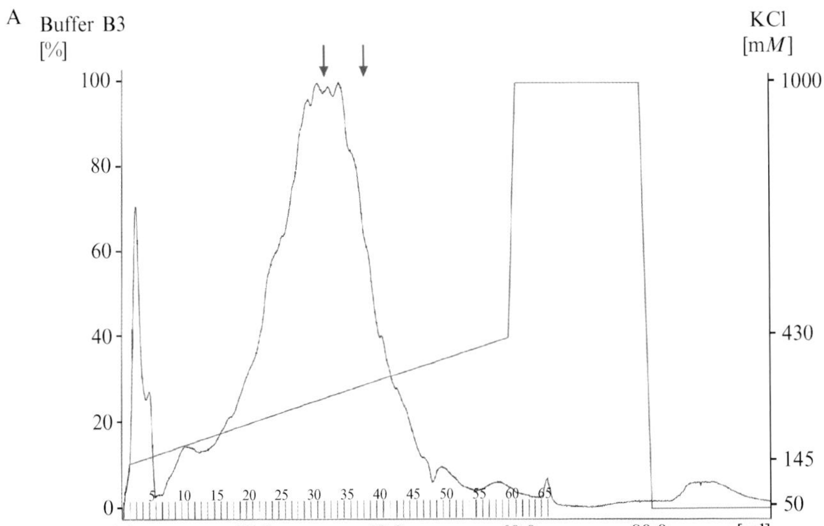


B

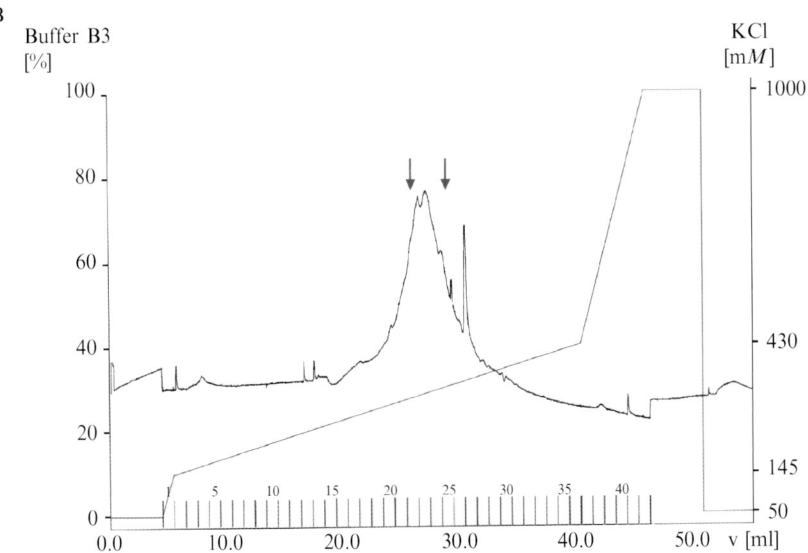


HETFELD *ET AL.*, CHAPTER 39, FIG. 3. Elution profile from FPLC columns. Resource Q column (A) and Mono Q column (B) with typical eluted protein profile and applied gradient. The red line shows the applied gradient from 10 to 40% buffer B3 (145–430 m*M* KCl) with subsequent regeneration of the column at 100% buffer B3 followed by buffer A. The black line indicates eluted protein concentration detected by an inline UV monitor (Monitor UV-M II, Amersham Biosciences). Numbers of collected 1-ml fractions are indicated above the *x* axis. Red arrows indicate fractions pooled for the next preparation step.

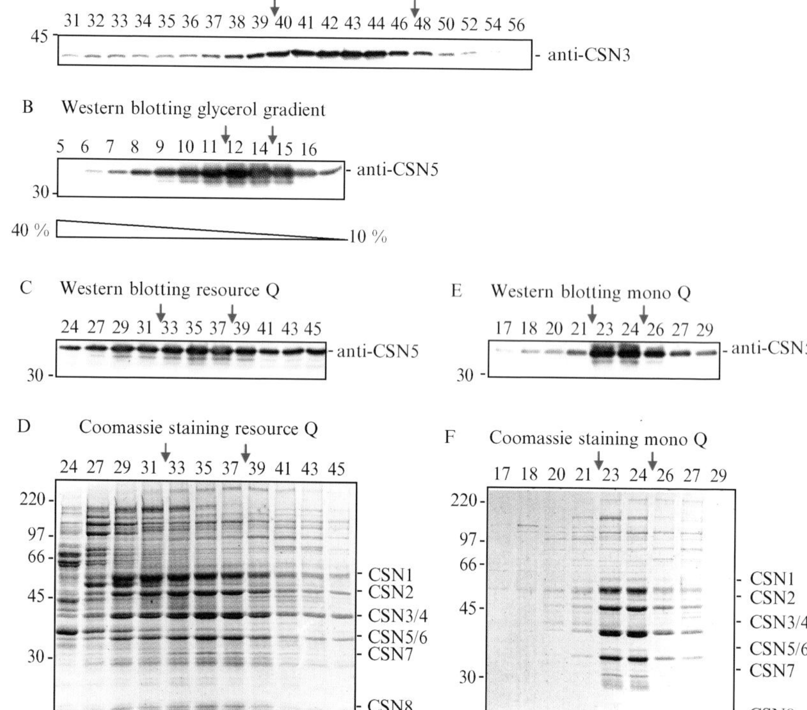


HETFELD *ET AL.*, CHAPTER 39, FIG. 4. Western blottings and Coomassie stainings at different purification steps. Red arrows indicate fractions pooled for next preparation step. (A) Western blotting of fractions 31–56 after DEAE column detected with anti-CSN3 antibody. (B) Western blotting of fractions 5–16 after density gradient centrifugation (10–40% glycerol) detected with anti-CSN5 antibody. (C and D) Western blotting and Coomassie staining of fractions 24–45 after Resource Q column. The CSN was detected with the anti-CSN5 antibody. (E and F) Western blotting and Coomassie staining of fractions 17–29 after Mono Q column. The CSN was detected with the anti-CSN5 antibody.

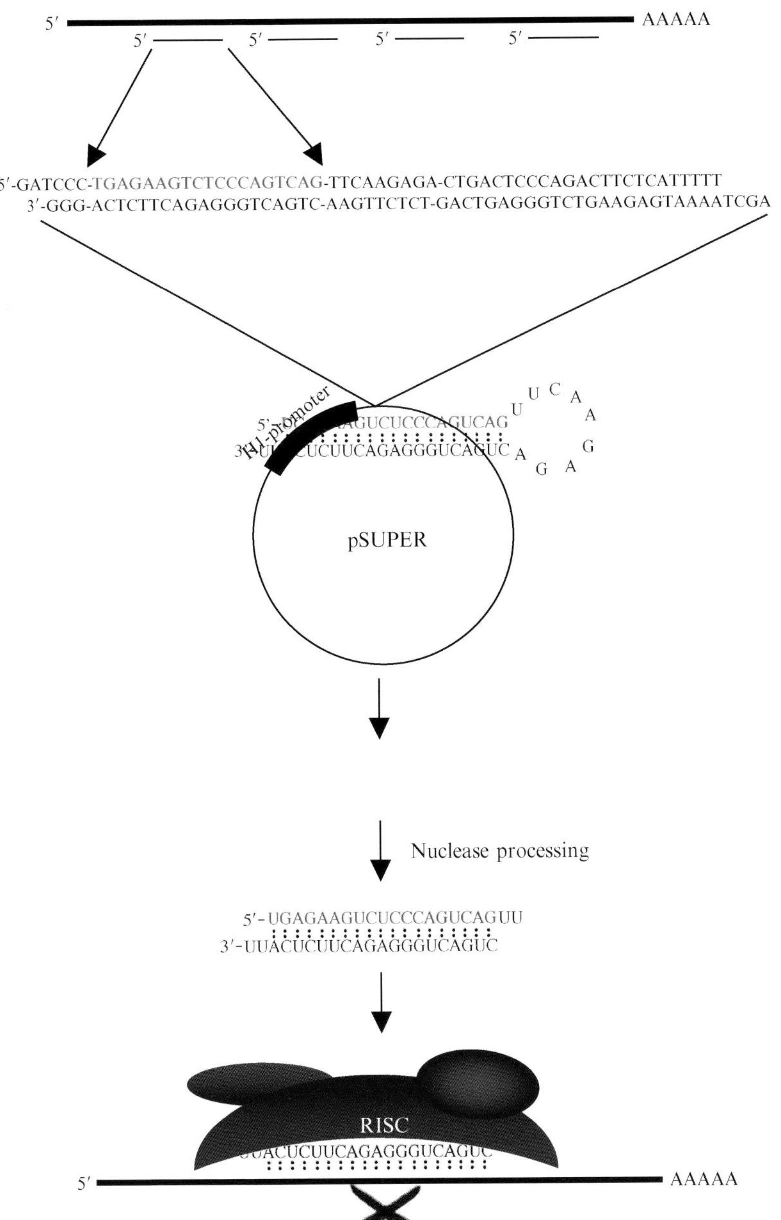


DIRAC *ET AL.*, CHAPTER 45, FIG. 1. Vector-based RNA interference. Four 19-mer targeting sequences are selected from each DUB transcript and two complementary 64-mer oligonucleotides are designed as indicated and cloned into pSUPER. Polymerase type III-driven transcription of the insert sequence generates a transcript having a stem–loop structure in which the 19-mer target sequence forms the stem. After cleavage by an endogenous nuclease, siRNAs are generated. The antisense sequence incorporated into the RISC interacts with the target mRNA, leading to cleavage of this target.

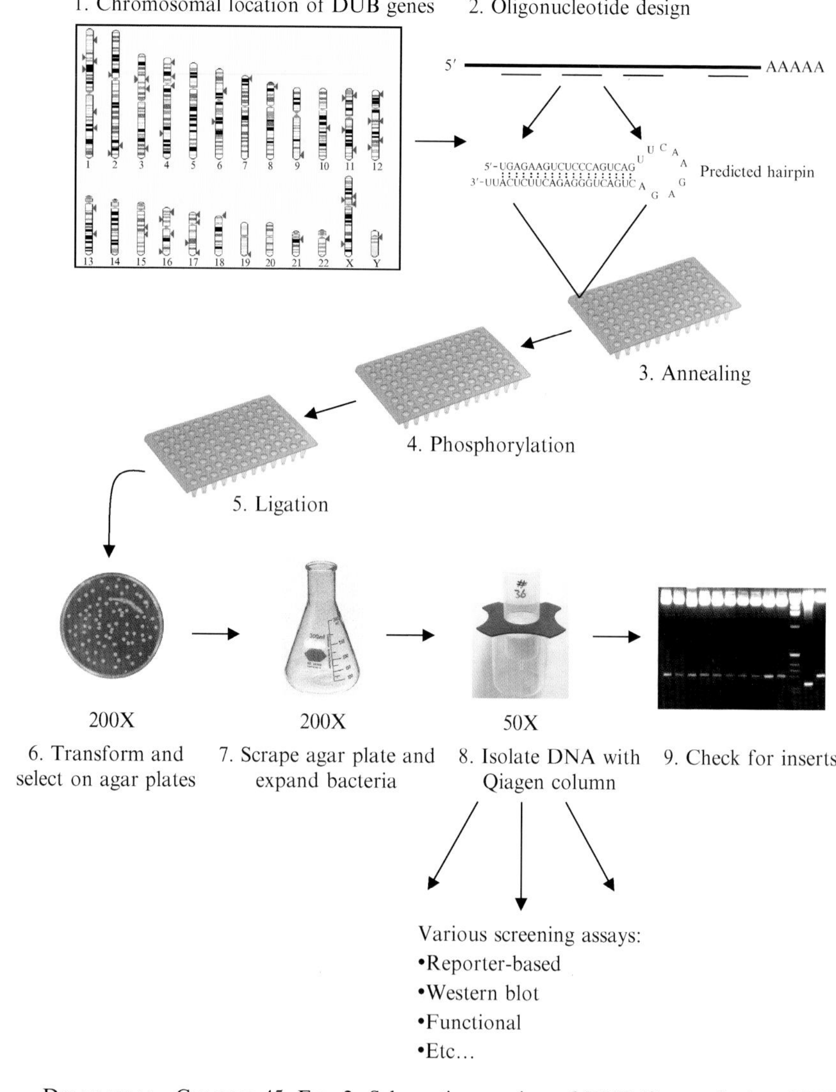


DIRAC *ET AL.*, CHAPTER 45, FIG. 2. Schematic overview of DUB library cloning. DUB sequences were selected, and four sets of oligonucleotides per transcript were designed. The 64-mers were annealed, phosphorylated, and ligated into pSUPER in a 96-well format. Ligated pSUPER plasmids were transformed into bacteria and selected on LB-ampicillin plates. Plates were scraped and bacteria were propagated in 50-ml TB cultures. Bacterial cultures were then pooled (four vectors targeting the same DUB per pool) and plasmid DNA was isolated using Qiagen columns. After checking the obtained DNA for inserts, various screening assays were performed.